Modern Strategies for Heterocycle Synthesis

Modern Strategies for Heterocycle Synthesis

Editor

Gianfranco Favi

MDPI • Basel • Beijing • Wuhan • Barcelona • Belgrade • Manchester • Tokyo • Cluj • Tianjin

Editor
Gianfranco Favi
University of Urbino "Carlo Bo"
Italy

Editorial Office
MDPI
St. Alban-Anlage 66
4052 Basel, Switzerland

This is a reprint of articles from the Special Issue published online in the open access journal *Molecules* (ISSN 1420-3049) (available at: https://www.mdpi.com/journal/molecules/special_issues/heterocycle_synthesis).

For citation purposes, cite each article independently as indicated on the article page online and as indicated below:

LastName, A.A.; LastName, B.B.; LastName, C.C. Article Title. *Journal Name* **Year**, *Volume Number*, Page Range.

ISBN 978-3-0365-0340-0 (Hbk)
ISBN 978-3-0365-0341-7 (PDF)

© 2021 by the authors. Articles in this book are Open Access and distributed under the Creative Commons Attribution (CC BY) license, which allows users to download, copy and build upon published articles, as long as the author and publisher are properly credited, which ensures maximum dissemination and a wider impact of our publications.

The book as a whole is distributed by MDPI under the terms and conditions of the Creative Commons license CC BY-NC-ND.

Contents

About the Editor vii

Gianfranco Favi
Modern Strategies for Heterocycle Synthesis
Reprinted from: *Molecules* **2020**, *25*, 2476, doi:10.3390/molecules25112476 1

Fengyan Jin, Tao Yang, Xian-Rong Song, Jiang Bai, Ruchun Yang, Haixin Ding and Qiang Xiao
TMSBr-Promoted Cascade Cyclization of *ortho*-Propynol Phenyl Azides for the Synthesis of 4-Bromo Quinolines and Its Applications
Reprinted from: *Molecules* **2019**, *24*, 3999, doi:10.3390/molecules24213999 5

Mara Tomassetti, Gabriele Lupidi, Pamela Piermattei, Federico V. Rossi, Samuele Lillini, Gianluca Bianchini, Andrea Aramini, Marco A. Ciufolini and Enrico Marcantoni
Catalyst-Free Synthesis of Polysubstituted 5-Acylamino-1,3-Thiazoles via Hantzsch Cyclization of α-Chloroglycinates
Reprinted from: *Molecules* **2019**, *24*, 3846, doi:10.3390/molecules24213846 21

Naoko Takenaga, Toshitaka Shoji, Takayuki Menjo, Akiko Hirai, Shohei Ueda, Kotaro Kikushima, Tomonori Hanasaki and Toshifumi Dohi
Nucleophilic Arylation of Halopurines Facilitated by Brønsted Acid in Fluoroalcohol
Reprinted from: *Molecules* **2019**, *24*, 3812, doi:10.3390/molecules24213812 37

Cecilia Ciccolini, Giacomo Mari, Gianfranco Favi, Fabio Mantellini, Lucia De Crescentini and Stefania Santeusanio
Sequential MCR via Staudinger/Aza-Wittig versus Cycloaddition Reaction to Access Diversely Functionalized 1-Amino-1*H*-Imidazole-2(3*H*)-Thiones
Reprinted from: *Molecules* **2019**, *24*, 3785, doi:10.3390/molecules24203785 47

Vasiliy M. Muzalevskiy, Zoia A. Sizova, Kseniya V. Belyaeva, Boris A. Trofimov and Valentine G. Nenajdenko
One-Pot Metal-Free Synthesis of 3-CF_3-1,3- Oxazinopyridines by Reaction of Pyridines with CF_3CO-Acetylenes
Reprinted from: *Molecules* **2019**, *24*, 3594, doi:10.3390/molecules24193594 61

Xabier del Corte, Edorta Martinez de Marigorta, Francisco Palacios and Javier Vicario
A Brønsted Acid-Catalyzed Multicomponent Reaction for the Synthesis of Highly Functionalized γ-Lactam Derivatives
Reprinted from: *Molecules* **2019**, *24*, 2951, doi:10.3390/molecules24162951 79

Xiaofei Chen, Liang Guo, Qin Ma, Wei Chen, Wenxi Fan and Jie Zhang
Design, Synthesis, and Biological Evaluation of Novel N-Acylhydrazone Bond Linked Heterobivalent β-Carbolines as Potential Anticancer Agents
Reprinted from: *Molecules* **2019**, *24*, 2950, doi:10.3390/molecules24162950 91

Wei Lin, Cangwei Zhuang, Xiuxiu Hu, Juanjuan Zhang and Juxian Wang
Alcohol Participates in the Synthesis of Functionalized Coumarin-Fused Pyrazolo[3,4-*b*]Pyridine from a One-Pot Three-Component Reaction
Reprinted from: *Molecules* **2019**, *24*, 2835, doi:10.3390/molecules24152835 111

Natalia A. Danilkina, Nina S. Bukhtiiarova, Anastasia I. Govdi, Anna A. Vasileva, Andrey M. Rumyantsev, Artemii A. Volkov, Nikita I. Sharaev, Alexey V. Povolotskiy, Irina A. Boyarskaya, Ilya V. Kornyakov, Polina V. Tokareva and Irina A. Balova
Synthesis and Properties of 6-Aryl-4-azidocinnolines and 6-Aryl-4-(1,2,3-1H-triazol-1-yl)cinnolines
Reprinted from: *Molecules* **2019**, *24*, 2386, doi:10.3390/molecules24132386 129

Anna Tripolszky, Krisztina Németh, Pál Tamás Szabó and Erika Bálint
Synthesis of (1,2,3-triazol-4-yl)methyl Phosphinates and (1,2,3-Triazol-4-yl)methyl Phosphates by Copper-Catalyzed Azide-Alkyne Cycloaddition
Reprinted from: *Molecules* **2019**, *24*, 2085, doi:10.3390/molecules24112085 155

Dalila Rocco, Catherine E. Housecroft and Edwin C. Constable
Synthesis of Terpyridines: Simple Reactions—What Could Possibly Go Wrong?
Reprinted from: *Molecules* **2019**, *24*, 1799, doi:10.3390/molecules24091799 169

Jin-ping Bao, Cui-lian Xu, Guo-yu Yang, Cai-xia Wang, Xin Zheng and Xin-xin Yuan
Novel 6a,12b-Dihydro-6H,7H-chromeno[3,4-c] chromen-6-ones: Synthesis, Structure and Antifungal Activity
Reprinted from: *Molecules* **2019**, *24*, 1745, doi:10.3390/molecules24091745 185

Eva Schütznerová, Anna Krchňáková and Viktor Krchňák
Traceless Solid-Phase Synthesis of Ketones via Acid-Labile Enol Ethers: Application in the Synthesis of Natural Products and Derivatives
Reprinted from: *Molecules* **2019**, *24*, 1406, doi:10.3390/molecules24071406 199

Xiuwen Jia, Pinyi Li, Xiaoyan Liu, Jiafu Lin, Yiwen Chu, Jinhai Yu, Jiang Wang, Hong Liu and Fei Zhao
Green and Facile Assembly of Diverse Fused N-Heterocycles Using Gold-Catalyzed Cascade Reactions in Water
Reprinted from: *Molecules* **2019**, *24*, 988, doi:10.3390/molecules24050988 209

Lucia Chiummiento, Rosarita D'Orsi, Maria Funicello and Paolo Lupattelli
Last Decade of Unconventional Methodologies for the Synthesis of Substituted Benzofurans
Reprinted from: *Molecules* **2020**, *25*, 2327, doi:10.3390/molecules25102327 239

Qili Lu, Dipesh S. Harmalkar, Yongseok Choi and Kyeong Lee
An Overview of Saturated Cyclic Ethers: Biological Profiles and Synthetic Strategies
Reprinted from: *Molecules* , *24*, 3778, doi:10.3390/molecules24203778 291

David Tejedor, Samuel Delgado-Hernández, Raquel Diana-Rivero, Abián Díaz-Díaz and Fernando García-Tellado
Recent Advances in the Synthesis of 2H-Pyrans
Reprinted from: *Molecules* **2019**, *24*, 2904, doi:10.3390/molecules24162904 313

Robert Pawlowski, Filip Stanek and Maciej Stodulski
Recent Advances on Metal-Free, Visible-Light- Induced Catalysis for Assembling Nitrogen- and Oxygen-Based Heterocyclic Scaffolds
Reprinted from: *Molecules* **2019**, *24*, 1533, doi:10.3390/molecules24081533 329

About the Editor

Gianfranco Favi received his Master's degree in Chemistry from the University of Bologna in 1999 and his Ph.D. in Chemistry and Pharmaceutical Sciences in 2005 from the University of Urbino working under the direction of Prof. O. A. Attanasi. Following postdoctoral studies, in 2016 he secured a temporary researcher position at the latter institution. In 2018, he spent a period of three months as a visiting scientist in the group of Prof. P. Melchiorre at the Institute of Chemical Research of Catalonia (ICIQ, Tarragona, Spain) where his studies centered on photochemical processes. After receiving the habilitation from the Italian Ministry for Research, in 2019 he was appointed to the position of Associate Professor in Organic Chemistry at the University of Urbino. His research interests include the design and development of new methodologies in organic synthesis, mainly aimed at obtaining novel mono- and polyheterocyclic systems with potential biological activities. He has authored more than 90 publications in international scientific journals.

Editorial

Modern Strategies for Heterocycle Synthesis

Gianfranco Favi

Department of Biomolecular Sciences, Section of Chemistry and Pharmaceutical Technologies, University of Urbino "Carlo Bo", Via I Maggetti 24, 61029 Urbino (PU), Italy; gianfranco.favi@uniurb.it

Received: 20 May 2020; Accepted: 25 May 2020; Published: 27 May 2020

Heterocycles constitute the largest and most diverse family of organic compounds that have received extensive interest owing to their popularity in many natural products, pharmaceuticals, and materials. It is estimated that of the over 50 million registered organic compounds in Chemical Abstracts [1], more than half are heterocycles, and the number is still increasing. Most frequently, nitrogen and oxygen heterocycles, or various positional combinations of nitrogen atoms, oxygen, and sulfur in five or six-membered rings, can be found. Due to the central role of heterocycles in chemistry and biology, new advances in synthetic methodologies that allow rapid access to a wide variety of functionalized heterocyclic compounds are of critical importance to the chemical community. For these reasons, a Special Issue collecting research on some aspects of innovative strategies for assembling heterocycles represents a good opportunity for dissemination of recent progress in this field. The vastness of this topic is documented by four review articles.

In their review, Stodulski and co-workers [2] highlight recent progress in the application of metal-free, visible-light-mediated catalysis for assembling five- and six-member heterocyclic scaffolds containing nitrogen and oxygen heteroatoms, particularly focused on the use of inexpensive organic dyes as an excellent alternative to the typical transition-metal complexes. García-Tellado's group [3] presents the nature of the different physicochemical factors affecting the valence isomerism between 2*H*-pyrans (2HPs) and 1-oxatrienes, and describes the most versatile synthetic methods reported in recent literature to access 2HPs. In more detail, a selection of the most transited routes able to generate these rings with a convenient amount of structural diversity, including the proper Knoevenagel reaction, the tandem propargyl Claisen rearrangement/*H*-shift reactions hosted by propargyl vinyl ethers [1,3], the cycloisomerization of diynes, and the Stille coupling of vinyl iodides and vinyl stannanes, is reported. The contribution of Lee and his group [4] provides an overview of the biological roles and synthetic strategies of saturated cyclic ethers, covering some of the most studied, and newly discovered, related natural products in recent years. This review also reports several promising and newly developed synthetic methods, emphasizing 3–7 membered rings. Chiummiento and co-workers [5] contribute to this Special Issue with a review describing recent developments in benzofurans synthesis. More specifically, new intramolecular and intermolecular C–C and/or C–O bond-forming processes, with transition-metal catalysis or metal-free are summarized.

In this Special Issue, Liu, Zhao and co-authors [6] report the first example of the generation of an indole/thiophene/pyrrole/pyridine/naphthalene/benzene-fused *N*-heterocycle library through an AuPPh$_3$Cl/AgSbF$_6$-catalyzed cascade reaction between amine nucleophiles and alkynoic acids in water. Low catalyst loading, good to excellent yields, high efficiency in bond formation (three new bonds together with two rings), excellent selectivity, great tolerance of functional groups, and extraordinarily broad substrate scope are features of this green cascade process.

The advantages of solid-phase synthesis in time-efficient and traceless preparation of ketones via acid-labile enol ethers are described in the article by Krchňák and co-authors [7]. The practicality of this synthetic strategy on the solid-phase construction of pyrrolidine-2,4-diones, which represent the core structure of several natural products, including tetramic acid, is also demonstrated.

A series of chromeno[3,4-c]chromen-6-one derivatives, a new type of dihydrocoumarins, are synthesized by Xu, Yang et al. [8] via Michael addition, transesterification and nucleophilic addition from the reaction of 3-trifluoroacetyl coumarins and phenols in the presence of an organic base. For these compounds, the in vitro antifungal activity is assessed against two fungal strains with the mycelial growth rate method.

In this Issue, a couple of articles concerning the copper(I)-catalyzed azide-alkyne cycloaddition (CuAAC) (the "click reaction") are presented. In this regard, Bálint and co-workers [9] describe a facile and efficient method for the synthesis of new (1,2,3-triazol-4-yl)methyl phosphinates/phosphates by the Cu(I)-catalyzed 1,3-dipolar (Huisgen) cycloaddition of organic azides and prop-2-ynyl phosphinate/phosphate. In the other article, Balova's group [10] first reports the synthesis of 6-aryl-4-azidocinnolines through the Richter-type cyclization of 2-ethynyl-4-aryltriazenes with the formation of 4-bromo-6-arylcinnolines and nucleophilic substitution of a bromine atom with an azide functional group, then their use in both CuAAC with terminal alkynes and SPAAC with diazacyclononyne, yielding 4-triazolylcinnolines.

This Special Issue presents three contributions of multicomponent reactions (MCRs), in which access to different molecular skeletons is realized. Participation of various alcohols in the construction of coumarin-fused pyrazolo[3,4-b]pyridines via a silica sulfuric acid (SSA)-catalyzed three-component domino reaction under microwave irradiation is presented for the first time by Lin, Wang and co-authors [11]. In another work, the group of Palacios and Vicario [12] reports a phosphoric acid-catalyzed MCR procedure for the synthesis of highly functionalized γ-lactam derivatives by the reaction of benzaldehyde, amines and acetylenedicarboxylate. Another article dealing with MCR strategy is described by our group [13]. Combining sequential azidation, Staudinger, and aza-Wittig reactions with CS_2 on α-halohydrazones in a one-pot protocol, variously substituted 1-amino-1H-imidazole-2(3H)-thiones are directly accessible in good yields and with complete control of regioselectivity. In addition, the concurrent presence of reactive appendages on the obtained scaffolds also ensures post-modifications toward N-bridgeheaded heterobicyclic structures.

Although the synthesis of terpyridines by one-pot reaction of acetylpyridines, aromatic aldehydes and ammonia is presented in the literature as an infallible synthetic method, there is ample precedent for the formation of a variety of alternative products. Based on this assumption, Constable et al. [14] provides another example of an unexpected product and a systematic survey of the products of such reactions.

Utilizing a pharmacophore hybridization approach, a novel series of 28 new heterobivalent β-carbolines bearing an acylhydrazone bond is reported by Zhang and co-workers [15]. The results of their in vitro antiproliferative activity using the MTT-based assay against five cancer cell lines (LLC, BGC-823, CT-26, Bel-7402, and MCF-7) contribute to the further elucidation of the biological regulatory role of these compounds, as well as providing helpful information on the development of vascular targeting antitumor drugs.

In this Issue, there are two articles describing the synthesis of heterocycles and their functionalization using metal-free approaches. In one of them, Trofimov and Nenajdenko's team [16] elaborates an efficient pathway towards trifluoromethylated oxazinopyridines on the base of a one-pot, metal-free 1:2 assembly of pyridines and CF_3-ynone. Target heterocycles are prepared in a stereoselective manner and up to quantitative yields. In the other work, through the direct arylation of halopurines with aromatic compounds, facilitated by the combination of triflic acid and fluoroalcohol, various aryl-substituted purine derivatives are synthesized by the Takenaga and Dohi group [17]. This metal-free method is complementary to conventional coupling reactions using metal catalysts and reagents for the syntheses of aryl-substituted purine analogues.

A publication on a TMSBr-promoted cascade cyclization of ortho-propynol phenyl azides for the synthesis of 4-bromo quinolines is reported by Xiao and his group [18]. Moreover, a variety of functionalized compounds with molecular diversity at C4 position of quinolines are obtained through the subsequent coupling or nucleophilic reactions.

Finally, a contribution of Tomassetti, Marcantoni et al. [19] deals with the synthesis of polysubstituted 5-acylamino-1,3-thiazoles via a Hantzsch heterocyclization reaction of α-chloroglycinates with thiobenzamides or thioureas. As result, the pharmaceutically relevant target compounds are obtained under mild conditions from readily available, inexpensive building blocks through an environmentally benign process that requires no stringent control of reaction parameters/atmosphere and no catalysts.

As guest editor, I hope that you find this Special Issue highlighting contributions in the vast field of heterocyclic chemistry both informative and inspiring.

Funding: The author declares no competing financial interest.

Acknowledgments: The guest editor wishes to thank all the authors for their contributions to this Special Issue, and all the reviewers for their work in evaluating the submitted articles. Special thanks are also given to the editorial staff of Molecules, especially Zack Li, as well as the other assistant editors of this journal, who have participated actively in compiling this Special Issue.

Conflicts of Interest: There are no conflict to declare.

References

1. Lipkus, A.H.; Yuan, Q.; Lucas, K.A.; Funk, S.A.; Bartelt, W.F.; Schenck, R.J.; Trippe, A.J. Structural Diversity of Organic Chemistry. A Scaffold Analysis of the CAS Registry. *J. Org. Chem.* **2008**, *73*, 4443. [CrossRef] [PubMed]
2. Pawlowski, R.; Stanek, F.; Stodulski, M. Recent Advances on Metal-Free, Visible-LightInduced Catalysis for Assembling Nitrogen- and Oxygen-Based Heterocyclic Scaffolds. *Molecules* **2019**, *24*, 1533. [CrossRef] [PubMed]
3. Tejedor, D.; Delgado-Hernández, S.; Diana-Rivero, R.; Díaz-Díaz, A.; García-Tellado, F. Recent Advances in the Synthesis of 2H-Pyrans. *Molecules* **2019**, *24*, 2904. [CrossRef] [PubMed]
4. Lu, Q.; Harmalkar, D.S.; Choi, Y.; Lee, K. An Overview of Saturated Cyclic Ethers: Biological Profiles and Synthetic Strategies. *Molecules* **2019**, *24*, 3778. [CrossRef] [PubMed]
5. Chiummiento, L.; D'Orsi, R.; Funicello, M.; Lupattelli, P. Last decade of unconventional methodologies for the synthesis of substituted benzofurans. *Molecules* **2020**, *25*, 2327. [CrossRef] [PubMed]
6. Jia, X.; Li, P.; Liu, X.; Lin, J.; Chu, Y.; Yu, J.; Wang, J.; Liu, H.; Zhao, F. Green and Facile Assembly of Diverse Fused N-Heterocycles Using Gold-Catalyzed Cascade Reactions in Water. *Molecules* **2019**, *24*, 988. [CrossRef] [PubMed]
7. Schütznerová, E.; Krchňáková, A.; Krchňák, V. Traceless Solid-Phase Synthesis of Ketones via Acid-Labile Enol Ethers: Application in the Synthesis of Natural Products and Derivatives. *Molecules* **2019**, *24*, 1406. [CrossRef] [PubMed]
8. Bao, J.-P.; Xu, C.-L.; Yang, G.-Y.; Wang, C.-X.; Zheng, X.; Yuan, X.-X. Novel 6a,12b-Dihydro-6H,7H-chromeno[3,4-c] chromen-6-ones: Synthesis, Structure and Antifungal Activity. *Molecules* **2019**, *24*, 1745. [CrossRef] [PubMed]
9. Tripolszky, A.; Németh, K.; Szabó, P.T.; Bálint, E. Synthesis of (1,2,3-triazol-4-yl)methyl Phosphinates and (1,2,3-Triazol-4-yl)methyl Phosphates by Copper-Catalyzed Azide-Alkyne Cycloaddition. *Molecules* **2019**, *24*, 2085. [CrossRef] [PubMed]
10. Danilkina, N.A.; Bukhtiiarova, N.S.; Govdi, A.I.; Vasileva, A.A.; Rumyantsev, A.M.; Volkov, A.A.; Sharaev, N.I.; Povolotskiy, A.V.; Boyarskaya, I.A.; Kornyakov, I.V.; et al. Synthesis and Properties of 6-Aryl-4-azidocinnolines and 6-Aryl-4-(1,2,3-1H-triazol-1-yl)cinnolines. *Molecules* **2019**, *24*, 2386. [CrossRef] [PubMed]
11. Lin, W.; Zhuang, C.; Hu, X.; Zhang, J.; Wang, J. Alcohol Participates in the Synthesis of Functionalized Coumarin-Fused Pyrazolo[3,4-b]Pyridine from a One-Pot Three-Component Reaction. *Molecules* **2019**, *24*, 2835. [CrossRef] [PubMed]
12. Del Corte, X.; de Marigorta, E.M.; Palacios, F.; Vicario, J. A Brønsted Acid-Catalyzed Multicomponent Reaction for the Synthesis of Highly Functionalized γ-Lactam Derivatives. *Molecules* **2019**, *24*, 2951. [CrossRef] [PubMed]
13. Ciccolini, C.; Mari, G.; Favi, G.; Mantellini, F.; De Crescentini, L.; Santeusanio, S. Sequential MCR via Staudinger/Aza-Wittig versus Cycloaddition Reaction to Access Diversely Functionalized 1-Amino-1H-Imidazole-2(3H)-Thiones. *Molecules* **2019**, *24*, 3785. [CrossRef] [PubMed]
14. Rocco, D.; Housecroft, C.E.; Constable, E.C. Synthesis of Terpyridines: Simple Reactions—What Could Possibly Go Wrong? *Molecules* **2019**, *24*, 1799. [CrossRef] [PubMed]

15. Chen, X.; Guo, L.; Ma, Q.; Chen, W.; Fan, W.; Zhang, J. Design, Synthesis, and Biological Evaluation of Novel N-Acylhydrazone Bond Linked Heterobivalent β-Carbolines as Potential Anticancer Agents. *Molecules* **2019**, *24*, 2950. [CrossRef] [PubMed]
16. Muzalevskiy, V.M.; Sizova, Z.A.; Belyaeva, K.V.; Trofimov, B.A.; Nenajdenko, V.G. One-Pot Metal-Free Synthesis of 3-CF$_3$-1,3- Oxazinopyridines by Reaction of Pyridines with CF$_3$CO-Acetylenes. *Molecules* **2019**, *24*, 3594. [CrossRef] [PubMed]
17. Takenaga, N.; Shoji, T.; Menjo, T.; Hirai, A.; Ueda, S.; Kikushima, K.; Hanasaki, T.; Dohi, T. Nucleophilic Arylation of Halopurines Facilitated by Brønsted Acid in Fluoroalcohol. *Molecules* **2019**, *24*, 3812. [CrossRef] [PubMed]
18. Jin, F.; Yang, T.; Song, X.-R.; Bai, J.; Yang, R.; Ding, H.; Xiao, Q. TMSBr-Promoted Cascade Cyclization of *ortho*-Propynol Phenyl Azides for the Synthesis of 4-Bromo Quinolines and Its Applications. *Molecules* **2019**, *24*, 3999. [CrossRef] [PubMed]
19. Tomassetti, M.; Lupidi, G.; Piermattei, P.; Rossi, F.V.; Lillini, S.; Bianchini, G.; Aramini, A.; Ciufolini, M.A.; Marcantoni, E. Catalyst-Free Synthesis of Polysubstituted 5-Acylamino-1,3-Thiazoles via Hantzsch Cyclization of α-Chloroglycinates. *Molecules* **2019**, *24*, 3846. [CrossRef] [PubMed]

© 2020 by the author. Licensee MDPI, Basel, Switzerland. This article is an open access article distributed under the terms and conditions of the Creative Commons Attribution (CC BY) license (http://creativecommons.org/licenses/by/4.0/).

Article

TMSBr-Promoted Cascade Cyclization of *ortho*-Propynol Phenyl Azides for the Synthesis of 4-Bromo Quinolines and Its Applications

Fengyan Jin [†], Tao Yang [†], Xian-Rong Song, Jiang Bai, Ruchun Yang, Haixin Ding and Qiang Xiao *

Institute of Organic Chemistry, Jiangxi Science & Technology Normal University, Key Laboratory of Organic Chemistry, Jiangxi Province, Nanchang 330013, China; jinfengyancg@163.com (F.J.); tao_yang2019@yeah.net (T.Y.); songxr2015@163.com (X.-R.S.); mtbaijiang@yeah.net (J.B.); ouyangruchun@yeah.net (R.Y.); dinghaixin0204@yeah.net (H.D.)
* Correspondence: xiaoqiang@tsinghua.org.cn; Tel.: + 86-1376-700-6775
† Co-first author.

Received: 30 September 2019; Accepted: 31 October 2019; Published: 5 November 2019

Abstract: Difficult-to-access 4-bromo quinolines are constructed directly from easily prepared *ortho*-propynol phenyl azides using TMSBr as acid-promoter. The cascade transformation performs smoothly to generate desired products in moderate to excellent yields with good functional groups compatibility. Notably, TMSBr not only acted as an acid-promoter to initiate the reaction, and also as a nucleophile. In addition, 4-bromo quinolines as key intermediates could further undergo the coupling reactions or nucleophilic reactions to provide a variety of functionalized compounds with molecular diversity at C4 position of quinolines.

Keywords: TMSBr; propargylic alcohols; azides; cascade cyclization; 4-bromo quinolines

1. Introduction

Quinolines are distinctive and significant frameworks which are widely existed in numerous pharmaceuticals, pesticide molecules, bioactive molecules, and natural products [1–8]. Moreover, such compounds using as ligands play crucial role in synthetic and catalysis chemistry [9–13]. Consequently, developing general and flexible approach towards these heterocycles has attracted much attention among synthetic chemists. Until now, despite significant achievements having been made in the construction of functionalized quinolines [14–22], methods for the direct synthesis of 4-halo quinolines are still limited [23–26]. 4-halo quinolines have been widely used as key synthetic intermediates for the construction of various bioactive molecules or drugs [27–29]. Therefore, the development of an efficient and versatile strategy towards 4-halo quinolines is highly desirable, especially through a cascade cyclization, because of the merits of efficiency and atomic economy.

Based on its distinctive bifunctional group characteristics, the cascade reaction of propynols is an important tactic in organic synthesis, which exerts a significant role in the construction of functionalized carbo- or heterocyclic compounds [30–34]. In the past few years, our group had developed various efficient methods to construct functionalized heterocyclics through the cascade cyclization of propargylic alcohols in the presence of acid-promoter [35–44]. For example, we recently reported an efficient approach for the construction of 4-chrolo quinolines via the cyclization of *ortho*-propynol phenyl azides with TMSCl as acid-promoter [45]. Taking into consideration that the coupling reaction of chloro-substituted compounds is more difficult than bromo- or iodo-substituted compounds, the further development of universal approach for the construction of 4-bromo quinolines is still desirable and necessary. Herein, we report a general TMSBr-promoted the cascade cyclization of *ortho*-propynol phenyl azides for constructing 4-bromo quinolines, which can further undergo the coupling reactions or nucleophilic reactions to provide a variety of functionalized compounds with molecular diversity at C4 position of quinolines (Scheme 1). Compared to the Shvartsberg's method [26], our developed strategy has the merits of good functional groups compatibility, easy preparation of the starting material, and simple operation.

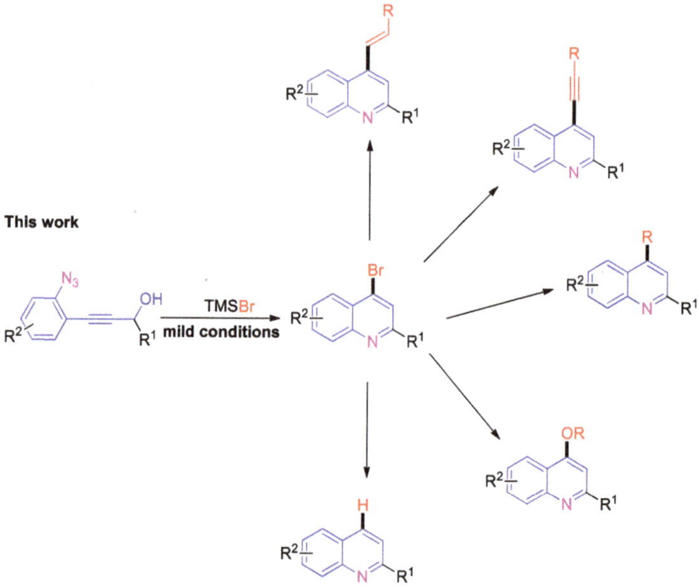

Scheme 1. Our strategy for the construction of 4-bromo quinolines and its applications.

2. Results and Discussion

Initially, the reaction conditions were optimized for cascade cyclization of *ortho*-propynol phenyl azides **1a** in the presence of TMSBr. Various solvents, temperatures, and TMSBr loading were investigated, and all cases were shown in Table 1. To our delight, with 2.5 equiv of TMSBr in different solvents—such as MeCN, CH_3NO_2, DCE, 1,4-dioxane, HOAc, and DCM—all reactions proceeded smoothly and cleanly to produce expected product 4-bromo-2-(4-methoxyphenyl)quinoline **2a** (Table 1, entries 1–5); CH_3NO_2 as solvent was most suitable for this transformation (73% yield). Encouraged by this preliminary result, further efforts were then directed toward improving the yield of desired product **2a** while suppressing the classical Meyer–Schuster rearrangement side reaction. Our studies on the loading of TMSBr with CH_3NO_2 as solvent showed that 3.5 equiv of TMSBr was the most efficient for this cascade transformation and could improve the yield of product **2a** to 81% (Table 1, entries 6–8). Subsequently, the examination of the reaction temperature indicated that the choice of reaction temperature was also an important in this transformation (entries 9, 10). Furthermore, no better yield was obtained when hydrobromic acid (HBr, 48 wt % in H_2O) was used instead of TMSBr as the acid promoter (entry 11). Therefore, we establish the reaction conditions as optimum: 0.2 mmol of 2-propynol phenyl azides, 3.5 equiv of TMSBr in CH_3NO_2 were stirred at 60 °C.

Table 1. Optimization of the reaction for the synthesis of 2a [a].

Entry	Solvent	TMSBr (x Equiv)	T [°C]	Yield [%]
1	DCE	2.5	60	45
2	MeCN	2.5	60	39
3	CH_2Cl_2	2.5	40	15
4	$MeNO_2$	2.5	60	73
5	HOAc	2.5	60	36
6	$MeNO_2$	3.5	60	81
7	$MeNO_2$	3.0	60	78
8	$MeNO_2$	2.0	60	67
9	$MeNO_2$	3.5	80	82
10	$MeNO_2$	3.5	rt	69
11 [b]	$MeNO_2$	3.5	60	75

[a] Unless otherwise noted, all reactions were performed with 0.2 mmol of **1a** in solvent (2.0 mL) for 1.0 h. [b] hydrobromic acid instead of TMSBr was used.

Then, we investigated the generality of the reaction with diverse substituted propynols **1** using TMSBr as acid-promoter and nucleophile, and the results are presented in Figure 1. Various substituents R^1 and R^2 on the aryl ring were well-tolerated under the optimal conditions, efficiently generating the corresponding products 4-bromo quinolines in favorable yields (up to 91% yield). Firstly, we investigated the influence of substituent electronic effects on this reaction, and the results indicated that substrates containing electron-donor groups (OMe, Me) gave better transformation than those containing electron-poor groups (F, Cl, Br). This might due to the fact that the reaction involved the carbocation intermediate (Intermediate **B**, see Scheme 4); and the electron-rich groups were good for the stabilization of carbocation intermediate. The corresponding products 4-bromo quinolines give the better yields compared to the synthesis of 4-chrolo quinolines bearing the electron-withdrawing groups. Substrates bearing *ortho*-position substituent provided slightly lower yields (**2j–2k**), indicating that the steric effect showed clear influence on this reaction. Importantly, the functionalities of halogen atoms such as fluorine, chlorine, and bromine were also tolerated for this transformation producing the target products. Such halogenated products could be converted into a variety of functionalized quinolines through cross-coupling reactions. Substrates containing two or three substituents attached to the benzene ring smoothly, and the target compounds were generated in good to excellent yields. Notably, the substrates with naphthyl or styryl group (**1m** and **1o**) were also compatible to generate the target products in good yields (**2k–2m**). Then we examined the effect of a substituent (R^2) on another aromatic ring on this transformation. Both electron-rich and electron-poor substituents were performed smoothly to produce the target compounds in 76–89% yields (**2m–2s**). It was noteworthy that the strong electron-deficient groups (CN and CF_3) in R^2 also proceeded well in this reaction and provided the target products in good yields. Unfortunately, no target product **2t** was generated when alkyl-substituted substrate **1t** was performed under the optimal conditions. Having successfully accomplished the direct formation of 4-bromo-quinolines, this cascade reaction was further extended to the construction of 4-iodo quinolines by using 2-propynol phenyl azides as starting materials with TMSI in CH_3NO_2 at 60 °C for 1.0 h under these circumstances. Some selected substrates (**1a, 1b, 1n**) were tolerated smoothly to the corresponding 4-iodo quinolines in moderate yields.

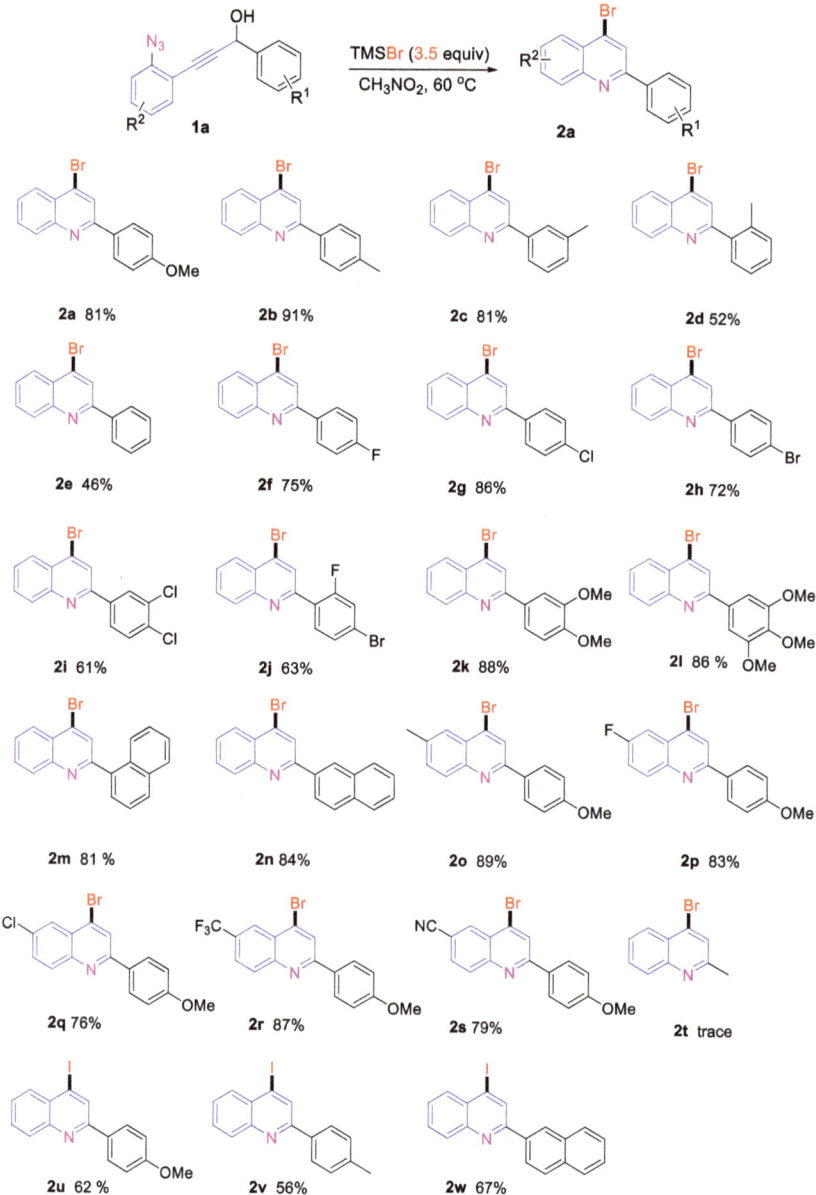

Figure 1. Transformation of *ortho*-propynol phenyl azides **1** to 4-bromo quinolines **2** [a]. [a] Unless otherwise noted, all reactions were performed with **1** (0.2 mmol) in CH$_3$NO$_2$ (2.0 mL) at 60 °C for 1 h. Isolated yield.

Furthermore, the synthetic utility of this TMSBr-promoted reaction of *ortho*-propynol azides was demonstrated by a gram-scale synthesis (Scheme 2-1). The yield of product **2a** was not obvious affected when a gram-scale (5 mmol, 1.40g) experiment of **1a** was performed under similar reaction conditions. Importantly, a bromine atom at the 4-position of obtained product quinolines moiety is useful and easily substituted by various functional hydrocarbon and heteroatomic groups, which persuades us to exploit synthetic transformation of 4-bromo quinolones [46–48]. As representative examples, the Suzuki coupling reaction of **2a** with arylboronic acids to 4-aryl quinolines **3a–3d** in good yield was achieved (Scheme 2-2) [46]. Notably, the corresponding product 4-vinyl quinoline **3e** was also generated when the reaction of **2a** with E-phenylethenylboronic acid. Furthermore, the Sonogashira coupling of **2a** with arylacetylene could smoothly proceed to produce the target products **4a–4b** in good yields (Scheme 2-3) [47]. More importantly, the classical reduction reaction of **2a** to the corresponding quinoline **5** was also investigated (Scheme 2-4). These results clearly demonstrate the usefulness of our obtained product 4-bromo quinolines as synthetic intermediates.

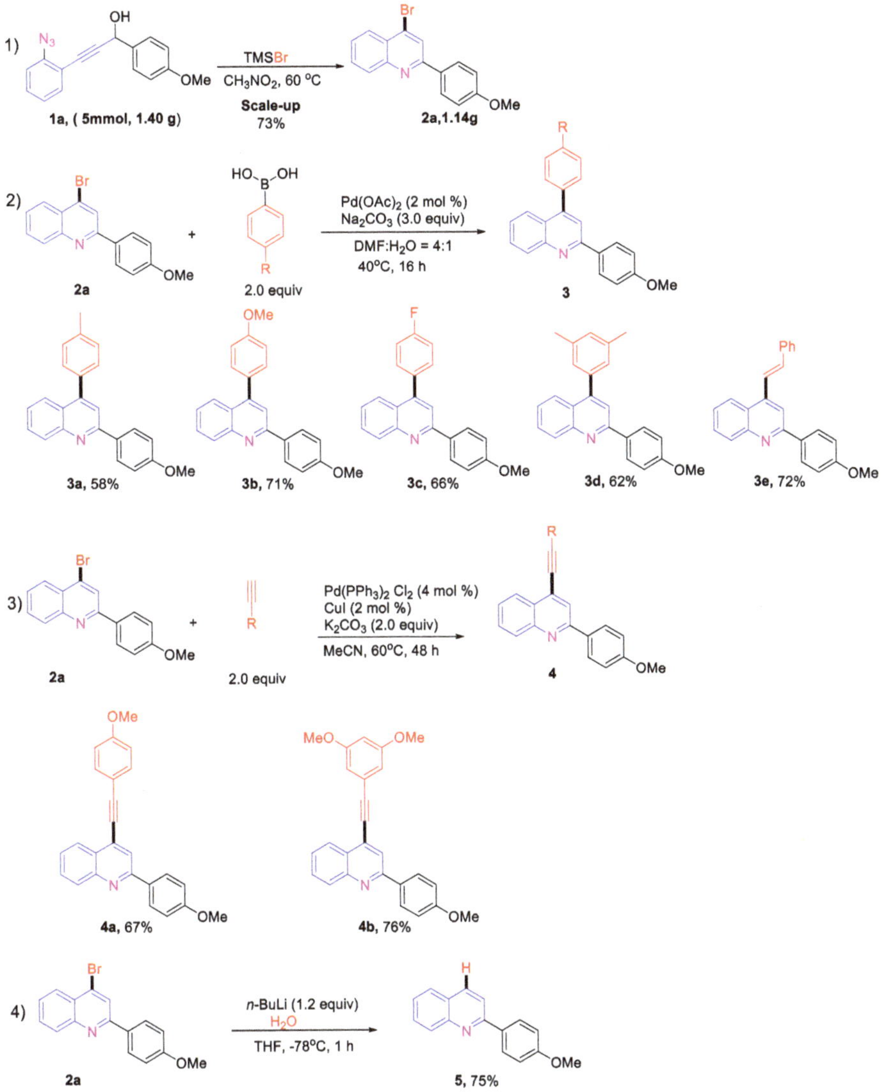

Scheme 2. Functionality elaboration of 4-bromo-quinolines.

As we all known, 4-aryloxy quinolines are significant structure frameworks which are existed widely in various bioactive molecules and natural products [49–52]. In this context, the synthesis of

4-aryloxy quinolines from 4-bromo quinolines is attractive because of the clean conversion and the mild reaction conditions. Therefore, the scope of the reactions was also investigated by varying the phenols. Some representative substituted 4-aryllkoxy quinolines **6a–6d** were generated in acceptable yields by choosing the appropriate nucleophilic reagents (Scheme 3).

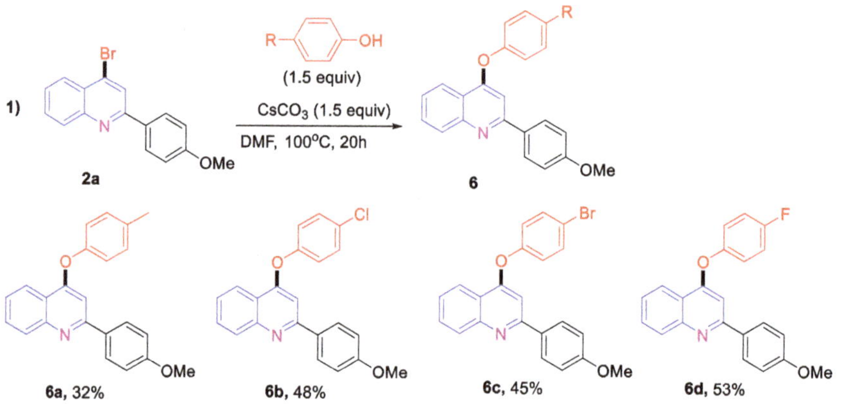

Scheme 3. Transformation of 4-bromo quinoline **2a** to 4-aryloxy quinolines **6**.

On the basis of the above experimental results and literature reports [45,53,54], we propose a plausible reaction mechanism for this reaction (Scheme 4). Firstly, a proargylic carbocation intermediate **A** was formed through the TMSBr-promoted the dehydration of propargylic alcohols **1**. Intermediate **A** could easily undergo tautomerization to generate allenic carbocation intermediate **B**, which could be attracted by nucleophile halide anion (Br⁻) to produce intermediate **C**. Subsequently, the 6-endo-trig cyclization of intermediate **C** in the presence of proton forms intermediate **D**. Finally, the target product **2** was generated through the aromatization of the intermediate **D** with the generation of a nitrogen gas and a proton.

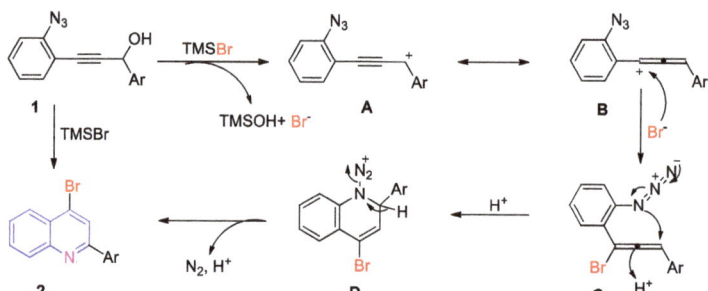

Scheme 4. Proposed reaction mechanism.

3. Materials and Methods

3.1. General Remarks

^{1}H-NMR spectra were recorded on 400 MHz in CDCl$_3$ and ^{13}C-NMR spectra were recorded on 100 MHz in CDCl$_3$. Chemical shifts (ppm) were recorded with tetramethylsilane (TMS) as the internal reference standard. Multiplicities are given as: s (singlet), d (doublet), t (triplet), dd (doublet of doublets), q (quartet), or m (multiplet). High-resolution mass spectrometry (HRMS) was performed on a TOF/Q–TOF mass spectrometer. Copies of the ^{1}H-NMR and ^{13}C-NMR spectra are provided in the Supporting Information. Commercially available reagents were used without further purification. All solvents were dried under standard method.

3.2. General Procedure for the Construction of 4-Bromo Quinolines 2

To a seal tube was added *ortho*-propynol phenyl azides (**1**) (0.2 mmol), TMSBr (0.7 mmol), in CH_3NO_2 at 60 °C. After 1.0 h, as monitored by TLC, the reaction mixture was concentrated in vacuum and purified by column chromatography to generate 4-bromo quinolines **2**.

4-Bromo-2-(4-Methoxyphenyl)quinoline (**2a**)

The title compound was prepared according to the 0.5, 130.8, 134.5, 148.7, 156.7, 161.1. general procedure and purified by column chromatography (silica gel, petroleum ether/ethyl acetate) to give a product **2a** (81%) [45]. ^{1}H-NMR (400 MHz, $CDCl_3$): δ 3. 79 (s, 3 H), 6.95 (dd, *J* = 2.0, 6.8 Hz, 2 H), 7.47–7.51 (m, 1 H), 7.63–7.67 (m, 1 H), 8.01–8.07 (m, 5 H). ^{13}C-NMR (100 MHz, $CDCl_3$): δ 55.4, 122.4, 126.3, 126.5, 127.0, 128.9, 129.8, 13.

4-Bromo-2-(p-tolyl)quinoline (**2b**)

The title compound was prepared according to the general procedure and purified by column chromatography (silica gel, petroleum ether/ethyl acetate) to give a product **2b** (91%). ^{1}H-NMR (400 MHz, $CDCl_3$): δ 2.34 (s, 3 H), 7.23 (d, *J* = 8.4 Hz, 2 H), 7.48–7.52 (m, 1 H), 7.64–7.68 (m, 1 H), 7.95 (d, *J* = 8.0 Hz, 2 H), 8.04–8.08 (m, 3 H). ^{13}C-NMR (100 MHz, $CDCl_3$): δ 21.3, 122.7, 126.5, 127.2, 127.4, 129.6, 130.0, 130.4, 134.5, 135.5, 139.9, 148.7, 157.1. HRMS (ESI, *m/z*): calcd for $C_{16}H_{12}BrN$: M + H = 298.0226; found: 298.0229.

4-Bromo-2-(m-tolyl)quinoline (**2c**)

The title compound was prepared according to the general procedure and purified by column chromatography (silica gel, petroleum ether/ethyl acetate) to give a product **2c** (81%). ^{1}H-NMR (400 MHz, $CDCl_3$): δ 2.39 (s, 3 H), 7.21 (d, *J* = 7.2 Hz, 1 H), 7.33 (t, *J* = 7.6 Hz, 1 H), 7.52 (t, *J* = 7.2 Hz, 1 H), 7.68 (t, *J* = 7.6 Hz, 1 H), 7.81 (d, *J* = 8.0 Hz, 1 H), 7.89 (s, 1 H), 8.07–8.10 (m, 3 H). ^{13}C-NMR (100 MHz, $CDCl_3$): δ 21.5, 123.0, 124.6, 126.5, 126.6, 127.4, 128.2, 128.8, 130.0, 130.5, 130.6, 134.5, 138.3, 138.6, 148.7, 157.4. HRMS (ESI, *m/z*): calcd for $C_{16}H_{12}BrN$: M + H = 298.0226; found: 298.0229.

4-Bromo-2-(o-tolyl)quinoline (**2d**)

The title compound was prepared according to the general procedure and purified by column chromatography (silica gel, petroleum ether/ethyl acetate) to give a product **2d** (52%). ^{1}H-NMR (400 MHz, $CDCl_3$): δ 2.35 (s, 3 H), 7.20–7.29 (m, 3 H), 7.41 (d, *J* = 6.8 Hz, 1 H), 7.58 (t, *J* = 7.6 Hz, 1 H), 7.70 (t, *J* = 7.6 Hz, 1 H), 7.78 (s, 1 H), 8.06–8.16 (m, 2 H). ^{13}C-NMR (100 MHz, $CDCl_3$): δ 20.3, 126.1, 126.1, 126.3, 126.6, 127.6, 128.9, 129.6, 130.0, 130.5, 131.1, 133.9, 136.1, 139.4, 148.4, 160.0. HRMS (ESI, *m/z*): calcd for $C_{16}H_{12}BrN$: M + H = 298.0226; found: 298.0227.

4-Bromo-2-phenylquinoline (**2e**)

The title compound was prepared according to the general procedure and purified by column chromatography (silica gel, petroleum ether/ethyl acetate) to give a product **2e** (46%). ^{1}H-NMR (400 MHz, $CDCl_3$): δ 7.40–7.48 (m, 3 H), 7.54–7.56 (m, 1 H), 7.67–7.71 (m, 1 H), 8.05–8.11 (m, 5 H). ^{13}C-NMR (100 MHz, $CDCl_3$): δ 122.9, 126.5, 126.7, 127.5, 127.5, 128.9, 129.8, 130.1, 130.5, 134.6, 138.4, 148.8, 157.2. HRMS (ESI, *m/z*): calcd for $C_{15}H_{10}BrN$: M + H = 284.0069; found: 284.0071.

4-Bromo-2-(4-fluorophenyl)quinoline (**2f**)

The title compound was prepared according to the general procedure and purified by column chromatography (silica gel, petroleum ether/ethyl acetate) to give a product **2f** (75%). ^{1}H-NMR (400 MHz, $CDCl_3$): δ 7.11–7.18 (m, 2 H), 7.52–7.56 (m, 1 H), 7.67–7.71 (m, 1 H), 8.05–8.11 (m, 5 H).

^{13}C-NMR (100 MHz, CDCl$_3$): δ 115.8, 116.0, 122.5, 126.6, 127.5, 129.4, 129.5, 130.0, 130.7, 134.5, 134.8, 148.7, 156.0, 162.8, 165.3. HRMS (ESI, *m/z*): calcd for C$_{15}$H$_9$BrFN: M + H = 301.9975; found: 301.9973.

4-Bromo-2-(4-chlorophenyl)quinoline (**2g**)

The title compound was prepared according to the general procedure and purified by column chromatography (silica gel, petroleum ether/ethyl acetate) to give a product **2g** (86%). ^{1}H-NMR (400 MHz, CDCl$_3$): δ 7.41 (d, *J* = 8.4 Hz, 2 H), 7.52–7.56 (m, 1 H), 7.67–7.71 (m, 1 H), 7.99–8.10 (m, 5 H). ^{13}C-NMR (100 MHz, CDCl$_3$): δ 122.5, 126.6, 126.7, 127.7, 128.7, 129.1, 130.0, 130.7, 134.8, 136.0, 136.7, 148.7, 155.8. HRMS (ESI, *m/z*): calcd for C$_{15}$H$_9$BrClN: M + H = 317.9680; found: 317.9682.

4-Bromo-2-(4-bromophenyl)quinoline (**2h**)

The title compound was prepared according to the general procedure and purified by column chromatography (silica gel, petroleum ether/ethyl acetate) to give a product **2h** (72%). ^{1}H-NMR (400 MHz, CDCl$_3$): δ 7.54–7.57 (m, 3 H), 7.68 (t, *J* = 8.4 Hz, 1 H), 7.94 (d, *J* = 8.4 Hz, 2 H), 8.04–8.10 (m, 3 H). ^{13}C-NMR (100 MHz, CDCl$_3$): δ 122.4, 124.5, 126.6, 126.7, 127.7, 129.0, 130.1, 130.7, 132.0, 134.8, 137.2, 148.7, 155.9. HRMS (ESI, *m/z*): calcd for C$_{15}$H$_9$Br$_2$N: M + H = 361.9175; found: 361.9179.

4-Bromo-2-(3,4-dichlorophenyl)quinoline (**2i**)

The title compound was prepared according to the general procedure and purified by column chromatography (silica gel, petroleum ether/ethyl acetate) to give a product **2i** (61%). ^{1}H-NMR (400 MHz, CDCl$_3$): δ 7.50 (d, *J* = 8.4 Hz, 1 H), 7.54–7.58 (m, 1 H), 7.69–7.73 (m, 1 H), 7.87–7.90 (m, 1 H), 8.02–8.11 (m, 3 H), 8.21 (d, *J* = 2.0 Hz, 1 H). ^{13}C-NMR (100 MHz, CDCl$_3$): δ 122.2, 126.4, 126.6, 126.8, 128.0, 129.3, 130.1, 130.8, 130.9, 133.3, 134.1, 135.0, 138.1, 148.6, 154.4. HRMS (ESI, *m/z*): calcd for C$_{15}$H$_8$BrCl$_2$N: M + H = 351.9290; found: 351.9291.

4-Bromo-2-(4-bromo-2-fluorophenyl)quinoline (**2j**)

The title compound was prepared according to the general procedure and purified by column chromatography (silica gel, petroleum ether/ethyl acetate) to give a product **2j** (63%). ^{1}H-NMR (400 MHz, CDCl$_3$): δ 7.34 (dd, *J* = 1.6, 6.8 Hz, 1 H), 7.40 (dd, *J* = 1.6, 8.4 Hz, 1 H) 7.59 (t, *J* = 7.2 Hz, 1 H), 7.72 (t, *J* = 7.2 Hz, 1 H), 7.97 (t, *J* = 7.6 Hz, 1 H), 8.07–8.14 (m, 3 H). ^{13}C-NMR (100 MHz, CDCl$_3$): δ 119.8, 120.0, 124.1, 124.2, 125.8, 125.9, 126.7, 126.9, 128.1, 128.2, 128.2, 130.0, 130.7, 132.5, 132.6, 134.4, 148.6, 152.6, 159.1, 161.8. HRMS (ESI, *m/z*): calcd for C$_{15}$H$_8$Br$_2$FN: M + H = 379.9080; found: 379.9084.

4-Bromo-2-(3,4-dimethoxyphenyl)quinoline (**2k**)

The title compound was prepared according to the general procedure and purified by column chromatography (silica gel, petroleum ether/ethyl acetate) to give a product **2k** (88%). ^{1}H-NMR (400 MHz, CDCl$_3$): δ 3.87 (s, 3 H), 3.97 (s, 3 H), 6.89 (d, *J* = 8.4 Hz, 1 H), 7.48–7.56 (m, 2 H), 7.64–7.68 (m, 1 H), 7.75 (d, *J* = 2.0 Hz, 1 H), 8.05–8.08 (m, 3 H). ^{13}C-NMR (100 MHz, CDCl$_3$): δ 55.9, 56.0, 110.2, 111.0, 120.3, 122.5, 126.4, 126.5, 127.1, 129.8, 130.5, 131.0, 134.5, 148.6, 149.4, 150.7, 156.6. HRMS (ESI, *m/z*): calcd for C$_{17}$H$_{14}$BrNO$_2$: M + H = 344.0281; found: 344.0283.

4-Bromo-2-(3,4,5-trimethoxyphenyl)quinoline (**2l**)

The title compound was prepared according to the general procedure and purified by column chromatography (silica gel, petroleum ether/ethyl acetate) to give a product **2l** (86%). ^{1}H-NMR (400 MHz, CDCl$_3$): δ 3.85 (s, 3 H), 3.93 (s, 6 H), 7.29 (s, 2 H), 7.53 (t, *J* = 7.2 Hz, 1 H), 7.69 (t, *J* = 7.6 Hz, 1 H), 8.04 (s, 1 H), 8.07–8.10 (m, 2 H). ^{13}C-NMR (100 MHz, CDCl$_3$): δ 56.3, 60.9, 104.8, 122.7, 126.5, 127.5, 129.9, 130.6, 133.8, 134.6, 139.8, 148.5, 153.6, 156.7. HRMS (ESI, *m/z*): calcd for C$_{18}$H$_{16}$BrNO$_3$: M + H = 374.0386; found: 374.0382.

4-Bromo-2-(naphthalen-1-yl)quinoline (**2m**)

The title compound was prepared according to the general procedure and purified by column chromatography (silica gel, petroleum ether/ethyl acetate) to give a product **2m** (81%). ^{1}H-NMR (400 MHz, CDCl$_3$): δ 7.41–7.48 (m, 2 H), 7.51–7.55 (m, 1 H), 7.61–7.66 (m, 2 H), 7.56 (t, *J* = 7.2 Hz, 1 H), 7.86–7.91 (m, 2 H), 7.96 (s, 1 H), 8.05 (d, *J* = 8.0 Hz, 1 H), 8.19 (dd, *J* = 8.8 Hz, 2 H). ^{13}C-NMR (100 MHz, CDCl$_3$): δ 125.3, 126.1, 126.6, 126.7, 126.9, 127.0, 127.9, 128.0, 128.5, 129.6, 129.9, 130.8, 131.0, 133.9, 134.4, 137.1, 148.4, 159.1. HRMS (ESI, *m/z*): calcd for C$_{19}$H$_{12}$BrN: M + H = 334.0226; found: 334.0227.

4-Bromo-2-(naphthalen-2-yl)quinoline (**2n**)

The title compound was prepared according to the general procedure and purified by column chromatography (silica gel, petroleum ether/ethyl acetate) to give a product **2n** (84%). ^{1}H-NMR (400 MHz, CDCl$_3$): δ 7.42–7.46 (m, 2 H), 7.50–7.54 (m, 1 H), 7.66–7.70 (m, 1 H), 7.78–7.81 (m, 1 H), 7.88 (dd, *J* = 2.8, 5.6 Hz, 2 H), 8.08–8.11 (m, 2 H), 8.22–8.24 (m, 2 H), 8.47 (d, *J* = 0.8 Hz, 1 H). ^{13}C-NMR (100 MHz, CDCl$_3$): δ 123.0, 124.7, 126.4, 126.6, 126.7, 126.9, 127.3, 127.5, 127.7, 128.7, 128.8, 130.1, 130.6, 133.3, 134.0, 134.6, 135.6, 148.8, 156.9. HRMS (ESI, *m/z*): calcd for C$_{19}$H$_{12}$BrN: M + H = 334.0226; found: 334.0227.

4-Bromo-2-(4-methoxyphenyl)-6-methylquinoline (**2o**)

The title compound was prepared according to the general procedure and purified by column chromatography (silica gel, petroleum ether/ethyl acetate) to give a product **2o** (89%). ^{1}H-NMR (400 MHz, CDCl$_3$): δ 2.48 (s, 3 H), 3.79 (s, 3 H), 6.94 (d, *J* = 8.8 Hz, 2 H), 7.47 (dd, *J* = 1.6, 8.4 Hz, 1 H), 7.80 (s, 1 H), 7.91 (d, *J* = 8.8 Hz, 1 H), 7.98–8.00 (m, 3 H). ^{13}C-NMR (100 MHz, CDCl$_3$): δ 21.7, 55.3, 114.2, 122.4, 125.3, 126.2, 128.7, 129.6, 131.1, 132.6, 133.7, 137.2, 147.3, 155.8, 160.9. HRMS (ESI, *m/z*): calcd for C$_{17}$H$_{14}$BrNO: M + H = 328.0332; found: 328.0331.

4-Bromo-6-fluoro-2-(4-methoxyphenyl)quinoline (**2p**)

The title compound was prepared according to the general procedure and purified by column chromatography (silica gel, petroleum ether/ethyl acetate) to give a product **2p** (83%). ^{1}H-NMR (400 MHz, CDCl$_3$): δ 3.81 (s, 3 H), 6.95 (d, *J* = 8.4 Hz, 2 H), 7.39–7.44 (m, 1 H), 7.70 (dd, *J* = 2.8, 9.6 Hz, 1 H), 7.99–8.05 (m, 4 H). ^{13}C-NMR (100 MHz, CDCl$_3$): δ 55.4, 110.2, 110.4, 114.3, 120.5, 120.7, 123.0, 127.2, 127.3, 128.8, 130.6, 132.4, 132.5, 133.3, 133.4, 145.8, 156.1, 156.2, 159.7, 161.2, 162.2. HRMS (ESI, *m/z*): calcd for C$_{16}$H$_{11}$BrFNO: M + H = 332.0081; found: 332.0081.

4-Bromo-6-chloro-2-(4-methoxyphenyl)quinoline (**2q**)

The title compound was prepared according to the general procedure and purified by column chromatography (silica gel, petroleum ether/ethyl acetate) to give a product **2q** (76%). ^{1}H-NMR (400 MHz, CDCl$_3$): δ 3.80 (s, 3 H), 6.95 (d, *J* = 8.8 Hz, 2 H), 7.57 (dd, *J* = 2.4, 8.8 Hz, 1 H), 7.94–8.05 (m, 5 H). ^{13}C-NMR (100 MHz, CDCl$_3$): δ 55.4, 114.3, 123.1, 125.5, 127.0, 128.8, 130.4, 131.4, 131.5, 133.0, 133.1, 147.1, 156.9, 161.3. HRMS (ESI, *m/z*): calcd for C$_{16}$H$_{11}$BrClNO: M + H = 347.9785; found: 347.9787.

4-Bromo-2-(4-methoxyphenyl)-6-(trifluoromethyl)quinoline (**2r**)

The title compound was prepared according to the general procedure and purified by column chromatography (silica gel, petroleum ether/ethyl acetate) to give a product **2r** (87%). ^{1}H-NMR (400 MHz, CDCl$_3$): δ 3.80 (s, 3 H), 6.94 (d, *J* = 8.8 Hz, 2 H), 7.79 (d, *J* = 8.8 Hz, 1 H), 8.03 (d, *J* = 8.4 Hz, 2 H), 8.10 (d, *J* = 8.4 Hz, 2 H), 8.34 (s, 1 H). ^{13}C-NMR (100 MHz, CDCl$_3$): δ 55.4, 114.4, 123.4, 124.6, 124.7, 125.5, 126.1, 126.1, 129.1, 130.0, 131.0, 135.0, 149.7, 158.6, 161.7. HRMS (ESI, *m/z*): calcd for C$_{17}$H$_{11}$BrF$_3$NO: M + H = 382.0049; found: 382.0045.

4-Bromo-2-(4-methoxyphenyl)quinoline-6-carbonitrile (**2s**)

The title compound was prepared according to the general procedure and purified by column chromatography (silica gel, petroleum ether/ethyl acetate) to give a product **2s** (79%). ^{1}H-NMR (400 MHz, CDCl$_3$): δ 3.83 (s, 3 H), 6.98 (d, *J* = 8.8 Hz, 2 H), 7.79 (dd, *J* = 1.6, 8.4 Hz, 1 H), 8.08 (dd,

J = 5.6, 8.8 Hz, 3 H), 8.16 (s, 1 H), 8.47 (d, J = 1.2 Hz, 1 H). ^{13}C-NMR (100 MHz, CDCl$_3$): δ 55.5, 110.4, 114.5, 118.5, 123.8, 126.0, 129.3, 129.7, 131.2, 131.3, 132.9, 134.5, 149.9, 159.4, 162.0. HRMS (ESI, m/z): calcd for C$_{17}$H$_{11}$BrN$_2$O: M + H = 339.0128; found: 339.0128.

4-Iodo-2-(4-methoxyphenyl)quinoline (**2u**)

The title compound was prepared according to the general procedure and purified by column chromatography (silica gel, petroleum ether/ethyl acetate) to give a product **2u** (62%) [45]. ^{1}H-NMR (400 MHz, CDCl$_3$): δ 3.89 (s, 3 H), 7.03–7.05 (m, 2 H), 7.54–7.58 (m, 1 H), 7.70–7.74 (m, 1 H), 7.98 (d, J = 8.4 Hz, 1 H), 8.05 (d, J = 8.4 Hz, 1 H), 8.09–8.12 (m, 2 H), 8.42 (s, 1 H). ^{13}C-NMR (100 MHz, CDCl$_3$): δ 66.4, 112.5, 114.3, 127.4, 128.9, 128.9, 130.1, 130.1, 130.5, 130.6, 131.4, 147.8, 156.7, 161.1.

4-Iodo-2-(*p*-tolyl)quinoline (**2v**)

The title compound was prepared according to the general procedure and purified by column chromatography (silica gel, petroleum ether/ethyl acetate) to give a product **2v** (56%). ^{1}H-NMR (400 MHz, CDCl$_3$): 2.44 (s, 3 H), 7.33 (d, J = 8.0 Hz, 2 H), 7.60 (t, J = 7.2 Hz, 1 H), 7.72–7.76 (m, 1 H), 7.99–8.08 (m, 4 H), 8.45 (s, 1 H). ^{13}C-NMR (100 MHz, CDCl$_3$): δ 21.3, 112.5, 127.4, 127.6, 129.0, 129.7, 130.2, 130.4, 130.5, 131.4, 135.2, 139.9, 147.8, 157.0. HRMS (ESI, m/z): calcd for C$_{16}$H$_{12}$IN: M + H = 346.0087; found: 346.0092.

4-Iodo-2-(naphthalen-2-yl)quinoline (**2w**)

The title compound was prepared according to the general procedure and purified by column chromatography (silica gel, petroleum ether/ethyl acetate) to give a product **2w** (67%). ^{1}H-NMR (400 MHz, CDCl$_3$): δ 7.54–7.57 (m, 2 H), 7.60–7.64 (m, 1 H), 7.75–7.80 (m, 1 H), 7.90–7.92 (m, 1 H), 7.99–8.05 (m, 3 H), 8.14 (d, J = 8.4 Hz, 1 H), 8.32–8.35 (m, 1 H), 8.59 (s, 1 H), 8.63 (s, 1 H). ^{13}C-NMR (100 MHz, CDCl$_3$): δ 112.6, 124.8, 126.5, 126.9, 127.3, 127.7, 127.9, 128.7, 128.8, 129.2, 130.3, 130.6, 130.7, 131.5, 133.4, 134.0, 135.3, 147.9, 156.9. HRMS (ESI, m/z): calcd for C$_{19}$H$_{12}$IN: M + H = 382.0087; found: 382.0089.

2-(4-Methoxyphenyl)-4-(p-tolyl)quinoline (**3a**)

The title compound was purified by column chromatography (silica gel, petroleum ether/ethyl acetate) to give a product **3a** (58%) [55]. ^{1}H-NMR (400 MHz, CDCl$_3$): δ 2.38 (s, 3 H), 3.78 (s, 3 H), 6.95 (d, J = 8.4 Hz, 2 H), 7.26 (d, J = 8.0 Hz, 2 H), 7.32–7.37 (m, 3 H), 7.59–7.63 (m, 1 H), 7.66 (s, 1 H), 7.81 (d, J = 8.4 Hz, 1 H), 8.05–8.12 (m, 3 H). ^{13}C-NMR (100 MHz, CDCl$_3$): δ 21.3, 55.3, 114.2, 118.8, 125.6, 125.7, 125.8, 128.9, 129.2, 129.3, 129.4, 129.8, 132.3, 135.6, 138.2, 148.8, 149.0, 156.4, 160.8.

2,4-bis(4-Methoxyphenyl)quinoline (**3b**)

The title compound was purified by column chromatography (silica gel, petroleum ether/ethyl acetate) to give a product **3b** (71%) [55]. ^{1}H-NMR (400 MHz, CDCl$_3$): δ 3.79 (s, 3 H), 3.81 (s, 3 H), 6.97 (dd, J = 8.8, 13.2 Hz, 4 H), 7.33–7.37 (m, 1 H), 7.41 (d, J = 8.8 Hz, 2 H), 7.59–7.63 (m, 1 H), 7.66 (s, 1 H), 7.82 (d, J = 8.4 Hz, 1 H), 8.06–8.12 (m, 3 H). ^{13}C-NMR (100 MHz, CDCl$_3$): δ 55.3, 55.4, 114.0, 114.2, 118.8, 125.6, 125.7, 125.8, 128.9, 129.3, 129.8, 130.7, 132.3, 148.6, 148.9, 156.4, 159.8, 160.8.

4-(4-Fluorophenyl)-2-(4-methoxyphenyl)quinoline (**3c**)

The title compound was purified by column chromatography (silica gel, petroleum ether/ethyl acetate) to give a product **3c** (66%) [55]. ^{1}H-NMR (400 MHz, CDCl$_3$): δ 3.80 (s, 3 H), 6.96 (d, J = 8.8 Hz, 2 H), 7.13–7.18 (m, 2 H), 7.35–7.39 (m, 1 H), 7.43–7.46 (m, 2 H), 7.61–7.65 (m, 2 H), 7.73 (d, J = 8.4 Hz, 1 H), 8.06–8.13 (m, 3 H). ^{13}C-NMR (100 MHz, CDCl$_3$): δ 55.4, 114.2, 115.5, 115.7, 118.9, 125.3, 125.5, 126.0, 128.9, 128.5, 130.0, 131.2, 131.3, 132.0, 134.4, 134.5, 147.9, 148.8, 156.4, 160.9, 161.6, 164.1.

4-(3,5-Dimethylphenyl)-2-(4-methoxyphenyl)quinoline (**3d**)

The title compound was purified by column chromatography (silica gel, petroleum ether/ethyl acetate) to give a product **3d** (62%). ^{1}H-NMR (400 MHz, CDCl$_3$): δ 2.34 (s, 6 H), 3.79 (s, 3 H), 6.95 (d, J = 8.8 Hz, 2 H), 7.05–7.08 (m, 3 H), 7.33–7.37 (m, 1 H), 7.59–7.63 (m, 1 H), 7.67 (s, 1 H), 7.81 (d, J = 8.4 Hz, 1 H), 8.06–8.12 (m, 3 H). ^{13}C-NMR (100 MHz, CDCl$_3$): δ 21.3, 55.3, 114.2, 118.7, 125.6, 125.7, 125.8, 127.3, 128.9, 129.3, 129.8, 129.9, 132.2, 138.1, 138.4, 148.7, 149.3, 156.4, 160.8. HRMS (ESI, *m/z*): calcd for: C$_{24}$H$_{21}$NO: M + H = 340.1696; found: M + H = 340.1692.

(*E*)-2-(4-Methoxyphenyl)-4-styrylquinoline (**3e**)

The title compound was purified by column chromatography (silica gel, petroleum ether/ethyl acetate) to give a product **3e** (72%) [56]. ^{1}H-NMR (400 MHz, CDCl$_3$): δ 3.83 (s, 3 H), 6.99 (d, J = 8.4 Hz, 2 H), 7.27–7.39 (m, 4 H), 7.44–7.49 (m, 1 H), 7.58 (d, J = 7.2 Hz, 2 H), 7.63–7.67 (m, 1 H), 7.77 (d, J = 16.0 Hz, 1 H), 7.95 (s, 1 H), 8.08–8.12 (m, 4 H). ^{13}C-NMR (100 MHz, CDCl$_3$): δ 55.4, 114.2, 114.7, 123.3, 123.6, 125.2, 125.9, 127.1, 128.7, 128.8, 128.9, 129.4, 130.2, 132.4, 134.9, 136.7, 143.5, 148.8, 156.8, 160.8.

2-(4-Methoxyphenyl)-4-((4-methoxyphenyl)ethynyl)quinoline (**4a**)

The title compound was purified by column chromatography (silica gel, petroleum ether/ethyl acetate) to give a product **4a** (67%) [45]. ^{1}H-NMR (400 MHz, CDCl$_3$): δ 3.86 (s, 3 H), 3.89 (s, 3 H), 6.95 (d, J = 8.8 Hz, 2 H), 7.05 (d, J = 8.8 Hz, 2 H), 7.55–7.59 (m, 1 H), 7.63 (d, J = 8.8 Hz, 2 H), 7.71–7.75 (m, 1 H), 8.00 (s, 1 H), 8.13–8.16 (m, 3 H), 8.33 (d, J = 7.6 Hz, 1 H). ^{13}C-NMR (100 MHz, CDCl$_3$): δ 55.4, 84.5, 98.2, 114.2, 114.2, 114.4, 120.9, 125.7, 126.3, 126.3, 128.8, 129.8, 129.9, 130.5, 131.8, 133.5, 148.2, 156.4, 160.4, 160.9.

4-((3,5-Dimethoxyphenyl)ethynyl)-2-(4-methoxyphenyl)quinoline (**4b**)

The title compound was purified by column chromatography (silica gel, petroleum ether/ethyl acetate) to give a product **4b** (76%). ^{1}H-NMR (400 MHz, CDCl$_3$): δ 3.78 (s, 6 H), 3.82 (s, 3 H), 6.48 (t, J = 2.0 Hz, 1 H), 6.76 (d, J = 2.4 Hz, 2 H), 6.98 (d, J = 8.8 Hz, 2 H), 7.51 (t, J = 7.6 Hz, 1 H), 7.67 (t, J = 8.0 Hz, 1 H), 7.96 (s, 1 H), 8.07–8.10 (m, 3 H), 8.25 (d, J = 8.0 Hz, 1 H). ^{13}C-NMR (100 MHz, CDCl$_3$): δ 55.5, 85.0, 97.8, 102.6, 109.7, 114.3, 121.3, 123.6, 125.6, 126.3, 126.5, 128.8, 129.9, 130.0, 130.0, 131.7, 148.2, 156.4, 160.7, 161.0. HRMS (ESI, *m/z*): calcd for: C$_{26}$H$_{21}$NO$_3$: M + H = 396.1594; found: 396.1596.

2-(4-Methoxyphenyl)quinoline (**5**)

The title compound was purified by column chromatography (silica gel, petroleum ether/ethyl acetate) to give a product **5** (75%) [57]. ^{1}H-NMR (400 MHz, CDCl$_3$): δ 3.79 (s, 3 H), 6.96 (d, J = 8.8 Hz, 2 H), 7.38–7.42 (m, 1 H), 7.59–7.64 (m, 1 H), 7.70–7.75 (m, 2 H), 8.04–8.09 (m, 4 H). ^{13}C-NMR (100 MHz, CDCl$_3$): δ 55.4, 114.2, 118.5, 125.8, 126.9, 127.4, 128.9, 129.5, 129.5, 132.2, 136.6, 148.3, 156.9, 160.8.

2-(4-Methoxyphenyl)-4-(p-tolyloxy)quinoline (**6a**)

The title compound was purified by column chromatography (silica gel, petroleum ether/ethyl acetate) to give a product **6a** (32%). ^{1}H-NMR (400 MHz, CDCl$_3$): δ 2.35 (s, 3 H), 3.77 (s, 3 H), 6.88–6.90 (m, 3 H), 7.04 (d, J = 8.4 Hz, 2 H), 7.18–7.21 (m, 2 H), 7.44 (t, J = 7.6 Hz, 1 H), 7.66 (t, J = 7.6 Hz, 1 H), 7.84 (d, J = 8.8 Hz, 2 H), 8.04 (d, J = 8.8 Hz, 1 H), 8.25 (d, J = 8.0 Hz, 1 H). ^{13}C-NMR (100 MHz, CDCl$_3$): δ 20.9, 55.4, 101.7, 114.0, 120.4, 120.7, 121.7, 125.4, 128.8, 129.1, 130.2, 130.7, 132.5, 135.1, 149.8, 152.3, 158.1, 160.7, 162.5. HRMS (ESI, *m/z*): calcd for: C$_{23}$H$_{19}$NO$_2$: M + H = 342.1489; found: 342.1495.

4-(4-Chlorophenoxy)-2-(4-methoxyphenyl)quinoline (**6b**)

The title compound was purified by column chromatography (silica gel, petroleum ether/ethyl acetate) to give a product **6b** (48%). ^{1}H-NMR (400 MHz, CDCl$_3$): δ 3.77 (s, 3 H), 6.89–6.92 (m, 3 H), 7.09 (d, J = 8.8 Hz, 2 H), 7.36 (d, J = 9.2 Hz, 2 H), 7.45 (t, J = 7.2 Hz, 1 H), 7.65–7.70 (m, 1 H), 7.86 (d, J = 8.8 Hz, 2 H), 8.05 (d, J = 8.4 Hz, 1 H), 8.19 (d, J = 7.6 Hz, 1 H). ^{13}C-NMR (100 MHz, CDCl$_3$): δ 55.5, 110.4, 114.5, 118.5, 123.8, 126.0, 129.3, 129.7, 131.2, 131.3, 132.9, 134.5, 149.9, 159.4, 162.0. HRMS (ESI, *m/z*): calcd for: C$_{22}$H$_{16}$ClNO$_2$: M + H = 362.0942; found: 362.0948.

4-(4-Bromophenoxy)-2-(4-methoxyphenyl)quinoline (6c)

The title compound was purified by column chromatography (silica gel, petroleum ether/ethyl acetate) to give a product **6c** (45%). ^{1}H-NMR (400 MHz, CDCl$_3$): δ 3.77 (s, 3 H), 6.89–6.93 (m, 3 H), 7.04 (d, J = 8.8 Hz, 2 H), 7.45 (t, J = 7.2 Hz, 1 H), 7.51 (d, J = 8.8 Hz, 2 H), 7.67 (t, J = 8.4 Hz, 1 H), 7.86 (d, J = 8.4 Hz, 2 H), 8.06 (d, J = 8.4 Hz, 1 H), 8.18 (d, J = 8.0Hz, 1 H). ^{13}C-NMR (100 MHz, CDCl$_3$): δ 55.4, 102.3, 114.1, 118.2, 120.2, 121.5, 122.5, 125.7, 128.8, 129.2, 130.4, 132.1, 133.3, 149.8, 153.9, 158.1, 160.9, 161.7. HRMS (ESI, m/z): calcd for: C$_{22}$H$_{16}$BrNO$_2$: M + H = 406.0437; found: 406.0431.

4-(4-Fluorophenoxy)-2-(4-methoxyphenyl)quinoline (6d)

The title compound was purified by column chromatography (silica gel, petroleum ether/ethyl acetate) to give a product **6d** (53%). ^{1}H-NMR (400 MHz, CDCl$_3$): δ 3.77 (s, 3 H), 6.89–6.93 (m, 3 H), 7.04 (d, J = 8.8 Hz, 2 H), 7.45 (t, J = 7.2 Hz, 1 H), 7.51 (d, J = 8.8 Hz, 2 H), 7.67 (t, J = 8.4 Hz, 1 H), 7.86 (d, J = 8.4 Hz, 2 H), 8.06 (d, J = 8.4 Hz, 1 H), 8.18 (d, J = 8.0Hz, 1 H). ^{13}C-NMR (100 MHz, CDCl$_3$): δ 55.4, 102.3, 114.1, 118.2, 120.2, 121.5, 122.5, 125.7, 128.8, 129.2, 130.4, 132.1, 133.3, 149.8, 153.9, 158.1, 160.9, 161.7. HRMS (ESI, m/z): calcd for: C$_{22}$H$_{16}$FNO$_2$: M + H = 346.1238; found: 346.1234.

4. Conclusions

In summary, we have developed an efficient and general approach for the synthesis of 4-bromo or 4-iodo quinolines through the TMSBr promoted the cascade cyclization of *ortho*-propynol phenyl azides. It is noteworthy that the obtained products 4-halo quinolines could be used as key intermediate for the construction of various bioactive molecules, natural products, and drugs. A variety of 4-halo quinolines were obtained in moderate to excellent yields under mild conditions. This process does not require the use of metal catalysts, additional oxidants; water and nitrogen gas are generated as the only side products.

Supplementary Materials: The following are available online at http://www.mdpi.com/1420-3049/24/21/3999/s1.

Author Contributions: Conceptualization, Q.X.; Data curation, J.B.; Formal analysis, F.J., T.Y., and X.-R.S.; Writing—original draft preparation, F.J.; Writing—review and editing, X.-R.S., R.Y., and H.D.

Funding: This work was funded by the National Science Foundation of China (21676131 and 21462019), the Science Foundation of Jiangxi Province (20181BAB203005 and 20143ACB20012), the Education Department of Jiangxi Province (GJJ180616), Jiangxi Science & Technology Normal University (2017QNBJRC004).

Conflicts of Interest: The authors declare no conflict of interest.

References

1. Joule, J.A.; Mills, K. *Heterocyclic Chemistry*, 4th ed.; Blackwell Science, Ltd.: Oxford, UK, 2000; p. 121.
2. Balasubraimanan, M.; Keay, J.G. *Comprehensive Heterocyclic Chemistry II*; Katritzky, A.R., Rees, C.W., Scriven, E.F.V., Eds.; Pergamon Press: Oxford, UK, 1996; Volume 5, p. 245.
3. Michael, J.P. Quinoline, quinazoline and acridone alkaloids. *Nat. Prod. Rep.* **2008**, *25*, 166–187. [CrossRef]
4. Boa, A.N.; Canavan, S.P.; Hirst, P.R.; Ramsey, C.; Stead, A.M.W.; McConkey, G.A. Synthesis of brequinar analogue inhibitors of malaria parasite dihydroorotate dehydrogenase. *Bioorg. Med. Chem.* **2005**, *13*, 1945–1967. [CrossRef]
5. Diao, Y.; Lu, W.; Jin, H.; Zhu, J.; Han, L.; Xu, M.; Gao, R.; Shen, X.; Zhao, Z.; Liu, X.; et al. Discovery of Diverse Human Dihydroorotate Dehydrogenase Inhibitors as Immunosuppressive Agents by Structure-Based Virtual Screening. *J. Med. Chem.* **2012**, *55*, 8341–8349. [CrossRef]
6. Munier-Lehmann, H.; Lucas-Hourani, M.; Guillou, S.; Helynck, O.; Zanghi, G.; Noel, A.; Tangy, F.; Vidalain, P.; Janin, Y.L. Original 2-(3-Alkoxy-1H-pyrazol-1-yl)pyrimidine Derivatives as Inhibitors of Human Dihydroorotate Dehydrogenase (DHODH). *J. Med. Chem.* **2015**, *58*, 860–877. [CrossRef]
7. Das, P.; Deng, X.; Zhang, L.; Roth, M.G.; Fontoura, B.M.A.; Phillips, M.A.; De Brabander, J.K. SAR-Based Optimization of a 4-Quinoline Carboxylic Acid Analogue with Potent Antiviral Activity. *ACS Med. Chem. Lett.* **2013**, *4*, 517–521. [CrossRef]

8. Strekowski, L.; Say, M.; Zegrocka, O.; Tanious, F.A.; Wilson, W.D.; Manzel, L.; Macfarlane, D.E. Bis-4-aminoquinolines: Novel triple-helix DNA intercalators and antagonists of immunostimulatory CpG-oligodeoxynucleotides. *Bioorg. Med. Chem.* **2003**, *11*, 1079–1085. [CrossRef]
9. Gogoi, S.; Zhao, C.-G. Organocatalyzed enantioselective synthesis of 6-amino-5-cyanodihydropyrano[2,3-c]pyrazoles. *Tetrahedron Lett.* **2009**, *50*, 2252–2255. [CrossRef]
10. Tan, B.; Shi, Z.; Chua, J.P.; Zhong, G. Control of Four Stereocenters in an Organocatalytic Domino Double Michael Reaction: Efficient Synthesis of Multisubstituted Cyclopentanes. *Org. Lett.* **2008**, *10*, 3425–3428. [CrossRef]
11. Wang, B.; Wu, F.; Wang, Y.; Liu, X.; Deng, L. Control of Diastereoselectivity in Tandem Asymmetric Reactions Generating Nonadjacent Stereocenters with Bifunctional Catalysis by Cinchona Alkaloids. *J. Am. Chem. Soc.* **2007**, *129*, 768–769. [CrossRef]
12. Biddle, M.M.; Lin, M.; Scheidt, K.A. Catalytic Enantioselective Synthesis of Flavanones and Chromanones. *J. Am. Chem. Soc.* **2007**, *129*, 3830–3831. [CrossRef]
13. Enders, D.; Grondal, C.; Hüttl, M.R.M. Asymmetric Organocatalytic Domino Reactions. *Angew. Chem. Int. Ed.* **2007**, *46*, 1570–1581. [CrossRef] [PubMed]
14. Bharate, J.B.; Vishwakarma, R.A.; Bharate, S.B. Metal-free domino one-pot protocols for quinoline synthesis. *RSC Adv.* **2015**, *5*, 42020–42053. [CrossRef]
15. Prajapati, S.M.; Patel, K.D.; Vekariya, R.H.; Panchal, S.N.; Patel, H.D. Recent advances in the synthesis of quinolines: A review. *RSC Adv.* **2014**, *4*, 24463–24476. [CrossRef]
16. Hassanin, H.M.; Ibrahim, M.A.; Gabr, Y.A.; Alnamer, Y.A. Synthesis and Chemical Reactivity of Pyrano[3,2-c]quinolinones. *J. Heterocycl. Chem.* **2012**, *49*, 1269–1289. [CrossRef]
17. Wang, Y.; Peng, C.; Liu, L.; Zhao, J.; Su, L.; Zhu, Q. Sulfuric acid promoted condensation cyclization of 2-(2-(trimethylsilyl) ethynyl)anilines with arylaldehydes in alcoholic solvents: An efficient one-pot synthesis of 4-alkoxy-2-arylquinolines. *Tetrahedron Lett.* **2009**, *50*, 2261–2265. [CrossRef]
18. Sridharan, V.; Avendano, C.; Menendez, J.C. CAN-catalyzed three-component reaction between anilines and alkyl vinyl ethers: Stereoselective synthesis of 2-methyl-1,2,3,4-tetrahydroquinolines and studies on their aromatization. *Tetrahedron* **2007**, *63*, 673–681. [CrossRef]
19. Gharpure, S.J.; Nanda, S.K.; Adate, P.A.; Shelke, Y.G. Lewis Acid Promoted Oxonium Ion Driven Carboamination of Alkynes for the Synthesis of 4-Alkoxy Quinolines. *J. Org. Chem.* **2017**, *82*, 2067–2080. [CrossRef]
20. Mao, X.-F.; Zhu, X.-P.; Li, D.-Y.; Liu, P.-N. Cu-Catalyzed Cascade Annulation of Alkynols with 2-Azidobenzaldehydes: Access to 6H-Isochromeno[4,3-c]quinoline. *J. Org. Chem.* **2017**, *82*, 7032–7039. [CrossRef]
21. Liu, Y.; Zhang, X.; Xi, C. MeOTf-induced annulation of arylisocyanates and arylalkynes leading to 4-methoxyl-2,3-diarylquinolines. *Tetrahedron Lett.* **2018**, *59*, 2440–2442. [CrossRef]
22. Yang, T.; Ding, H.; Li, R.; Jin, F.; Song, X.-R.; Chen, X.; Bai, J.; Xiao, Q.; Liang, Y.-M. para-TsOH-Promoted Cascade Reaction of ortho-Propynol Phenyl Azides for the Synthesis of 4-Methoxy Quinolines and Propargyl Methyl Ethers: Insight on Mechanism of Propargylic Alcohols. *Asian J. Org. Chem.* **2019**, *8*, 391–398. [CrossRef]
23. Abbiati, G.; Arcadi, A.; Marinelli, F.; Rossi, E. Domino [3 + 2] Cycloaddition/Annulation Reactions of β-(2-Aminophenyl)-α,β-ynones with Nitrile Oxides: Synthesis of Isoxazolo[4,5-c]quinolines. *Eur. J. Org. Chem.* **2003**, *2003*, 1423–1427. [CrossRef]
24. Strekowski, L.; Zegrocka, O.; Windham, C.; Czarny, A. Practical Synthesis of 4-Chloro-2-(2-naphthyl)quinoline, a Precursor to Triple-Helix DNA Intercalators. *Org. Process Res. Dev.* **1997**, *1*, 384–386. [CrossRef]
25. Akila, S.; Selvi, S.; Balasubramanian, K. The Vilsmeier cyclization of 2′-azido and 2′-aminochalcones—A mild one pot synthesis of 2-aryl-4-chloroquinoline and its N-formyl-1,2-dihydro derivatives. *Tetrahedron* **2001**, *57*, 3465–3469. [CrossRef]
26. Shvartsberg, M.S.; Kolodina, E.A. Synthesis of 4-haloquinolines and their fused polycyclic derivatives. *Mendeleev Commun.* **2008**, *18*, 109–111. [CrossRef]
27. O'Neill, P.M.; Bray, P.G.; Hawley, S.R.; Ward, S.A.; Park, B.K. 4-Aminoquinolines—Past, present, and future; A chemical perspective. *Pharmacol. Ther.* **1998**, *77*, 29–58. [CrossRef]

28. Blauer, G.; Akkawi, M.; Fleischhacker, W.; Hiessboeck, R. Synthesis and optical properties of the chloroquine enantiomers and their complexes with ferriprotoporphyrin IX in aqueous solution. *Chirality* **1998**, *10*, 556–563. [CrossRef]
29. Egan, T.J.; Hunter, R.; Kaschula, C.H.; Marques, H.M.; Misplon, A.; Walden, J. Structure–Function Relationships in Aminoquinolines: Effect of Amino and Chloro Groups on Quinoline–Hematin Complex Formation, Inhibition of β-Hematin Formation, and Antiplasmodial Activity. *J. Med. Chem.* **2000**, *43*, 283–291. [CrossRef]
30. Muzart, J. Gold-catalysed reactions of alcohols: Isomerisation, inter- and intramolecular reactions leading to C–C and C–heteroatom bonds. *Tetrahedron* **2008**, *64*, 5815–5849. [CrossRef]
31. Kabalka, G.W.; Yao, M.-L. Direct Propargylic Substitution of Hydroxyl Group in Propargylic Alcohols. *Curr. Org. Synth.* **2008**, *5*, 28–32. [CrossRef]
32. Zhang, L.; Fang, G.; Kumar, R.K.; Bi, X. Coinage-Metal-Catalyzed Reactions of Propargylic Alcohols. *Synthesis* **2015**, *47*, 2317–2346.
33. Zhu, Y.; Sun, L.; Lu, P.; Wang, Y. Recent Advances on the Lewis Acid-Catalyzed Cascade Rearrangements of Propargylic Alcohols and Their Derivatives. *ACS Catal.* **2014**, *4*, 1911–1925. [CrossRef]
34. Song, X.-R.; Qiu, Y.-F.; Liu, X.-Y.; Liang, Y.-M. Recent advances in the tandem reaction of azides with alkynes or alkynols. *Org. Biomol. Chem.* **2016**, *14*, 11317–11331. [CrossRef] [PubMed]
35. Han, Y.-P.; Li, X.-S.; Zhu, X.-Y.; Li, M.; Zhou, L.; Song, X.-R.; Liang, Y.-M. Lewis Acid Catalyzed Dehydrogenative Coupling of Tertiary Propargylic Alcohols with Quinoline N-Oxides. *J. Org. Chem.* **2017**, *82*, 1697–1704. [CrossRef] [PubMed]
36. Han, Y.-P.; Song, X.-R.; Qiu, Y.-F.; Li, X.-S.; Zhang, H.-R.; Zhu, X.-Y.; Liu, X.-Y.; Liang, Y.-M. Lewis Acid Catalyzed Cyclization of Propargylic Alcohols with 2-Vinylphenol. *Org. Lett.* **2016**, *18*, 3866–3869. [CrossRef] [PubMed]
37. Han, Y.-P.; Song, X.-R.; Qiu, Y.-F.; Zhang, H.-R.; Li, L.-H.; Jin, D.-P.; Sun, X.-Q.; Liu, X.-Y.; Liang, Y.-M. Lewis Acid Catalyzed [4 + 3] Cycloaddition of Propargylic Alcohols with Azides. *Org. Lett.* **2016**, *18*, 940–943. [CrossRef]
38. Qiu, Y.-F.; Song, X.-R.; Li, M.; Zhu, X.-Y.; Wang, A.-Q.; Yang, F.; Han, Y.-P.; Zhang, H.-R.; Jin, D.-P.; Li, Y.-X.; et al. BF$_3$·OEt$_2$-AgSCF$_3$ Mediated Trifluoromethylthiolation/Cascade Cyclization of Propynols: Synthesis of 4-((Trifluoromethyl)thio)-2H-chromene and 4-((Trifluoromethyl)thio)-1,2-dihydroquinoline Derivatives. *Org. Lett.* **2016**, *18*, 1514–1517. [CrossRef]
39. Li, R.; Jin, F.; Song, X.-R.; Yang, T.; Ding, H.; Yang, R.; Xiao, Q.; Liang, Y.-M. Acid-promoted cyclization of 2-propynolphenols leading to 4-tosyloxy-2H-chromenes. *Tetrahedron Lett.* **2019**, *60*, 331–334. [CrossRef]
40. Yang, T.; Song, X.-R.; Li, R.; Jin, F.; Zhang, Y.; Bai, J.; Yang, R.; Ding, H.; Xiao, Q. Metal-free and efficient approach to 4-thiocyanated 2H-chromenes via TFA-mediated cascade cyclization of 2-propynolphenols. *Tetrahedron Lett.* **2019**, *60*, 1248–1253. [CrossRef]
41. Li, R.; Song, X.-R.; Chen, X.; Ding, H.; Xiao, Q.; Liang, Y.-M. Copper-Catalyzed Cascade Cyclization of 2-Propynolphenols: Access to 4-Phosphorylated 2H-Chromenes. *Adv. Synth. Catal.* **2017**, *359*, 3962–3967. [CrossRef]
42. Song, X.-R.; Li, R.; Yang, T.; Chen, X.; Ding, H.; Xiao, Q.; Liang, Y.-M. Novel and Efficient Access to Flavones under Mild Conditions: Aqueous HI-Mediated Cascade Cyclization/Oxidative Radical Reaction of 2-Propynolphenols. *Eur. J. Org. Chem.* **2018**, *2018*, 5548–5552. [CrossRef]
43. Yang, T.; Kou, P.; Jin, F.; Song, X.-R.; Bai, J.; Ding, H.; Xiao, Q.; Liang, Y.-M. TFA-Promoted Sulfonation/Cascade Cyclization of 2-Propynolphenols with Sodium Sulfinates to 4-Sulfonyl 2H-Chromenes under Metal-free Conditions. *Org. Chem. Front.* **2019**, *6*, 3162–3166. [CrossRef]
44. Song, X.-R.; Li, R.; Ding, H.; Yang, R.; Xiao, Q.; Liang, Y.-M. Highly efficient access to 4-chloro-2H-chromenes and 1,2-dihydroquinolines under mild conditions: TMSCl-mediated cyclization of 2-propynolphenols/anilines. *Tetrahedron Lett.* **2016**, *57*, 4519–4524. [CrossRef]
45. Song, X.-R.; Li, R.; Ding, H.; Chen, X.; Yang, T.; Jiang, B.; Xiao, Q.; Liang, Y.-M. An efficient approach to 4-chloro quinolines via TMSCl-mediated cascade cyclization of ortho-propynol phenyl azides. *Org. Chem. Front.* **2018**, *5*, 1537–1541. [CrossRef]
46. Wada, T.; Iwasaki, M.; Kondoh, A.; Yorimitsu, H.; Hideki Yorimitsu Oshima, K. Palladium-Catalyzed Addition of Silyl-Substituted Chloroalkynes to Terminal Alkynes. *Chem. Eur. J.* **2010**, *16*, 10671–10674. [CrossRef]

47. Gopinath, V.S.; Pinjari, J.; Dere, R.T.; Verma, A.; Vishwakarma, P.; Shivahare, R.; Moger, M.; Goud, P.S.K.; Ramanathan, V.; Bose, P.; et al. Design, synthesis and biological evaluation of 2-substituted quinolines as potential antileishmanial agents. *Eur. J. Med. Chem.* **2013**, *69*, 52. [CrossRef] [PubMed]
48. Gopinath, V.S.; Rao, M.; Shivahare, R.; Vishwakarma, P.; Ghose, S.; Pradhan, A.; Hindupur, R.; Sarma, K.D.; Gupta, S.; Puri, S.K.; et al. Design, synthesis, ADME characterization and antileishmanial evaluation of novel substituted quinoline analogs. *Bioorg. Med. Chem. Lett.* **2014**, *24*, 2046–2052. [CrossRef]
49. Michael, J.P. Quinoline, quinazoline and acridone alkaloids. *Nat. Prod. Rep.* **2001**, *18*, 543–559. [CrossRef]
50. Michael, J.P. Quinoline, quinazoline and acridone alkaloids. *Nat. Prod. Rep.* **2003**, *20*, 476–493. [CrossRef]
51. Musiol, R.; Serda, M.; Bielowka, S.H.; Polanski, J. Quinoline-Based Antifungals. *Curr. Med. Chem.* **2010**, *17*, 1960–1973. [CrossRef]
52. Solomon, V.R.; Lee, H. Quinoline as a Privileged Scaffold in Cancer Drug Discovery. *Curr. Med. Chem.* **2011**, *18*, 1488–1508. [CrossRef]
53. Zhang, H.; Tanimoto, H.; Morimoto, T.; Nishiyama, Y.; Kakiuchi, K. Regioselective Rapid Synthesis of Fully Substituted 1,2,3-Triazoles Mediated by Propargyl Cations. *Org. Lett.* **2013**, *15*, 5222–5225. [CrossRef] [PubMed]
54. Zhu, Y.; Yin, G.; Hong, D.; Lu, P.; Wang, Y. Tandem Reaction of Propargylic Alcohol, Sulfonamide, and N-Iodosuccinimide: Synthesis of N-(2-Iodoinden-1-yl)arenesulfonamide. *Org. Lett.* **2011**, *13*, 1024–1027. [CrossRef] [PubMed]
55. Adeloye, A.O.; Mphahlele, M.J. 2,4-Diarylquinolines: Synthesis, absorption and emission properties. *J. Chem. Res.* **2011**, *35*, 254–259. [CrossRef]
56. Abbitati, G.; Arcadi, A.; Marinelli, F.; Rossi, E.; Verdecchia, M. Rh-Catalyzed Sequential Hydroarylation/Hydrovinylation–Heterocyclization of b-(2-Aminophenyl)-a,b-ynones with Organoboron Derivatives: A New Approach to Functionalized Quinolines. *Synlett* **2006**, *19*, 3218–3224.
57. Azizi, K.; Akrami, S.; Madsen, R. Manganese(III) Porphyrin-Catalyzed Dehydrogenation of Alcohols to form Imines, Tertiary Amines and Quinolines. *Chem. Eur. J.* **2019**, *25*, 6439–6446. [CrossRef] [PubMed]

Sample Availability: Samples of the final products are available from the authors.

© 2019 by the authors. Licensee MDPI, Basel, Switzerland. This article is an open access article distributed under the terms and conditions of the Creative Commons Attribution (CC BY) license (http://creativecommons.org/licenses/by/4.0/).

Article

Catalyst-Free Synthesis of Polysubstituted 5-Acylamino-1,3-Thiazoles via Hantzsch Cyclization of α-Chloroglycinates

Mara Tomassetti [1,*], Gabriele Lupidi [2], Pamela Piermattei [2], Federico V. Rossi [2], Samuele Lillini [1], Gianluca Bianchini [3], Andrea Aramini [1,3], Marco A. Ciufolini [4] and Enrico Marcantoni [2,*]

1. Dompé Farmaceutici S.p.A., Via Pietro Castellino, Napoli 80131, Italy; samuele.lillini@dompe.com (S.L.); andrea.aramini@dompe.com (A.A.)
2. School of Science and Technology, Chemistry Division, University of Camerino, Camerino 62032, Italy; gabriele.lupidi@unicam.it (G.L.); pamela.piermattei@unicam.it (P.P.); federico.rossi@unicam.it (F.V.R.)
3. Dompé Farmaceutici S.p.A., Via Campo di Pile, L'Aquila 67100, Italy; gianluca.bianchini@dompe.com
4. Department of Chemistry, University of British Columbia, 2036 Main Mall, Vancouver, BC V6T 1Z1, Canada; ciufi@chem.ubc.ca
* Correspondence: mara.tomassetti@dompe.com (M.T.); enrico.marcantoni@unicam.it (E.M.); Tel.: +39-(816)-132-220 (M.T.); Tel.: +39-(737)-402-255 (E.M.)

Academic Editor: Gianfranco Favi
Received: 30 September 2019; Accepted: 21 October 2019; Published: 25 October 2019

Abstract: A catalyst-free heterocyclization reaction of α-chloroglycinates with thiobenzamides or thioureas leading to 2,4-disubstituted-5-acylamino-1,3-thiazoles has been developed. The methodology provides straightforward access to valuable building blocks for pharmaceutically relevant compounds.

Keywords: α-chloroglycinates; 5-acylamino-1,3-thiazoles; Hantzsch reaction

1. Introduction

Heterocyclic compounds are an integral part of many biologically active small molecules. Indeed, many currently marketed drugs exhibit heterocycles as their core structures [1,2]. In particular, compounds based on a 1,3-thiazole display a wide range of activities [3]. Therefore, increasing attention has been devoted in recent years to the preparation of polysubstituted thiazoles, primarily for pharmaceutical applications [4–7], but also in connection with problems in material science [8]. Of special relevance in medicinal chemistry are aminothiazoles and their derivatives [9–16]. Such compounds show potential in oncology [17,18], in the treatment of inflammatory conditions [19,20] and neurological disorders [21]. Examples (Figure 1) include compound **1**, an experimental CDK5 inhibitor for the treatment of Alzheimer's disease [22], and avatrombopag **2**, approved in 2018 the treatment of adult thrombocytopenia [23].

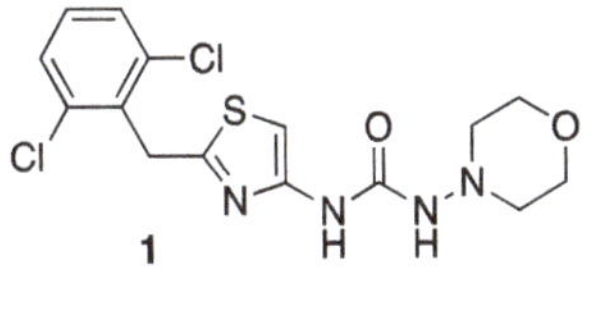

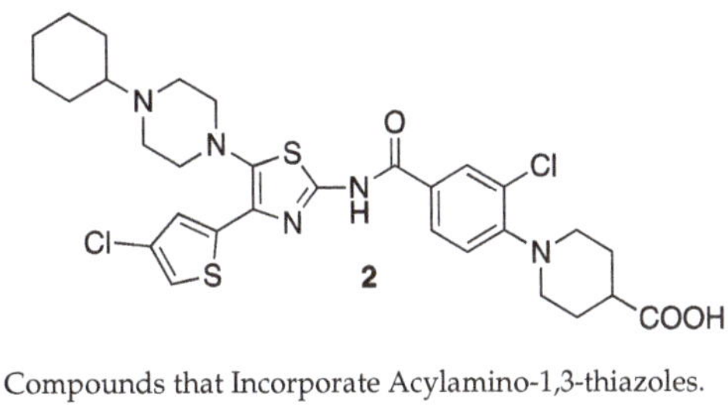

Figure 1. Examples of Biologically Active Compounds that Incorporate Acylamino-1,3-thiazoles.

The research described herein finds its genesis in Dompé Farmaceutici's identification of novel thiazole derivatives such as **3** (Figure 2), with proven efficacy in the urology and pain areas [4,24]. As a consequence of this discovery, congeners of **3** incorporating alkylamino-or acylamino substituents, i.e., substances **4–5**, became of special interest. Curiously, such thiazoles are scantly documented in the literature. For instance, the SciFinder database records only 47 compounds of the type **4**, described in 11 publications as of this writing [25,26]. Substances of general structure **5** are even rarer (11 compounds, 6 publications) [27,28]. Furthermore, good synthetic procedures that lead directly to compounds **4–5** are lacking. Possibly for these reasons, such heterocycles are quite uncommon in medicinal chemistry.

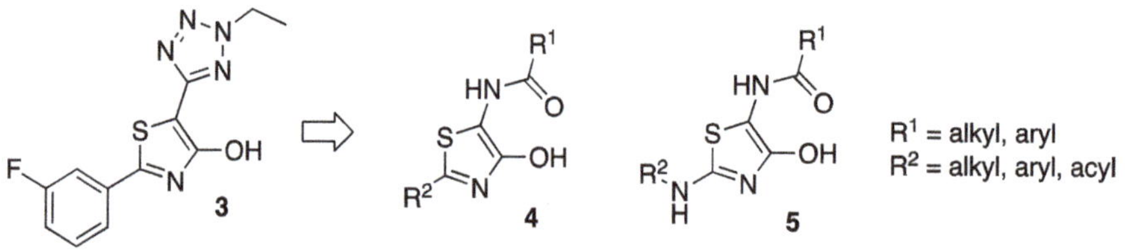

Figure 2. Examples of Biologically Active Compounds that Incorporate Acylamino-1,3-thiazoles.

Our interest in developing general methods for the synthesis of pharmaceutically relevant heterocyclic compounds [29–31] induced us to launch a program aiming to establish widely applicable procedures for the direct synthesis of the desired thiazoles. In drug discovery, the chemical modifications of thiazole ring moieties could be a useful tool in the discovery of new ways to make variations on existing drugs. But this approach is limited for organic chemists because there are only so many changes that can be made to a complex heterocyclic compound. The cyclization of polyfunctionalized acyclic precursors is much more advantageous for medical and biotechnological applications [32]. Taking into account a potential industrial development of the methodology, it was essential to avoid harsh reaction conditions, issues of regioselectivity that may result in the formation of multiple products, the need for costly catalysts, elaborate reaction protocols, and complex purification procedures.

2. Results

Among the numerous methods for thiazole synthesis [33–35], the venerable Hantzsch reaction [36] and its variants [37,38], i.e., the cyclocondensation of α-halocarbonyl compounds with thioamides or thioureas (Scheme 1, Equation (1)), remains especially popular. This transformation reliably produces 1,3-thiazoles having alkyl, aryl, or heterocyclic substituents in good to excellent yields. Furthermore, the reaction requires no metallic catalysts, expensive reagents, or stringent measures to exclude moisture and air: a significant advantage in terms of environmental impact and total cost of the synthetic procedure. It appeared that the target compounds **4–5** could be accessed by a Hatzsch-like reaction between an α-chloroglycinate, **8**, and a thioamide, **9**, or thiourea, **11** (Scheme 2). Compounds **8** are readily available starting with a Ben-Ishai addition of a primary amide, **6**, to, e.g., ethyl glyoxylate, followed by reaction of the resultant **7** with SOCl$_2$ [39–42]. They are perfectly isolable and fairly stable on storage at −20 °C with exclusion of moisture (two weeks at least) [43–47], even though the halogen atom is quite labile. Also, they are normally obtained is a state of good to excellent purity; therefore, it is generally expedient to use them directly. A caveat is that they are sensitive to the action of bases, which cause rapid formation of polymeric products [41]. A noteworthy illustration of this was provided in connection with their use in a useful oxazole synthesis: displacement of the chlorine with a poorly basic aluminum acetylide results in the efficient formation of polysubstituted oxazoles, but the action of basic alkali metal acetylides rapidly converts them into intractable mixtures of products [43–47]. On such grounds, it seemed plausible that poorly basic, but highly S-nucleophilic, thioamides/thioureas should combine with chlorogycinates **8** as desired.

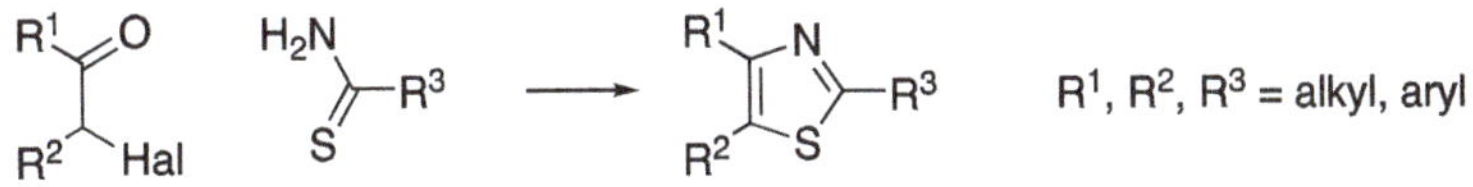

Scheme 1. The Hantzsch Thiazole Synthesis.

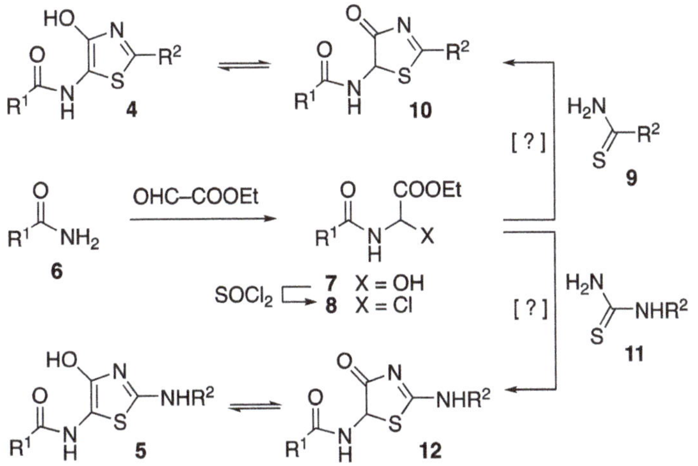

Scheme 2. Hypothetical Hantzsch-type Route to the Target Thiazoles 4–5.

The exploration of the new methodology started with a study of the reaction of N-benzoylchloroglycine ethyl ester, **8a**, with thiobenzamide, **9a**, (Scheme 3). When a solution of the reactants in THF was stirred at room temperature overnight, a precipitate appeared. This material consisted (NMR, MS) of a mixture of tautomers **10aa** and **4aa** of the expected product [48]. Unfortunately, the yield of product never exceeded 40%, regardless of solvent used (THF, DMF, and MeCN). Also, conduct of the reaction at higher temperatures (refluxing conditions) resulted in formation of complex mixtures. An HPLC-MS analysis of the reaction mixtures showed the presence of a dimer of tentative structure **13**, the formation of which is attributable to water contamination of the solvents. The formation of presumed **13** was accelerated substantially when hydroxyglycinate **7a**, R^1 = Ph, was exposed to the $CeCl_3 \cdot 7H_2O$-NaI system [49] in an attempt to effect conversion into the corresponding iodide. Fortunately, the use of freshly dried THF suppressed the formation of the dimeric product and greatly improved the yield of thiazoles. Furthermore, it transpired that it was best to allow the reaction to proceed at r.t. for only 2 h. In all cases, the workup procedure involved the removal of volatiles under vacuum, the resuspension of the solid residue in ether, and the recovery of the solid product by filtration. The thiazoles thus obtained were of excellent quality and required no further purification. Some of them existed in solution as mixtures of keto (**10**) and enol (**4**) tautomers (NMR). The keto form exhibited a diagnostic ^3J coupling between the C-5 and the NH protons (≈7.4 Hz), consistent with literature values in related systems [50]. The enol form may be the dominant/exclusive tautomer present in the solid state, as suggested by the broad OH signal observed in the FT-IR spectrum (see Supplementary Materials). Representative examples of the new transformation are shown in Table 1. It is apparent that the reaction tolerates both electron-donating and electron-withdrawing substituents on either reactant (entries 5, 8 and 10).

It is worthy of note that chloroglycinates derived from conjugated amides are good substrates for the present reaction (entry 4), even though they are quite poor for the oxazole-forming one [43–47]. It should also be stressed that the procedure is readily amenable to high-throughput chemical synthesis and that its scope was found to be considerably broader than the 12 examples of Table 1 suggest.

Thus, various points of diversification can be introduced to generate more complex molecules with interesting biological activities.

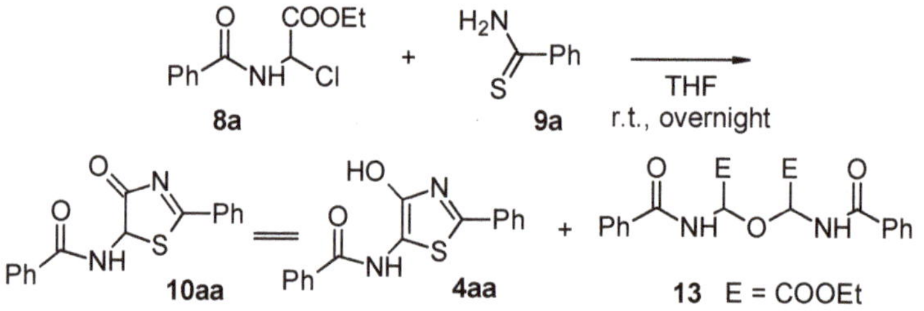

Scheme 3. Reaction of Ethyl N-Benzoylchloroglycinate with Thiobenzamide.

Table 1. Formation of 5-Acylamino-1,3-thiazoles from chloroglycinates and thiobenzamides [a].

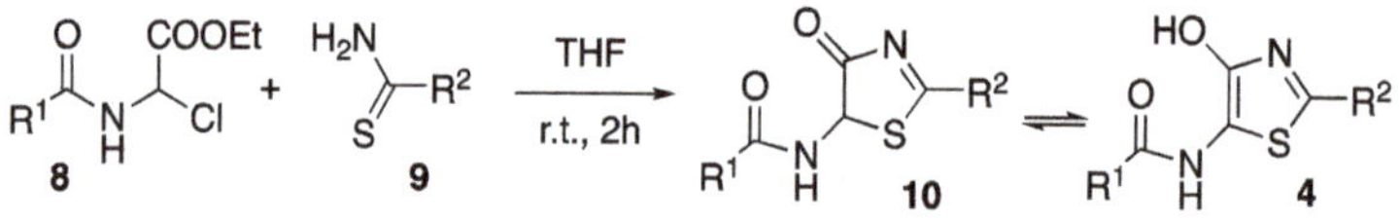

Entry	R^1	R^2	Product (10 + 4) [b]	Yield (%) [c]
1	Ph (8a)	Ph (9a)	10aa + 4aa [d]	88
2	Piperonyl (8b)	Ph (9a)	10ab + 4ab [d]	76
3	Et (8c)	Ph (9a)	10ac + 4ac [d]	94
4	PhCH=CH (8d)	Ph (9a)	10ad + 4ad [d]	81
5	Ph (8a)	4-NO$_2$-C$_6$H$_4$ (9b)	10ba + 4ba [d]	74
6	Piperonyl (8b)	4-NO$_2$-C$_6$H$_4$ (9b)	10bb + 4bb [d]	87
7	Et (8c)	4-NO$_2$-C$_6$H$_4$ (9b)	10bc + 4bc [d]	94
8	Ph (8a)	4-MeO-C$_6$H$_4$ (9c)	10ca + 4ca	94
9	Piperonyl (8b)	4-MeO-C$_6$H$_4$ (9c)	10cb + 4cb [d]	68
10	Ph (8a)	4-Cl-C$_6$H$_4$ (9d)	10da + 4da [d]	78
11	Piperonyl (8b)	4-Cl-C$_6$H$_4$ (9d)	10db + 4db [d]	95
12	Et (8c)	4-Cl-C$_6$H$_4$ (9d)	10dc + 4dc [d]	90

[a] Typical procedure: a thioamide (1.0 mmol) was added to a stirred solution of α-chloroglycinate (1.0 mmol) in dry THF under nitrogen. After 2 h, the solvent was removed under reduced pressure, and the residue was re-suspended in Et$_2$O and stirred for 1 h. The solid was collected by filtration to obtain a thiazole that required no further purification. [b] Equilibrium mixture of keto (10) and enol (4) form. [c] As a mixture of tautomers. [d] Predominant tautomer in DMSO-d6.

On a side note, substituted 2-thiazolinones/2-hydroxythiazoles are subject to acid-catalyzed ring mutation reactions [51–53]. Concerns about the possible sensitivity of 5-thiazolinones/5-hydroxythiazoles 10/4 to analogous isomerization processes were rapidly allayed by the observation that all such compounds remained unchanged upon storage for several weeks at low temperature.

The use of a thiourea in lieu of a thioamide in the reaction just described successfully led to the formation of compounds 5 in moderate to good yield (Table 2). No improvement in yields was observed when the reaction was carried out in the presence of 1,8-bis-(dimethylamino)naphthalene (proton sponge) [54]. The rate of product formation was also unaffected, providing additional evidence that the target thiazoles do not form by an initial dehydrohalogenation of 8 to an acylimine and subsequent nucleophilic addition thereto. Instead, they are likely to arise upon cyclization of intermediates 14 (Scheme 4, reaction pathway *a*), formed in turn by displacement of chlorine from 8 by the nucleophilic sulfur center of the thioamide. Interestingly, all attempts to detect 14 or other possible intermediates by ESI-MS techniques [55,56] met with failure (only reactants and products apparent in the spectra), indicating that the cyclization of 14 to 12/5 must be very fast. We note in passing that substance 14 could theoretically produce heterocycle **15** by a cyclization reaction involving the amide group (pathway *b*). However, no products of the type **15** were ever observed in our reactions, undoubtedly because of the weaker electrophilic reactivity of the amide relative to the ester and the lack of aromatic character in **15**.

Table 2. Formation of 5-Acylamino-2-amino-1,3-thiazole Derivatives from chloroglycinates and thioureas [a].

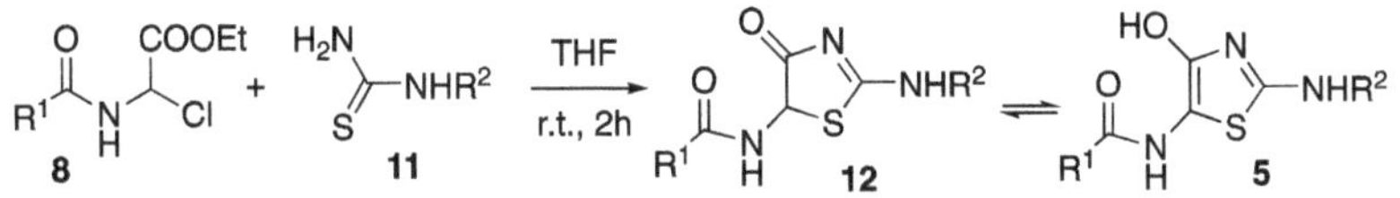

Entry	R^1	R^2	Product (12 + 5) [b]	Yield (%) [c]
1	Ph (**8a**)	H (**11a**)	**12aa** [d] + **5aa**	65
2	Ph (**8a**)	4-NO$_2$-C$_6$H$_4$ (**11b**)	**12ab** [d] + **5ab**	96
3	Ph (**8a**)	4-CH$_3$O-C$_6$H$_4$ (**11c**)	**12ac** [d] + **5ac**	97
4	Ph (**8a**)	4-CH$_3$CO-C$_6$H$_4$ (**11d**)	**12ad** [d] + **5ad**	77
5	Ph (**8a**)	CH$_3$CO (**11e**)	**12ae** [d] + **5ae**	62
6	Piperonyl (**8b**)	4-NO$_2$-C$_6$H$_4$ (**11b**)	**12bb** + **5bb** [d]	75
7	Piperonyl (**8b**)	4-CH$_3$CO-C$_6$H$_4$ (**11d**)	**12bd** + **5bd** [d]	76
8	Et (**8c**)	4-NO$_2$-C$_6$H$_4$ (**11b**)	**12cb** [d] + **5cb**	81
9	Et (**8c**)	4-CH$_3$CO-C$_6$H$_4$ (**11d**)	**12cd** [d] + **5cd**	80

[a] Typical procedure: a thiourea (1.0 mmol) was added to a stirred solution of α-chloroglycinate (1.0 mmol) in dry THF under nitrogen. After 2 h, the solvent was removed under reduced pressure, and the residue was re-suspended in Et$_2$O and stirred for 1 h. The solid was collected by filtration to obtain a thiazole that required no further purification. [b] Equilibrium mixture of keto (**12**) and enol (**5**) form. [c] As a mixture of tautomers. [d] Predominant tautomer in DMSO-d6.

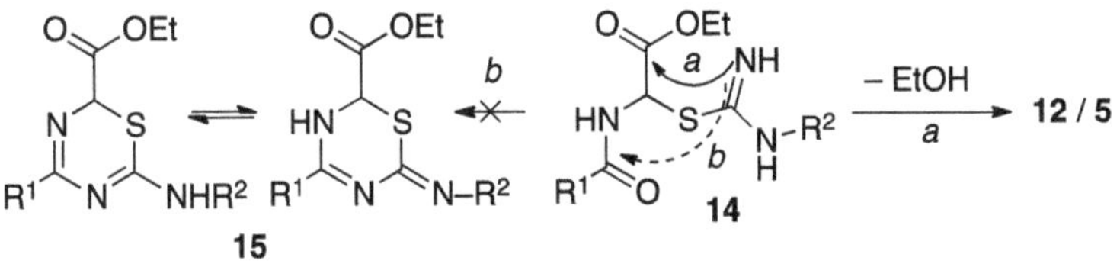

Scheme 4. Formation of Thiazoles **5** by Cyclization of Intermediates **14**.

On the other hand, ESI-MS monitoring of the reaction of compound **16** (prepared from propionamide and phenylglyoxal and subsequent treatment of the hydroxy derivative with SOCl$_2$) with thioamides **9a–b** did reveal the transient presence of hydroxy intermediates **17a–b**, in situ dehydration of which furnished 5-acylaminothiazoles **18a–b** in excellent yields (Scheme 5).

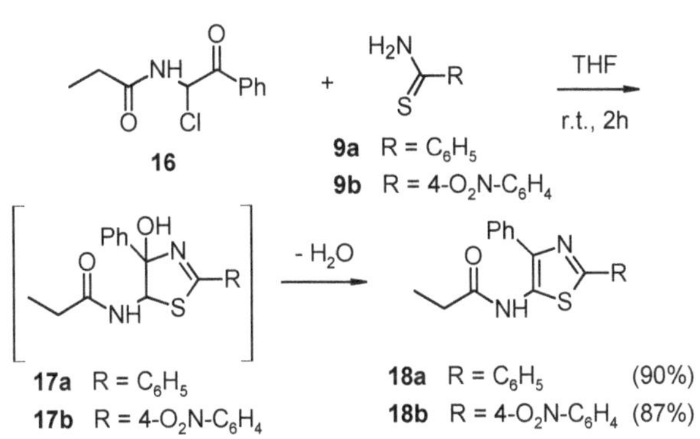

Scheme 5. Hantzsch cyclization of α-chloro carbonyl compounds and thiobenzamides.

In conclusion, a Hantzsch construction of thiazoles **4–5** and **18** through the reaction of α-chloroglycinate esters and congeners with thioamides or thioureas has been established. The target compounds are obtained under mild conditions from readily available, inexpensive building blocks through an environmentally benign process that requires no stringent control of reaction parameters/atmosphere and no catalysts. The medicinal chemistry of the products is being actively researched and pertinent results will be reported in due course.

3. Materials and Methods

3.1. General

All reagents and solvents were purchased from commercial suppliers and used without further purification, except THF (freshly distilled over metallic sodium) and DCM (freshly distilled over $CaCl_2$). All reactions were performed under nitrogen atmosphere. All glassware was oven dried at 100 °C for at least 2 h prior to use. Merck pre-coated TLC plates (silica gel 60 GF254 0.25mm) furnished by Merck KGaA (Darmstadt, Germany) were used for thin-layer chromatography (TLC). Compounds were visualized under UV light, or in an iodine, chamber, or by staining with phosphomolybdic acid solution. Proton (400 MHz), ^{13}C (100 MHz), and 135DEPT spectra were recorded on a Varian Mercury 400 (Varian, Inc., Palo Alto, CA, USA). Chemical shifts are reported in ppm from TMS and are referenced to solvent signals ($CDCl_3$: 7.26 ppm for the residual protio species in ^1H, 77.2 ppm in ^{13}C; DMSO-d_6: 2.50 ppm in ^1H and 39.5 ppm in ^{13}C). Coupling constants, J, are reported in hertz (Hz). Splitting patterns are described as s (singlet), d (doublet), t (triplet), q (quartet), m (multiplet). IR spectra (cm^{-1}) were recorded with a Perkin-Elmer FT-IR spectrometer Spectrum Two UATR (Perkin Elmer, Inc., Waltham, MA, USA). Low-resolution ESI/APCI mass spectra were recorded with an Agilent 1100 MSD ion-trap mass spectrometer (Agilent Technologies, Inc., Santa Clara, CA, USA) equipped with a standard ESI/APCI source. Nitrogen served both as the nebulizer gas and the dry gas. The analyte (10 mg) was dissolved in the appropriate mobile phase (1 mL) and introduced by direct infusion with a syringe pump. High-resolution mass spectra (HRMS) were obtained with a HPLC Ultimate 3000 (Thermofisher Scientific, MA, USA) coupled with a high-resolution Q Exactive Benchtop Quadrupole–Orbitrap (Thermofisher Scientific, MA, USA). The NMR spectra of compounds were provided in Supplementary Materials (Figures S1–S66).

3.2. General Procedure for the Synthesis of α-Hydroxyglycinates (**7**)

An amide (1.0 mmol) was added to a solution of ethyl glyoxylate (technical, 50% solution in toluene, 1.2 eq) in toluene (1 mL) and the reaction was stirred overnight at 70 °C. The next morning a white precipitate had appeared. The solvent was removed under reduced pressure and the residue was suspended in Et_2O. The precipitate of α-hydroxyglycinate ester was recovered by filtration and found to be pure enough for the next step. Yields were generally quantitative. The following compounds were thus prepared from appropriate amides:

Ethyl 2-benzamido-2-hydroxyacetate (**7a**) [57]: From benzamide. Yield: 98% as an amorphous white solid. FTIR (neat, cm^{-1}): 3380 (broad), 3307, 1750, 1646, 1536. ^1H-NMR (400 MHz, DMSO-d_6): δ 9.35 (d, J = 7.8 Hz, 1H), 7.93–7.84 (m, 2H), 7.58–7.52 (m, 3H), 6.57 (d, J = 6.46 Hz, 1H), 5.64 (t, J = 7.00 Hz, 1H), 4.15 (q, J = 7.1 Hz, 2H), 1.21 (t, J = 7.08 Hz, 3H). ^{13}C-NMR (100 MHz, $CDCl_3$): 170.41, 166.43, 133.98, 132.15, 128.81, 127.98, 72.38, 61.21, 14.50 HR-MS (ESI) calcd for $C_{11}H_{13}NO_4$: $[M + H]^+$ 224.0917, found 224.0913.

Ethyl 2-(2-(benzo[1,3]dioxol-5-yl)acetamido)-2-hydroxyacetate (**7b**): From 2-(benzo[d][1,3]dioxol-5-yl)acetamide. Yield: 98% as an amorphous white solid. FTIR (neat, cm^{-1}): 3407 (broad), 3326, 1727, 1650, 1540. ^1H-NMR (400 MHz, $CDCl_3$): δ 6.79 (d, J = 7.8 Hz, 1H), 6.75–6.68 (m, 3H), 5.96 (s, 2H), 5.50 (d, J = 7.4 Hz, 1H), 4.26 (q, J = 7.2 Hz, 2H), 3.52 (s, 2H), 1.30 (t, J = 7.2 Hz, 3H). ^{13}C-NMR (100 MHz, $CDCl_3$): δ 172.17,

169.35, 148.34, 147.28, 127.38, 122.84, 109.87, 108.89, 101.36, 72.45, 62.81, 43.16, 14.14. HR-MS (ESI) calcd for $C_{13}H_{15}NO_6$: $[M + H]^+$ 282.0972, found 282.0979.

Ethyl 2-hydroxy-2-propanamidoacetate (**7c**): From propanamide. Yield: 96% as an amorphous white solid. FTIR (neat, cm^{-1}): 3400 (broad), 3315, 1736, 1655, 1537. ^{1}H-NMR (400 MHz, CDCl$_3$): δ 6.98 (s, 1H), 5.60 (d, *J* = 7.7 Hz, 1H), 4.26 (q, *J* = 7.1 Hz, 2H), 2.27 (q, *J* = 7.5 Hz, 2H), 1.30 (t, *J* = 7.2 Hz, 3H), 1.14 (t, *J* = 7.5 Hz, 3H). ^{13}C-NMR (100 MHz, CDCl$_3$): δ 174.95, 169.73, 72.08, 62.60, 29.45, 14.13, 9.33. HR-MS (ESI) calcd for $C_7H_{13}NO_4$: $[M - H]^-$ 174.0771, found 174.0772.

Ethyl 2-cinnamamido-2-hydroxyacetate (**7d**): From cinnamamide. Yield: 95% an amorphous white solid. FTIR (neat, cm^{-1}): 3290 (broad), 3215, 1750, 1654, 1547. ^{1}H-NMR (400 MHz, CDCl$_3$): δ 7.68 (d, *J* = 15.6 Hz, 1H), 7.50 (dd, *J* = 6.7, 2.9 Hz, 2H), 7.40–7.28 (m, 3H), 7.11 (s, 1H), 6.46 (d, *J* = 15.6 Hz, 1H), 5.76 (d, *J* = 7.5 Hz, 1H), 4.31 (q, *J* = 7.1 Hz, 2H), 1.33 (t, *J* = 7.1 Hz, 3H). ^{13}C-NMR (100 MHz, CDCl$_3$): δ 169.37, 166.61, 143.33, 134.38, 130.42, 129.06, 128.19, 119.18, 72.72, 62.93, 14.21. HR-MS (ESI) calcd for $C_{13}H_{15}NO_4$: $[M + Na]^+$ 272.0893, found 272.0894.

3.3. General Procedure for the Synthesis of α-Chloroglycinates (**8**)

Thionyl chloride (10 eq) was added dropwise to a suspension of a hydroxyglycinate (**7**) (1 mmol) in dry DCM (1 mL) under nitrogen. The mixture was warmed to 40 °C and the progress of the reaction was periodically checked by ^{1}H-NMR. Full conversion typically required about 3 h. Excess thionyl chloride was removed under high vacuum and the residue of crude chloride, yellowish solid, was immediately used in subsequent coupling reactions without further purification to avoid degradation. Yields were essentially quantitative. Since the compounds are unstable in water solution it was not possible to perform an HPLC-MS analysis. The following compounds were thus prepared:

Ethyl 2-benzamido-2-chloroacetate (**8a**): From ethyl 2-benzamido-2-hydroxyacetate (**7a**). Yield 99% as an amorphous white solid. ^{1}H-NMR (400 MHz CDCl$_3$): δ 7.84–7.80 (m, 2H), 7.63–7.54 (m, 1H), 7.56–7.45 (m, 2H), 6.49 (d, *J* = 9.74, 1H), 4.38 (q, *J* = 7.10, 2H), 1.39 (t, *J* = 7.09, 3H) ^{13}C-NMR (400 MHz, CDCl$_3$) δ 166.63, 166.01, 132.80, 132.39, 128.84, 127.42, 63.32, 60.55, 13.91.

Ethyl 2-(2-(benzo[1,3]dioxol-5-yl)acetamido)-2-chloroacetate (**8b**): From ethyl 2-(2-(benzo[1,3]dioxol-5-yl)acetamido)-2-hydroxyacetate (**7b**). Yield: 99% as an amorphous yellow solid. ^{1}H-NMR (400 MHz, CDCl$_3$): δ 6.82–6.68 (m, 4H), 6.23 (d, *J* = 9.8 Hz, 1H), 5.98 (d, *J* = 0.7 Hz, 2H), 4.28 (m, 2H), 3.56 (s, 2H), 1.31 (t, *J* = 7.1 Hz, 3H). ^{13}C-NMR (100 MHz, CDCl$_3$): δ 170.21, 166.43, 148.45, 147.44, 126.82, 122.81, 109.80, 108.98, 101.41, 63.32, 59.95, 43.27, 13.97.

Ethyl 2-chloro-2-propanamidoacetate (**8c**): From ethyl 2-hydroxy-2-propanamidoacetate (**7c**). Yield: 99% as an amorphous pale yellow solid. ^{1}H-NMR (400 MHz CDCl$_3$): δ 7.07 (s, 1H), 6.27 (d, *J* = 9.6 Hz, 1H), 4.26 (q, *J* = 6.9 Hz, 2H), 2.31 (q, *J* = 7.0 Hz, 2H), 1.29 (t, *J* = 7.0 Hz, 3H), 1.13 (t, *J* = 7.0 Hz, 3H). ^{13}C-NMR (100 MHz CDCl$_3$): δ 173.04, 166.67, 63.27, 60.16, 29.60, 13.97, 9.11.

Ethyl 2-chloro-2-cinnamamidoacetate (**8d**): From *ethyl 2-cinnamamido-2-hydroxyacetate* (**7d**). Yield: 99% as an amorphous orange solid. ^{1}H-NMR (400 MHz, CDCl$_3$): δ 7.75 (d, *J* = 15.6 Hz, 1H), 7.56–7.51 (m, 2H), 7.42–7.37 (m, 3H), 6.90 (d, *J* = 9.7 Hz, 1H), 6.45 (m, 2H), 4.35 (q, *J* = 7.1 Hz, 2H), 1.37 (t, *J* = 7.1 Hz, 3H). ^{13}C-NMR (100 MHz, CDCl$_3$): δ 166.56, 164.61, 144.24, 134.10, 130.53, 128.98, 128.18, 118.60, 63.27, 60.43, 13.90.

3.4. General Procedure for the Synthesis of 5-Amido-4-Hydroxy Thiazoles **4** *and Their Keto Tautomers* **10**

A thioamide (1.0 mmol) was added to a solution of a chloroglycinate **8** (1.0 mmol) in dry THF (2 mL) under nitrogen and the reaction was stirred at room temperature for 2 h, whereupon a precipitate appeared. The solvent was removed under reduced pressure and the residue was resuspended in Et$_2$O.

The suspension was stirred for 1 h, then the solid product was collected by filtration. This material was of excellent quality and required no further purification unless otherwise specified. The following thiazoles were thus obtained:

N-(4-hydroxy-2-phenyl-1,3-thiazol-5-yl)benzamide (**4aa**): From ethyl 2-benzamido-2-chloroacetate **8a** and benzothioamide. Yield 88% as an amorphous yellow solid. FTIR (neat, cm^{-1}): 3380 (broad), 3252, 1655, 1634, 1521. ^{1}H-NMR (400 MHz, DMSO-d_6), Enol tautomer: δ 10.95 (bs, 1H), 10.62 (bs, 1H), 8.11–8.05 (m, 2H), 7.86–7.82 (m, 2H), 7.63–7.58 (m, 1H), 7.56–7.51 (m, 2H), 7.50–7.45 (m, 2H), 7.44–7.58 (m, 1H). ^{13}C-NMR (100 MHz, DMSO-d_6): δ 164.43, 154.79, 152.12, 134.08, 133.22, 132.43, 129.78, 129.62, 128.88, 128.42, 125.13, 108.59. HR-MS (ESI) calcd for $C_{16}H_{12}N_2O_2S$: [M + H]$^-$: 295.0546, found 295.0546.

N-(4-hydroxy-2-phenyl-1,3-thiazol-5-yl)-1,3-benzodioxole-5-carboxaamide (**4ab**): From ethyl 2-(2-(benzo[1,3]dioxol-5-yl)acetamido)-2-chloroacetate **8b** and benzothioamide. Yield 76% as an amorphous pale yellow solid. FTIR (neat, cm^{-1}): 3378 (broad), 3261, 1673, 1638, 1541. ^{1}H-NMR (400 MHz, DMSO-d_6), Enol tautomer: δ 10.78 (bs, 1H), 10.74 (bs, 1H), 7.8–7.74 (m, 2H), 7.46–7.40 (m, 2H), 7.39–7.34 (m, 1H), 7.91–7.88 (m, 1H), 6.87–6.83 (m, 1H), 6.80–6.86 (m, 1H), 5.97 (s, 2H), 3.64 (s, 2H). ^{13}C-NMR (100 MHz, DMSO-d_6): δ 168.16, 153.33, 150.72, 147.59, 146.42, 134.15, 129.80, 129.61, 129.52, 124.93, 122.60, 109.97, 108.72, 108.56, 101.27, 41.31. HR-MS (ESI) calcd for $C_{18}H_{14}N_2O_4S$: [M + H]$^+$: 355.0747, found 355.0748.

N-(4-hydroxy-2-phenyl-1,3-thiazol-5-yl)propanamide (**4ac**): From ethyl 2-chloro-2-proaonamidoacetate **8c** and benzothioamide. Yield: 94% as an amorphous pale yellow solid. FTIR (neat, cm^{-1}): 3393 (broad), 3277, 1649, 1636, 1527. ^{1}H-NMR (400 MHz, DMSO-d_6), Enol tautomer: δ 10.68 (s, 1H), 7.78 (m, 2H), 7.48–7.31 (m, 3H), 2.39 (q, *J* = 7.6 Hz, 2H), 1.06 (t, *J* = 7.6 Hz, 3H). ^{13}C-NMR (100 MHz, DMSO-d_6): δ 170.42, 152.56, 149.91, 133.82, 129.19, 129.04, 124.51, 108.50, 27.96, 9.83. HR-MS (ESI) calcd for $C_{12}H_{12}N_2O_2S$: [M + H]$^+$: 249.0692, found 249.0690.

(2E)-N-(4-hydroxy-2-phenyl-1,3-thiazol-5-yl)-3-phenylacrylamide (**4ad**): From ethyl 2-chloro-2-cinnammidoacetate **8d** and benzothioamide Yield: 81% as an amorphous yellow solid. FTIR (neat, cm^{-1}): 3200 (broad), 3108, 1638, 1628, 1525. ^{1}H-NMR (400 MHz, DMSO-d_6), Enol tautomer: δ 11.05 (s, 1H), 7.87–7.78 (m, 2H), 7.60 (m, 3H), 7.50–7.37 (m, 6H), 7.08 (d, *J* = 15.8 Hz, 1H). ^{13}C-NMR (100 MHz, DMSO-d_6): δ 161.63, 153.13, 150.39, 140.47, 134.76, 133.76, 129.88, 129.18, 129.09, 129.06, 127.74, 124.54, 120.39, 108.64. HR-MS (ESI) calcd for $C_{18}H_{14}N_2O_2S$: [M + H]$^+$: 323.0849, found 323.0848.

N-(4-hydroxy-2-(4-nitrophenyl)-1,3-thiazol-5-yl)benzamide (**4ba**): From ethyl 2-benzamido-2-chloroacetate **8a** and 4-nitrobenzothioamide. Yield 74% as an amorphous deep red solid. FTIR (neat, cm^{-1}): 3376 (broad), 3268, 1671, 1629, 1542. ^{1}H-NMR (400 MHz, DMSO-d_6), Enol tautomer: δ 10.91 (s, 1H), 8.32–8.25 (m, 2H), 8.09–8.05 (m, 2H), 8.04–7.98 (m, 2H), 7.65–7.59 (m, 1H), 7.57–7.51 (m, 2H). ^{13}C-NMR (100 MHz, DMSO-d_6): δ 164.66, 152.92, 151.00, 147.84, 139.84, 133.03, 132.59, 128.91, 128.49, 125.72, 125.04 112.05. HR-MS (ESI) calcd for $C_{16}H_{11}N_3O_4S$: [M − H]$^-$: 340.0397, found 340.0397.

N-[4-hydroxy-2-(4-nitrophenyl)-1,3-thiazol-5-yl]-1,3-benzodioxole-5-carboxamide (**4bb**): From ethyl 2-(2-(benzo[1,3]dioxol-5-yl)acetamido)-2-chloroacetate **8b** and 4-nitrobenzothioamide. The product existed in solution as a mixture of two tautomers. Yield: 87% as an amorphous red solid. FTIR (neat, cm^{-1}): 3340 (broad), 3231, 1670, 1629, 1538. ^{1}H-NMR (400 MHz, DMSO-d_6), Enol tautomer: δ 11.15 (s, 1H), 11.07 (s, 1H), 8.26 (d, *J* = 8.4 Hz, 2H), 8.00 (d, *J* = 8.7 Hz, 2H), 6.92–6.82 (m, 2H), 6.79 (s, 1H), 5.98 (s, 2H), 3.68 (s, 2H). ^{13}C-NMR (100 MHz, DMSO-d_6): δ 167.91, 151.07, 148.92, 147.17, 146.73, 146.02, 139.55, 129.16, 125.02, 124.64, 122.19, 111.95, 109.54, 108.15, 100.85, 42.10. HR-MS (ESI) calcd for $C_{18}H_{13}N_3O_6S$: [M − H]$^-$: 398.0452, found 398.0451.

N-[4-hydroxy-2-(4-nitrophenyl)-1,3-thiazol-5-yl)propanamide (**4bc**): From ethyl 2-chloro-2-propanamidoacetate **8c** and 4-nitrobenzothioamide. Yield: 94% as an amorphous red solid. FTIR (neat, cm^{-1}): 3400 (broad), 3403, 1650, 1641, 1576. ^{1}H-NMR (400 MHz, DMSO-d_6), Enol tautomer: δ 10.98 (s, 1H), 10.80 (s, 1H), 8.28–8.23 (m, 2H), 8.03–7.98 (m, 2H), 2.44 (d, J = 7.6 Hz, 2H), 1.08 (t, J = 7.6 Hz, 3H). ^{13}C-NMR (100 MHz, DMSO-d_6): δ 170.67, 150.77, 148.54, 146.64, 139.63, 124.93, 124.58, 112.19, 27.78, 9.64. HR-MS (ESI) calcd for C$_{12}$H$_{11}$N$_3$O$_4$S: [M − H]$^-$: 292.0397, found 292.0398.

N-(4-hydroxy-2-(4-methoxyphenyl)-1,3-thiazol-5-yl)benzamide (**4ca**) and *N-[2-(4-methoxylphenyl)-4-oxo-4,5-dihydro-1,3-thiazol-5-yl]benzamide* (**10ca**): From ethyl 2-benzamido-2-chloroacetate **8a** and 4-methoxybenzothioamide. The product existed in solution as a mixture of tautomers 4ca and 10ca. Yield 94% as an amorphous bright yellow solid. FTIR (neat, cm^{-1}): 3360 (broad), 3235, 1650, 1638, 1527, 1211. ^{1}H-NMR (400 MHz, DMSO-d_6): Enol tautomer **4ca**: δ 10.86 (s, 1H), 8.09–8.03 (m, 2H), 7.80–7.74 (m, 2H), 7.62–7.56 (m, 1H), 7.55–7.47 (m, 2H), 7.06–6.99 (m, 2H), 3.80 (s, 3H). Keto tautomer **10ca**: δ 9.84 (d, J = 7.41, 1H), 7.93–7.85 (m, 2H), 7.63–7.45 (m, 4H), 7.21–7.15 (m, 2H), 6.24 (d, J = 7.40; 1H), 3.90 (s, 3H). ^{13}C-NMR (100 MHz, DMSO-d_6): δ 192.84, 188.98, 166.47, 165.58, 164.45, 160.79, 155.31, 151.97, 133.30, 132.87, 132.68, 132.38, 131.20, 129.06, 128.87, 128.37, 127.94, 126.81, 126.75, 124.96, 115.36, 115.03, 107.24, 63.22, 56.36, 55.80. HR-MS (ESI) calcd for C$_{17}$H$_{14}$N$_2$O$_3$S: [M + H]$^+$: 327.0797, found 327.0796.

N-[4-hydroxy-2-(4-methoxyphenyl)-1,3-thiazol-5-yl)-1,3-benzodioxole-5-carboxamide (**4cb**): From ethyl 2-(2-(benzo[1,3]dioxol-5-yl)acetamido)-2-chloroacetate **8b** and 4-methoxybenzothioamide. Yield 68% as an amorphous yellow solid. FTIR (neat, cm^{-1}): 3366 (broad), 3255, 1668, 1641, 1546. ^{1}H-NMR (400 MHz, DMSO-d_6), Enol tautomer: δ 10.82 (s, 1H), 7.71 (d, J = 5.9 Hz, 2H), 7.10–6.63 (m, 5H), 5.98 (s, 2H), 3.79 (s, 3H), 3.62 (s, 2H). ^{13}C-NMR (100 MHz, CDCl$_3$): δ 167.49, 160.11, 153.32, 149.96, 147.14, 145.97, 129.43, 126.51, 126.03, 122.14, 114.61, 109.53, 108.12, 106.95, 100.84, 55.33, 40.97. HR-MS (ESI) calcd for C$_{19}$H$_{16}$N$_2$O$_5$S: [M + H]$^+$: 385.0853, found 385.0851.

Synthesis of *N-[2-(4-chlorophenyl)-4-hydroxy-1,3-thiazol-5-yl]benzamide* (**4da**): From ethyl 2-benzamido-2-chloroacetate **8a** and 4-chlorobenzothioamide. Yield 78% as an amorphous yellow solid. FTIR (neat, cm^{-1}): 3392 (broad), 3255, 1675, 1633, 1534. ^{1}H-NMR (400 MHz, DMSO-d_6), Enol tautomer: δ 10.94 (s, 1H), 8.08–8.03 (m, 2H), 7.87–7.81 (m, 2H), 7.63–7.57 (m, 1H), 7.56–7.49 (m, 4H). ^{13}C-NMR (100 MHz, DMSO-d_6): δ 164.47, 153.28, 152.22, 134.14, 133.15, 132.95, 132.48, 129.67, 128.90, 128.43, 126.78, 109.15. HR-MS (ESI) calcd for C$_{16}$H$_{11}$ClN$_2$O$_2$S: [M − H]$^-$: 329.0156, found 329.0158.

N-[2-(4-chlorophenyl)-4-hydroxy-1,3-thiazol-5-yl]-1,3-benzodioxole-5-carboxamide (**4db**): From ethyl 2-(2-(benzo[1,3]dioxol-5-yl)acetamido)-2-chloroacetate **8b** and 4-chlorobenzothioamide. Yield 95% as an amorphous yellow solid. FTIR (neat, cm^{-1}): 3355 (broad), 3267, 1663, 1638, 1534. ^{1}H-NMR (400 MHz, DMSO-d_6), Enol tautomer: δ 10.80 (s, 2H), 7.76 (d, J = 8.6 Hz, 2H), 7.47 (d, J = 8.5 Hz, 2H), 6.91–6.80 (m, 2H), 6.76 (m, 1H), 5.96 (s, 2H), 3.63 (s, 2H). ^{13}C-NMR (100 MHz, DMSO-d_6): δ 167.68, 151.28, 150.30, 147.11, 145.95, 133.36, 132.57, 129.16, 126.08, 122.11, 109.48, 108.92, 108.07, 100.78, 40.80. HR-MS (ESI) calcd for C$_{18}$H$_{13}$ClN$_2$O$_4$S: [M − H]$^-$: 387.0212, found 387.0214.

N-[2-(4-chlorophenyl)-4-hydroxy-1,3-thiazol-5-yl]propanamide (**4dc**): From ethyl 2-chloro-2-propanamidoacetate **8c** and 4-chlorobenzothioamide. The product existed as a mixture of two tautomers. Yield: 90% as an amorphous orange compound. FTIR (neat, cm^{-1}): 3450 (broad), 3285, 1650, 1635, 1525. ^{1}H-NMR (400 MHz, DMSO-d_6), Enol tautomer: δ 10.52 (s, 1H), 7.77 (d, J = 8.7 Hz, 2H), 7.48 (d, J = 8.7 Hz, 2H), 2.38 (q, J = 7.6 Hz, 2H), 1.05 (t, J = 7.6 Hz, 3H). ^{13}C-NMR (100 MHz, DMSO-d_6): δ 171.17, 151.63, 150.70, 133.95, 133.33, 129.86, 126.74, 109.76, 28.46, 10.41. HR-MS (ESI) calcd for C$_{12}$H$_{11}$ClN$_2$O$_2$S: [M − H]$^-$: 281.0157, found 281.0158.

3.5. General Procedure for the Synthesis of 5-Amido-2-Amino Thiazoles 5 and Their Keto Tautomers (12)

A thiourea (1 mmol) was added to a solution of a chloroglycinate 8 (1.0 mmol) in dry THF (2 mL) under nitrogen and the reaction was stirred at room temperature for 2 h, whereupon a precipitate appeared. The solvent was removed under reduced pressure and the residue was resuspended in Et$_2$O. The suspension was stirred for 1 h, then the solid thiazole was collected by filtration. This material was of excellent quality and required no further purification unless otherwise specified. The following thiazoles were thus obtained:

N-(2-amino-4-oxo-1,3-thiazol-5-yl)benzamide (**12aa**): From ethyl 2-benzamido-2-chloroacetate **8a** and thiourea. Yield: 65% as an amorphous yellow solid. FTIR (neat, cm^{-1}): 3351 (broad), 2869, 2521, 1776, 1667, 1619, 1563, 1484. ^{1}H-NMR (400 MHz, DMSO-d_6), Keto tautomer: δ 9.58 (d, J = 8.11 Hz, 1H), 9.17 (bss, 1H), 8.93 (bs, 1H), 7.92–7.85 (m, 2H,), 7.60–7.53 (m, 1H,), 7.53–7.46 (m, 2H), 6.08 (d, J = 8.09 Hz, 1H). ^{13}C-NMR (100 MHz, DMSO-d_6): δ 185,81, 181.13, 166,94, 133.34, 132.42, 128.92, 127.95, 64.19. ^{13}C-DEPT-135-NMR (100 MHz, DMSO-d_6): δ 132.42, 128.93, 127.95, 64.19 (CH). HR-MS (ESI) calcd for C$_{10}$H$_9$N$_3$O$_2$S: [M + H]$^+$: 236.0488, found 236.0489.

N-[2-(4-nitroanilino)-4-oxo-4,5-dihydro-1,3-thiazol-5-yl]benzamide (**12ab**): From ethyl 2-benzamido-2-hydroxyacetate **8a** and 4-nitrobenzothiourea. Yield 96%, as an amorphous yellow solid. FTIR (neat, cm^{-1}): 3370 (broad), 2854, 2508, 1783, 1672, 1621, 1532, 1492. ^{1}H-NMR (400 MHz, DMSO-d_6), Keto tautomer: δ12.22 (s, 1H), 9.72 (s, 1H), 8.24–8.22 (m, 2H), 7.87–7.85 (m, 2H), 7.58–7.48 (m, 3H), 7.14 (s, 1H), 6.17 (d, J = 7.7 Hz, 1H).^{13}C-NMR (100 MHz, DMSO-d_6): δ 166.77, 132.95, 132.70, 129.05, 128.81, 127.94, 125.66, 122.40. HR-MS (ESI) calcd for C$_{16}$H$_{12}$N$_4$O$_4$S: [M − H]$^-$: 355.0506, found 355.0502.

N-[2-(4-Methoxyanilino)-4-oxo-4,5-dihydro-1,3-thiazol-5-yl]benzamide (**12ac**): From ethyl 2-benzamido-2-hydroxyacetate **8a** and 4-methoxythiourea. Yield 97% as a yellow waxy solid. FTIR (neat, cm^{-1}): 3345 (broad), 2965, 2510, 1770, 1665, 1615, 1523. ^{1}H-NMR (400 MHz, DMSO-d_6) Keto tautomer: δ 11.78 (bs, 1H), 9.62 (d, J = 7.9 Hz, 1H), 7.90–7.78 (m, 2H), 7.66–7.47 (m, 4H), 7.02–6.89 (m, 3H), 6.15 (d, J = 8.0 Hz, 1H), 3.77 (s, 3H). ^{13}C-NMR (100 MHz, DMSO-d_6): δ 186.18, 175.72, 166.99, 156.92, 133.09, 132.70, 132.01, 129.05, 127.74, 122.59, 114.62, 62.81, 55.76. HR-MS (ESI) calcd for C$_{17}$H$_{15}$N$_3$O$_3$S: [M − H]$^-$: 340.0761, found 340.0760.

N-[2-(4-acetylanilino)-4-oxo-4,5-dihydro-1,3-thiazol-5-yl]benzamide (**12ad**): From ethyl 2-benzamido-2-hydroxyacetate **8a** and 1-(4-acetylphenyl)thiourea. Yield 77% as a yellow waxy solid. FTIR (neat, cm^{-1}): 3358 (broad), 2948, 2505, 1776, 1670, 1622, 1578, 1511.^{1}H-NMR (400 MHz, DMSO-d_6) Keto tautomer: δ 9.74 (s, 1H), 7.99–7.47 (m, 9H), 7.05 (s, 1H), 6.13 (d, J = 7.7 Hz, 1H), 2.54 (s, 3H).^{13}C-NMR (100 MHz, DMSO-d_6): δ 196.66, 166.29, 142.68, 132.90, 132.16, 130.30, 129.76, 128.53, 127.48, 121.21, 119.91, 26.56. M.W.: 353.4, ESI-MS: [M − H]$^-$ m/z = 352.0. HR-MS (ESI) calcd for C$_{18}$H$_{15}$N$_3$O$_3$S: [M − H]$^-$: 352.0761, found 352.0759.

N-(2-acetamido-4-oxo-4,5-dihydro-1,3-thiazol-5-yl)benzamid (**12ae**): From ethyl 2-benzamido-2-chloroacetate and N-carbamothioylacetamide. Yield 62% as an amorpohous off-white solid. FTIR (neat, cm^{-1}):3363 (broad), 2896, 2501, 1768, 1654, 1637, 1581. ^{1}H-NMR (400 MHz, DMSO-d_6), Keto tautomer: δ 9.57 (d, J = 7.5 Hz, NH), 7.90–7.85 (m, 2H), 7.62–7.55 (m, 1H), 7.54–7.45 (m, 2H), 5.81 (d, J = 7.5 Hz, 1H), 2.20 (s, 3H). ^{13}C-NMR (100 MHz, DMSO-d_6):δ185.70, 180.00, 173.40, 166.59, 133.13, 132.53, 129.00, 127.88, 59.90, 24.42.^{13}C-DEPT-135-NMR (100 MHz, DMSO-d_6): δ =132.54, 129.01, 127.88, 63.76, 59.91, 24.41. HR-MS (ESI) calcd for C$_{12}$H$_{11}$N$_3$O$_3$S: [M + H]$^+$: 278.0594, found 278.0594.

N-[4-hydroxy-2-(4-nitroanilino)-1,3-thiazol-5-yl]-2H-1,3-benzodioxole-5-carboxamide (**5bb**): From ethyl 2-(2-(benzo[1,3]dioxol-5-yl)acetamido)-2-chloroacetate **8b** and 4-nitrothiourea. The product existed as a mixture of two tautomers. Yield: 75% as an amorphous solid. FTIR (neat, cm^{-1}): 3360 (broad),

2867, 2517, 1778, 1679, 1630, 1523, 1501. ^{1}H-NMR (400 MHz, DMSO-d_6), Enol tautomer: δ 9.31 (d, J = 7.5 Hz, 1H), 8.11 (dd, J = 9.2 Hz, 3H), 6.86 (bs, 1H), 6.84–6.69 (m, 3H), 5.99 (s, 2H), 3.43 (s, 2H). ^{13}C-NMR (100 MHz, DMSO-d_6): δ 177.89, 171.62, 147.78, 146.62, 144.23, 143.07, 129.47, 125.85, 125.15, 122.82, 121.47, 113.04, 110.10, 108.78, 101.49, 41.83. HR-MS (ESI) calcd for $C_{18}H_{14}N_4O_6S$: [M − H]$^-$: 413.0561, found 413.0562.

N-[2-(4-Acetylanilino)-4-hydroxy-1,3-thiazol-5-yl]-2H-1,3-benzodioxole-5-carboxamide (**5bd**). From ethyl 2-(2-(benzo[1,3]dioxol-5-yl)acetamido)-2-chloroacetate **8b** and 4-acetophenylthiourea. The product existed as a mixture of two tautomers. Yield: 76% as an amorphous yellow solid. FTIR (neat, cm^{-1}): 3371 (broad), 2985, 2507, 1768, 1668, 1617, 1574, 1486. ^{1}H-NMR (400 MHz, DMSO-d_6) Enol tautomer: δ 9.26 (d, J = 7.5 Hz, 1H), 7.94 (m, 3H), 7.03–6.79 (m, 4H), 5.98 (s, 2H), 3.41 (s, 2H), 2.55 (s, 3H). ^{13}C-NMR (100 MHz, DMSO-d_6): δ 196.72, 170.90, 147.13, 145.96, 142.42, 132.96, 129.78, 129.08, 122.15, 121.19, 119.97, 109.44, 108.10, 100.83, 41.20, 26.60. HR-MS (ESI) calcd for $C_{20}H_{17}N_3O_5S$: [M − H]$^-$: 410.0816, found 410.0815.

N-[2-(4-nitroanilino)-4-oxo-4,5-dihydro-1,3-thiazol-5-yl]propanamide (**12cb**). From ethyl 2-chloro-2-propanamidoacetate **8c** and 4-nitrothiourea. Yield 81% as an amorphous yellow solid. FTIR (neat, cm^{-1}): 3367 (broad), 2875, 2512, 1783, 1665, 1618, 1561, 1497. ^{1}H-NMR (400 MHz, DMSO-d_6), Keto tautomer: δ 9.10 (bs, 1H), 8.41–7.60 (m, 4H), 7.13 (bs, 1H), 5.95 (d, J = 7.6 Hz, 1H), 2.16 (q, J = 7.3 Hz, 2H), 0.98 (t, J = 7.5 Hz, 3H). ^{13}C-NMR (100 MHz, DMSO-d_6): δ 173.57, 173.46, 173.08, 171.14, 143.56, 125.17, 122.09, 58.72, 27.93, 9.20. HR-MS (ESI) calcd for $C_{12}H_{12}N_4O_4S$: [M − H]$^-$: 307.0506, found 307.0503.

N-[2-(4-acetylanilino)-4-oxo-4,5-dihydro-1,3-thiazol-5-yl]propanamide (**12cd**). From ethyl 2-chloro-2-propanamidoacetate **8c** and 1-(4-acetylphenyl)thiourea. Yield 80% as an amorphous yellow solid. FTIR (neat, cm^{-1}): 3355 (broad), 2976, 2512, 1764, 1669, 1624, 1595, 1506. ^{1}H-NMR (400 MHz, DMSO-d_6), Keto tautomer: δ 12.04 (s, 1H), 11.48 (s, 1H), 9.03 (d, J = 8.2 Hz, 1H), 8.18–7.67 (m, 2H), 7.00 (s, 1H), 5.94 (d, J = 7.6 Hz, 1H), 2.54 (s, 3H), 2.15 (q, J = 15.1, 7.5 Hz, 2H), 0.97 (t, J = 7.6 Hz, 3H). ^{13}C-NMR (100 MHz, DMSO-d_6):δ 197.4, 174.2, 174.0, 133.6, 130.4, 121.8, 120.6, 28.6, 27.2, 10.0. HR-MS (ESI) calcd for $C_{14}H_{15}N_3O_3S$: [M − H]$^-$: 304.0761, found 304.0760.

N-(1-Chloro-2-oxo-2-phenylethyl)propenamide (**16**). Thionyl chloride (10 eq) was added dropwise to a suspension *N*-(1-hydroxy-2-oxo-2-phenylethyl)propionamide (1 mmol) in dry DCM (1 mL) under nitrogen. The mixture was stirred at 40 °C and the progress of the reaction was monitored by ^{1}H-NMR. Upon complete conversion (ca. 3h), excess thionyl chloride was removed under high vacuum to leave a yellowish solid residue of crude **15**, which was used without further purification. Yield: 99%. ^{1}H-NMR (400 MHz, CDCl$_3$): δ 8.10–8.06 (m, 2H), 7.69–7.63 (m, 1H), 7.56–7.50 (m, 2H), 7.18 (d, J = 9.2 Hz, 1H), 2.39 (q, J = 7.6 Hz, 2H), 1.22 (td, J = 7.6, 3.6 Hz, 3H). ^{13}C-NMR (100 MHz, CDCl$_3$): δ 194.67, 174.39, 134.39, 133.07, 129.51, 128.83, 72.32, 29.65, 9.40.

3.6. General Procedure for the Synthesis of 5-Amido-4-Phenyl Thiazoles (18)

A thioamide (1 mmol) was added to a solution of *N*-(1-chloro-2-oxo-2-phenylethyl)propionamide (**16**) (1.0 mmol) in dry THF (2 mL) under nitrogen, and the mixture was stirred at room temperature overnight. Upon complete conversion (no more **16** visible by TLC; eluent: DCM/MeOH 95/5) the solvent was removed undero reduced pressure. The residue was re-suspended in Et$_2$O and stirred for several hours. The solid 5-amido-4-phenylthiazole was collected by filtration. The following thiazoles were thus obtained:

N-(2,4-diphenyl-1,3-thiazol-5-yl)propanamide (**18a**): From *N*-(1-chloro-2-oxo-2-phenylethyl)propenamide (**16**) and benzothioamide. Yield: 90% as an amorphous yellow solid. FTIR (neat, cm^{-1}): 3226, 1650, 1595, 1536. ^{1}H-NMR (400 MHz, DMSO-d_6): δ 10.65 (s, 1H), 7.96–7.91 (m, 2H), 7.81 (d, J = 7.3 Hz, 2H),

7.50 (dt, J = 6.3, 5.4 Hz, 5H), 7.40 (s, 1H), 2.48–2.43 (m, 2H), 1.11 (t, J = 7.5 Hz, 3H). ^{13}C-NMR (100 MHz, DMSO-d_6): δ 172.37, 158.76, 141.32, 134.07, 133.43, 129.77, 129.22, 128.64, 128.14, 127.76, 125.60, 28.25, 9.54. HR-MS (ESI) calcd for $C_{18}H_{16}N_2OS$: [M − H]$^-$: 307.0911, found 307.0911.

Synthesis of N-[2-(4-nitrophenyl)-4-phenyl-1,3-thiazol-5-yl]propanamide (18b): From N-(1-chloro-2-oxo-2-phenylethyl)propanamide (16) and 4-nitrobenzothioamide. Yield: 87% as an amorphous brown solid. FTIR (neat, cm^{-1}): 3231, 1648, 1599, 1541. ^{1}H-NMR (400 MHz, DMSO-d_6): δ 10.89 (s, 1H), 8.34–8.28 (m, 2H), 8.20–8.14 (m, 2H), 7.82–7.76 (m, 2H), 7.56–7.49 (m, 2H), 7.45–7.40 (m, 1H), 2.52–2.50 (m, 2H), 1.11 (t, J = 7.5 Hz, 3H). ^{13}C-NMR (100 MHz, DMSO-d_6): δ 172.41, 155.30, 147.47, 141.54, 139.17, 133.73, 132.29, 128.70, 128.29, 128.01, 126.34, 124.54, 28.21, 9.44. HR-MS (ESI) calcd for $C_{18}H_{15}N_3O_3S$: [M − H]$^-$: 352.0761, found 352.0757.

Supplementary Materials: The following are available online at http://www.mdpi.com/1420-3049/24/21/3846/s1, Figures S1–S8: The NMR spectra of α-hydroxyglycinates 7(a–d), Figures S9–S16: The NMR spectra of α-chloroglycinates 8(a–d), Figures S17–S40: The NMR spectra of 5-amido-4-hydroxy thiazoles 4 and their keto tautomers 10, Figures S41–S60: The NMR spectra of 5-amido-2-amino thiazoles 5 and their keto tautomers 12, Figures S61–S66: The NMR spectra of 5-amido-4-phenyl thiazoles 16–18.

Author Contributions: The manuscript was written through contributions of all authors. All authors contributed equally, and all authors have given approval to the final version of the manuscript.

Acknowledgments: This work was carried out under the framework of the University Research Project 'Chemical swiss tools to treat tumors, metastases and infections (FAR2014-2015)' and 'FAR2018: Fondo di Ateneo per la Ricerca' supported by the University of Camerino. The portion of this work carried out at the University of British Columbia was supported through an NSERC grant to M.A.C.

Conflicts of Interest: The authors declare no competing financial interest.

References

1. Taylor, A.P.; Robinson, R.P.; Fobian, Y.M.; Blakemore, D.C.; Jones, L.H.; Fadeyi, O. Modern advances in heterocyclic chemistry in drug discovery. *Org. Biomol. Chem.* **2016**, *14*, 6611–6637. [CrossRef] [PubMed]
2. Li, J.J. *Heterocyclic Chemistry in Drug Discovery*; John Wiley and Sons: Hoboken, NJ, USA, 2013.
3. Pola, S. *Significance of Thiazole-Based Hetrocycles for Bioactive Systems, Scope of Selective Hetrocycles from Organic and Pharmaceutical Perspective*; Varala, R., Ed.; InTech: Rijeka, Croatia, 2016.
4. De Caro, C.; Russo, R.; Avagliano, C.; Cristiano, C.; Calignano, A.; Aramini, A.; Bianchini, G.; Allegretti, M.; Brandolini, L. Antinociceptive effect of two novel transient receptor potential melastatin 8 antagonists in acute and chronic pain models in rat. *Brit. J. Pharmac.* **2018**, *175*, 1691–1706. [CrossRef] [PubMed]
5. Moriconi, A.; Bianchini, G.; Cologioia, S.; Brandolini, L.; Aramini, A.; Liberati, C.; Bovolenta, S. TRPM8 Antagonists. Patent WO092711, June 2013.
6. Feng, M.; Tang, B.; Liang, S.H.; Jiang, X. Sulfur containing scaffolds in drugs: Synthesis and application in medicinal chemistry. *Curr. Top. Med. Chem.* **2016**, *16*, 1200–1216. [CrossRef] [PubMed]
7. Colella, M.; Musci, P.; Carlucci, C.; Lillini, S.; Tomassetti, M.; Aramini, A.; Degennaro, L.; Luisi, R. 1, 3-Dibromo-1,1-difluoro-2-propanone as a Useful Synthon for a Chemoselective Preparation of 4-Bromodifluoromethyl Thiazoles. *ACS Omega* **2018**, *3*, 14841–14848. [CrossRef]
8. Radhakrishnan, R.; Sreejalekshni, K.G.J. Computational Design, Synthesis, and Structure Property Evaluation of 1, 3-Thiazole-Based Color-Tunable Multi-heterocyclic Small Organic Fluorophores as Multifunctional Molecular Materials. *Org. Chem.* **2018**, *83*, 3453–3466. [CrossRef]
9. Patil, R.V.; Chavan, J.U.; Beldar, A.G. Synthesis of aminothiazoles: Polymer-supported approaches. *RSC Adv.* **2017**, *7*, 23765–23778. [CrossRef]
10. Das, D.; Sidkar, P.; Bairagi, M. Recent developments of 2-aminothiazoles in medicinal chemistry. *Eur. J. Med. Chem.* **2016**, *109*, 89–98. [CrossRef]
11. Yurttas, L.; Ciftci, G.A.; Temel, H.E.; Saglik, B.N.; Demir, B.; Levent, S. Biological Activity Evaluation of Novel 1, 2, 4-Triazine Derivatives Containing Thiazole/Benzothiazole Rings. *Anticancer Agents Med. Chem.* **2017**, *17*, 1846–1853. [CrossRef]
12. Budak, Y.; Kocyigit, U.M.; Gürdere, M.B.; Özcan, K.; Taslimi, P.; Gülçin, I.; Ceylan, M. Synthesis and investigation of antibacterial activities and carbonic anhydrase and acetyl cholinesterase inhibition profiles

of novel 4, 5-dihydropyrazol and pyrazolyl-thiazole derivatives containing methanoisoindol-1,3-dion unit. *Synth. Commun.* **2017**, *47*, 2313–2323. [CrossRef]
13. Li, L.; Zhang, C.L.; Song, H.R.; Tan, C.Y.; Ding, H.W.; Jiang, Y.Y. Discovery of novel dual inhibitors of VEGFR and PI3K kinases containing 2-ureidothiazole scaffold. *Chin. Chem. Lett.* **2016**, *27*, 1–6. [CrossRef]
14. Gaikwad, N.D.; Patil, S.V.; Bobade, V.D. Synthesis and biological evaluation of some novel thiazole substituted benzotriazole derivatives. *Bioorg. Med. Chem. Lett.* **2012**, *22*, 3449–3459. [CrossRef] [PubMed]
15. Gallardo-Godoy, A.; Gever, J.; Fife, K.L.; Silber, B.M.; Prusiner, S.B.; Renslo, A.R.J. 2-Aminothiazoles as therapeutic leads for prion diseases. *Med. Chem.* **2011**, *54*, 1010–1021. [CrossRef] [PubMed]
16. Pirotte, B.; Delarge, J.; Coyette, J.; Frere, J.M. Antibacterial activity of 5-acylaminothiazole derivatives, synthetic drugs related to beta-lactam antibiotics. *J. Antib.* **1991**, *44*, 844–853. [CrossRef] [PubMed]
17. Qin, J.; Ji, J.; Deng, R.; Tang, J.; Yang, F.; Feng, G.K.; Chen, W.D.; Wu, X.Q.; Qian, X.J.; Ding, K.; et al. DC120, a novel AKT inhibitor, preferentially suppresses nasopharyngeal carcinoma cancer stem-like cells by downregulating Sox2. *Oncotarget* **2015**, *6*, 6944–6958. [CrossRef]
18. Chang, S.; Zhang, Z.; Zhuang, X.; Luo, J.; Cao, X.; Li, H.; Tu, Z.; Lu, X.; Ren, X.; Ding, K. New thiazole carboxamides as potent inhibitors of Akt kinases. *Bioorg. Med. Chem. Lett.* **2012**, *22*, 1208–1212. [CrossRef]
19. Abdelazeem, A.H.; Habash, M.; Maghrabi, I.A.; Taha, M.O. Synthesis and evaluation of novel diphenylthiazole derivatives as potential anti-inflammatory agents. *Med. Chem. Res.* **2015**, *24*, 3681–3695. [CrossRef]
20. Shivaprasad, C.M.; Jagadish, S.; Swaroop, T.R.; Ashwini, N.; Harsha, K.B.; Rangappa, K.S. Synthesis, antibacterial, antioxidant and anti-inflammatory activities of new benzimidazole derivatives. *Asian J. Biochem. Pharmac. Res.* **2014**, *4*, 316–327.
21. Giles, K.; Berry, D.B.; Condello, C.; Hawley, R.C.; Gallardo-Godoy, A.; Bryant, C.; Oehler, A.; Elepano, M.; Bhardway, S.; Patel, S.; et al. Different 2-aminothiazole therapeutics produce distinct patterns of scrapie prion neuropathology in mouse brains. *Pharmacol. Exp. Ther.* **2015**, *355*, 2–12. [CrossRef]
22. Larsen, S.D.; Stachew, C.F.; Clare, P.M.; Cubbage, J.W.; Leach, K.L. A catch-and-release strategy for the combinatorial synthesis of 4-acylamino-1,3-thiazoles as potential CDK5 inhibitors. *Bioorg. Med. Chem. Lett.* **2003**, *13*, 3491–3495. [CrossRef]
23. Shih, A.; Nazi, I.; Kelton, J.G.; Arnold, D.M. Novel treatments for immune thrombocytopenia. *Press Med.* **2014**, *43*, e87–e95. [CrossRef]
24. Mistretta, F.A.; Russo, A.; Castiglione, F.; Battiga, A.; Calciago, G.; Montorsi, F.; Brandolini, L.; Bianchini, G.; Aramini, A.; Allegretti, M.; et al. DFL23448, A Novel Transient Receptor Potential Melastin 8–Selective Ion Channel Antagonist, Modifies Bladder Function and Reduces Bladder Overactivity in Awake Rats. *Pharmacol. Exp. Ther.* **2016**, *356*, 200–211. [CrossRef] [PubMed]
25. Baranak-Stojanovic, M.; Klaumunzer, U.; Markovic, R.; Kleinpeter, E. Structure, configuration, conformation and quantification of the push–pull effect of 2-alkylidene-4-thiazolidinones and 2-alkylidene-4, 5-fused bicyclic thiazolidine derivatives. *Tetrahedron* **2010**, *66*, 8958–8967. [CrossRef]
26. Baranac-Stojanovic, M.; Tatar, J.; Kleinpeter, E.; Markovic, R. High-yield synthesis of substituted and unsubstituted pyridinium salts containing a 4-oxothiazolidine moiety. *Synthesis* **2008**, *13*, 2117–2121. [CrossRef]
27. Stadelmann, B.; Scholl, S.; Muller, J.; Hemphill, A.J. Application of an in vitro drug screening assay based on the release of phosphoglucose isomerase to determine the structure–activity relationship of thiazolides against Echinococcus multilocularis metacestodes. *Antimicrob. Chemother.* **2010**, *65*, 512–519. [CrossRef]
28. Esposito, M.; Muller, N.; Hemphill, A. Structure–activity relationships from in vitro efficacies of the thiazolide series against the intracellular apicomplexan protozoan Neospora caninum. *Int. J. Parasitol.* **2007**, *37*, 183–190. [CrossRef]
29. Cimarelli, C.; Bordi, S.; Piermattei, P.; Pellei, M.; Del Bello, F.; Marcantoni, E. An efficient Lewis acid catalyzed Povarov reaction for the one-pot stereocontrolled synthesis of polyfunctionalized tetrahydroquinolines. *Synthesis* **2017**, *49*, 5387–5395. [CrossRef]
30. Cimarelli, C.; Di Nicola, M.; Diomedi, S.; Giovannini, R.; Hamprecht, D.; Properzi, R.; Sorana, F.; Marcantoni, E. An efficient one-pot two catalyst system in the construction of 2-substituted benzimidazoles: Synthesis of benzimidazo [1,2-c] quinazolines. *Org. Biomol. Chem.* **2015**, *13*, 11687–11695. [CrossRef]
31. Properzi, R.; Marcantoni, E. Construction of heterocyclic structures by trivalent cerium salts promoted bond forming reactions. *Chem. Soc. Rev.* **2014**, *43*, 779–791. [CrossRef]

32. Rehm, F.B.H.; Jackson, M.A.; De Geyter, E.; Yap, K.; Gilding, E.K.; Durek, T.; Craik, D.J. Papain-like cysteine proteases prepare plant cyclic peptide precursors for cyclization. *PNAS* **2019**, *116*, 7831–7836. [CrossRef]
33. Ambhaikar, N.B. Thiazoles and Benzothiazoles. In *Heterocyclic Chemistry in Drug Discovery*; Li, J.K., Ed.; Wiley: Hoboken, NJ, USA, 2013; pp. 283–322.
34. Metzger, J.V. Thiazoles and their Benzo Derivatives. In *Comprehensive Heterocyclic Chemistry*; Katritzky, A.R., Rees, C.W., Eds.; Pergamon Press: Oxford, UK, 1984; Volume 6, pp. 235–331.
35. Rajer, V.N.; Swaroop, T.R.; Anil, S.M.; Bommegowda, Y.K.; Rangappa, K.S.; Sadashiva, M.P. Base-Induced Cyclization of Active Methylene Isocyanides with Xanthate Esters: An Efficient Method for the Synthesis of 5-Alkoxy-4-(tosyl/ethoxycarbonyl)-1,3-thiazoles. *Synlett* **2017**, *28*, 2281–2284.
36. Hantzsch, A.; Weber, J.H. Ueber verbindungen des thiazols (pyridins der thiophenreihe). *Ber. Dtsch. Chem. Ges.* **1887**, *20*, 3118. [CrossRef]
37. Merritt, E.A.; Bagley, M.C. Holzapfel-Meyers-Nicolaou Modification of the Hantzsch Thiazole Synthesis. *Synthesis* **2007**, *22*, 3535–3541.
38. Facchinetti, V.; Avellar, M.N.; Nery, A.C.S.; Gomes, C.R.B.; Vasconcelos, T.R.A.; de Souza, M.V.N. An Eco-friendly, Hantzsch-Based, Solvent-Free Approach to 2-Aminothiazoles and 2-Aminoselenazoles. *Synthesis* **2016**, *48*, 437–440.
39. Ben-Ishai, D.; Sataty, I.; Bernstein, Z. A new synthesis of n-acyl aromatic α-amino acids—Amidoalkylation of aromatic and heterocyclic compounds with glyoxylic acid derivatives. *Tetrahedron* **1976**, *32*, 1571–1573. [CrossRef]
40. Zoller, U.; Ben-Ishai, D. Amidoalkylation of mercaptans with glyoxylic acid derivatives. *Tetrahedron* **1975**, *31*, 863–866. [CrossRef]
41. Bernstein, Z.; Ben-Ishai, D. Synthesis of N-substituted aziridine-2-carboxylates. *Tetrahedron* **1977**, *33*, 881–883. [CrossRef]
42. Samantha, S.S.; Roche, S.P.J. In Situ-Generated Glycinyl Chloroaminals for a One-Pot Synthesis of Non-proteinogenic α-Amino Esters. *Org. Chem.* **2017**, *82*, 8514–8526. [CrossRef]
43. Zhang, J.; Coqueron, P.Y.; Ciufolini, M.A. Development and applications of an oxazole-forming reaction. *Heterocycles* **2011**, *82*, 949–980.
44. Zhang, J.; Ciufolini, M.A. An approach to the bis-oxazole macrocycle of diazonamides. *Org. Lett.* **2011**, *13*, 390–393. [CrossRef]
45. Zhang, J.; Coqueron, P.Y.; Vors, J.P.; Ciufolini, M.A. Synthesis of 5-Amino-Oxazole-4-Carboxylates from α-Chloroglycinates. *Org. Lett.* **2010**, *12*, 3942–3945. [CrossRef]
46. Zhang, J.; Polishchuk, E.A.; Chen, J.; Ciufolini, M.A.J. Development of an oxazole conjunctive reagent and application to the total synthesis of siphonazoles. *Org. Chem.* **2009**, *74*, 9140–9151. [CrossRef] [PubMed]
47. Coqueron, P.Y.; Didier, C.; Ciufolini, M.A. Iterative Oxazole Assembly via α-Chloroglycinates: Total Synthesis of (−)-Muscoride A. *Angew. Chem. Int. Ed.* **2003**, *42*, 1411–1414. [CrossRef] [PubMed]
48. Raczynska, E.D.; Kosinska, W.; Osmialowski, B.; Gawinecki, R. Tautomeric equilibria in relation to pi-electron delocalization. *Chem. Rev.* **2005**, *105*, 3561–3621.
49. Di Deo, M.; Bartoli, G.; Bellucci, M.C.; Bosco, M.; Marcantoni, E.; Sambri, L.; Torregiani, E.J. A Simple, Efficient, and General Method for the Conversion of Alcohols into Alkyl Iodides by a CeCl$_3$ × 7H$_2$O/NaI System in Acetonitrile. *Org. Chem.* **2000**, *65*, 2830–2833. [CrossRef] [PubMed]
50. Lin, Y.; Andersen, K.K. On the Tautomerism of 2,4-Disubstituted Thiazolones. *Eur. J. Org. Chem.* **2002**, *3*, 557–563. [CrossRef]
51. Billi, K.; Cosimelli, B.; Leoni, A.; Spinelli, D.J. Ring-ring interconversions. Part 3†. On the effect of the substituents on the thiazole moiety in the ring-opening/ring-closing reactions of nitrosoimidazo[2,1-b][1,3]thiazoles with hydrochloric acid. *J. Heterocycl. Chem.* **2000**, *37*, 875–878. [CrossRef]
52. Pil'o, S.G.; Brovarets, V.S.; Vinogradova, T.K.; Golovchenko, A.V.; Drach, B.S. Synthesis of new 5-mercapto-1,3-oxazole derivatives on the basis of 2-acylamino-3,3-dichloroacrylonitriles and their analogs. *Russ. J. Gen. Chem.* **2002**, *72*, 1714–1723. [CrossRef]
53. Ishida, T.; Hirata, F.; Sato, H.; Kato, S.J. Molecular Theory of Solvent Effect on Keto–Enol Tautomers of Formamide in Aprotic Solvents: RISM-SCF Approach. *Phys. Chem. B* **1998**, *102*, 2045–2050. [CrossRef]
54. Dudding, T.; Hafez, A.M.; Taggi, A.E.; Wagerle, T.R.; Lectka, T. A catalyst that plays multiple roles: Asymmetric synthesis of β-substituted aspartic acid derivatives through a four-stage, one-pot procedure. *Org. Lett.* **2002**, *4*, 387–390. [CrossRef]

55. Eberlin, M.N. Electrospray ionization mass spectrometry: A major tool to investigate reaction mechanisms in both solution and the gas phase. *Eur. J. Mass Spectrom.* **2007**, *13*, 19–28. [CrossRef]
56. Silva Santos, L.; Knaack, L.; Metzger, J.O. Investigation of chemical reactions in solution using API-MS. *Int. J. Mass Spectrom.* **2005**, *246*, 84–104. [CrossRef]
57. Chau, J.; Zhang, J.; Ciufolini, M.A. A Peterson avenue to 5-alkenyloxazoles. *Tetrahedron Lett.* **2009**, *50*, 6163–6165. [CrossRef]

Sample Availability: Samples of the compounds are available from the authors, compounds (7a–d) and 7b, 7c, 7d, 4aa, 4ac, 4ad, 4bb, 4bc, 4cb, 12 ab, 12cd.

© 2019 by the authors. Licensee MDPI, Basel, Switzerland. This article is an open access article distributed under the terms and conditions of the Creative Commons Attribution (CC BY) license (http://creativecommons.org/licenses/by/4.0/).

Communication

Nucleophilic Arylation of Halopurines Facilitated by Brønsted Acid in Fluoroalcohol

Naoko Takenaga [1,*], Toshitaka Shoji [2], Takayuki Menjo [3], Akiko Hirai [2], Shohei Ueda [3], Kotaro Kikushima [2], Tomonori Hanasaki [3] and Toshifumi Dohi [2,*]

1. Faculty of Pharmacy, Meijo University, 150 Yagotoyama, Tempaku-ku, Nagoya 468-8503, Japan
2. College of Pharmaceutical Sciences, Ritsumeikan University, 1-1-1 Nojihigashi, Kusatsu, Shiga 525–8577, Japan; ph0088sh@ed.ritsumei.ac.jp (T.S.); akiko.xp@gmail.com (A.H.); kixy@fc.ritsumei.ac.jp (K.K.)
3. Department of Applied Chemistry, College of Life Sciences, Ritsumeikan University, 1-1-1 Nojihigashi, Kusatsu, Shiga 525–8577, Japan; sb0067xs@ed.ritsumei.ac.jp (T.M.); sc0029ev@ed.ritsumei.ac.jp (S.U.); hanasaki@sk.ritsumei.ac.jp (T.H.)
* Correspondence: ntakenag@meijo-u.ac.jp (N.T.); td1203@ph.ritsumei.ac.jp (T.D.); Tel.: +81-052-839-2706 (N.T.); +81-077-561-4908 (T.D.)

Academic Editor: Gianfranco Favi
Received: 8 October 2019; Accepted: 21 October 2019; Published: 23 October 2019

Abstract: Various aryl-substituted purine derivatives were synthesized through the direct arylation of halopurines with aromatic compounds, facilitated by the combination of triflic acid and fluoroalcohol. This metal-free method is complementary to conventional coupling reactions using metal catalysts and reagents for the syntheses of aryl-substituted purine analogues.

Keywords: purine; nucleobase; aromatic substitution; arylation; fluoroalcohol

---

1. Introduction

Biogenic purine bases are heteroaromatic compounds that constitute the basic subunits of DNA and RNA and play a crucial role in biological processes. In addition to these natural nucleosides, various chemically modified purine nucleosides have recently been discovered, and detailed analyses of their bioactivities have attracted much attention. Purine derivatives bearing an aryl substituent are of particular interest among these extensively studied classes of compounds, and their preparation has gained much attention owing to the promising biological properties of the derivatives, such as cytotoxicity and antitumor activity [1–3]. In addition, their applications as biological probes have also been consistent with the synthetic advances in artificial purine compounds [4].

The classical methods for the preparation of purines bearing aryl substituents are based on heterocyclization; however, the cyclization methodology usually requires multistep procedures. Consequently, the synthesis of the target aryl purines afforded only moderate to low yields [5]. The recent methods for the synthesis of aryl-substituted purines involve the transition metal-catalyzed cross-coupling reactions of aryl organometallics (Ar-M) with halopurines (Scheme 1A) [6,7]. For example, Suzuki–Miyaura [8–16], Stille [17,18], Negishi [19], and Kumada [20] coupling reactions have been frequently used for the preparation of aryl-substituted purines. Indeed, these approaches represent versatile and reliable synthetic methods; however, these coupling reactions require stoichiometric amounts of metallic reagents and the protection of the nucleophilic functional groups—such as the hydroxyl and amino groups—in the substrates. Hence, direct arylation of 6-chloropurines by electron-rich arenes using a three-fold excess of aluminum chloride ($AlCl_3$) was reported by Guo's group as an alternative method for preparing aryl purines in a short synthetic step (Scheme 1B) [21].

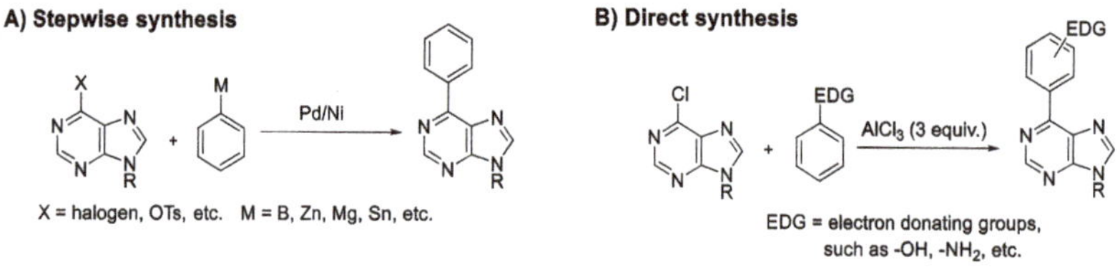

Scheme 1. General synthetic routes to obtain aryl-substituted purines from halopurines. (**A**) Stepwise synthesis. (**B**) Direct synthesis.

Despite these synthetic advances brought about by the alternative method, there unfortunately remain some limitations regarding the structural diversity of the obtainable aryl purines. To the best of our knowledge, the preparation of N-7-substituted 6-arylpurines has seldom been reported in the scientific literature [22–24]. To expand the synthetic scope for obtaining highly functionalized aryl purines with greater structural and steric diversities, developing a new practical method for preparing a variety of aryl-substituted purines is still necessary. In our continuous study on the development of a new method for the synthesis of functionalized nucleobases [25–27], we would like to report herein the metal-free arylation of purine derivatives facilitated by the combination of triflic acid and fluoroalcohol (Scheme 2).

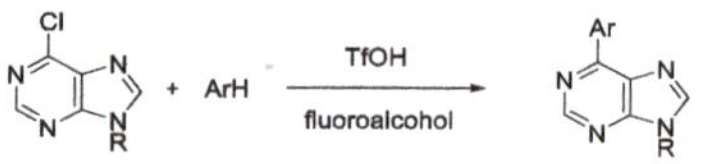

Scheme 2. Direct arylation of halopurines facilitated by Brønsted acid in fluoroalcohol. ArH: aryl nucleophile, TfOH: triflic acid.

2. Results and Discussions

Fluoroalcohols, such as 1,1,1,3,3,3-hexafluoro-2-propanol (HFIP) and 2,2,2-trifluoroethanol (TFE), possess specific properties that differentiate them from other non-fluorinated alcohols; they are highly polar [28] and weakly nucleophilic [29] and exhibit remarkable hydrogen-bond donor abilities [30]. Owing to their unique physical properties, these fluoroalcohols can dramatically direct the course of reactions; thus, as a means of developing new reactions, the authors utilized HFIP and TFE as attractive and distinctive alternatives to ordinary solvents in hypervalent iodine chemistry [31–33]. In these studies, we unexpectedly discovered the metal-free S_NAr-type arylation of heteroaromatic diaryliodonium salts by nucleophilic aromatic compounds facilitated by Lewis and Brønsted acids, i.e., boron trifluoride, trimethylsilyl triflate, and triflic acid (TfOH), in fluoroalcohols [34–36]. Golding's group also reported that the combination of trifluoroacetic acid (TFA) and TFE allowed the amination of halopurines by various substituted anilines under metal-free conditions; this method involves C–N bond formation [37–40]. Meanwhile, the metal-free nucleophilic arylation of halopurines involving C–C bond formation has not been reported.

In a pilot experiment, we first examined the S_NAr-type coupling reaction of 6-chloropurine **1a** initiated by a Brønsted acid [34,35], using methyl indole **2a** as an aromatic nucleophile in HFIP as the model case (Table 1). In order to optimize the coupling reaction, varying equivalents (Entries 1–4) of TfOH were used, and the desired arylation product of purine **3aa** was obtained in excellent yield when we used 0.5 to 1.0 equiv. of TfOH for the reactions (Entries 3 and 4). The usage of alternative Brønsted acids as additives, such as H_3PO_4, p-TsOH, and even TFA [37–40], was not as effective and provided inferior results in comparison with the use of TfOH. The fluoroalcohol HFIP plays an essential role in the reaction, and a solvent mixture of HFIP and 1,2-dichloroethane (DCE) did not smoothly produce the coupling product **3aa**. (Entry 5). Also, the replacement of HFIP with TFE and the use of methanol and acetonitrile as the solvent instead of HFIP yielded low or null amounts of product **3aa**.

Table 1. Optimization of the reaction conditions [a].

Entry	Solvent	TfOH	Yield (%) [b]
1	HFIP	0.1 equiv.	8
2	HFIP	0.2 equiv.	54
3	HFIP	0.5 equiv.	89
4	HFIP	1.0 equiv.	quant.
5	HFIP/DCE = 9:1	1.0 equiv.	60

[a] Reaction conditions: The reactions were performed using chloropurine **1a** (0.50 mmol), 1-methylindole **2a** (0.55 mmol), and TfOH in solvent (0.1 M). [b] Determined by ^{1}H-NMR using nitromethane as an internal standard. HFIP: hexafluoroisopropanol, DCE: 1,2-dichloroethane.

To evaluate the generality of the reaction system, the substrate scope under the optimized reaction conditions was examined (Table 2). The reaction of N-protected 9-benzyl-6-chloro-9H-purine **1b** cleanly favored the corresponding product **3ba** with good yield. When non-N-protected indoles **2b–d** were subjected to analogous reaction conditions, the desired products **3bb–bd** were also obtained in excellent yields. However, it was revealed that the reaction of indoles bearing electron-withdrawing groups, such as 5-nitroindole **2e**, did not proceed under these reaction conditions due to the deactivation of the aromatic nucleophile by hydrogen bonding with HFIP [41–43]. Furthermore, other electron-rich arenes were as compatible as the aromatic nucleophiles; similarly, good results were obtained from the coupling reactions with 1-naphthol **2f**, 1-methoxynaphthalene **2g**, several alkoxybenzenes **2h,i**, and resorcinol **2j**.

Table 2. Scope of substrates [a].

3aa: quant.	3ba: 82%	3bb: quant.	3bc: 96%
3bd: 90%	**3be: n.r**	**3af: 97%**	**3ag: 85%**
3bg: 92%	**3bh: quant.**	**3bi: quant.**	**3bj: quant.**

[a] All the reactions were performed using chloropurine **1** (0.50 mmol), indole or aromatic nucleophile **2** (0.55 mmol), and TfOH (0.5 mmol, 1.0 equiv.) in HFIP (0.1 M) at 60 °C. The yields after isolation are indicated.

One of the significant advantages of the present reaction system is the production of structural and sterically diverse N-7-substituted 6-arylpurines; these arylpurines are not easily accessible by other synthetic methods [22–24]. As a result, the proposed reaction conditions were also utilized for the coupling of 7-benzyl-6-chloro-7H-purine **1c** with indole **2k** and naphthalene nucleophile **2f** to afford the corresponding 6-arylated N-7-substituted purines **3ck** and **3cf** in good yields (Scheme 3).

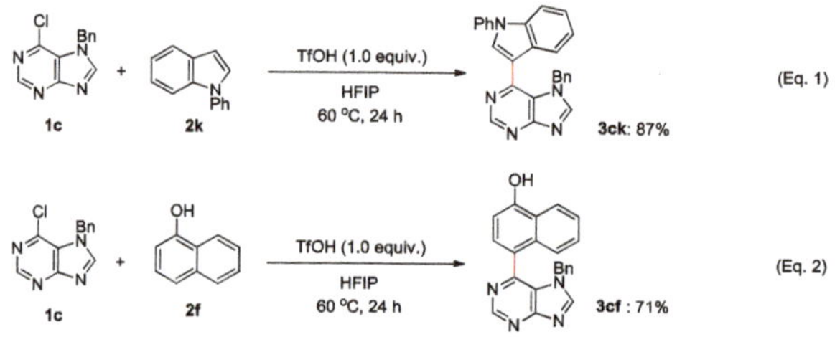

Scheme 3. Synthesis of 7-substituted 6-aryl purines.

When using *p*-anisidine **4** as a substrate, our reaction system with TfOH became valuable for the chemoselective N-arylation of halopurines at the 6 position under mild temperature (Scheme 4) [37]. We subjected 9H-chloropurine **1d** and aniline **4** to our optimized conditions at 60 °C, obtaining selective N-arylation that smoothly provided N-(4-methoxyphenyl)-9H-purine-6-amine **5** in 79% yield, without the formation of the C-arylated purine coupling product **5′**. On the other hand, Guo's group previously reported the reaction of purines and anilines or naphthylamines in the presence of a three-fold excess of AlCl₃ in DCE, which alternatively gave the C-arylated coupling products and likewise the biaryl **5′** [21]. Therefore, our reaction system is complementary to the AlCl₃-mediated coupling reaction [21] for the syntheses of C6-aryl-substituted purine derivatives in view of product selectivity.

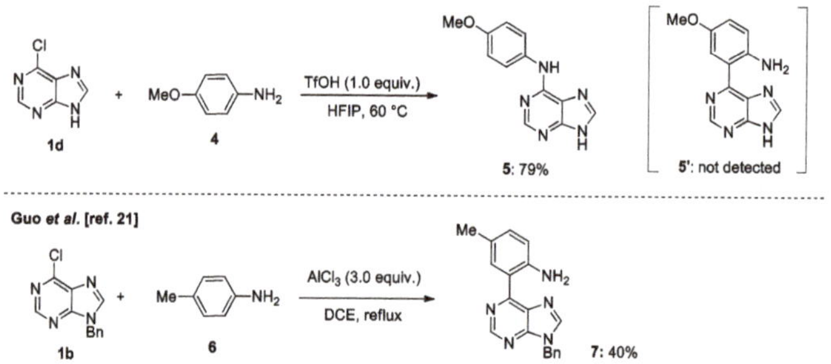

Scheme 4. Formation of different products using Brønsted acid and aluminum chloride as activators.

The success of the metal-free coupling reaction relies on the use of HFIP as the solvent. Although the precise role of HFIP [44–46] remains unclear, we presume that HFIP can increase the acidity of TfOH (Brønsted acid activation by H-bond donor) to enhance the reactivity of halopurine electrophiles through the purine nitrogen atoms [47–50]. Importantly, HFIP offers a means of improving the leaving group ability of the chloride atom in the purine substrates through hydrogen bonding as well as solvation [51]. Recently, such unique role of fluoroalcohol as the H-bond donor has been discussed in several Brønsted acid catalyzed reactions in regard to its ability to accelerate substitution processes [37,47–51]. Interestingly, these cases would involve intermediates activated by hydrogen bonding with fluoroalcohol, and, with our present system, the formation of a similar intermediate would also be expected to facilitate the aromatic substitution reactions.

3. Conclusions

In conclusion, we have developed a new metal-free coupling method of halopurines for the syntheses for diverse C6-aryl-substituted purine derivatives based on Brønsted acid activation. The combination of TfOH and HFIP is an efficient and practical methodology for the direct nucleophilic arylation of halopurines under mild conditions. We have elucidated that the unique properties of HFIP (hydrogen-bonding formation and weak nucleophilicity) could facilitate the direct arylation of halopurines by various nucleophilic arene molecules. Further investigations on the utilization of the obtained purine biaryls are currently underway in our research group.

4. Experimental Section

The melting points (mp) are uncorrected. The ^{1}H-NMR (and ^{13}C-NMR) spectra of the coupling products 3 and 5 were recorded by a JEOL JMN-400 spectrometer (JEOL Ltd., Tokyo, Japan) operating at 400 MHz (100 MHz for ^{13}C-NMR) in DMSO-d_6 at 25 °C with tetramethylsilane as the internal standard. The data are reported as follows: chemical shift in part per million (δ), multiplicity (s = singlet, d = doublet, t = triplet, q = quartet, br = broad singlet, m = multiplet), integration, and coupling constant (Hz). The infrared spectra (IR) were obtained using a Hitachi 270–50 spectrometer (Hitachi Ltd., Tokyo, Japan); absorptions are reported in reciprocal centimeters (cm^{-1}) for representative peaks. High-resolution mass spectra were measured with a Thermo Scientific Exactive Plus Orbitrap (Thermo Fisher Scientific., Inc., Waltham, MA, USA). All chemicals used in this study are commercially available and were used without further purification. Regarding fluoroalcohol, we used commercial water-containing hexafluoroisopropanol (HFIP) as supplied for the reactions.

4.1. General Procedure for Brønsted Acid Catalyzed Arylation of Halopurines in Fluoroalcohol (Table 2 and Scheme 3)

To a stirred solution of chloropurine 1 (0.50 mmol) in hexafluoroisopropanol (5 mL), aromatic nucleophile 2 (0.55 mmol, 1.1 equiv) and trifluoromethanesulfonic acid (TfOH, 44 µL, 0.5 mmol, 1 equiv) were successively added. The resulting mixture was stirred at 60 °C for 24 h. After completion of the reaction checked by TLC, the reaction mixture was poured into sat. NaHCO$_3$ aqueous. The resultant solution was extracted with ethyl acetate, dried with solid sodium sulfate, and then concentrated. The residue was purified by short-column chromatography on silica gel using hexane-ethyl acetate as the eluent to give the purine aromatic-linked compound 3 in the indicated yield in Table 2 or Scheme 3.

Compound 6-(1-methyl-1H-indol-3-yl)-7H-purine (**3aa**). A yellow powder, mp 346–350 °C. IR: 3647, 1732 cm^{-1}. ^{1}H-NMR (400 MHz, DMSO-d_6) δ 3.95 (s, 3H), 7.21–7.31 (m, 2H), 7.55 (d, J = 7.9 Hz, 1H), 8.49 (s, 1H), 8.81–8.84 (m, 2H), 8.97 (s, 1H), 13.4 (bs, 1H) ppm; ^{13}C-NMR (100 MHz, DMSO-d_6) δ 33.1, 110.4 (x 2), 121.0, 122.4, 122.9, 126.2, 128.0, 136.2, 137.2, 142.7, 151.3, 152.1, 152.3 ppm; HRMS (DART): Calcd. for C$_{14}$H$_{12}$N$_5$ [M + H]$^+$: 250.1087, found: 250.1087.

Compound 6-(1-methyl-1H-indol-3-yl)-9-phenylmethyl)-9H-purine (**3ba**) [21]. A yellow powder, mp 163–166 °C. IR: 3047, 2932, 1581, 1536, 1498, 1475 cm^{-1}. ^{1}H-NMR (400 MHz, DMSO-d_6) δ 3.96 (s, 3H), 5.51(s, 2H), 7.23–7.59 (m, 7H), 7.57 (d, 1H, J = 7.8 Hz), 8.68 (s, 1H), 8.80 (d, 1H, J = 7.8 Hz), 8.86 (s, 1H), 8.96 (s, 1H) ppm; ^{13}C-NMR (100 MHz, DMSO-d_6) δ 33.0, 54.9, 110.2, 110.3, 121.1, 122.5, 122.8, 126.2, 127.6, 127.8, 128.1, 128.7, 136.3, 136.8, 137.2, 144.4, 150.3, 152.3, 152.7 ppm.

Compound 6-(1H-indol-3-yl)-9-phenylmethyl-9H-purine (**3bb**) [21]. A yellow powder, mp 178–180 °C. IR: 3631, 1688 cm^{-1}. ^{1}H-NMR (400 MHz, DMSO-d_6) δ 5.51 (s, 2H), 7.17–7.41 (m, 7H), 7.52 (d, 1H, J = 6.8 Hz), 8.66 (s, 1H), 8.76–8.89 (m, 2H), 8.99 (s, 1H), 12.0 (s, 1H) ppm; ^{13}C-NMR (100 MHz, DMSO-d_6) δ 46.3, 111.2, 112.0, 120.9, 122.4, 122.7, 125.7, 127.8, 128.4, 128.8, 132.8, 136.6, 136.9, 144.5, 150.3, 152.3, 153.1 ppm.

Compound 6-(5-methyl-1H-indol-3-yl)-9-phenylmethyl-9H-purine (**3bc**). A brown liquid, IR: 3649, 1690, 1559, 1540 cm^{-1}. ^{1}H-NMR (400 MHz, DMSO-d_6) δ 2.36 (s, 3H), 5.45 (s, 2H), 7.08 (d, J = 8.3 Hz, 1H),

7.26–7.36 (m, 6H), 8.53 (s, 1H), 8.59 (s, 1H), 8.82 (s, 1H), 8.90 (s, 1H), 11.8 (s, 1H) ppm; ^{13}C-NMR (100 MHz, DMSO-d_6) δ 21.5, 46.0, 110.8, 111.7, 122.3, 124.0, 126.0, 127.6, 127.9, 128.3, 128.8, 129.5, 132.9, 135.0, 136.9, 144.3, 150.2, 152.3, 153.2 ppm; HRMS (DART): Calcd. for $C_{21}H_{18}N_5^+$ [M + H]$^+$: 340.1557, found: 340.1557.

Compound 6-(5-methoxy-1H-indol-3-yl)-9-phenylmethyl-9H-purine (**3bd**). A brown solid, mp 229–231 °C. IR: 3595, 1704, 1559, 1508, 1437 cm^{-1}. ^{1}H-NMR (400 MHz, DMSO-d_6) δ 3.89 (s, 3H), 5.57 (s, 2H), 6.92 (d, 1H, J = 8.6 Hz), 7.31–7.49 (m, 6H), 8.38 (s, 1H), 8.71 (s, 1H), 8.93 (s, 1H), 9.01 (s, 1H), 11.9 (s, 1H) ppm; ^{13}C-NMR (100 MHz, DMSO-d_6) δ 46.3, 55.4, 104.6, 111.0, 112.2, 112.7, 126.4, 127.6, 127.9, 128.2, 128.8, 131.6, 133.2, 136.9, 144.3, 150.2, 152.3, 153.2, 154.7 ppm; HRMS (DART): Calcd. for $C_{21}H_{18}N_5O^+$ [M + H]$^+$: 356.1506, found: 356.1507.

Compound 4-(7H-purin-6-yl)-naphthalene-1-ol (**3af**). A yellow powder, mp 204–208 °C. IR: 3650, 1541 cm^{-1}. ^{1}H-NMR (400 MHz, DMSO-d_6) δ 7.09 (d, 1H, J = 8.6 Hz), 7.51–7.57 (m, 2H), 8.06–8.63 (m, 4H), 9.06 (s, 1H), 10.8 (bs, 1H), 13.3 (bs, 1H) ppm; ^{13}C-NMR (100 MHz, DMSO-d_6) δ 107.8, 122.6, 123.5, 125.0, 125.3, 125.8, 127.3, 131.4, 132.2, 152.0, 155.5 ppm; HRMS (DART): Calcd. for $C_{15}H_{11}N_4O^+$ [M + H]$^+$: 263.0927, found: 263.0928.

Compound 6-(4-methoxynaphthalen-1-yl)-7H-purine (**3ag**). A yellow powder, mp 170–172 °C. IR: 3629, 1704, 1542, 1508 cm^{-1}. ^{1}H-NMR (400 MHz, DMSO-d_6) δ 4.06 (s, 3H), 7.16 (d, 1H, J = 8.6 Hz), 7.49–7.59 (m, 2H), 8.22–8.30 (m, 3H), 8.58 (bs, 1H), 9.02 (s, 1H), 13.6 (bs, 1H) ppm; ^{13}C-NMR (100 MHz, DMSO-d_6) δ 56.0, 103.9, 121.7, 124.8, 125.0, 125.6, 127.1, 131.5, 151.6, 156.3 ppm; HRMS (DART): Calcd. for $C_{16}H_{13}N_4O^+$ [M + H]$^+$: 277.1084, found: 277.1082.

Compound 6-(4-methoxynaphthalen-1-yl)-9-phenylmethyl-9H-purine (**3bg**) [21]. A white solid, mp 197–198 °C. IR: 3672, 2968, 1507 cm^{-1}. ^{1}H-NMR (400 MHz, DMSO-d_6) δ 4.02 (s, 3H), 5.55 (s, 2H), 7.12 (d, 1H, J = 8.0 Hz), 7.24–7.36 (m, 3H), 7.42 (d, 2H, J = 7.4 Hz), 7.48–7.55 (m, 2H), 8.13 (d, 1H, J = 8.6 Hz), 8.24–8.28 (m, 1H), 8.44–8.49 (m, 1H), 8.75 (s, 1H), 9.09 (s, 1H) ppm; ^{13}C-NMR (100 MHz, DMSO-d_6) δ 46.5, 55.8, 103.8, 121.7, 124.5, 125.0, 125.4, 125.9, 127.0, 127.8, 127.9, 128.7, 131.6, 131.8, 136.5, 146.1, 151.7, 151.8, 156.4, 156.5 ppm.

Compound 6-(1,3,5-trimethoxyphen-4-yl)-9-phenylmethyl-9H-purine (**3bh**) [21]. A white solid, mp 251–253 °C. IR: 3650, 1698 cm^{-1}. ^{1}H-NMR (400 MHz, DMSO-d_6) δ 3.58 (s, 6H), 3.84 (s, 3H), 5.49 (s, 2H), 6.35 (s, 2H), 7.27–7.47 (m, 5H), 8.61 (s, 1H), 8.90 (s, 1H) ppm; ^{13}C-NMR (100 MHz, DMSO-d_6) δ 46.6, 55.5, 55.7, 91.0, 106.6, 128.1, 128.8, 133.6, 136.6, 145.7, 150.8, 151.8, 154.2, 158.7, 162.0 ppm.

Compound 6-(1,3-dimethoxyphen-4-yl)-9-phenylmethyl-9H-purine (**3bi**) [52]. A white solid, mp 125–127 °C. IR: 3671, 1707 cm^{-1}. ^{1}H-NMR (400 MHz, DMSO-d_6) δ 3.74 (s, 3H), 3.83 (s, 3H), 5.50 (s, 2H), 6.66 (d, 1H, J = 8.8 Hz), 6.73 (s, 1H), 7.24–7.42 (m, 5H), 7.53 (d, 1H, J = 8.3 Hz), 8.66 (s, 1H), 8.93 (s, 1H) ppm; ^{13}C-NMR (100 MHz, DMSO-d_6) δ 46.5, 55.4, 55.7, 98.9, 105.2, 117.8, 127.8, 128.0, 128.8, 131.9, 132.5, 136.7, 145.6, 151.1, 151.8, 155.4, 158.8, 162.0 ppm.

Compound 4-(9-phenylmethyl-9H-purin-6-yl)-benzene-1,3-diol (**3bj**) [21]. A yellow solid, mp 250–253 °C. IR: 3691, 2983, 1686, 1507, 1318 cm^{-1}. ^{1}H-NMR (400 MHz, DMSO-d_6) δ 5.47 (s, 2H), 6.36 (s, 1H), 6.48 (d, 1H, J = 8.8 Hz), 7.20–7.34 (m, 5H), 8.72 (s, 1H), 8.79 (s, 1H), 9.21 (d, 1H, J = 8.8 Hz), 14.6 (s, 1H) ppm; ^{13}C-NMR (100 MHz, DMSO-d_6) δ 46.6, 103.3, 106.4, 108.2, 109.1, 127.8, 128.8, 133.8, 136.5, 145.6, 149.5, 151.1, 154.4, 158.6, 162.5, 163.5 ppm.

Compound 6-(1-phenyl-1H-indol-3-yl)-7-phenylmethyl-7H-purine (**3ck**). A yellow powder, mp 191–194 °C. IR: 1693, 1521 cm^{-1}. ^{1}H-NMR (400 MHz, DMSO-d_6) δ 5.54 (s, 2H), 7.26–7.39 (m, 7H), 7.49–7.74 (m, 6H), 8.72 (s, 1H), 8.91–8.95 (m, 2H), 9.14 (s, 1H) ppm; ^{13}C-NMR (100 MHz, DMSO-d_6) δ 46.4, 110.9, 112.8, 122.1, 123.2, 123.6, 124.5, 126.8, 127.6, 127.7, 127.9, 128.8, 130.1, 134.5, 136.0, 136.7, 138.2, 143.3, 145.0, 150.5, 152.0, 152.3 ppm. HRMS (DART): Calcd. for $C_{26}H_{20}N_5$ [M + H]$^+$: 402.1713, found: 402.1713.

Compound 4-(7-phenylmethyl-7H-purin-6-yl)-naphthalene-1-ol (**3cf**). A yellow powder, mp 259–263 °C. IR: 3613, 2980, 1697 cm^{-1}. ^{1}H-NMR (400 MHz, DMSO-d_6) δ 4.96-5.04 (m, 2H), 6.18 (d, 2H, J = 7.3 Hz), 6.77 (t, 2H, J = 7.3 Hz), 6.86 (t, 1H, J = 7.3 Hz), 6.93 (d, 1H, J = 7.8 Hz), 7.18 (d, 1H, J = 8.3 Hz), 7.26 (t, 2H, J = 5.4 Hz), 7.42 (t, 1H, J = 7.3 Hz), 8.22 (d, 1H, J = 8.3 Hz), 8.91 (s, 1H), 9.04 (s, 1H), 10.7 (bs, 1H) ppm; ^{13}C-NMR (100 MHz, DMSO-d_6) δ 49.8, 107.0, 122.2, 123.4, 123.6, 124.3, 124.7, 124.9, 125.7, 126.9, 127.3, 127.9, 128.8, 132.3, 135.6, 150.8, 151.7, 152.0, 154.9, 161.6 ppm; HRMS (DART): Calcd. for $C_{22}H_{17}N_4O^+$ [M + H]$^+$: 353.1397, found: 353.1395.

4.2. General Procedure for Brønsted Acid Catalyzed N-Coupling of Aniline Derivatives to Halopurines in Fluoroalcohol (Scheme 4)

To a stirred solution of 9H-chloropurine **1d** (77.3 mg, 0.50 mmol) in hexafluoroisopropanol (5 mL) *p*-methoxyaniline **4** (67.8 mg, 0.55 mmol, 1.1 equiv) and trifluoromethanesulfonic acid (TfOH, 44 µL, 0.5 mmol, 1 equiv) were successively added. The resulting mixture was stirred at 60 °C for 24 h. After completion of the reaction checked by TLC, the reaction mixture was poured into sat. NaHCO$_3$ aqueous. The resultant solution was extracted with ethyl acetate, dried with solid sodium sulfate, and then concentrated. The residue was purified by short-column chromatography on silica gel using hexane-ethyl acetate as the eluent to give *N*-(4-methoxyphenyl)-9H-purine-6-amine **5** in 79% yield (95.3 mg, 0.395 mmol) as a white powder.

Compound N-(4-methoxyphenyl)-9H-purine-6-amine (**5**) [53]. A white solid, mp 266–267 °C. IR: 3673, 3630 cm^{-1}. ^{1}H-NMR (400 MHz, DMSO-d_6) δ 3.33 (s, 3H), 6.89 (d, 2H, J = 9.3 Hz), 7.79 (d, 2H, J = 8.8 Hz), 8.22 (s, 1H), 8.29 (s, 1H), 9.61 (s, 1H), 13.1 (bs, 1H) ppm; ^{13}C-NMR (100 MHz, DMSO-d_6) δ 55.2, 113.6, 119.2, 122.4, 132.8, 139.5, 150.2, 151.9, 154.9, 159.7 ppm.

Author Contributions: N.T. and T.D. conceived and designed the experiments and directed the project; N.T., T.S., T.M., A.H., and S.U. performed the experiments; T.D. and K.K. analyzed the data and checked the experimental details; T.H. contributed to critical discussion and presentation of the results; N.T. and T.D. wrote the paper.

Funding: This work was supported by JSPS KAKENHI Grant Number 16K18854. T.D. acknowledges the support from JSPS KAKENHI (C) Grant Number 19K05466 and the Ritsumeikan Global Innovation Research Organization (R-GIRO) project.

Acknowledgments: We thank Central Glass Co., Ltd., for the generous gift of fluoroalcohol.

Conflicts of Interest: The authors declare no conflict of interest.

References

1. Legraverend, M.; Grierson, D.S. The purines: Potent and versatile small molecule inhibitors and modulators of key biological targets. *Bioorg. Med. Chem.* **2006**, *14*, 3987–4006. [CrossRef] [PubMed]
2. Bakkestuen, A.K.; Gundersen, L.L.; Utenova, B.T. Synthesis, biological activity, and SAR of antimycobacterial 9-aryl-, 9-arylsulfonyl-, and 9-benzyl-6-(2-furyl)purines. *J. Med. Chem.* **2005**, *48*, 2710–2723. [CrossRef] [PubMed]
3. Hocek, M.; Naus, P.; Pohl, R.; Votruba, I.; Furman, P.A.; Tharnish, P.M.; Otto, M.J. Cytostatic 6-arylpurine nucleosides. 6. SAR in anti-HCV and cytostatic activity of extended series of 6-hetarylpurine ribonucleosides. *J. Med. Chem.* **2005**, *48*, 5869–5873. [CrossRef] [PubMed]
4. Storr, T.E.; Strohmeier, J.A.; Baumann, C.G.; Fairlamb, I.J.S. A sequential direct arylation/Suzuki–Miyaura cross-coupling transformation of unprotected 20-deoxyadenosine affords a novel class of fluorescent analogues. *Chem. Commun.* **2010**, *46*, 6470–6472. [CrossRef] [PubMed]
5. Shaw, G. *Comprehensive Heterocyclic Chemistry*; Katritzky, A.R., Rees, C.W., Eds.; Pergamon Press: Oxford, UK, 1984; Volume 5, pp. 501–597.
6. Agrofoglio, L.A.; Gillaizeau, I.; Saito, Y. Palladium-assisted routes to nucleosides. *Chem. Rev.* **2003**, *103*, 1875–1916. [CrossRef]
7. Hocek, M. Synthesis of purines bearing carbon substituents in positions 2, 6 or 8 by metal- or organometal-mediated C–C bond-forming reaction. *Eur. J. Org. Chem.* **2003**, 245–254. [CrossRef]

8. Lakshman, M.K.; Hilmer, J.H.; Martin, J.Q.; Keeler, J.C.; Dinh, Y.Q.; Ngassa, F.N.; Russon, L.M. Palladium catalysis for the synthesis of hydrophobic C-6 and C-2 aryl 2′-deoxynucleosides. Comparison of C–C versus C–N bond formation as well as C-6 versus C-2 reactivity. *J. Am. Chem. Soc.* **2001**, *123*, 7779–7787. [CrossRef]
9. Hocek, M.; Holy, A.; Votruba, I.; Dvorakova, H. Synthesis and cytostatic activity of substituted 6-phenylpurine bases and nucleosides: Application of the Suzuki-Miyaura cross-coupling reactions of 6-chloropurine derivatives with phenylboronic acids. *J. Med. Chem.* **2000**, *43*, 1817–1825. [CrossRef]
10. Cerna, I.; Pohl, R.; Klepetarova, B.; Hocek, M. Synthesis of 6,8,9-tri- and 2,6,8,9-tetrasubstituted purines by a combination of the Suzuki cross-coupling, N-arylation, and direct C–H arylation reactions. *J. Org. Chem.* **2008**, *73*, 9048–9054. [CrossRef]
11. Lakshman, M.K.; Thomson, P.F.; Nuqui, M.A.; Hilmer, J.H.; Sevova, N.; Boggess, B. Facile Pd-catalyzed cross-coupling of 2′-deoxyguanosine O^6-arylsulfonates with arylboronic acids. *Org. Lett.* **2002**, *4*, 1479–1482. [CrossRef]
12. Gunda, P.; Russon, L.M.; Lakshman, M.K. Pd-catalyzed amination of nucleoside arylsulfonates to yield N6-aryl-2,6-diaminopurine nucleosides. *Angew. Chem. Int. Ed.* **2004**, *43*, 6372–6377. [CrossRef] [PubMed]
13. Lakshman, M.K.; Gunda, P.; Pradhan, P. Mild and room temperature C–C bond forming reactions of nucleoside C-6 arylsulfonates. *J. Org. Chem.* **2005**, *70*, 10329–10335. [CrossRef] [PubMed]
14. Liu, J.; Robins, M.J. Fluoro, alkylsulfanyl, and alkylsulfonyl leaving groups in suzuki cross-coupling reactions of purine 2′-deoxynucleosides and nucleosides. *Org. Lett.* **2005**, *7*, 1149–1151. [CrossRef] [PubMed]
15. Liu, J.; Robins, M.J. Azoles as Suzuki cross-coupling leaving groups: Syntheses of 6-arylpurine 2′-deoxynucleosides and nucleosides from 6-(imidazol-1-yl)- and 6-(1,2,4-triazol-4-yl)purine derivatives. *Org. Lett.* **2004**, *6*, 3421–3423. [CrossRef] [PubMed]
16. Kang, F.A.; Sui, Z.; Murray, W.V. Pd-catalyzed direct arylation of tautomerizable heterocycles with aryl boronic acids via C-OH bond activation using phosphonium salts. *J. Am. Chem. Soc.* **2008**, *130*, 11300–11302. [CrossRef]
17. Langli, G.; Gundersen, L.-L.; Rise, F. Regiochemistry in Stille couplings of 2,6-dihalopurines. *Tetrahedron* **1996**, *52*, 5625–5638. [CrossRef]
18. Havelková, M.; Dvořák, D.; Hocek, M. Covalent analogues of DNA base-pairs and triplets. Part 3: Synthesis of 1,4- and 1,3-bis(purin-6-yl)benzenes and 1-(1,3-dimethyluracil-5-yl)-3 or 4-(purin-9-yl)benzenes. *Tetrahedron* **2002**, *58*, 7431–7435. [CrossRef]
19. Gundersen, L.L.; Langli, G.; Rise, F. Regioselective Pd-mediated coupling between 2,6-dichloropurines and organometallic reagents. *Tetrahedron Lett.* **1995**, *36*, 1945–1948. [CrossRef]
20. Furstner, A.; Leitner, A.; Mendez, M.; Krause, H. Iron-catalyzed cross-coupling reactions. *J. Am. Chem. Soc.* **2002**, *124*, 13856–13863. [CrossRef]
21. Guo, H.-M.; Li, P.; Niu, H.-Y.; Wang, D.-C.; Qu, G.-R. Direct synthesis of 6-arylpurines by reaction of 6-chloropurines with activated aromatics. *J. Org. Chem.* **2010**, *75*, 6016–6018. [CrossRef]
22. Edenhofer, A. Novel route to imidazoles, and their use for the synthesis of purines and 4,6-dihydro-1,2-dimethyl-8-phenylimidazo[4,5-e]-1,4-diazepin-5(1H)-one. *Helv. Chim. Acta* **1975**, *58*, 2192–2209. [CrossRef]
23. Gundersen, L.-L.; Bakkestuen, A.K.; Aasen, J.; Overas, H.; Rise, F. 6-Halopurines in palladium-catalyzed coupling with organotin and organozinc reagents. *Tetrahedron* **1994**, *50*, 9743–9756. [CrossRef]
24. Havelkova, M.; Hocek, M.; Cesnek, M.; Dvorak, D. The Suzuki-Miyaura cross-coupling reactions of 6-halopurines with boronic acids leading to 6-aryl- and 6-alkenylpurines. *Synlett* **1999**, 1145–1147. [CrossRef]
25. Takenaga, N.; Ueda, S.; Hayashi, T.; Dohi, T.; Kitagaki, S. Facile synthesis of stable uracil-iodonium(III) salts with various counterions. *Heterocycles* **2018**, *97*, 1248–1256. [CrossRef]
26. Takenaga, N.; Ueda, S.; Hayashi, T.; Dohi, T.; Kitagaki, S. Vicinal functionalization of uracil heterocycles with base activation of iodonium(III) salts. *Heterocycles* **2019**, *99*, 865–874. [CrossRef]
27. Takenaga, N.; Hayashi, T.; Ueda, S.; Satake, H.; Yamada, Y.; Kodama, T.; Dohi, T. Synthesis of uracil-iodonium(III) salts for practical utilization as nucleobase synthetic modules. *Molecules* **2019**, *24*, 3034. [CrossRef]
28. Reichardt, C. Empirical parameters of solvent polarity as linear free-energy relationships. *Angew. Chem. Int. Ed. Engl.* **1979**, *18*, 98–110. [CrossRef]
29. Schadt, F.L.; Bentley, T.W.; Schleyer, P.V.R. The S_N2-S_N1 spectrum. 2. Quantitative treatments of nucleophilic solvent assistance. A scale of solvent nucleophilicities. *J. Am. Chem. Soc.* **1976**, *98*, 7667–7675. [CrossRef]

30. Kamlet, M.J.; Abbound, J.L.M.; Abraham, M.H.; Taft, R.W. Linear solvation energy relationships. 23. A comprehensive collection of the solvatochromic parameters, .pi*., alpha., and .beta., and some methods for simplifying the generalized solvatochromic equation. *J. Org. Chem.* **1983**, *48*, 2877–2887. [CrossRef]
31. Dohi, T.; Yamaoka, N.; Kita, Y. Fluoroalcohols: Versatile solvents in hypervalent iodine chemistry and syntheses of diaryliodonium(III) salts. *Tetrahedron* **2010**, *66*, 5775–5785. [CrossRef]
32. Kamitanaka, T.; Morimoto, K.; Tsuboshima, K.; Koseki, D.; Takamuro, H.; Dohi, T.; Kita, Y. Efficient coupling reaction of quinone monoacetal with phenols leading to phenol biaryls. *Angew. Chem. Int. Ed.* **2016**, *55*, 15535–15538. [CrossRef] [PubMed]
33. Dohi, T.; Ito, M.; Morimoto, K.; Minamitsuji, Y.; Takenaga, N.; Kita, Y. Versatile direct dehydrative approach for diaryliodonium(III) salts in fluoroalcohol media. *Chem. Commun.* **2007**, 4152–4154. [CrossRef] [PubMed]
34. Dohi, T.; Ito, M.; Yamaoka, N.; Morimoto, K.; Fujioka, H.; Kita, Y. Unusual *ipso* substitution of diaryliodonium bromides initiated by a single-electron-transfer oxidizing process. *Angew. Chem. Int. Ed.* **2010**, *49*, 3334–3337. [CrossRef] [PubMed]
35. Yamaoka, N.; Sumida, K.; Itani, I.; Kubo, H.; Ohnishi, Y.; Sekiguchi, S.; Dohi, T.; Kita, Y. Single-electron-transfer (SET)-induced oxidative biaryl coupling by polyalkoxybenzene-derived diaryliodonium(III) salts. *Chem. Eur. J.* **2013**, *19*, 15004–15011. [CrossRef]
36. Dohi, T.; Ueda, S.; Hirai, A.; Kojima, Y.; Morimoto, K.; Kita, Y. Selective aryl radical transfers into N-heteroaromatics from diaryliodonoium salts with trimethoxybenzene auxiliary. *Heterocycles* **2017**, *95*, 1272–1284. [CrossRef]
37. Carbin, B.; Coxon, C.R.; Lebraud, H.; Elliott, K.J.; Matheson, C.J.; Meschini, E.; Roberts, A.R.; Turner, D.M.; Wong, C.; Cano, C.; et al. Trifluoroacetic acid in 2,2,2-trifluoroethanol facilitates S_NAr reactions of heterocycles with arylamines. *Chem. Eur. J.* **2014**, *20*, 2311–2317. [CrossRef]
38. Whitfield, H.J.; Griffin, R.J.; Hardcastle, I.R.; Henderson, A.; Meneyrol, J.; Mesguiche, V.; Sayle, K.L.; Golding, B.T. Facilitation of addition–elimination reactions in pyrimidines and purines using trifluoroacetic acid in trifluoroethanol. *Chem. Commun.* **2003**, 2802–2803. [CrossRef]
39. Marchetti, F.; Cano, C.; Curtin, N.J.; Golding, B.T.; Griffin, R.J.; Haggerty, K.; Newell, D.R.; Parsons, R.J.; Payne, S.L.; Wang, L.Z.; et al. Synthesis and biological evaluation of 5-substituted O^4-alkylpyrimidines as CDK2 inhibitors. *Org. Biomol. Chem.* **2010**, *8*, 2397–2407. [CrossRef]
40. Wong, C.; Griffin, R.J.; Hardcastle, I.R.; Northen, J.S.; Wang, L.Z.; Golding, B.T. Synthesis of sulfonamide-based kinase inhibitors from sulfonates by exploiting the abrogated S_N2 reactivity of 2,2,2-trifluoroethoxysulfonates. *Org. Biomol. Chem.* **2010**, *8*, 2457–2464. [CrossRef]
41. Bégué, J.; Bonnet-delpon, D.; Crousse, B. Fluorinated alcohols: A new medium for selective and clean reaction. *Synlett* **2004**, 18–29. [CrossRef]
42. Shuklov, I.A.; Dubrovina, N.V.; Boerner, A. Fluorinated alcohols as solvents, cosolvents and additives in homogeneous catalysis. *Synthesis* **2007**, 2925–2943. [CrossRef]
43. Baeza, A.; Najera, C. Recent advances in the direct nucleophilic substitution of allylic alcohols through S_N1-type reactions. *Synthesis* **2014**, *46*, 25–34. [CrossRef]
44. Berkessel, A.; Adrio, J.A. Dramatic acceleration of olefin epoxidation in fluorinated alcohols: Activation of hydrogen peroxide by multiple H-bond networks. *J. Am. Chem. Soc.* **2006**, *128*, 13412–13420. [CrossRef] [PubMed]
45. Colomer, I.; Chamberlain, A.E.R.; Haughey, M.B.; Donohoe, T.J. Hexafluoroisopropanol as a highly versatile solvent. *Nat. Rev. Chem.* **2017**, *1*, 0088. [CrossRef]
46. Zhou, Z.; Cheng, Q.-Q.; Kürti, L. Aza-Rubottom oxidation: Synthetic access to primary α-aminoketones. *J. Am. Chem. Soc.* **2019**, *141*, 2242–2246. [CrossRef] [PubMed]
47. Liu, J.; Robins, M.J. S_NAr displacements with 6-(fluoro, chloro, bromo, iodo, and alkylsulfonyl)purine nucleosides: Synthesis, kinetics, and mechanism. *J. Am. Chem. Soc.* **2007**, *129*, 5962–5968. [CrossRef] [PubMed]
48. Vukovic, V.D.; Richmond, E.; Wolf, E.; Moran, J. Catalytic Friedel-Crafts reactions of highly electronically deactivated benzylic alcohols. *Angew. Chem. Int. Ed.* **2017**, *56*, 3085–3089. [CrossRef]
49. Liu, W.; Wang, H.; Li, C.-J. Metal-free markovnikov-type alkyne hydration under mild conditions. *Org. Lett.* **2016**, *18*, 2184–2187. [CrossRef]
50. Dohi, T.; Hu, Y.; Kamitanaka, T.; Washimi, N.; Kita, Y. [3 + 2] Coupling of quinone monoacetals by combined acid-hydrogen bond donor. *Org. Lett.* **2011**, *13*, 4814–4817. [CrossRef]

51. Lee, J.W.; Oliveira, M.T.; Jang, H.B.; Lee, S.; Chi, D.Y.; Kim, D.W.; Song, C.E. Hydrogen-bond promoted nucleophilic fluorination: Concept, mechanism and applications in positron emission tomography. *Chem. Soc. Rev.* **2016**, *45*, 4638–4650. [CrossRef]
52. Braendvang, M.; Gundersen, L.-L. Selective anti-tubercular purines: Synthesis and chemotherapeutic properties of 6-aryl- and 6-heteroaryl-9-benzylpurines. *Bioorg. Med. Chem.* **2005**, *13*, 6360–6373. [CrossRef]
53. Wang, X.; Han, C.; Wu, K.; Luo, L.; Wang, Y.; Du, X.; He, Q.; Ye, F. Design, synthesis and ability of non-gold complexed substituted purine derivatives to inhibit LPS-induced inflammatory response. *Eur. J. Med. Chem.* **2018**, *149*, 10–21. [CrossRef]

Sample Availability: Samples of the products are available from the authors.

© 2019 by the authors. Licensee MDPI, Basel, Switzerland. This article is an open access article distributed under the terms and conditions of the Creative Commons Attribution (CC BY) license (http://creativecommons.org/licenses/by/4.0/).

Article

Sequential MCR via Staudinger/Aza-Wittig versus Cycloaddition Reaction to Access Diversely Functionalized 1-Amino-1*H*-Imidazole-2(3*H*)-Thiones

Cecilia Ciccolini, Giacomo Mari, Gianfranco Favi *, Fabio Mantellini, Lucia De Crescentini and Stefania Santeusanio *

Department of Biomolecular Sciences, Section of Chemistry and Pharmaceutical Technologies, University of Urbino "Carlo Bo", Via I Maggetti 24, 61029 Urbino (PU), Italy; c.ciccolini@campus.uniurb.it (C.C.); giacomo.mari@uniurb.it (G.M.); fabio.mantellini@uniurb.it (F.M.); lucia.decrescentini@uniurb.it (L.D.C.)
* Correspondence: gianfranco.favi@uniurb.it (G.F.); stefania.santeusanio@uniurb.it (S.S.)

Academic Editor: Wim Dehaen
Received: 25 September 2019; Accepted: 15 October 2019; Published: 21 October 2019

Abstract: A multicomponent reaction (MCR) strategy, alternative to the known cycloaddition reaction, towards variously substituted 1-amino-1*H*-imidazole-2(3*H*)-thione derivatives has been successfully developed. The novel approach involves α-halohydrazones whose azidation process followed by tandem Staudinger/aza-Wittig reaction with CS_2 in a sequential MCR regioselectively leads to the target compounds avoiding the formation of the regioisomer iminothiazoline heterocycle. The approach can be applied to a range of differently substituted α-halohydrazones bearing also electron-withdrawing groups confirming the wide scope and the substituent tolerance of the process for the synthesis of the target compounds. Interestingly, the concurrent presence of reactive functionalities in the scaffolds so obtained ensures post-modifications in view of *N*-bridgeheaded heterobicyclic structures.

Keywords: multicomponent reaction; α-halohydrazones; Staudinger reaction; aza-Wittig; 1*H*-imidazole-2(3*H*)-thione; 2*H*-imidazo[2,1-*b*][1,3,4]thiadiazine

1. Introduction

Imidazoles belong to an important class of heterocyclic compounds that play a crucial role in various biochemical processes [1]. A lot of imidazole-based molecules have been shown bioactivities, [2] such as antifungal, antiinflammatory, antihystamine, antihelmintic, analgesic, antineoplastic, antihypertensive activity [3–7].

Among imidazole derivatives, imidazole-2-thiones have been associated to a special class of biologically relevant thiourea derivatives [8] endowed with antithyroid [9], antiproliferative [10], matrix metalloproteinases (MPP) inhibitory [11] properties and can be used as building blocks for the synthesis of *N*-aminoimidazole with antiretroviral activity [12].

To date, the most widespread method used for the synthesis of *N*-substituted 1-amino-1*H*-imidazol-2(3*H*)-thiones can be referred to the Schantl's protocol, which consists of reacting α-haloketones with potassium thiocyanate and monosubstituted arylhydrazines in weak acidic medium (Scheme 1) [13–19]. This multistep reaction is considered to proceed via the formation of conjugated azoalkenes, derived from α-thiocyanatohydrazones **D** (Scheme 2) and dipolarophile isothiocyanic acid intermediate that in turn undergo a [3+2] cycloaddition reaction providing substituted 1-arylamino-1*H*-imidazole-2(3*H*)-thione **I** scaffolds [20,21].

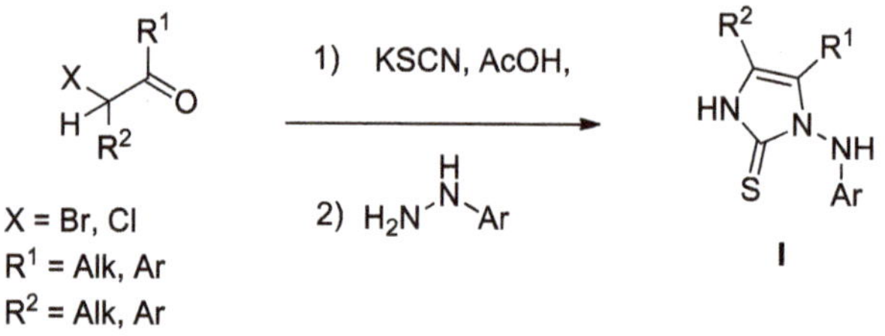

Scheme 1. Schantl's protocol for the synthesis of N-substituted 1-amino-1H-imidazole-2(3H)-thione derivatives **I**.

Even if this method appears robust, it seems to suffer of some limitations in terms of insertion of electron-withdrawing groups placed on the α-halohydrazone precursors of conjugated azoalkene intermediates. In this regard, for our research purposes, we tried to apply the Schantl's method reacting 2-chloro-N,N-dimethyl-3-oxobutanamide (**A**), potassiun thiocyanate (**B**) and *tert*-butyl hydrazinecarboxylate (**C**) in acetic acid to obtain the corresponding N-substituted 1-amino-1H-imidazole-2(3H)-thione derivative **I** but without success. As shown in Scheme 2, instead of the cycloaddition, a 5-exo-dig cyclization reaction leading to 2-iminothiazole **II** took place. This evidence is in agreement with the result obtained by Lagoja and coworkers where a pathway involving the key α-thiocyanatohydrazone intermediate **D** is invoked [12].

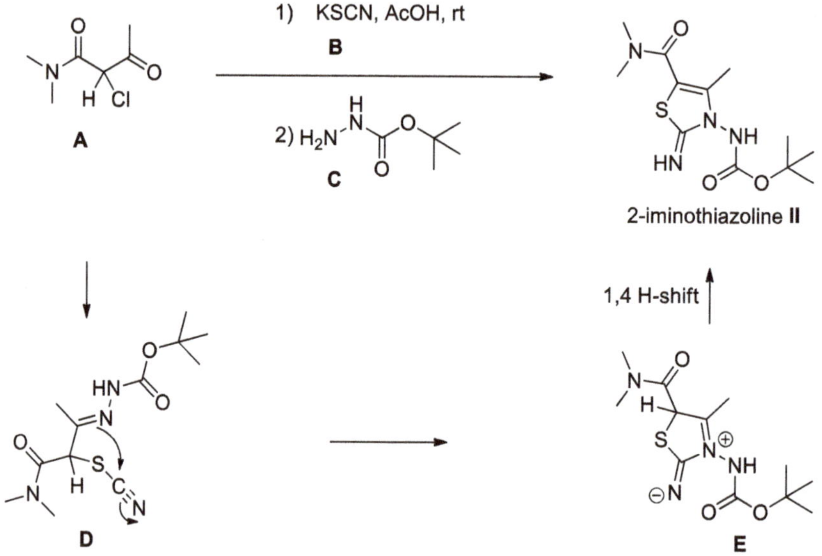

Scheme 2. Pathway for the formation of 2-iminothiazoline heterocycle **II**.

The structure of the iminothiazoline **II** was confirmed by comparison of the spectral data of the same compound obtained by means of a different procedure previously described by some of us that foresees the conjugated hydrothiocyanation of the pertinent conjugated azoalkene in acidic medium followed by intramolecular cyclization [22].

Inspired by our previous experience [23], and in order to perform a complete regioselective-oriented method for the desired 1-amino-1H-imidazole-2(3H)-thiones **I**, we have planned a different strategy that avoids the use of bidentate-nucleophilic reagents such as the potassium thiocyanate. In the construction of **I**, three strategic disconnections between the N1-C2, C2-N3 and N3-C4 were hypothesized (Scheme 3).

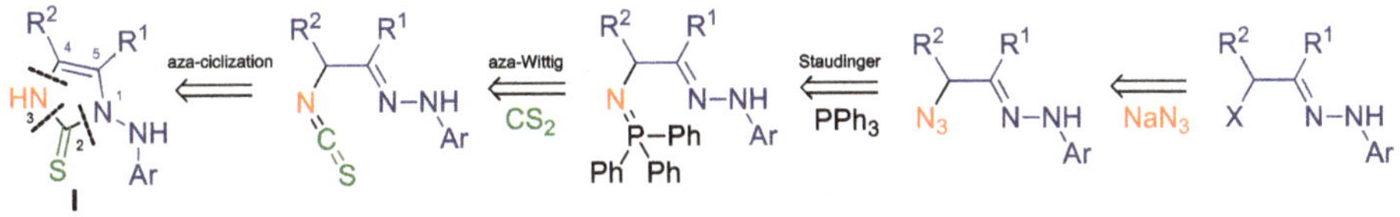

Scheme 3. Our hypothesized disconnection of 1-amino-1*H*-imidazole-2(3*H*)-thione **I** derivatives.

We reasoned that the azidation process of the pertinent α-halohydrazone derivative followed by tandem Staudinger/aza-Wittig reaction with CS$_2$ could have been a successful route [24,25].

2. Results and Discussion

To validate our hypothesis we began to explore the process step by step. Thus, α-chlorohydrazone derivative **1a** [26–30] (2.0 mmol) dissolved in THF (9.0 mL) subjected to α-azidation using an ice-cooled aqueous solution of NaN$_3$ [31] (2.0 mmol/1.0 mL) under magnetic stirring at room temperature. After the evaporation of the solvent and an appropriate extraction, the α-azidohydrazone derivative **2a** was obtained in 70% yield. In the next step, the addition of a stoichiometric amount of PPh$_3$ to **2a** (1.0 mmol) dissolved in CH$_2$Cl$_2$ (5.0 mL) furnished the iminophosphorane derivative **3a** by precipitation from the reaction medium (66%). Then, **3a** (0.65 mmol) was dissolved in 5.0 mL of THF/MeOH mixture (4:1) and treated with an excess of CS$_2$ at reflux to afford, after column chromatography purification, the corresponding *N*-substituted 1-amino-2,3-dihydro-1*H*-imidazole-2-thione derivative **5a** (53%) arising from intramolecular cyclization of the α-isothiocyanate hydrazone intermediate **4a** (Scheme 4).

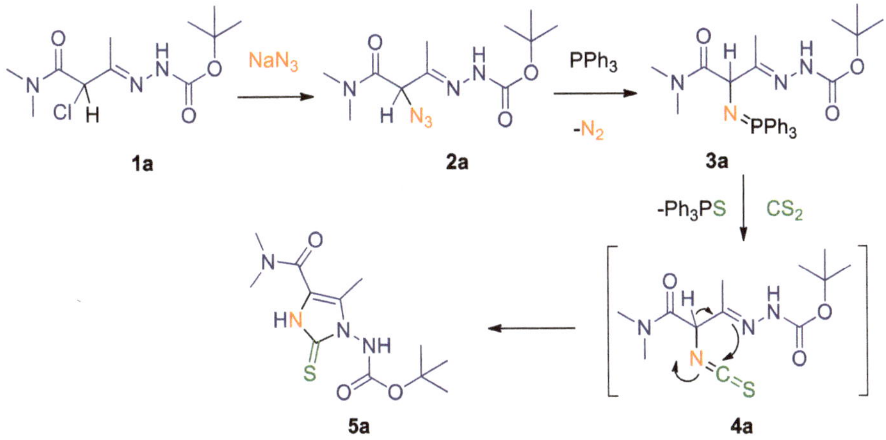

Scheme 4. Step-by-step synthetic pathway for *N*-substituted 1-amino-2,3-dihydro-1*H*-imidazole-2-thione derivative **5a**.

Motivated by this result, we aimed to develop a one-pot sequential multicomponent reaction (MCR) [32–35] as alternative method for regioselective synthesis of a new series of imidazole-2-thione-containing structures as suitable precursors for drug-like compounds [36].

Hence, our new approach to *N*-substituted 1-amino-1*H*-imidazole-2(3*H*)-thiones **5a–k** (53%–85%) is depicted in Scheme 5. The whole process that permits the formation of the desired heterocycle can be easily checked by the complete disappearance of the pertinent α-azidohydrazone derivative and by the observation of Ph$_3$P=S as byproduct (thin-layer chromatography (TLC) check, see Experimental Section). It is to be noted that for **5a**, the efficiency of the reaction benefits by this latter protocol increasing the overall yield from 25% (obtained employing the step-by-step procedure) to 79% (Table 1). Moreover, the implemented strategy broadens the substitution patterns at the amino-N1 and at C4 of the heterocycle skeleton with electron-withdrawing groups (5a–e) and tolerates the aromatic (amino-N1) and aliphatic (C4) groups, as for 5j [15,17,18] (Table 1).

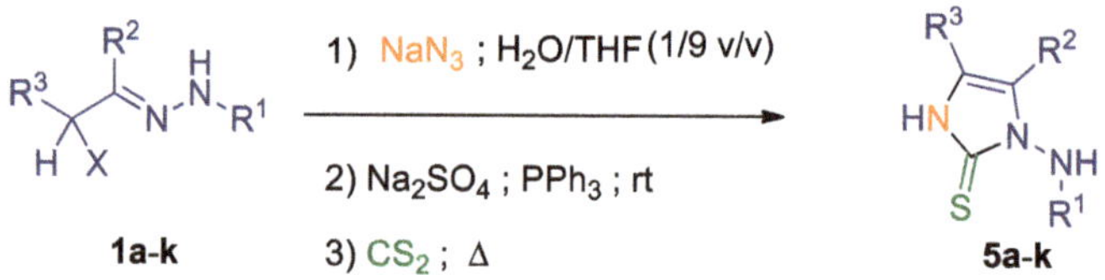

Scheme 5. New multicomponent reaction (MCR) method for N-substituted 1-amino-2,3-dihydro-1H-imidazole-2-thione derivatives 5a–k.

Table 1. Substrate scope of the MCR synthetic pathway for N-substituted 1-amino-2,3-dihydro-1H-imidazole-2-thione derivatives 5a–k.

Entry	α-Halohydrazone 1					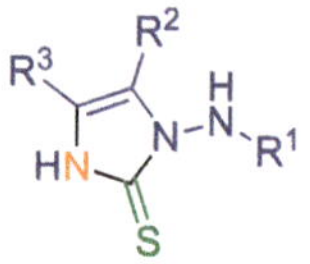 5	One-Pot MCR Yield (%) [a, b]
		R^1	R^2	R^3	X		
1	1a	CO_2Bu^t	Me	$CON(Me)_2$	Cl	5a	25 [a]; 79 [b]
2	1b	CONHPh	Me	$CON(Me)_2$	Cl	5b	53 [b]
3	1c	CO_2Bu^t	Me	$CON(Et)_2$	Br	5c	72 [b]
4	1d	CO_2Bu^t	Me	H	Cl	5d	69 [b]
5	1e	CO_2Bu^t	Me	$CONH_2$	Br	5e	58 [b]
6	1f	CO_2Bu^t	Me	CONHPh	Br	5f	67 [b]
7	1g	CONHPh	Me	H	Cl	5g	82 [b]
8	1h	COPh	Me	H	Cl	5h	59 [b]
9	1i	CONHPh	Me	Me	Cl	5i	85 [b]
10	1j	4-NO_2-Ph	Me	Me	Cl	5j	84 [b]
11	1k	CO_2Bu^t	Ph	H	Br	5k	66 [b]

[a] Overall yield of isolated product 5a from the step-by-step reaction based on 1a; [b] Overall yield of isolated products 5a–k from one-pot MCR based on 1a–k.

These results not only lie in the wide scenario of the heterocyclic scaffolds obtainable through tandem Staudinger/aza-Wittig sequence [24,25,37–41], but the concurrent presence of reactive functionalities in the target compounds 5a–k ensures post-modifications in view of heterobicyclic structures. In fact, the tautomerism thionoamide/thioloimide permits the introduction of a further element of diversity at the sulfur atom producing imidazole derivatives suitable to be combined with the useful 1-amino-Boc protected group [42] directly installed by this approach, as for 5a, 5c–f, 5k. Thus, as an example, 5c,d,f (1.0 mmol) solved in acetone (10.0 mL), were reacted with 2-bromo-1-phenylethanone (6a) (1.0 mmol), 1-chloropropan-2-one) (6b) (1.0 mmol), and ethyl 2-bromoacetate (6c) (1.0 mmol), respectively, in the presence of K_2CO_3 (1.0 mmol). After the removal of solvent followed by extraction, the corresponding α-(imidazol-2-ylthio) carbonyl compounds 7a–c were obtained as solid after column chromatography purification (84%–93%) (Scheme 6). The subsequent cleavage of the Boc-protecting group under homogeneous [43] or heterogeneous acidic conditions [44] was able to produce free amino function available to interact with the carbonyl appendage in 2-position of the ring, affording new 2H-imidazo [2,1-b][1,3,4]thiadiazine derivatives 8a,b by condensation or 2H-imidazo[2,1-b][1,3,4]thiadiazinone derivative 8c by acylic nucleophilic substitution process (Scheme 6, Table 2).

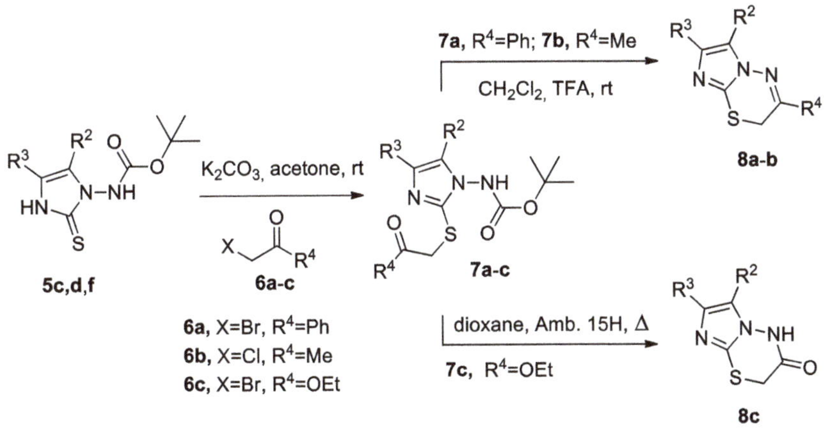

Scheme 6. Synthetic approach to 2H-imidazo[2,1-b][1,3,4]thiadiazine derivatives.

Table 2. Substrate scope of the reaction between 1-amino-2,3-dihydro-1H-imidazole-2-thione derivatives 5 with α-haloketones 6a,b or α-haloester 6c.

	5			6		7	Yield (%) [a]	8	Yield (%) [b]
	R^2	R^3		X	R^4				
5c	Me	CON(Et)$_2$	6a	Br	Ph	7a	84	8a	82
5d	Me	H	6b	Cl	Me	7b	93	8b	65
5f	Me	CONHPh	6c	Br	OEt	7c	92	8c	74

[a] Yield of isolated product 7a–c based on 6a–c; [b] Yield of isolated product 8a–c based on 7a–c.

It is worthwhile to note that the proposed synthetic pathway can offer an alternative method for obtaining 2H-imidazo[2,1-b][1,3,4]thiadiazine derivatives 8 with respect to the ring transformation of α-(oxazol-2-ylthio) ketones 9 on treatment with hydrazine hydrate 10 [45], together with the possibility of wide diversification of the substituents at the different positions of the N-bridgeheaded heterobicyclic structures. As depicted in Scheme 7, a different disconnection for the assembly of the 2H-imidazo[2,1-b][1,3,4]thiadiazine scaffold can be envisaged.

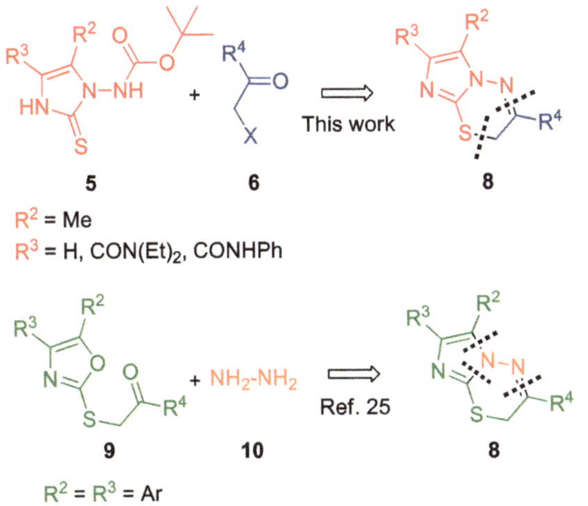

Scheme 7. Different synthetic approaches to 2H-imidazo[2,1-b][1,3,4]thiadiazine derivatives.

3. Experimental Section

3.1. General

All the commercially available reagents and solvents were used without further purification. α-Halohydrazones **1a–k** were synthesized by known procedures [26–30]. Chromatographic purification of compounds was carried out on silica gel (60–200 µm). Thin-layer chromatography (TLC) analysis was performed on pre-loaded (0.25 mm) glass supported silica gel plates (Silica gel 60, F254, Merck; Darmstadt, Germany); compounds were visualized by exposure to UV light. Melting points (Mp) were determined in open capillary tubes and are uncorrected.

All ^{1}H NMR and ^{13}C NMR spectra were recorded at 400 and 100 MHz, respectively at 25 °C on a Bruker Ultrashield 400 spectrometer (Bruker, Billerica, MA, USA). Proton and carbon spectra were referenced internally to residual solvent signals as follows: δ = 2.50 ppm for proton (middle peak) and δ = 39.50 ppm for carbon (middle peak) in DMSO-d_6 and δ = 7.27 ppm for proton and δ = 77.00 ppm for carbon (middle peak) in CDCl$_3$. The following abbreviations are used to describe peak patterns where appropriate: s = singlet, d = doublet, t = triplet q = quartet, m = multiplet and br = broad signal. All coupling constants (*J*) are given in Hz. Copies of ^{1}H-NMR and ^{13}C-NMR spectra of compounds **II**, **2a**, **3a**, **5a–k**, **7a–c**, and **8a–c** are in Supplementary Materials. FT-IR spectra were measured as Nujol mulls using a Nicolet Impact 400 (Thermo Scientific, Madison, WI, USA). Mass spectra were obtained by ESI-MS analyses performed on Thermo Scientific LCQ Fleet Ion Trap LC/MS and Xcalibur data System. High-resolution mass spectra (HRMS) were determined with ESI resource on a Waters Micromass QTOF instrument (Waters, Milford, MA, USA). Elemental analyses were within ±0.4 of the theoretical values (C, H, N).

3.2. Step-By-Step Synthetic Method for 5a

3.2.1. Synthesis of *tert*-butyl 2-(3-azido-4-(dimethylamino)-4-oxobutan-2-ylidene) hydrazinecarboxylate (**2a**)

To the α-halohydrazone **1a** (555.5 mg, 2.0 mmol) solved in THF (9.0 mL), an ice-cooled aqueous solution (1.0 mL, T = 4 °C) of NaN$_3$ (2.0 mmol, 130.02 mg) was added. The reaction mixture was stirred at room temperature until the disappearance of the starting **1a** (TLC check). THF was removed under reduced pressure and the residue was diluted with water and extracted with CH$_2$Cl$_2$ (3 × 15.0 mL). The combined organic layers were dried over anhydrous NaSO$_4$ and concentrated under reduced pressure. The crude reaction was purified by crystallization from Et$_2$O affording the α-azido derivative **2a**. Yield 70.0% (398.0 mg) as a white solid; Mp 120–124 °C (dec); ^{1}H-NMR, 400 MHz, DMSO-d_6) δ 1.44 (s, 9H, OBu^t), 1.84 (s, 3H, CH$_3$), 2.86 (s, 3H, NCH$_3$), 2.92 (s, 3H, NCH$_3$), 4.99 (s, 1H, CH), 9.82 (br s, 1H, NH, D$_2$O exch.); ^{13}C-NMR (100 MHz, DMSO-d_6) δ 13.8, 28.0, 35.5, 36.6, 64.6, 79.5, 146.2, 152.9, 166.5; IR (Nujol, ν, cm^{-1}): 3239, 3150, 2982, 2172, 2098, 1706, 1686, 1664; MS *m/z* (ESI): 285.07 (M + H)$^+$; anal. calcd. for C$_{11}$H$_{20}$N$_6$O$_3$ (284.31): 46.47; H, 7.09; N, 29.56; found: C, 46.36; H, 7.15; N, 29.65.

3.2.2. Synthesis of *tert*-butyl 2-(4-(dimethylamino)-4-oxo-3-((triphenylphosphoranylidene)amino) butan-2-ylidene)hydrazinecarboxylate (**3a**)

1.0 Mmol of **2a** (284.31 mg) was solved in CH$_2$Cl$_2$ (5.0 mL). The reaction flask was then immersed in an ice bath (T = 0 °C), and a cooled solution of PPh$_3$ (262.3 mg, 1.0 mmol) in CH$_2$Cl$_2$ (1.0 mL) was added dropwise. The reaction was brought back to room temperature and stirred until the disappearance of organic azide **2a** (monitored by TLC). The formation of phosphazene **3a** was accompanied by the development of N$_2$. After partial removal of the solvent under reduced pressure, **3a** was isolated by precipitation from a solution of CH$_2$Cl$_2$/EtOAc as white powder; yield 66% (342.3 mg); Mp 127–131 °C (dec.); ^{1}H-NMR (400 MHz, DMSO-d_6) δ 1.42 (s, 9H, OBu^t), 1.81 (s, 3H, CH$_3$), 2.59 (s, 3H, NCH$_3$), 2.73 (s, 3H, NCH$_3$), 4.62 (t, J_{H-P} = 9.2 Hz, 1H, CH), 7.57–7.90 (m, 15H, Ar), 9.64 (s, 1H, NH, D$_2$O exch.) ppm; ^{13}C-NMR (100 MHz, DMSO-d_6) δ 13.0, 28.0, 35.6, 36.1, 59.3, 79.5, 120.9 ($^1J_{C-P}$ = 102.0 Hz),

129.7 ($^2J_{C-P}$ = 14.0 Hz), 133.7 ($^3J_{C-P}$ = 11.0 Hz), 133.8 ($^3J_{-CP}$ = 12.0 Hz), 134.9 ($^4J_{C-P}$ = 2.0 Hz), 150.7, 167.0 ppm; IR (Nujol, ν, cm^{-1}): 3543, 3377, 3211, 1722, 1664; MS m/z (ESI): 519.31 (M + H)$^+$; anal. calcd. for C$_{29}$H$_{35}$N$_4$O$_3$P (518.59): C, 67.17; H, 6.80; N, 10.80; found: C, 67.31; H, 6.86; N, 10.72.

3.2.3. Synthesis of tert-butyl (4-(dimethylcarbamoyl)-5-methyl-2-thioxo-2,3-dihydro-1H-imidazol-1-yl)carbamate (5a)

0.65 Mmol of 3a (337.0 mg,) was solved in a mixture of THF:MeOH (4:1, 5.0 mL) heating. Then, 0.5 mL of CS$_2$ was added and the reaction was refluxed. The end of the reaction was defined (4.0 h) by the disappearance of 3a together with the formation of Ph$_3$P=S as byproduct (monitored by TLC). After removal of the reaction solvents under reduced pressure, a first crop of 5a was obtained as white powder from a solution of THF/light petroleum ether. A further amount was be gained by column chromatography eluting with CH$_2$Cl$_2$/EtOAc mixtures. White powder from THF/light petroleum ether; yield 53% (103.4 mg); Mp 172–173 °C (dec.); ^{1}H-NMR (400 MHz, DMSO-d_6,) δ 1.32 and 1.45 (2 s, 9H, OBu^t), 1.99 (s, 3H, CH$_3$), 2.94 [s, 6H, N(CH$_3$)$_2$], 9.69 and 10.15 (2 br s, 1H, NH, D$_2$O exch.), 12.50 (br s, 1H, NH, D$_2$O exch.) ppm; ^{13}C-NMR (100 MHz, DMSO-d_6) δ 8.9, 27.6, 27.8, 35.9, 80.8, 116.2, 128.2, 153.8, 160.2, 162.9 ppm; IR (Nujol, ν, cm^{-1}): 3188, 3115, 1741, 1645, 1607; MS m/z (ESI): 301.15 (M + H)$^+$; calcd. for C$_{12}$H$_{20}$N$_4$O$_3$S (300.38): C, 47.98; H, 6.71; N, 18.65; found: C, 48.11; H, 6.63; N, 18.57. The partition of some signals here, as well as in the following cases, is due to the N1-amide rotameric effect [46].

3.3. *Typical MCR Procedure for the Synthesis of N-Substituted 1-Amino-1H-Imidazole-2(3H)-Thione Derivatives 5a–k*

To a round flask equipped with a magnetic stirring bar containing ice-cooled solution of NaN$_3$ (1.0 mmol, 65.01 mg) dissolved in 0.5 mL of H$_2$O, the corresponding α-halohydrazone 1a–k (1.0 mmol) dissolved in THF (4.5 mL) was added. The mixture was stirred at room temperature until the disappearance of 1 (monitored by TLC). Upon completion, Na$_2$SO$_4$ (0.5 g), a solution of PPh$_3$ (1.1 mmol, 288.5 mg) in THF (1.0 mL) and CS$_2$ (1.0 mL) were added in sequence, and the mixture was refluxed for the appropriate reaction time (3.0–20.0 h). The formation of the final products 5a–k was revealed by the complete disappearance of the spot corresponding to the α-azidohydrazone 2a–k as well as the detection of the byproduct Ph$_3$P=S. The Na$_2$SO$_4$ was filtered in vacuo and washed with THF (10.0 mL). The filtrate was concentrated under reduced pressure and the residue was purified by crystallization and/or by chromatography eluting with cyclohexane:EtOAc or CH$_2$Cl$_2$:EtOAc mixtures. The resulting products 5a–k were isolated by crystallization from the specific solvents (see below). According to this procedure, 5a was obtained in 79% (237.3 mg).

N,N,5-trimethyl-1-(3-phenylureido)-2-thioxo-2,3-dihydro-1H-imidazole-4-carboxamide (5b): Yield 53% (169.3 mg), pink powder from CH$_2$Cl$_2$/Et$_2$O; Mp 247–248 °C (dec.); ^{1}H-NMR (400 MHz, DMSO-d_6) δ 2.06 (s, 3H, CH$_3$), 2.97 [s, 6H, N(CH$_3$)$_2$], 7.01 (t, J = 8.0 Hz, 1H, Ar), 7.29 (t, J = 8.0 Hz, 2H, Ar), 7.46 (d, J = 8.0 Hz, 2H, Ar), 9.00 (s, 1H, NH, D$_2$O exch.), 9.33 (br s, 1H, NH, D$_2$O exch.), 12.56 (s, 1H, NH, D$_2$O exch.) ppm; ^{13}C-NMR (100 MHz, DMSO-d_6) δ 9.3, 36.8, 116.1, 118.3, 122.3, 128.7, 129.0, 139.1, 153.6, 160.3, 162.6 ppm; IR (Nujol, ν, cm^{-1}): 3323, 3248, 3195, 3136, 1713, 1638, 1605; MS m/z (ESI): 320.40 (M + H)$^+$; calcd. for; C$_{14}$H$_{17}$N$_5$O$_2$S (319.38): C, 52.65; H, 5.37; N, 21.93; calcd. for; C$_{14}$H$_{17}$N$_5$O$_2$S (319.38): C, 52.65; H, 5.37; N, 21.93; found: C, 52.79; H, 5.44; N, 21.84.

tert-Butyl (4-(diethylcarbamoyl)-5-methyl-2-thioxo-2,3-dihydro-1H-imidazol-1-yl)carbamate (5c): Yield 72% (236.3 mg), white powder from EtOAc/THF/light petroleum ether; Mp 168–169 °C (dec.); ^{1}H-NMR (400 MHz, DMSO-d_6) δ 1.06–1.10 (m, 6H, 2xNCH$_2$CH$_3$), 1.32 and 1.45 (2s, 9H, OBu^t), 1.94 and 1.97 (2s, 3H, CH$_3$), 3.26–3.37 (m, 4H, 2xNCH$_2$CH$_3$), 9.68 and 10.07 (2 br s, 1H, NH, D$_2$O exch.), 12.49 (br s, 1H, NH, D$_2$O exch.) ppm; ^{13}C-NMR (100 MHz, DMSO-d_6) δ 8.6, 13.4, 27.5, 27.8, 34.8, 80.6, 117.0, 126.6, 153.8, 159.8, 162.7 ppm; IR (Nujol, ν, cm^{-1}): 3169, 3120, 1748, 1642, 1634; MS m/z (ESI): 329.23 (M + H)$^+$;calcd. for C$_{14}$H$_{24}$N$_4$O$_3$S (328.16): C, 51.20; H, 7.37; N, 17.06; found: C, 51.09; H, 7.42; N, 16.95.

tert-Butyl (5-methyl-2-thioxo-2,3-dihydro-1H-imidazol-1-yl)carbamate (**5d**): Yield 69% (158.1 mg), white powder from EtOAc/THF/light petroleum ether; Mp 168–169 °C (dec.); ^{1}H-NMR (400 MHz, DMSO-d_6) δ 1.32 and 1.45 (2s, 9H, OBu^t), 1.93 (s, 3H, CH$_3$), 6.60 (s, 1H, CH), 9.51 and 9.94 (2 br s, 1H, NH, D$_2$O exch.), 11.97 (br s, 1H, NH, D$_2$O exch.) ppm; ^{13}C-NMR (100 MHz, DMSO-d_6) δ 8.9, 27.9, 80.5, 108.9, 126.9, 153.9, 162.4 ppm; IR (Nujol, ν, cm^{-1}): 3271, 3144, 3098, 1744, 1732, 1640; MS *m/z* (ESI): 229.96 (M + H)$^+$; calcd. for C$_9$H$_{15}$N$_3$O$_2$S (229.09): C, 47.14; H, 6.59; N, 18.33; found: C, 47.01; H, 6.65; N, 18.41.

tert-Butyl (4-carbamoyl-5-methyl-2-thioxo-2,3-dihydro-1H-imidazol-1-yl)carbamate (**5e**): Yield 58% (157.8 mg), white powder from CH$_2$Cl$_2$/light petroleum ether; Mp 270 °C (dec.); ^{1}H-NMR (400 MHz, DMSO-d_6) δ 1.32 and 1.45 (2s, 9H, OBu^t), 2.23 and 2.26 (2s, 3H, CH$_3$), 7.23 and 7.53 (2 br s, 2H, NH$_2$, D$_2$O exch.), 9.71 and 10.17 (2s, 1H, NH, D$_2$O exch.), 12.42 (s, 1H, NH, D$_2$O exch.) ppm; ^{13}C NMR (100 MHz, DMSO-d_6) δ 8.9, 10.1, 27.6, 27.8, 80.9, 115.8, 133.1, 153.7, 159.6, 162.9 ppm; IR (Nujol, ν, cm^{-1}): 3395, 3354, 3182, 3137, 1754, 1717, 1676, 1594; MS *m/z* (ESI): 273.04 (M + H)$^+$; calcd. for C$_{10}$H$_{16}$N$_4$O$_3$S (272.09): C, 44.10; H, 5.92; N, 20.57; found: C, 44.23; H, 5.96; N, 20.45.

tert-Butyl (5-methyl-4-(phenylcarbamoyl)-2-thioxo-2,3-dihydro-1H-imidazol-1-yl)carbamate (**5f**): Yield 67% (233.2 mg), white powder from EtOAc; Mp 170–171 °C (dec.); ^{1}H-NMR (400 MHz, DMSO-d_6) δ 1.34 and 1.46 (2s, 9H, OBu^t), 2.28 (s, 3H, CH$_3$), 7.11 (t, *J* = 8.0 Hz, 1H, Ar), 7.35 (t, *J* = 8.0 Hz, 2H, Ar), 7.65 (d, *J* = 8.0 Hz, 2H, Ar), 9.68 (s, 1H, NH, D$_2$O exch.), 10.28 (s, 1H, NH, D$_2$O exch.), 12.69 (s, 1H, NH, D$_2$O exch.) ppm; ^{13}C-NMR (100 MHz, DMSO-d_6) δ 9.2, 27.8, 81.0, 116.0, 119.7, 123.8, 128.8, 133.9, 138.4, 153.7, 156.3, 163.2 ppm; IR (Nujol, ν, cm^{-1}): 3375, 3243, 3066, 1752, 1659, 1630, 1598, 1545; MS *m/z* (ESI): 349.22 (M + H)$^+$; calcd. for C$_{16}$H$_{20}$N$_4$O$_3$S (348.13): C, 55.16; H, 5.79; N, 16.08; found: C, 55.01; H, 5.72; N, 16.16.

1-(5-Methyl-2-thioxo-2,3-dihydro-1H-imidazol-1-yl)-3-phenylurea (**5g**): Yield 82% (203.4 mg), white powder from THF/EtOAc; Mp 245–248 °C (dec.); ^{1}H-NMR (400 MHz, DMSO-d_6) δ 2.01 (s, 3H, CH$_3$), 6.64 (s, 1H, CH), 6.99 (t, *J* = 8.0 Hz, 1H, Ar), 7.28 (t, *J* = 8.0 Hz, 2H, Ar), 7.46 (d, *J* = 8.0 Hz, 2H, Ar), 8.91 (s, 1H, NH, D$_2$O exch.), 9.25 (s, 1H, NH, D$_2$O exch.), 12.05 (s, 1H, NH, D$_2$O exch.) ppm; ^{13}C-NMR (100 MHz, DMSO-d_6) δ 9.1, 108.8, 118.3, 122.3, 127.6, 128.8, 139,1, 153.8, 161.9 ppm; IR (Nujol, ν, cm^{-1}): 3305, 3154, 3119, 3097, 1714, 1681, 1637, 1602; MS *m/z* (ESI): 249.07 (M + H)$^+$; calcd. for C$_{11}$H$_{12}$N$_4$OS (248.07): C, 53.21; H, 4.87; N, 22.56; found: C, 53.08; H, 4.94; N, 22.65.

N-(5-methyl-2-thioxoimidazolidin-1-yl)benzamide (**5h**): Yield 59% (137.6 mg) white powder from MeOH; Mp 240–242 °C (dec.); ^{1}H-NMR (400 MHz, DMSO-d_6) δ 1.97 (s, 3H, CH$_3$), 6.72 (s, 1H, CH), 7.56 (t, *J* = 8.0 Hz, 2H, Ar), 7.65 (t, *J* = 8.0 Hz, 1H, Ar), 7.99 (d, *J* = 8.0 Hz, 2H, Ar), 11.44 (s, 1H, NH, D$_2$O exch.), 12.15 (s, 1H, NH, D$_2$O exch.) ppm; ^{13}C-NMR (100 MHz, DMSO-d_6) δ 8.9, 109.3, 127.0, 127.7, 128.6, 131.5, 132.5, 162.0, 165.4 ppm; IR (Nujol, ν, cm^{-1}): 3168, 3106, 1666, 1631; MS *m/z* (ESI): 234.04 (M + H)$^+$; calcd. for C$_{11}$H$_{11}$N$_3$OS (233.29): C, 56.63; H, 4.75; N, 18.01; found: C, 56.76; H, 4.82; N, 17.89.

1-(4,5-Dimethyl-2.thioxo-2,3-dihydro-1H-imidazol-1-yl)-3-phenylurea (**5i**): Yield 85% (223.0 mg), white powder from THF/Et$_2$O; Mp 245–250 °C (dec.); ^{1}H-NMR (400 MHz, DMSO-d_6) δ 1.94 (s, 3H, CH$_3$), 1.99 (s, 3H, CH$_3$), 6.99 (t, *J* = 8.0 Hz, 2H, Ar), 7.28 (t, *J* = 8.0 Hz, 1H, Ar), 7.46 (d, *J* = 8.0 Hz, 2H, Ar), 8.89 (s, 1H, NH, D$_2$O exch.), 9.19 (s, 1H, NH, D$_2$O exch.), 12.00 (s, 1H, NH, D$_2$O exch.) ppm; ^{13}C-NMR (100 MHz, DMSO-d_6) δ 7.8, 8.9, 116.6, 118.3, 122.2, 122.5, 128.7, 139.1, 153.9, 160.7 ppm; IR (Nujol, ν, cm^{-1}): 3271, 3172, 3095, 1719, 1691, 1665, 1603; MS *m/z* (ESI): 263.11 (M + H)$^+$; calcd. for C$_{12}$H$_{14}$N$_4$OS (262.33): C, 54.94; H, 5.38; N, 21.36; found: C, 54.87; H, 5.46; N, 21.23.

4,5-Dimethyl-1-[(4-nitrophenyl)amino]-1H-imidazole-2(3H)-thione (**5j**): Yield 84% (222.0 mg), beige powder from THF/EtOAc/Et$_2$O; Mp 279–282 °C (dec.); ^{1}H-NMR (400 MHz, DMSO-d_6) δ 1.90 (s, 3H, CH$_3$), 2.03 (s, 3H, CH$_3$), 6.59 (d, *J* = 8.0 Hz, 2H, Ar), 8.10 (d, *J* = 8.0 Hz, 2H, Ar), 10.09 (s, 1H, NH, D$_2$O exch.), 12.19 (s, 1H, NH, D$_2$O exch.) ppm; ^{13}C-NMR (100 MHz, DMSO-d_6) δ 7.6, 9.0, 111.3, 117.8, 121.7, 125.8, 139.3, 153.0, 160.8 ppm; IR (Nujol, ν, cm^{-1}): 3199, 3094, 1673, 1594; HRMS *m/z* calcd. for [M + H]$^+$ C$_{11}$H$_{13}$N$_4$O$_2$S 265.0759; found 265.0774.

tert-Butyl (5-phenyl-2-thioxo-2,3-dihydro-1H-imidazol-1-yl)carbamate (**5k**): Yield 66% (192.3 mg), light yellow powder from THF/EtOAc/light petroleum ether; Mp 172–174 °C (dec.); ^{1}H-NMR (400 MHz, DMSO-d_6) δ 1.17 and 1.39 (2s, 9H, OBu^t), 7.18 (s, 1H, CH), 7.35–7.49 (m, 5H, Ar), 9.79 and 10.12 (2s, 1H, NH, D_2O exch.), 12.51 (br s, 1H, NH, D_2O exch.) ppm; ^{13}C-NMR (100 MHz, DMSO-d_6) δ 27.5, 27.9, 80.2, 80.5, 110.7, 110.8, 126.9, 127.0, 127.7, 127.9, 128.1, 128.5, 130.6, 130.8, 153.2, 153.9, 162.0, 164.2 ppm; IR (Nujol, ν, cm^{-1}): 3275, 3120, 3093, 1726, 1618, 1600; MS *m/z* (ESI): 292.18 (M + H)$^+$; calcd. for $C_{14}H_{17}N_3O_2S$ (291.37); C, 57.71; H, 5.88; N, 14.42; found: C, 57.83; H, 5.82; N, 14.37.

3.4. General Procedure for the Synthesis of α-(Imidazol-2-Ylthio) Carbonyl Compounds 7a–c.

To a suspension of the *N*-Boc-protected 1-amino-1*H*-imidazole-2(3*H*)-thione derivatives **5c,d,f** (1.0 mmol) and K_2CO_3 (1.0 mmol, 138 mg) in 10.0 mL of acetone, the corresponding α-halocarbonyl derivative **6a–c** (1.0 mmol) was added. The reaction mixture was kept under magnetic stirring at room temperature. Upon completion (monitored by TLC) the solvent was removed, and the crude reaction mixture was quenched to neutrality with a solution of HCl 1N and extracted with EtOAc (30.0 mL). The organic layer was washed with brine and dried over anhydrous Na_2SO_4. The solvent was removed in vacuum and the crude extract was purified by crystallization or by column chromatography eluting with cyclohexane:ethyl acetate mixtures to furnish **7a–c** derivatives in good yields (84%–93%).

tert-Butyl (4-(diethylcarbamoyl)-5-methyl-2-((2-oxo-2-phenylethyl)thio)-1H-imidazol-1-yl)carbamate (**7a**): Yield 84% (375.1 mg); white solid from Et_2O; Mp 123–126 °C; ^{1}H-NMR (400 MHz, CDCl$_3$) δ 1.14–1.23 (m, 6H, 2xNCH$_2$CH$_3$), 1.48 (s, 9H, OBu^t), 1.98 (s, 3H, CH$_3$), 3.37–3.64 (m, 4H, 2xNCH$_2$CH$_3$), 4.50 (br s, 1H, SCH$_a$H$_b$), 4.69 (br s, 1H, SCH$_a$H$_b$), 7.46 (t, *J* = 8.0 Hz, 2H, Ar), 7.58 (t, *J* = 8.0 Hz, 1H, Ar), 7.97 (d, *J* = 8.0 Hz, 2H, Ar), 9.45 (br s, 1H, NH, D_2O exch.) ppm; ^{13}C-NMR (100 MHz, CDCl$_3$) δ 8.7, 12.9, 14.5, 28.0, 40.6, 41.6, 43.4, 82.5, 128.4, 128.5, 128.7, 130.9, 133.7, 135.3, 135.4, 139.9, 153.8, 164.5, 193.7 ppm; IR (Nujol, ν, cm^{-1}): 3114, 3059, 1741, 1726, 1700, 1681, 1597, 1584; MS *m/z* (ESI): 447.35 (M + H)$^+$; calcd. for $C_{22}H_{30}N_4O_4S$ (446.56): C, 59.17; H, 6.77; N, 12.55; found: C, 59.02; H, 6.84; N, 11.67.

tert-Butyl (5-methyl-2-((2-oxopropyl)thio)-1H-imidazol-1-yl)carbamate (**7b**): Yield 93% (265.4 mg); ocher solid from EtOAc/cyclohexane; Mp 103–104 °C; ^{1}H-NMR (400 MHz, CDCl$_3$) δ 1.50 (s, 9H, OBu^t), 2.14 (s, 3H, CH$_3$), 2.24 (s, 3H, COCH$_3$), 3.81 (br s, 1H, SCH$_a$H$_b$), 3.94 (br s, 1H, SCH$_a$H$_b$), 6.77 (s, 1H, CH), 8.26 (br s, 1H, NH, D_2O exch.) ppm; ^{13}C-NMR (100 MHz, CDCl$_3$) δ 8.9, 28.0, 28.8, 44.8, 82.8, 125.1, 131.5, 139.5, 154.1, 203.5 ppm; IR (Nujol, ν, cm^{-1}): 3125, 1725, 1714; MS *m/z* (ESI): 286.17 (M + H)$^+$; calcd. for $C_{12}H_{19}N_3O_3S$ (285.36): C, 50.51; H, 6.71; N, 14.73; found: C, 50.40; H, 6.78; N, 14.86.

Ethyl 2-((1-((tert-Butoxycarbonyl)amino)-5-methyl-4-(phenylcarbamoyl)-1H-imidazol-2-yl)thio)acetate (**7c**): Yield 92% (399.7 mg), white solid from EtOAc/cyclohexane; Mp 134–136 °C; ^{1}H-NMR (400 MHz, CDCl$_3$) δ 1.27 (t, *J* = 8.0 Hz, 3H, OCH$_2$CH$_3$), 1.51 (s, 9H, OBu^t), 2.55 (s, 3H, CH$_3$), 3.66 (d, *J* = 16.0 Hz, 1H, SCH$_a$H$_b$), 3.90 (d, *J* = 16.0 Hz, 1H, SCH$_a$H$_b$), 4.17–4.24 (m, 2H, OCH$_2$CH$_3$), 7.10 (t, *J* = 8.0 Hz, 1H, Ar), 7.34 (t, *J* = 8.0 Hz, 2H, Ar), 7.67 (d, *J* = 8.0 Hz, 2H, Ar), 8.18 (br s, 1H, NH, D_2O exch.), 8.96 (s, 1H, NH, D_2O exch.) ppm; ^{13}C-NMR (100 MHz, CDCl$_3$): δ 9.4, 14.0, 28.0, 36.7, 62.7, 83.4, 119.5, 123.8, 128.9, 129.9, 137.3, 138.0, 139.0, 153.6, 160.6, 169.9 ppm; IR (Nujol, ν, cm^{-1}): 3315, 3182, 1733, 1647, 1601; MS *m/z* (ESI): 435.20 (M + H)$^+$; calcd. for $C_{20}H_{26}N_4O_5S$ (434.51): C, 55.28; H, 6.03; N, 12.89; found: C, 55.39; H, 5.97; N, 12.81.

3.5. General Procedure for the Synthesis of N-Bridgeheaded Heterobicyclic Derivatives 8a–c.

Derivative **7a,b** (1.0 mmol) was solved in 5.0 mL of a solution of trifluoroacetic acid (TFA) and CH_2Cl_2 (1:1). The reaction mixture was left at room temperature until the disappearance of the starting **7a,b** (TLC check). Then, the solvent was removed under reduced pressure and the crude reaction mixture was quenched to neutrality with a saturated solution of Na_2CO_3 and extracted with EtOAc (20.0 mL × 3). The combined organic layers were washed with brine and dried over anhydrous Na_2SO_4. After the removal of the solvent, the crude extract was purified by crystallization or by column chromatography eluting with cyclohexane/ethyl acetate mixtures to furnish **8a,b** derivatives.

For obtaining **8c**, the best condition found was to treat **7c** (1.0 mmol) with Amberlyst 15H (500 mg) in refluxing dioxane (15.0 mL) for 12.0 h. Upon completion (monitored by TLC) the resin was filtered off in vacuo and washed with THF (20.0 mL). The filtrate was evaporated under reduced pressure and the crude reaction mixture was purified by crystallization.

N,N-diethyl-6-methyl-3-phenyl-2H-imidazo[2,1-b][1,3,4]thiadiazine-7-carboxamide (**8a**): Yield 82% (269.3 mg) white powder from EtOAc/cyclohexane; Mp 125–127 °C; ^{1}H-NMR (400 MHz, CDCl$_3$) δ 1.22 (t, J = 8.0 Hz, 6H, 2xNCH$_2$CH$_3$), 2.56 (s, 3H, CH$_3$), 3.53 (br s, 2H, NCH$_2$CH$_3$), 3.74 (br s, 2H, NCH$_2$CH$_3$), 3.97 (s, 2H, SCH$_2$), 7.50–7.52 (m, 3H, Ar), 7.90 (d, J = 8.0 Hz, 2H, Ar) ppm; ^{13}C-NMR (100 MHz, CDCl$_3$) δ 9.7, 13.0, 14.5, 23.9, 40.3, 43.0, 127.0, 128.9, 129.7, 131.1, 131.3, 133.5, 134.2, 150.8, 164.1 ppm; IR (Nujol, ν, cm^{-1}): 1611, 1574, 1562, 1557; MS m/z (ESI): 329.28 (M + H)$^+$; calcd. for C$_{17}$H$_{20}$N$_4$OS (328.43): C, 62.17; H, 6.14; N, 17.06; found: C, 62.04; H, 6.19; N, 17.15.

3,6-Dimethyl-2H-imidazo[2,1-b][1,3,4]thiadiazine (**8b**): Yield 65% (108.7 mg); white needles from CHCl$_3$/cyclohexane; Mp 57–58 °C; ^{1}H-NMR (400 MHz, CDCl$_3$) δ 2.26 (s, 3H, CH$_3$), 2.32 (s, 3H, CH$_3$), 3.42 (s, 2H, SCH$_2$), 6.70 (s, 1H, CH), ppm; ^{13}C-NMR (100 MHz, CDCl$_3$) δ 8.8, 23.5, 26.1, 123.7, 128.6, 130.1, 152.0 ppm; IR (Nujol, ν, cm^{-1}): 1640. 1582; MS m/z (ESI): 168.06 (M + H)$^+$; calcd. for C$_7$H$_9$N$_3$S (167.23): C, 50.27; H, 5.42; N, 25.13; found: C, 50.39; H, 5.39 N, 25.06.

6-Methyl-3-oxo-N-phenyl-3,4-dihydro-2H-imidazo[2,1-b][1,3,4]thiadiazine-7-carboxamide (**8c**): Yield 74% (213.3 mg), light yellow powder from EtOAc/light petroleum ether; Mp 229–232 °C; ^{1}H-NMR (400 MHz, DMSO-d$_6$) δ 2.53 (s, 3H, CH$_3$), 3.81 (s, 2H, SCH$_2$), 7.04 (t, J = 8.0 Hz, 1H, Ar), 7.29 (t, J = 8.0 Hz, 2H, Ar), 7.81 (d, J = 8.0 Hz, 2H, Ar), 9.80 (s, 1H, NH, D$_2$O exch.), 12.27 (br s, 1H, NH, D$_2$O exch.) ppm; ^{13}C-NMR (100 MHz, DMSO-d$_6$): δ 9.2, 29.7, 119.8, 123.0, 128.3, 128.4, 130.6, 132.2, 138.8, 160.8, 164.6 ppm; IR (Nujol, ν, cm^{-1}): 1679, 1666, 1595, 1582; MS m/z (ESI): 288.97 (M + H)$^+$; calcd. for C$_{13}$H$_{12}$N$_4$O$_2$S (288.32): C, 54.15; H, 4.20; N, 19.43; found: C, 54.08; H, 4.27; N, 19.31.

4. Conclusions

In conclusion, combining sequential azidation, Staudinger, and aza-Wittig reactions with CS$_2$ on α-halohydrazones in a one-pot protocol, variously substituted 1-amino-1*H*-imidazole-2(3*H*)-thiones are directly accessible in good yields and with complete control of regioselectivity. The method is particularly attractive and advantageous for its mild conditions, operational simplicity, and its efficiency as well as its robustness (wide substrate scope and tolerance of various functional groups) and reliability. The concurrent presence of reactive appendages on the obtained scaffolds ensures post-modifications toward N-bridgeheaded heterobicyclic structures.

Supplementary Materials: The following are available online: procedure followed for obtaining **II**; copies of ^{1}H-NMR and ^{13}C-NMR spectra of **II** [22]; copies of ^{1}H-NMR and ^{13}C-MNR spectra of all newly synthesized compounds; copies of HMQC of compound **5a** and **8c**.

Author Contributions: C.C. performed all synthetic work in laboratory; G.M. designed the experiments; G.F.; supervised the project, F.M. funding acquisition; L.D.C. validation; S.S.* conceived the synthetic route and wrote the paper.

Funding: The authors declare no competing financial interest.

Acknowledgments: The authors gratefully thank Anna Maria Gioacchini and Samuele Lillini who competently performed the mass spectra and the Department of Biomolecular Science of the University of Urbino for the economical support.

Conflicts of Interest: There are no conflicts to declare.

References

1. Uçucu, O.; Karaburun, N.G.; Işikdağ, I. Synthesis and analgesic activity of some 1-benzyl-2-substituted-4,5-diphenyl-1*H*-imidazole derivatives. *Il Farmaco* **2001**, *56*, 285–290. [CrossRef]

2. Rani, N.; Sharma, A.; Gupta, G.K.; Singh, R. Imidazoles as potential antifungal agents: A review. *Mini-Rev. Med. Chem.* **2013**, *13*, 1626–1655. [CrossRef] [PubMed]
3. Khalafi-Nezhad, A.; Rad, M.N.S.; Mohabatkar, H.; Asrari, Z.; Hemmateenejad, B. Design, synthesis, antibacterial and QSAR studies of benzimidazole and imidazole chloroaryloxyalkyl derivatives. *Bioorg. Med. Chem.* **2005**, *13*, 1931–1938. [CrossRef]
4. Kerru, N.; Bhaskaruni, S.V.H.S.; Gummidi, L.; Maddila, S.N.; Maddila, S.; Jonnalagadda, S.B. Recent advances in heterogeneous catalysts for the synthesis of imidazole derivatives. *Synthetic Commun.* **2019**, *49*, 2437–2459. [CrossRef]
5. Ali, I.; Lone, M.N.; Aboul-Enein, H.Y. Imidazoles as potential anticancer agents. *MedChemComm* **2017**, *8*, 1742–1773. [CrossRef]
6. Sharma, A.; Kumar, V.; Kharb, R.; Kumar, S.; Sharma, P.C.; Pathak, D.P. Imidazole derivatives as potential therapeutic agents. *Curr. Pharm. Des.* **2016**, *22*, 3265–3301. [CrossRef]
7. Bellina, F.; Cauteruccio, S.; Rossi, R. Synthesis and biological activity of vicinal diaryl-substituted 1H-imidazoles. *Tetrahedron* **2007**, *63*, 4571–4624. [CrossRef]
8. Savjani, J.K.; Gajjar, A.K. Pharmaceutical importance and synthetic strategies for imidazolidine-2-thione and imidazole-2-thione derivatives. *Pak. J. Biol. Sci.* **2011**, *14*, 1076–1089. [CrossRef]
9. Isaia, F.; Aragoni, M.C.; Arca, M.; Demartin, F.; Devillanova, F.A.; Floris, G.; Garau, A.; Hursthouse, M.B.; Lippolis, V.; Medda, R.; et al. Interaction of methimazole with I2: X-ray crystal structure of the charge transfer complex methimazole-I2. implications for the mechanism of action of methimazole-based antithyroid drugs. *Med. Chem.* **2008**, *51*, 4050–4053. [CrossRef]
10. Cesarini, S.; Spallarossa, A.; Ranise, A.; Schenone, S.; Rosano, C.; La Colla, P.; Sanna, G.; Busonera, B.; Loddo, R. N-acylated and N,N'-diacylated imidazolidine-2-thione derivatives and N,N'-diacylated tetrahydropyrimidine-2(1H)-thione analogues: Synthesis and antiproliferative activity. *Eur. J. Med. Chem.* **2009**, *44*, 1106–1118. [CrossRef]
11. Sheppeck, J.; Gilmore, J.L. Substituted 1.3-dihydro-imidazol-2-one and 1.3-dihydro-imidazol-2-thione derivatives as inhibitors of matrix metallo proteinases and/or TNF-α converting enzyme (TACE). U.S. Patent 20050075384, 7 April 2005.
12. Lagoja, I.M.; Pannecouque, C.; Van Aerschot, A.; Witvrouw, M.; Debyser, Z.; Balzarini, J.; Herdewijn, P.; De Clercq, E.J. N-Aminoimidazole derivatives inhibiting retroviral replication via a yet unidentified mode of action. *Med. Chem.* **2003**, *46*, 1546–1553. [CrossRef] [PubMed]
13. Schantl, J.G.; Prean, M. Addition products of hydrazine derivatives to azo-alkenes, part V: The reaction of α-(1-phenylhydrazino)alkanone phyenylhydrazones with acids and acid derivatives. *Monatsh. Chem.* **1993**, *124*, 299–308. [CrossRef]
14. Schantl, J.G.; Kahlig, H.; Preans, M. 1-Arylamino-2,3-dihydro-1H-imidazole-2-thiones from the reaction of 1-[2-(2-arylhydrazono)alkyl]pyridinium iodides with potassium thiocyanate. *Heterocycles* **1994**, *37*, 1873–1878. [CrossRef]
15. Schantl, J.G.; Lagoia, I. Direct synthetic approach to N-Substituted 1-amino-2,3-dihydro-1H-imidazole-2-thiones. *Heterocycles* **1997**, *45*, 691–700. [CrossRef]
16. Schantl, J.; Nádenik, P. Tandem [3+2] cycloaddition reaction of azo-alkenes and thiocyanic acid: Extending the scope of the classical "criss-cross" cycloaddition reaction. *Synlett* **1998**, 786–788. [CrossRef]
17. Yurttaş, L.; Ertas, M.; Gulsen, A.C.; Temel, H.E.; Demirayak, Ş. Novel benzothiazole based imidazole derivatives as new cytotoxic agents against glioma (C6) and liver (HepG2) cancer cell lines. *Acta Pharm. Sci.* **2017**, *55*, 39–47. [CrossRef]
18. Neochoritis, C.; Tsoleridis, C.A.; Stephanidou-Stephanatou, J. 1-Arylaminoimidazole-2-thiones as intermediates in the synthesis of imidazo[2,1-b][1,3,4]thiadiazines. *Tetrahedron* **2008**, *64*, 3527–3533. [CrossRef]
19. Grimmett, M.R. Imidazoles. In *Science of Synthesis*; Neier, D., Bellus, D., Eds.; G. Thieme Verlag: Stuttgart, Germany, 2002; Volume 12, pp. 325–328. [CrossRef]
20. Schantl, J.G. Azomethine imines. In *Science of Synthesis*; Padwa, A., Ed.; G. Thieme Verlag: Stuttgart, Germany, 2004; Volume 27, pp. 731–738. [CrossRef]
21. Schantl, J.G. Cyclic azomethine imines from diazenes (azo compounds). In *Advances in Heterocyclic Chemistry*; Katritzky, A.R., Ed.; Academic Press: Oxford, UK; Volume 99, pp. 185–207. [CrossRef]
22. Attanasi, O.A.; Favi, G.; Filippone, P.; Perrulli, F.R.; Santeusanio, S. Direct access to variously substituted 2-imino-4-thiazolines. *Synlett* **2010**, 1859–1861. [CrossRef]

23. Attanasi, O.A.; Bartoccini, S.; Favi, G.; Filippone, P.; Perrulli, F.R.; Santeusanio, S. Tandem aza-Wittig/carbodiimide-mediated annulation applicable to 1,2-diaza-1,3-dienes for the one-pot synthesis of fully substituted 1,2-diaminoimidazoles. *J. Org. Chem.* **2012**, *77*, 9338–9343. [CrossRef]
24. Xie, H.; Liu, J.-C.; Wu, L.; Ding, M.-W. Unexpected synthesis of 2,4,5-trisubstituted oxazoles via a tandem aza-Wittig/Michael/isomerization reaction of vinyliminophosphorane. *Tetrahedron* **2012**, *68*, 7984–7990. [CrossRef]
25. Santhosh, L.; Durgamma, S.; Sureshbabu, V.V. Staudinger/aza-Wittig reaction to access N^β-protected amino alkyl isothiocyanates. *Org. Biomol. Chem.* **2018**, *16*, 4874–4880. [CrossRef] [PubMed]
26. Lopes, S.M.M.; Lemos, A.; Pinho e Melo, T.M.V.D. Reactivity of dipyrromethanes towards azoalkenes: Synthesis of functionalized dipyrromethanes, calix[4]pyrroles, and bilanes. *Eur. J. Org. Chem.* **2014**, 7039–7048. [CrossRef]
27. Attanasi, O.A.; De Crescentini, L.; Favi, G.; Filippone, P.; Mantellini, F.; Perrulli, F.R.; Santeusanio, S. Cultivating the passion to build heterocycles from 1,2-diaza-1,3-dienes: The force of imagination. *Eur. J. Org. Chem.* **2009**, 3109–3127. [CrossRef]
28. Attanasi, O.A.; De Crescentini, L.; Giorgi, R.; Perrone, A.; Santeusanio, S. Synthesis of 3-unsubstituted-1-aminopyrroles. *Heterocycles* **1996**, *43*, 1447–1457. [CrossRef]
29. Attanasi, O.A.; Favi, G.; Mantellini, F.; Mantenuto, S.; Moscatelli, G.; Nicolini, S. Regioselective formation of 5-methylene-6-methoxy-1,4,5,6-tetrahydropyridazines from the [4+2]-cycloaddition reaction of in situ generated 1,2-diaza-1,3-dienes with methoxyallene. *Synlett* **2015**, 193–196. [CrossRef]
30. Attanasi, O.A.; De Crescentini, L.; Favi, G.; Mantellini, F.; Mantenuto, S.; Nicolini, S. Interceptive [4+1] annulation of in situ generated 1,2-diaza-1,3-dienes with diazo esters: Direct access to substituted mono-, bi-, and tricyclic 4,5-dihydropyrazoles. *J. Org. Chem.* **2014**, *79*, 8331–8338. [CrossRef]
31. Attanasi, O.A.; Serra-Zanetti, F.; Zhiyuan, L. Easy one-pot conversion of 2-chlorohydrazone into 2-oxohydrazone derivatives via 2-azidohydrazone intermediates. *Tetrahedron* **1992**, *48*, 2785–2792. [CrossRef]
32. Herrera, R.P.; Marqués-Lopez, E. *Multicomponent Reactions: Concepts and Applications for Design and Synthesis*; John Wiley & Sons: Hoboken, NJ, USA, 2015.
33. Dömling, A. Recent developments in isocyanide based multicomponent reactions in applied chemistry. *Chem. Rev.* **2006**, *106*, 17–89. [CrossRef]
34. Dömling, A.; Wang, W.; Wang, K. Chemistry and biology of multicomponent reactions. *Chem. Rev.* **2012**, *112*, 3083–3135. [CrossRef]
35. Preeti; Singh, K.N. Multicomponent reactions: A sustainable tool to 1,2- and 1,3-azoles. *Org. Biomol. Chem.* **2018**, *16*, 9084–9116. [CrossRef] [PubMed]
36. Yurttaş, L.; Duran, M.; Demirayak, S.; Gençer, H.K.; Tunali, Y. Synthesis and initial biological evaluation of substituted 1-phenylamino-2-thio-4,5-dimethyl-1H-imidazole derivatives. *Bioorg. Med. Chem. Lett.* **2013**, *23*, 6764–6768. [CrossRef]
37. Palacios, F.; Alonso, C.; Aparicio, D.; Rubiales, G.; de los Santos, J.M. The aza-Wittig reaction: An efficient tool for the construction of carbon–nitrogen double bonds. *Tetrahedron* **2007**, *63*, 523–575. [CrossRef]
38. Eguchi, S. Recent progress in the synthesis of heterocyclic natural products by the Staudinger/intramolecular aza-Wittig reaction. *ARKIVOC* **2005**, *(ii)*, 98–119. [CrossRef]
39. Fresneda, P.M.; Molina, P. Application of iminophosphorane-based methodologies for the synthesis of natural products. *Synlett* **2004**, 1–17. [CrossRef]
40. Pavlova, A.S.; Ianova, O.A.; Chagarovskij, A.O.; Stebunov, N.S.; Orlov, N.V.; Shumsky, A.N.; Budynina, E.M.; RybaKov, V.B.; Trushkov, I.V. Domino Staudinger/aza-Wittig/Mannich reaction: An approach to diversity of di- and tetrahydropyrrole scaffolds. *Chem. Eur. J.* **2016**, *22*, 17967–17971. [CrossRef]
41. Xiong, J.; Wei, X.; Ding, M.-W. New Facile synthesis of 2-alkylthiopyrimidin-4(3H)-ones by tandem aza-Wittig reaction starting from the Baylis–Hillman adducts. *Synlett* **2017**, 1075–1078. [CrossRef]
42. Agami, C.; Couty, F. The reactivity of the N-Boc protecting group: An underrated feature. *Tetrahedron* **2002**, *58*, 2701–2724. [CrossRef]
43. Greene, T.W.; Wuts, P.G.M. *Protective Groups in Organic Synthesis*, 3rd ed.; Wiley: New York, NY, USA, 1999. [CrossRef]
44. Ballini, R.; Petrini, M. Amberlyst 15, a superior, mild, and selective catalyst for carbonyl regeneration from nitrogeneous derivatives. *J. Chem. Soc. Perkin Trans. 1* **1988**, 2563–2565. [CrossRef]

45. Sasaki, T.; Ito, E.; Shimizu, I. Ring transformations of oxazoles and their benzo analogues. New synthetic route for 2*H*-imidazo[2,1-*b*][1,3,4]thiadiazine and N-heteroaryl-o-aminophenol. *Heterocycles* **1982**, *19*, 2119–2129. [CrossRef]
46. Qi, L.-W.; Mao, J.-H.; Zhang, J.; Tan, B. Organocatalytic asymmetric arylation of indoles enabled by azo groups. *Nat. Chem.* **2018**, *10*, 58–64. [CrossRef] [PubMed]

Sample Availability: Samples of the compounds **5a–k** are available from the authors.

© 2019 by the authors. Licensee MDPI, Basel, Switzerland. This article is an open access article distributed under the terms and conditions of the Creative Commons Attribution (CC BY) license (http://creativecommons.org/licenses/by/4.0/).

Article

One-Pot Metal-Free Synthesis of 3-CF$_3$-1,3-Oxazinopyridines by Reaction of Pyridines with CF$_3$CO-Acetylenes

Vasiliy M. Muzalevskiy [1], Zoia A. Sizova [1], Kseniya V. Belyaeva [2], Boris A. Trofimov [2,*] and Valentine G. Nenajdenko [1,*]

1. M. V. Lomonosov Moscow State University, Department of Chemistry, Leninskie Gory 1, 119991 Moscow, Russia; muzvas@mail.ru (V.M.M); syzova@mail.ru (Z.A.S).
2. A. E. Favorsky Irkutsk Institute of Chemistry, Siberian Branch, Russian Academy of Sciences, 1 Favorsky Str., 664033 Irkutsk, Russia; belyaeva@irioch.irk.ru
* Correspondence: boris_trofimov@irioch.irk.ru (B.A.T.); nenajdenko@gmail.com (V.G.N)

Received: 18 September 2019; Accepted: 4 October 2019; Published: 6 October 2019

Abstract: The reaction of pyridines with trifluoroacetylated acetylenes was investigated. It was found that the reaction of various pyridines with two molecules of CF$_3$CO-acetylenes proceeds under mild metal-free conditions. As a result, efficient stereoselective synthesis of 3-arylethynyl-3-trifluoromethyl-1,3-oxazinopyridines was elaborated. Target heterocycles can be prepared in up to quantitative yields.

Keywords: pyridine; CF$_3$CO-acetylenes; 1,3-oxazines; fluorinated heterocycles

1. Introduction

Pyridine motif is the one of the most recognizable frameworks among heterocyclic molecules. A lot of attention has been paid to the chemistry of this class of heterocyclic compounds since the very beginning of its discovery. Nowadays the flow of the articles concerning pyridine is still far from the drying out. The high attractiveness of pyridine chemistry can be explained by high biological activity of pyridine derivatives both naturally occurred and prepared in the lab. Therefore, almost 300 alkaloids, having pyridine moiety (not including derivatives with fused pyridine ring, such as isoquinoline), were listed in "The Dictionary of Alkaloids" [1].

The pyridine scaffold is also a privileged structure for design of novel pharmaceuticals. Structural analysis of US FDA approved drugs showed that pyridine core is a consistent part of 62 marketed drugs (second place after piperidine) in the list of most frequent nitrogen heterocycles in structure of approved drugs [2,3]. One can also found 15 derivatives of pyridine among the "Top 200 Pharmaceutical Products by Retail Sales in 2018" which made together about $27 billion during 2018 alone [4]. Some pyridine-based drugs were approved by FDA in 2019 (for examples, see Figure 1).

On the other hand, investigation of organofluorine compounds is one of the most important trends in modern organic chemistry [5–9]. Due to unique physicochemical and biological properties, organofluorine compounds are widely used as construction materials, components of liquid crystalline compositions, agrochemicals and pharmaceuticals [10–15]. By some estimation, about 20–25% of currently used drugs [16–23] and agrochemicals [24–27] contain at least one fluorine atom. As for the year 2018, that value is even higher, because three out of ten drugs approved by the US FDA in 2018 contain this atom (18 out of 59 drugs) [28]. Heterocyclic compounds are also an important object for medicinal chemistry, which can be found among numerous drugs (about 59% of small-molecule drugs [2], approved by the US FDA before 2014). Last year, 35 out of 59 drugs contain any heterocyclic fragment, with 16 of them also having at least one fluorine atom, including six with fluorinated heterocyclic motif (Figure 1). It is not surprising that novel effective methodologies for the synthesis of fluorinated heterocycles have been in great demand in recent decades [29–35].

α,β-Unsaturated CF$_3$-ketones have been shown as versatile building blocks for the synthesis of various fluorinated heterocyclic compounds [36–40]. In a series of works, we have demonstrated a great potential of CF$_3$-ynones in different heterocyclizations to prepare fluorinated derivatives of diazepines [41], pyrimidines [42], thiophenes [43], triazoles [44], pyrazoles [45–47]. Recently, we focused our attention on the reactions of CF$_3$-ynones with azines. It was found that, depending on nature of azine and the acetylene–azine ratio, various products can be obtained very efficiently. The reaction with quinolines opened access to 1,3-oxazinoquinolines **6** [48,49] or **7** (Scheme 1) [50]. 1,3-Oxazine moiety has been experienced a growing interest in recent years [51,52] and became perspective targets for drug design [49,53,54]. For example, Dolutegravir (Tivicay®approved in 2013 [55] and in combination with Lamivudine as Dovato®approved in 2019) and Bictegravir (Biktarvy®approved in 2018) [56] are used for treatment of patients with HIV (Figure 1).

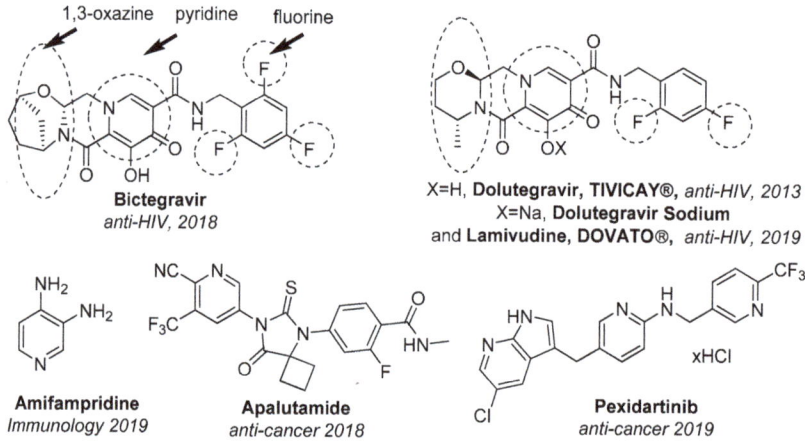

Figure 1. Selected FDA approved drugs in 2018 and 2019 containing pyridine moiety, fluorine atoms, 1,3-oxazine moiety.

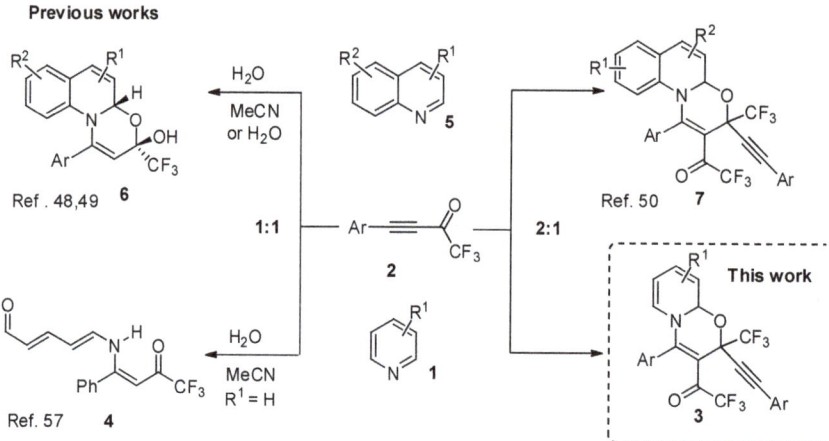

Scheme 1. CF$_3$-ynones in the reactions with quinolines and pyridines.

In contrast to the reaction with quinolines, our attempt to involve pyridines into 1,3-oxazine assembling reaction with CF$_3$-ynones has been less successful. The reaction of pyridine with equal amount of CF$_3$-ynone in wet acetonitrile afforded the corresponding ring opening product. Polyunsaturated 5-amino-2,4-pentadienal **7** has been isolated as a result of cascade transformation (Scheme 1) [57].

2. Results and Discussion

This study is devoted to the next step of our systematic study of the reactions of fluorinated acetylenes with azines. A simple and highly efficient approach towards 3-arylethynyl-3-trifluoromethyl-1,3-oxazinopyridines **3** is presented by the reaction of CF$_3$-ynones with pyridines in 2:1 ratio.

We assumed that using dry conditions and excess of ketone the reaction course could be redirected to formation of the corresponding trifluoromethylated 1,3-oxazines. Indeed, being mixed together without solvent, pyridine and CF$_3$-ynone **2a** 1:2 ratio new transformation was observed to form viscous mass in a few minutes.

Analysis of the reaction mixture by NMR showed clean formation of **3a** and unreacted starting materials. After addition of a small amount of MeCN to form homogeneous solution the reaction mixture was left overnight. As a result, oxazine **3a** was isolated in 94% yield in stereoselective manner. According to NMR a 90:10 mixture of 2S*,9aS* and 2R*,9aS* diastereomers was formed. Assignment of both diastereomers was maintained by careful comparison with 3-arylethynyl-3-trifluoromethyl-1,3-oxazinoquinolines **7** having similar structures (Figure 2) [50]. Therefore, values of δ(^{1}H-8), δ(^{1}H-9), δ(^{19}F-COCF$_3$) in 2S*,9aS*-**3a** are larger than in 2R*,9aS*-**3a'** while values of δ(^{1}H-9a) and δ(^{19}F-CF$_3$) are the other way around. The same regularity can be seen in the NMR of 3S*,4aS*- and 3R*,4aS*-diastereomers of **7a**.

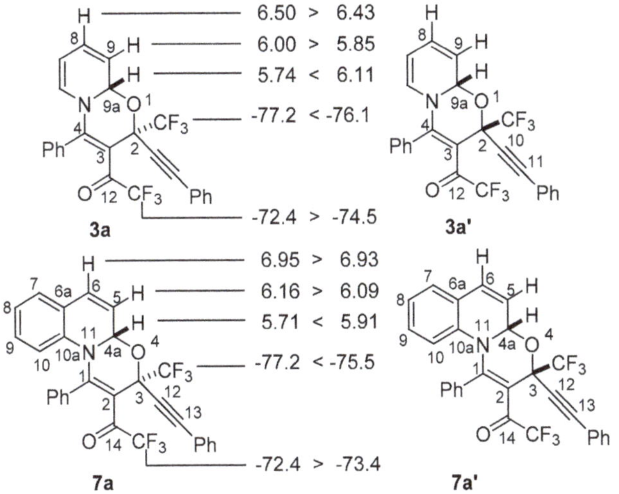

Figure 2. Comparison of characteristic values of chemical shifts of diastereomers of **3a** in ^{1}H- and ^{19}F-NMR spectra with the corresponding quinoline derivative **7a**.

Next, the reaction scope was studied. For this aim, the interaction of parent pyridine with various CF$_3$-ynones was investigated. To our delight, it was found that the reaction has no restrictions in terms of CF$_3$-ynones. The corresponding 1,3-oxazinoquinolines **3a–i** were isolated in 77–99% yield (Scheme 2). Similar stereoselectivity was observed for all these products. Compounds **3a–i** were formed as a mixture of 2S*,9aS* and 2R*,9aS* diastereomers in near 9:1 ratio in most cases.

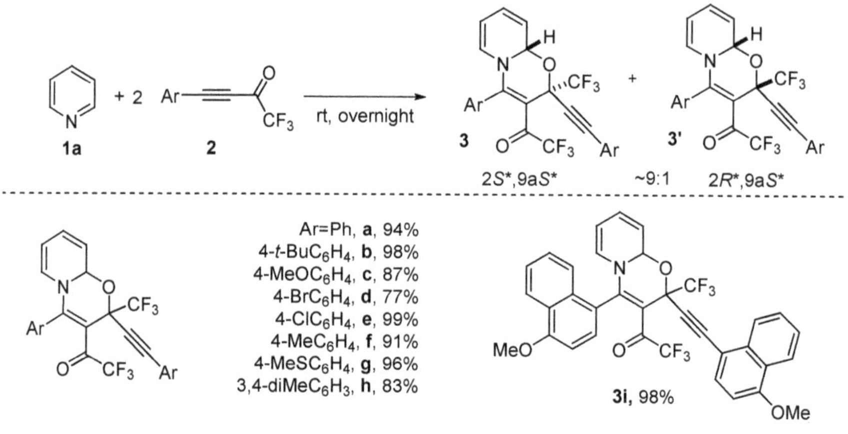

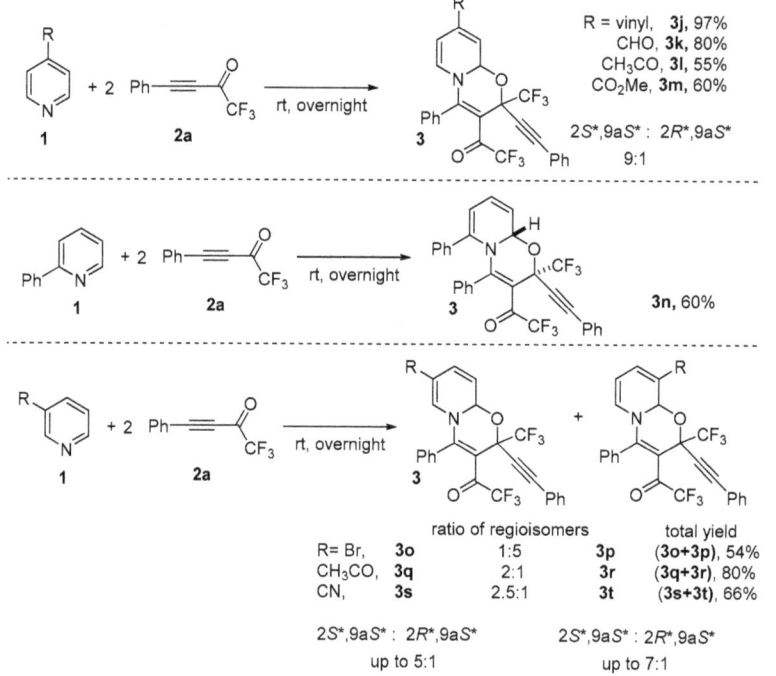

Scheme 2. Reactions of pyridine with CF$_3$-ynones **2a-i** to form 1,3-oxazinopyridines **3a–i**.

Next, the reaction of CF$_3$-ynone **2a** with several pyridines was studied in order to investigate the influence of nature of pyridine component of the reaction. A series of 4, 3 and 2-substituted pyridines was involved into reaction with **2a** (Scheme 3).

Scheme 3. Reactions of pyridine **1b–j** with CF$_3$-ynones **2a** to form 1,3-oxazinopyridines **3j–t**.

It was found that the reaction has broad scope in terms of pyridines and nature of substituents. However, the reaction is very sensitive to structure of starting pyridine. An especially important influence on the reaction is the nature of a substituent, pKa value of pyridine and its nucleophilicity, and the position of a substituent in the molecule of pyridine. Therefore, the reaction with 4-substituted pyridines afforded the corresponding oxazines **3j–m** in high yields. Again, a mixture of diastereomers in near to 8:1–9:1 ratio was formed in all cases (Scheme 3, compounds **3j–m**). In contrast, 2-substituted pyridines (2-phenylpyridine) reacted with CF$_3$-ynone **2a** 100% stereoselectively to form 2S*, 9aS* diastereomer exclusively (Scheme 3, compound **3n**). A more complex picture was observed for the reaction with 3-substituted pyridines. Due to the presence of two possible positions for cyclization in the pyridine framework, 7- and 9-isomers were formed in about 2:1 ratio for pyridines with electron-withdrawing acetyl- and cyano groups, having strong −M effect. In contrast mostly 9-isomer (in ratio 5:1 with 7-isomer) was formed in the reaction with 3-bromopyridine having bromine atom with slight +M effect (Scheme 3, compounds **3o–t**). It is noteworthy that both increase (30 °C) and decrease (7 °C) of the temperature did not change dramatically the regioselectivity of the reaction. However, the stereoselectivity of formation of compounds **3p–t** was again high to give 2S*,9aS-isomer as a major one in up to 7:1 ratio with minor 2R*,9aS*-isomer.

Some restrictions were also found. We observed that pKa of azine and therefore its nucleophilicity plays a decisive role in the possibility of the reaction to occur. Therefore, pyridines with pKa lower than ~1, 2-bromopyridine (0.79), 2-fluoropyridine (−0.43), 2-methoxy-5-bromopyridine (1.04) do not react with CF$_3$-ynone **2a**.

Based on our previous mechanistic rationalizations regarding interaction of CF$_3$-ynones with azines [48,49,58,59], the possible reaction mechanism can be proposed. The domino assembly of oxazinoquinolines **3** is initiated by the reversible formation of the intermediate zwitterion **A** resulted from the nitrogen nucleophilic addition to the triple bond. In contrast to 1:1 reaction, the carbanionic site of **A** is selectively attacked by the carbonyl group of the second molecule of **2a** to form anion **B**. Cyclization of **B** undergoes by the attacks of oxygen into alpha-position of the pyridine ring to give the corresponding 1,3-oxazine **3** (Scheme 4).

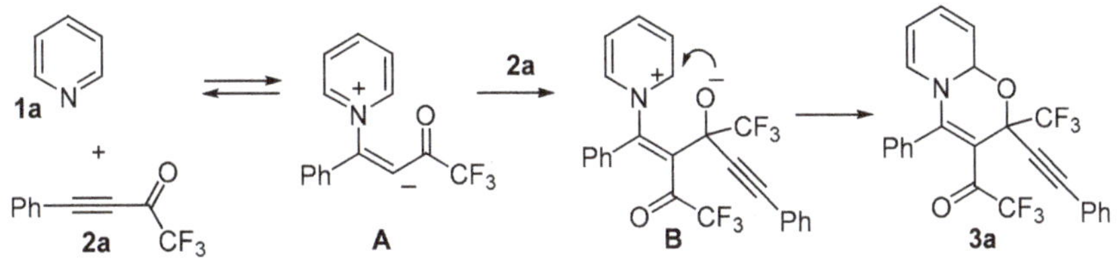

Scheme 4. Possible mechanism of the reaction of pyridines with CF$_3$-ynones.

3. Materials and Methods

3.1. General Details

^{1}H-, ^{13}C- and ^{19}F-NMR spectra were recorded on Bruker AVANCE 400 MHz spectrometer (Bruker Corp., Karlsruhe, Germany) in CD$_3$CN and CDCl$_3$ at 400.1, 100.6 and 376.3 MHz respectively (Supplementary Information). Chemical shifts (δ) in ppm were reported with the use of the residual CHD$_2$CN and chloroform signals (1.94 and 7.25 for ^{1}H and 77.0 for ^{13}C) as internal reference. The ^{19}F chemical shifts were referenced to C$_6$F$_6$, (−162.9 ppm). HRMS (ESI-TOF) spectra were measured with an Orbitrap Elite instrument (TermoFisher, Paisley, UK). TLC analysis was performed on "Merck 60 F$_{254}$" plates. Visualization was accomplished by UV light (254 nm) at Vilber Lourmat UV lamp. Silica gel (silica 60, 0.063–0.2 mm, 70–230 mesh), Screw neck vials (clear, flat bottom, 4 mL) and Screw caps were purchased at MACHEREY-NAGEL (Duren, Germany). All reagents were purchased at Sigma-Aldrich (Muenchen, Germany) and Acros companies (Geel, Belgium). The reagents were of reagent grade and were used as such or distilled prior to use. CF$_3$-ynones **2** were prepared as reported previously [46]. Melting points were determined on an Electrothermal 9100 apparatus.

3.2. Reaction of CF$_3$-Ynones and Pyridines (General Procedure)

A 4 mL vial with a screw cap was charged with CF$_3$-ynone **2** (1–1.05 mmol, 2–2.1 equiv.)* and then pyridine **1** (0.5 mmol, 1 equiv.) was added in one portion. After vigorous stirring for several minutes the reaction mixture became viscous due to crystallization of the product. At that moment MeCN (0.5 mL) was added to form homogeneous solution again and the reaction mixture was left overnight at stirring. Next volatiles were evaporated in vacuo, the residue was crystallized from appropriate amount of ether-hexane mixtures or purified via column chromatography on silica gel using mixtures of hexane with CH$_2$Cl$_2$. * In case of solid CF$_3$-ynones **2** MeCN (0.1–0.2 mL) was added to form clear solution.

2,2,2-Trifluoro-1-(4-phenyl-2-(phenylethynyl)-2-(trifluoromethyl)-2H,9aH-pyrido[2,1-b][1,3]oxazin-3-yl)ethan-1-one (**3a**). Obtained from pyridine **1a** (0.042 g, 0.53 mmol) and CF$_3$-ynone **2a** (0.212 g, 1.071 mmol). Yellow-brown powder, m.p. 109.4–111.8 °C (hexane), yield 0.238 g (94%). (2S*,9aS*):(2R*,9aS*)-isomers ratio is 90:10 (^{19}F-NMR). HRMS (ESI-TOF): m/z [M + H]$^+$ Calcd for C$_{25}$H$_{16}$F$_6$NO$_2$$^+$: 476.1080; found: 476.1085.

(2S*,9aS*)-**3a**: ^{1}H-NMR (400.1 MHz, CDCl$_3$): δ 7.68–7.39 (m, 7H), 7.37–7.27 (m, 3H), 6.50 (dd, $^3J_{8,9}$ = 9.7 Hz, $^3J_{7,8}$ = 6.1 Hz, 1H, H-8), 6.46 (d, $^3J_{6,7}$ = 7.7 Hz, 1H, H-6), 6.00 (dd, $^3J_{8,9}$ = 9.7 Hz, $^3J_{9a,9}$ = 3.9 Hz, 1H, H-9), 5.74 (d, $^3J_{9a,9}$ = 3.9 Hz, 1H, H-9a), 5.50 (pseudo-td, ^3J ~ 7 Hz, ^4J ~ 1 Hz, 1H, H-7). ^{13}C-NMR (100.6 MHz, CDCl$_3$): δ 180.7 (q, $^2J_{CF}$ = 35.0 Hz, C-12), 160.3 (C-4), 133.2, 132.1, 131.4 (C$_{i'}$ from Ar), 129.5, 129.3, 129.2, 128.3 (C-8), 126.2 (C-6), 125.8, 122.6 (q, $^1J_{CF}$ = 286.6 Hz, CF$_3$), 121.1 (C$_i$ from Ar), 116.5 (C-9), 115.5 [q, $^1J_{CF}$ = 292.7 Hz, C(O)CF$_3$], 104.0 (C-7), 88.4 (C-11), 81.3 (C-10), 79.1 (C-9a), 73.7 (q, $^2J_{CF}$ = 33.9 Hz, C-2). ^{19}F-NMR (376.3 MHz, CDCl$_3$): δ −72.5 [C(O)CF$_3$], −77.3 (CF$_3$).

(2R*,9aS*)-**3a'**: ^{1}H-NMR (400.1 MHz, CDCl$_3$): δ 6.43 (dd, $^3J_{8,9}$ = 9.8 Hz, $^3J_{7,8}$ = 6.1 Hz, 1H, H-8), 6.28 (d, $^3J_{6,7}$ = 7.6 Hz, 1H, H-6), 6.11 (d, $^3J_{9a,9}$ = 4.0 Hz, 1H, H-9a), 5.85 (dd, $^3J_{8,9}$ = 9.8 Hz, $^3J_{9a,9}$ = 4.0 Hz, 1H, H-9), 5.34 (pseudo-td, ^3J ~ 7 Hz, ^4J = 1 Hz, 1H, H-7). Other signals are overlapped with those of major isomer. ^{13}C-NMR (100.6 MHz, CDCl$_3$): δ 133.9, 132.3, 128.9, 126.5, 126.3, 115.0, 109.4, 102.2. Other signals are overlapped with those of major isomer or cannot be seen in the spectrum due to the low concentration of minor isomer. ^{19}F-NMR (376.3 MHz, CDCl$_3$): δ −74.6 [C(O)CF$_3$], −76.2 (CF$_3$).

1-(4-(4-(Tert-butyl)phenyl)-2-((4-(tert-butyl)phenyl)ethynyl)-2-(trifluoromethyl)-2H,9aH-pyrido[2,1-b][1,3]oxazin-3-yl)-2,2,2-trifluoroethan-1-one (**3b**). Obtained from pyridine **1a** (0.041 g, 0.518 mmol) and CF$_3$-ynone **2b** (0.267 g, 1.051 mmol). Yellow powder, m.p. 130.0–132.7 °C (hexane), yield 0.300 g (98%). (2S*,9aS*):(2R*,9aS*)-isomers ratio is 90:10 (^{19}F-NMR). HRMS (ESI-TOF): m/z [M + H]$^+$ Calcd for C$_{33}$H$_{32}$F$_6$NO$_2$$^+$: 588.2332; found: 588.2340.

(2S*,9aS*)-**3b**: ^{1}H-NMR (400.1 MHz, CDCl$_3$): δ 7.52 (d, ^3J = 8.4 Hz, 2H), 7.47–7.36 (m, 4H), 7.32 (d, ^3J = 8.4 Hz, 2H), 6.52 (d, $^3J_{6,7}$ = 7.8 Hz, 1H, H-6), 6.48 (dd, $^3J_{8,9}$ = 9.8 Hz, $^3J_{7,8}$ = 6.1 Hz, 1H, H-8), 5.99 (dd, $^3J_{8,9}$ = 9.8 Hz, $^3J_{9a,9}$ = 3.9 Hz, 1H, H-9), 5.72 (d, $^3J_{9a,9}$ = 3.9 Hz, 1H, H-9a), 5.49 (psedo-t, ^3J ~ 7 Hz, 1H, H-7), 1.35 (s, 9H, 3Me from t-Bu), 1.29 (s, 9H, 3Me from t-Bu). ^{13}C-NMR (100.6 MHz, CDCl$_3$): δ 180.8 (q, $^2J_{CF}$ = 35.0 Hz, C-12), 160.4 (C-4), 157.2 (C$_p$ from Ar), 152.7 (C$_{p'}$ from Ar), 131.9 (C$_{m,m'}$ from Ph), 128.6 (q, $^3J_{CF}$ = 2.2 Hz, C-3), 126.4, 126.3 (C-8), 126.1 (C-6), 125.3 (C$_{o,o'}$ from Ar), 122.6 (q, $^1J_{CF}$ = 286.6 Hz, CF$_3$), 118.1 (C$_{i'}$ from Ar), 116.5 (C-9), 115.7 [q, $^1J_{CF}$ = 292.5 Hz, C(O)CF$_3$], 103.6 (C-7), 88.5 (C-11), 80.9 (C-10), 79.0 (C-9a), 73.8 (q, $^2J_{CF}$ = 34.6 Hz, C-2), 35.2, 34.8, 31.1, 31.0. ^{19}F-NMR (376.3 MHz, CDCl$_3$): δ −72.5 [C(O)CF$_3$], −77.4 (CF$_3$).

(2R*,9aS*)-**3b'**: ^{1}H-NMR (400.1 MHz, CDCl$_3$): δ 6.43 (dd, $^3J_{8,9}$ = 9.8 Hz, $^3J_{7,8}$ = 6.0 Hz, 1H, H-8), 6.33 (d, $^3J_{6,7}$ = 7.6 Hz, 1H, H-6), 6.10 (d, $^3J_{9a,9}$ = 3.9 Hz, 1H, H-9a), 5.84 (dd, $^3J_{8,9}$ = 9.8 Hz, $^3J_{9a,9}$ = 3.9 Hz, 1H, H-9), 5.33 (pseudo-t, ^3J ~ 7 Hz, 1H, H-7), 1.33 (s, 9H, 3Me from t-Bu), 1.30 (s, 9H, 3Me from t-Bu). Other signals are overlapped with those of major isomer. ^{13}C-NMR (100.6 MHz, CDCl$_3$): δ 132.0, 126.6, 126.51, 126.49, 125.2, 109.2, 35.1, 31.2, 30.9. Other signals are overlapped with those of major isomer or cannot be seen in the spectrum due to the low concentration of minor isomer. ^{19}F-NMR (376.3 MHz, CDCl$_3$): δ −74.6 [C(O)CF$_3$], −76.2 (CF$_3$).

1-(4-(4-Methoxyphenyl)-2-((4-methoxyphenyl)ethynyl)-2-(trifluoromethyl)-2H,9aH-pyrido[2,1-b][1,3]oxazin-3-yl)-2,2,2-trifluoroethan-1-one (**3c**). Obtained from pyridine **1a** (0.041 g, 0.518 mmol) and CF$_3$-ynone **2c** (0.239 g, 1.048 mmol). Light brown powder, m.p. 117.3–118.7 °C (hexane), yield 0.242 g (87%). (2S*,9aS*):(2R*,9aS*)-isomers ratio is 90:10 (^{19}F-NMR). HRMS (ESI-TOF): *m/z* [M + H]$^+$ Calcd for C$_{27}$H$_{20}$F$_6$NO$_4$$^+$: 536.1291; found: 536.1296.

(2S*,9aS*)-**3c**: ^{1}H-NMR (400.1 MHz, CDCl$_3$): δ 7.48–7.30 (m, 4H), 7.07–6.91 (m, 2H), 6.82 (d, ^3J = 8.9 Hz, 2H), 6.49 (d, $^3J_{6,7}$ = 7.0 Hz, 1H, H-6), 6.48 (dd, $^3J_{8,9}$ = 9.8 Hz, $^3J_{7,8}$ = 6.1 Hz, 1H, H-8), 5.99 (dd, $^3J_{8,9}$ = 9.8 Hz, $^3J_{9a,9}$ = 4.1 Hz, 1H, H-9), 5.71 (d, $^3J_{9a,9}$ = 4.1 Hz, 1H, H-9a), 5.49 (pseudo-td, ^3J ~ 7 Hz, ^3J ~ 1 Hz, 1H, H-7), 3.88 (s, 3H, MeO), 3.79 (s, 3H, MeO). ^{13}C-NMR (100.6 MHz, CDCl$_3$): δ 180.6 (q, $^2J_{CF}$ = 34.7 Hz, C-12), 163.7, 160.4 (C-4), 160.3, 133.7, 126.3 (C-8), 126.0 (C-6), 123.7, 122.7 (q, $^1J_{CF}$ = 286.8 Hz, CF$_3$), 116.5 (C-9), 115.7 [q, $^1J_{CF}$ = 293.0 Hz, C(O)CF$_3$], 113.9, 113.2, 108.6, 103.8 (C-7), 88.3 (C-11), 80.3 (C-10), 78.9 (C-9a), 73.9 (q, $^2J_{CF}$ = 34.3 Hz, C-2), 55.6, 55.2. ^{19}F-NMR (376.3 MHz, CDCl$_3$): δ −72.4 [C(O)CF$_3$], −77.5 (CF$_3$).

(2R*,9aS*)-**3c'**: ^{1}H-NMR (400.1 MHz, CDCl$_3$): δ 7.52 (d, ^3J = 8.9 Hz, 2H), 6.42 (dd, $^3J_{8,9}$ = 9.8 Hz, $^3J_{7,8}$ = 6.1 Hz, 1H, H-8), 6.31 (d, $^3J_{6,7}$ = 7.5 Hz, 1H, H-6), 6.07 (d, $^3J_{9a,9}$ = 4.1 Hz, 1H, H-9a), 5.84 (dd, $^3J_{8,9}$ = 9.8 Hz, $^3J_{9a,9}$ = 4.1 Hz, 1H, H-9), 5.33 (pseudo-t, ^3J ~ 7 Hz, 1H, H-7), 3.86 (s, 3H, MeO), 3.81 (s, 3H, MeO). Other signals are overlapped with those of major isomer. ^{13}C-NMR (100.6 MHz, CDCl$_3$): δ 136.0, 133.8, 131.8, 126.51, 126.48, 114.4, 55.5. Other signals are overlapped with those of major isomer or cannot be seen in the spectrum due to the low concentration of minor isomer. ^{19}F-NMR (376.3 MHz, CDCl$_3$): δ −74.6 [C(O)CF$_3$], −76.2 (CF$_3$).

1-(4-(4-Bromophenyl)-2-((4-bromophenyl)ethynyl)-2-(trifluoromethyl)-2H,9aH-pyrido[2,1-b][1,3]oxazin-3-yl)-2,2,2-trifluoroethan-1-one (**3d**). Obtained from pyridine **1a** (0.0395 g, 0.5 mmol) and CF$_3$-ynone **2d** (0.292 g, 1.054 mmol). Yellow-brown powder, m.p. 83.9–86.7 °C (hexane), yield 0.244 g (77%). (2S*,9aS*):(2R*,9aS*)-isomers ratio is 89:11 (^{19}F-NMR). HRMS (ESI-TOF): *m/z* [M + H]$^+$ Calcd for C$_{25}$H$_{14}$Br$_2$F$_6$NO$_2$$^+$: 633.9270; found: 633.9282.

(2S*,9aS*)-**3d**: ^{1}H-NMR (400.1 MHz, CDCl$_3$): δ 7.68–7.29 (m, 8H), 6.49 (dd, $^3J_{8,9}$ = 9.8 Hz, $^3J_{7,8}$ = 6.1 Hz, 1H, H-8), 6.41 (d, $^3J_{6,7}$ = 7.6 Hz, 1H, H-6), 6.00 (dd, $^3J_{8,9}$ = 9.8 Hz, $^3J_{9a,9}$ = 3.9 Hz, 1H, H-9), 5.69 (d, $^3J_{9a,9}$ = 3.9 Hz, 1H, H-9a), 5.53 (pseudo-t, ^3J ~ 7 Hz, 1H, H-7). ^{13}C-NMR (100.6 MHz, CDCl$_3$): δ 180.4 (q, $^2J_{CF}$ = 35.4 Hz, C-12), 159.3 (C-4), 149.6, 133.5, 132.9, 131.6, 130.2, 128.5, 126.3 (C-8), 125.4 (C-6), 123.9, 122.4 (q, $^1J_{CF}$ = 286.8 Hz, CF$_3$), 119.9, 116.7 (C-9), 115.5 [q, $^1J_{CF}$ = 292.7 Hz, C(O)CF$_3$], 109.4, 104.5 (C-7), 87.4 (C-11), 82.2 (C-10), 79.2 (C-9a), 73.6 (q, $^2J_{CF}$ = 34.3 Hz, C-2). ^{19}F-NMR (376.3 MHz, CDCl$_3$): δ −72.3 [C(O)CF$_3$], −77.3 (CF$_3$).

(2R*,9aS*)-**3d'**: ^{1}H-NMR (400.1 MHz, CDCl$_3$): δ 6.45–6.42 (m, 1H, H-8), 6.22 (d, $^3J_{6,7}$ = 7.5 Hz, 1H, H-6), 6.07 (d, $^3J_{9a,9}$ = 4.0 Hz, 1H, H-9a), 5.84 (dd, $^3J_{8,9}$ = 9.8 Hz, $^3J_{9a,9}$ = 4.0 Hz, 1H, H-9), 5.36 (pseudo-t, ^3J = 6.8 Hz, 1H, H-7). Other signals are overlapped with those of major isomer. ^{13}C-NMR (100.6 MHz, CDCl$_3$): δ 136.1, 135.1, 133.7, 131.6, 126.5, 126.0, 123.8, 115.1, 102.6. Other signals are overlapped with those of major isomer or cannot be seen in the spectrum due to the low concentration of minor isomer. ^{19}F-NMR (376.3 MHz, CDCl$_3$): δ −74.5 [C(O)CF$_3$], −76.2 (CF$_3$).

1-(4-(4-Chlorophenyl)-2-((4-chlorophenyl)ethynyl)-2-(trifluoromethyl)-2H,9aH-pyrido[2,1-b][1,3]oxazin-3-yl)-2,2,2-trifluoroethan-1-one (**3e**). Obtained from pyridine **1a** (0.042 g, 0.53 mmol) and CF$_3$-ynone **2e** (0.254 g, 1.09 mmol). Yellow-brown powder, m.p. 68–70 °C (hexane), yield 0.286 g (99%). (2S*,9aS*):(2R*,9aS*)-isomers ratio is 89:11 (^{19}F-NMR). HRMS (ESI-TOF): *m/z* [M + H]$^+$ Calcd for C$_{25}$H$_{14}$Cl$_2$F$_6$NO$_2$$^+$: 544.0300; found: 544.0308.

(2S*,9aS*)-**3e**: ^{1}H-NMR (400.1 MHz, CDCl$_3$): δ 7.52–7.28 (m, 8H), 6.49 (dd, $^3J_{8,9}$ = 9.8 Hz, $^3J_{7,8}$ = 6.1 Hz, 1H, H-8), 6.41 (d, $^3J_{6,7}$ = 7.6 Hz, 1H, H-6), 6.00 (dd, $^3J_{8,9}$ = 9.8 Hz, $^3J_{9a,9}$ = 4.0 Hz, 1H, H-9), 5.69

(d, $^3J_{9a,9}$ = 4.0 Hz, 1H, H-9a), 5.53 (pseudo-t, ^3J ~ 7 Hz, 1H, H-7). ^{1}H-NMR (400.1 MHz, CD$_3$CN): δ 7.73–7.31 (m, 8H), 6.57–6.52 (m, 2H, H-8, H-6), 6.03 (dd, $^3J_{8,9}$ = 9.8 Hz, $^3J_{9a,9}$ = 3.8 Hz, 1H, H-9), 5.75 (d, $^3J_{9a,9}$ = 3.8 Hz, 1H, H-9a), 5.61 (pseudo-t, ^3J = 7.2 Hz, 1H, H-7). ^{13}C-NMR (100.6 MHz, CDCl$_3$): δ 180.3 (q, $^2J_{CF}$ = 34.7 Hz, C-12), 159.2 (C-4), 149.6, 140.0, 135.6, 133.3, 129.9, 128.7, 126.3 (C-8), 125.4 (C-6), 122.4 (q, $^1J_{CF}$ = 286.8 Hz, CF$_3$), 119.4, 116.7 (C-9), 115.5 [q, $^1J_{CF}$ = 292.7 Hz, C(O)CF$_3$], 109.4, 104.5 (C-7), 87.3 (C-11), 82.0 (C-10), 79.1 (C-9a), 73.6 (q, $^2J_{CF}$ = 34.3 Hz, C-2). ^{19}F-NMR (376.3 MHz, CD$_3$CN): δ −70.0 [C(O)CF$_3$], −75.4 (CF$_3$). ^{19}F-NMR (376.3 MHz, CDCl$_3$): δ −72.2 [C(O)CF$_3$], −77.3 (CF$_3$).

(2R*,9aS*)-3e': ^{1}H-NMR (400.1 MHz, CDCl$_3$): δ 6.22 (d, $^3J_{6,7}$ = 7.6 Hz, 1H, H-6), 6.07 (d, $^3J_{9a,9}$ = 4.0 Hz, 1H, H-9a), 5.85 (dd, $^3J_{8,9}$ = 9.8 Hz, $^3J_{9a,9}$ = 4.0 Hz, 1H, H-9), 5.36 (pseudo-t, ^3J ~ 7 Hz, 1H, H-7). Other signals are overlapped with those of major isomer. ^{1}H-NMR (400.1 MHz, CD$_3$CN): δ 6.48–6.42 (m, 1H, H-8), 6.32 (d, $^3J_{6,7}$ = 7.5 Hz, 1H, H-6), 5.89 (dd, $^3J_{8,9}$ = 9.8 Hz, $^3J_{9a,9}$ = 3.9 Hz, 1H, H-9), 5.42 (pseudo-t, ^3J = 6.8 Hz, 1H, H-7). Other signals are overlapped with those of major isomer. ^{13}C-NMR (100.6 MHz, CDCl$_3$): δ 136.1, 133.5, 129.7, 128.6, 126.0, 123.9, 115.1, 102.6. Other signals are overlapped with those of major isomer or cannot be seen in the spectrum due to the low concentration of minor isomer. ^{19}F-NMR (376.3 MHz, CD$_3$CN): δ −72.2 [C(O)CF$_3$], −74.1 (CF$_3$). ^{19}F-NMR (376.3 MHz, CDCl$_3$): δ −74.4 [C(O)CF$_3$], −76.1 (CF$_3$).

1-(4-(4-Methylphenyl)-2-((4-methylphenyl)ethynyl)-2-(trifluoromethyl)-2H,9aH-pyrido[2,1-b][1,3]oxazin-3-yl)-2,2,2-trifluoroethan-1-one (**3f**). Obtained from pyridine **1a** (0.044 g, 0.556 mmol) and CF$_3$-ynone **2f** (0.240 g, 1.13 mmol). Yellow-brown powder, m.p. 95.2–99.1 °C (hexane), yield 0.256 g (91%). (2S*,9aS*):(2R*,9aS*)-isomers ratio is 91:9 (^{19}F-NMR). HRMS (ESI-TOF): m/z [M + H]$^+$ Calcd for C$_{27}$H$_{20}$F$_6$NO$_2$$^+$: 504.1393; found: 504.1401.

(2S*,9aS*)-3f: ^{1}H-NMR (400.1 MHz, CDCl$_3$): δ 7.54–7.10 (m, 8H), 6.50–6.47 (m, 2H, H-8, H-6), 6.00 (dd, $^3J_{8,9}$ = 10.0 Hz, $^3J_{9a,9}$ = 3.8 Hz, 1H, H-9), 5.74 (d, $^3J_{9a,9}$ = 3.8 Hz, 1H, H-9a), 5.49 (pseudo-t, ^3J = 6.7 Hz, 1H, H-7), 2.44 (s, 3H, Me), 2.33 (s, 3H, Me). ^{13}C-NMR (100.6 MHz, CDCl$_3$): δ 180.7 (q, $^2J_{CF}$ = 35.0 Hz, C-12), 160.5 (C-4), 139.5, 134.0, 132.0, 130.2, 129.0, 128.7, 126.2 (C-8), 125.9 (C-6), 122.6 (q, $^1J_{CF}$ = 286.8 Hz, CF$_3$), 118.0, 116.4 (C-9), 115.6 [q, $^1J_{CF}$ = 293.4 Hz, C(O)CF$_3$], 109.0, 103.8 (C-7), 88.5 (C-11), 80.8 (C-10), 79.0 (C-9a), 73.8 (q, $^2J_{CF}$ = 34.3 Hz, C-2), 21.6, 21.5. ^{19}F-NMR (376.3 MHz, CDCl$_3$): δ −72.3 [C(O)CF$_3$], −77.2 (CF$_3$).

(2R*,9aS*)-3f': ^{1}H-NMR (400.1 MHz, CDCl$_3$): δ 6.47–6.41 (m, 1H, H-8), 6.30 (d, $^3J_{6,7}$ = 7.6 Hz, 1H, H-6), 6.10 (d, $^3J_{9a,9}$ = 3.8 Hz, 1H, H-9a), 5.84 (dd, $^3J_{8,9}$ = 9.6 Hz, $^3J_{9a,9}$ = 3.8 Hz, 1H, H-9), 5.33 (pseudo-t, ^3J ~ 7 Hz, 1H, H-7), 2.42 (s, 3H, Me), 2.36 (s, 3H, Me). Other signals are overlapped with those of major isomer. ^{13}C-NMR (100.6 MHz, CDCl$_3$): δ 143.2, 139.4, 132.1, 126.4, 118.4, 115.0, 102.0, 79.8. Other signals are overlapped with those of major isomer or can not be seen in the spectrum due to the low concentration of minor isomer. ^{19}F-NMR (376.3 MHz, CDCl$_3$): δ −74.4 [C(O)CF$_3$], −76.1 (CF$_3$).

1-(4-(4-Methylthiophenyl)-2-((4-methylthiophenyl)ethynyl)-2-(trifluoromethyl)-2H,9aH-pyrido[2,1-b][1,3]oxazin-3-yl)-2,2,2-trifluoroethan-1-one (**3g**). Obtained from pyridine **1a** (0.040 g, 0.506 mmol) and CF$_3$-ynone **2g** (0.256 g, 1.05 mmol). Brown powder, m.p. 120.5–123.2 °C (hexane), yield 0.274 g (96%). (2S*,9aS*):(2R*,9aS*)-isomers ratio is 92:8 (^{19}F-NMR). HRMS (ESI-TOF): m/z [M + H]$^+$ Calcd for C$_{27}$H$_{20}$F$_6$NO$_2$S$_2$$^+$: 568.0834; found: 568.0834.

(2S*,9aS*)-3g: ^{1}H-NMR (400.1 MHz, CDCl$_3$): δ 7.49–7.25 (m, 6H), 7.14 (d, 2H, ^3J = 8.5 Hz), 6.51–6.47 (m, 2H, H-8, H-6), 6.00 (dd, $^3J_{8,9}$ = 9.7 Hz, $^3J_{9a,9}$ = 3.8 Hz, 1H, H-9), 5.70 (d, $^3J_{9a,9}$ = 3.8 Hz, 1H, H-9a), 5.50 (pseudo-t, ^3J = 6.4 Hz, 1H, H-7), 2.53 (s, 3H, Me), 2.46 (s, 3H, Me). ^{1}H-NMR (400.1 MHz, CD$_3$CN): δ 7.58–7.30 (m, 6H), 7.24 (d, 2H, ^3J = 8.7 Hz), 6.57 (d, $^3J_{6,7}$ = 7.6 Hz, 1H, H-6), 6.54 (dd, $^3J_{8,9}$ = 9.8 Hz, $^3J_{7,8}$ = 6.0 Hz, 1H, H-8), 6.00 (dd, $^3J_{8,9}$ = 9.8 Hz, $^3J_{9a,9}$ = 4.0 Hz, 1H, H-9), 5.74 (d, $^3J_{9a,9}$ = 4.0 Hz, 1H, H-9a), 5.59 (pseudo-t, ^3J = 6.8 Hz, 1H, H-7), 2.53 (s, 3H, Me), 2.47 (s, 3H, Me). ^{13}C-NMR (100.6 MHz, CD$_3$CN): δ 180.8 (q, $^2J_{CF}$ = 34.1 Hz, C-12), 162.6 (C-4), 148.1, 142.6, 135.2, 132.9, 131.8, 129.8, 127.1, 126.8 (C-8), 126.5 (C-6), 123.8 (q, $^1J_{CF}$ = 285.8 Hz, CF$_3$), 117.8, 117.5, 116.7 [q, $^1J_{CF}$ = 292.2 Hz, C(O)CF$_3$], 108.7, 105.4 (C-7), 88.4 (C-11), 82.5 (C-10), 80.1 (C-9a), 74.5 (q, $^2J_{CF}$ = 33.7 Hz, C-2), 15.1, 14.7.

^{19}F-NMR (376.3 MHz, CDCl$_3$): δ −72.3 [C(O)CF$_3$], −77.4 (CF$_3$). ^{19}F-NMR (376.3 MHz, CD$_3$CN): δ −70.0 [C(O)CF$_3$], −75.5 (CF$_3$).

(2R*,9aS*)-**3g′**: ^{1}H-NMR (400.1 MHz, CDCl$_3$): δ 6.47–6.41 (m, 1H, H-8), 6.30 (d, $^3J_{6,7}$ = 7.5 Hz, 1H, H-6), 6.07 (d, $^3J_{9a,9}$ = 4.0 Hz, 1H, H-9a), 5.84 (dd, $^3J_{8,9}$ = 9.7 Hz, $^3J_{9a,9}$ = 4.0 Hz, 1H, H-9), 5.34 (pseudo-t, ^3J = 6.8 Hz, 1H, H-7), 2.51 (s, 3H, Me), 2.47 (s, 3H, Me). Other signals are overlapped with those of major isomer. ^{1}H-NMR (400.1 MHz, CD$_3$CN): δ 6.45–6.37 (m, 2H, H-8, H-9a), 6.20 (d, $^3J_{6,7}$ = 7.0 Hz, 1H, H-6), 5.87 (dd, $^3J_{8,9}$ = 9.8 Hz, $^3J_{9a,9}$ = 4.0 Hz, 1H, H-9), 5.42 (t, ^3J = 6.4 Hz, 1H, H-7), 2.52 (s, 3H, Me), 2.50 (s, 3H, Me). Other signals are overlapped with those of major isomer. ^{13}C-NMR (100.6 MHz, CD$_3$CN): δ 145.5, 142.8, 136.9, 133.1, 129.5, 129.0, 128.2, 126.7, 126.3, 104.4, 84.7, 15.1, 14.7. Other signals are overlapped with those of major isomer or cannot be seen in the spectrum due to the low concentration of minor isomer. ^{19}F-NMR (376.3 MHz, CDCl$_3$): δ −74.4 [s, 3F, C(O)CF$_3$], −76.1 (s, 3F, CF$_3$). ^{19}F-NMR (376.3 MHz, CD$_3$CN): δ −72.2 [C(O)CF$_3$], −74.0 (CF$_3$).

1-(4-(3,4-Dimethylphenyl)-2-((3,4-dimethylphenyl)ethynyl)-2-(trifluoromethyl)-2H,9aH-pyrido[2,1-b][1,3]oxazin-3-yl)-2,2,2-trifluoroethan-1-one (**3h**). Obtained from pyridine **1a** (0.039 g, 0.49 mmol) and CF$_3$-ynone **2h** (0.232 g, 1.027 mmol). Yellow-brown powder, m.p. 72.6–74.6 °C (hexane), yield 0.215 g (83%). (2S*,9aS*):(2R*,9aS*)-isomers ratio is 92:8 (^{19}F-NMR) HRMS (ESI-TOF): *m/z* [M + H]$^+$ Calcd for C$_{29}$H$_{24}$F$_6$NO$_2{}^+$: 532.1706; found: 532.1717.

(2S*,9aS*)-**3h**: ^{1}H-NMR (400.1 MHz, CDCl$_3$): δ 7.38–7.02 (m, 6H), 6.50–6.46 (m, 2H, H-8, H-6), 5.98 (dd, $^3J_{8,9}$ = 10.0 Hz, $^3J_{9a,9}$ = 3.9 Hz, 1H, H-9), 5.71 (d, $^3J_{9a,9}$ = 3.9 Hz, 1H, H-9a), 5.47 (pseudo-t, ^3J = 6.5 Hz, 1H, H-7), 2.34 (s, 3H, Me), 2.31 (s, 3H, Me), 2.24 (s, 3H, Me), 2.21 (s, 3H, Me). ^{13}C-NMR (100.6 MHz, CDCl$_3$): δ 180.8 (q, $^2J_{CF}$ = 35.0 Hz, C-12), 160.6 (C-4), 138.3, 136.6, 133.0, 130.6, 129.5, 129.1, 126.2 (C-8), 126.1 (C-6), 122.7 (q, $^1J_{CF}$ = 286.8 Hz, CF$_3$), 118.3, 116.4 (C-9), 115.6 [q, $^1J_{CF}$ = 293.0 Hz, C(O)CF$_3$], 109.0, 103.6 (C-7), 88.6 (C-11), 80.6 (C-10), 78.9 (C-9a), 73.8 (q, $^2J_{CF}$ = 34.3 Hz, C-2), 20.0, 19.7, 19.6 (br s), 19.4. ^{19}F-NMR (376.3 MHz, CDCl$_3$): δ −72.3 [C(O)CF$_3$], −77.3 (CF$_3$).

(2R*,9aS*)-**3h′**: ^{1}H-NMR (400.1 MHz, CDCl$_3$): δ 6.44–6.40 (m, 1H, H-8), 6.31 (d, $^3J_{6,7}$ = 7.6 Hz, 1H, H-6), 6.08 (d, $^3J_{9a,9}$ = 4.0 Hz, 1H, H-9a), 5.83 (dd, $^3J_{8,9}$ = 9.9 Hz, $^3J_{9a,9}$ = 4.0 Hz, 1H, H-9), 5.31 (pseudo-t, ^3J = 7.2 Hz, 1H, H-7), 2.32 (s, 3H, Me), 2.28 (s, 3H, Me), 2.25 (s, 3H, Me). Other signals are overlapped with those of major isomer. ^{13}C-NMR (100.6 MHz, CDCl$_3$): δ 142.7, 138.1, 134.8, 133.2, 131.7, 130.3, 129.7, 126.6, 126.5, 118.7, 114.9, 108.5, 101.8, 79.7, 20.2, 19.8. Other signals are overlapped with those of major isomer or cannot be seen in the spectrum due to the low concentration of minor isomer. ^{19}F-NMR (376.3 MHz, CDCl$_3$): δ −74.4 [C(O)CF$_3$], −77.3 (CF$_3$).

2,2,2-Trifluoro-1-(4-(4-methoxynaphthalen-1-yl)-2-((4-methoxynaphthalen-1-yl)ethynyl)-2-(trifluoromethyl)-2H,9aH-pyrido[2,1-b][1,3]oxazin-3-yl)ethan-1-one (**3i**). Obtained from pyridine **1a** (0.0405 g, 0.51 mmol) and CF$_3$-ynone **2i** (0.296 g, 1.06 mmol). Yellow-brown powder, m.p. 143.5–145.5 °C (hexane), yield 0.320 g (98%). (2S*,9aS*):(2R*,9aS*)-isomers ratio is 94:6. Rotamers ratio is (93:1):(4:2) (^{19}F-NMR). HRMS (ESI-TOF): *m/z* [M + H]$^+$ Calcd for C$_{35}$H$_{24}$F$_6$NO$_4{}^+$: 636.1604; found: 636.1608.

(2S*,9aS*)-**3i**: ^{1}H-NMR (400.1 MHz, CDCl$_3$): δ 8.39–8.36 (m, 1H), 8.25–8.22 (m, 2H), 7.70–7.47 (m, 7H), 6.87 (d, ^3J = 8.1 Hz, 1H), 6.75 (d, ^3J = 8.1 Hz, 1H), 6.49 (dd, $^3J_{8,9}$ = 9.7 Hz, $^3J_{7,8}$ = 6.1 Hz, 1H, H-8), 6.11–6.02 (m, 3H, H-6, H-9, H-9a), 5.36 (pseudo-t, ^3J = 7.2 Hz, 1H, H-7), 4.08 (s, 3H, MeO), 4.01 (s, 3H, MeO). ^{13}C-NMR (100.6 MHz, CDCl$_3$): δ 179.9 (q, $^2J_{CF}$ = 34.7 Hz, C-12), 160.8 (C-4), 160.2, 156.7, 137.4, 134.4, 132.3, 131.7, 128.8, 127.6, 126.3 (C-8), 126.2 (C-6), 125.93, 125.87, 125.8, 125.1, 124.8, 123.1 (q, $^1J_{CF}$ = 287.1 Hz, CF$_3$), 123.1, 122.1, 120.7, 117.0 (C-9), 115.7 [q, $^1J_{CF}$ = 292.9 Hz, C(O)CF$_3$], 111.0, 109.7, 104.0, 103.5, 103.3, 86.7 (C-11), 84.9 (C-10), 78.6 (C-9a), 74.2 (q, $^2J_{CF}$ = 33.9 Hz, C-2), 55.9, 55.6. ^{19}F-NMR (376.3 MHz, CDCl$_3$): δ major rotamer −71.6 [C(O)CF$_3$], −76.8 (CF$_3$); minor rotamer −73.2 [C(O)CF$_3$], −78.1 (CF$_3$).

(2R*,9aS*)-**3i′**: ^{1}H-NMR (400.1 MHz, CDCl$_3$): δ 8.32 (d, ^3J = 8.4 Hz, 1H), 7.96 (d, ^3J = 8.2 Hz, 1H), 6.86 (d, ^3J = 8.1 Hz, 1H), 6.81 (d, ^3J = 7.9 Hz, 1H), 4.04 (s, 3H, MeO), 4.02 (s, 3H, MeO). Other signals are overlapped with those of major isomer. ^{13}C-NMR (100.6 MHz, CDCl$_3$): δ 137.5, 135.0, 129.0, 126.6, 125.0, 122.9, 56.0, 55.8. Other signals are overlapped with those of major isomer or cannot be seen in the spectrum due to the low concentration of minor isomer. ^{19}F-NMR (376.3 MHz, CDCl$_3$): δ major rotamer −74.1 [C(O)CF$_3$], −75.9 (CF$_3$); minor rotamer −74.3 [C(O)CF$_3$], −76.2 (CF$_3$).

2,2,2-Trifluoro-1-(4-phenyl-2-(phenylethynyl)-2-(trifluoromethyl)-8-vinyl-2H,9aH-pyrido[2,1-b][1,3]oxazin-3-yl)ethan-1-one (**3j**). Obtained from pyridine **1b** (0.054 g, 0.51 mmol) and CF$_3$-ynone **2a** (0.206 g, 1.04 mmol). Brown powder, m.p. 80–83 °C (hexane), yield 0.249 g (97%). (2S*,9aS*):(2R*,9aS*)-isomers ratio is 89:11 (^{19}F-NMR). HRMS (ESI-TOF): *m/z* [M + H]$^+$ Calcd for C$_{27}$H$_{18}$F$_6$NO$_2$$^+$: 502.1246; found: 502.1246.

(2S*,9aS*)-**3j**: ^{1}H-NMR (400.1 MHz, CDCl$_3$): δ 7.67–7.44 (m, 7H), 7.37–7.28 (m, 3H), 6.49 (d, $^3J_{6,7}$ = 8.0 Hz, 1H, H-6), 6.44 (dd, ^3J = 17.6 Hz, ^3J = 11.0 Hz, 1H, C<u>H</u>=CH$_2$), 5.90 (d, $^3J_{9a,9}$ = 4.5 Hz, 1H, H-9), 5.78 (d, $^3J_{9a,9}$ = 4.5 Hz, 1H H-9a), 5.77 (dd, $^3J_{6,7}$ = 8.0 Hz, $^4J_{7,9}$ = 1.5 Hz, 1H, H-7), 5.56 (d, ^3J = 17.6 Hz, 1H, CH=C<u>H</u>$_2$), 5.33 (d, ^3J = 11.0 Hz, 1H, CH$_2$, CH=C<u>H</u>$_2$). ^{13}C-NMR (100.6 MHz, CDCl$_3$): δ 180.9 (q, $^2J_{CF}$ = 34.8 Hz, C-12), 160.2 (C-4), 135.2 (C, <u>C</u>H=CH$_2$), 134.3, 133.3, 132.1 (C$_{m,m'}$ from Ar), 131.4 (q, $^1J_{CF}$ = 1.7 Hz, C-3), 129.5 (br s), 129.3, 128.2 (C$_{o,o'}$ from Ar), 126.0 (C-6), 122.6, (q, $^1J_{CF}$ = 286.9 Hz, CF$_3$), 121.0, 116.6 (CH=<u>C</u>H$_2$), 114.3 (CH=<u>C</u>H$_2$), 115.5 [q, $^1J_{CF}$ = 292.8 Hz, C(O)CF$_3$], 109.3, 101.7 (C-7), 88.4 (C-11), 81.3 (C-10), 79.4 (C-9a), 73.9 (q, $^2J_{CF}$ = 34.1 Hz, C-2). ^{19}F-NMR (376.3 MHz, CDCl$_3$): δ −72.5 [C(O)CF$_3$], −77.2 (CF$_3$).

(2R*,9aS*)-**3j′**: ^{1}H-NMR (400.1 MHz, CDCl$_3$): δ 6.31 (d, $^3J_{6,7}$ = 7.9 Hz, 1H, H-6), 6.13–6.11 (m, 2H, H-9, H-9a), 5.60 (dd, ^3J = 7.9 Hz, ^3J = 1.6 Hz, 1H, H-7), 5.53 (d, ^3J = 17.4 Hz, 1H, CH=C<u>H</u>$_2$), 5.51 (d, ^3J = 17.2 Hz, 1H, CH=C<u>H</u>$_2$), 5.31 (d, ^3J = 10.0 Hz, 1H, CH=C<u>H</u>$_2$). Other signals are overlapped with those of major isomer. ^{13}C-NMR (100.6 MHz, CDCl$_3$): δ 137.1, 135.4, 132.4, 132.2, 129.2, 128.7, 128.2, 127.2, 126.6, 126.0, 117.6, 116.5, 113.0, 99.8. Other signals are overlapped with those of major isomer or can not be seen in the spectrum due to the low concentration of minor isomer. ^{19}F-NMR (376.3 MHz, CDCl$_3$): δ −74.6 [C(O)CF$_3$], −76.2 (CF$_3$).

4-Phenyl-2-(phenylethynyl)-3-(trifluoroacetyl)-2-(trifluoromethyl)-2H,9aH-pyrido[2,1-b][1,3]oxazine-8-carbaldehyde (**3k**). Obtained from pyridine **1c** (0.0475 g, 0.44 mmol) and CF$_3$-ynone **2a** (0.198 g, 1 mmol). Yellow powder, m.p. 77–79 °C (hexane), yield 0.178 g (80%). (2S*,9aS*):(2R*,9aS*)-isomers ratio is 89:11 (^{19}F-NMR). HRMS (ESI-TOF): *m/z* [M + H]$^+$ Calcd for C$_{26}$H$_{16}$F$_6$NO$_3$$^+$: 504.1029; found: 504.1035.

(2S*,9aS*)-**3k**: ^{1}H-NMR (400.1 MHz, CD$_3$CN): 9.70 (s, 1H, CHO), δ 7.75–7.25 (m, 10H), 6.81 (d, ^3J = 4.2 Hz, 1H, H-9), 6.66 (d, ^3J = 7.8 Hz, 1H, H-6), 6.06 (d, ^3J = 4.2 Hz, 1H, H-9a), 5.94 (dd, ^3J = 7.8 Hz, ^3J = 1.5 Hz, 1H, H-7). ^{13}C-NMR (100.6 MHz, CD$_3$CN): δ 192.0 (CHO), 181.7 (q, $^2J_{CF}$ = 34.8 Hz, C-12), 162.0 (C-4), 137.5, 134.9, 134.7, 134.0, 132.2 (q, $^3J_{CF}$ = 1.8 Hz), 130.9, 130.6 (br s), 130.2, 129.8, 129.6, 128.7, 123.7 (q, $^1J_{CF}$ = 286.0 Hz, CF$_3$), 121.5, 116.5 [q, $^1J_{CF}$ = 292.5 Hz, C(O)CF$_3$], 110.1, 99.0, 89.5 (C-11), 82.1 (C-10), 80.1 (C-9a), 75.2 (q, $^2J_{CF}$ = 34.3 Hz, C-2). ^{19}F-NMR (376.3 MHz, CD$_3$CN): δ −70.2 [C(O)CF$_3$], −75.2 (CF$_3$). ^{19}F-NMR (376.3 MHz, CDCl$_3$): δ −71.7 [C(O)CF$_3$], −76.0 (CF$_3$).

(2S*,9aS*)-**3k′**: ^{1}H-NMR (400.1 MHz, CD$_3$CN): 9.66 (s, 1H, CHO), 6.52–6.46 (m, 2H), 6.37 (d, ^3J = 4.2 Hz, 1H), 5.77 (dd, ^3J = 7.7 Hz, ^3J = 1.5 Hz, 1H). Other signals are overlapped with those of major isomer. ^{13}C-NMR (100.6 MHz, CD$_3$CN): δ 133.8, 132.9, 130.0, 129.7, 129.6, 129.4, 128.1, 115.1, 97.1, 83.8, 78.2. Other signals are overlapped with those of major isomer or cannot be seen in the spectrum due to the low concentration of minor isomer. ^{19}F-NMR (376.3 MHz, CD$_3$CN): δ −72.6 [C(O)CF$_3$], −74.4 (CF$_3$). ^{19}F-NMR (376.3 MHz, CDCl$_3$): δ −73.6 [C(O)CF$_3$], −75.2 (CF$_3$).

1-(8-Acetyl-4-phenyl-2-(phenylethynyl)-2-(trifluoromethyl)-2H,9aH-pyrido[2,1-b][1,3]oxazin-3-yl)-2,2,2-trifluoroethan-1-one (**3l**). Obtained from pyridine **1d** (0.030 g, 0.25 mmol) and CF$_3$-ynone **2a** (0.101 g, 0.51 mmol). Yellow powder, m.p. 122.8–124.2 °C (hexane), yield 0.071 g (55%). (2S*,9aS*):(2R*,9aS*)-isomers ratio is 87:13 (^{19}F-NMR). HRMS (ESI-TOF): *m/z* [M + H]$^+$ Calcd for C$_{27}$H$_{18}$F$_6$NO$_3$$^+$: 518.1185; found: 518.1214.

(2*S**,9a*S**)-3l: ^{1}H-NMR (400.1 MHz, CDCl$_3$): δ 7.66–7.45 (m, 7H), 7.38–7.29 (m, 3H), 6.72 (pseudo-d, $^3J_{9a,9}$ ~ 4 Hz, 1H, H-9), 6.54 (d, $^3J_{6,7}$ = 7.8 Hz, 1H, H-6), 6.04 (dd, $^3J_{6,7}$ = 7.8, Hz, $^3J_{7,9}$ = 1.6 Hz, 1H, H-7), 5.91 (d, $^3J_{9a,9}$ = 4.2 Hz, 1H, H-9a), 2.48 (s, 3H, Me). ^{13}C-NMR (100.6 MHz, CDCl$_3$): δ 195.6, 180.9 (q, $^2J_{CF}$ = 35.4 Hz, C-12), 159.3 (C-4), 136.5, 133.4, 132.1, 131.1 (q, $^1J_{CF}$ = 1.7 Hz, C-3), 129.6 (br s), 129.5, 128.3 (C$_{o,o'}$ from Ar), 126.7 (C-6), 122.4 (q, $^1J_{CF}$ = 286.6 Hz, CF$_3$), 121.2 (C-9), 120.8 (C$_i$ from Ar), 115.4 [q, $^1J_{CF}$ = 292.8 Hz, C(O)CF$_3$], 110.0, 100.5 (C-7), 89.1 (C-11), 80.9 (C-10), 78.8 (C-9a), 74.4 (q, $^2J_{CF}$ = 34.1 Hz, C-2), 25.3. ^{19}F-NMR (376.3 MHz, CDCl$_3$): δ −72.7 [C(O)CF$_3$], −77.2 (CF$_3$).

(2*R**,9a*S**)-3l′: ^{1}H-NMR (400.1 MHz, CDCl$_3$): δ 6.57 (pseudo-d, ^3J ~ 4 Hz, 1H, H-9), 6.37 (d, $^3J_{6,7}$ = 7.8 Hz, 1H, H-6), 6.28 (d, $^3J_{9a,9}$ = 4.3 Hz, 1H, H-9a), 5.88 (dd, $^3J_{6,7}$ = 7.8 Hz, $^3J_{7,9}$ = 1.5 Hz, 1H, H-7), 2.46 (s, 3H, Me). Other signals are overlapped with those of major isomer. ^{13}C-NMR (100.6 MHz, CDCl$_3$): δ 136.6, 132.6, 132.3, 129.4, 128.3, 127.3, 119.9, 98.5, 29.7. Other signals are overlapped with those of major isomer or cannot be seen in the spectrum due to the low concentration of minor isomer. ^{19}F-NMR (376.3 MHz, CDCl$_3$): δ −74.7 [C(O)CF$_3$], −76.3 (CF$_3$).

Methyl 4-phenyl-2-(phenylethynyl)-3-(2,2,2-trifluoroacetyl)-2-(trifluoromethyl)-2H,9aH- pyrido[2,1-b][1,3] oxazine-8-carboxylate (**3m**). Obtained from pyridine **1e** (0.048 g, 0.35 mmol) and CF$_3$-ynone **2a** (0.147 g, 0.74 mmol). Pale yellow powder, m.p. 115.4–116.5 °C (hexane), yield 0.112 g (60%). (2*S**,9a*S**):(2*R**,9a*S**)-isomers ratio is 87:13 (^{19}F-NMR). HRMS (ESI-TOF): *m/z* [M + H]$^+$ Calcd for C$_{27}$H$_{18}$F$_6$NO$_4^+$: 534.1135; found: 534.1140.

(2*S**,9a*S**)-3m: ^{1}H-NMR (400.1 MHz, CDCl$_3$): δ 7.66–7.44 (m, 7H), 7.37–7.29 (m, 3H), 6.89 (pseudo-d, ^3J ~ 4 Hz, 1H, H-9), 6.52 (d, $^3J_{6,7}$ = 7.8 Hz, 1H, H-6), 5.99 (dd, $^3J_{6,7}$ = 7.8 Hz, $^3J_{7,9}$ = 1.5 Hz, 1H, H-7), 5.87 (d, $^3J_{9a,9}$ = 4.2 Hz, 1H, H-9a), 3.85 (s, 3H, Me). ^{13}C-NMR (100.6 MHz, CDCl$_3$): δ 181.0 (q, $^2J_{CF}$ = 35.4 Hz, C-12), 164.6 (C-4), 159.2 (CO$_2$Me), 133.4, 132.1, 131.2 (q, $^1J_{CF}$ = 1.7 Hz, C-3), 130.1, 129.6 (br s), 129.5 (C-8), 128.3 (C$_{o,o'}$ from Ar), 126.5 (C-6), 122.4, (q, $^1J_{CF}$ = 287.3 Hz, CF$_3$), 121.5 (C-9), 120.8 (C$_i$ from Ar), 115.4 (q, $^1J_{CF}$ = 293.0 Hz, C(O)CF$_3$), 110.3, 101.5 (C-7), 89.0 (C-11), 80.9 (C-10), 78.8 (C-9a), 74.3 (q, $^2J_{CF}$ = 34.8 Hz, C-2), 52.5. ^{19}F-NMR (376.3 MHz, CDCl$_3$): δ −72.7 [C(O)CF$_3$], −77.2 (CF$_3$).

(2*R**,9a*S**)-3m′: ^{1}H-NMR (400.1 MHz, CDCl$_3$): δ 6.74 (pseudo-d, ^3J ~ 4 Hz, 1H, H-9), 6.35 (d, $^3J_{6,7}$ = 7.8 Hz, 1H, H-6), 6.24 (d, $^3J_{9a,9}$ = 4.3 Hz, 1H, H-9a), 5.83 (dd, $^3J_{6,7}$ = 7.8 Hz, $^3J_{7,9}$ = 1.5 Hz, 1H, H-7), 3.84 (s, 3H, Me). Other signals are overlapped with those of major isomer. ^{13}C-NMR (100.6 MHz, CDCl$_3$): δ 132.6, 132.3, 130.3, 129.4, 128.3, 127.1, 120.1, 99.8, 52.5. Other signals are overlapped with those of major isomer or cannot be seen in the spectrum due to the low concentration of minor isomer. ^{19}F-NMR (376.3 MHz, CDCl$_3$): δ −74.7 [C(O)CF$_3$], −76.4 (CF$_3$).

(2S*,9aS*)-1-(4,6-diPhenyl-2-(phenylethynyl)-2-(trifluoromethyl)-2H,9aH-pyrido[2,1-b][1,3]oxazin-3-yl)-2,2,2- trifluoroethan-1-one (**3n**). Obtained from pyridine **1f** (0.079 g, 0.51 mmol) and CF$_3$-ynone **2a** (0.204 g, 1.03 mmol). Orange powder, m.p. 90–91 °C (hexane), yield 0.168 g (60%). HRMS (ESI-TOF): *m/z* [M + H]$^+$ Calcd for C$_{31}$H$_{20}$F$_6$NO$_2^+$: 552.1393; found: 552.1393. (2*S**,9a*S**)-3o: ^{1}H-NMR (400.1 MHz, CDCl$_3$): δ 7.46–7.43 (m, 2H), 7.38–7.28 (m, 3H), 7.17–6.99 (m, 5H), 6.93 (br s, 5H), 6.61 (ddd, $^3J_{8,9}$ = 9.7 Hz, $^3J_{7,8}$ = 6.1 Hz, $^4J_{8,9a}$ = 0.8 Hz, 1H, H-8), 6.00 (ddd, $^3J_{8,9}$ = 9.7 Hz, $^3J_{9a,9}$ = 4.2 Hz, $^4J_{7,9}$ = 0.7 Hz, 1H, H-9), 5.84 (d, $^3J_{9a,9}$ = 4.2 Hz, 1H, H-9a), 5.43 (dd, $^3J_{7,8}$ = 6.1 Hz, $^3J_{7,9}$ = 0.7 Hz, 1H, H-7). ^{13}C-NMR (100.6 MHz, CDCl$_3$): δ 184.2 (q, $^2J_{CF}$ = 36.1 Hz, C-12), 157.3 (C-4), 139.2, 136.4, 134.7, 133.9, 132.5, 132.1, 131.4, 129.5, 128.9, 128.4, 128.3, 127.8, 127.7, 127.2, 122.8 (q, $^1J_{CF}$ = 285.7 Hz, CF$_3$), 120.7, 115.0 [q, $^1J_{CF}$ = 293.6 Hz, C(O)CF$_3$], 114.7, 106.7, 89.7 (C-11), 81.5 (C-9a), 81.2 (C-10), 74.5 (q, $^2J_{CF}$ = 34.8 Hz, C-2). ^{19}F-NMR (376.3 MHz, CDCl$_3$): δ −73.9 [C(O)CF$_3$], −76.0 (CF$_3$).

1-(9-Bromo-4-phenyl-2-(phenylethynyl)-2-(trifluoromethyl)-2H,9aH-pyrido[2,1-b][1,3]oxazin-3-yl)-2,2,2- trifluoroethan-1-one (**3p**). Major (9-Br)-regioisomer, obtained as a mixture (1:5) with minor (7-Br)-regioisomer (**3o**) from pyridine **1g** (0.082 g, 0.52 mmol) and CF$_3$-ynone **2a** (0.208 g, 1.05 mmol). Yellow powder, m.p. 76.0–77.8 °C (hexane), yield 0.153 g (53%). (2*S**,9a*S**):(2*R**,9a*S**)-isomers ratio of **3p** is 80:20 (^{19}F-NMR). HRMS (ESI-TOF) for the mixture of **3o** and **3p**: *m/z* [M + H]$^+$ Calcd for C$_{25}$H$_{15}$F$_6$BrNO$_2^+$: 554.0185; found: 554.0190.

(2*S**,9a*S**)-3p:** ^{1}H-NMR (400.1 MHz, CDCl$_3$): δ 7.65–7.30 (m, 10H), 6.80 (d, $^3J_{6,7}$ = 6.6 Hz, 1H, H-6), 6.45 (d, $^3J_{7,8}$ = 7.5 Hz, 1H, H-8), 5.79 (s, 1H, H-9a), 5.38 (pseudo-t, ^3J ~ 7 Hz, 1H, H-7). ^{13}C-NMR (100.6 MHz, CDCl$_3$): δ 181.1 (q, $^2J_{CF}$ = 35.6 Hz, C-12), 158.7 (C-4), 133.4, 132.3, 132.1, 131.1 (q, $^3J_{CF}$ = 1.8 Hz, C-3), 129.6 (br s), 129.4 (C-8), 128.7 (C-6), 128.3, 125.2, 122.4 (q, $^1J_{CF}$ = 286.4 Hz, CF$_3$), 121.0 (C$_i$ from Ar), 117.4 (C-9), 115.4 [q, $^1J_{CF}$ = 292.6 Hz, C(O)CF$_3$], 109.9, 103.3 (C-7), 89.2 (C-11), 82.9 (C-10), 80.5 (C-9a), 74.7 (q, $^2J_{CF}$ = 34.8 Hz, C-2). ^{19}F-NMR (376.3 MHz, CDCl$_3$): δ −72.7 [C(O)CF$_3$], −76.8 (CF$_3$).

(2*R,9a*S**)-3p′:** ^{1}H-NMR (400.1 MHz, CDCl$_3$): δ 6.75 (d, $^3J_{6,7}$ = 6.7 Hz, 1H, H-6), 6.30 (d, $^3J_{7,8}$ = 7.6 Hz, 1H, H-8), 6.16 (s, 1H, H-9a), 5.25 (pseudo-t, ^3J ~ 7 Hz, 1H, H-7). Other signals are overlapped with those of major isomer. ^{13}C-NMR of (2*R**,9a*S**)-3p′ and ^{13}C-NMR of (2*S**,9a*S**)-3o (100.6 MHz, CDCl$_3$): δ 158.6 (C-4), 152.3, 133.5, 132.6, 132.2, 130.8, 131.0 (q, $^3J_{CF}$ = 1.3 Hz, C-3), 130.5, 129.5, 129.3, 128.94, 128.91, 128.36, 128.31, 125.7, 121.5, 121.3, 120.8, 110.8, 101.8, 89.02 and 88.98 (C-11), 83.53 and 83.49 (C-10), 81.3 and 80.9 (C-9a), 77.8. Due to low concentration and equal amounts of (2*R**,9a*S**)-3p′ and ^{13}C-NMR of (2*S**,9a*S**)-3o assignment of their signals cannot be done. ^{13}C-NMR are reported together. Other signals are overlapped with those of major isomer 3p or cannot be seen in the spectrum due to the low concentration of minor isomers. ^{19}F-NMR (376.3 MHz, CDCl$_3$): δ −74.7 [C(O)CF$_3$], −75.9 (CF$_3$).

1-(7-Bromo-4-phenyl-2-(phenylethynyl)-2-(trifluoromethyl)-2H,9aH-pyrido[2,1-b][1,3]oxazin-3-yl)-2,2,2-trifluoroethan-1-one (**3p**). Minor (7-Br)-regioisomer, obtained as a mixture with major (9-Br)-regioisomer (**3q**) (see above). (2*S**,9a*S**):(2*R**,9a*S**)-isomers ratio is 78:22 (^{19}F-NMR). HRMS (ESI-TOF) for the mixture of **3o** and **3p**: *m/z* [M + H]$^+$ Calcd for C$_{25}$H$_{15}$F$_6$BrNO$_2$$^+$: 554.0185; found: 554.0190.

(2*S,9a*S**)-3o:** ^{1}H-NMR (400.1 MHz, CDCl$_3$): δ 6.65 (s, 1H, H-6), 6.55 (d, $^3J_{8,9}$ = 10.1 Hz, 1H, H-8), 5.98 (dd, $^3J_{8,9}$ = 10.1 Hz, $^3J_{9a,9}$ = 3.9 Hz, 1H, H-9), 5.71 (d, $^3J_{9,9a}$ = 3.9 Hz, 1H, H-9a). Other signals are overlapped with those of major isomer. ^{13}C-NMR (100.6 MHz, CDCl$_3$): See above in **3p** section. ^{19}F-NMR (376.3 MHz, CDCl$_3$): δ −72.7 [C(O)CF$_3$], −77.1 (CF$_3$).

(2*R,9a*S**)-3o′:** ^{1}H-NMR (400.1 MHz, CDCl$_3$): δ 6.62 (s, 1H, H-6), 6.40 (d, $^3J_{8,9}$ = 10.1 Hz, 1H, H-8), 5.84 (dd, $^3J_{8,9}$ = 10.1 Hz, $^3J_{9a,9}$ = 4.2 Hz, 1H, H-9), 6.06 (d, $^3J_{9,9a}$ = 4.2 Hz, 1H, H-9a). Other signals are overlapped with those of major isomer. ^{13}C-NMR (100.6 MHz, CDCl$_3$): cannot be seen in the spectrum due to the low concentration of minor isomer. ^{19}F-NMR (376.3 MHz, CDCl$_3$): δ −74.7 [C(O)CF$_3$], −75.6 (CF$_3$).

1-(7-Acetyl-4-phenyl-2-(phenylethynyl)-2-(trifluoromethyl)-2H,9aH-pyrido[2,1-b][1,3]oxazin-3-yl)-2,2,2-trifluoroethan-1-one (**3q**). Major (7-Ac)-regioisomer, obtained as a mixture (2:1) with minor (9-Ac)-regioisomer (**3r**) from pyridine **1h** (0.065 g, 0.54 mmol) and CF$_3$-ynone **2a** (0.214 g, 1.08 mmol). Yellow powder, m.p. 114.4–115.3 °C (hexane), yield 0.224 g (80%). (2*S**,9a*S**):(2*R**,9a*S**)-isomers ratio is 83:17 (^{1}H-NMR). HRMS (ESI-TOF) for the mixture of **3q** and **3r**: *m/z* [M + H]$^+$ Calcd for C$_{27}$H$_{18}$F$_6$NO$_3$$^+$: 518.1185; found: 518.1196.

(2*S,9a*S**)-3q:** ^{1}H-NMR (400.1 MHz, CDCl$_3$): δ 7.69–7.29 (m, 11H), 7.10 (d, $^3J_{8,9}$ = 10.1 Hz, 1H, H-8), 6.00 (dd, $^3J_{8,9}$ = 10.1 Hz, $^3J_{9a,9}$ = 3.6 Hz, 1H, H-9), 5.88 (d, $^3J_{9a,9}$ = 3.6 Hz, 1H, H-9a), 2.12 (s, 3H, Me). ^{13}C-NMR (100.6 MHz, CDCl$_3$): δ 193.1, 182.1 (q, $^2J_{CF}$ = 36.3 Hz, C-12), 156.2 (C-4), 133.7, 133.6, 132.1, 129.9, 129.7, 128.4 (C-8), 130.4 (q, $^3J_{CF}$ = 1.3 Hz, C-3), 124.2 (C-6), 122.2 (q, $^1J_{CF}$ = 286.0 Hz, CF$_3$), 120.4 (C$_i$ from Ar), 115.3 (C-9), 115.0 (q, $^1J_{CF}$ = 293.0 Hz, C(O)CF$_3$), 102.3, 90.1 (C-11), 80.3 (C-10), 79.1 (C-9a), 74.2 (q, $^2J_{CF}$ = 34.8 Hz, C-2), 25.0. ^{19}F-NMR (376.3 MHz, CDCl$_3$): δ −73.7 [C(O)CF$_3$], −76.8 (CF$_3$).

(2*R,9a*S**)-3q′:** ^{1}H-NMR (400.1 MHz, CDCl$_3$): δ 7.17 (s, 1H, H-6), 7.05 (d, $^3J_{8,9}$ = 10.2 Hz, 1H, H-8), 6.17 (d, $^3J_{9a,9}$ = 3.6 Hz, 1H, H-9a), 5.90 (dd, $^3J_{8,9}$ = 10.2 Hz, $^3J_{9a,9}$ = 3.6 Hz, 1H, H-9), 2.06 (s, 3H, Me). ^{13}C-NMR (100.6 MHz, CDCl$_3$): δ 193.0, 131.8, 128.3, 114.5, 24.9. Other signals are overlapped with those of major isomer or cannot be seen in the spectrum due to the low concentration of minor isomer. ^{19}F-NMR (376.3 MHz, CDCl$_3$): δ −74.9 [C(O)CF$_3$], −75.5 (CF$_3$).

1-(9-Acetyl-4-phenyl-2-(phenylethynyl)-2-(trifluoromethyl)-2H,9aH-pyrido[2,1-b][1,3]oxazin-3-yl)-2,2,2-trifluoroethan-1-one (**3r**). Minor (9-Ac)-regioisomer, obtained as a mixture with major (7-Ac)-regioisomer

(3q) (see above). (2S*,9aS*):(2R*,9aS*)-isomers ratio is 87:13 (^{1}H-NMR). HRMS (ESI-TOF) for the mixture of 3q and 3r: m/z [M + H]$^+$ Calcd for $C_{27}H_{18}F_6NO_3^+$: 518.1185; found: 518.1196.

(2S*,9aS*)-3r: ^{1}H-NMR (400.1 MHz, CDCl$_3$): δ 7.69–7.29 (m, 11H), 6.65 (d, $^3J_{6,7}$ = 7.3 Hz, 1H, H-6), 6.20 (s, 1H, H-9a), 5.62 (pseudo-t, ^3J ~ 7 Hz, 1H, H-7), 2.47 (s, 3H, Me). ^{13}C-NMR (100.6 MHz, CDCl$_3$): δ 194.6, 181.5 (q, $^2J_{CF}$ = 35.8 Hz, C-12), 158.0 (C-4), 134.2, 133.5, 132.1, 131.0 (q, $^3J_{CF}$ = 1.7 Hz, C-3), 129.4, 128.3, (C-8), 124.9 (C-6), 122.3 (q, $^1J_{CF}$ = 286.6 Hz, CF$_3$), 120.9 (C_i from Ar), 115.7 (C-9), 115.2 (q, $^1J_{CF}$ = 293.0 Hz, C(O)CF$_3$), 89.6 (C-11), 80.0 (C-10), 77.7 (C-9a), 74.4 (q, $^2J_{CF}$ = 34.5 Hz, C-2), 25.7. ^{19}F-NMR (376.3 MHz, CDCl$_3$): δ −73.0 [C(O)CF$_3$], −76.8 (CF$_3$).

(2R*,9aS*)-3r': ^{1}H-NMR (400.1 MHz, CDCl$_3$): δ 6.58 (s, 1H, H-9a), 6.51 (d, $^3J_{6,7}$ = 7.4 Hz, 1H, H-6), 5.62 (pseudo-t, ^3J ~ 7 Hz, 1H, H-7), 2.43 (s, 3H, Me). Other signals are overlapped with those of major isomer. ^{13}C-NMR (100.6 MHz, CDCl$_3$): δ 194.7, 134.3, 133.0, 132.2, 129.5, 114.9, 25.5. Other signals are overlapped with those of major isomer or cannot be seen in the spectrum due to the low concentration of minor isomer. ^{19}F-NMR (376.3 MHz, CDCl$_3$): δ −74.9 [C(O)CF$_3$], −76.4 (CF$_3$).

4-Phenyl-2-(phenylethynyl)-3-(2,2,2-trifluoroacetyl)-2-(trifluoromethyl)-2H,9aH-pyrido[2,1-b][1,3]oxazine-7-carbonitrile (3s). Major (7-CN)-regioisomer, obtained as a mixture (2.5:1) with minor (9-CN)-regioisomer (3t) from pyridine 1i (0.054 g, 0.5 mmol) and CF$_3$-ynone 2a (0.208 g, 1.05 mmol). Yellow powder, m.p. 95–96 °C (hexane), yield 0.165 g (66%). (2S*,9aS*):(2R*,9aS*)-isomers ratio is 76:24 (^{19}F-NMR). HRMS (ESI-TOF) for the mixture of 3s and 3t: m/z [M + H]$^+$ Calcd for $C_{26}H_{15}F_6N_2O_2^+$: 501.1032; found: 501.1055.

(2S*,9aS*)-3s: ^{1}H-NMR (400.1 MHz, CDCl$_3$): δ 7.69–7.28 (m, 10H), 6.98 (s, 1H, H-6), 6.49 (d, $^3J_{8,9}$ = 10.0 Hz, 1H, H-8), 6.00 (dd, $^3J_{8,9}$ = 10.0 Hz, $^3J_{9a,9}$ = 3.5 Hz, 1H, H-9), 5.92–5.89 (m, 1H, H-9a). ^{13}C-NMR (100.6 MHz, CDCl$_3$): δ 182.0 (q, $^2J_{CF}$ = 37.0 Hz, C-12), 154.5 (C-4), 136.0, 133.7, 132.0, 130.0, 129.9, 128.4, 130.3 (q, $^4J_{CF}$ = 1.7 Hz, C-3), 124.4 (C-6), 122.1 (q, $^1J_{CF}$ = 286.4 Hz, CF$_3$), 120.2 (C_i from Ar), 116.3 (C-9), 114.9 [q, $^1J_{CF}$ = 293.0 Hz, C(O)CF$_3$], 113.3 (CN), 101.4, 88.5 (C-11), 79.9 (C-10), 78.3 (C-9a), 74.3 (q, $^2J_{CF}$ = 34.7 Hz, C-2). ^{19}F-NMR (376.3 MHz, CDCl$_3$): δ −73.9 [C(O)CF$_3$], −76.8 (CF$_3$).

(2R*,9aS*)-3s': ^{1}H-NMR (400.1 MHz, CDCl$_3$): δ 6.87 (s, 1H, H-6), 6.44 (d, $^3J_{8,9}$ = 10.0 Hz, 1H, H-8), 6.20 (d, $^3J_{9a,9}$ = 3.6 Hz, 1H, H-9a). ^{13}C-NMR (100.6 MHz, CDCl$_3$): δ 149.8, 136.5, 133.1, 132.2, 120.8, 115.6, 112.4 (CN), 100.3, 86.8, 80.1, 78.8 (q, $^4J_{CF}$ = 3.7 Hz, C-10), 72.1 (q, $^2J_{CF}$ = 32.4 Hz, C-2). Other signals are overlapped with those of major isomer or cannot be seen in the spectrum due to the low concentration of minor isomer. ^{19}F-NMR (376.3 MHz, CDCl$_3$): δ −75.0 [C(O)CF$_3$], −76.5 (CF$_3$).

4-Phenyl-2-(phenylethynyl)-3-(2,2,2-trifluoroacetyl)-2-(trifluoromethyl)-2H,9aH-pyrido[2,1-b][1,3]oxazine-7-carbonitrile (3t). Minor (9-CN)-regioisomer, obtained as a mixture with major (7-CN)-regioisomer (3s) (see above). (2S*,9aS*):(2R*,9aS*)-isomers ratio is 86:14 (^{19}F-NMR). HRMS (ESI-TOF) for the mixture of 3s and 3t: m/z [M + H]$^+$ Calcd for $C_{26}H_{15}F_6N_2O_2^+$: 501.1032; found: 501.1055.

(2S*,9aS*)-3t: ^{1}H-NMR (400.1 MHz, CDCl$_3$): δ 7.69–7.28 (m, 10H), 7.14 (d, $^3J_{7,8}$ = 6.5 Hz, 1H, H-8), 6.64 (d, $^3J_{6,7}$ = 7.5 Hz, 1H, H-6), 5.92–5.89 (m, 1H, H-9a), 5.54 (pseudo-t, ^3J ~ 7 Hz, 1H, H-7). ^{13}C-NMR (100.6 MHz, CDCl$_3$): δ 182.5 (q, $^2J_{CF}$ = 37.2 Hz, C-12), 156.4 (C-4), 139.3, 133.6, 132.1, 131.3, 129.7, 129.6, 129.3, 128.3, 124.3 (C-6), 122.1 (q, $^1J_{CF}$ = 286.6 Hz, CF$_3$), 120.4 (C_i from Ar), 117.2 (C-9), 115.1 [q, $^1J_{CF}$ = 293.2 Hz, C(O)CF$_3$], 112.5 (CN), 98.7, 90.7 (C-11), 87.7, 79.8 (C-10), 78.2 (C-9a), 74.5 (q, $^2J_{CF}$ = 35.6 Hz, C-2). ^{19}F-NMR (376.3 MHz, CDCl$_3$): δ −73.5 [C(O)CF$_3$], −76.8 (CF$_3$).

(2R*,9aS*)-3t': ^{1}H-NMR (400.1 MHz, CDCl$_3$): δ 7.09 (d, $^3J_{7,8}$ = 6.4 Hz, 1H, H-8), 6.23 (s, 1H, H-9a). Other signals are overlapped with those of major isomer. ^{13}C-NMR (100.6 MHz, CDCl$_3$): δ 149.5, 132.9, 132.7, 97.8, 86.3, 90.4. Other signals are overlapped with those of major isomer or cannot be seen in the spectrum due to the low concentration of minor isomer. ^{19}F-NMR (376.3 MHz, CDCl$_3$): δ −74.9 [C(O)CF$_3$], −76.5 (CF$_3$).

4. Conclusions

In conclusion, a new efficient pathway towards to trifluoromethylated oxazinopyridines was elaborated on the base of a one-pot, metal-free 1:2 assembly of pyridines and CF_3-ynones. The reaction has a broad scope in terms of both pyridines and CF_3-ynones used. Therefore, pyridines with electron withdrawing as well as electron donating groups afforded corresponding products in up to 99% yield. Various CF_3-ynones including bulky ones can also be involved in the reaction. High stereoselectivity (up to 100% for 2-substitueted pyridines) is the advantage of the method. However, dramatic influence of the pKa values of pyridines on the reaction course was observed. Pyridines having pKa lower than ~1 do not react with CF_3-ynones. The possible mechanism of the reaction includes a cascade of ionic transformations triggered by attack of the nitrogen of pyridine molecule by electron-deficient triple bond of CF_3-ynone.

Supplementary Materials: Copy of all ^{1}H-, ^{13}C- and ^{19}F-NMR spectra are available online at http://www.mdpi.com/1420-3049/24/19/3594/s1.

Author Contributions: Conceptualization, V.M.M., B.A.T., and V.G.N.; investigation, V.M.M., Z.A.S., K.V.B.; writing—original draft preparation, V.M.M.; writing—review and editing, V.M.M., B.A.T., K.V.B.; V.G.N.; visualization, V.M.M.; funding acquisition, V.G.N.

Funding: This research was funded by Russian Science Foundation grant number 18-13-00136.

Acknowledgments: The authors acknowledge partial support in measuring oF-NMR from M.V. Lomonosov Moscow State University Program of Development. The authors acknowledge Thermo Fisher Scientific Inc., MS Analytica (Moscow, Russia), and personally to A. Makarov for providing mass spectrometry equipment for this work.

Conflicts of Interest: The authors declare no conflicts of interest.

References

1. Buckingham, J.; Baggaley, K.H.; Roberts, A.D.; Szabó, L.F. *Dictionary of Alkaloids*, 2nd ed.; CRC Press, Taylor and Francis Group: Boca Raton, FL, USA, 2010.
2. Vitaku, E.; Smith, D.T.; Njardarson, J.T. Analysis of the Structural Diversity, Substitution Patterns, and Frequency of Nitrogen Heterocycles among U.S. FDA Approved Pharmaceuticals. *J. Med. Chem.* **2014**, *57*, 10257–10274. [CrossRef] [PubMed]
3. Taylor, R.D.; MacCoss, M.; Lawson, A.D. Rings in drugs: Miniperspective. *Med. Chem.* **2014**, *57*, 5845–5859. [CrossRef] [PubMed]
4. McGrath, N.A.; Brichacek, M.; Njardarson, J.T. A Graphical Journey of Innovative Organic Architectures That Have Improved Our Lives. *J. Chem. Ed.* **2010**, *87*, 1348–1349. [CrossRef]
5. Liang, T.; Neumann, C.N.; Ritter, T. Introduction of fluorine and fluorine-containing functional groups. *Angew. Chem. Int. Ed.* **2013**, *52*, 8214–8264. [CrossRef] [PubMed]
6. Yang, X.; Wu, T.; Phipps, R.J.; Toste, F.D. Advances in catalytic enantioselective fluorination, mono-, di-, and trifluoromethylation, and trifluoromethylthiolation reactions. *Chem. Rev.* **2015**, *115*, 826–870. [CrossRef] [PubMed]
7. Ahrens, T.; Kohlmann, J.; Ahrens, M.; Braun, T. Functionalization of fluorinated molecules by transition metal mediated C–F bond activation to access fluorinated building blocks. *Chem. Rev.* **2015**, *115*, 931–972. [CrossRef]
8. Nenajdenko, V.G.; Muzalevskiy, V.M.; Shastin, A.V. Polyfluorinated ethanes as versatile fluorinated C2-building blocks for organic synthesis. *Chem. Rev.* **2015**, *115*, 973–1050. [CrossRef] [PubMed]
9. Yerien, D.E.; Barata-Vallejo, S.; Postigo, A. Difluoromethylation reactions of organic compounds. *Chem. Eur. J.* **2017**, *23*, 14676–14701. [CrossRef]
10. Kirsch, P. *Modern Fluoroorganic Chemistry: Synthesis, Reactivity, Applications*; Wiley-VCH: Weinheim, Germany, 2013.
11. Uneyama, K. *Organofluorine Chemistry*; Blackwell Publishing: Oxford, UK, 2006.
12. Theodoridis, G. Fluorine-containing agrochemicals: An overview of recent developments. In *Fluorine and the Environment: Agrochemicals, Archaeology, Green Chemistry & Water*; Tressaud, A., Ed.; Elsevier: Amsterdam, The Netherland, 2006; pp. 121–175.

13. Bégué, J.P.; Bonnet-Delpon, D. *Bioorganic and Medicinal Chemistry of Fluorine*; John Wiley & Sons: Hoboken, NJ, USA, 2008.
14. Tressaud, A.; Haufe, G. *Fluorine and Health: Molecular Imaging, Biomedical Materials and Pharmaceuticals*; Elsevier: Amsterdam, The Netherland, 2008; pp. 553–778.
15. Soloshonok, V.A.; Mikami, K.; Yamazaki, T.; Welch, J.T.; Honek, J.F. *Current Fluoroorganic Chemistry. New Synthetic Directions, Technologies, Materials, and Biological Applications*; ACS Symposium Series 949; American Chemical Society: Washington, DC, USA, 2007.
16. Meanwell, N.A. Fluorine and Fluorinated Motifs in the Design and Application of Bioisosteres for Drug Design. *J. Med. Chem.* **2018**, *61*, 5822–5880. [CrossRef]
17. Gillis, E.P.; Eastman, K.J.; Hill, M.D.; Donnelly, D.J.; Meanwell, N.A. Applications of fluorine in medicinal chemistry. *J. Med. Chem.* **2015**, *58*, 8315–8359. [CrossRef]
18. Zhu, W.; Wang, J.; Wang, S.; Gu, Z.; Aceña, J.L.; Izawa, K.; Liu, H.; Soloshonok, V.A. Recent advances in the trifluoromethylation methodology and new CF_3-containing drugs. *J. Fluorine Chem.* **2014**, *167*, 37–54. [CrossRef]
19. Purser, S.; Moore, P.R.; Swallow, S.; Gouverneur, V. Fluorine in medicinal chemistry. *Chem. Soc. Rev.* **2008**, *37*, 320–330. [CrossRef] [PubMed]
20. Hagmann, W.K. The many roles for fluorine in medicinal chemistry. *J. Med. Chem.* **2008**, *51*, 4359–4369. [CrossRef] [PubMed]
21. Zhou, Y.; Wang, J.; Gu, Z.; Wang, S.; Zhu, W.; Aceña, J.L.; Soloshonok, V.A.; Izawa, K.; Liu, H. Next generation of fluorine containing pharmaceuticals, compounds currently in phase II–III clinical trials of major pharmaceutical companies: New structural trends and therapeutic areas. *Chem. Rev.* **2016**, *116*, 422–518. [CrossRef] [PubMed]
22. Wang, J.; Sánchez-Roselló, M.; Aceña, J.L.; del Pozo, C.; Sorochinsky, A.E.; Fustero, S.; Soloshonok, V.A.; Liu, H. Fluorine in pharmaceutical industry: Fluorine-containing drugs introduced to the market in the last decade (2001–2011). *Chem. Rev.* **2014**, *114*, 2432–2506. [CrossRef] [PubMed]
23. Ilardi, E.A.; Vitaku, E.; Njardarson, J.T. Data-mining for sulfur and fluorine: An evaluation of pharmaceuticals to reveal opportunities for drug design and discovery. *J. Med. Chem.* **2014**, *57*, 2832–2842. [CrossRef]
24. Jeschke, P. The unique role of fluorine in the design of active ingredients for modern crop protection. *ChemBioChem* **2004**, *5*, 570–589. [CrossRef] [PubMed]
25. Jeschke, P. The unique role of halogen substituents in the design of modern agrochemicals. *Pest Manage. Sci.* **2010**, *66*, 10–27. [CrossRef]
26. Fujiwara, T.; O'Hagan, D. Successful fluorine-containing herbicide agrochemicals. *J. Fluorine Chem.* **2014**, *167*, 16–29. [CrossRef]
27. Jeschke, P. Latest generation of halogen-containing pesticides. *Pest Manage. Sci.* **2017**, *73*, 1053–1056. [CrossRef]
28. de la Torre, B.G.; Albericio, F. The Pharmaceutical Industry in 2018. An Analysis of FDA Drug Approvals from the Perspective of Molecules. *Molecules* **2019**, *24*, 809. [CrossRef] [PubMed]
29. Nenajdenko, V.G. *Fluorine in Heterocyclic Chemistry*; Springer: Berlin, Germany, 2014; pp. 681–760.
30. Petrov, V.A. (Ed.) *Fluorinated Heterocyclic Compounds: Synthesis, Chemistry, and Applications*; Wiley: Hoboken, NJ, USA, 2009.
31. Gakh, A.; Kirk, K.L. *Fluorinated Heterocycles*; Oxford University Press: Oxford, UK, 2008.
32. Muzalevskiy, V.M.; Nenajdenko, V.G.; Shastin, A.V.; Balenkova, E.S.; Haufe, G. Synthesis of Trifluoromethyl Pyrroles and Their Benzo Analogues. *Synthesis* **2009**, *23*, 3905–3929.
33. Serdyuk, O.V.; Abaev, V.T.; Butin, A.V.; Nenajdenko, V.G. Synthesis of Fluorinated Thiophenes and Their Analogues. *Synthesis* **2011**, *16*, 2505–2529. [CrossRef]
34. Serdyuk, O.V.; Muzalevskiy, V.M.; Nenajdenko, V.G. Synthesis and Properties of Fluoropyrroles and Their Analogues. *Synthesis* **2012**, *14*, 2115–2137.
35. Politanskaya, L.V.; Selivanova, G.A.; Panteleeva, E.V.; Tretyakov, E.V.; Platonov, V.E.; Nikul'shin, P.V.; Vinogradov, A.S.; Zonov, Y.V.; Karpov, V.M.; Mezhenkova, T.V.; et al. Organofluorine chemistry: Promising growth areas and challenges. *Rus. Chem. Rev.* **2019**, *88*, 425–569. [CrossRef]
36. Shimizu, M.; Hiyama, T. Modern Synthetic Methods for Fluorine-Substituted Target Molecules. *Angew. Chem.* **2005**, *44*, 214–231. [CrossRef]

37. Druzhinin, S.V.; Balenkova, E.S.; Nenajdenko, V.G. Recent advances in the chemistry of α,β-unsaturated trifluoromethylketones. *Tetrahedron* **2007**, *63*, 7753–7808. [CrossRef]
38. Nenajdenko, V.G.; Sanin, A.V.; Balenkova, E.S. Methods for the synthesis of α,β-unsaturated trifluoromethyl ketones and their use in organic synthesis. *Russ. Chem. Rev.* **1999**, *68*, 483–505. [CrossRef]
39. Nenajdenko, V.G.; Sanin, A.V.; Balenkova, E.S. Preparation of α,β-Unsaturated Ketones Bearing a Trifluoromethyl Group and Their Application in Organic Synthesis. *Molecules* **1997**, *12*, 186–232. [CrossRef]
40. Rulev, A.Y. The Wonderful Chemistry of Trifluoromethyl α-Haloalkenyl Ketones. *Eur. J. Org. Chem.* **2018**, *27–28*, 3609–3617. [CrossRef]
41. Romanov, A.R.; Rulev, A. Yu.; Ushakov, I.A.; Muzalevskiy, V.M.; Nenajdenko, V.G. Synthesis of trifluoromethylated [1,4]-diazepines based on cf3-ynones. *Mendeleev Commun.* **2014**, *24*, 269–271. [CrossRef]
42. Romanov, A.R.; Rulev, A.Y.; Ushakov, I.A.; Muzalevskiy, V.M.; Nenajdenko, V.G. One-Pot, Atom and Step Economy (PASE) Assembly of Trifluoromethylated Pyrimidines from CF_3-Ynones. *Eur. J. Org. Chem.* **2017**, *28*, 4121–4129. [CrossRef]
43. Muzalevskiy, V.M.; Iskandarov, A.A.; Nenajdenko, V.G. Reaction of CF_3-ynones with methyl thioglycolate. Regioselective synthesis of 3-CF_3-thiophene derivatives. *J. Fluorine Chem.* **2018**, *214*, 13–16. [CrossRef]
44. Muzalevskiy, V.M.; Mamedzade, M.N.; Chertkov, V.A.; Bakulev, V.A.; Nenajdenko, V.G. Reaction of CF_3-ynones with azides. An efficient regioselective and metal-free route to 4-trifluoroacetyl-1,2,3-triazoles. *Mendeleev Commun.* **2018**, *28*, 17–19. [CrossRef]
45. Muzalevskiy, V.M.; Iskandarov, A.A.; Nenajdenko, V.G. Synthesis of dibromo substituted cf3-enones and their reactions with n-nucleophiles. *Mendeleev Commun.* **2014**, *24*, 342–344. [CrossRef]
46. Muzalevskiy, V.M.; Rulev, A.Y.; Romanov, A.R.; Kondrashov, E.V.; Ushakov, I.A.; Chertkov, V.A.; Nenajdenko, V.G. Selective, Metal-Free Approach to 3- or 5-CF_3-Pyrazoles: Solvent Switchable Reaction of CF_3-Ynones with Hydrazines. *J. Org. Chem.* **2017**, *82*, 7200–7214. [CrossRef]
47. Topchiy, M.A.; Zharkova, D.A.; Asachenko, A.F.; Muzalevskiy, V.M.; Chertkov, V.A.; Nenajdenko, V.G.; Nechaev, M.S. Mild and Regioselective Synthesis of 3-CF_3-Pyrazoles by the AgOTf-Catalysed Reaction of CF_3-Ynones with Hydrazines. *Eur. J. Org. Chem.* **2018**, *27–28*, 3750–3755. [CrossRef]
48. Trofimov, B.A.; Belyaeva, K.V.; Nikitina, L.P.; Afonin, A.V.; Vashchenko, A.V.; Muzalevskiy, V.M.; Nenajdenko, V.G. Metal-free stereoselective annulation of quinolines with trifluoroacetylacetylenes and water: An access to fluorinated oxazinoquinolines. *Chem. Commun.* **2018**, *54*, 2268–2271. [CrossRef]
49. Muzalevskiy, V.M.; Trofimov, B.A.; Belyaeva, A.V.; Nenajdenko, V.G. Green, diastereoselective synthesis of CF_3-oxazinoquinolines in water. *Green Chem.* **2019**. Submitted (under revision).
50. Belyaeva, K.V.; Nikitina, L.P.; Afonin, A.V.; Vashchenko, A.V.; Muzalevskiy, V.M.; Nenajdenko, V.G.; Trofimov, B.A. Catalyst-free 1:2 annulation of quinolines with trifluoroacetylacetylenes: An access to functionalized oxazinoquinolines. *Org. Biomol. Chem.* **2018**, *16*, 8038–8041. [CrossRef]
51. Lazar, L.; Fulop, F. 1,3-Oxazines and Their Benzo Derivatives. In *Comprehensive Heterocyclic Chemistry III*; Katritzky, A.R., Ramsden, C.A., Scriven, E.F.V., Taylor, R.J.K., Eds.; Elsevier: Amsterdam, The Netherland, 2008; Volume 8, pp. 373–459.
52. Gaonkar, S.L.; Nagaraj, V.U.; Nayak, S. A Review on Current Synthetic Strategies of Oxazines. *Mini Rev. Org. Chem.* **2019**, *16*, 43–58.
53. Sindhu, T.J.; Sonia, D.A.; Girly, V.; Meena, C.; Bhat, A.R.; Krishnakumar, K. Biological Activities of Oxazine and Its Derivatives: A Review. *Int. J. Pharm. Sci. Res.* **2013**, *4*, 134–143.
54. Liu, P.; Lei, M.; Hu, L. Synthesis of benzo-annulated 1,3-oxazine derivatives through the multi-component reaction of arynes with N-heteroaromatics and aldehydes or ketones. *Tetrahedron* **2013**, *69*, 10405–10413. [CrossRef]
55. Min, S.; Song, I.; Borland, J.; Chen, S.; Lou, Y.; Fujiwara, T.; Piscitelli, S.C. Pharmacokinetics and safety of S/GSK1349572, a next-generation HIV integrase inhibitor, in healthy volunteers. *Antimicrob Agents Chemother.* **2010**, *54*, 254–258. [CrossRef] [PubMed]
56. Tsiang, M.; Jones, G.C.; Goldsmith, J.; Mulato, A.; Hansen, D.; Kan, E.; Tsai, L.; Bam, R.A.; Stepan, G.; Stray, K.M.; et al. Antiviral activity of bictegravir (GS-9883), a novel potent HIV1 integrase strand transfer inhibitor with an improved resistance profile. *Antimicrob Agents Chemother.* **2016**, *60*, 7086–7097. [PubMed]

57. Andriyankova, L.V.; Nikitina, L.P.; Belyaeva, K.V.; Mal'kina, A.G.; Afonin, A.V.; Muzalevskii, V.M.; Nenaidenko, V.G.; Trofimov, B.A. Opening of the pyridine ring in the system 1,1,1-trifluoro-4-phenylbut-3-yn-2-one–water. Stereoselective synthesis of 5-{[(1Z)-4,4,4-trifluoro-3-oxo-1-phenylbut-1-en-1-yl]amino}penta-2,4-dienal. *Rus. J. Org. Chem.* **2016**, *52*, 1857–1860. [CrossRef]
58. Gusarova, N.K.; Mikhaleva, A.I.; Schmidt, E. Yu.; Mal'kina, A.G. *Khimiya atsetilena. Novye glavy (The Chemistry of Acetylene. New Chapters)*; Egorov, M.P., Ed.; Nauka: Novosibirsk, Russia, 2013; Volume 92. (In Russian)
59. Trofimov, B.A.; Belyaeva, K.V.; Andriyankova, L.V.; Nikitina, L.P.; Mal'kina, A.G. Ring-opening of pyridines and imidazoles with electron-deficient acetylenes: En route to metal-free organic synthesis. *Mendeleev Commun.* **2017**, *27*, 109–115. [CrossRef]

Sample Availability: Samples of the compounds are available from the authors.

© 2019 by the authors. Licensee MDPI, Basel, Switzerland. This article is an open access article distributed under the terms and conditions of the Creative Commons Attribution (CC BY) license (http://creativecommons.org/licenses/by/4.0/).

Communication

A Brønsted Acid-Catalyzed Multicomponent Reaction for the Synthesis of Highly Functionalized γ-Lactam Derivatives

Xabier del Corte, Edorta Martinez de Marigorta, Francisco Palacios * and Javier Vicario *

Departamento de Química Orgánica I, Centro de Investigación y Estudios Avanzados "Lucio Lascaray"-Facultad de Farmacia, University of the Basque Country, UPV/EHU Paseo de la Universidad 7, 01006 Vitoria-Gasteiz, Spain
* Correspondence: francisco.palacios@ehu.eus (F.P.); javier.vicario@ehu.eus (J.V.);
  Tel.: +34-945-013103 (F.P.); +34-945-013087 (J.V.)

Academic Editor: Gianfranco Favi
Received: 24 July 2019; Accepted: 11 August 2019; Published: 14 August 2019

Abstract: Brønsted acids catalyze a multicomponent reaction of benzaldehyde with amines and diethyl acetylenedicarboxylate to afford highly functionalized γ-lactam derivatives. The reaction consists of a Mannich reaction of an enamine to an imine, both generated in situ, promoted by a phosphoric acid catalyst and a subsequent intramolecular cyclization. The hydrolysis of the cyclic enamine substrate can provide enol derivatives and, moreover, a second attack of the amine on the carboxylate can afford amide derivatives. An optimization of the reaction conditions is presented in order to obtain selectively cyclic enamines that can afford the enol species after selective hydrolysis.

Keywords: γ-lactam; pyrrolidones; multicomponent reactions; organocatalysis

---

1. Introduction

Multicomponent reactions (MCRs) [1,2] are valuable processes where three or more substrates, which are simultaneously (or almost) added, react in a single vessel to form a new structure that contains substantial portions of all the starting materials. Strecker, Hantzsch, Biginelli, Passerini, Gröbcke-Blackburn-Bienaymé, Kabachnik-Fields, or Ugi are some of the names of classical reactions that fit with this definition, and they are widely used in organic synthesis [1,2]. Due to the high degree of molecular diversity achieved in MCRs, they are now an essential tool in diversity-oriented synthesis [3,4], with huge potential in the field of medicinal chemistry [5,6]. Considering the relevance of the γ-lactam ring **I** (Figure 1) [7] and the increasing demand of potentially active compounds in medicinal sciences, MCR protocols were extensively used during the last decades for the synthesis of a wide number of densely functionalized γ-lactam derivatives [8,9]. In particular, 1,5-dihydro-2H-pyrrol-2-ones **II** (Figure 1) are conjugated unsaturated γ-lactam substrates with huge potential as intermediates in synthetic chemistry that also show assorted pharmacological activities [10–13].

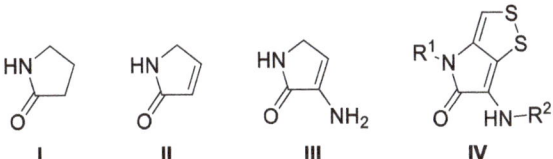

Figure 1. General structure of γ-lactams **I**, 1,5-dihydro-2H-pyrrol-2-ones **II**, 3-amino 1,5-dihydro-2H-pyrrol-2-ones **II**, and dithiopyrrolone antibiotics **IV**.

Within this family of compounds, the structure of their 3-amino substituted derivatives **III** (Figure 1) contains the enamine moiety and, in addition to their obvious applications as synthetic intermediates in organic synthesis [14,15], their skeleton is also present in many new bioactive ingredients such as antimicrobials with anti-biofilm activity, caspase-3 inhibitors, antipyretics, or analgesics [16–20]. Moreover, these cyclic α-dehydro α,β-diamino acid derivatives contain the essential structure of dithiopyrrolone antibiotics **IV** (Figure 1) [21] and are key intermediates for the synthesis of Amaryllidaceae and *Sceletium* alkaloids [22,23].

Several MCR procedures for the preparation of 3-amino 1,5-dihydro-2H-pyrrol-2-ones were reported to date [8]. In particular, some years ago, we reported a three-component reaction of ethyl pyruvate **1**, aldehydes **2**, and amines **3** mediated by sulfuric acid that yields very efficiently highly functionalized γ-lactam derivatives **7** [24]. In this reaction, an initial simultaneous condensation of amines **3** with both ethyl pyruvate **1** and aldehydes **2** leads to the formation of intermediate enamine **4** and aldimine **5** that undergo a subsequent Mannich reaction, followed by a cyclization reaction driven by the formation of an internal amide bond in the resulting adduct **6** (Scheme 1).

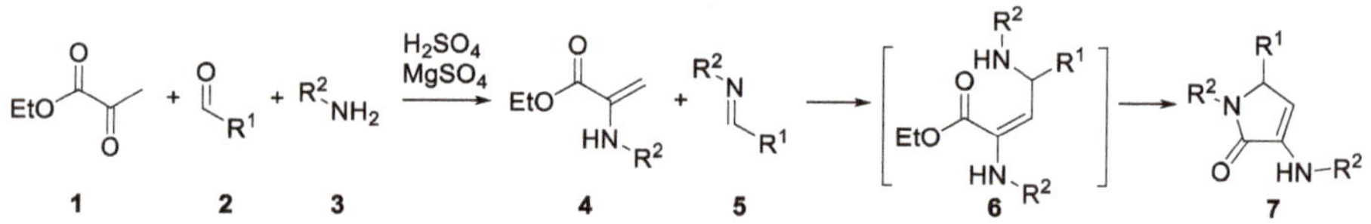

Scheme 1. Three-component reaction of ethyl pyruvate **1**, aldehydes **2**, and amines **3**.

Based on this report, some authors later described several modifications of this synthetic procedure, and the uncatalyzed [25] or solvent-free [26] reaction, and the use of recyclable catalysts [27] were reported in the last few years. Interestingly, it was also established that such reaction can be performed under organocatalysis [28] and, taking the advantage of this fact, very recently, we achieved a highly enantioselective version of this reaction using 1,1′-bi-2-naphthol (BINOL)-derived chiral phosphoric acids as catalysts [29].

A similar multicomponent process, where dialkyl acetylenedicarboxylates are used instead of ethyl pyruvate, was also reported for the synthesis of 3-amino 1,5-dihydro-2H-pyrrol-2-ones. In this case, the nucleophilic addition of aromatic amines to the activated alkyne gives rise to a deactivated enamine intermediate and 0.5 equivalents of benzoic acid are required in order to promote the subsequent Mannich reaction [30]. Activation of this process was also described by the use of molecular iodine [31] or graphene-oxide nanosheets under solvent-free conditions [32]. In this context, organocatalysis is identified to be at the heart of greening of chemistry, because this branch of science is found to reduce the environmental impact of chemical processes. Therefore, in view of the demonstrated ability of phosphoric acids to catalyze the nucleophilic addition of pyruvate-derived enamines **4** to imines **5**, we thought that this activation could be extended to the enamines derived from dialkyl acetylenedicarboxylates. Consequently, continuing with the interest of our research group in the synthesis of nitrogenated heterocycles [33–36] and amino-acid derivatives [37–40], we report here the use of phosphoric acids as catalysts in a three-component reaction of amines, benzaldehyde, and diethyl acetylenedicarboxylate to afford densely functionalized γ-lactam derivatives.

2. Results

Based on our previous experience in MCRs for the synthesis of 3-amino 1,5-dihydro-2H-pyrrol-2-ones [11,14], we firstly used BINOL-derived phosphoric acid **9** as a Brønsted acid catalyst in the three-component reaction of benzaldehyde **2**, p-toluidine **3** (R = p-MeC$_6$H$_4$), and diethyl acetylenedicarboxylate **8** using refluxing dichloromethane as solvent (Scheme 2). However, only the enamine and/or imine intermediates that result from the reaction of amine substrate **3** with benzaldehyde **2** or alkyne **8** were observed in the crude (Table 1, Entry 1).

Considering that our previously reported three-component reaction of ethyl pyruvate, benzaldehyde, and amines smoothly yields the corresponding 3-amino 1,5-dihydro-2H-pyrrol-2-ones, we thought that the increased steric hindrance, together with the additional deactivation present in the enamine intermediate when acetylenedicarboxylates 8 are used instead of pyruvate derivatives, may be the reason for the lack of reactivity in this case.

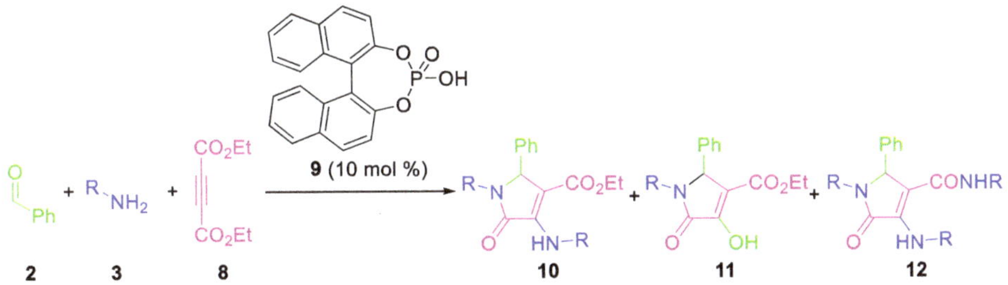

Scheme 2. Three-component reaction of diethyl acetylenedicarboxilate 8, benzaldehyde 2, and amines 3.

Table 1. Three-component reaction of ethyl pyruvate 1, benzaldehyde 2, and amines 3.

Entry	R	2/3/8	Solvent	T (°C)	Yield (%) [1]	10/11/12 [2]
1	p-MeC$_6$H$_4$	1/2/1	CH$_2$Cl$_2$	40	0	n.d.
2	p-MeC$_6$H$_4$	1/2/1	THF	65	0	n.d.
3	p-MeC$_6$H$_4$	1/2/1	DME	85	0	n.d.
4	p-MeC$_6$H$_4$	1/2/1	MTBE	55	72	40/60/0
5	p-MeC$_6$H$_4$	1/2/3	MTBE	55	0	n.d.
6	p-MeC$_6$H$_4$	1/2/1	Dioxane	101	81	80/0/20
7	p-MeC$_6$H$_4$	1/2/1	Toluene	110	77	95/0/5
8	p-MeOC$_6$H$_4$	1/2/1	Toluene	110	76	70/0/30
9	p-MeOC$_6$H$_4$	1/4/1	Toluene	110	76	70/0/30
10	Bn	1/2/1	Toluene	110	58	100/0/0

[1] Isolated total yield. [2] Determined by ^{1}H-NMR. n.d.—not determined.

Then, we tried to perform the reaction at higher temperature and, although the same results were observed using tetrahydrofurane (THF) or dimethoxyethane (DME) as solvents (Table 1, Entries 2 and 3), the reaction in refluxing methyl *tert*-butylether (MTBE) proceeded in full conversion in a few hours, affording the expected 3-amino 1,5-dihydro-2H-pyrrol-2-one 10a together with enol derivative 11a [41], which may result from the hydrolysis of enamine moiety in 10a (Table 1, Entry 4). The use of an excess of ethyl pyruvate in the parent MCR with benzaldehyde and amines proved to be very effective in reducing the reaction times and temperatures [29]; however, remarkably, when three equivalents of acetylene derivative 8 were used, no formation of γ-lactam derivatives 10a or 11a was observed due to the consumption of p-toluidine 3a (R = p-MeC$_6$H$_4$) by reaction with the excess of diethyl acetylenedicarboxylate 8 (Table 1, Entry 5).

Better selectivity was observed when the reaction was performed using hot dioxane as solvent. In this case, amino 1,5-dihydro-2H-pyrrol-2-one 10a was obtained together with a significant amount of amide derivative 12a, which presumably results from the nucleophilic attack of amine on the ethyl carboxylate moiety in compound 10a (Table 1, Entry 6). Finally, the selectivity of the reaction was further improved using toluene as the reaction solvent, and only a small amount (5%) of amide derivative 12a was obtained together with γ-lactam 10a (Table 1, Entry 7). Under the same conditions, the use of more nucleophilic p-anisidine 3b (R = p-MeOC$_6$H$_4$) in the reaction yielded 1,5-dihydro-2H-pyrrol-2-one 10b as the major product of the reaction although, in this case, together with a 30% of amide derivative 12b (Table 1, Entry 8). In order to obtain exclusively amide substrate 12b, four equivalents of amine were used under the same reaction conditions, but the same proportion of the products was observed (Table 1, Entry 9). However, the use of benzylamine 3c (R = Bn) afforded exclusively 1,5-dihydro-2H-pyrrol-2-one

10c, and no formation of enol 11c or amide 12c was observed (Table 1, Entry 10). The selectivity in this case could be explained by the lower steric crowding in the enamine moiety in benzylamine derivative 10c if compared to the aromatic derivatives 10a and 10b [42].

In view of the three compounds observed, the reaction mechanism could start with an initial concomitant addition of amines 3 to acetylene carboxylate 8 and benzaldehyde 2 that affords enamine 13 and aldimine 14. Both species 13 and 14 can be observed by ^{1}H-NMR. Then, a subsequent Mannich reaction leads to the formation of adduct 15, which undergoes an intramolecular cyclization by the formation of an internal amide bond between the amine and carboxylate moieties to afford enamine type γ-lactam 10. Due to the presence of water and some remaining amine 3, the γ-lactam 10 may afford enol type lactam 11 through hydrolysis of the enamine moiety or amide derivative 12, through the displacement of ethanol by the amine (Scheme 3). This is supported by the fact that, using high-boiling-point solvents, no enol derivative 11 is observed, which may be due to the instantaneous evaporation of water at high reaction temperatures.

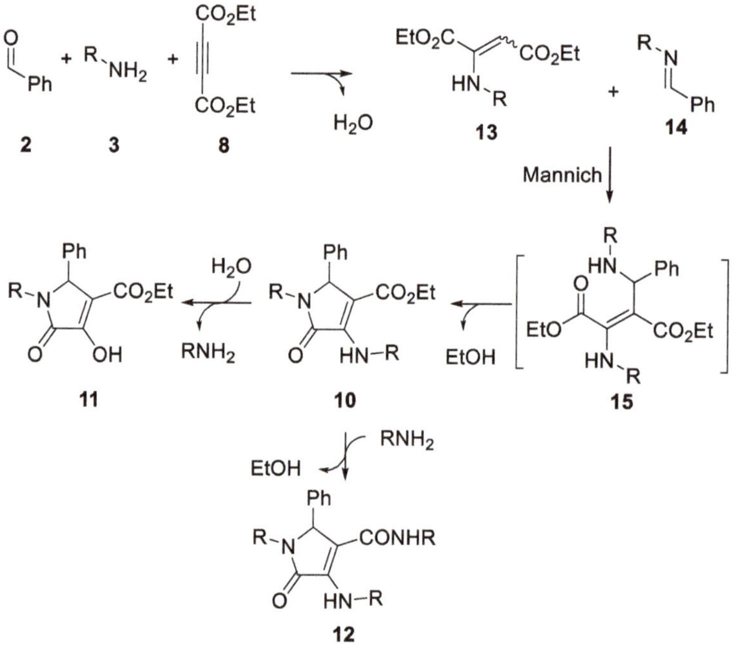

Scheme 3. Reaction mechanism for the three-component reaction of diethyl acetylenedicarboxylate 8, benzaldehyde 2, and amines 3.

In our case, the three resulting γ-lactam structures 10, 11, and 12 could be separated in all the cases by simple chromatography, and they were fully characterized on the basis of their spectroscopic data. However, due to the structural resemblance between all the lactam derivatives, in order to unambiguously determine the identity of the substrates of the reaction, a single crystal of enol 11a was prepared, and its X-ray diffraction structure was obtained (Figure 2). Key features of the crystal structure are the almost planar shape of the five-membered ring and the presence of a hydrogen bond between the enol hydrogen and the carboxylate group in a six-membered ring configuration rather than with the amide carboxylate, forming a five-membered ring.

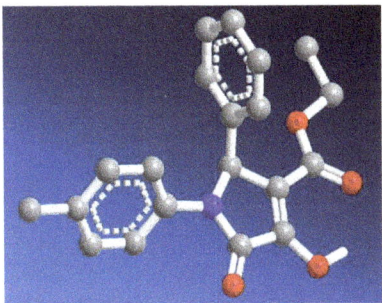

Figure 2. X-ray structure of **11a**. (blue ball, Nitrogen; gray ball, Carbon).

In order to set up the optimal conditions for the preparation of enol derivatives **11**, we proposed the corresponding reactions starting from their parent 1,5-dihydro-2H-pyrrol-2-ones **10** (Scheme 4). Therefore, the hydrolysis of enamine moiety in **10** was performed by treatment of 1,5-dihydro-2H-pyrrol-2-ones **10** in the presence of aqueous hydrochloric acid in refluxing THF. Despite the strong acidic conditions, no trace of the products derived from the hydrolysis of ester of amide groups are observed and enol derivatives **11** are obtained in quantitative yields.

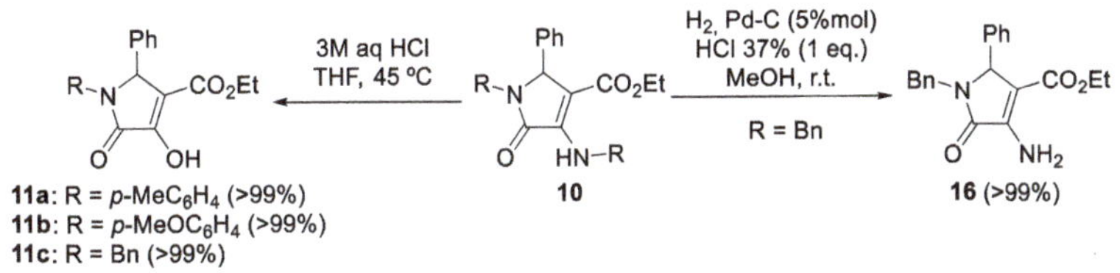

Scheme 4. Preparation of enol derivatives **11** and deprotection of benzylamine derivative **10**.

In addition, the treatment of benzylamine derivative γ-lactam **10** (R = Bn) with a catalytic amount of palladium under hydrogen atmosphere during several days led to the exclusive deprotection of the nitrogen at the enamine moiety in quantitative yield to afford lactam **16**. Remarkably, the benzyl group at the endocyclic nitrogen and the enamine double bond remained unaltered under those reaction conditions. Although the reaction times are very long, this process can be sped up by the addition of one equivalent of aqueous hydrochloric acid (Scheme 4).

Taking into account the typical activation accepted by phosphoric acid catalysts [43–45], we propose a tentative transition state for the key Mannich reaction, where a dual activation of imine and enamine species takes place by the simultaneous formation of two hydrogen bonds between the phosphoryl oxygen and the acidic proton of the phosphoric acid group with the enamine proton and the iminic nitrogen, respectively (Figure 3).

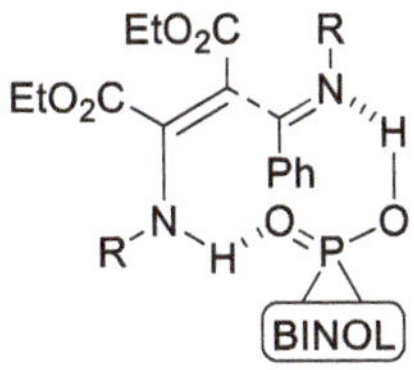

Figure 3. Transition state proposed for the Mannich reaction in the three-component reaction.

According to the transition state proposed, we may expect substantial enantiomeric excesses for this reaction. However, when enantiomerically pure chiral phosphoric acids were used as catalysts, very poor enantioselectivities were observed with a maximum enantiomeric excess of 5%. This may be attributable to the high temperatures required for the reaction conditions because of the steric hindrance present in the enamine substrate, together with the additional deactivation of the nucleophile due to the presence of two carboxylate groups.

In conclusion, we report a Brønsted acid-catalyzed MCR procedure for the preparation of 3-amino 1,5-dihydro-2H-pyrrol-2-ones where diethyl acetylenedicarboxylate, amines, and benzaldehyde are used as substrates. This is the first example of such a reaction using phosphoric acids as catalyst. Moreover, we present nine highly functionalized γ-lactam derivatives, adding some molecular diversity to the already published substrates. The hydrolysis process of 1,5-dihydro-2H-pyrrol-2-ones from enamine substrates to the enol derivatives **11** was not previously reported.

3. Materials and Methods

General. Solvents for extraction and chromatography were technical grade. All solvents used in reactions were freshly distilled from appropriate drying agents before use. All other reagents were recrystallized or distilled as necessary. All reactions were performed under an atmosphere of dry nitrogen. Analytical thin layer chromatography (TLC) was performed with silica gel 60 F_{254} plates. Visualization was accomplished by ultraviolet (UV) light. ^{1}H-, ^{13}C-, and ^{31}P-NMR spectra were recorded on a Varian Unity Plus (at 300 MHz, 75 MHz, and 120 MHz, respectively, Advanced Research Facilities (SGIker), by the University of the Basque Country, Vitoria-Gasteiz, Spain) and on a Bruker Avance 400 (at 400 MHz, 100 MHz, and 160 MHz, respectively, Advanced Research Facilities (SGIker), by the University of the Basque Country, Vitoria-Gasteiz, Spain). Chemical shifts (δ) are reported in ppm relative to residual CHCl$_3$ (δ = 7.26 ppm for ^{1}H, and δ = 77.16 ppm for ^{13}C-NMR). Coupling constants (J) are reported in Hertz. Data for ^{1}H-NMR spectra are reported as follows: chemical shift, multiplicity, coupling constant, integration. Multiplicity abbreviations are as follows: s = singlet, d = doublet, t = triplet, q = quartet, m = multiplet, b = broad. ^{13}C-NMR peak assignments were supported by distortionless enhanced polarization transfer (DEPT). High-resolution mass spectra (HRMS) were obtained by positive-ion electrospray ionization (ESI). Data are reported in the form *m/z* (intensity relative to base = 100). Infrared spectra (IR) were taken in a Nicolet iS10 Termo Scientific spectrometer as neat solids. Peaks are reported in cm^{-1}. Copies of ^{1}H- and ^{13}C {^{1}H} NMR spectra for γ-lactams **10**, **11**, **12**, and **16** are in Supplementary Materials.

Crystal structure determination for compound **11a**. Intensity data were collected on an Agilent Technologies Super-Nova diffractometer (Advanced Research Facilities (SGIker), by the University of the Basque Country, Leioa, Spain), which was equipped with monochromated Cu kα radiation (λ = 1.54184 Å) and Atlas CCD detector. Measurement was carried out at 150.00 (10) K with the help of an Oxford Cryostream 700 PLUS temperature device (Advanced Research Facilities (SGIker), by the University of the Basque Country, Leioa, Spain). Data frames were processed (united cell determination, analytical absorption correction with face indexing, intensity data integration, and correction for Lorentz and polarization effects) using the Crysalis software package (Version 1.171.37.31, release 14-01-2014 CryAlis171.NET, compiled Jan 14 2014, 18:38:05, Advanced Research Facilities (SGIker), by the University of the Basque Country, Leioa, Spain). The structure was solved using ShelXS (Sheldrick, 2008, Advanced Research Facilities (SGIker), by the University of the Basque Country, Leioa, Spain) [46] and refined by full-matrix least-squares with SHELXL-97 (Sheldrick, 2008, Advanced Research Facilities (SGIker), by the University of the Basque Country, Leioa, Spain) [47]. Final geometrical calculations were carried out with Mercury [48] and PLATON [49,50] as integrated in WinGX [51].

General procedure for the synthesis of 3-amino-1,5-dihydro-2H-pyrrol-2-ones **10**. A solution of benzaldehyde **2** (0.1 mL, 1 mmol), diethyl acetylenedicarboxylate **8** (0.16 mL, 1 mmol), amine **3** (2 mmol), phosphoric acid catalyst **9** (34.8 mg, 0.1 mmol), and anhydrous MgSO$_4$ was stirred in toluene

(5 mL) at 110 °C for 48 h. The volatiles were dried off at reduced pressure, and the crude residue was purified by column chromatography (AcOEt/hexanes) to afford pure lactams **10**.

Ethyl 5-oxo-2-phenyl-1-(p-tolyl)-4-(p-tolylamino)-2,5-dihydro-1H-pyrrole-3-carboxylate (**10a**). The general procedure was followed, using p-toluidine (0.21 g, 2 mmol), affording 0.311 g (73%) of **10a** as a white solid. Melting point (m.p.) (Et$_2$O) 154–155 °C. ^{1}H-NMR (400 MHz, CDCl$_3$): δ 8.17 (bs, 1H, NH), 7.34 (d, $^3J_{HH}$ = 8.5 Hz, 2H, 2× CHar), 7.26–7.21 (m, 5H, 5× CHar), 7.12 (d, $^3J_{HH}$ = 8.3 Hz, 2H, 2× CHar), 7.08 (d, J = 8.5 Hz, 2H, 2× CHar), 7.03 (d, $^3J_{HH}$ = 8.3 Hz, 2H, 2× CHar), 5.77 (s, 1H, CHN), 4.01 (q, $^3J_{HH}$ = 7.1 Hz, 2H, CH$_2$ OEt), 2.33 (s, 3H, CH$_3$), 2.23 (s, 3H, CH$_3$), 1.01 (t, $^3J_{HH}$ = 7.1 Hz, 3H, CH$_3$ OEt). ^{13}C {^{1}H} NMR (101 MHz, CDCl$_3$) δ 164.7 (C=O ester), 164.1 (C=O amide), 142.7 (=C$_{quat}$), 137.2 (C$_{quat}$), 136.1 (C$_{quat}$), 135.5 (C$_{quat}$), 134.6 (C$_{quat}$), 134.2 (C$_{quat}$), 129.5 (2× CHar), 129.1 (2× CHar), 128.4 (2× CHar), 128.1 (CHar), 127.83 (2× CHar), 123.2 (2× CHar), 122.8 (2× CHar), 108.9 (=C$_{quat}$), 63.3 (CHN), 60.2 (CH$_2$ OEt), 21.1 (CH$_3$), 21.0 (CH$_3$), 14.0 (CH$_3$ OEt). Fourier-transform IR (FTIR) (neat) ν$_{max}$: 3289 (N–H), 1701 (C=O), 1679 (C=O), 1632 (C=C). HRMS (Q-TOF) m/z calculated for C$_{27}$H$_{26}$N$_2$O$_3$ [M]$^+$ 426.1943, found 426.1950.

Ethyl 1-(4-methoxyphenyl)-4-((4-methoxyphenyl)amino)-5-oxo-2-phenyl-2,5-dihydro-1H-pyrrole-3-carboxylate (**10b**). The general procedure was followed, using p-anisidine (0.25 g, 2 mmol), affording 0.284 g (63%) of **10b** as a yellow solid. m.p. (Et$_2$O) 116–117 °C. ^{1}H-NMR (300 MHz, CDCl$_3$) δ 8.20 (bs, 1H, NH), 7.29 (d, $^3J_{HH}$ = 9.1, 2H, 2× CHar), 7.24–7.18 (m, 5H, 5× CHar), 7.15 (d, $^3J_{HH}$ = 8.9 Hz, 2H, 2× CHar), 6.85 (d, $^3J_{HH}$ = 8.9 Hz, 2H, 2× CHar), 6.74 (d, $^3J_{HH}$ = 9.1, 2H, 2× CHar), 5.69 (bs, 1H, CHN), 4.01 (q, $^3J_{HH}$ = 7.1, 2H, CH$_2$ OEt), 3.80 (s, 3H, CH$_3$O), 3.71 (s, 3H, CH$_3$O), 1.02 (t, $^3J_{HH}$ = 7.1, 3H, CH$_3$ OEt). ^{13}C {^{1}H} NMR (75 MHz, CDCl$_3$) δ 164.9 (C=O ester), 163.9 (C=O amide, 157.5 (C$_{quat}$), 157.3 (C$_{quat}$), 143.4 (=C$_{quat}$), 137.3 (C$_{quat}$), 131.6 (C$_{quat}$), 129.7 (C$_{quat}$), 128.4 (2× CHar), 128.1 (CHar), 127.9 (2× CHar), 125.1 (2× CHar), 124.8 (2× CHar), 114.1 (2× CHar), 113.8 (2× CHar), 107.9 (=C$_{quat}$), 63.6 (CHN), 60.1 (CH$_2$ OEt), 55.6 (CH$_3$), 55.5(CH$_3$), 14.1 (CH$_3$ OEt). FTIR (neat) ν$_{max}$: 3436 (N–H), 1704 (C=O), 1672 (C=O), 1629 (C=C). HRMS (Q-TOF) m/z calculated for C$_{22}$H$_{15}$Br$_2$N$_3$O$_3$ [M]$^+$ 458.1842, found 458.1844.

Ethyl 1-benzyl-4-(benzylamino)-5-oxo-2-phenyl-2,5-dihydro-1H-pyrrole-3-carboxylate (**10c**). The general procedure was followed, using benzylamine (0.21 g, 2 mmol), affording 0.234 g (58%) of **10c** as a white solid. m.p. (Et$_2$O) 106–108 °C. ^{1}H-NMR (400 MHz, DMSO-d_6) δ 7.36 (m, 4H, 4× CHar), 7.34–7.21 (m, 8H, 7× Char + NH), 7.08 (m, 4H, 4× CHar), 5.09 (d, $^3J_{HH}$ = 6.8 Hz, 2H, CH$_2$ Bn), 4.95 (s, 1H, CHN), 4.86 (d, $^3J_{HH}$ = 15.1 Hz, 1H, CH$_2$ Bn), 3.96–3.81 (m, 2H, CH$_2$ OEt), 3.65 (d, $^3J_{HH}$ = 15.1 Hz, 1H, CH$_2$ Bn), 0.91 (t, $^3J_{HH}$ = 7.1 Hz, 3H, CH$_3$ OEt). ^{13}C {^{1}H} NMR (101 MHz, DMSO-d_6) δ 164.6 (C=O ester), 163.5 (C=O amide, 145.3 (=C$_{quat}$), 139.8 (C$_{quat}$), 137.0 (C$_{quat}$), 136.2 (C$_{quat}$), 128.0 (2× CHar), 127.9 (2× CHar), 127.8 (2× CHar), 127.5 (CHar), 127.3 (2× CHar), 127.1 (2× CHar), 126.8 (CHar), 126.7 (2× CHar), 126.4 (CHar), 103.4 (=C$_{quat}$), 60.8 (CHN), 58.4 (CH$_2$ OEt), 45.3 (CH$_2$ Bn), 43.4 (CH$_2$ Bn), 13.3 (CH$_3$ OEt). FTIR (neat) ν$_{max}$: 3430 (N–H), 1691 (C=O), 1665 (C=O) 1624 (C=C). HRMS (Q-TOF) m/z calculated for C$_{22}$H$_{15}$F$_2$N$_3$O$_3$ [M]$^+$ 426.1943, found 426.1942.

General procedure for the hydrolysis of compounds **10**. To 10 mL of a 3 M HCl/THF (1:1) solution, compound **10** (0.5 mmol) was added; the mixture was heated to 75 °C and stirred overnight. The reaction was monitored by TLC and, once it was finished, the mixture was concentrated under reduced pressure to eliminate the THF, washed with 3 M NaOH (2× 5 mL) and H$_2$O (2× 5mL), and extracted with ethyl acetate. The combined organic phases were dried with anhydrous Mg$_2$SO$_4$, and the crude residue was crystalized in Et$_2$O: pentane.

Ethyl 4-hydroxy-5-oxo-2-phenyl-1-(p-tolyl)-2,5-dihydro-1H-pyrrole-3-carboxylate (**11a**). The general procedure was followed, affording 0.161 g (95%) of **11a** as a white solid. m.p. (Et$_2$O) 170–172 °C. ^{1}H-NMR (300 MHz, CDCl$_3$) δ 9.19 (bs, 1H, OH), 7.38 (d, $^3J_{HH}$ = 8.2 Hz, 2H, 2× CHar), 7.32–7.25 (m, 5H, 5× CHar), 7.09 (d, $^3J_{HH}$ = 8.2 Hz, 2H, 2× CHar), 5.74 (s, 1H, CHN), 4.20 (q, $^3J_{HH}$ =, 7.1 Hz, 2H, CH$_2$ OEt) 2.26 (s, 3H), 1.20 (t, $^3J_{HH}$ = 7.1Hz, 3H, CH$_3$ OEt). ^{13}C {^{1}H} NMR (75 MHz, CDCl$_3$) δ 165.0 (C=O ester), 162.9 (C=O amide), 156.4 (=C$_{quat}$), 135.7 (C$_{quat}$), 135.3 (C$_{quat}$), 133.7 (C$_{quat}$), 129.6 (2× CHar), 128.6 (2× CHar), 128.5 (CHar), 127.6 (2× CHar), 122.4 (2× CHar), 113.1 (=C$_{quat}$), 61.8

(CHN), 61.2 (CH$_2$ OEt), 20.9 (CH$_3$), 14.0 (CH$_3$ OEt). FTIR (neat) ν_{max}: 3425 (O–H), 1704 (C=O), 1675 (C=O), 1643 (C=C). HRMS (Q-TOF) m/z calculated for C$_{27}$H$_{26}$N$_2$O$_3$ [M]$^+$ 337.1314, found 337.1319.

Ethyl 4-hydroxy-1-(4-methoxyphenyl)-5-oxo-2-phenyl-2,5-dihydro-1H-pyrrole-3-carboxylate (11b). The general procedure was followed, affording 0.162 g (92%) of 11b as a white solid. m.p. (Et$_2$O) 182 °C (dec.). ^{1}H-NMR (300 MHz, CDCl$_3$) δ 9.05 (bs, 1H, OH), 7.30 (d, $^3J_{HH}$ = 8.9 Hz, 2H, 2× CHar), 7.24–7.16 (m, 5H, 5× CHar), 6.79 (d, $^3J_{HH}$ = 8.8 Hz, 2H, 2× CHar), 5.63 (s, 1H, CHN), 4.17 (q, $^3J_{HH}$ = 7.1 Hz, 2H, CH$_2$ OEt), 3.72 (s, 3H, CH$_3$), 1.16 (t, $^3J_{HH}$ = 7.1 Hz, 3H, CH$_2$ OEt). ^{13}C {^{1}H} NMR δ 165.4 (C=O ester), 162.8 (C=O amide), 157.7 (C$_{quat}$), 157.1 (=C$_{quat}$), 135.3 (C$_{quat}$), 129.3 (C$_{quat}$), 128.7 (2× CHar), 128.6 (2× CHar), 127.7 (2× CHar), 124.5 (2× CHar), 120.5 (CHar), 114.4 (2× CHar), 113.0 (=C$_{quat}$), 62.2 (CHN), 61.3 (CH$_2$ OEt), 55.5 (CH$_3$), 14.1 (CH$_3$ OEt). FTIR (neat) ν_{max}: 3431 (O–H), 1711 (C=O), 1677 (C=O), 1653 (C=CH). HRMS (Q-TOF) m/z calculated for C$_{27}$H$_{26}$N$_2$O$_3$ [M]$^+$ 353.1263, found 353.1268.

Ethyl 1-benzyl-4-hydroxy-5-oxo-2-phenyl-2,5-dihydro-1H-pyrrole-3-carboxylate (11c). The general procedure was followed, affording 0.157 g (94%) of 11c as a white solid. m.p. (Et$_2$O) 178–179 °C. ^{1}H-NMR (300 MHz, CDCl$_3$) δ 9.11 (bs, 1H, OH), 7.39–7.33 (m, 3H, 3× CHar), 7.32–7.27 (m, 3H, 3× CHar), 7.15–7.08 (m, 4H, 4× CHar), 5.20 (d, $^3J_{HH}$ = 14.8 Hz, 1H, CH$_2$ Bn), 4.88 (s, 1H, CHN), 4.08 (q, $^3J_{HH}$ = 7.2, 2H, , CH$_2$ OEt), 3.55 (d, $^3J_{HH}$ = 14.8 Hz, 1H, CH$_2$ Bn), 1.06 (t, $^3J_{HH}$ = 7.1 Hz, 3H, CH$_3$ OEt). ^{13}C {^{1}H} NMR (75 MHz, CDCl$_3$) δ 165.59 (C$_{quat}$), 163.59 (C$_{quat}$), 157.91 (C$_{quat}$), 136.43 (C$_{quat}$), 134.68 (C$_{quat}$), 128.97 (CH), 128.68 (CH), 128.02 (CH), 127.98 (CH), 113.37 (C$_{quat}$), 61.14 (CH$_2$), 59.75 (CH), 44.11 (CH$_2$), 13.97 (CH$_3$). FTIR (neat) ν_{max}: 3450 (N–H), 1735 (C=O), 1675 (C=O), 1632 (C=C). HRMS (Q-TOF) m/z calculated for C$_{27}$H$_{26}$N$_2$O$_3$ [M]$^+$ 337.1314, found 337.1333.

General procedure for the isolation of amides 12. A solution of benzaldehyde 2 (0.1 mL, 1 mmol), diethyl acetylenedicarboxylate 8 (0.16 mL, 1 mmol), amine 3 (2 mmol), phosphoric acid catalyst 9 (34.8 mg, 0.1 mmol), and anhydrous MgSO$_4$ was stirred in toluene (5 mL) at 110 °C for 48 h. The volatiles were dried off at reduced pressure, and the crude residue was purified by column chromatography (AcOEt/hexanes) to afford pure lactams 12.

5-oxo-2-phenyl-N,1-di-p-tolyl-4-(p-tolylamino)-2,5-dihydro-1H-pyrrole-3-carboxamide (12a). The general procedure was followed, affording 0.02 g (4%) of 12a as a white solid. m.p. (Et$_2$O) 226 °C (dec.). (300 MHz, CDCl$_3$) δ 8.31 (bs, 1H, NH), 7.38–7.28 (m, 6H, 6× CHar), 7.11–7.04 (m, 7H, 7× CHar), 6.96 (d, $^3J_{HH}$ = 8.5 Hz, 2H, 2× CHar), 6.84 (d, $^3J_{HH}$ = 8.5 Hz, 2H, 2× CHar), 6.63 (bs, 1H, NH), 5.85 (s, 1H, CHN), 2.28 (s, 3H,CH$_3$), 2.25 (s, 3H, CH$_3$), 2.24 (s, 3H, CH$_3$). ^{13}C {^{1}H} NMR (75 MHz, CDCl$_3$) (75 MHz, CDCl$_3$) δ 164.75 (C=O), 162.12 (C=O), 139.1 (=C$_{quat}$), 136.6 (C$_{quat}$), 136.1 (C$_{quat}$), 135.8 (C$_{quat}$), 134.8 (C$_{quat}$), 134.6 (C$_{quat}$), 133.9 (C$_{quat}$), 133.8 (C$_{quat}$), 129.7 (4× CHar), 129.5 (2× CHar), 129.4 (2× CHar), 129.3 (CHar), 128.0 (2× CHar), 123.3 (2× CHar), 122.5 (2× CHar), 119.8 (2× CHar), 112.4 (=C$_{quat}$), 63.8 (CHN), 21.1 (CH$_3$), 21.0 (CH$_3$), 21.0 (CH$_3$). FTIR (neat) ν_{max}: 3309 (N–H), 3251 (N–H), 1685 (C=O), 1632 (C=C). HRMS (Q-TOF) m/z calculated for C$_{27}$H$_{26}$N$_2$O$_3$ [M]$^+$ 487.22598, found 487.2255.

N,1-bis(4-methoxyphenyl)-4-((4-methoxyphenyl)amino)-5-oxo-2-phenyl-2,5-dihydro-1H-pyrrole-3-carboxamide (12b). The general procedure was followed, affording 0.07 g (13%) of 12b as a white solid. m.p. (Et$_2$O) 228–229 °C. ^{1}H-NMR (400 MHz, CDCl$_3$) δ 8.46 (bs, 1H, NH), 7.37–7.28 (m, 5H, 5× CHar), 7.26–7.22 (m, 2H, 2× CHar), 7.17 (d, $^3J_{HH}$ = 8.8 Hz, 2H, 2× CHar), 6.90 (d, $^3J_{HH}$ = 9.1 Hz, 2H, 2× CHar), 6.82 (d, $^3J_{HH}$ = 8.9 Hz, 2H, 2× CHar), 6.78 (d, $^3J_{HH}$ = 9.1 Hz, 2H, 2× CHar), 6.71 (d, $^3J_{HH}$ = 9.1 Hz, 2H, 2× CHar), 6.56 (bs, 1H, NH), 5.76 (s, 1H, CHN), 3.74 (s, 3H, CH$_3$), 3.73 (s, 3H, CH$_3$), 3.72 (s, 3H, CH$_3$). ^{13}C {^{1}H} NMR (101 MHz, CDCl$_3$) δ 164.6 (C=O), 162.4 (C=O), 157.8 (C$_{quat}$), 157.2 (C$_{quat}$), 156.5 (C$_{quat}$), 140.2 (=C$_{quat}$), 136.7 (C$_{quat}$), 131.6 (C$_{quat}$), 130.5 (C$_{quat}$), 129.5 (2× CHar), 129.3 (CHar), 128.0 (2× CHar), 125.4 (2× CHar), 124.5 (2× CHar), 121.5 (2× CHar), 114.3 (2× CHar), 114.3 (2× CHar), 114.1 (2× CHar), 110.9 (=C$_{quat}$), 64.1 (CHN), 55.6 (CH$_3$), 55.6 (CH$_3$), 55.5 (CH$_3$). FTIR (neat) ν_{max}: 3344 (N–H), 3286 (N–H), 1662 (C=O), 1682 (C=O), 1632 (C=C). HRMS (Q-TOF) m/z calculated for C$_{27}$H$_{26}$N$_2$O$_3$ [M]$^+$ 535.2107, found 535.2105.

Ethyl 4-amino-1-benzyl-5-oxo-2-phenyl-2,5-dihydro-1*H*-pyrrole-3-carboxylate (**16**). A mixture of **10c** (21.3 mg, 0.5 mmol), 10% palladium on carbon (276 mg, 0.025 mmol), and 37% HCl (0.05 mL, 0.5 mmol) in methanol (30 mL) was stirred for 10 h under hydrogen pressure at 70 psi. The reaction mixture was filtered through Celite, and the filtered solution was treated with $NaHCO_3$ until neutral and extracted with dichloromethane (3 × 15 mL). The combined organic fractions were dried with anhydrous $MgSO_4$, and distilled off at reduced pressure; the residue was crystallized in Et_2O/pentane (1:2) to afford 0.163 g (97%) of **11c** as a white solid. m.p. (Et_2O) 139–142 °C. ^{1}H-NMR (400 MHz, $CDCl_3$) δ 7.37–7.27 (m, 6H, 6× CHar), 7.18–7.05 (m, 4H, 4× CHar), 5.74 (bs, 2H, NH), 5.13 (d, $^3J_{HH}$ = 14.8 Hz, 1H, CH_2 Bn), 4.89 (s, 1H, CHN), 4.10–3.87 (m, 2H, CH_2 OEt), 3.57 (d, $^3J_{HH}$ = 14.8 Hz, 1H, CH_2 Bn), 1.05 (t, $^3J_{HH}$ = 7.1 Hz, 3H, CH_3 OEt). ^{13}C {^{1}H} NMR (75 MHz, $CDCl_3$) δ 165.5 (C=O ester), 165.0 (C=O amide), 145.9 (=C_{quat}), 136.6 (C_{quat}), 136.6 (C_{quat}), 128.9 (2× CHar), 128.7 (2× CHar), 128.5 (2× CHar), 128.4 (CHar), 128.0 (2× CHar), 127.9 (CHar), 104.8 (=C_{quat}), 61.5 (CHN) , 59.8 (CH_2 OEt), 44.2 (CH_2 Bn), 14.2 (CH_3 OEt). FTIR (neat) ν_{max}: 3450 and 3319 (N–H_2), 1685 (C=O), 1654 (C=O), 1643 (C=C). HRMS (Q-TOF) m/z calculated for $C_{27}H_{26}N_2O_3$ [M]$^+$ 336.1474, found 336.1476.

Supplementary Materials: Copies of ^{1}H- and ^{13}C {^{1}H} NMR spectra for γ-lactams **10**, **11**, **12**, and **16** are available online. CCDC 1938640 contains the supplementary crystallographic data for this paper (compound **11a**). The data can be obtained free of charge from the Cambridge Crystallographic Data Center via www.ccdc.cam.ac.uk/structures

Author Contributions: Conceptualization, X.C., E.M.M, F.P. and J.V.; methodology, X.C.; software, X.C.; validation, E.M.M and J.V.; formal analysis, X.C; investigation, X.C; resources, E.M.M., F.P. J.V.; data curation, X.C.; writing—original draft preparation, J.V.; writing—review and editing, X.C, E.M.M., F.P., J.V.; visualization, E.M.M., F.P., J.V..; supervision, E.M.M. and J.V.; project administration, E.M.M. and J.V.; funding acquisition, F.P.

Funding: Financial support was provided by the Ministerio de Ciencia, Innovación, y Universidades (RTI2018-101818-B-I00, MCIU/AEI/FEDER, UE), and Gobierno Vasco (GV, IT 992-16) is gratefully acknowledged. X. del Corte thanks Gobierno Vasco for a predoctoral grant.

Acknowledgments: The authors thank the technical and human support provided by SGIker (UPV/EHU/ ERDF, EU).

Conflicts of Interest: The authors declare no conflicts of interest. The funders had no role in the design of the study; in the collection, analyses, or interpretation of data; in the writing of the manuscript, or in the decision to publish the results.

References

1. Zhu, J.; Wang, Q.; Wang, M.-X. (Eds.) *Multicomponent Reactions in Organic Synthesis*; Wiley-VCH: Weinheim, Germnay, 2015.
2. Müller, T.J. *J in Science of Synthesis, Multicomponent Reactions, Vol 1 and 2*; Thieme: Stutgart, Germnay, 2014.
3. Knapp, J.M.; Kurth, M.J.; Shaw, J.T.; Younai, A. *Diversity-Oriented Synthesis*; Trabocchi, A., Ed.; Willey: New York, NY, USA, 2013; Volume 41, pp. 29–57.
4. Schreiber, S.L. Target-oriented and diversity-oriented organic synthesis in drug discovery. *Science* **2000**, *287*, 1964–1969. [CrossRef] [PubMed]
5. de Moliner, F.; Kielland, N.; Lavilla, R.; Vendrell, M. Modern Synthetic Avenues for the Preparation of Functional Fluorophores. *Angew. Chem. Int. Ed.* **2017**, *56*, 3758–3769. [CrossRef] [PubMed]
6. Hall, D.G.; Rybak, T.; Verdelet, T. Multicomponent Hetero-[4 + 2] Cycloaddition/Allylboration Reaction: From Natural Product Synthesis to Drug Discovery. *Acc. Chem. Res.* **2016**, *49*, 2489–2500. [CrossRef] [PubMed]
7. Caruano, J.; Muccioli, G.G.; Robiette, R. Biologically active γ-lactams: Synthesis and natural sources. *Org. Biomol. Chem.* **2016**, *14*, 10134–10156. [CrossRef] [PubMed]
8. Martinez de Marigorta, E.; de los Santos, J.; Ochoa de Retana, A.M.; Vicario, J.; Palacios, F. Multicomponent Reactions in the Synthesis of γ –Lactams. *Synthesis* **2018**, *50*, 4539–4554.
9. Martinez de Marigorta, E.; de los Santos, J.; Ochoa de Retana, A.M.; Vicario, J.; Palacios, F. Multicomponent reactions (MCRs): A useful access to the synthesis of benzo-fused γ-lactams. *Beilstein J. Org. Chem.* **2019**, *150*, 1065–1085. [CrossRef] [PubMed]
10. Kirpotina, L.N.; Schepetkin, I.A.; Khlebnikov, A.I.; Ruban, O.I.; Ge, Y.; Ye, R.D.; Kominsky, D.J.; Quinn, M.T. 4-Aroyl-3-hydroxy-5-phenyl-1H-pyrrol-2(5H)-ones as N-formyl peptide receptor 1 (FPR1) antagonists. *Biochem. Pharmacol.* **2017**, *142*, 120–132. [CrossRef]

11. Ma, K.; Wang, P.; Fu, W.; Wan, X.; Zhou, L.; Chu, Y.; Ye, D. Rational design of 2-pyrrolinones as inhibitors of HIV-1 integrase. *Bioorg. Med. Chem. Lett.* **2011**, *21*, 6724–6727. [CrossRef]
12. Zhuang, C.; Miao, Z.; Zhu, L.; Dong, G.; Guo, Z.; Wang, S.; Zhang, Y.; Wu, Y.; Yao, J.; Sheng, C.; et al. Discovery, Synthesis, and Biological Evaluation of Orally Active Pyrrolidone Derivatives as Novel Inhibitors of p53–MDM2 Protein–Protein Interaction. *J. Med. Chem.* **2012**, *55*, 9630–9642. [CrossRef]
13. Peifer, C.; Selig, R.; Kinkel, K.; Ott, D.; Totzke, F.; Schaechtele, C.; Heidenreich, R.; Roecken, M.; Schollmeyer, D.; Laufer, S. Design, Synthesis, and Biological Evaluation of Novel 3-Aryl-4-(1H-indole-3yl)-1,5-dihydro-2H-pyrrole-2-ones as Vascular Endothelial Growth Factor Receptor (VEGF-R) Inhibitors. *J. Med. Chem.* **2008**, *51*, 3814–3824. [CrossRef]
14. Liu, Q.-J.; Wang, L.; Kang, Q.-K.; Zhang, X.P.; Tang, Y. Cy-SaBOX/Copper(II)-Catalyzed Highly Diastereo- and Enantioselective Synthesis of Bicyclic N,O Acetals. *Angew. Chem. Int. Ed.* **2016**, *55*, 9220–9223. [CrossRef] [PubMed]
15. Bures, J.; Armstrong, A.; Blackmond, D. Explaining Anomalies in Enamine Catalysis: "Downstream Species as a New Paradigm for Stereocontrol G. *Acc. Chem. Res.* **2016**, *49*, 214–222. [CrossRef] [PubMed]
16. Khalaf, A.I.; Waigh, R.D.; Drummond, A.J.; Pringle, B.; McGroarty, I.; Skellern, G.G.; Suckling, C.J. Distamycin Analogues with Enhanced Lipophilicity: Synthesis and Antimicrobial Activity. *J. Med. Chem.* **2004**, *47*, 2133–2156. [CrossRef] [PubMed]
17. Ye, Y.; Fang, F.; Li, Y. Synthesis and anti-biofilm activities of dihydro-pyrrol-2-one derivatives on *Pseudomonas aeruginosa*. *Bioorg. Med. Chem. Lett.* **2015**, *25*, 597–601. [CrossRef] [PubMed]
18. Zhu, Q.; Gao, L.; Chen, Z.; Zheng, S.; Shu, H.; Li, J.; Jiang, H.; Liu, S. A novel class of small-molecule caspase-3 inhibitors prepared by multicomponent reactions. *Eur. J. Med. Chem.* **2012**, *54*, 232–238. [CrossRef] [PubMed]
19. Gein, V.L.; Popov, A.V.; Kolla, V.E.; Popova, N.A.; Potemkin, K.D. Synthesis and biological activity of 1,5-diaryl-3-arylamino-4-carboxymethyl-2,5-dihydro-2-pyrrolones and 1,5-diaryl-4-carboxymethyltetrahydropyrrole-2, 3-diones. *Pharm. Chem. J.* **1993**, *27*, 343–346. [CrossRef]
20. Gein, V.L.; Popov, A.V.; Kolla, V.E.; Popova, N.A. Synthesis and biological activity of 1,5-diaryl-3-alkylamino-4-carboxymethyl-2,5-dihydropyrrol-2-ones and 1,5-diaryl-4-carboxymethyl-tetrahydropyrrol-2,3-diones. *Pharmazie* **1993**, *8*, 107–109.
21. Li, B.; Wever, W.J.; Walsh, C.T.; Bowers, A.A. Dithiolopyrrolones: Biosynthesis, synthesis, and activity of a unique class of disulfide-containing antibiotics. *Nat. Prod. Rep.* **2014**, *31*, 905–923. [CrossRef] [PubMed]
22. Rigby, J.H.; Hughes, R.C.; Heeg, M.J.J. Endo-Selective Cyclization Pathways in the Intramolecular Heck Reaction. *Am. Chem. Soc.* **1995**, *117*, 7834–7835. [CrossRef]
23. Lewis, J.R. Amaryllidaceae and Sceletium alkaloids. *Nat. Prod. Rep.* **1994**, *11*, 329–332. [CrossRef]
24. Palacios, F.; Vicario, J.; Aparicio, D. An efficient synthesis of achiral and chiral cyclic dehydro-α-amino acid derivatives through nucleophilic addition of Amines to β,γ-unsaturated α-keto esters. *Eur. J. Org. Chem.* **2006**, 2843–2850. [CrossRef]
25. Shaterian, H.R.; Ranjbar, M. Uncatalyzed synthesis of 3-amino-1,5-dihydro-2H-pyrrol-2-ones. *Res. Chem. Intermed.* **2014**, *40*, 2059–2074. [CrossRef]
26. Niknam, K.; Mojikhalifeh, S. Synthesis of new 1,5-diaryl-3-(arylamino)-1H-pyrrol-2(5H)-ones under catalyst-free and solvent-free conditions. *Mol. Diver.* **2014**, *18*, 111–117. [CrossRef] [PubMed]
27. Quian, J.; Yi, W.; Cai, C. Recyclable fluorous organocatalysts promoted three-component reactions of pyruvate, aldehyde and amine at room temperature. *Tetrahedron Lett.* **2013**, *54*, 7100–7102. [CrossRef]
28. Li, X.; Deng, H.; Luo, S.; Cheng, J.-P. Organocatalytic Three-Component Reactions of Pyruvate, Aldehyde and Aniline by Hydrogen-Bonding Catalysts. *Eur. J. Org. Chem.* **2008**, 4350–4356. [CrossRef]
29. del Corte, X.; Maestro, A.; Vicario, J.; Martinez de Marigorta, E.; Palacios, F. Brönsted-Acid-Catalyzed Asymmetric Three-Component Reaction of Amines, Aldehydes, and Pyruvate Derivatives. Enantioselective Synthesis of Highly Functionalized γ-Lactam Derivatives. *Org. Lett.* **2018**, *20*, 317–320. [CrossRef] [PubMed]
30. Gao, H.; Sun, J.; Yan, C.-G. Synthesis of functionalized 2-pyrrolidinones via domino reactions of arylamines, ethyl glyoxylate and acetylenedicarboxylates. *Tetrahedron* **2013**, *69*, 589–594. [CrossRef]
31. Khan, A.T.; Gosh, A.; Khan, M.M. One-pot four-component domino reaction for the synthesis of substituted dihydro-2-oxypyrrole catalyzed by molecular iodine. *Tetrahedron Lett.* **2012**, *53*, 2622–2626. [CrossRef]

32. Saha, M.; Das, A.R. Access of Diverse 2-Pyrrolidinone, 3,4,5-Substituted Furanone and 2-Oxo-dihydropyrroles Applying Graphene Oxide Nanosheet: Unraveling of Solvent Selectivity. *ChemistrySelect* **2017**, *2*, 10249–10260. [CrossRef]
33. Palacios, F.; Aparicio, D.; García, J.; Vicario, J.; Ezpeleta, J.M. Regioselective alkylation reactions of enamines derived from phosphane oxides - Synthesis of phosphorus substituted enamino esters, δ-amino-phosphonates, pyridone derivatives and pyrroles. *Eur. J. Org. Chem.* **2001**, 3357–3365. [CrossRef]
34. Palacios, F.; Aparicio, D.; Vicario, J. Synthesis of quinolinylphosphane oxides and -phosphonates from N-arylimines derived from phosphane oxides and phosphonates. *Eur. J. Org. Chem.* **2002**, 4131–4136.
35. Palacios, F.; Vicario, J.; Aparicio, D. Aza-Diels-Alder reaction of α,β-unsaturated sulfinylimines derived from α-amino acids with enolethers and enamines. *Tetrahedron Lett.* **2007**, *48*, 6747–6750. [CrossRef]
36. Palacios, F.; Vicario, J.; Aparicio, D. A diastereoselective aza-Diels-Alder reaction of N-aryl-1-azadienes derived from α-amino acids with enamines. *Tetrahedron Lett.* **2011**, *52*, 4109–4111.
37. Palacios, F.; Vicario, J.; Aparicio, D. Efficient synthesis of 1-azadienes derived from α-aminoesters. Regioselective preparation of α-dehydroamino acids, vinylglycines, and α-amino acids. *J. Org. Chem.* **2006**, *71*, 7690–7696. [CrossRef] [PubMed]
38. Vicario, J.; Ortiz, P.; Ezpeleta, J.M.; Palacios, F. Asymmetric Synthesis of Functionalized Tetrasubstituted α-Aminophosphonates through Enantioselective Aza-Henry Reaction of Phosphorylated Ketimines. *J. Org. Chem.* **2015**, *80*, 156–164. [CrossRef] [PubMed]
39. Vicario, J.; Ezpeleta, J.M.; Palacios, F. Asymmetric Cyanation of α-Ketiminophosphonates Catalyzed by Cinchona Alkaloids: Enantioselective Synthesis of Tetrasubstituted α-Aminophosphonic Acid Derivatives from Trisubstituted α-Aminophosphonates. *Adv. Synth. & Catal.* **2012**, *354*, 2641–2647.
40. Maestro, A.; Martinez de Marigorta, E.; Palacios, F.; Vicario, J. Enantioselective alpha-Aminophosphonate Functionalization of Indole Ring through an Organocatalyzed Friedel-Crafts Reaction. *J. Org. Chem.* **2019**, *84*, 1094–1102. [CrossRef] [PubMed]
41. Metten, B.; Kostermans, M.; Van Baelen, G.; Smet, M.; Dehaen, W. Synthesis of 5-aryl-2-oxopyrrole derivatives as synthons for highly substituted pyrroles. *Tetrahedron* **2006**, *62*, 6018. [CrossRef]
42. Winstein-Holnes values: A_{Bn} = 1.68 kcal.mol-1 and A_{Ph} = 2.80 kcal.mol-1. In *Stereochemistry of Organic Compounds*; Eliel, E.L.; Wilen, S.H.; Mander, L.N. (Eds.) Wiley: New York, NY, USA, 1994.
43. Ávila, E.P.; Justo, R.M.S.; Gonçalves, V.P.; Pereira, A.A.; Diniz, R.; Amarante, G.W. Chiral Brønsted Acid-Catalyzed Stereoselective Mannich-Type Reaction of Azlactones with Aldimines. *J. Org. Chem.* **2015**, *80*, 590–594. [CrossRef] [PubMed]
44. Shaoa, Y.-D.; Chengb, D.-J. Catalytic Asymmetric 1,2-Difunctionalization of Indolenines with α-(Benzothiazol-2-ylsulfonyl) Carbonyl Compounds. *Adv. Synth. Catal.* **2017**, *359*, 2549–2556. [CrossRef]
45. Huang, Q.; Cheng, Y.; Yuan, H.; Chang, X.; Li, P.; Li, W. Organocatalytic enantioselective Mannich-type addition of 5H-thiazol-4-ones to isatin-derived imines: Access to 3-substituted 3-amino-2-oxindoles featured by vicinal sulfur-containing tetrasubstituted stereocenters. *Org. Chem. Front.* **2018**, *5*, 3226–3230. [CrossRef]
46. Sheldrick, G.M. A short history of SHELX. *Acta Cryst.* **2008**, *A64*, 112–122. [CrossRef] [PubMed]
47. Sheldrick, G.M. Crystal Structure Refinement with SHELXL. *Acta Cryst.* **2015**, *C71*, 3–8.
48. Macrae, C.F.; Bruno, I.J.; Chisholm, J.A.; Edgington, P.R.; McCabe, P.; Pidcock, E.; Rodriguez-Monge, L.; Taylor, R.; van de Streek, J.; Wood, P.A. Mercury CSD 2.0 – new features for the visualization and investigation of crystal structures. *J. Appl. Cryst.* **2008**, *41*, 466–470. [CrossRef]
49. Spek, A.L. *PLATON, A Multipurpose Crystallographic Tool*; Utrecht University: Utrecht, The Netherlands, 2010.
50. Spek, A.L. Single-crystal structure validation with the program. *PLATON J. Appl. Cryst.* **2003**, *36*, 7–13. [CrossRef]
51. Farrugia, L.J. WinGX suite for small-molecule single-crystal crystallography. *J. Appl. Cryst.* **1999**, *32*, 837–838. [CrossRef]

Sample Availability: Samples of the compounds **10**, **11**, **12**, and **16** are available from the authors.

© 2019 by the authors. Licensee MDPI, Basel, Switzerland. This article is an open access article distributed under the terms and conditions of the Creative Commons Attribution (CC BY) license (http://creativecommons.org/licenses/by/4.0/).

Article

Design, Synthesis, and Biological Evaluation of Novel N-Acylhydrazone Bond Linked Heterobivalent β-Carbolines as Potential Anticancer Agents

Xiaofei Chen [1,†], Liang Guo [1,†], Qin Ma [2], Wei Chen [2], Wenxi Fan [2] and Jie Zhang [1,*]

1. School of Chemistry and Chemical Engineering, Key Laboratory for Green Processing of Chemical Engineering of Xinjiang Bingtuan, Shihezi University, Shihezi 832003, China
2. Xinjiang Huashidan Pharmaceutical Research Co. Ltd., 175 He Nan East Road, Urumqi 830011, China
* Correspondence: zhangjie-xj@163.com; Tel.: +86-993-205-7215
† These authors contributed equally to the work.

Received: 21 July 2019; Accepted: 8 August 2019; Published: 14 August 2019

Abstract: Utilizing a pharmacophore hybridization approach, we have designed and synthesized a novel series of 28 new heterobivalent β-carbolines. The in vitro cytotoxic potential of each compound was evaluated against the five cancer cell lines (LLC, BGC-823, CT-26, Bel-7402, and MCF-7) of different origin—murine and human, with the aim of determining the potency and selectivity of the compounds. Compound **8z** showed antitumor activities with half-maximal inhibitory concentration (IC_{50}) values of 9.9 ± 0.9, 8.6 ± 1.4, 6.2 ± 2.5, 9.9 ± 0.5, and 5.7 ± 1.2 μM against the tested five cancer cell lines. Moreover, the effect of compound **8z** on the angiogenesis process was investigated using a chicken chorioallantoic membrane (CAM) in vivo model. At a concentration of 5 μM, compound **8z** showed a positive effect on angiogenesis. The results of this study contribute to the further elucidation of the biological regulatory role of heterobivalent β-carbolines and provide helpful information on the development of vascular targeting antitumor drugs.

Keywords: asymmetric dimeric β-carboline; acylhydrazone group; cytotoxic; antitumor; structure–activity relationship

---

1. Introduction

Cancer remains a leading cause of death in developed and developing countries, although much significant progress has been achieved recently [1]. Cancer resistance to therapy is becoming a common phenomenon that threatens the current strategies against this disease. For that reason, we need to discover new anticancer agents. One of the successful and effective methods for the discovery of new anticancer drugs from natural products is synthesis of novel compounds through chemical structural modifications on the basis of leading compounds.

β-Carbolines are a large group of heterocyclic compounds with a 9H-pyrido[3,4-b]indole structural unit. They compose a class of alkaloids that are widely distributed in nature, including plants, foodstuffs, marine creatures, insects, mammals, human tissues, and body fluids [2]. In the last few decades, there have been intense research efforts in the design and development of β-carbolines as a new class of antitumor agents. A large number of β-carboline derivatives have been prepared in search of more potent antitumor agents. The structure–activity relationships (SARs) of these β-carbolines have been extensively investigated [3–10]. Research has indicated that this class of compounds exert their antitumor effects through multiple mechanisms of action, including intercalating into DNA [11–13] and inhibiting topoisomerases I and II [14,15], cyclin-dependent kinase (CDK) [16,17], polo-like kinase 1 (PLK1) [18], kinesin-like protein Eg5 [19], and IκB kinases [20].

Among frequently studied novel bioactive chemical entities, the acylhydrazone scaffold (–CONHN=) has attracted considerable attention for decades due to its broad applications ranging from medicinal agents to agrochemicals to functional materials. Many compounds containing this moiety have been reported, and many reports demonstrate that the introduction of this pharmacophore may have high potential for antitumor activity [21–26]. The acylhydrazone moiety is able to act as pharmacophore or auxophore subunit in different pharmaceutic classes, with a variety of action profiles, depending on the other functionalities present in the molecular structure [27]. For example, Mylotarg^{TM} (gemtuzumab ozogamicin; Pfizer) [28] (Figure 1) is a humanized anti-CD33 monoclonal antibody linked covalently to the cytotoxic agent N-acetyl gamma calicheamicin. Peterson reported PAC-1 (Figure 1), another N-acylhydrazone small-molecule, induces apoptotic death in cancer cells via the chelation of inhibitory zinc from procaspase-3, which leads to autocatalytic activation and subsequent generation of caspase-3 [29]. Carbazochrome (Figure 1), a semicarbazone-related compound, has been used as a hemostatic agent and is specifically indicated for capillary and parenchymal hemorrhage [30].

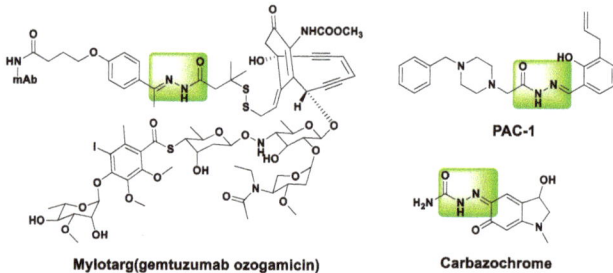

Figure 1. Structures of Mylotarg, PAC-1, and carbazochrome.

Our research group [31–34] has focused on incorporating substituents into positions 1, 2, 3, 7, and 9 of the β-carboline nucleus as antitumor agents. Structure–activity relationship (SAR) analysis indicated that (1) the β-carboline moiety was associated with their potential antitumor activities, and (2) the introduction of appropriate substituents into positions 1, 3, and 9 of the β-carboline nucleus enhanced their antitumor potencies. Previous research has shown that some antitumor agents when dimerized via an appropriate linker can lead to significantly improved antitumor effects, giving 100- to 500-fold improvement over the corresponding monomers [35–38].

So our group reported the synthesis, in vitro evaluation, in vivo efficacies, and SARs of the new homobivalent β-carbolines and heterobivalent β-carbolines with alkyl or alkylamino spacers in positions 1, 3, 7, and 9 of the β-carboline nucleus (Figure 2) [39–42]. In these homobivalent β-carbolines, 1-Methyl-9-[4-(1-methyl-β-carboline-9-yl)butyl]-β-carboline (B-9-3) [43,44] exhibited potent antitumor activity. The pharmacological mechanisms showed that B-9-3 selectively induces apoptosis of endothelial cells, in part through disruption of VEGF-A/VEGFR2 signaling [45], and also acts on the TGF-β signaling pathway [46]. Compounds B-3 [39] and B-4 [40] exhibited significant angiogenesis inhibitory effects in chicken chorioallantoic membrane (CAM) assay, and the anti-angiogenetic potency was comparable or more potent with the drug Endostar.

Continuing our studies to develop effective cytotoxic agents, the objective of this study was to synthesize potential anticancer compounds that are hybrids of β-carboline and acylhydrazone fragments (Figure 3). We have evaluated their cytotoxic activities for the first time, and the study also includes an investigation of the mechanism of action of these compounds for angiogenesis inhibition. These findings as well as our study of the SARs of the new compounds are discussed.

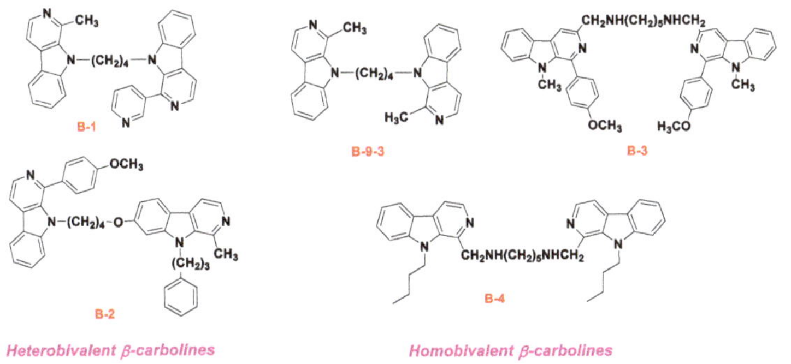

Molecules **2019**, 24, 2950

[Figure 2 illustration omitted]

Figure 2. The chemical structure of the representative reported diremic β-carbolines.

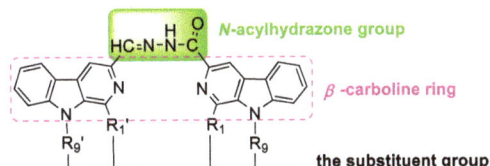

Figure 3. Hybrids of β-carboline and acylhydrazone fragments giving the target heterobivalent β-carbolines.

2. Results and Discussion

2.1. Chemistry

The syntheses of compounds **8a–ab** are depicted in Schemes 1–3. Monovalent β-carbolines **6a–l** and **7a–l** were synthesized according to previously published methods [39,47,48]. Using L-tryptophan as starting material, the tetrahydro-β-carboline skeleton (**2a–g**) was constructed via Pictet–Spengler cyclization. Then, the obtained carboxylic acid **2** reacted with thionyl chloride and ethanol to form ethyl ester **3**, which subsequently reacted with sulfur in xylene to afford compounds **4a–g**. Then compounds **4a–g** were reduced to their corresponding alcohols by lithium borohydride (LiBH$_4$) in dry THF to provide compounds **5a–g**, and further oxidized by MnO$_2$ in CH$_3$CN to afford the key intermediates, the 3-carboxaldehyde derivatives **6a–g** [39]. Alternatively, refluxing of compounds **4a–g** with 80% hydrazine hydrate in ethanol gave the other key intermediates, hydrazides **7a–g** [39] (see Scheme 1).

The N^9-alkylated derivative of compound **4a** was prepared by the action of sodium hydride (NaH) in anhydrous *N*,*N*-dimethylformamide (DMF) followed by the addition of alkyl halide to afford compounds **4h–l**, and following this, the intermediates **6h–l** and **7h–l** were prepared according to the same method for compounds **6a–g** and **7a–g** (see Scheme 2). Finally, the synthesis of compounds **8a–ac** (see Scheme 3) containing the acylhydrazone fragment was accomplished by the condensation of compounds **7a–l** with the corresponding aldehydes **6a–l**. We obtained the products easily in moderate to good yields when the aldehyde was 1 equiv and when the reactions were carried out in ethanol under reflux conditions. The structures of all compounds were confirmed by ^{1}H-NMR, ^{13}C-NMR (see Supplementary Materials), and high-resolution mass spectra (HRMS).

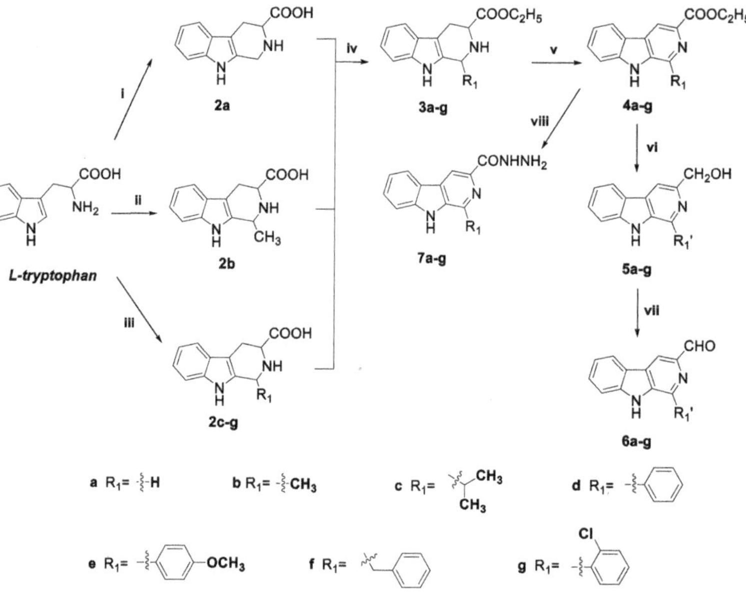

Scheme 1. Synthesis of the key intermediates **6a–g, 7a–g**. Reagents and conditions: (i) NaOH, H$_2$O, formaldehyde, reflux, 3 h; (ii) H$_2$SO$_4$, H$_2$O, acetaldehyde, room temperature, 3 h; (iii) acetic acid, R$_1$CHO, reflux, 3 h; (iv) ethanol, SOCl$_2$, reflux, 4 h; (v) xylene, S$_8$, reflux, 8 h; (vi) THF, LiBH$_4$, stirred at RT; (vii) CH$_3$CN, MnO$_2$, reflux, 2 h. (viii) hydrazine hydrate, ethanol, reflux, 4 h.

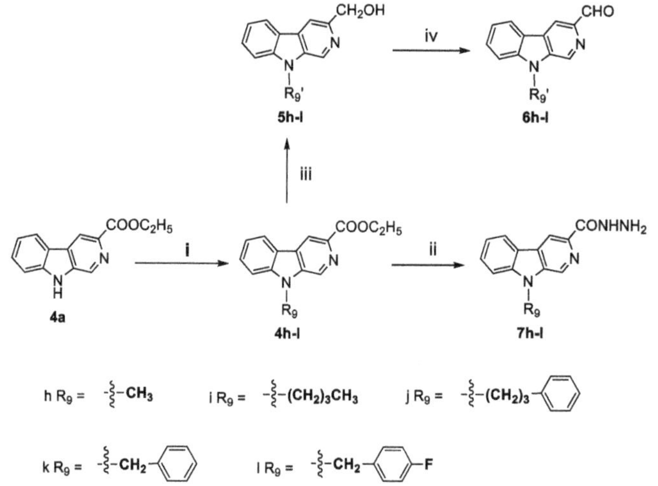

Scheme 2. Synthesis of the key intermediates **6h–l, 7h–l**. Reagents and conditions: (i) DMF, NaH, alkyl halogenide, stirred at RT; (ii) hydrazine hydrate, ethanol, reflux, 4 h. (iii) THF, LiBH$_4$, stirred at RT; (iv) CH$_3$CN, MnO$_2$, reflux, 2 h.

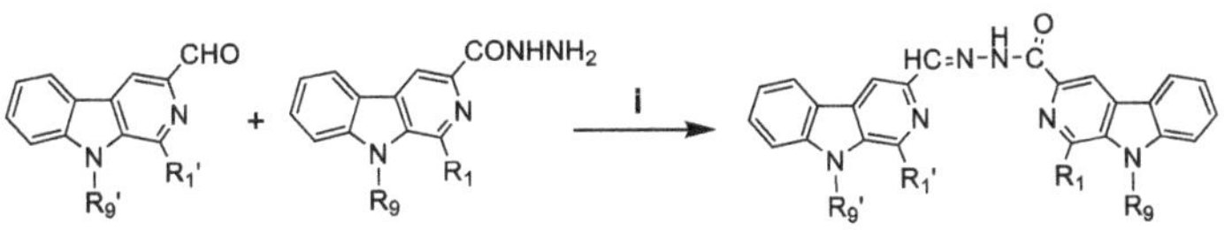

6a-l   7a-l   8a-ab

8a $R_1' = R_9' = R_1 = R_9 = H$
8b $R_1' = CH_3$ $R_9' = R_1 = R_9 = H$
8c $R_1' = CH(CH_3)_2$ $R_9' = R_1 = R_9 = H$
8d $R_1' = $ phenyl $R_9' = R_1 = R_9 = H$
8e $R_1' = $ 4-methoxyphenyl $R_9' = R_1 = R_9 = H$
8f $R_1' = $ benzyl $R_9' = R_1 = R_9 = H$
8g $R_1' = $ 2-chlorophenyl $R_9' = R_1 = R_9 = H$
8h $R_9' = CH_3$ $R_1' = R_1 = R_9 = H$
8i $R_9' = $ n-butyl $R_1' = R_1 = R_9 = H$
8j $R_9' = $ 3-phenylpropyl $R_1' = R_1 = R_9 = H$
8k $R_9' = $ benzyl $R_1' = R_1 = R_9 = H$
8l $R_9' = $ 4-fluorobenzyl $R_1' = R_1 = R_9 = H$
8m $R_1 = CH_3$ $R_1' = R_9' = R_9 = H$
8n $R_1 = CH(CH_3)_2$ $R_1' = R_9' = R_9 = H$

8o $R_1 = $ phenyl $R_1' = R_9' = R_9 = H$
8p $R_1 = $ 4-methoxyphenyl $R_1' = R_9' = R_9 = H$
8q $R_1 = $ benzyl $R_1' = R_9' = R_9 = H$
8r $R_9 = CH_3$ $R_1' = R_9' = R_1 = H$
8s $R_9 = $ n-butyl $R_1' = R_9' = R_1 = H$
8t $R_9 = $ 3-phenylpropyl $R_1' = R_9' = R_1 = H$
8u $R_9 = $ benzyl $R_1' = R_9' = R_1 = H$
8v $R_9 = $ 4-fluorobenzyl $R_1' = R_9' = R_1 = H$
8w $R_1' = R_1 = H$ $R_9' = R_9 = $ n-butyl
8x $R_1' = H$ $R_1 = CH_3$ $R_9' = R_9 = $ n-butyl
8y $R_1' = H$ $R_1 = CH_3$ $R_9' = $ n-butyl $R_9 = $ benzyl
8z $R_1' = CH_3$ $R_1 = H$ $R_9' = $ benzyl $R_9 = $ n-butyl
8aa $R_1' = R_1 = CH_3$ $R_9' = $ benzyl $R_9 = $ n-butyl
8ab $R_1' = R_1 = CH_3$ $R_9' = R_9 = $ benzyl

Scheme 3. Synthesis of the asymmetric dimeric β-carboline derivatives **8a–ab**. Reagents and conditions: (i) methanol, reflux, 4–6 h.

2.2. MTT Assay of Compounds 8a–8ab

From the synthetic route mentioned above, we obtained a series of novel heterobivalent β-carbolines. All of the target compounds were assayed for anticancer activity in various cancer cell lines including LLC (Lewis lung carcinoma), BGC-823 (gastric carcinoma), CT-26 (murine colon carcinoma), Bel-7402 (liver carcinoma), and MCF-7 (breast carcinoma), using the MTT method. The half-maximal inhibitory concentration (IC_{50}) values for each compound with respect to the five cancer cell lines were calculated, and the results are summarized in Table 1. These values represent the concentrations at which a 50% decrease in cell growth is observed after 72 h of incubation in the presence of the drug compared with control cells treated with DMSO or positive control Cisplatin (DDP) under similar conditions.

For the first experiment, we examined the influence of the substituents in position 1 of the β-carboline core on cytotoxic activities. In order to enhance the range of substituents, we designed 11 novel compounds with methyl and isopropyl substitutions and different patterns of aryl rings substituted by electron withdrawing (Cl) and donating (OCH_3) groups in the C-1 position of β-carboline. Of these 11 compounds, most of them showed medium or marginal cytotoxic activities in all cell lines. Interestingly, compounds **8b**, **8c**, **8d**, and **8o** were selectively active against BGC823 cells with IC_{50} values lower than 10 µM. In particular, compound **8b** was more potent against BGC823 cells than against the four other cell lines, with potencies in the double-digit µM range. Compounds **8m** ($R_1 = CH_3$) and **8q** ($R_1 = $ benzyl) were exceptional; they showed no distinct difference between each other, and their IC_{50} values were in the ranges of 13.8–24.7 µM and 13.3–24.5 µM, respectively. Next, we examined the influence of the substituents in position 9 of the β-carboline ring on antiproliferative effects. Compound **8s**, with an *n*-butyl group, was found to be the most potent agent among the heterobivalent β-carbolines, with IC_{50} values of 2.4 ± 0.6 µM (for BGC823) and 3.1 ± 1.2 µM (for CT26). Introduction of benzyl to the R_9 position on β-carboline yielded compound **8u**, and it demonstrated higher cytotoxic activity than other compounds against all tested tumor cell lines, except for the BGC823 cell line.

Table 1. Cytotoxic activity of acylhydrazone linked heterobivalent β-carbolines **8a–ab** in vitro [a] (IC$_{50}$, μM [b]).

Comp.	R$_1'$	R$_9'$	R$_1$	R$_9$	LLC[c]	BGC823	CT-26	Bel-7402	MCF-7
8a	H	H	H	H	91.9 ± 6.8	68.4 ± 6.2	63.6 ± 7.5	86.3 ± 9.4	56.6 ± 3.5
8b	CH$_3$	H	H	H	81.2 ± 7.5	3.2 ± 0.7	94.7 ± 11.8	45.7 ± 3.2	40.6 ± 5.4
8c	CH(CH$_3$)$_2$	H	H	H	76.7 ± 6.3	9.7 ± 0.9	38.5 ± 10.4	14.6 ± 3.2	15.3 ± 2.7
8d	-C$_6$H$_5$	H	H	H	61.3 ± 5.5	6.3 ± 1.6	57.2 ± 4.1	20.7 ± 3.3	11.5 ± 2.1
8e	-C$_6$H$_4$-OCH$_3$	H	H	H	58.5 ± 5.7	10.8 ± 1.4	68.4 ± 4.8	54.3 ± 6.9	34.2 ± 5.2
8f	-C$_6$H$_4$-Cl	H	H	H	70.3 ± 6.8	10.2 ± 2.3	22.2 ± 3.2	11.7 ± 0.9	17.6 ± 2.6
8g	-C$_6$H$_4$-Cl	H	H	H	48.8 ± 3.8	61.4 ± 7.6	57.4 ± 4.5	52.6 ± 5.8	30.9 ± 4.2
8h	H	CH$_3$	H	H	71.3 ± 10.8	78.6 ± 6.5	43.4 ± 4.2	41.2 ± 3.1	21.4 ± 5.3
8i	H	n-butyl	H	H	30.7 ± 2.9	27.4 ± 3.2	57.1 ± 5.6	15.3 ± 3.5	14.0 ± 3.5
8j	H	-(CH$_2$)$_3$-	H	H	40.7 ± 4.7	25.6 ± 2.1	23.4 ± 2.8	68.9 ± 5.4	21.6 ± 3.7
8k	H	-C$_6$H$_5$	H	H	17.4 ± 4.6	52.9 ± 5.8	15.6 ± 4.1	16.3 ± 4.7	18.5 ± 5.1
8l	H	-C$_6$H$_4$-F	H	H	29.2 ± 5.1	69.9 ± 8.4	15.8 ± 3.2	15.5 ± 2.9	17.3 ± 3.4
8m	H	H	CH$_3$	H	24.7 ± 3.9	17.5 ± 2.1	13.8 ± 2.7	15.6 ± 2.9	16.0 ± 4.3
8n	H	H	CH(CH$_3$)$_2$	H	38.2 ± 5.7	19.9 ± 2.9	22.0 ± 3.7	41.1 ± 7.8	14.5 ± 2.1
8o	H	H	-C$_6$H$_5$	H	72.1 ± 6.5	9.1 ± 1.4	56.3 ± 4.2	73.7 ± 9.8	14.4 ± 4.2
8p	H	H	-C$_6$H$_4$-OCH$_3$	H	68.4 ± 9.4	76.6 ± 5.2	69.5 ± 11.7	70.2 ± 8.5	>100
8q	H	H	H	H	20.3 ± 1.3	24.5 ± 5.1	13.4 ± 3.7	13.3 ± 2.1	14.7 ± 4.2
8r	H	H	H	CH$_3$	88.7 ± 5.6	34.8 ± 4.6	14.0 ± 3.3	12.4 ± 4.2	68.9 ± 7.5
8s	H	H	H	n-butyl	86.5 ± 10.7	2.4 ± 0.6	3.1 ± 1.2	21.5 ± 4.2	16.8 ± 1.9
8t	H	H	H	-(CH$_2$)$_3$-C$_6$H$_5$	13.3 ± 2.3	14.7 ± 2.1	12.6 ± 1.7	20.1 ± 5.2	2.5 ± 0.4

Table 1. Cont.

Comp.	R_1'	R_9'	R_1	R_9	IC$_{50}$(μM) ± SD				
					LLC[c]	BGC823	CT-26	Bel-7402	MCF-7
8u	H	H	H	phenethyl	83.2 ± 14.6	5.3 ± 0.9	2.7 ± 0.3	8.7 ± 1.5	2.8 ± 0.7
8v	H	H	H	fluorophenethyl	41.5 ± 5.1	43.4 ± 3.6	33.5 ± 2.7	46.7 ± 8.3	76.8 ± 5.2
8w	H	n-butyl	H	n-butyl	58.6 ± 4.4	28.2 ± 2.4	33.3 ± 6.2	25.3 ± 7.5	73.0 ± 12.9
8x	H	n-butyl	CH$_3$	n-butyl	70.2 ± 6.4	66.2 ± 5.8	55.8 ± 4.7	72.2 ± 11.4	49.5 ± 9.8
8y	H	n-butyl	CH$_3$	phenethyl	57.1 ± 4.6	39.2 ± 3.4	41.8 ± 5.3	60.4 ± 4.2	>100
8z	CH$_3$	phenethyl	H	n-butyl	9.9 ± 0.9	8.6 ± 1.4	6.2 ± 2.5	9.9 ± 0.5	5.7 ± 1.2
8aa	CH$_3$	phenethyl	H	n-butyl	44.5 ± 4.9	54.8 ± 3.2	44.3 ± 5.5	58.6 ± 4.3	28.2 ± 3.2
8ab	CH$_3$	phenethyl	CH$_3$	phenethyl	18.4 ± 2.7	61.6 ± 4.3	19.9 ± 3.5	44.7 ± 5.7	30.1 ± 4.2
DDP					21.3 ± 1.1	8.4 ± 0.7	4.2 ± 0.7	15.4 ± 1.9	10.5 ± 2.3

[a] Data represent the mean values of three independent determinations; [b] cytotoxicity as IC$_{50}$ for each cell line is the concentration of compound which reduced the optical density of treated cells by 50% with respect to untreated cells using the MTT assay. The data represent the mean values ± SD of at least three independent determinations. Values > 100 μM indicate less than 50% growth inhibition at > 100 μM; [c] cell lines include Lewis lung carcinoma (LLC), gastric carcinoma (BGC), murine colon carcinoma (CT-26), liver carcinoma (Bel-7402), and breast carcinoma (MCF-7).

Among all these novel molecules, the cytotoxic potencies of most heterobivalent β-carbolines (**8a–ab**) showed no distinct differences, and the IC_{50} values of this class of compounds ranged from 10 to 100 μM. **8a** and **8p** had poor inhibitory activities with IC_{50} values above 50 μM. Compound **8z** exhibited the most potent anticancer activity against the LLC, BGC-823, CT-26, Bel-7402, and MCF-7 cell lines, with IC_{50} values of 9.9 ± 0.9, 8.6 ± 1.4, 6.2 ± 2.5, 9.9 ± 0.5, and 5.7 ± 1.2 μM, respectively.

2.3. Inhibition of Angiogenesis in the Chicken Chorioallantoic Membrane (CAM) Assay

The CAM assay was deployed to assess the inhibitory effect of compound **8z** on neovascularization. In this experiment, we used Combretastatin A4 disodium phosphate (CA4P) as a positive control. The inhibitory effects of compound **8z** on angiogenesis in CAM are shown in Figure 4A. At the dose 0.5 μM, the reference anti-angiogenic drug CA4P elicited 21% inhibition of angiogenesis, and compound **8z** did not show a significant anti-angiogenic activity at this concentration (13% inhibition). The anti-angiogenetic activity of compound **8z** was comparable with CA4P in an in vivo CAM assay at the 5 μM level. In this assay, **8z** inhibited blood vessel formation by 45%, compared to 47% inhibition induced by CA4P ($p < 0.05$). At 50 μM, CA4P significantly inhibited blood vessel formation, eliciting 78% inhibition (Figure 4B).

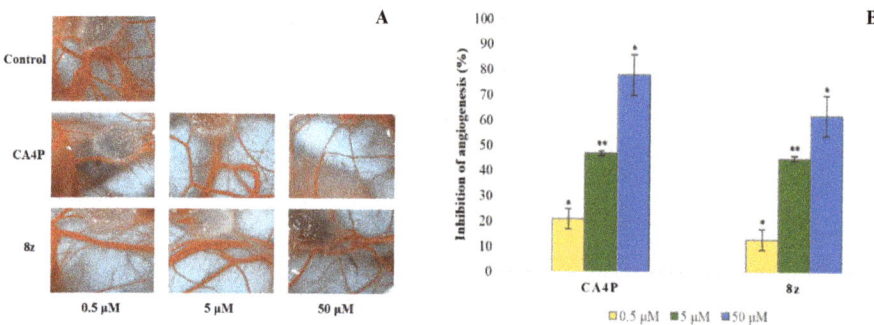

Figure 4. In vivo anti-angiogenic effect of compounds **8z** in CAM assay. (**A**) The representative photos of the experiments. (**B**) The anti-angiogenetic activity of compound **8z** was semi-quantitatively analyzed using Graph Pad Prism 5.0. The data represent the mean values ± SD of at least three independent determinations. (**, $p < 0.05$; *, $p < 0.5$).

3. Materials and Methods

3.1. Reagents and General Methods

MTT was obtained from Sigma-Aldrich (Darmstadt, Germany) and Cisplatin from Qilu pharmaceutical (Jinan, China). Other commercially available starting materials and solvents were reagent grade and were purchased from Adamas-beta and used without further purification. Reactions and products were routinely monitored by thin-layer chromatography (TLC) on silica gel F254 plates (Qingdao Haiyang Inc., Qingdao, China). ^{1}H-NMR and ^{13}C-NMR spectra were recorded at room temperature on a Bruker Avance III HD 400 instrument (Bruker Company, Bremen, Gemany) using tetramethylsilane as the internal reference. Chemical shifts (δ) were reported in ppm relative to the residual solvent peak, and the multiplicity of each signal was designated by the following abbreviations—s, singlet; d, doublet; t, triplet; q, quartet; m, multiplet. Coupling constants (*J*) were quoted in Hz. HRMS were recorded on a Bruker ultrafleXtreme MALDI-TOF/TOF-MS and Thermo Scientific LTQ Orbitrap XL (Thermo Fisher Scientific Inc, Waltham, USA). Column chromatography was performed with silica gel (200–300 mesh, Qingdao Haiyang Inc., Qingdao, China).

3.2. General Procedure for the Preparation of 6a–1c

A mixture of compound **5a** (1.98 g, 10 mmol) and activated MnO$_2$ (30 mmol) in CH$_3$CN (60 mL) was stirred under reflux for 2 h. After completion of the reaction (monitored by TLC), the products were cooled to room temperature and filtered through Celite. The filtrate was passed through silica gel and washed with dichloromethane, and the solvent was removed under reduced pressure. The residue was crystallized from acetone or acetone–petroleum ether to give the corresponding compound **6a**. Products **6b–l** were prepared according to the same method as **6a**.

1-benzyl-β-carboline-3-carbaldehyde (**6f**): The compound was obtained as a white solid with 81% yield. ^{1}H-NMR (400 MHz, CDCl$_3$) δ 10.24 (s, 1H, CHO), 8.68 (s, 1H, ArH), 8.64 (s, 1H, ArH), 8.14 (d, J = 8.0 Hz, 1H, ArH), 7.51 (d, J = 8.0 Hz, 1H, ArH), 7.43 (d, J = 8.4 Hz, 1H, ArH), 7.35–7.30 (m, 4H, ArH), 7.25–7.23 (m, 2H, ArH), 4.61 (s, 2H, ArCH$_2$). ^{13}C-NMR (100 MHz, CDCl$_3$) δ 193.41, 144.21, 143.97, 137.68, 129.04, 129.00, 128.94, 128.81, 128.77, 127.08, 127.01, 121.92, 121.80, 121.20, 114.09, 112.11, 41.56.

1-(2-chlorophenyl)-β-carboline-3-carbaldehyde (**6g**): The compound was obtained as a light yellow solid with 72% yield. ^{1}H-NMR (400 MHz, CDCl$_3$) δ 10.28 (s, 1H, CHO), 8.79 (s, 1H, ArH), 8.39 (s, 1H, ArH), 8.25 (d, J = 8.0 Hz, 1H, ArH), 7.69–7.66 (m, 1H, ArH), 7.64–7.59 (m, 2H, ArH), 7.56–7.49 (m, 3H, ArH), 7.40 (t, J = 7.6 Hz, 1H, ArH). ^{13}C-NMR (100 MHz, CDCl$_3$) δ 193.33, 144.43, 141.54, 140.61, 136.39, 136.02, 132.89, 132.03, 130.79, 130.32, 129.56, 129.42, 127.71, 122.22, 122.16, 121.48, 114.17, 111.97.

3.3. General Procedure for the Preparation of Compounds 7a–l

We added 85% hydrazine hydrate (10 mL) to a solution of compound **4a** (2.40 g, 10 mmol) in ethanol (100 mL), and then the mixture was refluxed for 8 h. Following the completion of reaction (as demonstrated by TLC), the resulting mixture was cooled to 5 °C and the precipitate was collected by filtration. The crude product was further purified first by washing with ethanol and then by recrystallization in ethanol to obtain compound **7a** with a yield of 85%. Product **7b–l** was prepared according to the same method as **7a**.

1-isopropyl-β-carboline-3-carbohydrazide (**7c**): The compound was obtained as a white solid with 91% yield. ^{1}H-NMR (400 MHz, DMSO-d_6) δ 9.47 (s, 1H, NH), 8.65 (s, 1H, ArH), 8.35 (d, J = 7.6 Hz, 1H, ArH), 7.65 (d, J = 8.0 Hz, 1H, ArH), 7.60–7.55 (m, 1H, ArH), 7.30–7.26 (m, 1H, ArH), 4.58 (s, 2H, NH$_2$), 3.75–3.64 (m, 1H, CH), 1.43 (d, J = 6.8 Hz, 6H, CH$_3$). ^{13}C-NMR (100 MHz, DMSO-d_6) δ 164.65, 149.72, 141.16, 138.99, 134.81, 128.67, 128.22, 122.44, 121.92, 120.27, 112.62, 112.19, 31.21, 21.81.

1-benzyl-β-carboline-3-carbohydrazide (**7f**): The compound was obtained as a white solid with 94% yield. ^{1}H-NMR (400 MHz, DMSO-d_6) δ 9.54 (s, 1H, NH), 8.69 (s, 1H, ArH), 8.37 (d, J = 8.0 Hz, 1H, ArH), 7.67 (d, J = 8.4 Hz, 1H, ArH), 7.63–7.57 (m, 1H, ArH), 7.51 (d, J = 7.2 Hz, 2H, ArH), 7.34–7.25 (m, 4H, ArH), 7.20–7.15 (m, 1H, ArH), 4.60 (s, 2H, NH$_2$), 4.51 (s, 2H, ArCH$_2$). ^{13}C-NMR (100 MHz, DMSO-d_6) δ 164.43, 143.63, 141.29, 139.36, 139.33, 135.69, 129.36, 129.05, 128.92, 128.82, 128.76, 126.65, 122.60, 121.88, 120.44, 112.68, 112.59, 61.00.

9-n-butyl-β-carboline-3-carbohydrazide (**7i**): The compound was obtained as a light yellow solid with 80% yield. ^{1}H-NMR (400 MHz, DMSO-d_6) δ 9.70 (s, 1H, NH), 9.05 (d, J = 1.2 Hz, 1H, ArH), 8.84 (d, J = 0.8 Hz, 1H, ArH), 8.44 (d, J = 8.0 Hz, 1H, ArH), 7.77 (d, J = 8.4 Hz, 1H, ArH), 7.68–7.63 (m, 1H, ArH), 7.36–7.31 (m, 1H, ArH), 4.59–4.54 (m, 4H, NH$_2$, –CH$_2$CH$_2$CH$_2$CH$_3$), 1.85–1.76 (m, 2H, –CH$_2$CH$_2$CH$_2$CH$_3$), 1.35–1.24 (m, 2H, –CH$_2$CH$_2$CH$_2$CH$_3$), 0.88 (t, J = 7.2 Hz, 3H, –CH$_2$CH$_2$CH$_2$CH$_3$). ^{13}C-NMR (100 MHz, DMSO-d_6) δ 164.37, 141.62, 140.00, 137.58, 131.70, 129.16, 128.14, 122.87, 121.16, 120.55, 114.14, 110.97, 43.05, 31.38, 20.20, 14.14.

9-(3-phenylpropyl)-β-carboline-3-carbohydrazide (**7j**): The compound was obtained as a gray powder with 86% yield. ^{1}H-NMR (400 MHz, DMSO-d_6) δ 9.69 (s, 1H, NH), 9.00 (d, J = 1.2 Hz, 1H, ArH), 8.83 (d, J = 0.8 Hz, 1H, ArH), 8.44 (d, J = 8.0 Hz, 1H, ArH), 7.72 (d, J = 8.4 Hz, 1H, ArH), 7.68–7.62 (m, 1H, ArH), 7.37–7.31 (m, 1H, ArH), 7.29–7.22 (m, 2H, ArH), 7.21–7.13 (m, 3H, ArH), 4.60 (t, J = 7.2 Hz,

2H, ArCH$_2$CH$_2$CH$_2$), 4.56 (s, 2H, NH$_2$), 2.67 (t, J = 7.6 Hz, 2H, ArCH$_2$CH$_2$CH$_2$), 2.19–2.10 (m, 2H, ArCH$_2$CH$_2$CH$_2$). ^{13}C-NMR (101 MHz, DMSO-d_6) δ 164.34, 141.59, 141.55, 140.08, 137.52, 131.62, 129.20, 128.80, 128.58, 128.24, 126.37, 122.92, 121.22, 120.62, 114.16, 110.89, 43.03, 32.85, 30.84.

9-(4-fluorobenzyl)-β-carboline-3-carbohydrazide (**7l**): The compound was obtained as a white solid with 96% yield. ^{1}H-NMR (400 MHz, DMSO-d_6) δ 9.76 (s, 1H, NH), 9.11 (d, J = 1.2 Hz, 1H, ArH), 8.89 (d, J = 1.2 Hz, 1H, ArH), 8.48 (d, J = 7.6 Hz, 1H, ArH), 7.83 (d, J = 8.4 Hz, 1H, ArH), 7.68–7.63 (m, 1H, ArH), 7.39–7.34 (m, 1H, ArH), 7.33–7.29 (m, 2H, ArH), 7.17–7.11 (m, 2H, ArH), 5.85 (s, 2H, ArCH$_2$), 4.61 (s, 2H, NH$_2$). ^{13}C-NMR (100 MHz, DMSO-d_6) δ 164.30, 162.52 (d, J = 242.1 Hz), 141.65, 140.50, 137.58, 133.86 (d, J = 3.2 Hz), 131.90, 129.51, 129.43, 129.39, 128.59, 122.98, 121.17 (d, J = 46.3 Hz), 115.99 (d, J = 21.4 Hz), 114.25, 111.18, 45.88.

*3.4. General Procedure for the Preparation of Heterobivalent β-Carbolines **8a–ab***

We added β-carboline-3-carbaldehyde **6a–l** (1 mmol) to a solution of β-carboline-3-carbohydrazide **7a–l** (1 mmol) in EtOH (50 mL), and then the reaction mixture was refluxed for 5 h. The solution was allowed to cool to room temperature. Then, the precipitates formed and were filtered, and the crude product was recrystallized with ethanol to afford compounds **8a–ab**.

N'-((9H-pyrido[3,4-b]indol-3-yl)methylene)-9H-pyrido[3,4-b]indole-3-carbohydrazide (**8a**): The compound was obtained as a yellow solid with 93% yield. ^{1}H-NMR (400 MHz, DMSO-d_6) δ 12.13 (s, 1H, NH), 12.11 (s, 1H, NH), 9.28 (s, 1H, ArH), 9.16 (d, J = 0.8 Hz, 1H, ArH), 9.02 (s, 1H, ArH), 8.63 (s, 1H, ArH), 8.46 (d, J = 8.0 Hz, 1H, CH), 8.34 (d, J = 8.0 Hz, 1H, ArH), 7.80 (s, 1H, ArH), 7.73 (d, J = 8.0 Hz, 1H, ArH), 7.70 (d, J = 8.0 Hz, 1H, ArH), 7.68–7.61 (m, 2H, ArH), 7.39–7.32 (m, 2H, ArH). ^{13}C-NMR (100 MHz, DMSO-d_6) δ 162.68, 142.36, 141.65, 141.50, 140.77, 139.78, 137.85, 135.54, 133.55, 133.31, 129.49, 129.16, 128.80, 128.66, 122.81, 122.53, 121.47, 121.24, 120.74, 120.59, 118.99, 115.46, 112.99, 112.78. HRMS m/z calculated for C$_{24}$H$_{17}$N$_6$O$^+$ (M + H)$^+$: 405.1458; found 405.1459.

N'-((1-methyl-9H-pyrido[3,4-b]indol-3-yl)methylene)-9H-pyrido[3,4-b]indole-3-carbohydrazide (**8b**): The compound was obtained as a yellow solid with 84% yield. ^{1}H-NMR (400 MHz, DMSO-d_6) δ 12.15 (s, 1H, NH), 12.02 (s, 1H, NH), 9.08 (d, J = 0.8 Hz, 1H, ArH), 9.03 (s, 1H, ArH), 8.46 (d, J = 8.0 Hz, 1H, CH), 8.44 (s, 1H, ArH), 8.29 (d, J = 8.0 Hz, 1H, ArH), 7.74 (s, 1H, ArH), 7.73–7.69 (m, 2H, ArH), 7.66–7.60 (m, 2H, ArH), 7.34 (t, J = 7.6 Hz, 2H, ArH), 3.20 (s, 3H, CH$_3$). ^{13}C-NMR (100 MHz, DMSO-d_6) δ 162.54, 141.96, 141.86, 141.48, 141.36, 140.92, 139.96, 137.89, 134.27, 133.02, 129.17, 129.13, 128.75, 128.02, 122.80, 122.45, 121.71, 121.48, 120.66, 120.60, 117.15, 115.48, 112.88, 112.81, 20.96. HRMS m/z calculated for C$_{25}$H$_{19}$N$_6$O$^+$ (M + H)$^+$: 419.1615; found 419.1620.

N'-((1-isopropyl-9H-pyrido[3,4-b]indol-3-yl)methylene)-9H-pyrido[3,4-b]indole-3-carbohydrazide (**8c**): The compound was obtained as a yellow solid with 89% yield. ^{1}H-NMR (400 MHz, DMSO-d_6) δ 15.64 (s, 1H, CONH), 12.10 (s, 1H, NH), 12.05 (s, 1H, NH), 9.07 (s, 1H, ArH), 9.02 (s, 1H, ArH), 8.49 (d, J = 8.0 Hz, 1H, CH), 8.46 (s, 1H, ArH), 8.30 (d, J = 8.0 Hz, 1H, ArH), 7.78 (s, 1H, ArH), 7.75–7.69 (m, 2H, ArH), 7.67–7.62 (m, 2H, ArH), 7.38–7.33 (m, 2H, ArH), 4.02–3.91 (m, 1H, CH(CH$_3$)$_2$), 1.76 (d, J = 7.2 Hz, 6H, CH(CH$_3$)$_2$). ^{13}C-NMR (100 MHz, DMSO-d_6) δ 162.76, 151.03, 142.16, 141.51, 141.38, 141.17, 139.54, 138.00, 132.90, 132.63, 129.25, 129.20, 128.88, 128.60, 122.87, 122.35, 121.72, 121.45, 120.66, 120.62, 117.59, 115.71, 112.87, 112.82, 31.57, 21.63. HRMS m/z calculated for C$_{27}$H$_{23}$N$_6$O$^+$ (M + H)$^+$: 447.1928; found 447.1932.

N'-((1-phenyl-9H-pyrido[3,4-b]indol-3-yl)methylene)-9H-pyrido[3,4-b]indole-3-carbohydrazide (**8d**): The compound was obtained as a yellow solid with 96% yield. ^{1}H-NMR (400 MHz, DMSO-d_6) δ 15.58 (s, 1H, CONH), 12.06 (s, 1H, NH), 11.95 (s, 1H, NH), 8.99 (s, 1H, ArH), 8.65 (s, 1H, ArH), 8.44 (d, J = 8.4 Hz, 2H, CH, ArH), 8.35 (d, J = 8.0 Hz, 1H, ArH), 8.26–8.23 (m, 1H, ArH), 7.85 (s, 1H, ArH), 7.82–7.73 (m, 4H, ArH), 7.69–7.59 (m, 3H, ArH), 7.41–7.35 (m, 1H, ArH), 7.35–7.29 (m, 1H, ArH). ^{13}C-NMR (100 MHz, DMSO-d_6) δ 162.83, 142.56, 142.51, 142.15, 141.43, 140.85, 139.47, 137.75, 137.64, 133.10, 132.78, 130.10,

129.75, 129.47, 129.44, 129.32, 129.16, 128.59, 122.81, 122.36, 121.46, 121.41, 120.91, 120.57, 118.56, 115.52, 113.36, 112.74. HRMS m/z calculated for $C_{30}H_{21}N_6O^+$ (M + H)$^+$: 481.1771; found 481.1777.

N'-((1-(4-methoxyphenyl)-9H-pyrido[3,4-b]indol-3-yl)methylene)-9H-pyrido[3,4-b]indole-3-carbohydrazide (**8e**): The compound was obtained as a yellow solid with 91% yield. ^{1}H-NMR (400 MHz, DMSO-d_6) δ 15.53 (s, 1H, CONH), 12.12 (s, 1H, NH), 11.90 (s, 1H, NH), 9.01 (s, 1H, ArH), 8.59–8.57 (m, 2H, ArH), 8.45 (d, J = 7.6 Hz, 1H, CH), 8.33 (d, J = 7.6 Hz, 1H, ArH), 8.23 (d, J = 8.8 Hz, 2H, ArH), 7.83 (s, 1H, ArH), 7.74 (d, J = 8.4 Hz, 1H, ArH), 7.70 (d, J = 8.4 Hz, 1H, ArH), 7.68–7.60 (m, 2H, ArH), 7.39–7.31 (m, 4H, ArH), 4.02 (s, 3H, OCH$_3$). ^{13}C-NMR (100 MHz, DMSO-d_6) δ 162.86, 160.53, 142.53, 142.45, 142.07, 141.48, 140.97, 139.51, 137.76, 132.92, 132.80, 131.10, 130.00, 129.88, 129.30, 129.18, 128.65, 122.82, 122.29, 121.50, 121.42, 120.84, 120.58, 118.19, 115.63, 114.72, 113.36, 112.75, 56.05. HRMS m/z calculated for $C_{31}H_{23}N_6O_2^+$ (M + H)$^+$: 511.1877; found 511.1878.

N'-((1-benzyl-9H-pyrido[3,4-b]indol-3-yl)methylene)-9H-pyrido[3,4-b]indole-3-carbohydrazide (**8f**): The compound was obtained as a yellow solid with 82% yield. ^{1}H-NMR (400 MHz, DMSO-d_6) δ 12.19 (s, 1H, NH), 11.95 (s, 1H, NH), 9.06 (s, 1H, ArH), 8.88 (s, 1H, ArH), 8.50 (s, 1H, ArH), 8.47 (d, J = 7.6 Hz, 1H, CH), 8.29 (d, J = 7.6 Hz, 1H, ArH), 7.78–7.72 (m, 4H, ArH), 7.69–7.59 (m, 4H, ArH), 7.37–7.32 (m, 2H, ArH), 7.23 (t, J = 7.6 Hz, 2H, ArH), 7.13 (t, J = 7.2 Hz, 1H, ArH), 4.90 (s, 2H, ArCH$_2$). ^{13}C-NMR (100 MHz, DMSO-d_6) δ 162.61, 144.08, 142.15, 141.48, 141.44, 140.83, 139.92, 139.26, 137.88, 133.58, 132.82, 131.50129.39, 129.23, 129.11, 128.95, 128.82, 126.84, 122.85, 122.44, 121.68, 121.45, 120.80, 120.64, 117.69, 115.64, 112.96, 112.78. HRMS m/z calculated for $C_{31}H_{23}N_6O^+$ (M + H)$^+$: 495.1928; found 495.1932.

N'-((1-(2-chlorophenyl)-9H-pyrido[3,4-b]indol-3-yl)methylene)-9H-pyrido[3,4-b]indole-3-carbohydrazide (**8g**): The compound was obtained as a yellow solid with 81% yield. ^{1}H-NMR (400 MHz, DMSO-d_6) δ 12.04 (s, 1H, NH), 11.85 (s, 1H, NH), 8.92 (s, 1H, ArH), 8.70 (s, 1H, ArH), 8.41 (d, J = 8.0 Hz, 1H, CH), 8.37 (d, J = 8.0 Hz, 1H, ArH), 7.96 (d, J = 1.2 Hz, 1H, ArH), 7.90–7.85 (m, 3H, ArH), 7.84–7.79 (m, 1H, ArH), 7.77–7.72 (m, 1H, ArH), 7.70–7.55 (m, 5H, ArH), 7.40–7.35 (m, 1H, ArH), 7.31 (t, J = 7.2 Hz, 1H, ArH). ^{13}C-NMR (100 MHz, DMSO-d_6) δ 162.71, 142.01, 141.37, 140.58, 140.53, 139.50, 137.61, 136.59, 133.76, 133.32, 132.75, 132.47, 131.09, 130.30, 129.58, 129.55, 129.09, 128.46, 127.95, 122.75, 122.50, 121.40, 121.38, 120.86, 120.52, 118.77, 115.25, 113.09, 112.70. HRMS m/z calculated for $C_{30}H_{20}ClN_6O^+$ (M + H)$^+$: 515.1382; found 515.1386.

N'-((9-methyl-9H-pyrido[3,4-b]indol-3-yl)methylene)-9H-pyrido[3,4-b]indole-3-carbohydrazide (**8h**): The compound was obtained as an ivory solid with 84% yield. ^{1}H-NMR (400 MHz, DMSO-d_6) δ 12.14 (s, 1H, NH), 9.45 (s, 1H, ArH), 9.15 (s, 1H, ArH), 9.01 (s, 1H, ArH), 8.64 (s, 1H, ArH), 8.46 (d, J = 8.0 Hz, 1H, CH), 8.36 (d, J = 8.0 Hz, 1H, ArH), 7.83 (d, J = 8.4 Hz, 1H, ArH), 7.80 (s, 1H, ArH), 7.74 (d, J = 7.6 Hz, 1H, ArH), 7.70 (d, J = 8.4 Hz, 1H, ArH), 7.63 (t, J = 7.6 Hz, 1H, ArH), 7.41 (t, J = 7.6 Hz, 1H, ArH), 7.34 (t, J = 7.4 Hz, 1H, ArH), 4.15 (s, 3H, CH$_3$). ^{13}C-NMR (100 MHz, DMSO-d_6) δ 162.68, 142.51, 142.47, 141.50, 140.63, 139.79, 137.86, 136.16, 133.45, 132.20, 129.61, 129.18, 128.68, 128.32, 122.82, 122.60, 121.47, 120.92, 120.60, 118.70, 115.47, 112.77, 111.11, 30.24. HRMS calculated for $C_{25}H_{19}N_6O$ [M + H]$^+$: 419.1615; found 419.1615.

N'-((9-n-butyl-9H-pyrido[3,4-b]indol-3-yl)methylene)-9H-pyrido[3,4-b]indole-3-carbohydrazide (**8i**): The compound was obtained as an ivory solid with 79% yield. ^{1}H-NMR (400 MHz, DMSO-d_6) δ 12.12 (s, 1H, NH), 9.46 (s, 1H, ArH), 9.17 (s, 1H, ArH), 9.02 (s, 1H, ArH), 8.64 (s, 1H, ArH), 8.47 (d, J = 7.6 Hz, 1H, CH), 8.36 (d, J = 8.0 Hz, 1H, ArH), 7.84 (d, J = 8.4 Hz, 1H, ArH), 7.80 (s, 1H, ArH), 7.75–7.68 (m, 2H, ArH), 7.63 (t, J = 7.6 Hz, 1H, ArH), 7.40 (t, J = 7.6 Hz, 1H, ArH), 7.34 (t, J = 7.6 Hz, 1H, ArH), 4.67 (t, J = 7.2 Hz, 2H, –CH$_2$CH$_2$CH$_2$CH$_3$), 1.92–1.84 (m, 2H, –CH$_2$CH$_2$CH$_2$CH$_3$), 1.42–1.31 (m, 2H, –CH$_2$CH$_2$CH$_2$CH$_3$), 0.92 (t, J = 7.6 Hz, 3H, –CH$_2$CH$_2$CH$_2$CH$_3$). ^{13}C-NMR (100 MHz, DMSO-d_6) δ 162.70, 142.48, 141.84, 141.51, 140.60, 139.77, 137.85, 135.67, 133.51, 132.16, 129.61, 129.19, 128.69, 128.42, 122.82, 122.69, 121.48, 121.00, 120.90, 120.60, 118.83, 115.47, 112.78, 111.30, 43.26, 31.50, 20.27, 14.19. HRMS calculated for $C_{28}H_{25}N_6O$ [M + H]$^+$: 461.2084; found 461.2088.

N'-((9-(3-phenylpropyl)-9H-pyrido[3,4-b]indol-3-yl)methylene)-9H-pyrido[3,4-b]indole-3-carbohydrazide (**8j**): The compound was obtained as a yellow solid with 84% yield. ^{1}H-NMR (400 MHz, DMSO-d_6) δ 12.15 (s, 1H, NH), 9.35 (s, 1H, ArH), 9.15 (d, *J* = 1.2 Hz, 1H, ArH), 9.03 (s, 1H, ArH), 8.64 (s, 1H, ArH), 8.47 (d, *J* = 7.6 Hz, 1H, CH), 8.36 (d, *J* = 8.0 Hz, 1H, ArH), 7.81–7.78 (m, 2H, ArH), 7.73–7.68 (m, 2H, ArH), 7.66–7.63 (m, 1H, ArH), 7.42–7.38 (m, 1H, ArH), 7.37–7.34 (m, 1H, ArH), 7.32–7.27 (m, 3H, ArH), 7.25–7.20 (m, 3H, ArH), 4.69 (t, *J* = 7.2 Hz, 2H, ArCH$_2$CH$_2$CH$_2$), 2.74 (t, *J* = 7.2 Hz, 2H, ArCH$_2$CH$_2$CH$_2$), 2.27–2.18 (m, 2H, ArCH$_2$CH$_2$CH$_2$). ^{13}C-NMR (100 MHz, DMSO-d_6) δ 162.70, 142.55, 141.79, 141.57, 141.51, 140.59, 139.77, 137.85, 135.59, 133.44, 131.99, 129.63, 129.20, 128.84, 128.82, 128.69, 128.62, 126.43, 122.83, 122.72, 121.48, 121.06, 120.96, 120.61, 118.85, 115.50, 112.78, 111.19, 43.12, 32.84, 30.87. HRMS calculated for C$_{33}$H$_{27}$N$_6$O [M + H]$^+$: 523.2241; found 523.2242.

N'-((9-benzyl-9H-pyrido[3,4-b]indol-3-yl)methylene)-9H-pyrido[3,4-b]indole-3-carbohydrazide (**8k**): The compound was obtained as a yellow solid with 86% yield. ^{1}H-NMR (400 MHz, DMSO-d_6) δ 12.32 (s, 1H, NH), 12.06 (s, 1H, NH), 9.13 (d, *J* = 0.8 Hz, 1H, ArH), 9.04 (s, 1H, ArH), 9.01 (d, *J* = 1.2 Hz, 1H, ArH), 8.92 (s, 1H, ArH), 8.83 (d, *J* = 0.8 Hz, 1H, ArH), 8.48 (t, *J* = 8.0 Hz, 2H, ArH, CH), 7.79 (d, *J* = 8.4 Hz, 1H, ArH), 7.71 (d, *J* = 8.4 Hz, 1H, ArH), 7.67–7.63 (m, 1H, ArH), 7.38–7.23 (m, 8H, ArH), 5.83 (s, 2H, ArCH$_2$). ^{13}C-NMR (100 MHz, DMSO-d_6) δ 161.86, 150.06, 144.04, 141.98, 141.53, 139.51, 137.86, 137.69, 137.05, 132.81, 129.36, 129.26, 129.19, 128.79, 128.76, 128.00, 127.35, 127.31, 122.96, 122.90, 121.43, 121.30, 120.74, 120.64, 115.62, 112.80, 111.71, 111.15, 46.53. HRMS calculated for C$_{31}$H$_{23}$N$_6$O [M + H]$^+$: 495.1928; found 495.1922.

N'-((9-(4-fluorobenzyl)-9H-pyrido[3,4-b]indol-3-yl)methylene)-9H-pyrido[3,4-b]indole-3-carbohydrazide (**8l**): The compound was obtained as a light gray solid with 92% yield. ^{1}H-NMR (400 MHz, DMSO-d_6) δ 12.31 (s, 1H, NH), 12.04 (s, 1H, NH), 9.15 (d, *J* = 0.8 Hz, 1H, ArH), 9.03 (s, 1H, ArH), 9.00 (d, *J* = 1.2 Hz, 1H, ArH), 8.91 (s, 1H, ArH), 8.82 (d, *J* = 1.2 Hz, 1H, ArH), 8.48 (dd, *J* = 8.0, 4.8 Hz, 2H, CH, ArH), 7.81 (d, *J* = 8.4 Hz, 1H, ArH), 7.71 (d, *J* = 8.4 Hz, 1H, ArH), 7.68–7.61 (m, 2H, ArH), 7.38–7.30 (m, 4H, ArH), 7.17–7.11 (m, 2H, ArH), 5.82 (s, 2H, ArCH$_2$). ^{13}C-NMR (100 MHz, DMSO-d_6) δ 161.96 (d, *J* = 241.8 Hz), 161.84, 150.05, 144.12, 141.86, 141.53, 139.51, 137.85, 136.94, 133.92 (d, *J* = 3.1 Hz), 132.80, 129.53, 129.45, 129.40, 129.24, 128.81, 128.79, 123.00 (d, *J* = 9.6 Hz), 121.43, 121.34, 120.80, 120.64, 116.11 (d, *J* = 21.3 Hz), 115.62, 112.79, 111.70, 111.12, 45.79. HRMS calculated for C$_{31}$H$_{22}$FN$_6$O [M + H]$^+$: 513.1834; found 513.1832.

N'-((9H-pyrido[3,4-b]indol-3-yl)methylene)-1-methyl-9H-pyrido[3,4-b]indole-3-carbohydrazide (**8m**): The compound was obtained as a yellow solid with 85% yield. ^{1}H-NMR (400 MHz, DMSO-d_6) δ 12.05 (s, 1H, NH), 12.02 (s, 1H, NH), 11.83 (s, 1H, NH), 8.95 (d, *J* = 0.8 Hz, 1H, ArH), 8.90 (s, 1H, ArH), 8.85 (s, 1H, ArH), 8.78 (s, 1H, ArH), 8.42 (t, *J* = 8.4 Hz, 2H, CH, ArH), 7.69–7.57 (m, 4H, ArH), 7.35–7.28 (m, 2H, ArH), 2.94 (s, 3H, CH$_3$). ^{13}C-NMR (100 MHz, DMSO-d_6) δ 161.84, 150.09, 143.33, 141.66, 141.61, 141.28, 138.85, 136.87, 136.60, 133.94, 129.05, 128.88, 128.76, 127.95, 122.77, 122.70, 121.90, 121.33, 120.55, 120.26, 113.68, 112.71, 112.63, 111.84, 20.87. HRMS calculated for C$_{25}$H$_{19}$N$_6$O [M + H]$^+$: 419.1615; found 419.1617.

N'-((9H-pyrido[3,4-b]indol-3-yl)methylene)-1-isopropyl-9H-pyrido[3,4-b]indole-3-carbohydrazide (**8n**): The compound was obtained as a yellow solid with 91% yield. ^{1}H-NMR (400 MHz, DMSO-d_6) δ 15.65 (s, 1H, CONH), 12.17 (s, 1H, NH), 12.06 (s, 1H, NH), 9.22 (s, 1H, ArH), 8.88 (s, 1H, ArH), 8.65 (s, 1H, ArH), 8.42 (d, *J* = 7.6 Hz, 1H, CH), 8.34 (d, *J* = 7.6 Hz, 1H, ArH), 7.80 (s, 1H, ArH), 7.74–7.60 (m, 4H, ArH), 7.40–7.30 (m, 2H, ArH), 3.92–3.82 (m, 1H, CH(CH$_3$)$_2$), 1.68 (d, *J* = 6.8 Hz, 6H, CH(CH$_3$)$_2$). ^{13}C-NMR (100 MHz, DMSO-d_6) δ 162.96, 149.74, 142.52, 141.59, 141.29, 140.56, 139.03, 135.63, 135.24, 132.99, 129.52, 128.88, 128.79, 128.53, 122.59, 121.98, 121.19, 120.75, 120.50, 119.37, 113.80, 112.96, 112.71, 31.12, 22.22. HRMS calculated for C$_{27}$H$_{23}$N$_6$O [M + H]$^+$: 447.1928; found 447.1926.

N'-((9H-pyrido[3,4-b]indol-3-yl)methylene)-1-phenyl-9H-pyrido[3,4-b]indole-3-carbohydrazide (**8o**): The compound was obtained as a yellow solid with 91% yield. ^{1}H-NMR (400 MHz, DMSO-d_6) δ 15.97 (s, 1H, CONH), 12.27 (s, 1H, NH), 12.03 (s, 1H, NH), 9.05 (d, *J* = 5.2 Hz, 2H, ArH), 8.63 (s, 1H, ArH), 8.50

(d, *J* = 8.0 Hz, 1H, CH), 8.39 (d, *J* = 6.8 Hz, 2H, ArH), 8.33 (d, *J* = 7.6 Hz, 1H, ArH), 7.91 (t, *J* = 7.6 Hz, 2H, ArH), 7.81 (s, 1H, ArH), 7.79–7.72 (m, 3H, ArH), 7.68–7.62 (m, 2H, ArH), 7.39–7.34 (m, 2H, ArH). ^{13}C-NMR (100 MHz, DMSO-d_6) δ 162.62, 142.40, 142.07, 141.63, 141.12, 140.82, 139.66, 138.23, 135.57, 134.93, 132.94, 130.58, 129.75, 129.51, 129.20, 128.74, 122.61, 122.56, 121.71, 121.18, 120.84, 120.72, 119.28, 114.68, 113.22, 112.93. HRMS calculated for $C_{30}H_{21}N_6O$ [M + H]$^+$: 481.1771; found 481.1772.

N'-((9H-pyrido[3,4-b]indol-3-yl)methylene)-1-(4-methoxyphenyl)-9H-pyrido[3,4-b]indole-3-carbohydrazide (**8p**): The compound was obtained as a yellow solid with 87% yield. ^{1}H-NMR (400 MHz, DMSO-d_6) δ 15.89 (s, 1H, CONH), 12.27 (s, 1H, NH), 11.96 (s, 1H, NH), 9.07 (s, 1H, ArH), 9.00 (s, 1H, ArH), 8.64 (s, 1H, ArH), 8.47 (d, *J* = 7.6 Hz, 1H, CH), 8.37–8.32 (m, 2H, ArH), 7.81 (s, 1H, ArH), 7.75 (dd, *J* = 8.0, 4.4 Hz, 2H, ArH), 7.69–7.61 (m, 2H, ArH), 7.44 (d, *J* = 8.8 Hz, 2H, ArH), 7.39–7.32 (m, 2H, ArH), 4.04 (s, 3H, OCH$_3$). ^{13}C-NMR (100 MHz, DMSO-d_6) δ 162.69, 160.72, 142.41, 142.01, 141.63, 141.10, 140.77, 139.53, 135.60, 134.70, 133.00, 130.57, 130.49, 130.34, 129.52, 129.08, 128.77, 122.58, 122.54, 121.75, 121.20, 120.78, 120.73, 119.33, 114.92, 114.17, 113.22, 112.94, 56.01. HRMS calculated for $C_{31}H_{23}N_6O_2$ [M + H]$^+$: 511.1877; found 511.1874.

N'-((9H-pyrido[3,4-b]indol-3-yl)methylene)-1-benzyl-9H-pyrido[3,4-b]indole-3-carbohydrazide (**8q**): The compound was obtained as an ivory solid with 83% yield. ^{1}H-NMR (400 MHz, DMSO-d_6) δ 12.14 (s, 1H, CONH), 11.99 (s, 1H, NH), 11.86 (s, 1H, NH), 8.98 (s, 1H, ArH), 8.91 (d, *J* = 2.4 Hz, 2H, ArH), 8.81 (s, 1H, ArH), 8.45 (d, *J* = 8.0 Hz, 1H, CH), 8.40 (d, *J* = 7.6 Hz, 1H, ArH), 7.72–7.66 (m, 2H, ArH), 7.65–7.55 (m, 4H, ArH), 7.35–7.28 (m, 4H, ArH), 7.21 (t, *J* = 7.2 Hz, 1H, ArH), 4.65 (s, 2H, ArCH$_2$). ^{13}C-NMR (100 MHz, DMSO-d_6) δ 161.65, 150.29, 143.62, 143.18, 141.67, 141.39, 139.26, 138.89, 136.90, 136.13, 133.99, 129.39, 129.08, 129.04, 128.96, 128.83, 128.78, 126.71, 122.77, 122.70, 121.89, 121.32, 120.66, 120.27, 113.96, 112.77, 112.64, 111.94. HRMS calculated for $C_{31}H_{23}N_6O$ [M + H]$^+$: 495.1928; found 495.1929.

N'-((9H-pyrido[3,4-b]indol-3-yl)methylene)-9-methyl-9H-pyrido[3,4-b]indole-3-carbohydrazide (**8r**): The compound was obtained as a light yellow solid with 81% yield. ^{1}H-NMR (400 MHz, DMSO-d_6) δ 12.31 (s, 1H, CONH), 11.82 (s, 1H, NH), 9.14 (s, 1H, ArH), 9.03 (s, 1H, ArH), 8.94 (d, *J* = 0.8 Hz, 1H, ArH), 8.92 (s, 1H, ArH), 8.78 (s, 1H, ArH), 8.52 (d, *J* = 7.6 Hz, 1H, CH), 8.41 (d, *J* = 8.0 Hz, 1H, ArH), 7.79 (d, *J* = 8.4 Hz, 1H, ArH), 7.73–7.68 (m, 1H, ArH), 7.64 (d, *J* = 8.4 Hz, 1H, ArH), 7.62–7.57 (m, 1H, ArH), 7.41 – 7.36 (m, 1H, ArH), 7.32–7.28 (m, 1H, ArH), 4.10 (s, 3H, CH$_3$). ^{13}C-NMR (100 MHz, DMSO-d_6) δ 161.77, 150.35, 143.37, 142.39, 141.65, 139.73, 138.38, 136.86, 133.88, 131.51, 129.35, 129.04, 128.75, 128.36, 122.96, 122.70, 121.34, 121.09, 120.81, 120.25, 115.37, 112.62, 111.79, 110.90, 30.18. HRMS calculated for $C_{25}H_{19}N_6O$ [M + H]$^+$: 419.1615; found 419.1618.

N'-((9H-pyrido[3,4-b]indol-3-yl)methylene)-9-n-butyl-9H-pyrido[3,4-b]indole-3-carbohydrazide (**8s**): The compound was obtained as a light yellow solid with 92% yield. ^{1}H-NMR (400 MHz, DMSO-d_6) δ 12.28 (s, 1H, CONH), 11.82 (s, 1H, NH), 9.15 (d, *J* = 0.8 Hz, 1H, ArH), 9.04 (d, *J* = 0.8 Hz, 1H, ArH), 8.95 (d, *J* = 0.8 Hz, 1H, ArH), 8.92 (s, 1H, ArH), 8.79 (s, 1H, ArH), 8.52 (d, *J* = 7.6 Hz, 1H, CH), 8.41 (d, *J* = 8.0 Hz, 1H, ArH), 7.81 (d, *J* = 8.4 Hz, 1H, ArH), 7.71–7.69 (m, 1H, ArH), 7.65 (d, *J* = 8.0 Hz, 1H, ArH), 7.62–7.57 (m, 1H, ArH), 7.40–7.35 (m, 1H, ArH), 7.32–7.28 (m, 1H, ArH), 4.63 (t, *J* = 6.8 Hz, 2H, –CH$_2$CH$_2$CH$_2$CH$_3$), 1.88–1.80 (m, 2H, –CH$_2$CH$_2$CH$_2$CH$_3$), 1.36–1.26 (m, 2H, –CH$_2$CH$_2$CH$_2$CH$_3$), 0.89 (t, *J* = 7.2 Hz, 3H, –CH$_2$CH$_2$CH$_2$CH$_3$). ^{13}C-NMR (100 MHz, DMSO-d_6) δ 161.76, 150.35, 143.36, 141.75, 141.65, 139.72, 137.94, 136.87, 133.89, 131.55, 129.35, 129.04, 128.75, 128.44, 123.07, 122.69, 121.34, 121.19, 120.80, 120.25, 115.45, 112.62, 111.80, 111.11, 49.07, 31.41, 20.21, 14.17. HRMS calculated for $C_{28}H_{25}N_6O$ [M + H]$^+$: 461.2084; found 461.2085.

N'-((9H-pyrido[3,4-b]indol-3-yl)methylene)-9-(3-phenylpropyl)-9H-pyrido[3,4-b]indole-3-carbohydrazide (**8t**): The compound was obtained as a yellow solid with 88% yield. ^{1}H-NMR (400 MHz, DMSO-d_6) δ 12.32 (s, 1H, CONH), 11.85 (s, 1H, NH), 9.10 (s, 1H, ArH), 9.06 (s, 1H, ArH), 8.97 (s, 1H, ArH), 8.94 (s, 1H, ArH), 8.80 (s, 1H, ArH), 8.52 (d, *J* = 7.6 Hz, 1H, CH), 8.39 (d, *J* = 8.0 Hz, 1H, ArH), 7.76 (d, *J* = 8.0 Hz, 1H, ArH), 7.71–7.64 (m, 2H, ArH), 7.62–7.58 (m, 1H, ArH), 7.38 (t, *J* = 7.6 Hz, 1H, ArH), 7.32–7.25 (m, 3H, ArH), 7.19–7.16 (m, 3H, ArH), 4.65 (t, *J* = 7.2 Hz, 2H, ArCH$_2$CH$_2$CH$_2$), 2.69 (t, *J* = 7.6 Hz, 2H,

ArCH$_2$CH$_2$CH$_2$), 2.22–2.14 (m, 2H, ArCH$_2$CH$_2$CH$_2$). ^{13}C-NMR (100 MHz, DMSO-d_6) δ 161.81, 150.40, 143.37, 141.70, 141.64, 141.50, 139.79, 137.86, 136.88, 133.94, 131.43, 129.37, 129.03, 128.81, 128.73, 128.58, 128.55, 126.39, 123.09, 122.67, 121.34, 121.25, 120.85, 120.24, 115.49, 112.63, 111.84, 110.99, 43.06, 32.85, 30.82. HRMS calculated for C$_{33}$H$_{27}$N$_6$O [M + H]$^+$: 523.2241; found 523.2242.

N'-((9H-pyrido[3,4-b]indol-3-yl)methylene)-9-benzyl-9H-pyrido[3,4-b]indole-3-carbohydrazide (**8u**): The compound was obtained as a light yellow solid with 80% yield. ^{1}H-NMR (400 MHz, DMSO-d_6) δ 12.28 (s, 1H, CONH), 11.82 (s, 1H, NH), 9.19 (s, 1H, ArH), 9.07 (s, 1H, ArH), 8.95 (s, 1H, ArH), 8.90 (s, 1H, ArH), 8.78 (s, 1H, ArH), 8.55 (d, *J* = 8.0 Hz, 1H, CH), 8.41 (d, *J* = 8.0 Hz, 1H, ArH), 7.84 (d, *J* = 8.0 Hz, 1H, ArH), 7.70–7.63 (m, 2H, ArH, NH), 7.62–7.57 (m, 1H, ArH), 7.39 (t, *J* = 7.5 Hz, 1H, ArH), 7.34–7.24 (m, 6H, ArH), 5.91 (s, 2H, ArCH$_2$). ^{13}C-NMR (100 MHz, DMSO-d_6) δ 161.70, 150.39, 143.33, 141.89, 141.64, 140.14, 138.04, 137.60, 136.87, 133.89, 131.84, 129.54, 129.23, 129.04, 128.84, 128.75, 128.09, 127.41, 123.17, 122.70, 121.40, 121.34, 121.11, 120.25, 115.52, 112.63, 111.80, 111.35, 46.73. HRMS calculated for C$_{31}$H$_{23}$N$_6$O [M + H]$^+$: 495.1928; found 495.1927.

N'-((9H-pyrido[3,4-b]indol-3-yl)methylene)-9-(4-fluorobenzyl)-9H-pyrido[3,4-b]indole-3-carbohydrazide (**8v**): The compound was obtained as a light yellow solid with 90% yield. ^{1}H-NMR (400 MHz, DMSO-d_6) δ 12.27 (s, 1H, CONH), 11.82 (s, 1H, NH), 9.20 (d, *J* = 1.2 Hz, 1H, ArH), 9.06 (d, *J* = 1.2 Hz, 1H, ArH), 8.94 (d, *J* = 1.2 Hz, 1H, ArH), 8.90 (s, 1H, ArH), 8.78 (s, 1H, ArH), 8.55 (d, *J* = 7.6 Hz, 1H, CH), 8.41 (d, *J* = 8.0 Hz, 1H, ArH), 7.86 (d, *J* = 8.4 Hz, 1H, ArH), 7.71–7.57 (m, 3H, ArH), 7.39 (t, *J* = 7.6 Hz, 1H, ArH), 7.35–7.28 (m, 3H, ArH), 7.19–7.12 (m, 2H, ArH), 5.90 (s, 2H, ArCH$_2$). ^{13}C-NMR (100 MHz, DMSO-d_6) δ 162.00 (d, *J* = 242.1 Hz), 161.68, 150.43, 143.34, 141.77, 141.64, 140.21, 137.94, 136.88, 133.90, 133.82 (d, *J* = 3 Hz), 131.81, 129.59, 129.51, 129.04, 128.89, 128.74, 123.20, 122.70, 121.39 (d, *J* = 10.2 Hz), 121.17, 120.25, 116.05 (d, *J* = 21.3 Hz), 115.52, 112.62, 111.80, 111.32, 49.07. HRMS calculated for C$_{31}$H$_{22}$FN$_6$O [M + H]$^+$: 513.1834; found 513.1835.

9-*n*-butyl-*N'*-((9-*n*-butyl-9H-pyrido[3,4-b]indol-3-yl)methylene)-9H-pyrido[3,4-b]indole-3-carbohydrazide (**8w**): The compound was obtained as a yellow solid with 91% yield. ^{1}H-NMR (400 MHz, DMSO-d_6) δ 12.28 (s, 1H, CONH), 9.15 (s, 1H, ArH), 9.10 (s, 1H, ArH), 9.03 (s, 1H, ArH), 8.91 (s, 1H, ArH), 8.80 (s, 1H, ArH), 8.52 (d, *J* = 8.0 Hz, 1H, CH), 8.45 (d, *J* = 8.0 Hz, 1H, ArH), 7.82 (d, *J* = 8.4 Hz, 1H, ArH), 7.76 (d, *J* = 8.4 Hz, 1H, ArH), 7.71–7.64 (m, 2H, ArH), 7.40–7.32 (m, 2H, ArH), 4.63 (t, *J* = 6.8 Hz, 2H, –CH$_2$CH$_2$CH$_2$CH$_3$), 4.55 (t, *J* = 6.8 Hz, 2H, –CH$_2$CH$_2$CH$_2$CH$_3$), 1.88–1.79 (m, 4H, –CH$_2$CH$_2$CH$_2$CH$_3$), 1.37–1.27 (m, 4H, –CH$_2$CH$_2$CH$_2$CH$_3$), 0.92–0.88 (m, 6H, –CH$_2$CH$_2$CH$_2$CH$_3$). ^{13}C-NMR (100 MHz, DMSO-d_6) δ 161.75, 150.25, 143.59, 141.85, 141.75, 139.70, 137.95, 136.96, 132.61, 131.57, 129.36, 129.17, 128.44, 128.40, 123.08, 122.88, 121.19, 121.09, 120.80, 120.40, 115.45, 111.64, 111.12, 110.90, 43.12, 42.99, 31.42, 31.34, 20.24, 20.22, 14.18. HRMS calculated for C$_{32}$H$_{33}$N$_6$O [M + H]$^+$: 517.2710; found 517.2706.

9-*n*-butyl-*N'*-((9-*n*-butyl-9H-pyrido[3,4-b]indol-3-yl)methylene)-1-methyl-9H-pyrido[3,4-b]indole-3-carbohydrazide (**8x**): The compound was obtained as a yellow solid with 88% yield. ^{1}H-NMR (400 MHz, CDCl$_3$) δ 11.32 (s, 1H, CONH), 9.04 (s, 1H, ArH), 8.92 (s, 1H, ArH), 8.84 (s, 1H, ArH), 8.59 (s, 1H, CH), 8.25 (d, *J* = 8.0 Hz, 1H, ArH), 8.22 (d, *J* = 8.0 Hz, 1H, ArH), 7.64–7.59 (m, 2H, ArH), 7.47 (dd, *J* = 8.4, 2.8 Hz, 2H, ArH), 7.35 (t, *J* = 7.6 Hz, 2H, ArH), 4.54 (t, *J* = 7.6 Hz, 2H, –CH$_2$CH$_2$CH$_2$CH$_3$), 4.36 (t, *J* = 7.2 Hz, 2H, –CH$_2$CH$_2$CH$_2$CH$_3$), 3.09 (s, 3H, CH$_3$), 1.94–1.81 (m, 4H, –CH$_2$CH$_2$CH$_2$CH$_3$), 1.53–1.36 (m, 4H, –CH$_2$CH$_2$CH$_2$CH$_3$), 1.00 (t, *J* = 7.6 Hz, 3H, –CH$_2$CH$_2$CH$_2$CH$_3$), 0.97 (t, *J* = 7.2 Hz, 3H, –CH$_2$CH$_2$CH$_2$CH$_3$). ^{13}C-NMR (100 MHz, CDCl$_3$) δ 161.29, 148.92, 142.53, 141.84, 141.81, 139.67, 137.77, 137.00, 136.59, 130.81, 129.58, 129.19, 128.72, 128.53, 122.59, 121.98, 121.69, 121.48, 120.56, 120.27, 113.75, 113.00, 109.97, 109.60, 44.84, 43.34, 33.02, 31.27, 23.71, 20.52, 20.21, 13.87, 13.81. HRMS calculated for C$_{33}$H$_{35}$N$_6$O [M + H]$^+$: 531.2867; found 531.2873.

9-benzyl-*N'*-((9-*n*-butyl-9H-pyrido[3,4-b]indol-3-yl)methylene)-1-methyl-9H-pyrido[3,4-b]indole-3-carbohydrazide (**8y**): The compound was obtained as a yellow solid with 84% yield. ^{1}H-NMR (400 MHz, CDCl$_3$) δ 11.27 (s, 1H, CONH), 9.03 (s, 1H, ArH), 8.99 (s, 1H, ArH), 8.84 (s, 1H, ArH), 8.55 (s, 1H, CH), 8.26 (dd, *J* = 10.4, 8.0 Hz, 2H, ArH), 7.64–7.55 (m, 2H, ArH), 7.47 (d, *J* = 8.4 Hz, 1H, ArH), 7.41–7.34 (m,

3H, ArH), 7.33–7.23 (m, 3H, ArH), 7.01–6.95 (m, 2H, ArH), 5.83 (s, 2H, ArCH$_2$), 4.37 (t, J = 7.2 Hz, 2H, –CH$_2$CH$_2$CH$_2$CH$_3$), 2.93 (s, 3H, CH$_3$), 1.94–1.87 (m, 2H, –CH$_2$CH$_2$CH$_2$CH$_3$), 1.47–1.37 (m, 2H, –CH$_2$CH$_2$CH$_2$CH$_3$), 0.96 (t, J = 7.2 Hz, 3H, –CH$_2$CH$_2$CH$_2$CH$_3$). ^{13}C-NMR (100 MHz, CDCl$_3$) δ 161.19, 149.29, 142.63, 142.26, 141.78, 140.15, 138.33, 137.49, 137.06, 131.05, 129.80, 129.13, 129.07, 128.91, 128.64, 127.72, 125.32, 122.56, 122.05, 121.75, 121.51, 121.03, 120.23, 113.81, 113.00, 110.11, 109.59, 48.30, 43.32, 31.27, 23.32, 20.52, 13.82. HRMS calculated for C$_{36}$H$_{33}$N$_6$O [M + H]$^+$: 565.2710; found 565.2715.

N'-((9-benzyl-1-methyl-9H-pyrido[3,4-b]indol-3-yl)methylene)-9-n-butyl-9H-pyrido[3,4-b]indole-3-carbohydrazide (**8z**): The compound was obtained as a yellow solid with 85% yield. ^{1}H-NMR (400 MHz, CDCl$_3$) δ 11.22 (s, 1H, CONH), 9.06 (s, 1H, ArH), 8.92 (s, 1H, ArH), 8.78 (s, 1H, ArH), 8.50 (s, 1H, CH), 8.25 (t, J = 7.2 Hz, 2H, ArH), 7.65–7.61 (m, 1H, ArH), 7.56–7.48 (m, 2H, ArH), 7.39–7.32 (m, 3H, ArH), 7.31–7.23 (m, 3H, ArH), 7.00 (d, J = 6.4 Hz, 2H, ArH), 5.76 (s, 2H, ArCH$_2$), 4.39 (t, J = 7.2 Hz, 2H, –CH$_2$CH$_2$CH$_2$CH$_3$), 2.88 (s, 3H, CH$_3$), 1.94–1.87 (m, 2H, –CH$_2$CH$_2$CH$_2$CH$_3$), 1.46–1.36 (m, 2H, –CH$_2$CH$_2$CH$_2$CH$_3$), 0.97 (t, J = 7.2 Hz, 3H, –CH$_2$CH$_2$CH$_2$CH$_3$). ^{13}C-NMR (100 MHz, CDCl$_3$) δ 161.36, 149.11, 142.47, 142.18, 141.57, 140.82, 138.81, 137.87, 137.70, 136.08, 130.00, 129.88, 129.05, 128.95, 128.81, 128.75, 127.59, 125.40, 122.41, 122.16, 121.76, 121.48, 120.69, 120.57, 115.31, 111.46, 109.91, 109.81, 48.26, 43.48, 31.32, 23.00, 20.52, 13.81. HRMS calculated for C$_{36}$H$_{33}$N$_6$O [M + H]$^+$: 565.2710; found 565.2703.

N'-((9-benzyl-1-methyl-9H-pyrido[3,4-b]indol-3-yl)methylene)-9-n-butyl-1-methyl-9H-pyrido[3,4-b]indole-3-carbohydrazide (**8aa**): The compound was obtained as a yellow solid with 83% yield. ^{1}H-NMR (400 MHz, CDCl$_3$) δ 11.30 (s, 1H, CONH), 8.90 (d, J = 4.4 Hz, 2H, ArH), 8.51 (s, 1H, CH, ArH), 8.26 (d, J = 7.6 Hz, 1H, ArH), 8.22 (d, J = 7.6 Hz, 1H, ArH), 7.63–7.59 (m, 1H, ArH), 7.56–7.52 (m, 1H, ArH), 7.46 (d, J = 8.4 Hz, 1H, ArH), 7.38–7.29 (m, 4H, ArH), 7.29–7.22 (m, 2H, ArH), 7.01–6.98 (m, 2H, ArH), 5.74 (s, 2H, ArCH$_2$), 4.51 (t, J = 7.6 Hz, 2H, –CH$_2$CH$_2$CH$_2$CH$_3$), 3.06 (s, 3H, CH$_3$), 2.88 (s, 3H, CH$_3$), 1.88–1.80 (m, 2H, –CH$_2$CH$_2$CH$_2$CH$_3$), 1.51–1.42 (m, 2H, –CH$_2$CH$_2$CH$_2$CH$_3$), 1.00 (t, J = 7.6 Hz, 3H, –CH$_2$CH$_2$CH$_2$CH$_3$). ^{13}C-NMR (100 MHz, CDCl$_3$) δ 161.26, 142.48, 142.10, 141.79, 140.81, 139.64, 137.73, 137.68, 136.55, 136.02, 129.90, 129.54, 129.06, 128.78, 128.52, 127.60, 125.40, 125.32, 122.17, 121.97, 121.73, 121.68, 120.71, 120.55, 113.67, 111.45, 109.97, 109.91, 48.24, 44.82, 33.01, 23.71, 22.93, 20.21, 13.86. HRMS calculated for C$_{37}$H$_{35}$N$_6$O [M + H]$^+$: 579.2867; found 579.2861.

9-benzyl-N'-((9-benzyl-1-methyl-9H-pyrido[3,4-b]indol-3-yl)methylene)-1-methyl-9H-pyrido[3,4-b]indole-3-carbohydrazide (**8ab**): The compound was obtained as a yellow solid with 85% yield. ^{1}H-NMR (400 MHz, CDCl$_3$) δ 11.25 (s, 1H, CONH), 8.96 (s, 1H, ArH), 8.90 (s, 1H, ArH), 8.48 (s, 1H, ArH), 8.28–8.25 (m, 2H, ArH, CH), 7.58–7.52 (m, 2H, ArH), 7.39–7.32 (m, 4H, ArH), 7.30–7.26 (m, 3H, ArH), 7.25–7.21 (m, 2H, ArH), 7.00–6.95 (m, 5H, ArH), 5.78 (s, 2H, ArCH$_2$), 5.74 (s, 2H, ArCH$_2$), 2.88 (s, 3H, CH$_3$), 2.86 (s, 3H, CH$_3$). ^{13}C-NMR (100 MHz, CDCl$_3$) δ 161.08, 142.44, 142.22, 142.14, 140.82, 140.09, 138.28, 137.70, 137.49, 136.98, 136.05, 129.83, 129.73, 129.11, 129.04, 128.88, 128.72, 127.70, 127.58, 125.39, 125.30, 122.14, 122.01, 121.75, 121.72, 120.99, 120.68, 113.67, 111.34, 110.09, 109.90, 48.24, 23.31. HRMS calculated for C$_{40}$H$_{33}$N$_6$O [M + H]$^+$: 613.2710; found 613.2719.

3.5. MTT Assay

Target compounds were assayed by the MTT method for determining cytotoxic activity as described previously [41]. The panel of cell lines included the human umbilical vein cell line EA.HY926, Lewis lung carcinoma (LLC), gastric carcinoma (BGC-823), murine colon carcinoma (CT-26), liver carcinoma (Bel-7402), and breast carcinoma (MCF-7). Cell lines were obtained from Shanghai Cell Institute, Chinese Academy of Science. Growth inhibition rates were calculated with the following equitation: Inhibition ratio (%) = $\frac{OD_{compd} - OD_{blank}}{OD_{DMSO} - OD_{blank}} \times 100\%$. The half-maximal inhibitory concentration (IC$_{50}$) of each compound was calculated using GraphPad Prism software (version 6.0).

3.6. CAM Assay in Vivo

To determine in vivo anti-angiogenic activity of heterobivalent β-carbolines, a CAM assay was performed as previously described [39]. In brief, five-day-old fertilized chicken eggs were purchased from a local hatchery. All the eggs were incubated at 37 °C in an incubator. We injected 0.5 mL of saline, and the eggs were incubated horizontally to allow the CAM to detach from the shell to make a bogus chamber. Compound **8z** was prepared in gelatin sponge discs at concentrations of 0.5, 5.0, and 50 μM/disc. CA4P was used as a positive control drug. Discs containing the vehicle only (DMSO) were used as negative controls. A small window opening was made in the shell, and the discs were directly applied onto the CAM. The square opening was covered with sterilized surgical tape, and the embryos were incubated for 48 h at 38.5 °C. The CAMs were photographed under a dissecting microscope, and blood vessels in each CAM were counted. The results are presented as a mean percentage of inhibition compared to the control ± SD, $n = 3$.

4. Conclusions

On the basis of our previous work, this study has focused on the synthesis of a series of heterobivalent β-carbolines bearing an acylhydrazone bond (**8a–ab**). All of the target compounds were investigated for their in vitro antiproliferative activity using the MTT-based assay against five cancer cell lines (LLC, BGC-823, CT-26, Bel-7402, and MCF-7). Most compounds showed medium antiproliferative activities against the tested cancer cell lines. In particular, compound **8z** showed antitumor activities with IC_{50} values of 9.9, 8.6, 6.2, 9.9, and 5.7 μM against the LLC, BGC-823, CT-26, Bel-7402, and MCF-7 cell lines, respectively. The anti-angiogenic activity of compound **8z** was comparable with CA4P in an in vivo CAM assay at the 5 μM level.

Supplementary Materials: The following are available online, ^{1}H and ^{13}C-NMR spectra for the target compounds are available online.

Author Contributions: Conceptualization, J.Z.; methodology, L.G., X.C., Q.M., and W.F.; formal analysis, L.G. and W.C.; investigation, X.C. and L.G.; writing—original draft preparation, X.C. and L.G.; writing—review and editing, J.Z. and L.G.; project administration, J.Z.

Funding: We gratefully thank the Scientific Research Innovation Project in Shihezi University (No. SHYL-YB201804), the Xinjiang Science and Technology Major Project (No. 2016A03005-1), the Program for Changjiang Scholars and Innovative Research Team in University (No. IRT15R46), and Xinjiang Huashidan Pharmaceutical Research Co. Ltd.

Conflicts of Interest: The authors declare no conflict of interest.

References

1. Song, M. Progress in discovery of KIF5B-RET kinase inhibitors for the treatment of non-small-cell lung cancer. *J. Med. Chem.* **2015**, *58*, 3672–3681. [CrossRef] [PubMed]
2. Cao, R.H.; Peng, W.L.; Wang, Z.H.; Xu, A.L. β-Carboline alkaloids: Biochemical and pharmacological functions. *Curr. Med. Chem.* **2007**, *14*, 479–500. [CrossRef] [PubMed]
3. Sathish, M.; Kavitha, B.; Nayak, V.L.; Tangella, Y.; Ajitha, A.; Nekkanti, S.; Alarifi, A.; Shankaraiah, N.; Nagesh, N.; Kamal, A. Synthesis of podophyllotoxin linked β-carboline congeners as potential anticancer agents and DNA topoisomerase II inhibitors. *Eur. J. Med. Chem.* **2018**, *144*, 557–571. [CrossRef] [PubMed]
4. Tokala, R.; Thatikonda, S.; Vanteddu, U.S.; Sana, S.; Godugu, C.; Shankaraiah, N. Design and synthesis of DNA-interactive β-carboline-oxindole hybrids as cytotoxic and apoptosis-inducing agents. *ChemMedChem* **2018**, *13*, 1909–1922. [CrossRef] [PubMed]
5. Ling, Y.; Guo, J.; Yang, Q.X.; Zhu, P.; Miao, J.F.; Gao, W.J.; Peng, Y.F.; Yang, J.Y.; Xu, K.; Xiong, B.; et al. Development of novel β-carboline-based hydroxamate derivatives as HDAC inhibitors with antiproliferative and antimetastatic activities in human cancer cells. *Eur. J. Med. Chem.* **2018**, *144*, 398–409. [CrossRef] [PubMed]

6. Patil, S.A.; Addo, J.K.; Deokar, H.; Sun, S.; Wang, J.; Li, W.; Suttle, D.P.; Wang, W.; Zhang, R.; Buolamwini, J.K. Synthesis, biological evaluation and modeling studies of new pyrido[3,4-b]indole derivatives as broad-spectrum potent anticancer agents. *Drug Des.* **2017**, *6*, 143. [CrossRef] [PubMed]
7. Kumar, S.; Singh, A.; Kumar, K.; Kumar, V. Recent insights into synthetic β-carbolines with anti-cancer activities. *Eur. J. Med. Chem.* **2017**, *142*, 48–73. [CrossRef]
8. Wu, J.; Zhao, M.; Qian, K.; Lee, K.H.; Morris-Natschke, S.; Peng, S.Q. Novel N-(3-carboxyl-9-benzyl-β-carboline-1-yl) ethylamino acids: Synthesis, anti-tumor evaluation, intercalating determination, 3D QSAR analysis and docking investigation. *Eur. J. Med. Chem.* **2009**, *44*, 4153–4161. [CrossRef]
9. Ikeda, R.; Kurosawa, M.; Okabayashi, T.; Takei, A.; Yoshiwara, M.; Kumakura, T.; Sakai, N.; Morita, O.A.; Ikekita, M.; Nakaike, Y.; et al. 3-(3-Phenoxybenzyl) amino-β-carboline: A novel antitumor drug targeting α-tubulin. *Bioorg. Med. Chem. Lett.* **2011**, *21*, 4784–4787. [CrossRef]
10. Yang, M.; Kuo, P.; Hwang, T.; Chiou, W.; Qian, K.; Lai, C.; Lee, K.; Wu, T. Synthesis, in vitro anti-inflammatory and cytotoxic evaluation, and mechanism of action studies of 1-benzoyl-β-carboline and 1-benzoyl-3-carboxy-β-carboline derivatives. *Bioorg. Med. Chem.* **2011**, *19*, 1674–1682. [CrossRef]
11. Shankaraiah, N.; Siraj, K.P.; Nekkanti, S.; Srinivasulu, V.; Sharma, P.; Senwar, K.R.; Sathish, M.; Vishnuvardhan, M.V.P.S.; Ramakrishna, S.; Jadala, C.; et al. DNA-binding affinity and anticancer activity of β-carboline–chalcone conjugates as potential DNA intercalators: Molecular modelling and synthesis. *Bioorg. Chem.* **2015**, *59*, 130–139. [CrossRef]
12. Taira, Z.; Kanzawas, S.; Dohara, C.; Ishida, S.; Matsumoto, M.; Sakiya, Y. Intercalation of six beta-carboline derivatives into DNA. *Jpn. J. Toxicol. Environ. Health* **1997**, *43*, 83–91. [CrossRef]
13. Cao, R.H.; Peng, W.L.; Chen, H.S.; Ma, Y.; Liu, X.D.; Hou, X.R.; Guan, H.J.; Xu, A.L. DNA binding properties of 9-substituted harmine derivatives. *Biochem. Biophys. Res. Commun.* **2005**, *338*, 1557–1563. [CrossRef]
14. Kamal, A.; Sathish, M.; Nayak, V.L.; Srinivasulu, V.; Kavitha, B.; Tangella, Y.; Thummuri, D.; Bagul, C.; Shankaraiah, N.; Nagesh, N. Design and synthesis of dithiocarbamate linked β-carboline derivatives: DNA topoisomerase II inhibition with DNA binding and apoptosis inducing ability. *Bioorg. Med. Chem.* **2015**, *23*, 5511–5526. [CrossRef]
15. Figueiredo, P.O.; Perdomo, R.T.; Garcez, F.R.; Matos, M.F.C.; Carvalho, J.E.; Garcez, W.S. Further constituents of Galianthe thalictroides (Rubiaceae) and inhibition of DNA topoisomerases I and IIα by its cytotoxic β-carboline alkaloids. *Bioorg. Med. Chem. Lett.* **2014**, *24*, 1358–1361. [CrossRef]
16. Song, Y.; Kesuma, D.; Wang, J.; Deng, Y.; Duan, J.; Wang, J.H.; Qi, R.Z. Specific inhibition of cyclin-dependent kinases and cell proliferation by harmine. *Biochem. Biophys. Res. Commun.* **2004**, *317*, 128–132. [CrossRef]
17. Li, Y.; Liang, F.; Jiang, W.; Yu, F.S.; Cao, R.H.; Ma, Q.H.; Dai, X.Y.; Jiang, J.D.; Wang, Y.C.; Si, S.Y. DH334, a β-carboline anticancer drug, inhibits the CDK activity of budding yeast. *Cancer Biol. Ther.* **2007**, *6*, 1193–1199. [CrossRef]
18. Zhang, J.; Li, Y.; Guo, L.; Cao, R.H.; Zhao, P.; Jiang, W.; Ma, Q.; Yi, H.; Li, Z.; Jiang, J.D.; et al. DH166, a beta-carboline derivative, inhibits the kinase activity of PLK1. *Cancer Biol. Ther.* **2009**, *8*, 2374–2383. [CrossRef]
19. Barsanti, P.A.; Wang, W.; Ni, Z.; Duhl, D.; Brammeier, N.; Martin, E. The discovery of tetrahydro-β-carbolines as inhibitors of the kinesin Eg5. *Bioorg. Med. Chem. Lett.* **2010**, *20*, 157–160. [CrossRef]
20. Castro, A.C.; Dang, L.C.; Soucy, F.; Grenier, L.; Mazdiyasni, H.; Hottelet, M.; Parent, L.; Pien, C.; Palombella, V.; Adams, J. Novel IKK inhibitors: β-carbolines. *Bioorg. Med. Chem. Lett.* **2003**, *13*, 2419–2422. [CrossRef]
21. Barbosa, V.A.; Barea, P.; Mazia, R.S.; Ueda-Nakamura, T.; Costa, W.F.D.; Foglio, M.A.; Goes Ruiz, A.L.T.; Carvalho, J.E.; Vendramini-Costa, D.B.; Nakamura, C.V.; et al. Synthesis and evaluation of novel hybrids beta-carboline-4-thiazolidinones as potential antitumor and antiviral 766 agents. *Eur. J. Med. Chem.* **2016**, *124*, 1093–1104. [CrossRef]
22. Misra, S.; Ghatak, S.; Patil, N.; Dandawate, P.; Ambike, V.; Adsule, S.; Unni, D.; Venkateswara Swamy, K.; Padhye, S. Novel dual cyclooxygenase and lipoxygenase inhibitors targeting hyaluronan-CD44v6 pathway and inducing cytotoxicity in colon cancer cells. *Bioorg. Med. Chem.* **2013**, *21*, 2551–2559. [CrossRef]
23. Cardoso, L.N.F.; Nogueira, T.C.M.; Rodrigues, F.A.R.; Oliveira, A.C.A.; Luciano, M.C.S.; Pessoa, C.; de Souza, M.V.N. N-acylhydrazones containing thiophene nucleus: A new anticancer class. *Med. Chem. Res.* **2017**, *26*, 1605–1608. [CrossRef]

24. Sun, K.; Peng, J.D.; Suo, F.Z.; Zhang, T.; Fu, Y.D.; Zheng, Y.C.; Liu, H.M. Discovery of tranylcypromine analogs with an acylhydrazone substituent as LSD1 inactivators: Design, synthesis and their biological evaluation. *Bioorg. Med. Chem. Lett.* **2017**, *27*, 5036–5039. [CrossRef]
25. Zheng, Y.W.; Ren, J.; Wu, Y.Q.; Meng, X.T.; Zhao, Y.B.; Wu, C.L. Proteolytic unlocking of ultrastable twin-acylhydrazone linkers for lysosomal acid-triggered release of anticancer drugs. *Bioconjug. Chem.* **2017**, *28*, 2620–2626. [CrossRef]
26. Li, F.Y.; Wang, X.; Duan, W.G.; Lin, G.S. Synthesis and in vitro anticancer activity of novel dehydroabietic acid-based acylhydrazones. *Molecules* **2017**, *22*, 1087. [CrossRef]
27. Rodrigues, A.P.; Costa, L.M.; Santos, B.L.; Maia, R.C.; Miranda, A.L.; Barreiro, E.J.; Fraga, C.A. Novel furfurylidene N-acylhydrazones derived from natural safrole: Discovery of LASSBio-1215, a new potent antiplatelet prototype. *J. Enzym Inhib. Med. Chem.* **2012**, *27*, 101–109. [CrossRef]
28. Norsworthy, K.J.; Ko, C.W.; Lee, J.E.; Liu, J.; John, C.S.; Przepiorka, D.; Farrell, A.T.; Pazdur, R. FDA approval summary: Mylotarg for treatment of patients with relapsed or refractory CD33-positive acute myeloid leukemia. *Oncologist* **2018**, *23*, 1103–1108. [CrossRef]
29. Peterson, Q.P.; Goode, D.R.; West, D.C.; Ramsey, K.N.; Lee, J.J.; Hergenrother, P.J. PAC-1 activates procaspase-3 in vitro through relief of zinc-mediated inhibition. *J. Mol. Biol.* **2009**, *388*, 144–158. [CrossRef]
30. Misra, M.C. Drug treatment of haemorrhoids. *Drugs* **2005**, *65*, 1481–1491. [CrossRef]
31. Guo, L.; Cao, R.H.; Fan, W.X.; Ma, Q. Synthesis and biological evaluation of 1,2,7,9-tetrasubstituted harmine derivatives as potential antitumor agents. *Chem. J. Chin. Univ.* **2014**, *35*, 518–523.
32. Guo, L.; Fan, W.X.; Chen, X.M.; Ma, Q.; Cao, R.H. Synthesis and antitumor activities of β-carboline derivatives. *Chin. J. Org. Chem.* **2013**, *33*, 332–338. [CrossRef]
33. Guo, L.; Fan, W.X.; Gan, Z.Y.; Chen, W.; Ma, Q.; Cao, R.H. Design and synthesis of 1-substituted-β-carboline derivatives as potential anticancer agents. *J. Chin. Pharm. Sci.* **2015**, *24*, 801–808.
34. Zhang, G.X.; Cao, R.H.; Guo, L.; Ma, Q.; Fan, W.X.; Chen, X.M.; Li, J.R.; Shao, G.; Qiu, L.Q.; Ren, Z.H. Synthesis and structureeactivity relationships of N2-alkylated quaternary β-carbolines as novel antitumor agents. *Eur. J. Med. Chem.* **2013**, *65*, 21–31. [CrossRef]
35. Posner, G.H.; D'Angelo, J.; O'Neill, P.M.; Mercer, A. Anticancer activity of artemisinin-derived trioxanes. *Expert Opin. Ther. Pat.* **2006**, *16*, 1665–1672. [CrossRef]
36. Alagbala, A.A.; McRiner, A.J.; Borstnik, K.; Labonte, T.; Chang, W.; D'Angelo, J.G.; Posner, G.H.; Foster, B.A. Biological mechanisms of action of novel C-10 non-acetal trioxane dimers in prostate cancer cell lines. *J. Med. Chem.* **2006**, *49*, 7836–7842. [CrossRef]
37. Posner, G.H.; McRiner, A.J.; Paik, I.H.; Sur, S.; Borstnik, K.; Xie, S.; Shapiro, T.A.; Alagbala, A.A.; Foster, B. Anticancer and antimalarial efficacy and safety of artemisininderived trioxane dimers in rodents. *J. Med. Chem.* **2004**, *47*, 1299–1301. [CrossRef]
38. Jung, M.; Lee, S.; Ham, J.; Lee, K.; Kim, H.; Kim, S.K. Antitumor activity of novel deoxoartemisinin monomers, dimers, and trimer. *J. Med. Chem.* **2003**, *46*, 987–994. [CrossRef]
39. Guo, L.; Chen, W.; Fan, W.X.; Ma, Q.; Sun, R.Q.; Shao, G.; Cao, R.H. Synthesis and preliminary evaluation of novel alkyl diamine linked bivalent β-carbolines as angiogenesis inhibitors. *Med. Chem. Commun.* **2016**, *7*, 2177–2183. [CrossRef]
40. Chen, W.; Zhang, G.X.; Guo, L.; Fan, W.X.; Ma, Q.; Zhang, X.D.; Du, R.L.; Cao, R.H. Synthesis and biological evaluation of novel alkyl diamine linked bivalent β-carbolines as angiogenesis inhibitors. *Eur. J. Med. Chem.* **2016**, *124*, 249–261. [CrossRef]
41. Guo, L.; Chen, W.; Cao, R.H.; Fan, W.X.; Ma, Q.; Zhang, J.; Dai, B. Synthesis and structure-activity relationships of asymmetric dimeric β-carboline derivatives as potential antitumor agents. *Eur. J. Med. Chem.* **2018**, *147*, 253–265. [CrossRef]
42. Guo, L.; Ma, Q.; Chen, W.; Fan, W.X.; Zhang, J.; Dai, B. Synthesis and biological evaluation of novel N^9-heterobivalent β-carbolines as angiogenesis inhibitors. *J. Enzyme Inhib. Med. Chem.* **2019**, *34*, 375–387. [CrossRef]
43. Shi, B.X.; Cao, R.H.; Fan, W.X.; Guo, L.; Ma, Q.; Chen, X.M.; Zhang, G.X.; Qiu, L.Q.; Song, H.C. Design, synthesis and in vitro and in vivo antitumor activities of novel bivalent β-carbolines. *Eur. J. Med. Chem.* **2013**, *60*, 10–22. [CrossRef]

44. Daoud, A.; Song, J.; Xiao, F.; Shang, J. B-9-3, a novel β-carboline derivative exhibits anti-cancer activity via induction of apoptosis and inhibition of cell migration in vitro. *Eur. J. Pharmacol.* **2014**, *724*, 219–230. [CrossRef]
45. Ma, Q.; Chen, W.; Chen, W. Anti-tumor angiogenesis effect of a new compound: B-9-3 through interference with VEGFR2 signaling. *Tumor Biol.* **2016**, *37*, 6107–6116. [CrossRef]
46. Zhong, H.; Daoud, A.; Han, J.; An, X.; Qiao, C.; Duan, L.; Wang, Y.; Chen, Z.; Zhou, J.; Shang, J. A small β-carboline derivative "B-9-3" modulates TGF-β signaling pathway causing tumor regression in vivo. *Front. Pharmacol.* **2018**, *9*, 788. [CrossRef]
47. Guo, L.; Xie, J.W.; Fan, W.X.; Chen, W.; Dai, B.; Ma, Q. Synthesis and antitumor activities of novel bivalent 1-Heterocyclic-β-carbolines linked by alkylamino spacer. *Chin. J. Org. Chem.* **2017**, *37*, 1741–1747. [CrossRef]
48. Cao, R.H.; Chen, H.S.; Peng, W.L.; Ma, Y.; Hou, X.R.; Guan, H.J.; Liu, X.; Xu, A.L. Design, synthesis and in vitro and in vivo antitumor activities of novel β-carboline derivatives. *Eur. J. Med. Chem.* **2005**, *40*, 991–1001. [CrossRef]

Sample Availability: Samples of the compounds **8a–ab** are available from the authors.

© 2019 by the authors. Licensee MDPI, Basel, Switzerland. This article is an open access article distributed under the terms and conditions of the Creative Commons Attribution (CC BY) license (http://creativecommons.org/licenses/by/4.0/).

Article

Alcohol Participates in the Synthesis of Functionalized Coumarin-Fused Pyrazolo[3,4-*b*]Pyridine from a One-Pot Three-Component Reaction

Wei Lin [1],*, Cangwei Zhuang [1], Xiuxiu Hu [1], Juanjuan Zhang [2] and Juxian Wang [3],*

[1] School of Chemistry and Environmental Engineering, Jiangsu University of Technology, Changzhou 213001, China
[2] State Key Laboratory of Pharmaceutical Biotechnology, Nanjing University, Nanjing 210093, China
[3] Institute of Medicinal Biotechnology, Chinese Academy of Medical Science and Peking Union Medical College, Beijing 100050, China
* Correspondence: linwei@jsut.edu.cn (W.L.); imbjxwang@163.com (J.-W.)

Academic Editors: Derek J. McPhee and Roman Dembinski
Received: 10 July 2019; Accepted: 2 August 2019; Published: 4 August 2019

Abstract: A concise and efficient approach to synthesizing coumarin-fused pyrazolo[3,4-*b*]pyridine via silica sulfuric acid (SSA) catalyzed three-component domino reaction under microwave irradiation has been demonstrated. Participation of various alcohols in construction of coumarin derivatives has been described for the first time. Short reaction time, high yields, one-pot procedure, usage of eco-friendly catalyst, and solvent are the key features of this method.

Keywords: coumarin; pyrazolo[3,4-*b*]pyridine; synthesis; silica sulfuric acid

1. Introduction

As one of the most important heterocyclic compounds, coumarin was widely found in nature products [1,2], and several synthetic coumarins [3] with a variety of pharmacophoric groups at C-3, C-4, and C-7 positions have been intensively screened for various biological activities like AChE inhibitors [4–6], anticancer [7–9], anticoagulant [10,11], anti-HIV [12–14], antitubercular [15,16], anti-inflammatory [17,18], antioxidant [19], antibacterial [20], antihypertensive [21], anticonvulsant [22], antifungal [23], and antihyperglycemic [24]. A recent literature survey suggests quite a few coumarin derivatives have been patented for their biological properties (Figure 1). Besides the high biological activity, coumarin is also considered to be a functional material [25,26] such as receptors [27–29], signaling units in sensors and biosensors, as well as in advanced photophysical systems [30,31].

Among various nitrogen-containing heterocyclic compounds, pyrazolo[3,4-*b*]pyridine is recognized as important drug molecular skeleton in recent years due to a wide varieties of biological activities (Figure 2), such as antimicrobial [32,33], anti-inflammatory [34,35], anti-proliferative [36,37], and many other [38,39] important effects.

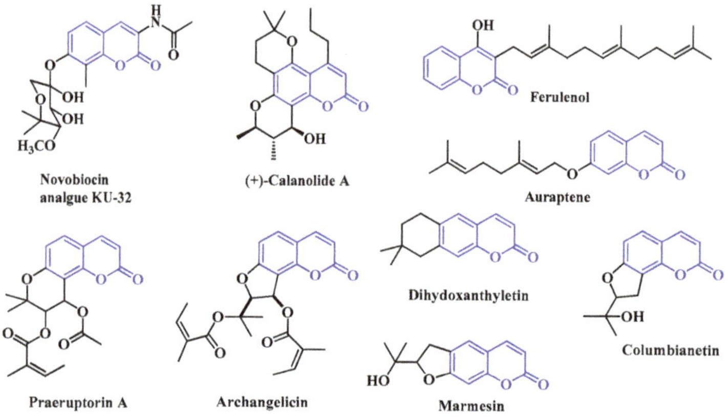

Figure 1. General structures of coumarin molecules possessing biological activity.

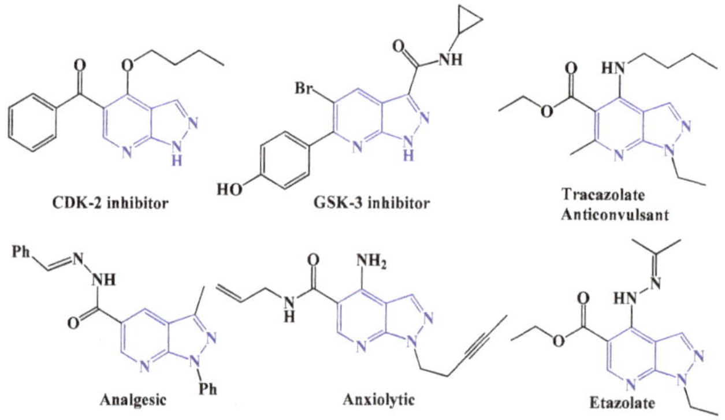

Figure 2. Biologically active compounds having pyrazolo[3,4-b]pyridine unit.

Therefore, development and introduction of a convenient, efficient method for the synthesis of coumarin-fused pyrazolo[3,4-b]pyridine is highly desirable for their immense pharmacological potential. As a part of our research on the synthesis of novel functionalized heterocyclic derivatives [40–46], in the current paper, we report a novel three-component domino reaction for the synthesis of functionalized coumarin-fused pyrazolo[3,4-b]pyridine derivatives using silica sulfuric acid as the catalyst. It worth mentioning that participation of alcohols in construction of coumarin derivatives is described for the first time.

2. Results and Discussion

In the early literature reports of our group [44], the coumarino[4,3-d]pyrazolo[3,4-b]pyridine derivative (**3a**) was synthesized by the reaction of 3-acylcoumarin (**1a**) with 5-aminopyrazole (**2a**) catalyzed by silica sulfuric acid (SSA) in EtOH at 90 °C for 20 min under microwave irradiation (Scheme 1).

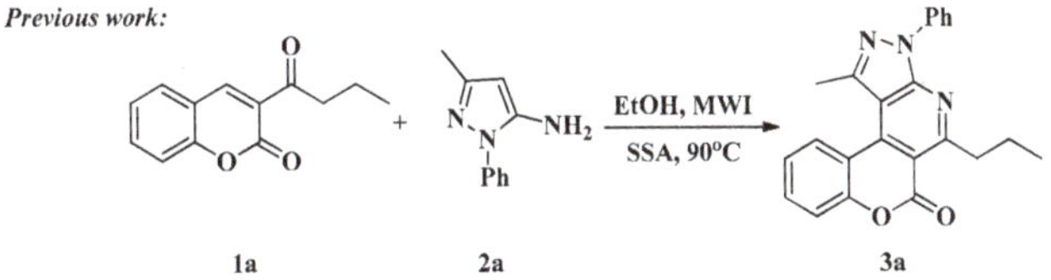

Previous work:

Scheme 1. Synthesis of coumarino[4,3-*d*]pyrazolo[3,4-*b*]pyridine derivative **3a**.

According to our previously reported synthetic procedure, we speculate that the coumarin derivative **6a** could be obtained from the 2-butyryl-3*H*-benzo[*f*]chromen-3-one (**4a**) and 3-methyl-1-phenyl-1*H*-pyrazol-5-amine (**2a**) used as the starting materials. However, product **6a** was not available as expect (Scheme 2-1). Considering the steric hindrance effect of the reaction, when ethanol and ethylene glycol (EG) as mixed solvent (volume ratio of EG/EtOH = 1:1) was added to the reaction, and further increasing the temperature (120 °C), a new product **7a** formed unexpectedly (Scheme 2-2), which was identified by ^{1}H-NMR, ^{13}C-NMR, HRMS analysis. Moreover, we also obtained the single crystal of **7a** suitable for X-ray analysis (Figure 3) [47]. To our surprise, the solvent ethanol also participated in this reaction and a novel coumarin derivative was constructed.

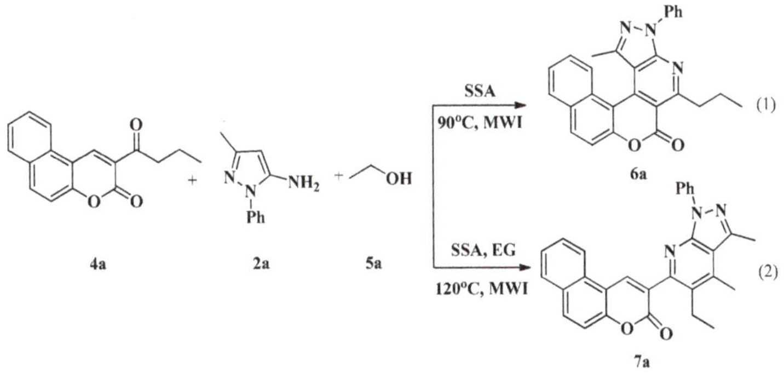

Scheme 2. New multicomponent domino reactions.

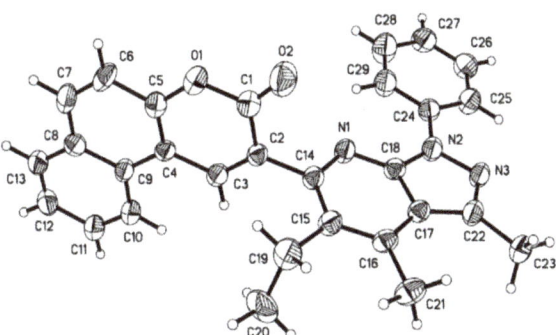

Figure 3. Crystal structure of **7a**.

In order to achieve the optimal conditions of three-component reaction, a series of catalysts, solvents, and temperature were screened, as shown in Table 1. Some other acid catalysts such as *p*-TsOH, HClO$_3$S, H$_2$SO$_4$, SiO$_2$-H$_2$SO$_4$ (Table 1. entries 1, 3–5) and base catalysts such as K$_2$CO$_3$, NaOH, Cs$_2$CO$_3$ (Table 1, entries 6–8) were tested. However, none of them gave better results, lead to the identification of SSA as the most effective catalyst (Table 1. entry 2). To further increase the yield of desired product **7a**, different solvents were evaluated. The results revealed that EtOH and EG as mixed solvents greatly improved the transformation, in control to EtOH, PEG, glycerol, and DMF as a single solvent (Table 1, entries 2, 9–12). When the volume ratio of EG/EtOH = 3:1, the yield of **7a** could further increase to 68% (Table 1, entry 15). Much to our delight, we observed that increasing of the temperature to 140 °C resulted in affording **7a** in 84% yield (Table 1, entry 20).

Table 1. Optimizing the reaction conditions for the synthesis of **7a** under microwave [a].

Entry	Catalyst	Solvent (v/v)	Temperature (°C)	Yield (%) [b]
1	p-TsOH (20 mol%)	EG/EtOH=1:1	120	trace
2	SSA (0.25 g)	EG/EtOH=1:1	120	58
3	HClO$_3$S (5 mol%)	EG/EtOH=1:1	120	36
4	SiO$_2$-H$_2$SO$_4$ (0.25 g)	EG/EtOH=1:1	120	-
5	H$_2$SO$_4$ (20 mol%)	EG/EtOH=1:1	120	-
6	K$_2$CO$_3$ (20 mol%)	EG/EtOH=1:1	120	-
7	NaOH (20 mol%)	EG/EtOH=1:1	120	-
8	Cs$_2$CO$_3$ (20 mol%)	EG/EtOH=1:1	120	-
9	SSA (0.25 g)	EtOH	110	20
10	SSA (0.25 g)	PEG/EtOH = 1:1	120	45
11	SSA (0.25 g)	Glycerol/EtOH = 1:1	120	32
12	SSA (0.25 g)	DMF/EtOH = 1:1	120	24
13	SSA (0.25 g)	EG/EtOH=1:2	120	21
14	SSA (0.25 g)	EG/EtOH = 2:1	120	55
15	SSA (0.25 g)	EG/EtOH = 3:1	120	68
16	SSA (0.25 g)	EG/EtOH = 4:1	120	57
17	SSA (0.25 g)	EG/EtOH = 3:1	100	trace
18	SSA (0.25 g)	EG/EtOH = 3:1	110	trace
19	SSA (0.25 g)	EG/EtOH = 3:1	130	78
20	SSA (0.25 g)	EG/EtOH = 3:1	140	84
21	SSA (0.25 g)	EG/EtOH = 3:1	150	76

[a] Reaction conditions: **4a** (0.5 mmol), **2a** (0.5 mmol), **5a** (1.0 mL), 45 min; [b] GC yield of **7a** determined using tridecane as internal standard.

With optimal conditions in hand, the corresponding novel coumarin-fused pyrazolo[3,4-b]pyridine derivatives **7** were synthesized (Scheme 3).

As illustrated in Scheme 3, the substrate scope of the transformation was examined using arylbenzo[f]chromen-3-one **4**, enaminone **2**, and alkyl alcohol **5** as staring materials. Notably, electronic effects had an important impact on this reaction. When the substituent R^3 was electron-donating group, such as Me, the desired products could not be obtained at all (**7e, 7f**).

To further expand the scope of substrates, aryl alcohols (**8**) instead of alkyl alcohols (**5**) were also tested. It was found that aryl alcohols were well tolerated under the optimal reaction conditions, the corresponding products were afforded in moderate to good yields. When substituent R^3 was electron-withdrawing groups (Ph), the yields were good and no more than 1 h cost (Table 2, entries 1–19). However, the substituents R^3 was electron-donating groups (CH$_3$) (Table 2, entries 20–22), the yields were lower and the reaction time was longer. Unfortunately, When R^3 and R^4 was electron rich group, such as Me, the reaction could not proceed successfully (Table 2, entry 23).

Table 2. Synthesis of coumarin-fused pyrazolo[3,4-b]pyridine derivatives 9 [a].

Entry	Product	R^1	R^2	R^3	R^4	Ar	Time (h)	Isolated yield (%)
1	9a	CH_2CH_3	H	Ph	CH_3	C_6H_5	1	69
2	9b	CH_2CH_3	H	Ph	CH_3	$4\text{-}CH_3C_6H_4$	1	74
3	9c	CH_2CH_3	H	Ph	CH_3	$4\text{-}OCH_3C_6H_4$	0.75	77
4	9d	CH_2CH_3	H	Ph	CH_3	$3\text{-}OCH_3C_6H_4$	0.75	76
5	9e	CH_2CH_3	H	Ph	CH_3	$4\text{-}BrC_6H_4$	0.75	71
6	9f	CH_2CH_3	H	Ph	CH_3	Pyridine-4-yl	1	70
7	9g	CH_2CH_3	H	Ph	CH_3	Furan-2-yl	1	76
8	9h	CH_2CH_3	OCH_3	Ph	CH_3	C_6H_5	1	62
9	9i	CH_2CH_3	OCH_3	Ph	CH_3	$4\text{-}CH_3C_6H_4$	1	70
10	9j	CH_2CH_3	OCH_3	Ph	CH_3	$4\text{-}OCH_3C_6H_4$	1	72
11	9k	CH_3	H	Ph	CH_3	C_6H_5	1	68
12	9l	CH_3	H	Ph	CH_3	$4\text{-}CH_3C_6H_4$	1.25	70
13	9m	CH_3	H	Ph	CH_3	$4\text{-}OCH_3C_6H_4$	1.25	74
14	9n	CH_3	H	Ph	CH_3	$3\text{-}OCH_3C_6H_4$	1.25	72
15	9o	CH_3	OCH_3	Ph	CH_3	$4\text{-}CH_3C_6H_4$	1.25	67
16	9p	CH_3	OCH_3	Ph	CH_3	$4\text{-}OCH_3C_6H_4$	1.25	70
17	9q	CH_3	OCH_3	Ph	CH_3	C_6H_5	1.25	60
18	9r	H	H	Ph	CH_3	C_6H_5	1.5	56
19	9s	H	H	Ph	CH_3	$4\text{-}OCH_3C_6H_4$	1.5	58
20	9t	CH_2CH_3	H	CH_3	Ph	C_6H_5	2	55
21	9u	CH_3	OCH_3	CH_3	Ph	C_6H_5	2	50
22	9v	CH_3	OCH_3	CH_3	Ph	$4\text{-}OCH_3C_6H_4$	2	45
23	9w	CH_2CH_3	H	CH_3	CH_3	$4\text{-}OCH_3C_6H_4$	2.5	trace

[a] Reaction conditions: arylbenzo[f]chromen-3-one **4** (0.5 mmol), enaminone **2** (0.5 mmol), aryl alcohols **8** (1.0 mL), EG (3 mL) and SSA (0.25 g), 140 °C.

To gain insight into the mechanism of this one-spot three-component reaction process, some additional experiments were performed. When benzaldehyde (**10**) was added to the reaction instead of phenylmethanol (**8a**) under standard conditions, 73% yield of desired product (**9a**) could be obtained, and reaction time reduced from 1 h to 15 min (Scheme 4A), and when butyraldehyde (**11**) was added to the reaction 50% yield of desired product (**7c**) could be obtained (Scheme 4B). The reaction did not proceed successfully without SSA catalyzed. Just phenylmethanol (**8a**) was heated to 140 °C, directly with the catalyst of SSA, benzaldehyde (**10**) and benzoic acid (**12**) could be detected by GC-MS (Scheme 4C). We speculated that the benzaldehyde was most likely the key intermediate in this protocol.

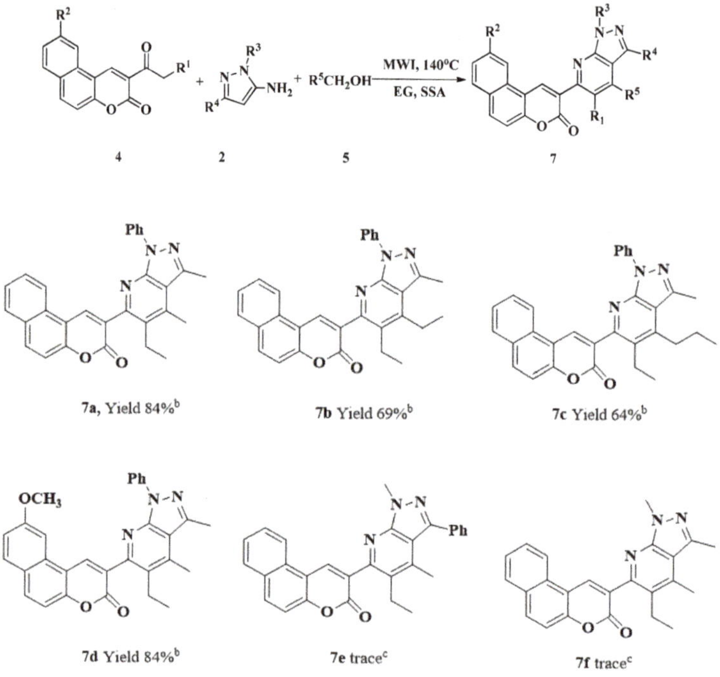

Scheme 3. Synthesis of coumarin-fused pyrazolo[3,4-b]pyridine derivatives **7** [a]. [a] Reaction conditions: arylbenzo[f]chromen-3-one **4** (0.5 mmol), enaminone **2** (0.5 mmol), alkyl alcohol **5** (1.0 mL), EG (3 mL) and SSA (0.25 g), 140 °C, 45 min; [b] Isolated yield; [c] 2 h.

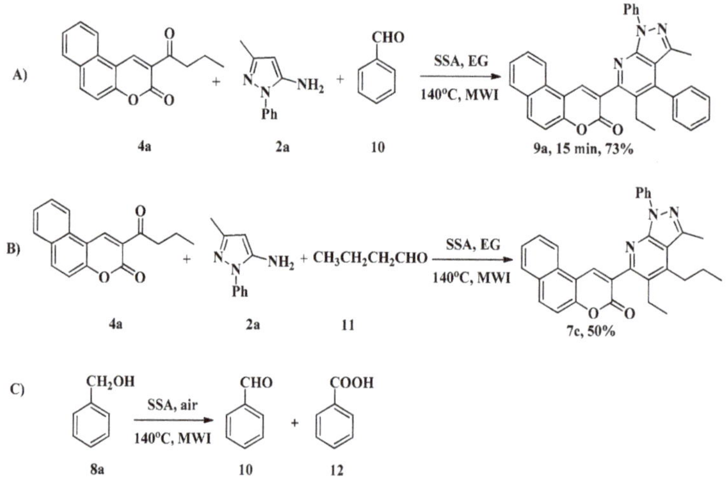

Scheme 4. Preliminary mechanistic studies. (**A**) Synthesis of coumarin-fused pyrazolo[3,4-b]pyridine derivatives **9a**. (**B**) Synthesis of coumarin-fused pyrazolo[3,4-b]pyridine derivatives **7c**. (**C**) Reaction of phenylmethanol with the catalyst of SSA.

Herein, we propose the following mechanism for the reaction (Scheme 5). SSA catalyzed alkyl alcohol **5** to afford the corresponding aldehyde, then the intermediate **A** is formed by means of a

Knoevenagel condensation of aldehyde and arylbenzo[f]chromen-3-one (**4**). The intermediate **A** is activated by SSA, which subsequently undergoes Michael addition with enaminone (**2**) via attack of the nucleophilic C-4 of the intermediate **A** to give intermediate **B**, which transformed to more-stable intermediate **C**. Then, intermediate **C** tautomerizes to intermediate **D**, which undergoes intramolecular nucleophilic addition to form intermediate **E**. In the last step, loss of H_2O affords the desired product.

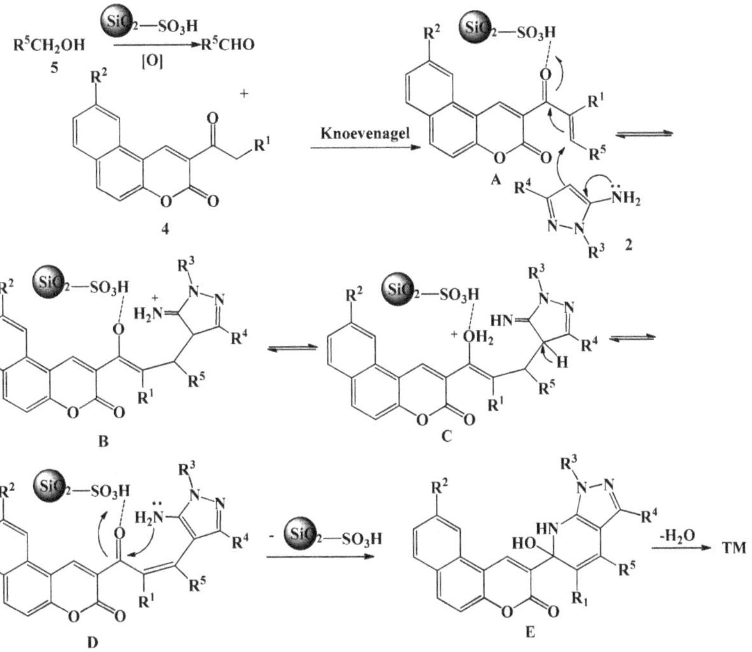

Scheme 5. Proposed mechanism for this reaction.

3. Conclusions

In conclusion, we have developed a protocol for the facile synthesis of various potentially biologically active coumarin-fused pyrazolo[3,4-b]pyridine derivatives, based on a novel three-component domino reaction under microwave irradiation. Using this method, coumarin derivatives could be rapidly constructed in moderate-to-good yields with short reaction time. Further study to deeply understand the reaction mechanism is currently underway in our lab.

4. Experimental Section

4.1. General

All reagents were purchased from commercial suppliers (Aladdin, Shanghai, China) and used without further purification. Microwave irradiation was carried out with Initiator 2.5 Microwave Synthesizers from Biotage, Uppsala, Sweden. The reaction temperatures were measured by infrared detector during microwave heating. Melting points are uncorrected. IR spectra were recorded on a Tensor 27 spectrometer (Bruker Corp., Karlsruhe, Germany) in KBr with absorptions in cm^{-1}. ^{1}H-NMR (400 MHz) and ^{13}C-NMR (75 MHz or 100 MHz) spectra were recorded on a Varian Inova-400 MHz or Varian Inova-300 MHz (Varian, CA, America) in CDCl$_3$, DMSO-d_6 or CF$_3$COOD as solution. J values are in hertz. Chemical shifts are expressed in parts per million downfield from interal standard TMS. High-resolution mass spectra (HRMS) for all the compounds were determined on Bruker MicrOTOF-QII mass spectrometer (Bruker Corp., Karlsruhe, Germany) with ESI resource. X-ray diffraction analysis was recorded on a Smart-1000 diffractometer (PANalytical B.V., Holland).

4.2. General Procedure for the Synthesis of Products 4 Are Represented as Follows

Typically, 2-hydroxy-1-naphthaldehyde (5 mmol), ethyl 3-oxopentanoate or ethyl 3-oxohexanoate or ethyl acetoacetate (5 mmol) and piperidine (0.5 mmol) were introduced in a 20 mL vial with ethanol (10 mL) as solution. Subsequently, the reaction vial was closed and then prestirred for 10 s. The mixture was irradiated at 90 °C for 10 min. After the completion, the reaction mixture was then cooled to room temperature and concentrated in vacuo to remove the solvent. The residue was then washed with water, filtered, dried, and the precipitate was purified by recrystallization from 95% EtOH to give the products of **4**. The analytical data for represent compounds are shown below. ^{1}H-NMR and ^{13}C-NMR spectra of compounds **4** in Supplementary Materials.

4.2.1. 2-Butyryl-3H-benzo[f]chromen-3-one (**4a**)

Yellow solid; yield 89%; m.p.: 127–129 °C; IR (KBr): v 1734, 1626, 1557, 1513, 1383, 1109, 864 cm^{-1}; ^{1}H-NMR (CDCl$_3$, 400 MHz) δ (ppm): 9.21 (s, 1H, ArH), 8.91 (s, 1H, ArH), 8.06 (d, J = 8.8 Hz, 1H, ArH), 7.83 (d, J = 8.8 Hz, 1H, ArH), 7.57–7.56 (m, 1H, ArH), 7.20 (d, J = 9.2 Hz, 1H, ArH), 7.12 (dd, J_1 = 8.8 Hz, J_2 = 2.0 Hz, 1H, ArH), 3.00 (t, J = 7.2 Hz, 2H, CH$_2$), 1.63–1.58 (m, 2H, CH$_2$), 0.93 (t, J = 7.2 Hz, 3H, CH$_3$); ^{13}C-NMR (100 MHz, DMSO-d_6) δ (ppm): 197.5, 159.1, 158.9, 156.4, 142.8, 136.6, 132.1, 131.5, 124.6, 121.9, 118.6, 113.0, 111.4, 104.8, 43.9, 17.3, 14.1;

4.2.2. 2-Butyryl-9-methoxy-3H-benzo[f]chromen-3-one (**4b**)

Yellow solid, yield 88%; m.p.: 125–128 °C; IR (KBr) v: 1730, 1667, 1601, 1556, 1513, 1386, 1365, 1196, 948, 836 cm^{-1}; ^{1}H-NMR (CDCl$_3$, 400 MHz) δ (ppm): 9.01 (s, 1H, ArH), 7.86 (d, J = 8.8 Hz, 1H, ArH), 7.68 (d, J = 8.8 Hz, 1H, ArH), 7.40 (s, 1H, ArH), 7.14 (t, J = 8.4 Hz, 2H, ArH), 3.94 (s, 3H, CH$_3$O), 3.12 (t, J = 8.4 Hz, 2H, CH$_2$); 1.76–1.70 (m, 2H, CH$_2$), 1.00 (t, J = 8.4 Hz, 3H, CH$_3$); ^{13}C-NMR (100 MHz, CDCl$_3$) δ (ppm): 197.6, 159.9, 158.7, 156.0, 142.5, 135.3, 131.2, 130.2, 124.7, 121.1, 117.9, 113.1, 111.4, 100.7, 55.5, 44.0, 16.9, 13.3.

4.2.3. 2-Propionyl-3H-benzo[f]chromen-3-one (**4c**)

Yellow solid, yield 87%; m.p.: 134–136 °C; IR (KBr): v 1732, 1662, 1601, 1556, 1524, 1387, 1365, 1196, 945, 823 cm^{-1}; ^{1}H-NMR (CDCl$_3$, 400 MHz) δ (ppm): 9.15 (s, 1H, ArH), 8.74 (s, 1H, ArH), 7.90 (d, J = 8.8 Hz, 1H, ArH), 7.66 (d, J = 8.8 Hz, 1H, ArH), 7.41–7.40 (m, 1H, ArH), 7.04 (d, J = 8.8 Hz, 1H, ArH), 6.95 (dd, J_1 = 8.8 Hz, J_2 = 2.0 Hz, 1H, ArH), 3.14–3.08 (m, 2H, CH$_2$), 1.08 (t, J = 7.2 Hz, 3H, CH$_3$); ^{13}C-NMR (100 MHz, CDCl$_3$) δ (ppm): 198.4, 159.9, 159.4, 156.7, 143.3, 135.9, 131.9, 130.8, 125.3, 121.8, 118.7, 113.7, 112.1, 102.0, 35.2, 10.7.

4.2.4. 9-Methoxy-2-propionyl-3H-benzo[f]chromen-3-one (**4d**)

Yellow solid, yield 87%; m.p.: 125–128 °C; IR (KBr): v 1730, 1667, 1601, 1556, 1513, 1386, 1365, 1196, 948, 836 cm^{-1}; ^{1}H-NMR (DMSO-d_6, 400 MHz) δ (ppm): 9.19 (s, 1H, ArH), 8.14 (d, J = 9.2 Hz, 1H, ArH), 7.90 (d, J = 9.2 Hz, 1H, ArH), 7.79 (s, 1H, ArH), 7.32 (t, J = 8.8 Hz, 1H, ArH), 7.21 (dd, J_1 = 8.8 Hz, J_2 = 2.0 Hz, 1H, ArH), 3.98 (s, 3H, CH$_3$O), 3.10–3.05 (m, 2H, CH$_2$), 1.09 (t, J = 7.2 Hz, 3H, CH$_3$); ^{13}C-NMR (100 MHz, DMSO-d_6) δ (ppm): 198.7, 160.4, 158.8, 156.1, 143.0, 136.2, 131.9, 131.2, 125.4, 122.8, 118.6, 113.9, 112.1, 102.4, 56.2, 35.4, 8.4.

4.2.5. 2-Acetyl-3H-benzo[f]chromen-3-one (**4e**)

Yellow solid, yield 88%; m.p.: 189–190 °C [48]; IR (KBr): v 2959, 1696, 1622, 1562, 1384, 1227, 1206, 857 cm^{-1}.

4.3. General Procedure for the Synthesis of Products 7 and 9 Are Represented as Follows

Typically, benzo[f]chromen-3-one **4** (0.5 mmol), enaminone **2** (0.5 mmol), alkyl alcohol **5** (1.0 mL) or aryl alcohols **8** (1.0 mL) and SSA (0.25 g) were introduced in a 5 mL vial with ethylene glycol (3 mL) as solution. Subsequently, the reaction vial was closed and then prestirred for 10 s. The mixture was irradiated at 140 °C. The reaction was monitored by TLC. After the completion, the reaction mixture was then cooled to room temperature and diluted with cold water (30 mL), and extracted with CH_2Cl_2 (3 × 30 mL). The extracts were washed with water (3 × 50 mL) and dried over anhydrous Na_2SO_4. After evaporation of the solvent under reduced pressure, the precipitate was collected and purified by recrystallization from 95% EtOH or by flash column chromatography (petroleum ether:ethyl acetate = 8:1) to give the products **7** or **9**. The analytical data for represent compounds are shown below. ^{1}H-NMR and ^{13}C-NMR spectra of compounds **7** and **9** in Supplementary Materials.

4.3.1. 2-(5-Ethyl-3,4-dimethyl-1-phenyl-1H-pyrazolo[3,4-b]pyridin-6-yl)-3H-benzo[f]chromen-3-one (7a)

White solid, m.p.: 258–260 °C; IR (KBr, cm^{-1}) ν: 2960, 1722, 1629, 1572, 1507, 1415, 1387, 1315, 1290, 1248, 1211, 1096, 989, 906, 815, 787, 713, 691, 605; ^{1}H-NMR (400 MHz, DMSO-d_6) δ (ppm): 9.07 (s, 1H, ArH), 8.58 (d, J = 8.0 Hz, 1H, ArH), 8.23 (t, J = 8.0 Hz, 3H, ArH), 8.08 (d, J = 8.0 Hz, 1H, ArH), 7.69–7.61 (m, 3H, ArH), 7.44 (t, J = 8.0 Hz, 2H, ArH), 7.20 (t, J = 7.2 Hz, 1H, ArH), 2.78–2.66 (m, 8H, 2 × CH_3 + CH_2), 1.05 (t, J = 7.2 Hz, 3H, CH_3); ^{13}C-NMR (75 MHz, CF_3COOD) δ (ppm): 156.0, 148.8, 146.2, 145.4, 140.5, 139.1, 135.6, 134.1, 132.6, 132.0, 131.2, 130.8, 130.2, 129.4, 128.5, 126.5, 122.3, 121.3, 116.7, 116.2, 22.7, 17.0, 14.3, 13.4; HRMS: m/z cacld. for $C_{29}H_{24}N_3O_2$ [M + H]$^+$ 446.1869, Found 446.1853.

4.3.2. 2-(4,5-Diethyl-3-methyl-1-phenyl-1H-pyrazolo[3,4-b]pyridin-6-yl)-3H-benzo[f]chromen-3-one (7b)

White solid, m.p.: >300 °C; IR (KBr, cm^{-1}) ν: 2974, 1719, 1688, 1656, 1628, 1596, 1628, 1571, 1546, 1506, 1413, 1357, 1204, 1071, 909, 817, 752, 694, 676, 589; ^{1}H-NMR (400 MHz, DMSO-d_6) δ (ppm): 9.16 (s, 1H, ArH), 8.66 (d, J = 8.4 Hz, 1H, ArH), 8.29 (d, J = 9.2 Hz, 1H, ArH), 8.21 (d, J = 7.6 Hz, 2H, ArH), 8.11 (d, J = 8.0 Hz, 1H, ArH), 7.73–7.63 (m, 3H, ArH), 7.47 (t, J = 8.0 Hz, 2H, ArH), 7.23 (t, J = 7.6 Hz, 1H, ArH), 3.51–3.48 (m, 2H, CH_2), 3.17–3.14 (m, 2H, CH_2), 2.79 (s, 3H, CH_3), 1.35 (t, J = 7.2 Hz, 3H, CH_3), 1.09 (t, J = 7.6 Hz, 3H, CH_3); ^{13}C-NMR (75 MHz, DMSO-d_6) δ (ppm): 160.2, 154.3, 153.8, 149.4, 148.1, 142.4, 139.6, 134.1, 130.5, 130.3, 129.6, 129.0, 128.3, 126.8, 125.7, 123.2, 120.5, 117.2, 115.9, 113.5, 100.0, 22.2, 21.6, 16.6, 16.2, 15.5; HRMS: m/z cacld. for $C_{30}H_{25}N_3O_2$ (M)$^+$ 459.1947, Found 459.1946.

4.3.3. 2-(5-Ethyl-3-methyl-1-phenyl-4-propyl-1H-pyrazolo[3,4-b]pyridin-6-yl)-3H-benzo[f]chromen-3-one (7c)

White solid, m.p.: 242–245 °C; IR (KBr, cm^{-1}) ν: 2974, 2880, 2703, 2545, 1789, 1722, 1665, 1573, 1503, 1439, 1414, 1389, 1359, 1320, 1288, 1248, 1217, 1155, 1091, 915, 858, 813, 792, 745, 695, 641, 610; ^{1}H-NMR (400 MHz, DMSO-d_6) δ (ppm): 9.14 (s, 1H, ArH), 8.63 (d, J = 8.8 Hz, 1H, ArH), 8.26 (d, J = 8.8 Hz, 1H, ArH), 8.22 (d, J = 8.0 Hz, 1H, ArH), 8.09 (d, J = 8.0 Hz, 1H, ArH), 7.70–7.63 (m, 3H, ArH), 7.46 (t, J = 7.6 Hz, 2H, ArH), 7.22 (t, J = 7.6 Hz, 1H, ArH), 3.07–3.06 (m, 2H, CH_2), 2.76–2.73 (m, 5H, CH_3 + CH_2), 1.72–1.69 (m, 2H, CH_2), 1.13 (s, 3H, CH_3), 1.07 (s, 3H, CH_3); ^{13}C-NMR (75 MHz, CF_3COOD) δ (ppm): 165.7, 155.9, 148.1, 146.2, 145.9, 140.8, 139.1, 135.0, 134.1, 132.6, 131.9, 131.2, 130.7, 130.2, 129.3, 128.5, 126.5, 121.6, 121.3, 116.3, 33.4, 26.2, 22.2, 14.5, 13.7; HRMS: m/z cacld. for $C_{31}H_{28}N_3O_2$ [M + H]$^+$ 474.2182, Found 474.2210.

4.3.4. 2-(5-Ethyl-3,4-dimethyl-1-phenyl-1H-pyrazolo[3,4-b]pyridin-6-yl)-9-methoxy-3H-benzo[f]chromen-3-one (7d)

White solid, m.p.: >300 °C; IR (KBr, cm^{-1}) ν: 2975, 2026, 1795, 1728, 1628, 1574, 1509, 1230, 1091, 989, 917, 840, 794, 751, 686, 610; ^{1}H-NMR (400 MHz, DMSO-d_6) δ (ppm): 9.21 (s, 1H, ArH), 8.24–8.17 (m, 3H, ArH), 8.01–7.96 (m, 2H, ArH), 7.50–7.45 (m, 3H, ArH), 7.27–7.21 (m, 2H, ArH), 3.90 (s, 3H, OCH_3), 2.80–2.79 (m, 8H, CH_2 + 2 × CH_3), 1.07 (t, J = 7.2 Hz, 3H, CH_3); ^{13}C-NMR (75 MHz, DMSO-d_6) δ (ppm): 160.3, 160.1, 154.4, 154.0, 149.0, 143.1, 142.4, 140.0, 139.7, 133.8, 131.4, 131.0, 129.5, 127.5, 125.6, 125.5,

120.3, 118.7, 116.9, 114.3, 102.6, 56.3, 22.4, 16.1, 15.4, 15.0; HRMS: m/z cacld. for $C_{30}H_{26}N_3O_3$ [M + H]$^+$ 476.1974, Found 476.1980.

4.3.5. 2-(5-Ethyl-3-methyl-1,4-diphenyl-1H-pyrazolo[3,4-b]pyridin-6-yl)-3H-benzo[f]chromen-3-one (9a)

Yellow solid, m.p.: >300 °C; IR (KBr, cm^{-1}) ν: 3032, 2978, 2888, 2763, 1725, 1049, 958, 815, 756, 699, 679, 588; ^{1}H-NMR (400 MHz, CF$_3$COOD) δ (ppm): 10.16 (s, 1H, ArH), 9.15–9.14 (m, 2H, ArH), 8.88–8.87 (m, 1H, ArH), 8.65–8.64 (m, 1H, ArH), 8.59–8.55 (m, 4H, ArH), 8.51–8.47 (m, 6H, ArH), 8.38–8.37 (m, 2H, ArH), 3.77–3.76 (m, 2H, CH$_2$), 3.05 (s, 3H, CH$_3$), 1.91 (s, 3H, CH$_3$); ^{13}C-NMR (75 MHz, CF$_3$COOD) δ (ppm): 163.7, 156.1, 149.4, 146.8, 146.4, 140.8, 139.1, 135.8, 134.3, 132.8, 132.6, 132.0, 131.6, 131.2, 130.8, 130.2, 129.9, 129.5, 128.5, 127.8, 126.5, 121.9, 121.4, 116.8, 116.3, 113.6, 22.9, 14.3, 12.5; HRMS: m/z cacld. for $C_{34}H_{26}N_3O_2$ [M + H]$^+$ 508.2025, Found 508.2025.

4.3.6. 2-(5-Ethyl-3-methyl-1-phenyl-4-(p-tolyl)-1H-pyrazolo[3,4-b]pyridin-6-yl)-3H-benzo[f]chromen-3-one (9b)

Yellow solid, m.p.: >300 °C; IR (KBr, cm^{-1}) ν: 2968, 1972, 1779, 1572, 1505, 1413, 1360, 1207, 1088, 961, 898, 806, 758, 728, 690, 642; ^{1}H-NMR (400 MHz, CF$_3$COOD) δ (ppm): 10.13 (s, 1H, ArH), 9.15–9.09 (m, 2H, ArH), 8.87–8.84 (m, 1H, ArH), 8.63–8.60 (m, 1H, ArH), 8.54–8.40 (m, 9H, ArH), 8.25–8.24 (m, 2H, ArH), 3.76–3.74 (m, 2H, CH$_2$), 3.37 (s, 3H, CH$_3$), 3.06 (s, 3H, CH$_3$), 1.88–1.87 (m, 3H, CH$_3$); ^{13}C-NMR (75 MHz, CF$_3$COOD) δ (ppm): 162.8, 155.1, 148.6, 145.6, 145.4, 142.0, 139.8, 138.2, 135.0, 133.3, 131.7, 131.1, 130.3, 129.9, 129.6, 129.3, 128.7, 128.5, 127.6, 126.9, 125.6, 121.0, 120.4, 115.9, 115.4, 112.7, 21.9, 19.5, 13.4, 11.7; HRMS: m/z cacld. for $C_{35}H_{28}N_3O_2$ [M + H]$^+$ 522.2182, Found 522.2180.

4.3.7. 2-(5-Ethyl-4-(4-methoxyphenyl)-3-methyl-1-phenyl-1H-pyrazolo[3,4-b]pyridin-6-yl)-3H-benzo[f]chromen-3-one (9c)

Yellow solid, m.p.: >300 °C; IR (KBr, cm^{-1}) ν: 2967, 1711, 1597, 1571, 1505, 1412, 1286, 1249, 1211, 1048, 982, 897, 849, 806, 758, 690, 641, 587; ^{1}H-NMR (400 MHz, CF$_3$COOD) δ (ppm): 9.29 (s, 1H, ArH), 8.28–8.23 (m, 2H, ArH), 7.99 (d, J = 8.4 Hz, 1H, ArH), 7.76 (t, J = 7.6 Hz, 1H, ArH), 7.69–7.56 (m, 7H, ArH), 7.50 (d, J = 8.4 Hz, 2H, ArH), 7.35 (d, J = 8.4 Hz, 2H, ArH), 4.06 (s, 3H, OCH$_3$), 2.93–2.88 (m, 2H, CH$_2$), 2.23 (s, 3H, CH$_3$), 1.02 (t, J = 7.2 Hz, 3H, CH$_3$); ^{13}C-NMR (75 MHz, CF$_3$COOD) δ (ppm): 162.9, 160.9, 160.5, 155.2, 148.4, 145.8, 145.7, 140.0, 138.3, 135.3, 133.5, 131.8, 131.2, 130.4, 130.0, 129.4, 129.2, 128.7, 127.7, 125.8, 125.3, 121.3, 120.6, 116.0, 115.5, 114.9, 112.8, 55.1, 22.1, 13.4, 12.0; HRMS: m/z cacld. for $C_{35}H_{28}N_3O_3$ [M + H]$^+$ 538.2131, Found 538.2111.

4.3.8. 2-(5-Ethyl-4-(3-methoxyphenyl)-3-methyl-1-phenyl-1H-pyrazolo[3,4-b]pyridin-6-yl)-3H-benzo[f]chromen-3-one (9d)

Yellow solid, m.p.: >300 °C;.IR (KBr, cm^{-1}) ν: 2965, 2023, 1785, 1712, 1573, 1504, 1382, 1357, 1285, 1158, 1136, 1046, 782, 759, 712, 689, 588; ^{1}H-NMR (400 MHz, DMSO-d$_6$) δ (ppm): 9.23 (s, 1H, ArH), 8.66 (d, J = 8.4 Hz, 1H, ArH), 8.29–8.23 (m, 3H, ArH), 8.10 (d, J = 8.0 Hz, 1H, ArH), 7.73–7.62 (m, 3H, ArH), 7.54–7.47 (m, 3H, ArH), 7.25 (t, J = 7.6 Hz, 1H, ArH), 7.14–7.12 (m, 1H, ArH), 7.03–7.01 (m, 2H, ArH), 3.84 (s, 3H, OCH$_3$), 2.58–2.56 (m, 2H, CH$_2$), 1.96 (s, 3H, CH$_3$), 0.89 (t, J = 7.2 Hz, 3H, CH$_3$); ^{13}C-NMR (75 MHz, DMSO-d$_6$) δ (ppm): 160.2, 159.6, 154.2, 153.9, 148.7, 145.3, 142.9, 140.1, 139.6, 137.3, 134.3, 130.5, 130.4, 130.1, 129.6, 129.5, 129.4, 129.0, 127.8, 126.8, 125.9, 123.1, 121.4, 120.6, 117.2, 115.8, 114.6, 113.5, 55.8, 22.5, 16.0, 14.2; HRMS: m/z cacld. for $C_{35}H_{27}N_3O_3$ (M)$^+$ 537.2052, Found 537.2053.

4.3.9. 2-(4-(4-Bromophenyl)-5-ethyl-3-methyl-1-phenyl-1H-pyrazolo[3,4-b]pyridin-6-yl)-3H-benzo[f]chromen-3-one (9e)

Yellow solid, m.p.: >300 °C;.IR (KBr, cm^{-1}) ν: 2968, 2032, 1775, 1721, 1574, 1385, 1357, 1285, 1166, 1047, 782, 759, 712, 681, 588; ^{1}H-NMR (400 MHz, DMSO-d$_6$) δ (ppm): 10.19 (s, 1H, ArH), 8.57–8.53 (m, 2H, ArH), 8.43 (d, J = 9.2 Hz, 1H, ArH), 8.06 (d, J = 8.0 Hz, 1H, ArH), 7.86 (t, J = 7.6 Hz, 1H, ArH), 7.78–7.23 (m, 10H, ArH), 2.79 (s, 2H, CH$_2$), 2.54 (s, 3H, CH$_3$), 1.35 (t, J = 7.2 Hz, 3H, CH$_3$); ^{13}C-NMR (75 MHz,

DMSO-d_6) δ (ppm): 165.7, 159.3, 157.7, 152.3, 151.4, 144.7, 141.6, 138.9, 134.8, 132.3, 131.0, 129.7, 126.7, 125.2, 121.4, 118.1, 113.1, 111.4, 111.3, 109.6, 107.5, 21.9, 17.0, 14.6; HRMS: *m/z* cacld. for $C_{34}H_{24}BrN_3O_2$ (M)$^+$ 585.1052, Found 585.1057.

4.3.10. *2-(5-Ethyl-3-methyl-1-phenyl-4-(pyridin-4-yl)-1H-pyrazolo[3,4-b]pyridin-6-yl)-3H-benzo[f]chromen-3-one* (**9f**):

Yellow solid, m.p.: >300 °C; IR (KBr, cm^{-1}) ν: 2965, 1972, 1783, 1573, 1505, 1413, 1362, 1089, 961, 898, 805, 758, 693, 642; ^{1}H-NMR (400 MHz, CF$_3$COOD) δ (ppm): 10.09 (s, 1H, ArH), 9.11–9.05 (m, 2H, ArH), 8.80 (d, *J* = 8.0 Hz, 1H, ArH), 8.60–8.35 (m, 10H, ArH), 8.21–8.19 (m, 2H, ArH), 3.72–3.70 (m, 2H, CH$_2$), 3.01 (s, 3H, CH$_3$), 1.83 (t, *J* = 6.8 Hz, 3H, CH$_3$);^{13}C-NMR (75 MHz, CF$_3$COOD) δ (ppm): 162.8, 155.0, 148.5, 145.5, 145.4, 142.0, 139.7, 138.1, 135.0, 133.3, 131.6, 131.0, 130.2, 129.8, 129.5, 129.2, 128.7, 128.5, 127.5, 126.8, 125.5, 121.0, 120.4, 115.8, 115.3, 112.6, 21.8, 13.3, 11.6; HRMS: *m/z* cacld. for $C_{33}H_{25}N_4O_2$ [M + H]$^+$ 509.1978, Found 509.1963.

4.3.11. *2-(5-Ethyl-4-(furan-2-yl)-3-methyl-1-phenyl-1H-pyrazolo[3,4-b]pyridin-6-yl)-3H-benzo[f]chromen-3-one* (**9g**):

Yellow solid, m.p.: >300 °C; IR (KBr, cm^{-1}) ν: 2966, 1720, 1629, 1566, 1412, 1383, 1264, 1084, 959, 852, 797, 766, 724, 691, 640, 617; ^{1}H-NMR (400 MHz, CF$_3$COOD) δ (ppm): 9.24 (s, 1H, ArH), 8.21 (d, *J* = 9.2 Hz, 1H, ArH), 7.95 (d, *J* = 8.8 Hz, 1H, ArH), 7.65–7.63 (m, 2H, ArH), 7.62–7.55 (m, 6H, ArH), 7.46 (d, *J* = 9.2 Hz, 1H, ArH), 7.39 (m, 3H, ArH), 2.93–2.87 (m, 2H, CH$_2$), 2.21 (s, 3H, CH$_3$), 1.03 (t, *J* = 7.6 Hz, 3H, CH$_3$); ^{13}C-NMR (75 MHz, CF$_3$COOD) δ (ppm): 163.0, 160.2, 156.0, 148.7, 145.7, 142.1, 140.0, 137.9, 135.2, 133.5, 131.9, 131.5, 130.7, 130.5, 129.8, 129.0, 127.1, 126.9, 125.8, 121.2, 118.0, 115.4, 113.7, 112.2, 22.1, 13.6, 11.8; HRMS: *m/z* cacld. for $C_{32}H_{24}N_3O_3$ [M + H]$^+$ 498.1818, Found 498.1831.

4.3.12. *2-(5-Ethyl-3-methyl-1,4-diphenyl-1H-pyrazolo[3,4-b]pyridin-6-yl)-9-methoxy-3H-benzo[f]chromen-3-one* (**9h**)

White solid, m.p.: 248–250 °C; IR (KBr, cm^{-1}) ν: 2968, 1724, 1631, 1573, 1507, 1434, 1414, 1384, 1354, 1281, 1241, 1135, 1105, 980, 960, 905, 827, 789, 758, 705, 692, 636; ^{1}H-NMR (400 MHz, DMSO-d_6) δ (ppm): 9.32 (s, 1H, ArH), 8.24 (d, *J* = 8.0 Hz, 2H, ArH), 8.17 (d, *J* = 9.2 Hz, 1H, ArH), 8.00–7.97 (m, 2H, ArH), 7.61–7.57 (m, 3H, ArH), 7.51–7.46 (m, 5H, ArH), 7.27–7.24 (m, 2H, ArH), 3.91 (s, 3H, OCH$_3$), 2.54–2.53 (m, 2H, CH$_2$), 1.89 (s, 3H, CH$_3$), 0.86 (t, *J* = 7.6 Hz, 3H, CH$_3$); ^{13}C-NMR (75 MHz, DMSO-d_6) δ (ppm): 160.3, 160.2, 154.5, 154.4, 148.7, 145.5, 142.8, 140.6, 139.5, 135.9, 134.0, 131.5, 131.0, 130.5, 129.6, 129.1, 129.0, 128.9, 127.0, 125.9, 125.7, 120.6, 118.7, 115.8, 114.4, 112.8, 102.7, 56.3, 22.5, 15.8, 14.2; HRMS: *m/z* cacld. for $C_{35}H_{28}N_3O_3$ [M + H]$^+$ 538.2131, Found 538.2122.

4.3.13. *2-(5-Ethyl-3-methyl-1-phenyl-4-(p-tolyl)-1H-pyrazolo[3,4-b]pyridin-6-yl)-9-methoxy-3H-benzo[f]chromen-3-one* (**9i**)

Yellow solid, m.p.: >300 °C; IR (KBr, cm^{-1}) ν: 2966, 1720, 1628, 1570, 1417, 1383, 1264, 1084, 959, 904, 832, 796, 761, 725, 691, 678, 640, 602; ^{1}H-NMR (400 MHz, CF$_3$COOD) δ (ppm): 9.24 (s, 1H, ArH), 8.20 (d, *J* = 9.2 Hz, 1H, ArH), 7.95 (d, *J* = 8.8 Hz, 1H, ArH), 7.65–7.54 (m, 8H, ArH), 7.46 (d, *J* = 9.2 Hz, 1H, ArH), 7.39 (d, *J* = 7.6 Hz, 3H, ArH), 4.04 (s, 3H, OCH$_3$), 2.92–2.87 (m, 2H, CH$_2$), 2.52 (s, 3H, CH$_3$), 2.20 (s, 3H, CH$_3$), 1.02 (t, *J* = 7.6 Hz, 3H, CH$_3$); ^{13}C-NMR (75 MHz, CF$_3$COOD) δ (ppm): 162.9, 160.1, 155.9, 148.6, 145.6, 145.5, 142.0, 139.8, 137.8, 135.1, 133.4, 131.8, 131.4, 130.6, 130.4, 129.6, 128.9, 127.0, 126.8, 125.7, 121.1, 117.9, 115.3, 113.6, 112.1, 55.3, 22.0, 19.6, 13.5, 11.8; HRMS: *m/z* cacld. for $C_{36}H_{30}N_3O_3$ [M + H]$^+$ 552.2287, Found 552.2246.

4.3.14. *2-(5-Ethyl-4-(4-methoxyphenyl)-3-methyl-1-phenyl-1H-pyrazolo[3,4-b]pyridin-6-yl)-9-methoxy-3H-benzo[f]chromen-3-one* (**9j**)

White solid, m.p.: 256–258 °C; IR (KBr, cm^{-1}) ν: 2965, 2145, 1735, 1717, 1629, 1572, 1463, 1381, 1286, 1227, 1077, 960, 887, 884, 805, 691, 604, 567; ^{1}H-NMR (400 MHz, DMSO-d_6) δ (ppm): 9.33 (s, 1H, ArH),

8.23 (d, *J* = 8.0 Hz, 2H, ArH), 8.19 (d, *J* = 8.8 Hz, 1H, ArH), 8.01–7.99 (m, 2H, ArH), 7.52–7.48 (m, 3H, ArH), 7.39–7.38 (m, 2H, ArH), 7.28–7.25 (m, 2H, ArH), 7.16 (d, *J* = 8.8 Hz, 2H, ArH), 3.92 (s, 3H, OCH$_3$), 2.87 (s, 3H, OCH$_3$), 2.56–2.55 (m, 2H, CH$_2$), 1.95 (s, 3H, CH$_3$), 0.87 (t, *J* = 7.2 Hz, 3H, CH$_3$); ^{13}C-NMR (75 MHz, DMSO-d_6) δ (ppm): 160.3, 160.2, 159.7, 154.6, 154.4, 148.8, 145.5, 142.9, 140.6, 139.6, 134.0, 131.5, 131.0, 130.9, 130.4, 129.7, 127.8, 127.1, 125.9, 125.7, 120.6, 118.8, 116.2, 114.3, 112.8, 102.7, 56.4, 55.7, 22.5, 15.8, 14.5; HRMS: *m/z* cacld. for C$_{36}$H$_{30}$N$_3$O$_4$ [M + H]$^+$ 568.2236, Found 568.2248.

4.3.15. 2-(3,5-Dimethyl-1,4-diphenyl-1H-pyrazolo[3,4-b]pyridin-6-yl)-3H-benzo[f]chromen-3-one (**9k**)

Yellow solid, m.p.: >300 °C; IR (KBr, cm^{-1}) ν: 2934, 2173, 1710, 1598, 1572, 1438, 1278, 965, 909, 820, 791, 692, 651, 633, 585; ^{1}H-NMR (400 MHz, CF$_3$COOD) δ (ppm): 9.33 (s, 1H, ArH), 8.29–8.24 (m, 2H, ArH), 8.00–7.97 (m, 1H, ArH), 7.72–7.61 (m, 11H, ArH), 7.48–7.47 (m, 2H, ArH), 2.42 (s, 3H, CH$_3$), 2.22 (s, 3H, CH$_3$); ^{13}C-NMR (75 MHz, CF$_3$COOD) δ (ppm): 155.3, 148.3, 146.2, 145.8, 139.9, 138.5, 133.5, 132.4, 131.8, 131.1, 130.8, 130.4, 130.0, 129.4, 129.3, 129.0, 128.7, 127.7, 126.9, 125.7, 120.6, 120.4, 115.9, 115.5, 112.9, 14.5, 12.0; HRMS: *m/z* cacld. for C$_{33}$H$_{24}$N$_3$O$_2$ [M + H]$^+$ 494.1869, Found 494.1887.

4.3.16. 2-(3,5-Dimethyl-1-phenyl-4-(p-tolyl)-1H-pyrazolo[3,4-b]pyridin-6-yl)-3H-benzo[f]chromen-3-one (**9l**)

Yellow solid, m.p.: 286–290 °C; IR (KBr, cm^{-1}) ν: 3078, 2187, 1719, 1626, 1606, 1575, 1507, 1447, 1380, 1212, 1093, 963, 813, 790, 741, 685; ^{1}H-NMR (400 MHz, CF$_3$COOD) δ (ppm): 9.35 (s, 1H, ArH), 8.31 (d, *J* = 9.2 Hz, 1H, ArH), 8.27 (d, *J* = 7.6 Hz, 1H, ArH), 8.02 (d, *J* = 8.4 Hz, 1H, ArH), 7.79 (t, *J* = 7.2 Hz, 1H, ArH), 7.72–7.57 (m, 9H, ArH), 7.39 (d, *J* = 7.6 Hz, 2H, ArH), 2.55 (s, 3H, CH$_3$), 2.46 (s, 3H, CH$_3$), 2.29 (s, 3H, CH$_3$); ^{13}C-NMR (75 MHz, CF$_3$COOD) δ (ppm): 154.3, 147.5, 145.2, 144.5, 141.3, 138.9, 137.5, 132.5, 130.8, 130.2, 129.4, 129.0, 128.8, 128.4, 128.3, 128.1, 127.7, 126.7, 126.0, 124.7, 119.6, 119.5, 114.9, 114.4, 18.6, 14.0, 10.9; HRMS: *m/z* cacld. for C$_{34}$H$_{26}$N$_3$O$_2$ [M + H]$^+$ 508.2025, Found 508.2020.

4.3.17. 2-(4-(4-Methoxyphenyl)-3,5-dimethyl-1-phenyl-1H-pyrazolo[3,4-b]pyridin-6-yl)-3H-benzo[f]chromen-3-one (**9m**)

White solid, m.p.: 258–260 °C; IR (KBr, cm^{-1}) ν: 2904, 2342, 1735, 1631, 1574, 1427, 1367, 1240, 1200, 1158, 1103, 1061, 849, 818, 759, 712, 668, 589; ^{1}H-NMR (400 MHz, DMSO-d_6) δ (ppm): 9.20 (s, 1H, ArH), 8.66 (d, *J* = 8.4 Hz, 1H, ArH), 8.28 (t, *J* = 7.6 Hz, 3H, ArH), 8.11 (d, *J* = 8.0 Hz, 1H, ArH), 7.75–7.63 (m, 3H, ArH), 7.50 (t, *J* = 8.0 Hz, 2H, ArH), 7.37 (d, *J* = 8.4 Hz, 2H, ArH), 7.26 (t, *J* = 7.2 Hz, 1H, ArH), 7.16 (d, *J* = 8.4 Hz, 2H, ArH), 3.87 (s, 3H, OCH$_3$), 2.13 (s, 3H, CH$_3$), 2.02 (s, 3H, CH$_3$); ^{13}C-NMR (75 MHz, DMSO-d_6) δ (ppm): 159.8, 159.6, 154.2, 154.0, 149.0, 145.3, 142.7, 140.3, 139.6, 134.3, 130.6, 130.5, 129.6, 129.5, 128.1, 126.7, 125.8, 124.9, 123.1, 120.6, 117.1, 115.9, 114.4, 113.6, 55.7, 16.3, 14.7; HRMS: *m/z* cacld. for C$_{34}$H$_{26}$N$_3$O$_3$ [M + H]$^+$ 524.1974, Found 524.1978.

4.3.18. 2-(4-(3-Methoxyphenyl)-3,5-dimethyl-1-phenyl-1H-pyrazolo[3,4-b]pyridin-6-yl)-3H-benzo[f]chromen-3-one (**9n**)

White solid, m.p.: 260–263 °C; IR (KBr, cm^{-1}) ν: 2970, 2372, 1718, 1573, 1505, 1410, 1362, 1279, 1239, 1142, 1054, 1019, 988, 970, 877, 815, 786, 744, 714, 670, 586; ^{1}H-NMR (400 MHz, DMSO-d_6) δ (ppm): 9.19 (s, 1H, ArH), 8.66 (d, *J* = 8.4 Hz, 1H, ArH), 8.29–8.25 (m, 3H, ArH), 8.10 (d, *J* = 8.0 Hz, 1H, ArH), 7.75–7.63 (m, 3H, ArH), 7.52–7.48 (m, 3H, ArH), 7.25 (t, *J* = 7.6 Hz, 1H, ArH), 7.13–7.11 (m, 1H, ArH), 7.00–6.98 (m, 2H, ArH), 3.84 (s, 3H, OCH$_3$), 2.13 (s, 3H, CH$_3$), 2.00 (s, 3H, CH$_3$); ^{13}C-NMR (75 MHz, DMSO-d_6) δ (ppm): 159.7, 159.6, 154.3, 154.0, 148.9, 145.2, 142.7, 140.4, 139.6, 137.6, 134.3, 130.5, 130.3, 129.6, 129.5, 129.4, 129.0, 128.1, 126.8, 125.9, 124.5, 123.1, 121.3, 120.6, 117.1, 115.5, 114.7, 114.6, 113.6, 55.8, 16.2, 14.4; HRMS: *m/z* cacld. for C$_{34}$H$_{26}$N$_3$O$_3$ [M + H]$^+$ 524.1974, Found 524.1978.

4.3.19. 2-(3,5-Dimethyl-1-phenyl-4-(p-tolyl)-1H-pyrazolo[3,4-b]pyridin-6-yl)-9-methoxy-3H-benzo[f]chromen-3-one (**9o**)

Yellow solid, m.p.: 288–290 °C; IR (KBr, cm^{-1}) ν: 2929, 1718, 1631, 1600, 1346, 1239, 1204, 1173, 1149, 1125, 1019, 852, 827, 795, 749, 690, 643, 606; ^{1}H-NMR (400 MHz, CF$_3$COOD) δ (ppm): 9.28 (s, 1H, ArH),

8.22 (d, J = 8.8 Hz, 1H, ArH), 7.96 (d, J = 8.8 Hz, 1H, ArH), 7.65–7.62 (m, 6H, ArH), 7.56–7.54 (m, 2H, ArH), 7.46 (d, J = 8.8 Hz, 1H, ArH), 7.41–7.35 (m, 3H, ArH), 4.04 (s, 3H, OCH$_3$), 2.52 (s, 3H, CH$_3$), 2.44 (s, 3H, CH$_3$), 2.26 (s, 3H, CH$_3$); ^{13}C-NMR (75 MHz, CF$_3$COOD) δ (ppm): 159.0, 158.9, 155.0, 147.4, 145.2, 144.4, 141.2, 138.7, 137.0, 132.4, 130.7, 130.4, 129.6, 129.4, 128.8, 128.2, 128.1, 126.0, 125.8, 124.6, 119.4, 116.6, 115.2, 114.1, 112.5, 54.2, 18.5, 14.0, 10.9; HRMS: m/z cacld. for C$_{35}$H$_{28}$N$_3$O$_3$ [M + H]$^+$ 538.2131, Found 538.2130.

4.3.20. *9-Methoxy-2-(4-(4-methoxyphenyl)-3,5-dimethyl-1-phenyl-1H-pyrazolo[3,4-b]pyridin-6-yl)-3H-benzo[f]chromen-3-one* (**9p**)

Yellow solid, m.p.: 287–289 °C; IR (KBr, cm^{-1}) v: 1716, 1630, 1611, 1571, 1513, 1464, 1385, 1246, 1107, 1033, 960, 832, 795, 754, 691; ^{1}H-NMR (400 MHz, CF$_3$COOD) δ (ppm): 9.29 (s, 1H, ArH), 8.23 (d, J = 8.8 Hz, 1H, ArH), 7.56 (d, J = 8.8 Hz, 1H, ArH), 7.65–7.63 (m, 6H, ArH), 7.48–7.42 (m, 3H, ArH), 7.39–7.35 (m, 3H, ArH), 4.07–4.05 (m, 6H, 2 × OCH$_3$), 2.46 (s, 3H, CH$_3$), 2.30 (s, 3H, CH$_3$); ^{13}C-NMR (75 MHz, CF$_3$COOD) δ (ppm): 159.6, 159.0, 155.0, 147.2, 145.1, 144.6, 138.9, 137.0, 132.4, 130.7, 130.4, 129.6, 129.3, 128.3, 128.1, 125.8, 124.6, 116.6, 114.0, 112.5, 54.3, 54.0, 14.0, 11.1; HRMS: m/z cacld. for C$_{35}$H$_{28}$N$_3$O$_4$ [M + H]$^+$ 554.2080, Found 554.2093.

4.3.21. *2-(3,5-Dimethyl-1,4-diphenyl-1H-pyrazolo[3,4-b]pyridin-6-yl)-9-methoxy-3H-benzo[f]chromen-3-one* (**9q**)

Yellow solid, m.p.: 252–254 °C; IR (KBr, cm^{-1}) v: 2961, 1725, 1629, 1582, 1557, 1435, 1397, 1335, 1290, 1250, 1219, 1196, 999, 906, 819, 797, 753, 695, 625; ^{1}H-NMR (400 MHz, DMSO-d_6) δ (ppm): 9.29 (s, 1H, ArH), 8.28–8.26 (m, 2H, ArH), 8.17 (d, J = 9.2 Hz, 1H, ArH), 8.01–7.98 (m, 2H, ArH), 7.62–7.57 (m, 3H, ArH), 7.52–7.43 (m, 5H, ArH), 7.28–7.25 (m, 2H, ArH), 3.92 (s, 3H, OCH$_3$), 2.11 (s, 3H, CH$_3$), 1.95 (s, 3H, CH$_3$); ^{13}C-NMR (75 MHz, DMSO-d_6) δ (ppm): 160.2, 159.7, 154.6, 154.5, 148.9, 145.3, 142.6, 140.8, 139.6, 136.2, 134.0, 131.5, 131.1, 129.7, 129.2, 129.1, 127.4, 125.9, 125.7, 124.5, 120.7, 118.7, 115.5, 114.3, 112.9, 102.8, 56.3, 16.2, 14.5; HRMS: m/z cacld. for C$_{34}$H$_{26}$N$_3$O$_3$ [M + H]$^+$ 524.1974, Found 524.1988.

4.3.22. *2-(3-Methyl-1,4-diphenyl-1H-pyrazolo[3,4-b]pyridin-6-yl)-3H-benzo[f]chromen-3-one* (**9r**)

Yellow solid, m.p.: 268–270 °C; IR (KBr, cm^{-1}) v: 2935, 2355, 1729, 1667, 1553, 1092, 891, 818, 746, 694, 657, 631, 585; ^{1}H-NMR (400 MHz, CF$_3$COOD) δ (ppm): 10.22 (s, 1H, ArH), 8.61–8.57 (m, 2H, ArH), 8.46 (d, J = 9.2 Hz, 1H, ArH), 8.09 (d, J = 8.0 Hz, 1H, ArH), 7.90 (t, J = 7.6 Hz, 1H, ArH), 7.82–7.76 (m, 11H, ArH), 7.70 (d, J = 8.8 Hz, 1H, ArH), 2.58 (s, 3H, CH$_3$); ^{13}C-NMR (75 MHz, CF$_3$COOD) δ (ppm): 163.5, 159.1, 155.0, 147.5, 145.0, 144.8, 140.3, 138.7, 132.9, 132.4, 131.1, 130.4, 130.3, 130.0, 129.8, 128.8, 128.3, 127.9, 127.4, 127.3, 122.6, 120.1, 114.3, 114.1, 113.5, 11.9; HRMS: m/z cacld. for C$_{32}$H$_{22}$N$_3$O$_2$ [M + H]$^+$ 480.1712, Found 480.1726.

4.3.23. *2-(4-(4-Methoxyphenyl)-3-methyl-1-phenyl-1H-pyrazolo[3,4-b]pyridin-6-yl)-3H-benzo[f]chromen-3-one* (**9s**)

Yellow solid, m.p.: >300 °C; IR (KBr, cm^{-1}) v: 2988, 2355, 1987, 1730, 1512, 1089, 1066, 959, 810, 809, 788, 765, 689, 654, 633, 599; ^{1}H-NMR (400 MHz, CF$_3$COOD) δ (ppm): 11.03 (s, 1H, ArH), 9.43 (d, J = 8.4 Hz, 1H, ArH), 9.34–9.30 (m, 2H, ArH), 8.94 (d, J = 8.0 Hz, 1H, ArH), 8.75 (t, J = 7.6 Hz, 1H, ArH), 7.66–7.63 (m, 8H, ArH), 8.54 (t, J = 9.2 Hz, 1H, ArH), 8.23 (d, J = 8.8 Hz, 2H, ArH), 4.95 (s, 3H, OCH$_3$), 3.50 (s, 3H, CH$_3$); ^{13}C-NMR (75 MHz, CF$_3$COOD) δ (ppm): 164.3, 159.2, 155.9, 148.2, 145.7, 145.3, 141.1, 139.8, 133.6, 131.3, 130.9, 130.6, 129.7, 128.8, 128.2, 126.6, 123.5, 120.9, 117.1, 115.1, 114.6, 55.1, 13.1; HRMS: m/z cacld. for C$_{33}$H$_{24}$N$_3$O$_3$ [M + H]$^+$ 510.1818, Found 510.1835.

4.3.24. *2-(5-Ethyl-1-methyl-3,4-diphenyl-1H-pyrazolo[3,4-b]pyridin-6-yl)-3H-benzo[f]chromen-3-one* (**9t**)

Yellow solid, m.p.: 285–288 °C; IR (KBr, cm^{-1}) v: 2396, 1732, 1574, 1353, 1099, 1515, 1088, 1076, 959, 810, 803, 704; ^{1}H-NMR (400 MHz, CF$_3$COOD) δ (ppm): 10.26 (s, 1H, ArH), 9.25 (d, J = 8.8 Hz, 2H, ArH), 8.95 (d, J = 8.0 Hz, 1H, ArH), 8.73 (t, J = 7.6 Hz, 1H, ArH), 8.63 (t, J = 7.6 Hz, 1H, ArH), 8.46 (d,

J = 7.6 Hz, 1H, ArH), 8.26 (t, J = 7.6 Hz, 1H, ArH), 8.21–8.14 (m, 3H, ArH), 8.09–8.02 (m, 4H, ArH), 7.94 (d, J = 7.6 Hz, 2H, ArH), 5.31 (s, 3H, CH$_3$), 3.87–3.85 (m, 2H, CH$_2$), 1.84 (t, J = 7.2 Hz, 3H, CH$_3$); ^{13}C-NMR (75 MHz, CF$_3$COOD) δ (ppm): 162.9, 155.3, 150.3, 146.0, 145.4, 140.3, 138.2, 134.3, 131.2, 131.1, 130.2, 129.9, 129.5, 129.3, 128.5, 128.3, 127.9, 127.8, 127.6, 127.1, 120.5, 119.3, 116.3, 115.5, 34.9, 21.8, 13.1; HRMS: m/z cacld. for C$_{34}$H$_{26}$N$_3$O$_2$ [M + H]$^+$ 508.2025, Found 508.2027.

4.3.25. 2-(1,5-Dimethyl-3,4-diphenyl-1H-pyrazolo[3,4-b]pyridin-6-yl)-9-methoxy-3H-benzo[f]chromen-3-one (**9u**)

Yellow solid, m.p.: 260–262 °C; IR (KBr, cm^{-1}) ν: 2697, 2551, 1783, 1708, 1628, 1567, 1511, 1469, 1441, 1387, 1330, 1218, 1149, 1017, 976, 898, 845, 796, 756, 725, 702, 601; ^{1}H-NMR (400 MHz, CF$_3$COOD) δ (ppm): 10.10 (s, 1H, ArH), 9.03 (d, J = 9.2 Hz, 1H, ArH), 8.76 (d, J = 8.8 Hz, 1H, ArH), 8.66–8.65 (m, 1H, ArH), 8.28 (d, J = 8.8 Hz, 1H, ArH), 8.20 (d, J = 9.2 Hz, 1H, ArH), 8.12 (d, J = 7.6 Hz, 1H, ArH), 8.05–7.99 (m, 3H, ArH), 7.91–7.87 (m, 4H, ArH), 7.81 (d, J = 7.6 Hz, 2H, ArH), 5.18 (s, 3H, OCH$_3$), 4.85 (s, 3H, CH$_3$), 3.21 (s, 3H, CH$_3$). ^{13}C-NMR (75 MHz, CF$_3$COOD) δ (ppm): 162.3, 160.0, 156.0, 150.1, 146.1, 145.8, 140.3, 138.0, 131.5, 131.4, 130.6, 130.3, 129.5, 128.5, 128.3, 128.2, 128.0, 127.9, 127.2, 126.7, 117.5, 113.5, 112.2, 112.1, 103.0, 55.2, 34.8, 15.1. HRMS: m/z cacld. for C$_{34}$H$_{26}$N$_3$O$_3$ [M + H]$^+$ 554.2080, Found 554.2093.

4.3.26. 9-Methoxy-2-(4-(4-methoxyphenyl)-1,5-dimethyl-3-phenyl-1H-pyrazolo[3,4-b]pyridin-6-yl)-3H-benzo[f]chromen-3-one (**9v**)

Yellow solid, m.p.: 240–244 °C; IR (KBr, cm^{-1}) ν: 2932, 1720, 1624, 1608, 1564, 1512, 1463, 1383, 1353, 1289, 1249, 1208, 1173, 1025, 970, 902, 836, 801, 698, 664, 607; ^{1}H-NMR (400 MHz, CF$_3$COOD) δ (ppm): 10.15 (s, 1H, ArH), 9.09 (d, J = 9.2 Hz, 1H, ArH), 8.82 (d, J = 9.2 Hz, 1H, ArH), 8.72–8.71 (m, 1H, ArH), 8.34 (d, J = 9.2 Hz, 1H, ArH), 8.26 (d, J = 9.2 Hz, 1H, ArH), 8.17–8.13 (m, 1H, ArH), 8.02–7.95 (m, 4H, ArH), 7.89–7.87 (m, 2H, ArH), 7.69–7.66 (m, 2H, ArH), 5.23 (s, 3H, OCH$_3$), 4.90 (s, 3H, OCH$_3$), 4.72 (s, 3H, CH$_3$), 3.31 (s, 3H, CH$_3$); ^{13}C-NMR (75 MHz, CF$_3$COOD) δ (ppm): 162.4, 160.3, 156.1, 150.0, 146.0, 145.9, 140.3, 138.1, 131.5, 130.4, 129.4, 128.7, 128.1, 128.0, 127.6, 126.8, 125.1, 117.5, 114.2, 113.5, 55.2, 55.1, 34.9, 15.1; HRMS: m/z cacld. for C$_{35}$H$_{28}$N$_3$O$_4$ [M + H]$^+$ 524.1974, Found 524.1972.

Supplementary Materials: The following are available online, Crystal date of compound 7a [47], ^{1}H NMR and ^{13}C NMR Spectra of all compounds and GC-MS spectra of Scheme 4B.

Author Contributions: W.L. conceived the synthetic route, supervised the project and wrote the paper; J.Z. and J.W. designed the experiments; C.Z. and X.H. performed all synthetic work in the laboratory.

Funding: This research was supported financially by the Natural Science Foundation of China (no. 21502074), Qing Lan Project of Jiangsu Province, Postgraduate Research & Practice Innovation Program of Jiangsu Province (SJCX18_0985 and SJCX19_0766) and CAMS Initiative for Innovative Medicine (2016-I2M-3-014).

Conflicts of Interest: The authors declare no conflict of interest.

References and Note

1. Revankar, H.M.; Bukhari, S.N.A.; Kumar, G.B.; Qin, H.L. Coumarins scaffolds as COX inhibitors. *Bioorg. Chem.* **2017**, *71*, 146–159. [CrossRef] [PubMed]
2. Dandriyal, J.; Singla, R.; Kumar, M.; Jaitak, V. Recent developments of C-4 substituted coumarin derivatives as anticancer agents. *Eur. J. Med. Chem.* **2016**, *119*, 141–168. [CrossRef] [PubMed]
3. Ibrar, A.; Shehzadi, S.A.; Saeed, F.; Khan, I. Developing hybrid molecule therapeutics for diverse enzyme inhibitory action: Active role of coumarin-based structural leads in drug discovery. *Bioorg. Med. Chem.* **2018**, *26*, 3731–3762. [CrossRef] [PubMed]
4. Singla, S.; Piplani, P. Coumarin derivatives as potential inhibitors of acetylcholinesterase: Synthesis, molecular docking and biological studies. *Bioorg. Med. Chem.* **2016**, *24*, 4587–4599. [CrossRef] [PubMed]
5. Hamulakova, S.; Janovec, L.; Soukup, O.; Jun, D.; Kuca, K. Synthesis, in vitro acetylcholinesterase inhibitory activity and molecular docking of new acridine-coumarin hybrids. *Int. J. Biol. Macromol.* **2017**, *104*, 333–338. [CrossRef]

6. Lan, J.S.; Ding, Y.; Liu, Y.; Kang, P.; Hou, J.W.; Zhang, X.Y.; Xie, S.S.; Zhang, T. Design, synthesis and biological evaluation of novel coumarin-N-benzyl pyridinium hybrids as multi-target agents for the treatment of Alzheimer's disease. *Eur. J. Med. Chem.* **2017**, *139*, 48–59. [CrossRef] [PubMed]
7. Emami, S.; Dadashpour, S. Current developments of coumarin-based anti-cancer agents in medicinal chemistry. *Eur. J. Med. Chem.* **2015**, *102*, 611–630. [CrossRef]
8. Kaur, M.; Kohli, S.; Sandhu, S.; Bansal, Y.; Bansal, G. Coumarin: A promising scaffold for anticancer agents. *Anti Cancer Agents Med. Chem.* **2015**, *15*, 1032–1048. [CrossRef]
9. Klenkar, J.; Molnar, M. Natural and synthetic coumarins as potential anticancer agents. *J. Chem. Pharm. Res.* **2015**, *7*, 1223–1238.
10. Poole, S.K.; Poole, C.F. Thin-layer chromatographic method for the determination of the principal polar aromatic flavour compounds of the cinnamons of commerce. *Analyst* **1994**, *119*, 113–120. [CrossRef]
11. Riveiro, M.E.; De Kimpe, N.; Moglioni, A.; Vazquez, R.; Monczor, F.; Shayo, C.; Davio, C. Coumarins: Old compounds with novel promising therapeutic perspectives. *Curr. Med. Chem.* **2010**, *17*, 1325–1338. [CrossRef] [PubMed]
12. Patil, A.D.; Freyer, A.J.; Eggleston, D.S.; Haltiwanger, R.C.; Bean, M.F.; Taylor, P.B.; Caranfa, M.J.; Breen, A.L.; Bartus, H.R. The inophyllums, novel inhibitors of HIV-1 reverse transcriptase isolated from the malaysian tree, calophyllum inophyllum linn. *J. Med. Chem.* **1993**, *36*, 4131–4138. [CrossRef] [PubMed]
13. Spino, C.; Dodier, M.; Sotheeswaran, S. Anti-HIV coumarins from calophyllum seed oil. *Bioorg. Med. Chem. Lett.* **1998**, *8*, 3475–3478. [CrossRef]
14. Kostova, I.; Mojzis, J. Biologically active coumarins as inhibitors of HIV-1. *Future HIV Ther.* **2007**, *1*, 315–329. [CrossRef]
15. Shin, E.; Choi, K.M.; Yoo, H.S.; Lee, C.K.; Hwang, B.Y.; Lee, M.K. Inhibitory effects of coumarins from the stem barks of Fraxinus rhynchophylla on adipocyte differentiation in 3T3-L1 cells. *Biol. Pharm. Bull.* **2010**, *33*, 1610–1614. [CrossRef] [PubMed]
16. Keri, R.S.; Sasidhar, B.S.; Nagaraja, B.M.; Santos, M.A. Recent progress in the drug development of coumarin derivatives as potent antituberculosis agents. *Eur. J. Med. Chem.* **2015**, *100*, 257–269. [CrossRef] [PubMed]
17. Piller, N.B. A comparison of the effectiveness of some anti-inflammatory drugs on thermal oedema. *Br. J. Exp. Pathol.* **1975**, *56*, 554–560. [PubMed]
18. Bansal, Y.; Sethi, P.; Bansal, G. Coumarin: A potential nucleus for anti-inflammatory molecules. *Med. Chem. Res.* **2013**, *22*, 3049–3060. [CrossRef]
19. Whang, W.K.; Park, H.S.; Ham, I.; Oh, M.; Namkoong, H.; Kim, H.K.; Hwang, D.W.; Hur, S.Y.; Kim, T.E.; Park, Y.G. Natural compounds, fraxin and chemicals structurally related to fraxin protect cells from oxidative stress. *Exp. Mol. Med.* **2005**, *37*, 436–446. [CrossRef]
20. Rosselli, S.; Maggio, A.M.; Faraone, N.; Spadaro, V.; Morris-Natschke, S.L.; Bastow, K.F.; Lee, K.H.; Bruno, M. The cytotoxic properties of natural coumarins isolated from roots of ferulago campestris (Apiaceae) and of synthetic ester derivatives of aegelinol. *Nat. Prod. Commun.* **2009**, *4*, 1701–1706. [CrossRef]
21. Crichton, E.G.; Waterman, P.G. Dihydromammea C/OB: A new coumarin from the seed of mammea Africana. *Phytochemistry* **1978**, *17*, 1783–1786. [CrossRef]
22. Baek, N.I.; Ahn, E.M.; Kim, H.Y.; Park, Y.D. Furanocoumarins from the root of Angelica dahurica. *Arch. Pharm. Res.* **2000**, *23*, 467–470. [CrossRef] [PubMed]
23. Teng, M.C.; Lin, H.; Ko, F.N.; Wu, T.S.; Huang, T.F. The relaxant action of osthole isolated from Angelica pubescens in guinea-pig trachea. *Naunyn Schmiedeberg's. Arch. Pharmacol.* **1994**, *349*, 202–208. [CrossRef] [PubMed]
24. Fort, D.; Rao, K.; Jolad, S.; Luo, J.; Carlson, T.; King, S. Antihyperglycemic activity of teramnus labialis (fabaceae). *Phytomedicine* **2000**, *6*, 465–467. [CrossRef]
25. Liu, Y.H.; Xu, J.R.; Wang, Q.; Li, M. Coupling coumarin to gold nanoparticles by DNA chains for sensitive detection of DNase I. *Anal. Biochem.* **2018**, *555*, 50–54. [CrossRef] [PubMed]
26. Chesterman, J.P.; Hughes, T.C.; Amsden, B.G. Reversibly photo-crosslinkable aliphatic polycarbonates functionalized with coumarin. *Eur. Polym. J.* **2018**, *105*, 186–193. [CrossRef]
27. Sarkar, N.; Datta, A.; Das, S.; Bhattacharyya, K. Solvation dynamics of coumarin 480 in micelles. *J. Phys. Chem.* **1996**, *100*, 15483–15486. [CrossRef]
28. Arzhantsev, S.; Ito, N.; Heitz, M.; Maroncelli, M. Solvation dynamics of coumarin 153 in several classes of ionic liquids: Cation dependence of the ultrafast component. *Chem. Phys. Lett.* **2003**, *381*, 278–286. [CrossRef]

29. Lang, B.; Angulo, G.; Vauthey, E. Ultrafast solvation dynamics of coumarin 153 in imidazolium-based ionic liquids. *J. Phys. Chem. A* **2006**, *110*, 7028–7034. [CrossRef]
30. Birau, M.M.; Wang, Z.Y. A dual-mode molecular switch based on a chiral binaphthol-coumarin compound. *Tetrahedron Lett.* **2000**, *41*, 4025–4028. [CrossRef]
31. Deng, G.W.; Xu, H.J.; Kuang, L.; He, C.C.; Li, B.K.; Yang, M.; Zhang, X.L.; Li, Z.H.; Liu, J.L. Novel nonlinear optical chromophores based on coumarin: Synthesis and properties studies. *Opt. Mater.* **2019**, *88*, 218–222. [CrossRef]
32. Goda, F.E.; Abdel-Azizb, A.A.M.; Attef, O.A. Synthesis, antimicrobial activity and conformational analysis of novel substituted pyridines: BF_3-promoted reaction of hydrazine with 2-alkoxy pyridines. *Bioorg. Med. Chem.* **2004**, *12*, 1845–1852. [CrossRef] [PubMed]
33. Foks, H.; Pancechowska-Ksepko, D.; Kędzia, A.; Zwolska, Z.; Janowiec, M.; Augustinowicz-Kopeć, E. Synthesis and antibacterial activity of 1H-pyrazolo[3,4-b] pyrazine and-pyridine derivatives. *Farmaco* **2005**, *60*, 513–517. [CrossRef] [PubMed]
34. Bharate, S.B.; Mahajan, T.R.; Gole, Y.R.; Nambiar, M.; Matan, T.T.; Kulkarni-Almeida, A.; Balachandran, S.; Junjappa, H.; Balakrishnan, A.; Vishwakarma, R.A. Synthesis and evaluation of pyrazolo[3,4-b] pyridines and its structural analogues as TNF-α and IL-6 inhibitors. *Bioorg. Med. Chem.* **2008**, *16*, 7167–7176. [CrossRef] [PubMed]
35. De Mello, H.; Echevarria, A.; Bernardino, A.M.; CantoCavalheiro, M.; Leon, L.L. Antileishmanial pyrazolopyridine derivatives: Synthesis and structure-activty relationship analysis. *J. Med. Chem.* **2004**, *47*, 5427–5432. [CrossRef] [PubMed]
36. Misra, R.N.; Rawlins, D.B.; Xiao, H.; Shan, W.; Bursuker, I.; Kellar, K.A.; Mulheron, J.G.; Sack, J.S.; Tokarski, J.S.; Kimball, S.D.; et al. 1H-Pyrazolo[3,4-b] pyridine inhibitors of cyclin-dependent kinases: Highly potent 2,6-difluorophenacyl analogues. *Bioorg. Med. Chem. Lett.* **2003**, *13*, 2405–2408. [CrossRef]
37. Lin, R.; Connolly, P.J.; Lu, Y.; Chiu, G.; Li, S.; Yu, Y.; Huang, S.; Li, X.; Emanuel, S.L.; Middleton, S.A.; et al. Synthesis and evaluation of pyrazolo[3,4-b] pyridine CDK1 inhibitors as anti-tumor agents. *Bioorg. Med. Chem. Lett.* **2007**, *17*, 4297–4302. [CrossRef] [PubMed]
38. Parker, W.B. Enzymology of purine and pyrimidine antimetabolites used in the treatment of cancer. *Chem. Rev.* **2009**, *109*, 2880–2893. [CrossRef]
39. Miliutina, M.; Janke, J.; Hassan, S.; Zaib, S.; Iqbal, J.; Lecka, J.; Sévigny, J.; Villinger, A.; Friedrich, A.; Lochbrunner, S.; et al. A domino reaction of 3-chlorochromones with aminoheterocycles. Synthesis of pyrazolopyridines and benzofuropyridines and their optical and ecto-5′-nucleotidase inhibitory effects. *Org. Biomol. Chem.* **2018**, *16*, 717–732. [CrossRef]
40. Lin, W.; Hu, X.X.; Song, S.; Cai, Q.; Wang, Y.; Shi, D.Q. Microwave-assisted synthesis of novel hetero[5] helicene-like molecules and coumarin derivatives. *Org. Biomol. Chem.* **2017**, *15*, 7909–7916. [CrossRef]
41. Wang, H.Y.; Liu, X.C.; Feng, X.; Huang, Z.B.; Shi, D.Q. GAP chemistry for pyrrolyl coumarin derivatives: A highly efficient one-pot synthesis under catalyst-free conditions. *Green Chem.* **2013**, *15*, 3307–3311. [CrossRef]
42. Wang, H.Y.; Shi, D.Q. Efficient synthesis of functionalized dihydro-1H-indol-4(5H)-ones via one-pot three-component reaction under catalyst-free conditions. *ACS Comb. Sci.* **2013**, *15*, 261–266. [CrossRef] [PubMed]
43. Liu, X.C.; Lin, W.; Wang, H.Y.; Huang, Z.B.; Shi, D.Q. Improved and efficient synthesis of chromeno [4,3-d]pyrazolo[3,4-b]pyridine-6(3H)-ones and their fluorescence properties. *J. Heterocycl. Chem.* **2014**, *51*, 1036–1044. [CrossRef]
44. Wang, J.X.; Lin, W.; Liu, H.T.; Hu, M.H.; Feng, X.; Ren, J.F.; Huang, Z.B.; Shi, D.Q. An efficient synthesis of coumarino[4,3-d]pyrazolo[3,4-b]-pyridine derivatives catalyzed by silica sulfuric acid under microwave irradiation. *Chin. J. Org. Chem.* **2015**, *35*, 927–933. [CrossRef]
45. Lin, W.; Cai, Q.; Zheng, C.Z.; Zheng, Y.X.; Shi, D.Q. Synthesis of functionalized coumarino[4,3-d]pyrazolo [3,4-b]pyridine derivatives and their selective recognition for Zn^{2+}. *Chin. J. Org. Chem.* **2017**, *37*, 2392–2398. [CrossRef]
46. Lin, W.; Zheng, Y.X.; Wang, Y.Z.; Shi, D.Q. An efficient synthesis of functionalized chromeno[4,3-d]pyrazolo [3,4-b]pyridine derivatives. *Heterocycles* **2016**, *92*, 2235–2243.

47. Crystallographic Data for **7a** Have been Deposited at the Cambridge Crystallographic Data Centre (CCDC 1881623). Available online: www.ccdc.cam.ac.uk/conts/retrieving.html (accessed on 30 July 2019).
48. Bogdal, D. Coumarins: Fast synthesis by knoevenagel condensation under microwave irradiation. *J. Chem. Res. Synop.* **1998**, *12*, 468–469. [CrossRef]

Sample Availability: Samples of the compounds **4** and **7** are available from the authors.

© 2019 by the authors. Licensee MDPI, Basel, Switzerland. This article is an open access article distributed under the terms and conditions of the Creative Commons Attribution (CC BY) license (http://creativecommons.org/licenses/by/4.0/).

Article

Synthesis and Properties of 6-Aryl-4-azidocinnolines and 6-Aryl-4-(1,2,3-1*H*-triazol-1-yl)cinnolines

Natalia A. Danilkina [1], Nina S. Bukhtiiarova [1], Anastasia I. Govdi [1], Anna A. Vasileva [1], Andrey M. Rumyantsev [2], Artemii A. Volkov [2], Nikita I. Sharaev [2], Alexey V. Povolotskiy [1], Irina A. Boyarskaya [1], Ilya V. Kornyakov [1,3], Polina V. Tokareva [1] and Irina A. Balova [1,*]

[1] Institute of Chemistry, Saint Petersburg State University (SPSU), Universitetskaya nab. 7/9, Saint Petersburg 199034, Russia
[2] Department of Genetics and Biotechnology, Saint Petersburg State University (SPSU), Universitetskaya nab. 7/9, Saint Petersburg 199034, Russia
[3] Laboratory of Nature-Inspired Technologies and Environmental Safety of the Arctic, Kola Science Centre, Russian Academy of Sciences, Fersmana 14, Apatity 184209, Russia
* Correspondence: i.balova@spbu.ru; Tel.: +7-812-428-6733

Received: 11 June 2019; Accepted: 24 June 2019; Published: 27 June 2019

Abstract: An efficient approach towards the synthesis of 6-aryl-4-azidocinnolines was developed with the aim of exploring the photophysical properties of 6-aryl-4-azidocinnolines and their click reaction products with alkynes, 6-aryl-4-(1,2,3-1*H*-triazol-1-yl)cinnolines. The synthetic route is based on the Richter-type cyclization of 2-ethynyl-4-aryltriazenes with the formation of 4-bromo-6-arylcinnolines and nucleophilic substitution of a bromine atom with an azide functional group. The developed synthetic approach is tolerant to variations of functional groups on the aryl moiety. The resulting azidocinnolines were found to be reactive in both CuAAC with terminal alkynes and SPAAC with diazacyclononyne, yielding 4-triazolylcinnolines. It was found that 4-azido-6-arylcinnolines possess weak fluorescent properties, while conversion of the azido function into a triazole ring led to complete fluorescence quenching. The lack of fluorescence in triazoles could be explained by the non-planar structure of triazolylcinnolines and a possible photoinduced electron transfer (PET) mechanism. Among the series of 4-triazolylcinnoline derivatives a compound bearing hydroxyalkyl substituent at triazole ring was found to be cytotoxic to HeLa cells.

Keywords: azides; cinnolines; triazoles; CuAAC; alkynes; cycloalkynes; Richter cyclization; Suzuki coupling; fluorescence; cytotoxicity

1. Introduction

During last decade azido-substituted heterocycles have been broadly studied in several areas. The first one is photoaffinity labeling [1] which is an important biological tool for the investigation of different biological processes by specific modification of proteins using their interaction with singlet nitrenes generated from azidoheterocyles under UV irradiation (Figure 1A) [2–4]. The second approach involves use of azidoheterocycles in the synthesis of triazolylheterocycles, mainly by Cu-catalyzed alkyne-azide cycloaddition (CuAAC) [5,6]. This methodology allowed a great variety of biologically active triazolyl substituted heterocycles to be synthesized and tested in numerous biological assays (Figure 1B) [7–10]. The third main field where azidoheterocycles are in great demand is as azide-based fluorophore tags for biological imaging (Figure 1C) [11–16].

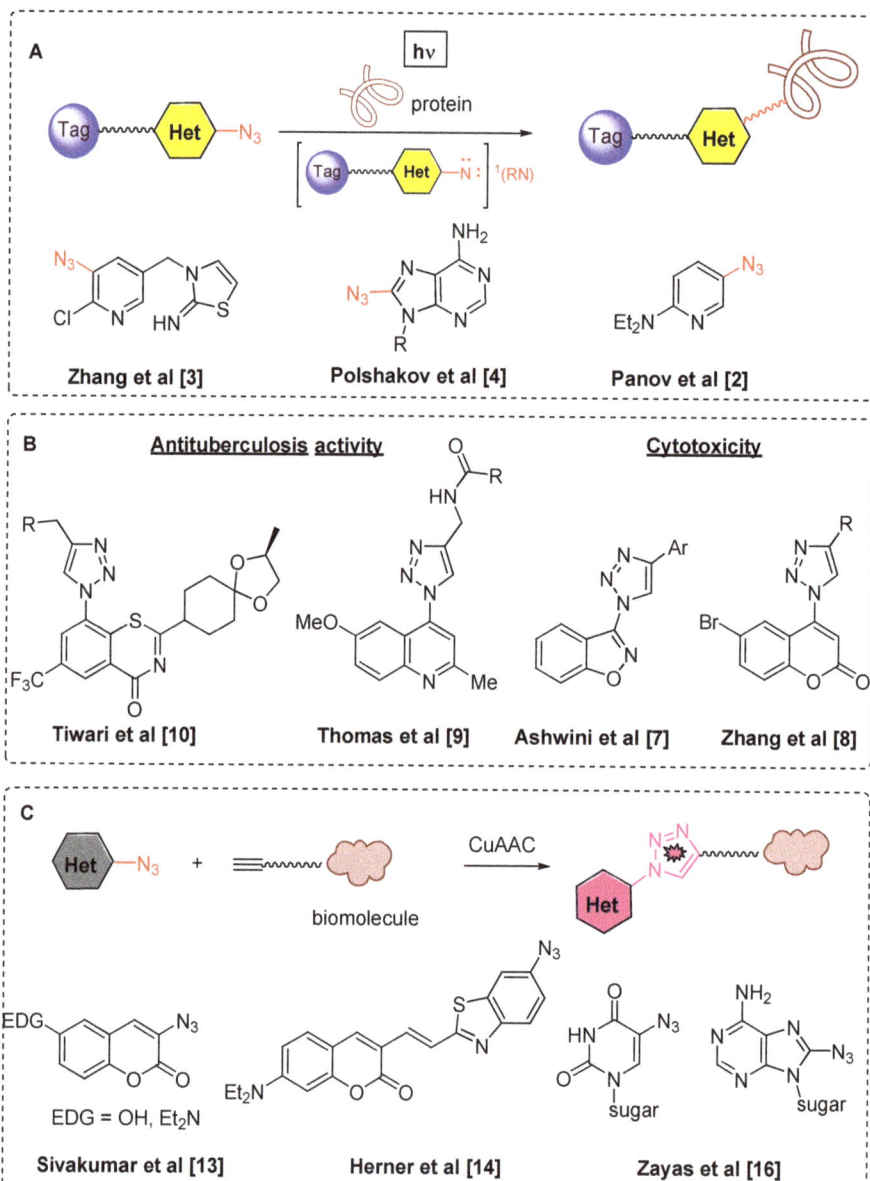

Figure 1. Recent areas of azidoheterocycles and triazolylheterocycles application–photoaffinity labeling (**A**); biologically-active compounds (**B**); fluorogenic azides (**C**).

Among azidoheterocycles, azidoquinolines (Figure 2A) [17–23] and azidopyridazines (Figure 2B) [24–28] are well known, however only one known example of an azidocinnloline has been reported to date (Figure 2C) [29], while triazolylcinnolines still remain unknown. Taking into account that the cinnoline core can be found both in biologically active compounds with different types of activities [30–34] and in fluorescent materials [35–40], the development of a synthetic route towards azidocinnolines and triazolylcinnolines could lead to new biologically active molecules and interesting fluorescent probes.

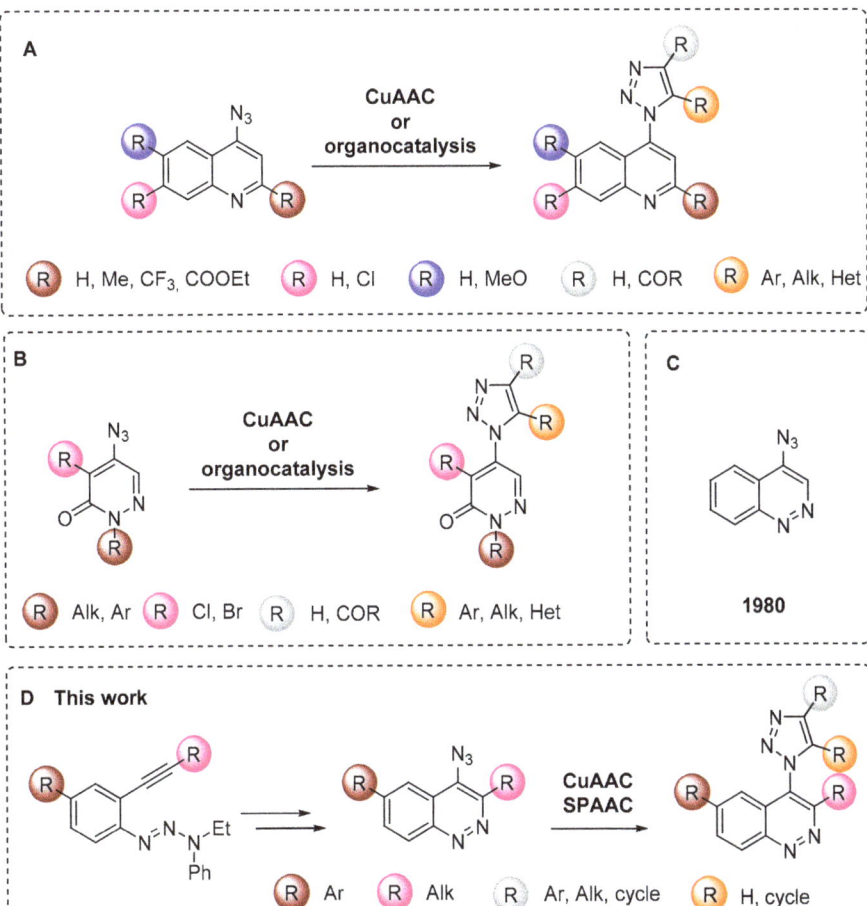

Figure 2. Known examples of azidoquinolines, azidopyridazines and the corresponding triazoles compared to azidocinnoline derivatives (**A**–**D**).

2. Results and Discussion

2.1. Synthesis of 6-aryl-4-azidocinnolines

Several routes have been established for introduction of an azido group into a heterocyclic ring [41]. The most common techniques are azidodediazoniation of arenediazonium salts, synthesis of azides from the corresponding heteroarylboronic acids and nucleophilic substitution of activated halogens. In most cases involving 4-azidoquinolines nucleophilic substitution of a chlorine atom at the C4 position of a quinoline ring has been employed [42–45]. A few examples are known of bromine substitution [46,47] along with azidodediazoniation [48] and conversion of the corresponding quinolones to azidoquinolines employing diphenylphospharylazide [49].

The synthetic routes towards cinnolines include different cyclization techniques and named reactions [33,50,51]. One of them, which allows 4-halocinnolines to be obtained in high yields, is the Richter-type cyclization [52] of *ortho*-ethynylarenediazonium salts [53] and *ortho*-ethynylaryltriazenes [54–56]. The halogen atom at C4 position can be substituted with various nucleophilic reagents, i.e., water [53], alcoholates [57–59], sulfides [56,58,60] and amines [58,61–64]. The only mentioned example of a 4-azidocinnoline molecule was also obtained from the corresponding chlorocinnoline by nucleophilic substitution with sodium azide in ethanol [29].

To reach the target 4-azido-6-arylcinnolines compounds we decided to use 4-bromocinnolines, which are synthetically accessible by the Richter cyclization in higher yields compared to corresponding chloro derivatives [54], while the bromine atom at the C4 position should be still reactive towards nucleophilic substitution. Firstly, two synthetic routes (A and B) have been proposed for target molecules (Scheme 1).

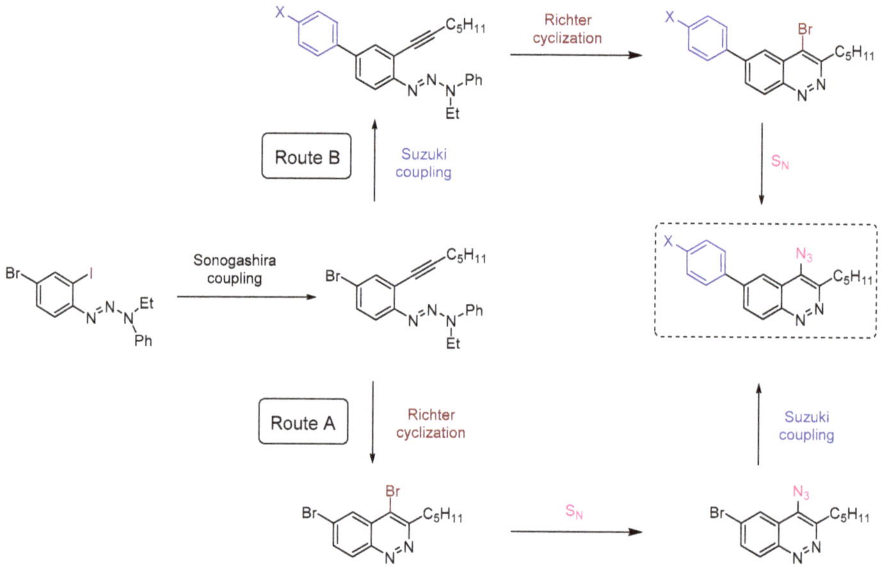

Scheme 1. Proposed synthetic "Routes A" and "B" towards 6-aryl-4-azidicinnolines.

The first step for both routes is the chemoselective Sonogashira coupling of the C-I group of 4-bromo-2-iodophenyltriazene with hept-1-yne (Scheme 1), which has been used previously for the synthesis of poly(arylene ethynylene)s [40].

"Route A" seems to be more divergent and rational than "Route B", because it allows for variations of the aromatic substituent during the last synthetic step. Therefore we chose "Route A" to start with. The Richter-type cyclization followed by the regioselective substitution of bromine atom with sodium azide in absolute DMF at the activated C4 position proceeded smoothly and did not affect the bromine atom at C6. Hence 4-azido-6-bromocinnoline **5** was obtained in good yield (Scheme 2). Next we tried to carry out the Suzuki coupling. Despite the fact that this transformation of substrates bearing aromatic azido groups is known, yields usually are low, because the whole process is accompanied by several side reactions that are common for both Suzuki coupling (homocoupling, reductive dehalogenation) and for azides (denitrogenative decomposition through nitrene intermediates and through the Staudinger reaction if phosphine ligands are present in the reaction mixture) [65]. Two types of conditions tested differed in the base (Na_2CO_3 or K_3PO_4) and solvents (toluene/dioxane/H_2O or only dioxane, respectively) used (Scheme 2). The formation of the desired 4-azido-6-phenylcinnoline **7a** as the main product was observed in both cases. However the isolated yields of the product **7a** were found to be between low (conditions *d*, Scheme 2) to moderate (conditions *e*, Scheme 2). Moreover the Suzuki coupling proceeded with the formation of many byproducts that required time-consuming purification of the target azide by column chromatography. Therefore we turned to "Route B".

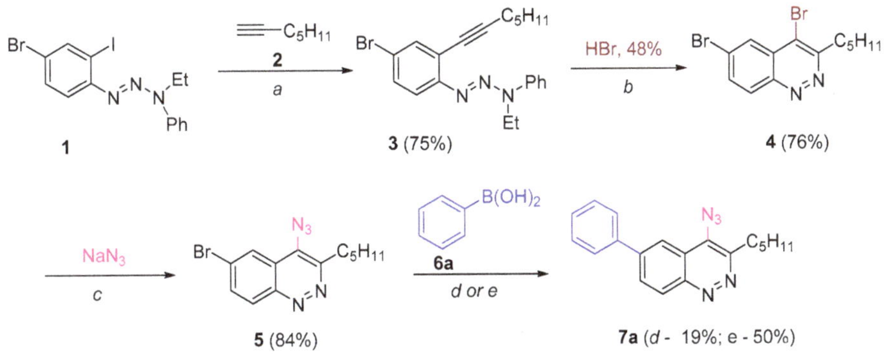

Scheme 2. Synthesis 6-aryl-4-azidocinnolines through "Route A". Reagents and conditions: (a) Pd(PPh$_3$)$_4$, CuI, Et$_3$N, THF, 40 °C, 5 h; (b) HBr (20 eqiuv), acetone, 0.01 M, rt, 10 min; (c) NaN$_3$ (5 equiv), 50 °C, 24 h; (d, Method A) – Pd(PPh$_3$)$_4$, Na$_2$CO$_3$, toluene/dioxane/H$_2$O, 80 °C, 1 h, (e, Method B) Pd(PPh$_3$)$_4$, K$_3$PO$_4$, dioxane, 80 °C, 1 h.

The Suzuki coupling of aromatic halotriazenes has been employed recently in the initial steps of the synthesis of hexahydrotribenzo[12]annulene [66]. Similar conditions (Na$_2$CO$_3$, toluene/dioxane/H$_2$O, 100 °C) for our substrates gave 4-phenyltriazene **8a** in 72% yield. Applying next the Richter-type cyclization and nucleophilic substitution of bromine with sodium azide under the same conditions as for "Route A" enabled us to produce the final 4-azido-6-phenylcinnoline **7a** in better overall yield (42%, "Route B") (Scheme 3) compared to "Route A" (24%) (Scheme 2). Same routes?

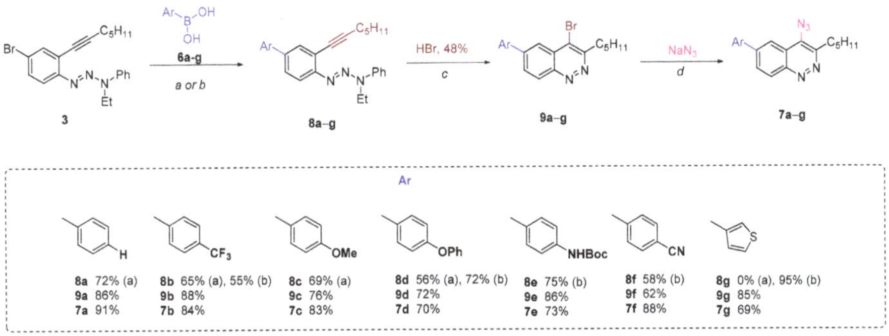

8a 72% (a)	**8b** 65% (a), 55% (b)	**8c** 69% (a)	**8d** 56% (a), 72% (b)	**8e** 75% (b)	**8f** 58% (b)	**8g** 0% (a), 95% (b)	
9a 86%	**9b** 88%	**9c** 76%	**9d** 72%	**9e** 86%	**9f** 62%	**9g** 85%	
7a 91%	**7b** 84%	**7c** 83%	**7d** 70%	**7e** 73%	**7f** 88%	**7g** 69%	

Scheme 3. Synthesis of 6-aryl-4-azidicinnolines **7a–g** through "Route B". Reagents and conditions: (a, Method A) Pd(PPh$_3$)$_4$, Na$_2$CO$_3$, toluene/dioxane/H$_2$O, 100 °C, 24 h; (b, Method B) Pd(PPh$_3$)$_4$, K$_3$PO$_4$, dioxane, 100 °C, 2–5 h; (c) HBr (20 eqiuv), acetone, 0.01M, rt, 10 min; (d) NaN$_3$ (5 equiv), 50 °C, 24 h.

The scope of "Route B" was then checked with different aryl- and heteroarylboronic acids **6b–g**. "Conditions a" worked well for the CF$_3$ (**b**), MeO (**c**) and PhO (**d**) series. However, these Suzuki coupling conditions failed when applied to other boronic acids bearing NHBoc (**6e**), CN (**6f**) and 3-thienyl (**6g**) groups. Thus, either complete or partial decomposition of the starting triazenes occurred when a mixture of organic solvent with water in the presence of sodium carbonate as a base was used. Excluding water from the reaction mixture and changing the base to potassium phosphate allowed all three triazenes **8e–g** to be obtained in high yields. The Richter cyclization and the subsequent nucleophilic substitution of bromine by an azido group for all compounds proceeded smoothly under the same conditions as for the unsubstituted phenyltriazene **8a**, providing a series of key 6-aryl(heteroaryl)-4-azidocinnolines **7** in high yields (Scheme 3).

2.2. Study of 6-aryl-4-azidocinnoline's Reactivity in the Sythesis of 6-aryl-4-triazolylcinnolines

All azidocinnolines were found to be stable crystalline compounds when stored at −18 °C and slowly decomposed in solution. Thus complete decomposition of azidocinnoline **7a** in acetone-d_6 was detected after a week. Despite this fact, all azides were stable under conditions used for CuAAC. Thus carrying out CuAAC of azidocinnolines **7a–g** both with terminal aromatic alkynes bearing EWG, EDG and aliphatic alkynes in the mixture of THF/H$_2$O using the copper (II) sulfate/sodium ascorbate catalytic system afforded the corresponding 4-triazolylcinnolines mostly in good yields (Scheme 4).

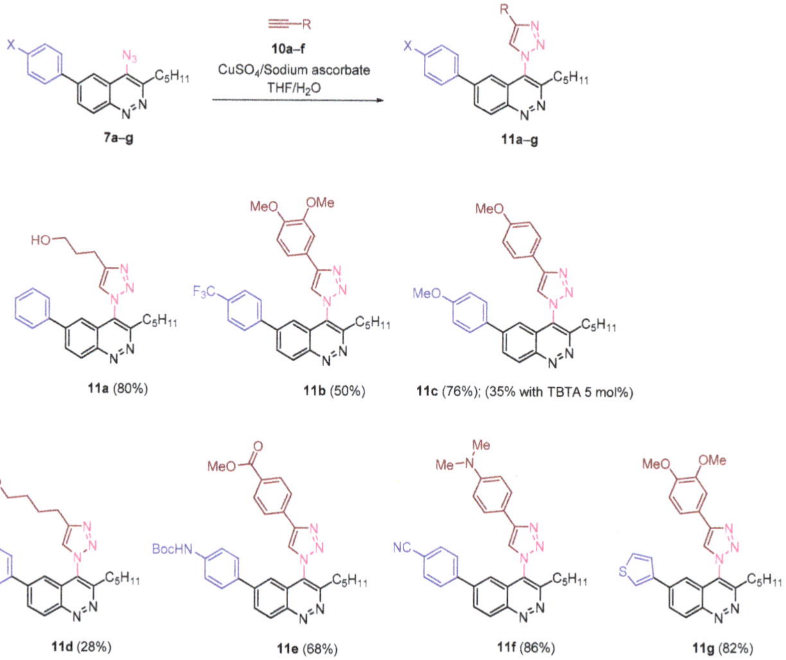

Scheme 4. Reactivity of 4-azidocinnolines **7a–g** in CuAAC.

One modification of the well-studied CuAAC is to carry out the reaction in the presence of tris(triazolyl) ligands, which improves the yields, shortens the reaction time and improves the synthetic accessibility of triazole derivatives that are not available when the common Cu(II)/ascorbate catalytic system is used [67]. Surprisingly, when the CuAAC of azidocinnoline **7c** with 3,4-dimethoxyphenylacetylene was run in the presence of the TBTA ligand (tris[(1-benzyl-1H-1,2,3-triazol-4-yl)methyl]amine), the reaction proceeded very slowly and full conversion was not achieved even after two days. Therefore triazolylcinnoline **11c** was isolated in only 35 % yield (the yield of **11c** without TBTA ligand was 76%).

Next we checked the reactivity of azidocinnolines synthesized in reactions other than CuAAC used for the synthesis of 1H-1,2,3-triazole derivatives, like the enol-mediated organocatalytic synthesis of triazoles and strain-promoted alkyne-azide cycloaddition (SPAAC). Unfortunately, enol-mediated reaction [68] of azide **7b** with propionic aldehyde catalysed by 1,8-diazabicyclo-[5.4.0]undec-7-ene (DBU) did not go as expected. The only isolated product was 4-aminocinnoline, which was presumably formed from the corresponding nitrene as an intermediate of azide decomposition (Scheme 5).

SPAAC of thienylazidocinnoline **7g** with tosylated diazacyclononyne [69] proceeded smoothly at 30 °C and afforded corresponding triazole in excellent yield (95%, Scheme 6). Thus synthesized

6-aryl-4-azidocinnolines were found to be suitable substrates for the synthesis of 6-aryl-4-triazolylcinnolines through both CuAAC and SPAAC.

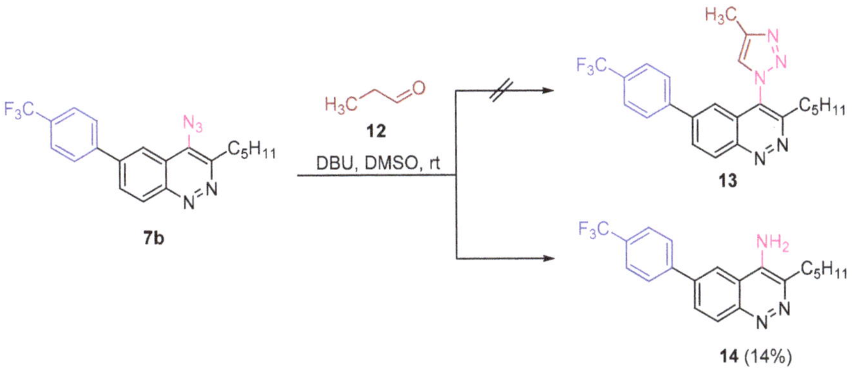

Scheme 5. Reactivity of azidocinnoline 7b in enol-mediated organocatalytic triazole synthesis.

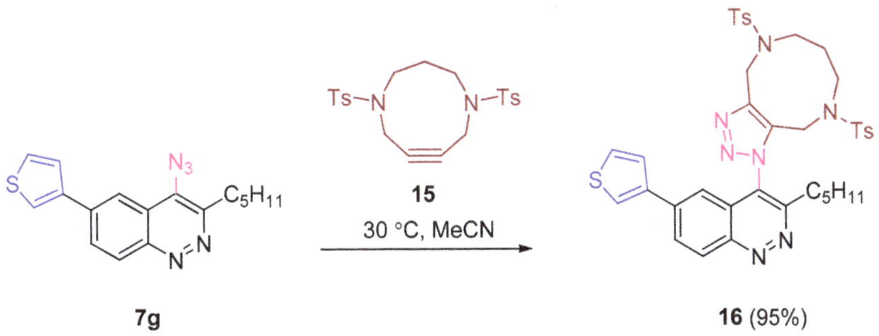

Scheme 6. Reactivity of 4-azidocinnoline 7g in SPAAC.

2.3. Photophysical Properties of 6-aryl-4-azidocinnolines and 6-aryl-4-triazolylcinnolines

Having both azidocinnolines and triazolylcinnolines in hand, we next studied their photophysical properties. Fluorogenic azides represent a class of fluorescent dyes which are only weakly fluorescent without conversion into triazoles. However, when a triazole moiety is introduced into the molecule, it imparts fluorescent properties to the entire molecular scaffold. Fluorogenic azides are of special interest for bioimaging, because they allow avoiding a series of steps to wash the excess of dye from the labeled materials and thus they eliminate the problem of background fluorescence. Syntheses and applications of various fluorogenic heterocyclic azides have been reported recently and discussed in the Introduction section [11] (Figure 1). Taking into account that the aryl substituent, azide group and azo-group of a cinnoline core are conjugated, we decided to investigate whether 6-aryl-4-azidocinnolines could be used as fluorogenic azide dyes.

Firstly, the absorption and emission spectra of azidocinnoline solutions in THF were measured (Figure 3). The obtained data revealed that both EWG and EDG groups at the *para*-position of an aryl ring attached to C6 atom provided an increase in fluorescence brightness. To quantify the observed fluorescence, the absolute quantum yields of fluorescence (QY) of azidocinnolines 7 were measured (Table 1).

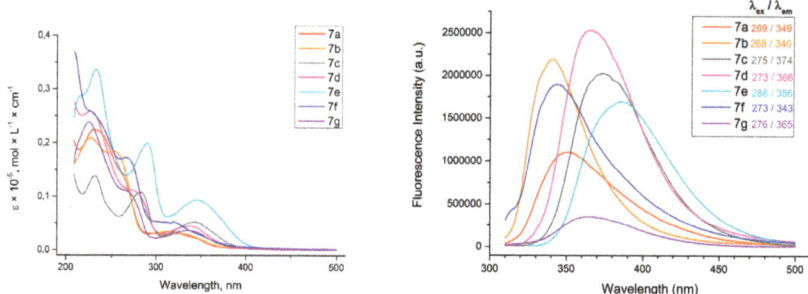

Figure 3. Absorption spectra of 4-azidocinnolines **7a–g** solutions in THF, C = 1 × 10^{-5} mol/L (left) and emission spectra of 4-azidocinnolines **7a–g** solutions in THF, C = 0.33 × 10^{-5} mol/L (right).

Table 1. Absolute QY for THF solutions of 6-aryl-4-azifocinnolines **7a–g**.

Entry	Azidocinnoline 7 (R)	Quantum yield, %
1	7a (H)	0.8
2	7b (CF$_3$)	0.1
3	7c (OMe)	0.9
4	7d (OPh)	0.3
5	7e (NHBoc)	0.8
6	7f (CN)	0.2
7	7g (3-thienyl) [1]	0.0

[1] Attached directly to C6 of cinnoline core.

The obtained data revealed that all 6-aryl-4-azidocinnolines **7** possess weak fluorescence with quantum yield values of less than 1%, with the highest QY being observed for 4-azido-6-(4-methoxyphenyl)cinnoline. Hoping to observe fluorescent properties in triazole derivatives **11, 16** the absorption and emission spectra for the series of triazolylcinnolines **11a–c,e–g, 16** were measured (Figures 4 and 5).

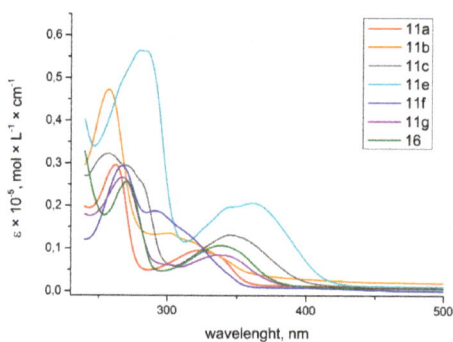

Figure 4. Absorption spectra of 4-triazolylcinnolines **11a–c, e–g, 16** solutions in THF, C = 1 × 10^{-5} mol/L.

Unfortunately, all triazoles were found to be devoid of any fluorescence. Emission spectra of azide/triazole pairs with MeO group **7c/11c** and 3-thienyl group **7f/11f** are presented on Figure 5 (for the spectra of the whole series see Appendix A–Figures A1 and A2).

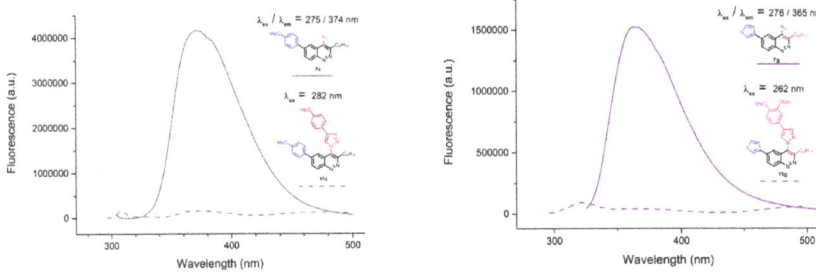

Figure 5. Emission spectra of 4-azidocinnolines/4-triazolylcinnolnes **7c/11c** (left) and **7g/11g** (right) solutions in THF, C = 1 × 10^{-5} mol/L.

To explain the observed loss of fluorescence for triazoles, quantum chemical calculations for S^0 states of the azide/triazoles pairs **7c/11c** and **7f/11f** were carried out. Geometry optimization was achieved using DFT calculations (B3LYP 6-311 + G(2d,2p)). The obtained data revealed that triazoles **11c** and **11f** are non-planar molecules: the triazole rings of both compounds lay out of the corresponding cinnoline plane (Table 2). X-Ray studies confirmed the non-planar geometry of triazole **11f** in the solid state (Figure 6). The dihedral angle between the triazole and cinnoline rings was found to be 64.8°. Therefore, despite the extension of the conjugated chain, the non-planar geometry of 4-triazolylcinnolines could be one of reasons for the observed absence of fluorescence.

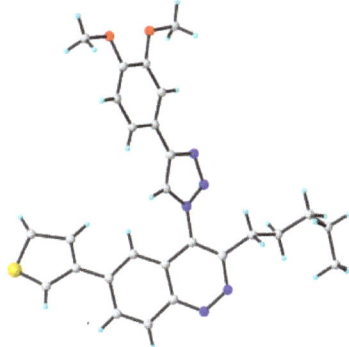

Figure 6. Molecular structures of triazolylcinnoline **11g** obtained from X-Ray analysis.

Analysis of the frontier molecular orbitals of triazolylcinnolines **11c** and **11f** showed that the HOMO is associated with the donor part of the molecule (triazole) whereas the LUMO is concentrated in the acceptor cinnoline part. On the other hand the frontier molecular orbitals of azidocinnolines **7c** and **7f** are evenly distributed over the entire molecule (Table 2).

Therefore, another reason for the lack of fluorescence of triazolylcinnolines compared to azidocinnolines could be intramolecular photoinduced electron transfer (PET) that is known to be responsible for fluorescence quenching [70]. For example, 1,4-bis(2-hydrohyphenyl)-1,2,3-triazole with a similar frontier molecular orbital separation between both 2-hydrohyphenyl substituents exhibited very weak fluorescence [71].

Table 2. Optimized geometries of azidocinnolines, triazolylcinnolines and their frontier molecular orbitals.

Entry	Compound [1]	Torsion Angle [2]	HOMO (E, ev)	LUMO (E, ev)
1	7c	–	−6.23	−2.43
2	11c	106.3°	−5.93	−2.48
3	7f	–	−6.41	−2.45
4	11f	100.0°	−5.95	−2.58

[1] Geometry optimization (B3LYP 6-311 + G(2d,2p)); [2] Torsion angle between triazole and cinnoline rings in optimized strucures.

2.4. Biological Studies of 6-aryl-4-triazolylcinnolines

Next we turned to biological activity screening tests. Both the triazole fragment and cinnoline moiety can be found in variety of compounds with antibacterial, antifungal and anticancer properties. Therefore we studied the antibacterial and antifungal activities of triazoles **11a,c,e−g, 16** against *Escherichia coli* and *Saccharomyces cerevisiae*, respectively. The cytotoxicity of triazolylcinnolines and their activity as DNA-cleavage agents were also tested.

The obtained data revealed that the studied triazolylcinnolines **11a,c,e−g, 16** are inactive towards the Gram-negative bacterium *Escherichia coli* and fungus *Saccharomyces cerevisiae*. On the other hand the MTT test for screening of cytotoxicity allowed identifying triazolylcinnoline **11a** as being active. Despite the fact that other triazolylcinnolines did not display reduced HeLA cell viability, the IC_{50} for compound **11a** bearing hydroxyalkyl substituent was found to be 56.5 µM (Figure 7). Triazolylcinnoline **11a** possess a flat heteroaromatic moiety–6-arylcinnoline, which might have DNA intercalating activity. Therefore we tested the ability of triazolylcinnolines to cleave DNA. None of the compounds affected the DNA plasmid pBR322, indicating a different mechanism of cytotoxicity.

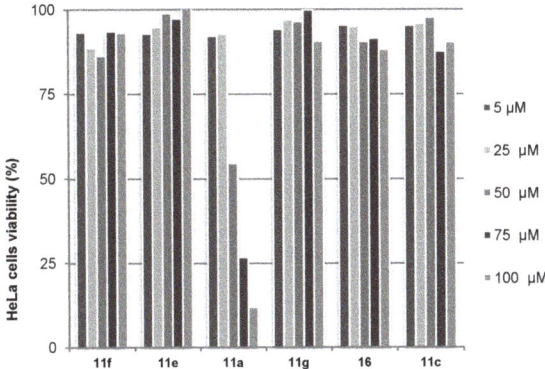

Figure 7. MTT test results with HeLa cell line for triazolylcinnolines **11,16**.

3. Materials and Methods

3.1. General Information

Solvents and reagents used for reactions were purchased from commercial suppliers. Catalyst Pd(PPh$_3$)$_4$ was purchased from Sigma-Aldrich (München, Germany). Solvents were dried under standard conditions; chemicals were used without further purification. 4-Bromo-2-iodoaniline [72], 1-(4-bromo-2-iodophenyl)-3-ethyl-3-phenyltriaz-1-ene (**1**) [40], Co$_2$(CO)$_6$-complex of diazacyclononyne **15** [69] and TBTA [67] were synthesized by known procedures without any modifications. Evaporation of solvents and concentration of reaction mixtures were performed in vacuum at 35 °C on a rotary evaporator. Thin-layer chromatography (TLC) was carried out on silica gel plates (Silica gel 60, F254, Merck (Darmstadt, Germany) with detection by UV or staining with a basic aqueous solution of KMnO$_4$. Melting points (mp) determined are uncorrected. ^{1}H and ^{13}CNMR spectra were recorded at 400 and 100 MHz, respectively, at 25 °C in acetone-d_6 without the internal standard using a 400 MHz Avance spectrometer (Bruker, Billerica, MA, USA). The ^{1}H-NMR data are reported as chemical shifts (δ), multiplicity (s, singlet; d, doublet; t, triplet; q, quartet; m, multiplet; br, broad), coupling constants (J, given in Hz), and number of protons. The ^{13}C NMR data are reported as the chemical shifts (δ) with coupling constant J(C–F) for F-containing compounds. Chemical shifts for ^{1}H and ^{13}C are reported as δ values (ppm) and referenced to residual solvent (δ = 2.05 ppm for ^{1}H; δ = 29.84 for ^{13}C–for spectra in acetone-d_6, δ = 7.26 ppm for ^{1}H; δ = 77.16 ppm for ^{13}C–for spectra in CDCl$_3$ and δ = 2.50 ppm for ^{1}H; δ = 39.52 ppm for ^{13}C–for spectra in DMSO-d_6). High resolution mass spectra (HRMS) were determined using electrospray ionization (ESI) in the mode of positive ion registration with a Bruker microTOF mass analyzer (Billerica, MA, USA). UV–vis spectra for solutions of all compounds in THF were recorded on a UV-1800 spectrophotometer (Shimadzu, Kyoto, Japan) at 20 °C. Fluorescence spectra for the same solutions were recorded on a FluoroMax-4 spectrofluorometer (Horiba Scientific, Glasgow, Scotland) at 20 °C. IR spectra were measured using a Nicolet 8700 spectrometer (Thermo Scientific, Madison, WI, USA) equipped with a Thermo Scientific Smart iTR™ as an Attenuated Total Reflectance (ATR) sampling accessory. Data for **11f** were collected using an XtaLAB SuperNova diffractometer (Rigaku Oxford Diffraction, Tokio, Japan) equipped with an HyPix3000 CCD area detector operated with monochromated microfocused CuKα radiation (λ[CuKα] = 1.54184 Å). All the data were integrated and corrected for background, Lorentz, and polarization effects by means of the CrysAlisPro (Tokyo, Japan) [73] program complex. Absorption correction was applied using the empirical spherical model within the CrysAlisPro program complex using spherical harmonics, implemented in the SCALE3 ABSPACK scaling algorithm. The unit-cell parameters were refined by the least-squares techniques. The structures were solved by direct methods and refined using the SHELX [74] program incorporated in the OLEX2 [75] program package.

3.2. Synthetic Methods

3.2.1. The Richter-Type Cyclization Protocol of Triazenes **3, 8**

To a solution of triazene **3** or **8** (1 equiv) in acetone (C = 0.1 M) HBr (48% aqueous solution, 20 equiv) was quickly added dropwise while maintaining the reaction mixture temperature of 20 °C by cooling the reaction mixture with a water bath. The resulting mixture was stirred for 10 minutes. Upon completion of the reaction, the reaction mixture was diluted with an aqueous solution of triethylamine (21 equiv). The resulting mixture was extracted with ethyl acetate, the combined organic layers were washed with water, brine and dried over anhydrous Na_2SO_4. The solvent was removed in vacuum and the crude 4-bromocinnoline was purified by column chromatography.

3.2.2. General Procure for the Suzuki Coupling (Method A)

To a solution of triazene **3** or azidocinnoline **5** (1 equiv) in a mixture of toluene/1,4-dioxane/water (1:2:2) (C = 0.1 M) in a vial was added arylboronic acid **6** (1.5 equiv), $Pd(PPh_3)_4$ (5 mol%) and Na_2CO_3 (2 equiv). The vial was sealed; the reaction mixture was evacuated and flushed with Ar several times. The vial with the reaction mixture was placed in a preheated oil bath (80–100 °C) and stirred for 1–24 h (TLC control). After completion the reaction, the mixture was cooled and poured into a saturated aqueous solution of NH_4Cl and extracted with ethyl acetate. The combined organic layers were washed with saturated aqueous solutions of NH_4Cl and brine, dried over anhydrous Na_2SO_4, and concentrated under reduced pressure to yield the crude product, which was purified by column chromatography on silica gel.

3.2.3. General Procure for the Suzuki Coupling (Method B)

Triazene **3** or azidocinnoline **5** (1 equiv), $ArB(OH)_2$ **6** (1.5 equiv), K_3PO_4 (2 equiv), and $Pd(PPh_3)_4$ (5 mol %) were placed in a vial. The vial was sealed, and the mixture was evacuated and flushed with Ar several times. 1,4-Dioxane (C = 0.1 M) was added, and the vial with the reaction mixture was placed in a preheated oil bath (80–100 °C) and stirred for 1–20 h (TLC control). After cooling to rt, the reaction mixture was filtered through a pad silica gel and washed with ethyl acetate. Solvents were removed under reduced pressure, and the crude product was purified by column chromatography on silica gel.

3.2.4. General Procure for the Nucleophilic Substitution

Sodium azide (2–5 equiv) was added to a solution of bromocinnoline (1 equiv) in absolute DMF (C = 0.1 M). The mixture was degassed and stirred under argon at 50 °C for 24 h (TLC control). Upon completion the reaction, the reaction mixture was poured into water, extracted with ethyl acetate; the combined organic layers were washed three times with water and two times with brine, dried over anhydrous Na_2SO_4. The solvent was removed in vacuum to yield the crude product, which was purified by column chromatography on silica gel.

3.2.5. General Procure for CuAAC Synthesis of 4-triazolylcinnolines

To a stirred mixture of a terminal alkynes (1 equiv) and 4-azidocinnolines (1 equiv) in a mixture of water/THF (1:1) were added sodium ascorbate (10 mol%) and $CuSO_4 \times 5H_2O$ (5 mol%). The reaction mixture was vigorously stirred at 30 °C until the completion of the reaction (TLC control). The resulting mixture was diluted with saturated aqueous solution of NH_4Cl (10 mL) and extracted with ethyl acetate (3 × 15 mL). The combined organic layers were dried over anhydrous Na_2SO_4 and concentrated under reduced pressure to yield the crude product, which was purified by column chromatography on silica gel.

3.2.6. Detailed Synthetic Procedures

1-[4-Bromo-2-(hept-1-yn-1-yl)phenyl]-3-ethyl-3-phenyl-triaz-1-ene (**3**). To a solution of 1-(4-bromo-2-iodophenyl)-3-ethyl-3-phenyltriaz-1-ene (**1**) (1 equiv, 860 mg, 2 mmol) in a mixture of triethylamine (8 mL) and THF (4 mL) degassed by freeze-pump-thaw cycling were added Pd(PPh$_3$)$_4$ (5 mol.%, 116 mg, 0.1 mmol), CuI (15 mol.%, 57.0 mg, 0.3 mmol). The reaction mixture was degassed once again. Then hept-1-yne **2** (1.5 equiv, 289 mg, 0.39 mL, 3 mmol) was added via a syringe and the reaction mixture was stirred at 40 °C for 5 hours (TLC control). Upon completion of the reaction, the reaction mixture was poured into a saturated aqueous solution of NH$_4$Cl, extracted with ethyl acetate (3 x50 mL); the combined organic layers were washed with saturated aqueous solutions of NH$_4$Cl, brine and dried over anhydrous Na$_2$SO$_4$. The solvent was removed under reduced pressure and the crude product was purified by column chromatography, using hexane/EtOAc/Et$_3$N (90:1:0.01) as the eluent to give **3** (600 mg, 75%) as an orange oil. ^{1}H-NMR (acetone-d_6) δ 7.58 – 7.60 (m, 3 H), 7.52 – 7.41 (m, 4 H), 7.17 (t, J = 7.4 Hz, 1H), 4.43 (q, J = 7.0 Hz, 2H), 2.48 (t, J = 7.0 Hz, 2H), 1.62 (p, J = 7.1 Hz, 2H), 1.51 – 1.44 (m, 2H), 1.41 – 1.32 (m, 5H), 0.91 (t, J = 7.3 Hz, 3H). ^{13}C-NMR (acetone-d_6) 151.6, 144.9, 135.9, 132.2, 130.2, 124.8, 123.0, 120.3, 119.4, 118.0, 97.3, 77.9, 41.2, 31.9, 29.2, 22.9, 20.1, 14.3, 11.2. HRMS (ESI): calcd. for C$_{21}$H$_{25}$BrN$_3$ [M + H]$^+$, 398.1232; found 398.1240.

4,6-Dibromo-3-pentylcinnoline (**4**). This compound was synthesized according to the general procedure for the Richter cyclization from triazene **3** (219 mg, 0.55 mmol). Purification of the crude product by column chromatography using hexane/EtOAc (40:1) as the eluent gave 150 mg (76%) of the title compound. ^{1}H-NMR (acetone-d_6) δ 8.42 (d, J = 9.0 Hz, 1H), 8.36 (d, J = 1.9 Hz, 1H), 8.08 (dd, J = 9.0, 2.0 Hz, 1H), 3.45 – 3.36 (m, 2H), 1.98 – 1.86 (m, 2H), 1.52 – 1.38 (m, 4H), 0.92 (t, J = 7.1 Hz, 3H). ^{13}C-NMR (acetone-d_6) δ 158.6, 149.0, 135.0, 132.9, 128.8, 128.4, 128.1, 125.5, 36.8, 32.3, 29.4, 23.1, 14.3. HRMS (ESI): calcd. for C$_{13}$H$_{14}$N$_2$Br$_2$Na [M + Na]$^+$, 378.9416; found 378.9426.

4-Azido-6-bromo-3-pentylcinnoline (**5**). This compound was synthesized according to the general procedure for the nucleophilic substitution from bromocinnoline **4** (804 mg, 2.25 mmol) and sodium azide (730 mg, 11.25 mmol) in absolute DMF (22.3 mL). Purification of the crude product by column chromatography using hexane/EtOAc (10:1) as the eluent gave azidocinnoline **5** (603 mg, 84%). ^{1}H-NMR (CDCl$_3$) δ 8.36 (d, J = 9.1 Hz, 1H), 8.32 (d, J = 2.1 Hz, 1H), 7.87 (dd, J = 9.0, 2.0 Hz, 1H), 3.40–3.32 (m, 2H), 1.96–1.84 (m, 2H), 1.54–1.35 (m, 4H), 0.93 (t, J = 7.1 Hz, 3H). ^{13}C-NMR (CDCl$_3$) δ 151.4, 148.4, 134.2, 131.7, 131.4, 126.3, 123.9, 121.8, 32.5, 31.7, 29.7, 22.6, 14.1. HRMS (ESI): calcd. for C$_{13}$H$_{14}$N$_5$BrNa [M + Na]$^+$, 342.0325; found 342.0336.

3-Ethyl-1-[3-(hept-1-yn-1-yl)-[1,1'-biphenyl]-4-yl]-3-phenyltriaz-1-ene (**8a**). (A) This compound was synthesized according to the general procedure for the Suzuki coupling (Method A) from triazene **3** (230 mg, 0.57 mmol) and phenylboronic acid **6a** (105 mg, 0.86 mmol) at 100 °C; reaction time – 24. Purification of the crude product by column chromatography using hexane/EtOAc (70:1) as the eluent gave **8a** (166 mg, 73%) as an orange oil. ^{1}H-NMR (acetone-d_6) δ 7.73–7.60 (m, 7H), 7.50–7.36 (m, 5 H), 7.5 –7.37 (m, 5H), 7.16 (t, J = 7.4 Hz, 1H), 4.46 (q, J = 7.0 Hz, 2H), 2.50 (t, J = 7.0 Hz, 2H), 1.6–1.61 (m, 2H), 1.53–1.46 (m, 2H), 1.42–1.33 (m, 5H), 0.92 (t, J = 7.3 Hz, 3H). ^{13}C-NMR (acetone-d_6) δ 150.6, 144.1, 139.8, 138.8, 131.0, 129.3, 128.9, 127.5, 126.8, 126.6, 123.6, 120.7, 118.1, 116.8, 94.8, 78.4, 40.0, 31.0, 28.5, 22.0, 19.2, 13.4, 10.4. HRMS (ESI): calcd. for C$_{27}$H$_{30}$N$_3$ [M + H]$^+$, 396.2434; found: 396.2433.

3-Ethyl-1-[3-(hept-1-yn-1-yl)-4'-(trifluoromethyl)-[1,1'-biphenyl]-4-yl]-3-phenyltriaz-1-ene (**8b**). (A) This compound was synthesized by the general procedure for the Suzuki coupling (Method A) from triazene **3** (100 mg, 0.25 mmol) and 4-trifluoromethylphenylboronic acid **6b** (72 mg, 0.38 mmol) at 100 °C; reaction time–24 h. Purification of the crude product by column chromatography using hexane/EtOAc (80:1) as the eluent gave **8b** (75.0 mg, 65%) as an orange oil.

(B) This compound was synthesized by the general procedure for the Suzuki coupling (method B) from triazene **3** (300 mg, 0.754 mmol) and 4-trifluoromethylphenylboronic acid **6b** (72 mg, 0.38 mmol) **6b** (190 mg, 1.50 mmol) at 100 °C; reaction time – 24 h. Purification of the crude product by column chromatography using hexane/EtOAc (80:1) as the eluent gave **8b** (205 mg, 55%) as an orange oil. ^{1}H-NMR (acetone-d_6) δ 7.93 (d, J = 8.2 Hz, 2H), 7.84–7.80 (m, 3H), 7.75–7.65 (m, 2H), 7.64–7.58 (m, 2H), 7.49–7.40 (m, 2H), 7.22–7.13 (m, 1H), 4.47 (q, J = 7.0 Hz, 2H), 2.51 (t, J = 7.0 Hz, 2H), 1.71 – 1.59 (m, 2H), 1.56–1.44 (m, 2H), 1.43–1.31 (m, 5H), 0.92 (t, J = 7.3 Hz, 3H). ^{13}C-NMR (acetone-d_6) δ 152.3, 145.0, 144.6, 138.0, 132.3, 130.2, 129.7 (q, $^2J_{C-F}$ = 32.3 Hz), 128.2, 128.0, 126.7 (q, $^3J_{C-F}$ = 3.8 Hz), 125.5 (q, $^1J_{C-F}$ = 270.5 Hz), 124.7, 121.8, 119.3, 117.9, 96.1, 79.1, 41.1, 31.9, 29.4, 22.9, 20.1, 14.3, 11.3. HRMS (ESI): calcd. for $C_{28}H_{29}F_3N_3$ [M + H]$^+$, 464.2313; found: 464.2308.

3-Ethyl-1-[3-(hept-1-yn-1-yl)-4'-methoxy-[1,1'-biphenyl]-4-yl]-3-phenyltriaz-1-ene (**8c**). This compound was synthesized by the general procedure for the Suzuki coupling (method A) from triazene **3** (100 mg, 0.25 mmol) and 4-methoxyphenylboronic acid **6c** (58 mg, 0.38 mmol) at 100 °C; reaction time – 24 h. Purification of the crude product by column chromatography using hexane/EtOAc (80:1) as the eluent gave **8c** (66 mg, 62%) as a brown oil. ^{1}H-NMR (acetone-d_6) δ 7.68 (d, J = 2.0 Hz, 1H), 7.66–7.54 (m, 6H), 7.48–7.38 (m, 2H), 7.17–7.13 (m, 1H), 7.07–6.99 (m, 2H), 4.45 (q, J = 7.0 Hz, 2H), 3.85 (s, 3H), 2.50 (t, J = 7.0 Hz, 2H), 1.70–1.58 (m, 2H), 1.55–1.43 (m, 2H), 1.43 – 1.30 (m, 5H), 0.91 (t, J = 7.3 Hz, 3H). ^{13}C-NMR (acetone-d_6) δ 160.5, 150.9, 145.1, 139.5, 133.0, 131.4, 130.2, 128.6, 127.3, 124.4, 121.6, 119.0, 117.7, 115.2, 95.5, 79.4, 55.7, 40.8, 31.9, 29.4, 22.9, 20.2, 14.3, 11.3. HRMS (ESI): calcd. for $C_{28}H_{32}N_3O$ [M + H]$^+$, 426.2540; found 426.2525.

3-Ethyl-1-[3-(hept-1-yn-1-yl)-4'-phenoxy-[1,1'-biphenyl]-4-yl]-3-phenyltriaz-1-ene (**8d**). (A) This compound was synthesized by the general procedure for the Suzuki coupling (method A) from triazene **3** (100 mg, 0.25 mmol) and 4-phenoxyphenylboronic acid **6d** (81 mg, 0.38 mmol) at 100 °C; reaction time – 24 h. Purification of the crude product by column chromatography using hexane/EtOAc (70:1) as the eluent gave **8d** (68 mg, 56%) as a brown oil.

(B) This compound was synthesized by the general procedure for the Suzuki coupling (method B) from triazene **3** (300 mg, 0.75 mmol) and 4-phenoxyphenylboronic acid **6d** (323 mg, 1.51 mmol) at 100 °C; reaction time – 4 h. Purification of the crude product by column chromatography using hexane/EtOAc (80:1) as the eluent gave **8d** (265 mg, 72%) as a brown oil. ^{1}H-NMR (acetone-d_6) δ 7.76–7.68 (m, 3H), 7.67–7.58 (m, 4H), 7.48–7.38 (m, 4H), 7.22–7.14 (m, 2H), 7.12–7.04 (m, 4H), 4.46 (q, J = 7.0 Hz, 2H), 2.50 (t, J = 7.0 Hz, 2H), 1.70–1.59 (m, 2H), 1.56–1.44 (m, 2H), 1.42–1.33 (m, 5H), 0.91 (t, J = 7.3 Hz, 3H). ^{13}C-NMR (acetone-d_6) δ 158.1, 158.0, 151.3, 145.1, 139.1, 135.8, 131.7, 130.9, 130.2, 129.1, 127.5, 124.5, 124.5, 121.7, 119.9, 119.9, 119.1, 117.7, 95.7, 79.3, 40.9, 31.9, 23.0, 20.2, 14.3, 11.3. HRMS (ESI): calcd. for $C_{33}H_{33}N_3ONa$ [M + Na]$^+$, 510.2516; found 510.2521.

tret-Butyl [4'-(3-ethyl-3-phenyltriaz-1-en-1-yl)-3'-(hept-1-yn-1-yl)-[1,1'-biphenyl]-4-yl]carbamate (**8e**). This compound was synthesized by the general procedure for the Suzuki coupling (method B) from triazene **3** (360 mg, 0.905 mmol) and {4-[(*tert*-butoxycarbonyl)amino]phenyl}boronic acid **6e** (430 mg, 1.8 mmol) at 100 °C; reaction time – 2.5 h. Purification of the crude product by column chromatography using hexane/EtOAc (15:1) as the eluent gave **8e** (345 mg, 75%) as an yellow oil. ^{1}H-NMR (acetone-d_6) δ 8.49 (s, 1H), 7.72–7.56 (m, 9H), 7.48–7.39 (m, 2H), 7.20–7.10 (m, 1H), 4.45 (q, J = 7.0 Hz, 2H), 2.50 (t, J = 7.0 Hz, 2H), 1.69–1.59 (m, 2H), 1.54–1.45 (m, 11H), 1.42–1.32 (m, 5H), 0.91 (t, J = 7.3 Hz, 3H). ^{13}C-NMR (acetone-d_6) δ 153.7, 151.0, 145.1, 140.3, 139.4, 134.5, 131.4, 130.2, 127.8, 127.2, 124.4, 121.6, 119.5, 119.0, 117.7, 95.5, 80.1, 79.4, 40.8, 31.9, 29.4, 28.6, 22.9, 20.1, 14.3, 11.3. HRMS (ESI): calcd. for $C_{32}H_{38}N_4O_2$ [M + H]$^+$, 511.3068; found: 511.3086.

4'-(3-Ethyl-3-phenyltriaz-1-en-1-yl)-3'-(hept-1-yn-1-yl)-[1,1'-biphenyl]-4-carbonitrile (**8f**). This compound was synthesized by the general procedure for the Suzuki coupling (method B) from triazene **3** (340 mg, 0.854 mmol) and 4-cyanophenylboronic acid **6f** (251 mg, 1.7 mmol) at 80 °C; reaction time – 10 h. Purification of the crude product by column chromatography using hexane/EtOAc (30:1) as the eluent gave **8f** (190 mg, 53%) as an yellow oil. ^{1}H-NMR (acetone-d_6) δ 7.96–7.83 (m, 4H), 7.79 (d, J = 2.0 Hz, 1H), 7.76–7.63 (m, 2H), 7.62–7.56 (m, 2H), 7.50–7.38 (m, 2H), 7.21–7.15 (m, 1H), 4.45 (q, J = 7.1 Hz, 2H), 2.49 (t, J = 7.0 Hz, 2H), 1.66–1.59 (m, 2H), 1.55–1.41 (m, 2H), 1.40–1.31 (m, 5H), 0.89 (t, J = 7.3 Hz, 3H).

^{13}C-NMR (acetone-d_6) δ 152.5, 145.1, 145.0, 137.6, 133.6, 132.4, 130.2, 128.4, 128.0, 124.8, 121.9, 119.4, 119.3, 117.9, 111.8, 96.2, 79.0, 41.1, 31.9, 22.9, 20.1, 14.3, 11.3. HRMS (ESI): calcd. for $C_{28}H_{28}N_4Na$ [M + Na]$^+$, 443.2206; found 443.2191.

3-Ethyl-1-(2-(hept-1-yn-1-yl)-4-(thiophen-3-yl)phenyl)-3-phenyltriaz-1-ene (**8g**). This compound was synthesized by the general procedure for the Suzuki coupling (method B) from triazene 3 (330 mg, 0.83 mmol) and thiophen-3-ylboronic acid **6g** (286 mg, 1.66 mmol) at 100 °C; reaction time – 4 h. Purification of the crude product by column chromatography using hexane/EtOAc (30:1) as the eluent gave **8g** (316 mg, 95%) as a yellow oil. ^{1}H-NMR (acetone-d_6) δ 7.79–7.78 (m, 2H), 7.68 (dd, J = 8.5, 2.1 Hz, 1H), 7.64–7.51 (m, 5H), 7.46–7.37 (m, 2H), 7.17–7.13 (m, 1H), 4.44 (q, J = 7.0 Hz, 2H), 2.49 (t, J = 7.0 Hz, 2H), 1.70–1.58 (m, 2H), 1.56–1.43 (m, 2H), 1.41–1.32 (m, 5H), 0.91 (t, J = 7.3 Hz, 3H). ^{13}C-NMR (acetone-d_6) δ 151.2, 145.0, 141.9, 134.7, 131.2, 130.2, 127.5, 127.1, 126.9, 124.5, 121.7, 121.5, 119.0, 117.7, 95.7, 79.3, 40.9, 31.9, 29.4, 22.9, 20.2, 14.3, 11.3. HRMS (ESI): calcd. for $C_{25}H_{27}N_3SCu$ [M + Cu]$^+$, 464.1216; found 464.1194.

4-Bromo-3-pentyl-6-phenylcinnoline (**9a**). This compound was synthesized by the general method for the Richter cyclization from triazene **8a** (200 mg, 0.505 mmol). Purification of the crude product by column chromatography using hexane/EtOAc (10:1→7:1) as the eluent gave **9a** (153 mg, 86%) as a yellow oil. ^{1}H-NMR (acetone-d_6) δ 8.51 (d, J = 8.8 Hz, 1H), 8.30–8.21 (m, 2H), 7.89–7.84 (m, 2H), 7.60–7.47 (m, 3H), 3.40–3.35 (m, 2H), 1.95–1.87 (m, 2H), 1.51–1.36 (m, 4H), 0.92 (t, J = 7.1 Hz, 3H). ^{13}C-NMR (acetone-d_6) δ 158.0, 149.8, 145.6, 139.8, 131.4, 131.1, 130.2, 129.8, 128.6, 127.5, 127.1, 123.5, 36.8, 32.3, 29.4, 23.1, 14.3. HRMS (ESI): calcd. for $C_{19}H_{20}N_2Br$ [M + H]$^+$, 355.0804; found 355.0802.

4-Bromo-3-pentyl-6-[4-(trifluoromethyl)phenyl]cinnoline (**9b**). This compound was synthesized by the general method for the Richter cyclization from triazene **8b** (50 mg, 0.108 mmol). Purification of the crude product by column chromatography using hexane/EtOAc (20:1) as the eluent gave **9b** (40 mg, 88%), m.p. 88–89 °C. ^{1}H-NMR (acetone-d_6) δ 8.58 (d, J = 8.8 Hz, 1H), 8.39 (d, J = 1.6 Hz, 1H), 8.30 (dd, J = 8.8, 1.9 Hz, 1H), 8.12 (d, J = 8.1 Hz, 2H), 7.92 (d, J = 8.2 Hz, 2H), 3.42–3.39 (m, 2H), 2.00–1.86 (m, 2H), 1.53–1.36 (m, 4H), 0.93 (t, J = 7.1 Hz, 3H). ^{13}C-NMR (100 MHz, acetone-d_6) δ 158.3, 149.9, 144.1, 143.8, 131.7, 131.2 (q, $^2J_{C-F}$ = 32.3 Hz), 131.0, 129.5, 127.4, 127.3, 127.0 (q, $^3J_{C-F}$ = 3.4 Hz), 126.7 (q, $^1J_{C-F}$ = 271.5 Hz), 124.6, 36.9, 32.3, 29.4, 23.1, 14.3. HRMS (ESI): calcd. for $C_{20}H_{18}BrF_3N_2Na$ [M + Na]$^+$, 445.0498; found 445.0480.

4-Bromo-6-(4-methoxyphenyl)-3-pentylcinnoline (**9c**). This compound was synthesized by the general method for the Richter cyclization from triazene **8c** (228 mg, 0.536 mmol). Purification of the crude product by column chromatography using hexane/EtOAc (10:1→5:1) as the eluent gave **9c** (157 mg, 76%), m.p. 85–87 °C. ^{1}H-NMR (acetone-d_6) δ 8.52–8.45 (m, 1H), 8.25–8.21 (m, 2H), 7.88–7.82 (m, 2H), 7.17–7.09 (m, 2H), 3.90 (s, 3H), 3.43–3.30 (m, 2H), 1.99–1.86 (m, 2H), 1.54–1.36 (m, 4H), 0.93 (t, J = 7.1 Hz, 3H). ^{13}C-NMR (acetone-d_6) δ 161.7, 157.9, 149.7, 145.3, 132.0, 131.3, 130.9, 129.9, 127.7, 126.9, 122.3, 115.7, 55.8, 36.9, 32.3, 23.2, 14.3. HRMS (ESI): calcd. for $C_{20}H_{21}BrN_2ONa$ [M + Na]$^+$, 407.0729; found 407.0708.

4-Bromo-3-pentyl-6-(4-phenoxyphenyl)cinnoline (**9d**). This compound was synthesized by the general method for the Richter cyclization from triazene **8d** (180 mg, 0.37 mmol). Purification of the crude product by column chromatography using hexane/EtOAc (10:1→5:1) as the eluent gave **9d** (121 mg, 72%), m.p. 77–78 °C. ^{1}H-NMR (400 MHz, acetone-d_6) δ = 8.51 (dd, J = 8.8, 0.7 Hz, 1H), 8.31–8.21 (m, 2H), 7.95–7.87 (m, 2H), 7.50–7.40 (m, 2H), 7.26–7.08 (m, 5H), 3.43–3.34 (m, 2H), 1.99–1.86 (m, 2H), 1.54–1.35 (m, 4H), 0.92 (t, J = 7.1 Hz, 3H). ^{13}C-NMR (100 MHz, acetone-d_6) δ 159.5, 158.0, 157.5, 149.7, 144.9, 134.6, 131.4, 131.0, 130.9, 130.3, 127.6, 127.0, 124.9, 122.9, 120.3, 119.8, 36.8, 32.3, 23.1, 14.3. HRMS (ESI): calcd. for $C_{25}H_{23}BrN_2ONa$ [M + Na]$^+$, 469.0886; found 469.0897.

tert-Butyl [4-(4-bromo-3-pentylcinnolin-6-yl)phenyl]carbamate (**9e**). This compound was synthesized by the general method for the Richter cyclization from triazene **8e** (170 mg, 0.333 mmol). Purification of the crude product by column chromatography using hexane/EtOAc (5:1) as the eluent gave **9e** (135 mg, 86%). ^{1}H-NMR (acetone-d_6) δ 8.66 (s, 1H), 8.53–8.45 (m, 1H), 8.31–8.22 (m, 2H), 7.89–7.74 (m, 4H), 3.43–3.34 (m, 2H), 1.99–1.86 (m, 2H), 1.52 (s, 9H), 1.51–1.37 (m, 5H), 0.93 (t, J = 7.0 Hz, 3H). ^{13}C NMR (acetone-d_6) δ 158.0, 153.7, 149.8, 145.3, 141.8, 133.3, 131.3, 130.8, 129.0, 127.7, 126.9, 122.3, 119.6, 80.4, 36.8, 32.3, 28.5, 23.1, 14.3. HRMS (ESI): calcd. for $C_{24}H_{28}BrN_3O_2Na$ [M + Na]$^+$, 492.1257; found 492.1262.

4-(4-Bromo-3-pentylcinnolin-6-yl)benzonitrile (**9f**). This compound was synthesized by the general method for Richter cyclization from triazene **8f** (190 mg, 0.452 mmol). Purification of the crude product by recrystallization from hexane/EtOAc (1:1) gave **9f** (107 mg, 62%) as a brown solid. ^{1}H-NMR (acetone-d_6) δ 8.60 (d, J = 8.8 Hz, 1H), 8.44–8.40 (m, 1H), 8.32 (dd, J = 8.8, 1.9 Hz, 1H), 8.15–8.10 (m, 2 H), 8.02–7.97 (m, 2H), 3.45–3.38 (m, 2H), 1.98–1.87 (m, 2H), 1.54–1.34 (m, 4H), 0.93 (t, J = 7.0 Hz, 3H).^{13}C-NMR (acetone-d_6) δ 152.4, 149.9, 144.2, 143.8, 133.9, 133.7, 131.8, 130.9, 129.6, 127.3, 124.9, 119.3, 113.4, 36.8, 32.3, 29.4, 23.1, 14.3. HRMS (ESI): calcd. for $C_{20}H_{18}BrN_3Na$ [M + Na]$^+$, 402.0576; found 402.0583.

4-Bromo-3-pentyl-6-(thiophen-3-yl)cinnoline (**9g**). This compound was synthesized by the general method for the Richter cyclization from triazene **8g** (280 mg, 0.698 mmol). Purification of the crude product by column chromatography using hexane/EtOAc (10:1) as the eluent gave **9g** (214 mg, 85%). ^{1}H-NMR (acetone-d_6) δ 8.47–8.40 (m, 1H), 8.33–8.25 (m, 2H), 8.14 (dd, J = 2.9, 1.4 Hz, 1H), 7.74 (dd, J = 5.1, 1.4 Hz, 1H), 7.68 (dd, J = 5.1, 2.9 Hz, 1H), 3.39–3.31 (m, 2H), 1.95–1.85 (m, 2H), 1.50–1.36 (m, 4H), 0.91 (t, J = 7.1 Hz, 3H). ^{13}C-NMR (acetone-d_6) δ 158.0, 149.7, 141.0, 140.1, 131.3, 130.6, 128.5, 127.8, 127.1, 126.9, 125.2, 122.0, 36.8, 32.3, 29.4, 23.1, 14.3. HRMS (ESI): calcd. for $C_{17}H_{17}BrN_2SNa$ [M + Na]$^+$, 383.0188; found 383.0188.

4-Azido-3-pentyl-6-phenylcinoline (**7a**). (A) This compound was synthesized by the general procedure for the Suzuki coupling (method A) from 4-azido-6-bromocinnoline **5** (50 mg, 0.156 mmol) and phenylboronic acid **6a** (38 mg, 0.312 mmol) at 80 °C; reaction time – 1 h. Purification of the crude product by column chromatography using hexane/EtOAc (5:1→2:1→1:2) as the eluent gave **7a** (9,5 mg, 19%). (B) Azidocinnoline **7a** was obtained by the general procedure for the Suzuki coupling (method B) from 4-azido-6-bromcinnoline **5** (50 mg, 0.156 mmol), phenylboronic acid **6a** (38 mg, 0.312 mmol) and K_3PO_4 (99 mg, 0.468 mmol, 3 equiv) at 80 °C; reaction time–1 h. Purification of the crude product by column chromatography using hexane/EtOAc (5:1→2:1→1:2) as the eluent gave **7a** (25 mg, 50%). C) Azidocinnoline **7a** was prepared by the general procure for the nucleophilic substitution from bromocinnoline **9a** (650 mg, 1.83 mmol) and sodium azide (594 mg, 9.15 mmol). Purification of the crude product by column chromatography using hexane/EtOAc (8:1) as the eluent gave **7a** (528 mg, 91%). ^{1}H-NMR (acetone-d_6) δ 8.49 (d, J = 8.9 Hz, 1H), 8.41 (d, J = 1.6 Hz, 1H), 8.23 (dd, J = 8.9, 1.6 Hz, 1H), 7.91–7.84 (m, 2H), 7.63–7.53 (m, 2H), 7.52–7.48 (m, 1H), 3.41–3.28 (m, 2H), 1.97–1.90 (m, 2H), 1.57–1.35 (m, 4H), 0.93 (t, J = 7.1 Hz, 3H). ^{13}C-NMR (acetone-d_6) δ 151.6, 150.3, 144.1, 140.2, 133.2, 131.2, 130.8, 130.2, 129.7, 128.5, 121.8, 119.3, 32.9, 32.4, 23.2, 14.3. HRMS (ESI): calcd. for $C_{19}H_{20}N_5$ [M + H]$^+$, 318.1713; found 318.1714. IR, cm^{-1}, ν: 2117 (N_3).

4-Azido-3-pentyl-6-(4-(trifluoromethyl)phenyl)cinnoline (**7b**). This compound was synthesized by the general procure for the nucleophilic substitution from bromocinnoline **9b** (130 mg, 0.308 mmol) and sodium azide (100 mg, 1.54 mmol). Purification of the crude product by column chromatography using hexane/EtOAc (3:1) as the eluent gave **7b** (100 mg, 84%). ^{1}H-NMR (acetone-d_6) δ 8.57–8.48 (m, 2H), 8.27 (dd, J = 8.8 Hz, 2.0 Hz, 1H), 8.11 (d, J = 8.1 Hz, 2H), 7.92 (d, J = 8.1 Hz, 2H), 3.42–3.33 (m, 2H), 2.00–1.88 (m, 2H), 1.54–1.38 (m, 4H), 0.93 (t, J = 7.0 Hz, 3H). ^{13}C-NMR (acetone-d_6) δ 151.8, 150.3, 144.1, 142.5, 133.4, 131.5, 130.9 (q, $^2J_{C-F}$ = 32.4 Hz), 130.6, 129.3, 127.0 (q, $^3J_{C-F}$ = 3.8 Hz), 125.4 (q, $^1J_{C-F}$ = 271.1 Hz), 121.6, 120.5, 32.9, 32.4, 30.0, 23.2, 14.3 ppm. HRMS (ESI): calcd. for $C_{20}H_{18}F_3N_5Na$ [M + Na]$^+$, 408.1407; found 408.1403. IR, cm^{-1}, ν: 2122 (N_3).

4-Azido-6-(4-methoxyphenyl) -3-pentylcinnoline (**7c**). This compound was synthesized by the general procure for the nucleophilic substitution from bromocinnoline **9c** (155 mg, 0.404 mmol) and sodium azide (65.7 mg, 1.01 mmol). Purification of the crude product by column chromatography using hexane/EtOAc (5:1→3:1) as the eluent gave **7c** (116 mg, 83%). ^{1}H-NMR (acetone-d_6) δ 8.44 (d, J = 8.9 Hz, 1H), 8.33 (d, J = 2.0 Hz, 1H), 8.20 (dd, J = 8.9, 1.9 Hz, 1H), 7.88–7.77 (m, 2H), 7.22–7.08 (m, 2H), 3.89 (s, 3H), 3.38–3.21 (m, 2H), 1.96–1.88 (m, 2H), 1.53–1.40 (m, 4H), 0.93 (t, J = 7.1 Hz, 3H). ^{13}C-NMR (acetone-d_6) δ 161.6, 151.5, 150.2, 143.8, 132.9, 132.2, 131.1, 130.5, 129.7, 121.9, 118.0, 115.6, 55.8, 32.9, 32.4, 30.0, 23.2, 14.3. HRMS (ESI): calcd. for $C_{20}H_{21}N_5ONa$ [M + Na]$^+$, 370.1638; found 370.1626. IR, cm^{-1}, ν: 2116 (N$_3$).

4-Azido-3-pentyl-6-(4-phenoxyphenyl)cinnoline (**7d**). This compound was synthesized by the general procure for the nucleophilic substitution from bromocinnoline **9d** (120 mg, 0.269 mmol) and sodium azide (87 mg, 1.35 mmol). Purification of the crude product by column chromatography using hexane/EtOAc (5:1→ 3:1) as the eluent gave **7d** (77 mg, 70%). ^{1}H-NMR (acetone-d_6) δ 8.47 (d, J = 8.9 Hz, 1H), 8.38 (d, J = 1.8 Hz, 1H), 8.22 (dd, J = 8.9, 1.9 Hz, 1H), 7.94–7.86 (m, 2H), 7.49 – 7.40 (m, 2H), 7.25–7.13 (m, 3H), 7.16–7.08 (m, 2H), 3.39–3.30 (m, 2H), 1.99–1.87 (m, 2H), 1.53–1.39 (m, 4H), 0.93 (t, J = 7.0 Hz, 3H). ^{13}C-NMR (acetone-d_6) δ 159.4, 157.6, 151.6, 150.2, 143.4, 134.9, 133.1, 131.2, 131.0, 130.9, 130.6, 130.1, 124.9, 121.8, 120.3, 119.8, 119.7, 118.7, 32.9, 32.4, 30.0, 23.2, 14.3. HRMS (ESI): calcd. for $C_{25}H_{23}N_5ONa$ [M + Na]$^+$, 432.1795; found 432.1797.

tert-Butyl (4-(4-azido-3-pentylcinnolin-6-yl)phenyl)carbamate (**7e**). This compound was synthesized by the general procure for the nucleophilic substitution from bromocinnoline **9e** (88 mg, 0.188 mmol) and sodium azide (61 mg, 0.94 mmol). Purification of the crude product by column chromatography using hexane/EtOAc (2:1→ 1:1) as the eluent gave **7e** (59 mg, 73%). ^{1}H-NMR (acetone-d_6) δ 8.65 (s, 1H), 8.46 (d, J = 8.9 Hz, 1H), 8.38 (d, J = 1.7 Hz, 1H), 8.23 (dd, J = 8.9, 1.8 Hz, 1H), 7.86–7.73 (m, 4H), 3.39–3.29 (m, 2H), 1.98–1.87 (m, 2H), 1.52 (s, 9H), 1.49–1.40 (m, 4H), 0.93 (t, J = 7.0 Hz, 3H). ^{13}C-NMR (acetone-d_6) δ = 153.7, 153.6, 151.5, 150.2, 143.7, 141.7, 133.6, 133.0, 131.1, 130.5, 128.9, 121.9, 119.6, 119.5, 118.1, 80.4, 32.9, 32.4, 29.8, 28.5, 23.2, 14.3. HRMS (ESI): calcd. for $C_{24}H_{29}N_6O_2$ [M + H]$^+$, 433.2347; found 433.2342.

4-(4-Azido-3-pentylcinnolin-6-yl)benzonitrile (**7f**). This compound was synthesized by the general procure for the nucleophilic substitution from bromocinnoline **9f** (100 mg, 0.264 mmol) and sodium azide (86 mg, 1.32 mmol). Purification of the crude product by column chromatography using hexane/EtOAc (5:1) as the eluent gave **7f** (80 mg, 88%). ^{1}H-NMR (acetone-d_6) δ 8.57–8.47 (m, 2H), 8.27 (dd, J = 8.9, 2.0 Hz, 1H), 8.15–8.04 (m, 2H), 8.02–7.92 (m, 2H), 3.46–3.22 (m, 2H), 2.00–1.88 (m, 2H), 1.56–1.36 (m, 4H), 0.93 (t, J = 7.1 Hz, 3H). ^{13}C-NMR (acetone-d_6) δ 151.9, 150.3, 144.5, 142.1, 133.9, 133.5, 131.5, 130.4, 129.5, 121.6, 120.7, 119.1, 113.2, 32.9, 32.4, 29.1, 23.2, 14.3. HRMS (ESI): calcd. for $C_{20}H_{18}N_6Na$ [M + Na]$^+$, 365.1485; found 365.1472.

4-Azido-3-pentyl-6-(thiophen-3-yl)cinnoline (**7g**). This compound was synthesized by the general procure for the nucleophilic substitution from bromocinnoline **9g** (185 mg, 0.514 mmol) and sodium azide (167 mg, 2.57 mmol). Purification of the crude product by column chromatography using hexane/EtOAc (7:1→3:1) as the eluent gave **7g** (115 mg, 69%), m.p. 100–102 °C. ^{1}H-NMR (acetone-d_6) δ 8.46–8.39 (m, 2H), 8.29 (dd, J = 8.9, 1.9 Hz, 1H), 8.13 (dd, J = 2.9, 1.4 Hz, 1H), 7.76 (dd, J = 5.1, 1.4 Hz, 1H), 7.70 (dd, J = 5.1, 2.9 Hz, 1H), 3.37–3.29 (m, 2H), 1.98–1.86 (m, 2H), 1.54–1.37 (m, 4H), 0.93 (t, J = 7.1 Hz, 3H). ^{13}C-NMR (acetone-d_6) δ 151.6, 150.2, 141.3, 138.7, 133.0, 131.2, 130.3, 128.4, 127.1, 124.7, 122.0, 117.9, 32.9, 32.4, 30.0, 23.2, 14.3. HRMS (ESI): calcd. for $C_{17}H_{17}N_5SNa$ [M + Na]$^+$, 346.1097; found 346.1085.

3-[1-(3-Pentyl-6-phenylcinnolin-4-yl)-1H-1,2,3-triazol-4-yl]propan-1-ol (**11a**). This compound was synthesized in accordance with the general procedure from pent-4-yn-1-ol **10a** (47.8 mg, 0.568 mmol), 4-azidocinnoline **7a** (90.0 mg, 0.284 mmol), CuSO$_4$·5H$_2$O (3.5 mg, 0.014 mmol), and sodium ascorbate (5.60 mg, 0.028 mmol) in a mixture of THF/H$_2$O (5.6 mL); reaction time – 5 h. The crude product was purified by column chromatography (eluent: hexane/EtOAc = 2:1) to afford triazole **11a** as a yellowish solid (100 mg, 88%), m.p. 89–91 °C. ^{1}H-NMR (acetone-d_6) δ 8.67 (d, J = 8.9 Hz, 1H), 8.35 (s, 1H), 8.30 (dd, J = 8.9, 1.9 Hz, 1H), 7.75–7.68 (m, 2H), 7.56–7.42 (m, 4H), 3.74–3.67 (m, 3H), 3.04–2.95 (m, 4H), 2.03–1.98 (m, 2H), 1.83–1.71 (m, 2H), 1.34–1.24 (m, 4H), 0.88–0.80 (m, 3H). ^{13}C-NMR (acetone-d_6) δ

154.3, 150.8, 149.2, 146.0, 139.7, 131.3, 131.2, 130.4, 130.1, 129.9, 128.5, 125.8, 123.6, 119.1, 61.6, 33.3, 32.5, 32.2, 30.0, 22.9, 22.7, 14.2. HRMS ESI: [M + H]$^+$ calcd. for $C_{24}H_{27}N_5O^+$: 402.2288; found: 402.2293.

4-[4-(3,4-Dimethoxyphenyl)-1H-1,2,3-triazol-1-yl]-3-pentyl-6-[4-(trifluoromethyl)phenyl]cinnoline (**11b**). This compound was synthesized in accordance with the general procedure from 4-ethynyl-1,2-dimethoxybenzene **10b** (21.0 mg, 0.130 mmol), 4-azidocinnoline **7b** (50.0 mg, 0.130 mmol), CuSO$_4$·5H$_2$O (1.7 mg, 0.007 mmol), and sodium ascorbate (2.6 mg, 0.013 mmol) in a mixture of THF/H$_2$O (5 mL); reaction time − 3 h. The crude product was purified by column chromatography (eluent: hexane/EtOAc = 2:1) to afford triazole **11b** as a yellowish solid (36 mg, 50%). ^{1}H-NMR (acetone-d_6) δ 8.92 (s, 1H), 8.76 (d, J = 8.9 Hz, 1H), 8.37 (dd, J = 8.9, 1.7 Hz, 1H), 7.98 (d, J = 8.5 Hz, 2H), 7.86–7.79 (m, 3H), 7.66 (d, J = 2.0 Hz, 1H), 7.58 (dd, J = 8.3, 2.0 Hz, 1H), 7.08 (d, J = 8.3 Hz, 1H), 3.90 (s, 3H), 3.87 (s, 3H), 3.11–3.03 (m, 2H), 1.83 (p, J = 7.2 Hz, 2H), 1.36–1.23 (m, 4H), 0.83 (t, J = 7.0 Hz, 3H). ^{13}C-NMR (acetone-d_6) δ 154.5, 150.88, 150.87, 150.82, 149.0, 144.6, 143.6 (q, $^5J_{C-F}$ = 1.2 Hz), 131.6, 131.32, 131.1 (q, $^2J_{C-F}$ = 32.4 Hz), 130.3, 129.5, 126.9 (q, $^3J_{C-F}$ = 3.8 Hz), 125.2 (q, $^1J_{C-F}$ = 271.4 Hz), 124.06, 124.05, 123.3, 120.4, 119.4, 113.2, 110.6, 56.25, 56.24, 32.6, 32.2, 30.0, 22.9, 14.2. HRMS (ESI): calcd. for $C_{30}H_{28}N_5O_2F_3Na$ [M + Na]$^+$; 570.2087; found: 570.2081.

6-(4-Methoxyphenyl)-4-[4-(4-methoxyphenyl)-1H-1,2,3-triazol-1-yl]-3-pentylcinnoline (**11c**). (A) This compound was prepared in accordance with the general procedure (method A) from 4-ethynyl-1-methoxybenzene **10c** (13.8 mg, 0.104 mmol), 4-azidocinnoline **7c** (40.0 mg, 0.104 mmol), CuSO$_4$·5H$_2$O (1.3 mg, 0.005 mmol), and sodium ascorbate (2.10 mg, 0.010 mmol) in a mixture of THF/H$_2$O (4 mL); reaction time − 2 h. The crude product was purified by column chromatography (eluent: hexane/EtOAc = 3:1) to afford triazole **11c** as a yellowish solid (35 mg, 76%). (B) This compound was prepared in accordance with the general procedure (method A) from mixture 4-ethynyl-1-methoxybenzene **10c** (24.0 mg, 0.177 mmol), 4-azidocinnoline **7c** (68.0 mg, 0.177 mmol), CuSO$_4$·5H$_2$O (2.5 mg, 0.009 mmol), and sodium ascorbate (3.6 mg, 0.009 mmol) in the mixture of THF/H$_2$O (4 mL) with the addition of TBTA (5.0 mg, 0.009 mmol); reaction time − 50 h. The crude product was purified by column chromatography (eluent: hexane/EtOAc = 3:1) to afford triazole **11c** as a solid (30 mg, 35%). ^{1}H-NMR (CDCl$_3$) δ 8.68 (d, J = 8.9 Hz, 1H), 8.11 (dd, J = 8.9, 1.9 Hz, 1H), 8.01 (s, 1H), 7.94–7.88 (m, 2H), 7.58–7.52 (m, 2H), 7.44 (d, J = 1.8 Hz, 1H), 7.07–7.03 (m, 2H), 7.00–6.95 (m, 2H), 3.89 (s, 3H), 3.84 (s, 3H), 3.07–2.98 (m, 2H), 1.90–1.82 (m, 2H), 1.35–1.28 (m, 4H), 0.85 (t, J = 6.8 Hz, 4H). ^{13}C-NMR (CDCl$_3$) δ 160.8, 160.3, 153.7, 150.0, 148.5, 145.6, 131.2, 130.7, 130.5, 129.1, 127.5, 123.2, 122.4, 121.5, 117.0, 114.9, 114.7, 55.6, 32.0, 31.7, 29.8, 22.4, 14.0. HRMS (ESI): calcd. for $C_{29}H_{29}N_5O_2Na$ [M + Na]$^+$; 502.2213; found: 502.2227.

4{1-[3-Pentyl-6-(4-phenoxyphenyl)cinnolin-4-yl]-1H-1,2,3-triazol-4-yl}butan-1-ol (**11d**). This compound was prepared in accordance with the general procedure from hex-5-yn-1-ol **10d** (3.60 mg, 0.037 mmol), 4-azido-3-pentyl-6-(4-phenoxyphenyl)cinnoline **7d** (15.0 mg, 0.037 mmol), CuSO$_4$·5H$_2$O (0.46 mg, 0.0018 mmol), and sodium ascorbate (0.67 mg, 0.0037 mmol) in a mixture of THF/H$_2$O (1.5 mL); reaction time − 21 h. The crude product was purified by column chromatography (eluent: hexane/acetone = 2:1) to afford triazole **11d** as a brown oil (5 mg, 28%). ^{1}H-NMR (DMSO-d_6) δ 8.71 (d, J = 8.9 Hz, 1H), 8.55 (s, 1H), 8.35 (dd, J = 8.9, 1.9 Hz, 1H), 7.78–7.72 (m, 2H), 7.48–7.40 (m, 2H), 7.38 (d, J = 1.9 Hz, 1H), 7.23–7.17 (m, 1H), 7.15–7.04 (m, 4H), 4.41 (t, J = 5.1 Hz, 1H, OH), 3.45 (td, J = 6.4, 5.1 Hz, 2H), 2.94–2.90 (m, 2H), 2.84 (t, J = 7.4 Hz, 2H), 1.81–1.72 (m, 2H), 1.64 (m, 2H), 1.53 (m, 2H), 1.22 (m, 4H), 0.81 (m, 3H). ^{13}C-NMR (DMSO-d_6) δ 158.0, 155.9, 153.0, 149.3, 147.9, 143.9, 132.8, 130.5, 130.2, 130.1, 129.3, 129.2, 125.4, 124.1, 122.2, 119.2, 118.9, 117.1, 60.3, 31.8, 31.2, 30.8, 28.5, 25.4, 24.7, 21.6, 13.6. HRMS (ESI): calcd. for $C_{31}H_{33}N_5O_2Na$ [M + Na]$^+$; 530.2526; found: 530.2510.

Methyl 4-{1-[6-(4-{[tert-butoxycarbonyl]amino}phenyl)-3-pentylcinnolin-4-yl]-1H-1,2,3-triazol-4-yl} benzoate (**11e**). This compound was prepared in accordance with the general procedure from methyl 4-ethynylbenzoate **10e** (10.7 mg, 0.067 mmol), 4-azidocinnoline **7e** (29.0 mg, 0.067 mmol), CuSO$_4$·5H$_2$O (0.84 mg, 0.0034 mmol), and sodium ascorbate (1.22 mg, 0.067 mmol) in a mixture of THF/H$_2$O

(3 mL); reaction time – 2 h. The crude product was purified by column chromatography (eluent: hexane/acetone = 3:1) to afford triazole **11e** as a yellow solid (27 mg, 68%). ^{1}H-NMR (DMSO-d_6) δ 9.56 (s, 1H), 9.40 (s, 1 H), 8.70 (d, J = 9.0 Hz, 1H), 8.36 (dd, J = 9.0, 1.9 Hz, 1H), 8.19–8.15 (m, 2H), 8.13–8.09 (m, 2H), 7.72–7.67 (m, 2H), 7.62–7.56 (m, 2H), 7.54 (d, J = 1.9 Hz, 1H), 3.89 (s, 3H), 2.97–2.93 (m, 2H), 1.77–1.64 (m, 2H), 1.47 (s, 9H), 1.26–1.15 (m, 4H), 0.80–0.72 (m, 3H). ^{13}C-NMR (DMSO-d_6) δ 165.9, 152.8, 152.6, 149.4, 146.2, 144.4, 140.5, 134.4, 131.1, 130.5, 130.0, 129.2, 128.7, 128.1, 126.0, 125.7, 122.0, 118.4, 116.3, 79.4, 52.2, 31.2, 30.7, 28.5, 28.0, 21.6, 13.6. HRMS (ESI): calcd. for $C_{34}H_{36}N_6O_4Na$ [M + Na]$^+$; 615.2690; found: 615.2683.

4-{4-[4-(4-{Dimethylamino}phenyl)-1H-1,2,3-triazol-1-yl]-3-pentylcinnolin-6-yl}benzonitrile (**11f**). This compound was prepared in accordance with the general procedure from 4-ethynyl-N,N-dimethylaniline **10f** (25.4 mg, 0.175 mmol), 4-azidocinnoline **7f** (60.0 mg, 0.175 mmol), CuSO$_4$·5H$_2$O (2.19 mg, 0.0088 mmol), and sodium ascorbate (3.19 mg, 0.017 mmol) in the mixture of THF/H$_2$O (6 mL); reaction time – 4 h. The crude product was purified by column chromatography (eluent: hexane/acetone = 3:1) to afford triazole **11f** as a yellow solid (74 mg, 86%). ^{1}H-NMR (DMSO-d_6) δ 9.04 (s, 1H), 8.79–8.76 (m, 1H), 8.41–8.37 (m, 1H), 7.96–7.95 (m, 4H), 7.85–7.75 (m, 2H), 7.69–7.68 (m, 1H), 6.84 (d, J = 8.4 Hz, 2H), 3.00 (t, J = 7.6 Hz, 2H), 2.96 (s, 6H), 1.82–1.57 (m, 2H), 1.23–1.17 (m, 4H), 0.86–0.57 (m, 3H). ^{13}C-NMR (DMSO-d_6) δ 153.1, 150.4, 149.5, 147.9, 142.8, 142.5, 133.1, 130.6, 129.2, 128.6, 126.5, 122.8, 121.7, 119.4, 118.5, 117.4, 112.3, 111.6, 31.3, 30.7, 28.5, 21.6, 13.6. HRMS (ESI): calcd. for $C_{30}H_{30}N_7$ [M + H]$^+$; 488.2557; found: 488.2558.

4-(4-(3,4-Dimethoxyphenyl)-1H-1,2,3-triazol-1-yl)-3-pentyl-6-(thiophen-3-yl)cinnoline (**11g**). This compound was prepared in accordance with the general procedure from 4-ethynyl-1,2-dimethoxybenzene **10b** (27.6 mg, 0.170 mmol), 4-azidocinnoline **7g** (55.0 mg, 0.170 mmol), CuSO$_4$·5H$_2$O (2.12 mg, 0.0085 mmol), and sodium ascorbate (3.70 mg, 0.017 mmol) in the mixture of THF/H$_2$O (6 mL); reaction time – 4 h. The crude product was purified by column chromatography (eluent: hexane/EtOAc = 2:1) to afford triazole **11g** as a yellow solid (68 mg, 82%). ^{1}H-NMR (DMSO-d_6) δ 9.17 (s, 1H), 8.68 (dd, J = 9.0, 0.6 Hz, 1H), 8.44 (dd, J = 9.0, 1.9 Hz, 1H), 8.23 (dd, J = 2.9, 1.4 Hz, 1H), 7.69 (dd, J = 5.1, 2.9 Hz, 1H), 7.61–7.60 (m, 3H), 7.56 (dd, J = 8.2, 2.0 Hz, 1H), 7.11 (d, J = 8.4 Hz, 1H), 3.86 (s, 3H), 3.82 (s, 3H), 2.97–2.93 (m, 2H), 1.74–1.67 (m, 2H), 1.24–1.20 (m, 4H), 0.82–0.74 (m, 3H). ^{13}C-NMR (DMSO-d_6) δ 153.0, 149.4, 149.2, 149.1, 147.4, 139.4, 139.3, 130.4, 130.1, 128.9, 128.2, 126.4, 125.6, 124.0, 122.6, 122.3, 118.2, 116.2, 112.2, 109.3, 55.61, 55.60, 31.2, 30.8, 28.6, 21.6, 13.6. HRMS (ESI): calcd. for $C_{27}H_{27}N_5O_2SNa$ [M + Na]$^+$; 508.1778; found: 508.1790. Single crystals of **11g** were grown from the solution in chloroform by the vapor exchange with hexane. Crystal data for **11g** $C_{27}H_{27}N_5O_2S$ (M = 485.59 g/mol): triclinic, space group P$\bar{1}$ (no. 2), a = 8.2511(5) Å, b = 10.9441(6) Å, c = 14.6090(6) Å, α = 105.898(5)°, β = 94.236(4)°, γ = 107.263(5)° V = 1194.17(12) Å^3, Z = 2, T = 115(3) K, μ(CuKα) = 1.489 mm^{-1}, D$_{calc}$ = 1.350 g/cm^3, 9946 reflections measured (6.382° ≤ 2Θ ≤ 139.978°), 4507 unique (Rint = 0.0378, Rsigma = 0.0505) which were used in all calculations. The final R$_1$ was 0.0516 (I > 2σ(I)) and wR$_2$ was 0.1510. Crystallographic data (excluding structure factors) for the structures reported in this work have been deposited with the Cambridge Crystallographic Data Centre as supplementary publication no. CCDC 1902675 (Supplementary Materials).

3-Pentyl-6-(4-(trifluoromethyl)phenyl)cinnolin-4-amine (**14**). To a stirred solution of 4-azidocinnolines **7b** (50.0 mg, 0.13 mmol) in DMSO (0.3 mL) at room temperature were propanal **12** (22.0 mg, 30.0 μL, 0.39 mmol) followed by 1,8-diazabicyclo[5.4.0]undec-7-ene (DBU) (6.0 mg, 5.8 μL, 0.039 mmol). The reaction mixture was stirred at room temperature for 3 hours (TLC control). The resulting mixture was diluted with water (10 mL) and extracted with ethyl acetate (3 × 8 mL). The combined organic layers were dried over anhydrous Na$_2$SO$_4$ and concentrated under reduced pressure to yield the crude product, which was purified by column chromatography on silica gel. ^{1}H-NMR (acetone-d_6) δ 8.60 (d, J = 1.9 Hz, 1H), 8.24 (d, J = 8.9 Hz, 1H), 8.08 (d, J = 8.7 Hz, 3H), 7.85 (d, J = 8.2 Hz, 2H), 6.51 (s, 2H), 3.23–3.04 (m, 2H), 1.93–1.79 (m, 2H), 1.49–1.35 (m, 4H), 0.90 (t, J = 7.0 Hz, 3H). ^{13}C-NMR (acetone-d_6) δ 149.0, 144.9 ($^5J_{C-F}$ = 1.2 Hz), 143.4, 139.5, 138.7, 130.7, 130.1 ($^2J_{C-F}$ = 32.3 Hz), 129.0, 128.9, 126.7

($^3J_{C-F}$ = 3.8 Hz), 125.4 ($^1J_{C-F}$ = 271.2 Hz), 120.6, 116.1, 32.6, 32.3, 28.3, 23.3, 14.4. HRMS (ESI): calcd. for $C_{20}H_{21}F_3N_3$ [M + H]$^+$; 360.1682; found: 360.1674.

1,5-Ditosyl-7,8-didehydro-2,3,4,5,6,9-hexahydro-1H-1,5-diazonine (**15**). To a stirred at 0 °C solution of $Co_2(CO)_6$-complex of **15** (100.0 mg, 0.139 mmol) [69] in acetone (3 mL) tetrabutylammonium fluoride hydrate (195 mg, 0.696 mmol, calculated for TBAF monohydrate) was added. The reaction mixture was allowed to warm to room temperature and then was stirred at this temperature for 1.5 hours (TLC control). After completion of the reaction the mixture was diluted with water (10 mL) and extracted with ethyl acetate (3 × 15 mL). The combined organic layers were dried over anhydrous Na_2SO_4 and concentrated under reduced pressure to yield the crude product, which was purified by column chromatography on silica gel (eluent: hexane/acetone = 3:1) to afford diazacyclononyne **15** as a white solid (56 mg, 93%). Spectral data were identical to those reported previously [69]. ^{1}H-NMR (CDCl$_3$) δ 7.69–7.62 (m, 4H), 7.37–7.30 (m, 4H), 3.79 (s, 4H), 3.31–3.24 (m, 4H), 2.44 (s, 6H), 2.14–2.09 (m, 2H).

1-[3-Pentyl-6-(thiophen-3-yl)cinnolin-4-yl]-5,9-ditosyl-1,4,5,6,7,8,9,10-octahydro-[1,2,3]triazolo-[4,5-g][1,5]-diazonine (**16**). A mixture of diazacyclononyne **15** (58 mg, 0.133 mmol) and 4-azidocinnoline **7g** (43.0 mg, 0.133 mmol) in acetonitrile (2 mL) was stirred at 30 °C for 5 h. The solvent was removed under reduced pressure and the crude product was purified by column chromatography on silicagel usin hexane/acetone (3:1) as the eluent to give triazole **16** as a yellowish solid (95 mg, 95%). ^{1}H-NMR (DMSO-d_6) δ 8.71 (d, J = 9.0 Hz, 1H), 8.48 (dd, J = 9.0, 1.9 Hz, 1H), 8.22 (dd, J = 2.9, 1.4 Hz, 1H), 7.84 – 7.77 (m, 2H), 7.69 (dd, J = 5.2, 2.9 Hz, 1H), 7.59 (dd, J = 5.1, 1.4 Hz, 1H), 7.49 (d, J = 8.0 Hz, 2H), 7.32 (d, J = 2.0 Hz, 1H), 7.18–7.09 (m, 4H), 4.78 (d, J = 16.2 Hz, 1H), 4.62 (d, J = 16.3 Hz, 1H), 4.18 (s, 2H), 3.24 (dt, J = 7.3, 3.6 Hz, 2H), 3.22–3.11 (m, 1H), 2.98–2.90 (m, 1H), 2.68–2.61 (m, 1H), 2.43 (s, 3H), 2.27 (s, 3H), 1.91–1.74 (m, 4H), 1.30–1.22 (m, 4H), 0.86–0.78 (m, 3H). ^{13}C-NMR (DMSO-d_6) δ 154.2, 149.3, 143.8, 143.7, 143.2, 139.6, 139.2, 134.6, 134.5, 133.9, 130.4, 130.3, 130.0, 129.7, 128.2, 127.3, 126.8, 126.6, 126.2, 125.6, 122.8, 115.6, 48.7, 48.6, 45.3, 42.1, 31.0, 31.0, 29.4, 28.5, 21.7, 21.0, 20.8, 13.7. HRMS (ESI): calcd. for $C_{38}H_{42}N_7O_4S_3$ [M + H]$^+$: 756.2455; found: 756.2466.

3.3. DFT Calculations

All computations were carried out at the DFT/HF hybrid level of theory using Becke's three-parameter hybrid exchange functional in combination with the gradient-corrected correlation functional of Lee, Yang, and Parr (B3LYP) by using the GAUSSIAN 2003 program packages [76]. The geometries optimization was performed using the 6-311+G(2d,2p) basis set (standard 6–311 basis set added with polarization (d, p) and diffuse functions). Optimizations were performed on all degrees of freedom, and optimized structures were verified as true minima with no imaginary frequencies. The Hessian matrix was calculated analytically for the optimized structures in order to prove the location of correct minima and to estimate the thermodynamic parameters.

3.4. The Absolute Fluorescence Quantum Yield Measurements

The absolute fluorescence quantum yield was measured on a Horiba Fluorolog-3 spectrometer (Edison, NJ, USA) equipped with an integrating sphere. A xenon lamp coupled to a double monochromator was used as excitation light source. The sample (1 cm quartz cuvette cell with molecular solution in THF) or blank (pure THF) were directly illuminated in the center of the integrating sphere. The optical density of all investigated sample solutions in THF did not exceed 0.1 at the luminescence excitation wavelength. Under the same conditions (e.g., excitation wavelength, spectral resolution, temperature), the luminescence spectrum of the sample E_c, the luminescence spectrum of the blank E_a, the Rayleigh scattering spectrum of the sample L_c, and the Rayleigh scattering spectrum of the solvent L_a were measured. The absolute fluorescence quantum yield was determined according to the formula:

$$\Phi = (E_c - E_a)/(L_a - L_c) \tag{1}$$

3.5. Cell Culture Cultivation and Cytotoxicity Studies

The HeLa cells were obtained from the Russian Collection of Cell cultures (Institute of Cytology RAS, Saint Petersburg, Russia). Cells were cultivated in DMEM medium containing 10% FBS (fetal bovine serum) at 37 °C and 5% CO_2. Compounds were diluted with DMSO to concentrations of 7.5; 5; 2.5; 0.5 and 0.1 mM. These solutions were diluted 100 x times with culture medium to the final concentrations of 100, 75, 50, 25, 5 and 1 µM/L (1% DMSO). Cytotoxicity assays were made using 3-(4,5-dimethyl-2-thiazolyl)-2,5-diphenyl-2H-tetrazolium bromide (MTT) [77]. Cells were placed in 96-well plates (5000 of cells per well in 100 µL of DMEM + FBS without compounds) and preincubated for 24 hours. Subsequently culture medium was aspirated and 100 µL volumes of media with compounds were added to the wells. After 24 h of drug exposure, cytotoxicity assays were performed using MTT. Twenty µL of MTT solution (5 mg/mL) was added to the cell cultures. During the following 3 h of incubation mitochondrial dehydrogenase activity of vital cells caused the production of formazan crystals from MTT. Culture medium was aspirated and formazan was dissolved in 100 µL of DMSO. The optical density at 570 nm, which is directly proportional to the number of surviving cells, was measured using microplate reader Uniplan AIFR-01 (Pikon, Moscow, Russia).

4. Conclusions

In summary, a convenient synthetic route towards 4-azido-6-arylcinnolines based on Suzuki coupling, Richter-type cyclization and nucleophilic substitution of activated aromatic bromine atom with azido function was developed. The crucial point in the synthetic sequences is to carry out the Suzuki coupling as the initial step and the introduction of an azido group as the final step. Despite the possibility of chemoselective substitution of the bromine atom at C4 in 4,6-dibromocinnoline with the formation of 4-azido-6-bromocinnoline, the azide moiety at C4 of cinnoline ring was found to be unstable under Suzuki reaction conditions. The synthetic route developed here is functional group tolerant to various substituents on the aryl moiety. All 4-azidocinnolines underwent CuAAC with terminal alkynes providing 4-triazolylcinnolines in moderate to high yields. The reactivity of 4-azidocinnolines in SPAAC was also confirmed. Whereas 4-azidocinnolines display fluorescent properties under excitation at their low wavelength absorption maximum, the corresponding 4-triazolylcinnolines were devoid of any fluorescence, probably due to their non-planar geometry and PET mechanism. Among the series of 4-triazolylcinnolines synthesized the compound **11a** bearing a hydroxyalkyl-substituted triazole ring was found to be cytotoxic toward HeLa cells at micromolar concentrations.

Supplementary Materials: The following are available online, copies of ^{1}H, ^{13}C{^{1}H}, and DEPT NMR spectra for all newly synthesized compounds; Cartesian coordinates and energies and isosurfaces of MO for optimized geometries of **7c,f** and **11c,f**; cif file with X-ray data for compound **11g**.

Author Contributions: Conceptualization, N.A.D. and I.A.B.*; methodology, N.A.D.; validation, A.A.V, A.I.G.; formal analysis, I.V.K., A.V.P., I.A.B; investigation, N.S.B., A.I.G., A.A.V., A.M.R., A.A.V., N.I.S., I.V.K., A.V.P., I.A.B, P.V.T.; writing—N.A.D., A.I.G, A.M.R., A.V.P.; writing—review and editing, N.A.D., I.A.B.*; visualization, N.A.D., A.I.G., I.A.B.; supervision, N.A.D.; project administration, N.A.D.; I.A.B.*; funding acquisition, A.A.V.

Funding: This research was funded by Russian Foundation for Basic Research (RFBR), grant number 18-33-01265.

Acknowledgments: We are grateful to all staff members of the Research Park of SPSU, who carried out analysis for current research. The research was carried out by using the equipment of the SPSU Resource Centres: Magnetic Resonance Research Centre, Chemical Analysis and Materials Research Centre, Centre for Optical and Laser Materials Research, Centre for X-ray Diffraction Studies, Chemistry Educational Centre. The authors are grateful to N. Samusik (Stanford University) for the helpful comments and additions to the text.

Conflicts of Interest: The authors declare no conflict of interest. The funders had no role in the design of the study; in the collection, analyses, or interpretation of data; in the writing of the manuscript, or in the decision to publish the results.

Appendix A

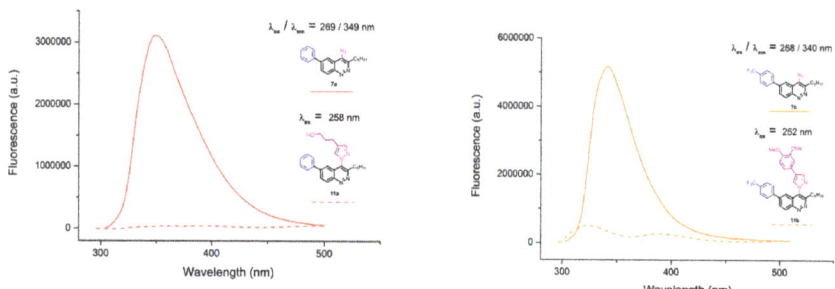

Figure A1. Emission spectra of 4-azidocinnolines/4-triazolylcinnolnes **7a/11a; 7b/11b; 7e/11e; 7f/11f**.

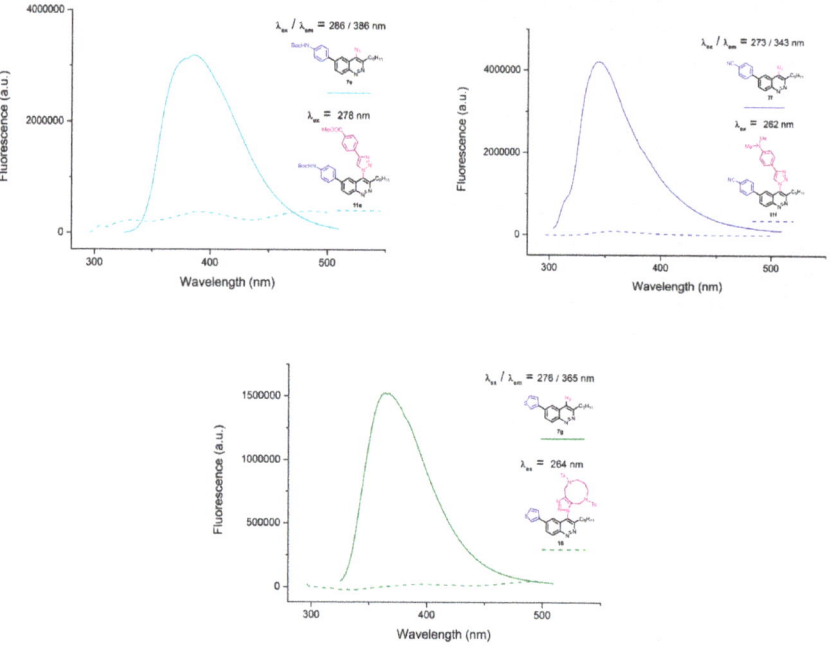

Figure A2. Emission spectra of 4-azidocinnolines/4-triazolylcinnolnes **7a/11a; 7b/11b; 7e/11e; 7f/11f; 7g/16** solutions in THF, C = 1 × 10^{-5} mol/L.

References

1. Vodovozova, E.L. Photoaffinity labeling and its application in structural biology. *Biochemisty* **2007**, *72*, 1–20. [CrossRef]
2. Panov, M.S.; Voskresenska, V.D.; Ryazantsev, M.N.; Tarnovsky, A.N.; Wilson, R.M. 5-azido-2-aminopyridine, a new nitrene/nitrenium ion photoaffinity labeling agent that exhibits reversible intersystem crossing between singlet and triplet nitrenes. *J. Am. Chem. Soc.* **2013**, *135*, 19167–19179. [CrossRef] [PubMed]
3. Zhang, N.; Tomizawa, M.; Casida, J.E. Structural features of azidopyridinyl neonicotinoid probes conferring high affinity and selectivity for mammalian α4β2 and Drosophila nicotinic receptors. *J. Med. Chem.* **2002**, *45*, 2832–2840. [CrossRef] [PubMed]

4. Polshakov, D.; Rai, S.; Wilson, R.M.; Mack, E.T.; Vogel, M.; Krause, J.A.; Burdzinski, G.; Platz, M.S. Photoaffinity labeling with 8-azidoadenosine and its derivatives: Chemistry of closed and opened adenosine diazaquinodimethanes. *Biochemistry* **2005**, *44*, 11241–11253. [CrossRef] [PubMed]
5. Rostovtsev, V.V.; Green, L.G.; Fokin, V.V.; Sharpless, K.B. A stepwise Huisgen cycloaddition process: copper(I)-catalyzed regioselective "ligation" of azides and terminal alkynes. *Angew. Chem. Int. Ed.* **2002**, *41*, 2596–2599. [CrossRef]
6. Tornøe, C.W.; Christensen, C.; Meldal, M. Peptidotriazoles on solid phase: [1,2,3]-Triazoles by regiospecific copper(I)-catalyzed 1,3-dipolar cycloadditions of terminal alkynes to azides. *J. Org. Chem.* **2002**, *67*, 3057–3064. [CrossRef] [PubMed]
7. Ashwini, N.; Garg, M.; Mohan, C.D.; Fuchs, J.E.; Rangappa, S.; Anusha, S.; Swaroop, T.R.; Rakesh, K.S.; Kanojia, D.; Madan, V.; et al. Synthesis of 1,2-benzisoxazole tethered 1,2,3-triazoles that exhibit anticancer activity in acute myeloid leukemia cell lines by inhibiting histone deacetylases, and inducing p21 and tubulin acetylation. *Bioorg. Med. Chem.* **2015**, *23*, 6157–6165. [CrossRef]
8. Zhang, W.; Li, Z.; Zhou, M.; Wu, F.; Hou, X.; Luo, H.; Liu, H.; Han, X.; Yan, G.; Ding, Z.; et al. Synthesis and biological evaluation of 4-(1,2,3-triazol-1-yl)coumarin derivatives as potential antitumor agents. *Bioorg. Med. Chem. Lett.* **2014**, *24*, 799–807. [CrossRef]
9. Thomas, K.D.; Adhikari, A.V.; Chowdhury, I.H.; Sumesh, E.; Pal, N.K. New quinolin-4-yl-1,2,3-triazoles carrying amides, sulphonamides and amidopiperazines as potential antitubercular agents. *Eur. J. Med. Chem.* **2011**, *46*, 2503–2512. [CrossRef]
10. Tiwari, R.; Miller, P.A.; Chiarelli, L.R.; Mori, G.; Šarkan, M.; Centárová, I.; Cho, S.; Mikušová, K.; Franzblau, S.G.; Oliver, A.G.; et al. Design, syntheses, and anti-TB activity of 1,3-benzothiazinone azide and click chemistry products inspired by BTZ043. *ACS Med. Chem. Lett.* **2016**, *7*, 266–270. [CrossRef]
11. Cserép, G.B.; Herner, A.; Kele, P. Bioorthogonal fluorescent labels: A review on combined forces. *Methods Appl. Fluoresc.* **2015**, *3*, 042001. [CrossRef] [PubMed]
12. Beatty, K.E.; Liu, J.C.; Xie, F.; Dieterich, D.C.; Schuman, E.M.; Wang, Q.; Tirrell, D.A. Fluorescence visualization of newly synthesized proteins in mammalian cells. *Angew. Chem. Int. Ed.* **2006**, *45*, 7364–7367. [CrossRef] [PubMed]
13. Sivakumar, K.; Xie, F.; Cash, B.M.; Long, S.; Barnhill, H.N.; Wang, Q. A fluorogenic 1,3-dipolar cycloaddition reaction of 3-azidocoumarins and acetylenes. *Org. Lett.* **2004**, *6*, 4603–4606. [CrossRef]
14. Herner, A.; Estrada Girona, G.; Nikić, I.; Kállay, M.; Lemke, E.A.; Kele, P. New generation of bioorthogonally applicable fluorogenic dyes with visible excitations and large stokes shifts. *Bioconjugate Chem.* **2014**, *25*, 1370–1374. [CrossRef] [PubMed]
15. Neef, A.B.; Schultz, C. Selective fluorescence labeling of lipids in living cells. *Angew. Chem. Int. Ed.* **2009**, *48*, 1498–1500. [CrossRef] [PubMed]
16. Zayas, J.; Annoual, M.; Das, J.K.; Felty, Q.; Gonzalez, W.G.; Miksovska, J.; Sharifai, N.; Chiba, A.; Wnuk, S.F. Strain promoted click chemistry of 2- or 8-azidopurine and 5-azidopyrimidine nucleosides and 8-azidoadenosine triphosphate with cyclooctynes. Application to living cell fluorescent imaging. *Bioconjugate Chem.* **2015**, *26*, 1519–1532. [CrossRef] [PubMed]
17. Brandão, G.C.; Rocha Missias, F.C.; Arantes, L.M.; Soares, L.F.; Roy, K.K.; Doerksen, R.J.; Braga de Oliveira, A.; Pereira, G.R. Antimalarial naphthoquinones. Synthesis via click chemistry, in vitro activity, docking to Pf DHODH and SAR of lapachol-based compounds. *Eur. J. Med. Chem.* **2018**, *145*, 191–205. [CrossRef]
18. Ellanki, A.R.; Islam, A.; Rama, V.S.; Pulipati, R.P.; Rambabu, D.; Rama Krishna, G.; Malla Reddy, C.; Mukkanti, K.; Vanaja, G.R.; Kalle, A.M.; et al. Solvent effect on copper-catalyzed azide–alkyne cycloaddition (CuAAC): Synthesis of novel triazolyl substituted quinolines as potential anticancer agents. *Bioorg. Med. Chem. Lett.* **2012**, *22*, 3455–3459. [CrossRef]
19. Leite, D.I.; de Fontes, F.V.; Bastos, M.M.; Hoelz, L.V.B.; Bianco, M.D.C.A.D.; de Oliveira, A.P.; da Silva, P.B.; da Silva, C.F.; Batista, D.D.G.J.; da Gama, A.N.S.; et al. New 1,2,3-triazole-based analogues of benznidazole for use against Trypanosoma cruzi infection: In vitro and in vivo evaluations. *Chem. Biol. Drug Des.* **2018**, *92*, 1670–1682. [CrossRef]
20. Chopra, R.; Chibale, K.; Singh, K. Pyrimidine-chloroquinoline hybrids: Synthesis and antiplasmodial activity. *Eur. J. Med. Chem.* **2018**, *148*, 39–53. [CrossRef]

21. Holla, B.S.; Mahalinga, M.; Karthikeyan, M.S.; Poojary, B.; Akberali, P.M.; Kumari, N.S. Synthesis, characterization and antimicrobial activity of some substituted 1,2,3-triazoles. *Eur. J. Med. Chem.* **2005**, *40*, 1173–1178. [CrossRef] [PubMed]
22. Li, J.; Zhang, X.; Jin, H.; Fan, J.; Flores, H.; Perlmutter, J.S.; Tu, Z. Synthesis of fluorine-containing phosphodiesterase 10A (PDE10A) inhibitors and the in vivo evaluation of F-18 labeled PDE10A PET tracers in rodent and nonhuman primate. *J. Med. Chem.* **2015**, *58*, 8584–8600. [CrossRef] [PubMed]
23. Hamann, A.R.; De Kock, C.; Smith, P.J.; Van Otterlo, W.A.L.; Blackie, M.A.L. Synthesis of novel triazole-linked mefloquine derivatives: Biological evaluation against Plasmodium falciparum. *Bioorg. Med. Chem. Lett.* **2014**, *24*, 5466–5469. [CrossRef] [PubMed]
24. Brooke, D.G.; van Dam, E.M.; Watts, C.K.W.; Khoury, A.; Dziadek, M.A.; Brooks, H.; Graham, L.K.; Flanagan, J.U.; Denny, W.A. Targeting the Warburg Effect in cancer; relationships for 2-arylpyridazinones as inhibitors of the key glycolytic enzyme 6-phosphofructo-2-kinase/2,6-bisphosphatase 3 (PFKFB3). *Bioorg. Med. Chem.* **2014**, *22*, 1029–1039. [CrossRef] [PubMed]
25. Janosi, T.Z.; Makkai, G.; Kegl, T.; Matyus, P.; Kollar, L.; Erostyak, J. Light-enhanced fluorescence of multi-level cavitands possessing pyridazine upper rim. *J. Fluoresc.* **2016**, *26*, 679–688. [CrossRef] [PubMed]
26. Qian, W.; Wang, H.; Bartberger, M.D. Accelerating effect of triazolyl and related heteroaryl substituents on SNAr reactions: Evidence of hydrogen-bond stabilized transition states. *J. Am. Chem. Soc.* **2015**, *137*, 12261–12268. [CrossRef]
27. Qian, W.; Winternheimer, D.; Allen, J. A "Click and Activate" approach in one-pot synthesis of a triazolyl-pyridazinone library. *Org. Lett.* **2011**, *13*, 1682–1685. [CrossRef]
28. Qian, W.; Winternheimer, D.; Amegadzie, A.; Allen, J. One-pot synthesis of [1,2,3]triazole-fused pyrazinopyridazindione tricycles by a 'click and activate' approach. *Tetrahedron Lett.* **2012**, *53*, 271–274. [CrossRef]
29. Kamiya, S.; Sueyoshi, S.; Miayhara, M.; Yanagimachi, K.; Nakashima, T. Synthesis of 4-azidoquinoline 1-oxides and related compounds. *Chem. Pharm. Bull. (Tokyo)* **1980**, *28*, 1485–1490. [CrossRef]
30. Hu, E.; Kunz, R.K.; Rumfelt, S.; Chen, N.; Bürli, R.; Li, C.; Andrews, K.L.; Zhang, J.; Chmait, S.; Kogan, J.; et al. Discovery of potent, selective, and metabolically stable 4-(pyridin-3-yl)cinnolines as novel phosphodiesterase 10A (PDE10A) inhibitors. *Bioorg. Med. Chem. Lett.* **2012**, *22*, 2262–2265. [CrossRef]
31. Awad, E.D.; El-Abadelah, M.M.; Matar, S.; Zihlif, M.A.; Naffa, R.G.; Al-Momani, E.Q.; Mubarak, M.S. Synthesis and biological activity of some 3-(4-(substituted)-piperazin-1-yl)cinnolines. *Molecules* **2011**, *17*, 227–239. [CrossRef] [PubMed]
32. Mondal, C.; Halder, A.K.; Adhikari, N.; Jha, T. Structural findings of cinnolines as anti-schizophrenic PDE10A inhibitors through comparative chemometric modeling. *Mol. Divers.* **2014**, *18*, 655–671. [CrossRef] [PubMed]
33. Han, Y.T.; Jung, J.-W.; Kim, N.-J. Recent advances in the synthesis of biologically active cinnoline, phthalazine and quinoxaline derivatives. *Curr. Org. Chem.* **2017**, *21*, 1265–1291. [CrossRef]
34. Al-Qtaitat, M.A.; El-Abadelah, M.M.; Sabri, S.S.; Matar, S.A.; Hammad, H.M.; Mubarak, M.S. Synthesis, Characterization, and Bioactivity of Novel Bicinnolines Having 1-Piperazinyl Moieties. *J. Heterocycl. Chem.* **2019**, *56*, 158–164. [CrossRef]
35. Mitsumori, T.; Bendikov, M.; Sedó, J.; Wudl, F. Synthesis and properties of novel highly fluorescent pyrrolopyridazine derivatives. *Chem. Mater.* **2003**, *15*, 3759–3768. [CrossRef]
36. Shen, Y.; Shang, Z.; Yang, Y.; Zhu, S.; Qian, X.; Shi, P.; Zheng, J.; Yang, Y. Structurally rigid 9-amino-benzo[c]cinnoliniums make up a class of compact and large stokes-shift fluorescent dyes for cell-based imaging applications. *J. Org. Chem.* **2015**, *80*, 5906–5911. [CrossRef]
37. Galenko, E.E.; Galenko, A.V.; Khlebnikov, A.F.; Novikov, M.S.; Shakirova, J.R. Synthesis and intramolecular azo coupling of 4-diazopyrrole-2-carboxylates: Selective approach to benzo and hetero [c]-fused 6H-pyrrolo[3,4-c]pyridazine-5-carboxylates. *J. Org. Chem.* **2016**, *81*, 8495–8507. [CrossRef]
38. Mayakrishnan, S.; Arun, Y.; Balachandran, C.; Emi, N.; Muralidharan, D.; Perumal, P.T. Synthesis of cinnolines via Rh(III)-catalysed dehydrogenative C–H/N–H functionalization: Aggregation induced emission and cell imaging. *Org. Biomol. Chem.* **2016**, *14*, 1958–1968. [CrossRef] [PubMed]
39. Wang, N.; Yang, J.-C.; Chen, L.-D.; Li, J.; An, Y.; Lü, C.-W.; Tian, Y.-Q. Efficient synthesis of diethyl benzo[c]cinoline-3,8-dicarboxylate for fluorescence quenching materials. *New J. Chem.* **2017**, *41*, 2786–2792. [CrossRef]

40. Danilkina, N.A.; Vlasov, P.S.; Vodianik, S.M.; Kruchinin, A.A.; Vlasov, Y.G.; Balova, I.A. Synthesis and chemosensing properties of cinnoline-containing poly(arylene ethynylene)s. *Beilstein J. Org. Chem.* **2015**, *11*, 373–384. [CrossRef]
41. Pinho e Melo, T.M.V.D. Synthesis of Azides. In *Organic Azides*; Bräse, S., Banert, K., Eds.; John Wiley & Sons, Ltd.: Chichester, UK, 2010; Volume 75, pp. 53–94.
42. De Souza, M.V.N.; Pais, K.C.; Kaiser, C.R.; Peralta, M.A.; de, L.; Ferreira, M.; Lourenço, M.C.S. Synthesis and in vitro antitubercular activity of a series of quinoline derivatives. *Bioorg. Med. Chem.* **2009**, *17*, 1474–1480. [CrossRef] [PubMed]
43. Guantai, E.M.; Ncokazi, K.; Egan, T.J.; Gut, J.; Rosenthal, P.J.; Smith, P.J.; Chibale, K. Design, synthesis and in vitro antimalarial evaluation of triazole-linked chalcone and dienone hybrid compounds. *Bioorg. Med. Chem.* **2010**, *18*, 8243–8256. [CrossRef] [PubMed]
44. Mantoani, S.; Chierrito, T.; Vilela, A.; Cardoso, C.; Martínez, A.; Carvalho, I. Novel Triazole-Quinoline Derivatives as Selective Dual Binding Site Acetylcholinesterase Inhibitors. *Molecules* **2016**, *21*, 193. [CrossRef] [PubMed]
45. Wu, G.; Gao, Y.; Kang, D.; Huang, B.; Huo, Z.; Liu, H.; Poongavanam, V.; Zhan, P.; Liu, X. Design, synthesis and biological evaluation of tacrine-1,2,3-triazole derivatives as potent cholinesterase inhibitors. *MedChemComm* **2018**, *9*, 149–159. [CrossRef] [PubMed]
46. Sumangala, V.; Poojary, B.; Chidananda, N.; Fernandes, J.; Kumari, N.S. Synthesis and antimicrobial activity of 1,2,3-triazoles containing quinoline moiety. *Arch. Pharmacal Res.* **2010**, *33*, 1911–1918. [CrossRef] [PubMed]
47. Ismail, M.M.; Abass, M.; Hassan, M.M. Chemistry of substituted quinolinones. v. synthesis and use of quinolinylphosphazenes in amination of 8-methylquinoline. *Phosphorus Sulfur Silicon Relat. Elem.* **2000**, *167*, 275–288. [CrossRef]
48. Stadlbauer, W. Methoden zur Darstellung von 4-Azido-2(1H)-chinolonen. *Monatsh. Chem.* **1986**, *117*, 1305–1323. [CrossRef]
49. Aizikovich, A.; Kuznetsov, V.; Gorohovsky, S.; Levy, A.; Meir, S.; Byk, G.; Gellerman, G. A new application of diphenylphosphorylazide (DPPA) reagent: convenient transformations of quinolin-4-one, pyridin-4-one and quinazolin-4-one derivatives into the 4-azido and 4-amino counterparts. *Tetrahedron Lett.* **2004**, *45*, 4241–4243. [CrossRef]
50. Vinogradova, O.V.; Balova, I.A. Methods for the synthesis of cinnolines (review). *Chem. Heterocycl. Compd.* **2008**, *44*, 501–522. [CrossRef]
51. Mathew, T.; Papp, A.Á.; Paknia, F.; Fustero, S.; Surya Prakash, G.K. Benzodiazines: Recent synthetic advances. *Chem. Soc. Rev.* **2017**, *46*, 3060–3094. [CrossRef]
52. Richter, V. Ueber Cinnoliderivate. *Ber. Dtsch. Chem. Ges.* **1883**, *16*, 677–683. [CrossRef]
53. Vasilevsky, S.F.; Tretyakov, E.V. Cinnolines and pyrazolopyridazines.—Novel synthetic and mechanistic aspects of the Richter reaction. *Liebigs Ann.* **1995**, 775–779. [CrossRef]
54. Bräse, S.; Dahmen, S.; Heuts, J. Solid-phase synthesis of substituted cinnolines by a Richter type cleavage protocol. *Tetrahedron Lett.* **1999**, *40*, 6201–6203. [CrossRef]
55. Goeminne, A.; Scammells, P.J.; Devine, S.M.; Flynn, B.L. Richter cyclization and co-cyclization reactions of triazene-masked diazonium ions. *Tetrahedron Lett.* **2010**, *51*, 6882–6885. [CrossRef]
56. Vinogradova, O.V.; Sorokoumov, V.N.; Balova, I.A. A short route to 3-alkynyl-4-bromo(chloro)cinnolines by Richter-type cyclization of ortho-(dodeca-1,3-diynyl)aryltriaz-1-enes. *Tetrahedron Lett.* **2009**, *50*, 6358–6360. [CrossRef]
57. Holzer, W. On the Synthesis and Reactivity of 4-(oxiran-2-ylmethoxy)cinnoline: Targeting a cinnoline analogue of propranolol. *Sci. Pharm.* **2008**, *76*, 19–32. [CrossRef]
58. Holzer, W.; Eller, G.A.; Schönberger, S. On the tautomerism of cinnolin-4-ol, cinnolin-4-thiol, and cinnolin-4-amine. *Heterocycles* **2008**, *75*, 77–86. [CrossRef]
59. Hoegenauer, K.; Soldermann, N.; Hebach, C.; Hollingworth, G.J.; Lewis, I.; von Matt, A.; Smith, A.B.; Wolf, R.M.; Wilcken, R.; Haasen, D.; et al. Discovery of novel pyrrolidineoxy-substituted heteroaromatics as potent and selective PI3K delta inhibitors with improved physicochemical properties. *Bioorg. Med. Chem. Lett.* **2016**, *26*, 5657–5662. [CrossRef]
60. Danilkina, N.A.; Gorbunova, E.G.; Sorokoumov, V.N.; Balova, I.A. Study of cyclyzation of o-(1-Alkynyl)- and o-(1,3-Butadiynyl)aryltriazenes under the action of acids. *Russ. J. Org. Chem.* **2012**, *48*, 1424–1434. [CrossRef]

61. Satyanarayana, M.; Feng, W.; Cheng, L.; Liu, A.A.; Tsai, Y.-C.; Liu, L.F.; LaVoie, E.J. Syntheses and biological evaluation of topoisomerase I-targeting agents related to 11-[2-(N,N-dimethylamino)ethyl]-2,3-dimethoxy-8,9-methylenedioxy-11H-isoquino[4,3-c]cinnolin-12-one (ARC-31). *Bioorg. Med. Chem.* **2008**, *16*, 7824–7831. [CrossRef]
62. Devine, W.; Woodring, J.L.; Swaminathan, U.; Amata, E.; Patel, G.; Erath, J.; Roncal, N.E.; Lee, P.J.; Leed, S.E.; Rodriguez, A.; et al. Protozoan parasite growth inhibitors discovered by cross-screening yield potent scaffolds for lead discovery. *J. Med. Chem.* **2015**, *58*, 5522–5537. [CrossRef] [PubMed]
63. Barlaam, B.; Cadogan, E.; Campbell, A.; Colclough, N.; Dishington, A.; Durant, S.; Goldberg, K.; Hassall, L.A.; Hughes, G.D.; MacFaul, P.A.; et al. Discovery of a series of 3-cinnoline carboxamides as orally bioavailable, highly potent, and selective ATM inhibitors. *ACS Med. Chem. Lett.* **2018**, *9*, 809–814. [CrossRef] [PubMed]
64. Scott, D.A.; Dakin, L.A.; Del Valle, D.J.; Bruce Diebold, R.; Drew, L.; Gero, T.W.; Ogoe, C.A.; Omer, C.A.; Repik, G.; Thakur, K.; et al. 3-Amido-4-anilinocinnolines as a novel class of CSF-1R inhibitor. *Bioorg. Med. Chem. Lett.* **2011**, *21*, 1382–1384. [CrossRef] [PubMed]
65. Mphahlele, M.J.; Mphahlele, M.M. Direct one-pot synthesis of primary 4-amino-2,3-diarylquinolines via Suzuki-Miyaura cross-coupling of 2-aryl-4-azido-3-iodoquinolines with arylboronic acids. *Molecules* **2011**, *16*, 8958–8972. [CrossRef] [PubMed]
66. Hisaki, I.; Senga, H.; Shigemitsu, H.; Tohnai, N.; Miyata, M. Construction of 1D π-stacked superstructures with inclusion channels through symmetry-decreasing crystallization of discotic molecules of C3 cymmetry. *Chem. Eur. J.* **2011**, *17*, 14348–14353. [CrossRef] [PubMed]
67. Hein, J.E.; Krasnova, L.B.; Iwasaki, M.; Fokin, V.V. Cu-Catalyzed azide-alkyne aycloaddition: preparation of tris((1-benzyl-1H-1,2,3-triazolyl)methyl)amine. *Org. Synth.* **2011**, *88*, 238–246.
68. Ramachary, D.B.; Shashank, A.B.; Karthik, S. An organocatalytic azide-aldehyde [3 + 2] cycloaddition: high-yielding regioselective synthesis of 1,4-disubstituted 1,2,3-triazoles. *Angew. Chem. Int. Ed.* **2014**, *53*, 10420–10424. [CrossRef] [PubMed]
69. Ni, R.; Mitsuda, N.; Kashiwagi, T.; Igawa, K.; Tomooka, K. Heteroatom-embedded Medium-Sized Cycloalkynes: Concise Synthesis, Structural Analysis, and Reactions. *Angew. Chem. Int. Ed.* **2015**, *54*, 1190–1194. [CrossRef]
70. Batat, P.; Vives, G.; Bofinger, R.; Chang, R.-W.; Kauffmann, B.; Oda, R.; Jonusauskas, G.; McClenaghan, N.D. Dynamics of ion-regulated photoinduced electron transfer in BODIPY-BAPTA conjugates. *Photochem. Photobiol. Sci.* **2012**, *11*, 1666–1674. [CrossRef]
71. Meisner, Q.J.; Accardo, J.V.; Hu, G.; Clark, R.J.; Jiang, D.E.; Zhu, L. Fluorescence of hydroxyphenyl-substituted "click" triazoles. *J. Phys. Chem. A* **2018**, *122*, 2956–2973. [CrossRef]
72. Fra, L.; Millán, A.; Souto, J.A.; Muñiz, K. Indole synthesis based on a modified koser reagent. *Angew. Chem. Int. Ed.* **2014**, *53*, 7349–7353. [CrossRef] [PubMed]
73. CrysAlisPro; Version 1.171.39.35a; Rigaku Oxford Diffraction; Rigaku Corporation: Tokyo, Japan, 2015.
74. Sheldrick, G.M. Crystal structure refinement with SHELXL. *Acta Crystallogr. Sect. C Struct. Chem.* **2015**, *71*, 3–8. [CrossRef] [PubMed]
75. Dolomanov, O.V.; Bourhis, L.J.; Gildea, R.J.; Howard, J.A.K.; Puschmann, H. OLEX2: A complete structure solution, refinement and analysis program. *J. Appl. Crystallogr.* **2009**, *42*, 339–341. [CrossRef]
76. Frisch, M.J.; Trucks, G.W.; Schlegel, H.B.; Scuseria, G.E.; Robb, M.A.; Cheeseman, J.R.; Montgomery, J.A., Jr.; Vreven, T.; Kudin, K.N.; Burant, J.C.; et al. *Gaussian 03, Revision D.02*; Gaussian, Inc.: Wallingford, CT, USA, 2004.
77. Van Meerloo, J.; Kaspers, G.J.L.; Cloos, J. Cancer Cell Culture. In *Cancer Cell Culture: Methods and Protocols, Second Edition, Methods in Molecular Biology*; Langdon, S.P., Ed.; Humana Press: Totowa, NJ, USA, 2003; Volume 88, pp. 237–245.

© 2019 by the authors. Licensee MDPI, Basel, Switzerland. This article is an open access article distributed under the terms and conditions of the Creative Commons Attribution (CC BY) license (http://creativecommons.org/licenses/by/4.0/).

Article

Synthesis of (1,2,3-triazol-4-yl)methyl Phosphinates and (1,2,3-Triazol-4-yl)methyl Phosphates by Copper-Catalyzed Azide-Alkyne Cycloaddition

Anna Tripolszky [1], Krisztina Németh [2], Pál Tamás Szabó [2] and Erika Bálint [1,*]

1. Department of Organic Chemistry and Technology, Budapest University of Technology and Economics, 1521 Budapest, Hungary; tripolszky.anna@mail.bme.hu
2. MS Metabolomics Laboratory, Instrumentation Center, Research Centre for Natural Sciences, Hungarian Academy of Sciences, Magyar Tudósok Krt. 2., H-1117 Budapest, Hungary; nemeth.krisztina.94@ttk.mta.hu (K.N.); szabo.pal@ttk.mta.hu (P.T.S.)
* Correspondence: ebalint@mail.bme.hu; Tel.: +36-1-463-3653

Received: 21 May 2019; Accepted: 30 May 2019; Published: 31 May 2019

Abstract: An efficient and practical method was developed for the synthesis of new (1,2,3-triazol-4-yl)methyl phosphinates and (1,2,3-triazol-4-yl)methyl phosphates by the copper(I)-catalyzed azide-alkyne cycloaddition (CuAAC) of organic azides and prop-2-ynyl phosphinate or prop-2-ynyl phosphate. The synthesis of (1-benzyl-1H-1,2,3-triazol-4-yl)methyl diphenylphosphinate was optimized with respect to the reaction parameters, such as the temperature, reaction time, and catalyst loading. The approach was applied to a range of organic azides, which confirmed the wide scope and the substituent tolerance of the process. The method elaborated represents a novel approach for the synthesis of the target compounds.

Keywords: 1,2,3-triazol; triazolylmethyl phosphinate; triazolylmethyl phosphate; copper-catalyzed azide-alkyne cycloaddition; click reaction

1. Introduction

The family of 1,2,3-triazoles has attracted considerable attention in the last decades due to the wide range of their applications in medicine and biochemistry, as well as in materials science [1–3]. Several 1,2,3-triazoles proved to be effective antibacterial, antifungal, anticancer, antiviral, or anti-inflammatory agents [4]. The 1,4-disubstituted 1H-1,2,3-triazoles also have important applications as agrochemicals, photostabilizers, dyes, or anticorrosives [5–8].

Organophosphorus compounds also present significant importance as biologically active agents [9]. The 1,2,3-Triazolyl phosphonates combine the advantages of the triazole and the phosphonate moieties [10]. Additionally, several triazolylphosphonate derivatives were found to be suitable for bioconjugation [11,12] or showed an anti-HIV effect [13].

The most convenient synthetic method for the preparation of 1,4-disubstituted 1H-1,2,3-triazoles is the Cu(I)-catalyzed 1,3-dipolar (Huisgen) cycloaddition of azides with alkynes (CuAAC) (the "click reaction"), which was developed by Meldal and Sharpless [14,15]. The 1,2,3-Triazolyl phosphonates may be synthesized by the reaction of azides with phosphorus-containing acetylenes [12, 13,16–20] or by the cycloaddition of azides incorporating a phosphonate moiety with alkynes [21–24]. In this paper, reports of the former method will be presented in detail.

The 1,3-dipolar cycloaddition of benzyl azide and ethyl ethynylphosphonate (**1**) was carried out in the absence of any catalyst at reflux temperature in toluene (Scheme 1) [16]. The reaction was not selective, two regioisomers, 1,2,3-triazolyl-4-phosphonate (**2a**) and 1,2,3-triazolyl-5-phosphonate (**2b**) were formed.

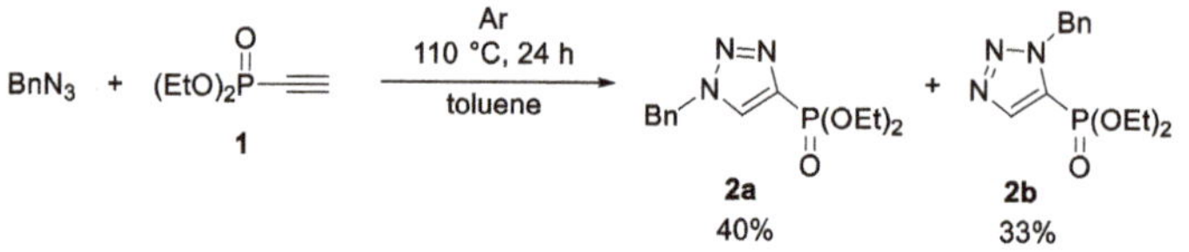

Scheme 1. Cycloaddition of benzyl azide and diethyl ethynylphosphonate (**1**).

Bisphosphonates incorporating a triazole ring (**4**) were prepared by the CuAAC reaction of a propargyl-substituted bisphosphonate (**3**) and organic azides (Scheme 2) [17,18]. The reactions were performed using different methods. According to method A, copper iodide was applied as a catalyst in the presence of N,N-diisopropylethylamine (DIPEA) as a base in tetrahydrofuran (THF) [17]. In another case, the Cu(I) catalyst was formed in situ from copper(II) sulfate pentahydrate and sodium ascorbate in the mixture of *tert*-butanol and water [18].

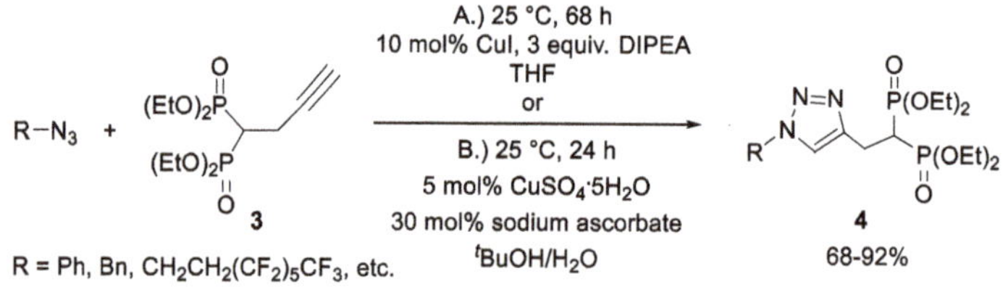

Scheme 2. Synthesis of bisphosphonates bearing a triazole ring (**4**) via CuAAc reaction.

Röschenthaler and co-workers performed the synthesis of triazole-containing α-CF$_3$-α-aminophosphonates (**6**) by the Cu(I)-catalyzed cycloaddition of azides and ethynyl- or propargyl-substituted phosphonate (**5**) (Scheme 3) [19].

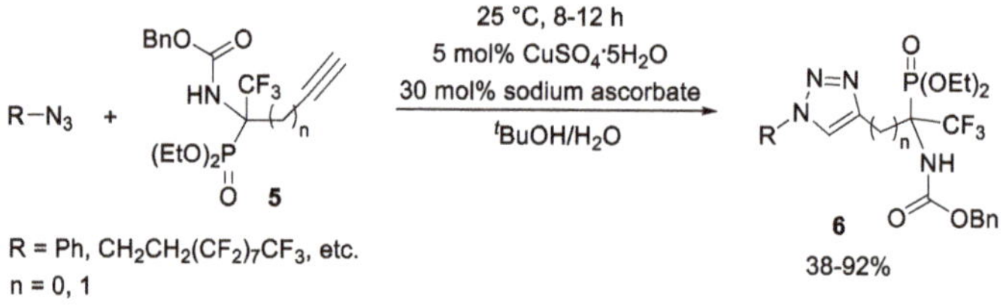

Scheme 3. Synthesis of triazole-containing α-CF$_3$-α-aminophosphonates (**6**).

The click reaction of 2-azidoethanol and triprop-2-ynyl phosphate (**7**) was also reported, in which a triazole-functionalized phosphate flame-retardant monomer (**8**) was synthesized (Scheme 4) [20].

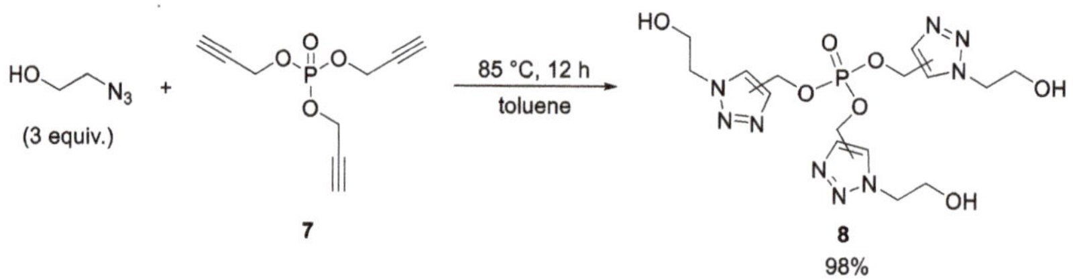

Scheme 4. Synthesis of a triazole-functionalized phosphate monomer (**8**).

To the best of our knowledge, there is no example for the azide-alkyne cycloaddition of simple azides and prop-2-ynyl phosphinate or diethyl prop-2-ynyl phosphate. The reaction of prop-2-ynyl diphenylphosphinate was reported only with steroidal azides [25]. Hence, we set a sight on building a phosphinate or phosphate side-chain on position 4 of the triazole ring by the click reaction with simple azides.

In this paper, we report on an efficient and fast synthesis of (1,2,3-triazol-4-yl)methyl phosphinates and (1,2,3-triazol-4-yl)methyl phosphates by the copper(I)-catalyzed 1,3-dipolar (Huisgen) cycloaddition of organic azides and prop-2-ynyl phosphinates or diethyl prop-2-ynyl phosphate.

2. Results and Discussion

At first, the starting materials of the cycloadditions were prepared. The synthesis of azides was carried out based on the literature data [26,27] (Scheme 5). The benzyl or substituted benzyl bromides were reacted with 1.5 equivalents of sodium azide at room temperature for 24 h in a mixture of acetone/water in a ratio of 4:1, and the corresponding azides (9a–f) were obtained in yields of 80–93% (Scheme 5/I, Method A). The synthesis of octyl-, *i*-octyl- and cyclohexyl azides (9g–i) was performed using 1.2 equivalents of sodium azide at 70 °C in dimethylformamide (DMF) (Scheme 5/I, Method B). For the preparation of phenyl azide (11), aniline was reacted with sodium nitrite in HCl/H_2O solution at 0 °C for 15 min, and the diazonium salt (10) formed was further reacted with sodium azide at ambient temperature (Scheme 5/II). After workup, phenyl azide (11) was obtained in a yield of 65%.

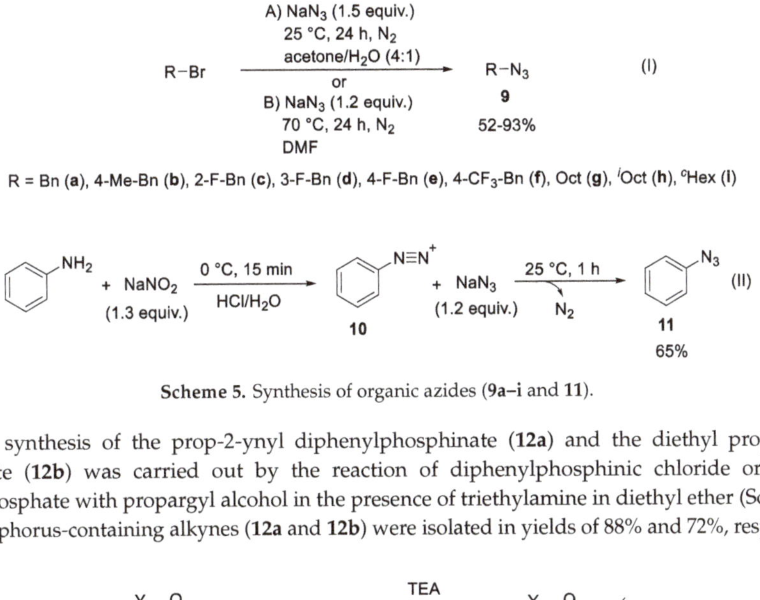

Scheme 5. Synthesis of organic azides (9a–i and 11).

The synthesis of the prop-2-ynyl diphenylphosphinate (12a) and the diethyl prop-2-ynyl phosphate (12b) was carried out by the reaction of diphenylphosphinic chloride or diethyl chlorophosphate with propargyl alcohol in the presence of triethylamine in diethyl ether (Scheme 6). The phosphorus-containing alkynes (12a and 12b) were isolated in yields of 88% and 72%, respectively.

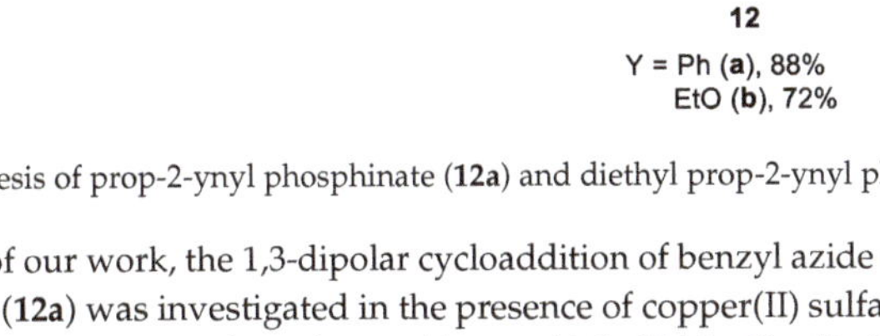

Scheme 6. Synthesis of prop-2-ynyl phosphinate (12a) and diethyl prop-2-ynyl phosphate (12d).

As the next step of our work, the 1,3-dipolar cycloaddition of benzyl azide (9a) and prop-2-ynyl diphenylphosphinate (12a) was investigated in the presence of copper(II) sulfate pentahydrate and sodium ascorbate in the mixture of *tert*-butanol/water (4:1) (Table 1). Performing the reaction at room temperature using 5 mol% of $CuSO_4·5H_2O$ and 30 mol% of sodium ascorbate, it was complete after 1 h and, after column chromatography, the desired (1-benzyl-1*H*-1,2,3-triazol-4-yl)methyl diphenylphosphinate (**13a**) was obtained in a yield of 84% (Table 1, Entry 1). Decreasing the amount of the reducing agent (sodium ascorbate) to 10 mol%, the conversion was only 78% under the conditions applied before (25 °C, 60 min) (Table 1, Entry 2). In this case, complete conversion could be reached after 3 h (Table 1, Entry 3). Carrying out the reaction at 60 °C for 5 min under conventional heating, the triazolylmethyl phosphinate (**13a**) was formed in a conversion of 93% (Table 1, Entry 4). Applying microwave (MW) irradiation, the result obtained was similar (Table 1, Entry 5), thus further experiments were performed in an oil bath. Increasing the reaction time to 10 min, the cycloaddition was complete and product **13a** was isolated in a yield of 89% (Table 1, Entry 6). In the next step, the effect of the catalyst loading was studied (Table 1/Entries 7–9). Using 3% of $CuSO_4 \cdot 5H_2O$ and 5 mol% of sodium ascorbate, the reaction was similar to the previous experiment (Table 1, Entries 6 and 7). Upon decreasing the amount of the Cu(II) catalyst to 2%, the conversion was lower (90%) (Table 1, Entry 7). Based on the results obtained, a temperature of 60 °C, a reaction time of 10 min and application of 3 mol% of $CuSO_4 \cdot 5H_2O$, as well as 5 mol% of sodium ascorbate, were found to be the optimum parameters (Table 1, Entry 7).

Table 1. Optimization of the reaction of benzyl azide (**9a**) and propynyl diphenylphosphinate (**11a**).

Entry	Mode of Heating	T [°C]	t [min]	Catalyst [mol%]		Conversion [%] [a]	Yield [%] [b]
				$CuSO_4 \cdot 5H_2O$	Sodium Ascorbate		
1	-	25	60	5	30	100	84
2	-	25	60	5	10	78	-
3	-	25	180	5	10	100	89
4	Δ	60	5	5	10	93	-
5	MW	60	5	5	10	92	-
6	Δ	60	10	5	10	100	89
7	Δ	60	10	3	5	100	91
8	Δ	60	10	2	5	90	-

[a] Based on ^{31}P NMR (phosphorus-31 nuclear magnetic resonance); [b] Isolated yield; MW (microwave).

In the next series of experiments, the cycloaddition of prop-2-ynyl diphenylphosphinate (**12a**) with a wide range of organic azides (**9** and **11**) was studied under the optimized conditions (Scheme 7). The reactions were complete in all cases. Using 4-methylbenzyl azide (**9b**), the [1-(4-methylbenzyl)-1H-1,2,3-triazol-4-yl)]methyl diphenylphosphinate (**13b**) was isolated in a yield of 86%. Changing for fluoro-substituted benzyl azides (2-, 3- or 4-fluorobenzyl azide) (**9c–e**), the desired triazolylmethyl phosphinates (**13c–e**) were obtained in yields of 81–83% after column chromatography. Carrying out the reaction starting from 4-(trifluoromethyl)benzyl azide (**9f**), the product **13f** was prepared in a yield of 88%. Applying octyl azide (**9g**), the corresponding (1-octyl-1H-1,2,3-triazol-4-yl)methyl diphenylphosphinate (**13g**) was obtained in a yield of 89%, while using iso-octyl azide (**9h**), product **13h** was isolated in a yield of 77%. The cyclohexyl azide (**9i**) was also tried out as the azide component; however, the triazolylmethyl phosphinate (**13i**) could be obtained in a slightly lower yield (62%) due to the steric effects of the cyclohexyl group. Finally, the click reaction of phenyl azide (**11**) was performed, and the desired triazolylmethyl phosphinate (**13j**) was synthesized in a yield of 82%.

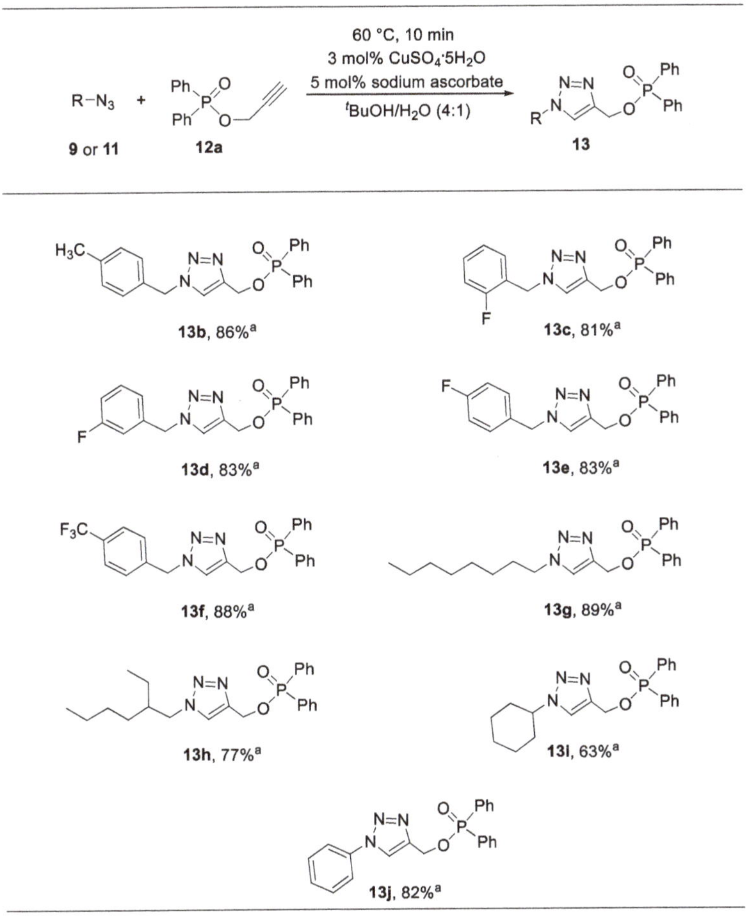

Scheme 7. Synthesis of (1H-1,2,3-triazol-4-yl)methyl diphenylphosphinates (**13**).

[a] Isolated yield.

The cycloaddition of benzyl azide (**9a**) and diethyl prop-2-ynyl phosphate (**12b**) was also investigated (Table 2). Using the optimized conditions (3 mol% of CuSO$_4$·5H$_2$O and 5 mol% of sodium ascorbate, 60 °C, 10 min), the reaction was incomplete, and the (1-benzyl-1H-1,2,3-triazol-4-yl)methyl diethyl phosphate (**14a**) was formed in a conversion of only 48% (Table 2, Entry 1). Increasing the reaction time to 20 min, the conversion increased to 59%; however, after 30 min, the reaction was complete, and the triazolylmethyl phosphate (**14a**) was obtained in a yield of 75% (Table 2, Entries 2 and 3). The diethyl prop-2-ynyl phosphate (**12b**) proved to be somewhat less reactive in the click reaction as compared to the prop-2-ynyl diphenylphosphinate (**12a**).

In the next round, the reaction of diethyl prop-2-ynyl phosphate (**12b**) was also carried out with a wide range of organic azides (**9** and **11**) in the presence of 3 mol% of CuSO$_4$·5H$_2$O and 5 mol% of sodium ascorbate at 60 °C for 30 min (Scheme 8). Applying substituted benzyl azides (4-methyl-, 2-, 3- or 4-fluorobenzylazide and 4-(trifluoromethyl)benzyl azide) (**11b–f**), the corresponding (1-benzyl-1H-1,2,3-triazol-4-yl)methyl diethyl phosphate derivatives (**14b–f**) were prepared in yields of 54–69%. Performing the cycloaddition starting from octyl or iso-octyl azide (**11g** or **11h**), the desired products (**14g** or **14h**) were isolated in yields of 60% and 54%, respectively. The reaction of cyclohexyl azide (**11i**) and diethyl prop-2-ynyl phosphate (**12b**) was also carried out, and the product **14i** was obtained in a yield of 51%. Using aromatic azide, such as phenyl azide (**11**), the triazolylmethyl phosphate (**14j**) was synthesized in a yield of 77%.

Table 2. Reaction of benzyl azide (**9a**) and diethyl prop-2-ynyl phosphate (**12b**).

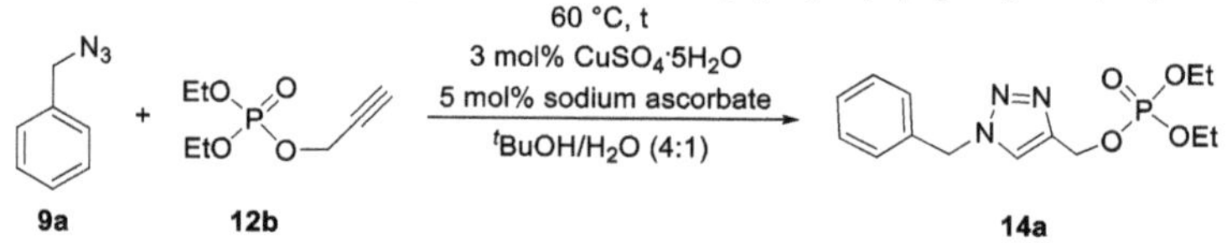

Entry	t (min)	Conversion (%) [a]	Yield (%) [b]
1	10	40	–
2	20	59	–
3	30	98 [c]	75

[a] Based on ^{31}P NMR. [b] Isolated yield. [c] No change for longer reaction time.

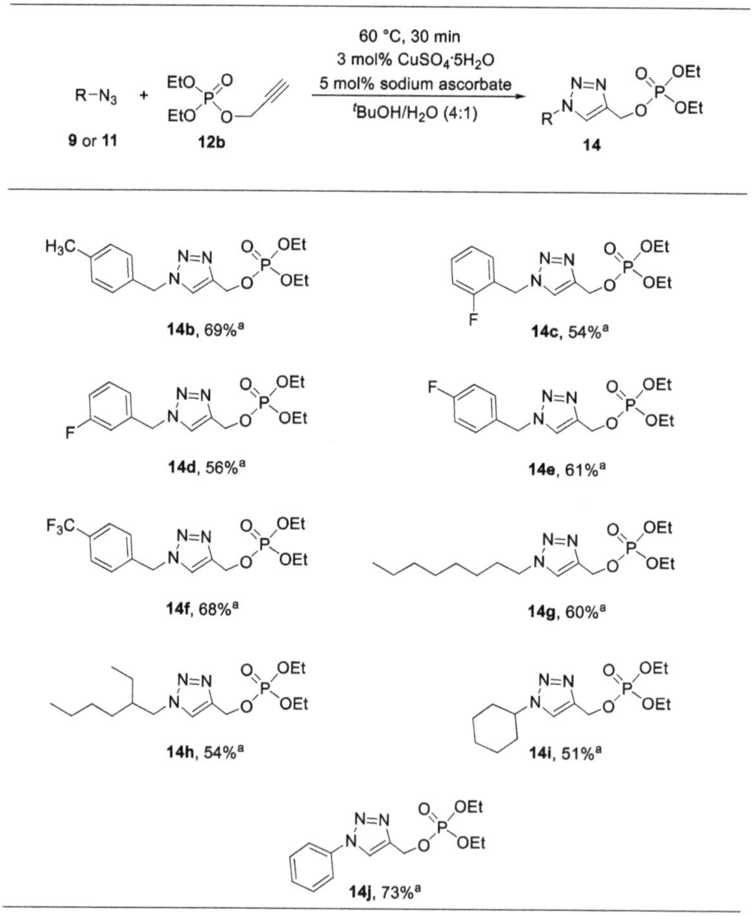

14b, 69% [a]

14c, 54% [a]

14d, 56% [a]

14e, 61% [a]

14f, 68% [a]

14g, 60% [a]

14h, 54% [a]

14i, 51% [a]

14j, 73% [a]

[a] Isolated yield.

Scheme 8. Synthesis of (1*H*-1,2,3-triazol-4-yl)methyl diethyl phosphates (**14**).

3. Materials and Methods

3.1. General

The reactions under conventional heating were carried out in an oil bath. The microwave-assisted experiments were performed in a 300 W CEM Discover focused microwave reactor (CEM Microwave Technology Ltd., Buckingham, UK) equipped with a pressure controller using 5–10 W irradiation under isothermal conditions.

High-performance liquid chromatography-mass spectrometry (HPLC–MS) measurements were performed with an Agilent 1200 liquid chromatography system coupled with a 6130 quadrupole mass spectrometer equipped with an ESI ion source (Agilent Technologies, Palo Alto, CA, USA). Analysis was performed at 40 °C on a Gemini C18 column (150 mm × 4.6 mm, 3 µm; Phenomenex, Torrance, CA, USA) with a mobile phase flow rate of 0.6 mL/min. The composition of eluent A was 0.1% (NH_4)(HCOO) in water; eluent B was 0.1% (NH_4)(HCOO) and 8% water in acetonitrile, 0–3 min 5% B, 3–13 min gradient, 13–20 min 100% B. The injection volume was 5 µL. The chromatographic profile was registered at 254 nm. The MSD operating parameters were as follows: positive ionization mode, scan spectra from m/z 120 to 1200, drying gas temperature 300 °C, nitrogen flow rate 10 L/min, nebulizer pressure 60 psi, capillary voltage 4000 V.

High-resolution mass spectrometric measurements were performed using a Sciex 5600+ Q-TOF (time-of-flight) mass spectrometer in positive electrospray mode.

The ^{31}P, ^{1}H, ^{13}C, NMR spectra were taken in $CDCl_3$ solution on a Bruker AV-300 spectrometer (Bruker AXS GmBH, Karlsruhe, Germany) operating at 121.5, 75.5, and 300 MHz, respectively. Chemical shifts are downfield relative to 85% H_3PO_4 and TMS (spectra for all compounds synthesized can be found in Supplementary).

3.2. General Procedure for the Synthesis of Benzyl Azides (Method A)

To a stirred solution of 10.0 mmol alkyl halides (1.19 mL of benzyl bromide, 1.85 g of 4-methylbenzyl bromide, 1.21 mL of 2-fluorobenzyl bromide, 1.23 mL of 3-fluorobenzyl bromide, 1.25 mL of 4-fluorobenzyl bromide, or 1.55 mL of 4-(trifluoromethyl)benzyl bromide) in 100 mL of acetone/H_2O 4:1 (*v/v*) was added 15.0 mmol (0.98 g) of sodium azide. The reaction mixture was stirred at room temperature for 24 h. After, the reaction was extracted with Et_2O (3 × 50 mL), dried over Na_2SO_4, and concentrated under reduced pressure to give benzyl azides as pale yellow oils. The following azides were thus prepared (Table 3):

Table 3. MS Data for benzyl azides (**9a–f**).

Compound	Yield	[M+H]$^+$ found	[M+H]$^+$ requires
Benzyl azide (**9a**) [26]	93% (1.24 g)	134.0725	134.0718
4-Methylbenzyl azide (**9b**) [28]	80% (1.18 g)	148.0880	148.0874
2-Fluorobenzyl azide (**9c**) [29]	68% (1.02 g)	152.0632	152.0624
3-Fluorobenzyl azide (**9d**) [30]	76% (1.14 g)	152.0633	152.0624
4-Fluorobenzyl azide (**9e**) [28]	83% (1.25 g)	152.0632	152.0624
4-(Trifluoromethyl)benzyl azide (**9f**) [31]	86% (1.72 g)	202.0601	202.0592

3.3. General Procedure for the Synthesis of Alkyl Azides (Method B)

To a stirred solution of 10.0 mmol alkyl halides (1.76 mL of octyl bromide, 1.78 mL of *iso*-octyl bromide, or 1.23 mL of bromocyclohexane) in 20 mL of DMF, 12.0 mmol (0.78 g) of sodium azide was added. The reaction mixture was stirred at 70 °C for 24 h in an oil bath. After, the reaction was extracted with Et_2O (3 × 50 mL), dried over Na_2SO_4, and concentrated under reduced pressure to give alkyl azides as pale yellow oils. The following azides were thus prepared (Table 4):

Table 4. MS Data for alkyl azides (9g–i).

Compound	Yield	[M+H]$^+$ found	[M+H]$^+$ requires
Octyl azide (9g) [32]	67% (1.04 g)	156.1514	156.1501
Iso-octyl azide (9h) [33]	55% (0.85 g)	156.1514	156.1501
Cyclohexyl azide (9i) [27]	52% (0.65 g)	126.1038	126.1031

3.4. General Procedure for the Synthesis of Phenyl Azide

To a stirred solution of 0.46 mL aniline (5.0 mmol) in 25 mL 17% HCl solution at 0 °C, 0.51 g (7.5 mmol) of sodium nitrite in water (3 mL) was added. After stirring for 15 min, a solution of 0.32 g sodium azide (7.5 mmol) in water (3 mL) was carefully added. The reaction was left to stir for 1 h, followed by extraction with Et$_2$O (3 × 30 mL). The combined organic layers were dried over Na$_2$SO$_4$, and carefully concentrated under reduced pressure to give phenyl azide as an orange oil.

Phenyl azide (**11**) [26]: Yield: 65% (0.77 g) of compound **11** as orange oil; [M+H]$^+$ found = 120.0567, C$_6$H$_6$N$_3$ requires 120.0562.

3.5. General Procedure for the Synthesis of Prop-2-ynyl Diphenylphosphinate and Diethyl Prop-2-ynyl Phosphate

To a stirred solution of 10 mmol of diphenylphosphinic chloride (1.91 mL) or diethyl chlorophosphate (1.44 mL) in 10 mL of Et$_2$O, 1.39 mL (10 mmol) of Et$_3$N and 0.50 mL (10.0 mmol) of propargyl alcohol at 0 °C were added under a nitrogen atmosphere. The solution was left stirring at room temperature for 3–6 h and the reaction mixture obtained was passed through a 1 cm silica gel layer using Et$_2$O. After evaporating the solvent, the products were obtained as white crystals (**12a**) or colorless oil (**12b**). The following products were thus prepared (Table 5):

Table 5. ^{31}P NMR and MS Data for prop-2-ynyl diphenylphosphinate (**12a**) and diethyl prop-2-ynyl phosphate (**12b**).

Compound	Yield	δp in CDCl$_3$	δp [lit.] in CDCl$_3$	[M+H]$^+$ found	[M+H]$^+$ requires
12a	88% (2.25 g)	34.3	34.2 [25]	257.0735	257.0731
12b	72% (1.38 g)	−0.4	−0.7 [34]	193.0639	193.0629

3.6. General Procedure for the Synthesis of (1H-1,2,3-Triazol-4-yl)methyl Phosphinates or Diethyl Phosphates

The 1.0 mmol organic azide (0.13 g of benzyl azide, 0.15 g of 4-methylbenzyl azide, 0.15 g of 2-fluorobenzyl azide, 0.15 g of 3-fluorobenzyl azide, 0.15 g of 4-fluorobenzyl azide, 0.20 g of 4-(trifluoromethyl)benzyl azide, 0.16 g of octyl azide, 0.16 g of iso-octyl azide, 0.13 g of cyclohexyl azide or 0.12 g of phenyl azide) and 1.0 mmol acetylene (0.26 g of prop-2-ynyl diphenylphosphinate or 0.19 g of diethyl prop-2-ynyl phosphate) were suspended in a mixture of ^{t}BuOH/H$_2$O (4:1) (2 mL). To this 7.5 mg (0.03 mmol) of CuSO$_4$·5H$_2$O and 20 mg (0.1 mmol) of sodium ascorbate were added. The mixture was stirred at 60 °C for 10 min. The resulting solution was extracted with ethyl acetate (3 × 30 mL) and the combined organic layers were dried over Na$_2$SO$_4$. After evaporating the solvent, the crude product was purified by column chromatography using silica gel and dichloromethane/methanol 97:3 as the eluent. The following products were thus prepared:

(1-Benzyl-1H-1,2,3-triazol-4-yl)methyl diphenylphosphinate (**13a**): Yield: 91% (0.35 g), white crystals; Mp: 91-92 °C; ^{31}P NMR (CDCl$_3$) δ 33.5; ^{1}H NMR (CDCl$_3$) δ 5.18 (d, $^3J_{HP}$ = 9.1, 2H, CH$_2$O), 5.46 (s, 2H, CH$_2$Ph), 7.20–7.27 (m, 2H, C$_2$H), 7.32–7.37 (m, 3H, C$_3$H, C$_4$H), 7.38–7.44 (m, 4H, C$_{3'}$H), 7.47–7.54 (m, 2H, C$_{4'}$H), 7.52 (s, 1H, CH), 7.79 (dd, $^3J_{HP}$ = 11.8, $^3J_{HH}$ = 7.4, 4H, C$_{2'}$H); ^{13}C NMR (CDCl$_3$) δ 54.3 (CH$_2$Ph), 58.2 (d, $^2J_{CP}$ = 5.3, CH$_2$O), 123.8 (CH=), 128.3 (C$_2$), 128.7 (d, $^3J_{CP}$ = 13.3, C$_{3'}$), 128.8 (C$_4$), 129.3 (C$_3$), 131.2 (d, $^1J_{CP}$ = 137.0, C$_{1'}$), 131.8 (d, $^2J_{CP}$ = 10.3, C$_{2'}$), 132.5 (d, J_{CP} = 2.8, C$_{4'}$), 134.5 (C$_1$), 144.0 (d, $^3J_{CP}$ = 6.7, C=); [M+H]$^+$ found = 390.1363., C$_{22}$H$_{21}$N$_3$O$_2$P requires 390.1371.

[1-(4-Methylbenzyl)-1H-1,2,3-triazol-4-yl]methyl diphenylphosphinate (**13b**): Yield: 86% (0.34 g), white crystals; Mp: 119-121 °C; ^{31}P NMR (CDCl$_3$) δ 33.3; ^{1}H NMR (CDCl$_3$) δ 2.35 (s, 3H, CH$_3$Ph), 5.18 (d, $^3J_{HP}$ = 9.0, 2H, CH$_2$O), 5.41 (s, 2H, CH$_2$Ph), 7.07–7.24 (m, 4H, C$_{3'}$H), 7.32–7.60 (m, 7H, C$_2$H, C$_3$H, C$_{4'}$H), 7.79 (s, 1H, CH), 7.80 (dd, $^3J_{HP}$ = 12.4, $^3J_{HH}$ = 7.4, 4H, C$_{2'}$H); ^{13}C NMR (CDCl$_3$) δ 21.3 (CH$_3$Ph), 54.2 (CH$_2$Ph), 58.2 (d, $^2J_{CP}$ = 5.3, CH$_2$O), 123.6 (CH=), 128.3 (C$_2$), 128.7 (d, $^3J_{CP}$ = 13.2, C$_{3'}$), 129.9 (C$_3$), 131.3 (d, $^1J_{CP}$ = 136.6, C$_{1'}$), 131.4 (C$_1$), 131.8 (d, $^2J_{CP}$ = 10.3, C$_{2'}$), 132.4 (d, J_{CP} = 2.8, C$_{4'}$), 138.9 (C$_4$), 144.0 (d, $^3J_{CP}$ = 6.7, C=); [M+H]$^+$$_{found}$ = 404.1519, C$_{23}$H$_{23}$N$_3$O$_2$P requires 404.1527.

[1-(2-Fluorobenzyl)-1H-1,2,3-triazol-4-yl]methyl diphenylphosphinate (**13c**): Yield: 81% (0.33 g), pale yellow crystals; Mp: 91-93 °C; ^{31}P NMR (CDCl$_3$) δ 33.5; ^{1}H NMR (CDCl$_3$) δ 5.18 (d, $^3J_{HP}$ = 8.9, 2H, CH$_2$O), 5.52 (s, 2H, CH$_2$Ph), 7.03–7.17 (m, 2H, C$_3$H, C$_5$H), 7.18–7.26 (m, 2H, C$_6$H), 7.28–7.53 (m, 7H, C$_4$H, C$_{3'}$H, C$_{4'}$H), 7.60 (s, 1H, CH), 7.79 (dd, $^3J_{HP}$ = 12.3, $^3J_{HH}$ = 6.9, 4H, C$_{2'}$H); ^{13}C NMR (CDCl$_3$) δ 47.8 (d, $^3J_{CF}$ = 4.4, CH$_2$Ph), 58.1 (d, $^2J_{CP}$ = 5.3, CH$_2$O), 115.9 (d, $^2J_{CF}$ = 21.1, C$_3$), 121.8 (d, $^2J_{CF}$ = 14.6, C$_1$), 124.0 (CH=), 124.9 (d, J_{CF} = 3.8, C$_5$), 128.7 (d, $^3J_{CP}$ = 13.3, C$_{3'}$), 130.7 (d, $^3J_{CF}$ = 3.2, C$_4$), 131.1 (d, $^3J_{CF}$ = 8.2, C$_6$), 131.1 (d, $^1J_{CP}$ = 136.8, C$_{1'}$), 131.8 (d, $^2J_{CP}$ = 10.4, C$_{2'}$), 132.4 (d, J_{CP} = 2.9, C$_{4'}$), 144.0 (d, $^3J_{CP}$ = 6.6, C=), 160.0 (d, $^1J_{CF}$ = 248.1, C$_2$); [M+H]$^+$$_{found}$ = 408.1268, C$_{22}$H$_{20}$N$_3$O$_2$FP requires 408.1277.

[1-(3-Fluorobenzyl)-1H-1,2,3-triazol-4-yl]methyl diphenylphosphinate (**13d**): Yield: 83% (0.34 g), pale yellow crystals; Mp: 93-94 °C; ^{31}P NMR (CDCl$_3$) δ 33.7; ^{1}H NMR (CDCl$_3$) δ 5.20 (d, $^3J_{HP}$ = 9.3, 2H, CH$_2$O), 5.46 (s, 2H, CH$_2$Ph), 6.87–6.96 (m, 1H, C$_4$H), 6.89–7.11 (m, 2H, C$_2$H, C$_6$H), 7.26–7.36 (m, 1H, C$_5$H), 7.37–7.55 (m, 6H, C$_{3'}$H, C$_{4'}$H), 7.59 (s, 1H, CH), 7.79 (dd, $^3J_{HP}$ = 12.4, $^3J_{HH}$ = 6.8, 4H, C$_{2'}$H); ^{13}C NMR (CDCl$_3$) δ 53.6 (d, J_{CF} = 2.0, CH$_2$Ph), 58.2 (d, $^2J_{CP}$ = 5.3, CH$_2$O), 115.2 (d, $^2J_{CF}$ = 22.2, C$_4$), 116.0 (d, $^2J_{CF}$ = 21.0, C$_2$), 123.8 (d, J_{CF} = 3.1, C$_6$), 123.9 (CH=), 128.7 (d, $^3J_{CP}$ = 13.3, C$_{3'}$), 130.9 (d, $^3J_{CF}$ = 8.3, C$_5$), 131.1 (d, $^1J_{CP}$ = 136.7, C$_{1'}$), 131.8 (d, $^2J_{CP}$ = 10.3, C$_{2'}$), 132.5 (d, J_{CP} = 2.8, C$_{4'}$), 136.8 (d, $^3J_{CF}$ = 7.4, C$_1$), 144.2 (d, $^3J_{CP}$ = 6.5, C=), 163.1 (d, $^1J_{CF}$ = 248.0, C$_3$); [M+H]$^+$$_{found}$ = 408.1269, C$_{22}$H$_{20}$N$_3$O$_2$FP requires 408.1277.

[1-(4-Fluorobenzyl)-1H-1,2,3-triazol-4-yl]methyl diphenylphosphinate (**13e**): Yield: 83% (0.34 g), pale yellow crystals; Mp: 95-97 °C; ^{31}P NMR (CDCl$_3$) δ 33.5; ^{1}H NMR (CDCl$_3$) δ 5.18 (d, $^3J_{HP}$ = 9.2, 2H, CH$_2$O), 5.43 (s, 2H, CH$_2$Ph), 7.04 (t, J_{HF} = $^3J_{HH}$ = 8.6, 2H, C$_2$H), 7.23 (dd, $^3J_{HF}$ = 8.4, $^3J_{HH}$ = 5.1, 2H, C$_3$H), 7.33–7.54 (m, 6H, C$_{3'}$H, C$_{4'}$H), 7.55 (s, 1H, CH), 7.79 (dd, $^3J_{HP}$ = 12.4, $^3J_{HH}$ = 7.4, 4H, C$_{2'}$H); ^{13}C NMR (CDCl$_3$) δ 53.5 (CH$_2$Ph), 58.2 (d, $^2J_{CP}$ = 5.3, CH$_2$O), 116.2 (d, $^2J_{CF}$ = 21.9, C$_3$), 123.8 (CH=), 128.7 (d, $^3J_{CP}$ = 13.3, C$_{3'}$), 130.1 (d, $^3J_{CF}$ = 8.4, C$_2$), 130.4 (d, J_{CF} = 3.3, C$_1$), 131.2 (d, $^1J_{CP}$ = 136.7, C$_{1'}$), 131.8 (d, $^2J_{CP}$ = 10.3, C$_{2'}$), 132.5 (d, J_{CP} = 2.9, C$_{4'}$), 144.2 (d, $^3J_{CP}$ = 6.2, C=), 163.0 (d, $^1J_{CF}$ = 248.3, C$_4$); [M+H]$^+$$_{found}$ = 408.1268, C$_{22}$H$_{20}$N$_3$O$_2$FP requires 408.1277.

[1-(4-Trifluoromethyl)benzyl)-1H-1,2,3-triazol-4-yl]methyl diphenylphosphinate (**13f**): Yield: 88% (0.40 g), white crystals; Mp: 124-125 °C; ^{31}P NMR (CDCl$_3$) δ 33.6; ^{1}H NMR (CDCl$_3$) δ 5.20 (d, $^3J_{HP}$ = 9.5, 2H, CH$_2$O), 5.43 (s, 2H, CH$_2$Ph), 7.33 (d, $^3J_{HH}$ = 8.0, 2H, C$_2$H), 7.39–7.46 (m, 4H, C$_{3'}$H), 7.48–7.53 (m, 2H, C$_{4'}$H), 7.61 (d, $^3J_{HH}$ = 8.0, 2H, C$_3$H), 7.63 (s, 1H, CH), 7.79 (dd, $^3J_{HP}$ = 12.4, $^3J_{HH}$ = 6.9, 4H, C$_{2'}$H); ^{13}C NMR (CDCl$_3$) δ 53.6 (CH$_2$Ph), 58.2 (d, $^2J_{CP}$ = 5.4, CH$_2$O), 122.8 (CH=), 123.9 (q, $^1J_{CF}$ = 272.1, CF$_3$), 126.2 (q, $^3J_{CF}$ = 3.7, C$_3$), 128.4 (C$_2$), 128.7 (d, $^3J_{CP}$ = 13.3, C$_{3'}$), 131.1 (q, $^2J_{CF}$ =32.9, C$_4$), 131.2 (d, $^1J_{CP}$ = 136.7, C$_{1'}$), 131.7 (d, $^2J_{CP}$ = 10.3, C$_{2'}$), 132.5 (d, J_{CP} = 2.9, C$_{4'}$), 138.5 (C$_1$), 144.4 (d, $^3J_{CP}$ = 6.0, C=); [M+H]$^+$$_{found}$ = 458.1242, C$_{23}$H$_{20}$N$_3$O$_2$F$_3$P requires 458.1245.

(1-Octyl-1H-1,2,3-triazol-4-yl)methyl diphenylphosphinate (**13g**): Yield: 89% (0.37 g), yellow oil; ^{31}P NMR (CDCl$_3$) δ 33.2; ^{1}H NMR (CDCl$_3$) δ 0.87 (t, $^3J_{HH}$ = 6.5, 3H, CH$_3$), 1.10–1.41 (m, 10H, CH$_2$CH$_2$(CH$_2$)$_5$CH$_3$), 1.72–1.97 (m, 2H, CH$_2$CH$_2$(CH$_2$)$_5$CH$_3$), 4.28 (t, 2H, $^3J_{HH}$ = 7.3, CH$_2$CH$_2$(CH$_2$)$_5$CH$_3$), 5.21 (d, $^3J_{HP}$ = 9.1, 2H, CH$_2$O), 7.34–7.57 (m, 6H, C$_3$H, C$_4$H), 7.60 (s, 1H, CH), 7.82 (dd, $^3J_{HP}$ = 12.4, $^3J_{HH}$ = 7.3, 4H, C$_2$H); ^{13}C NMR (CDCl$_3$) δ 14.2 (CH$_3$), 22.7 (CH$_2$CH$_3$), 26.6 (CH$_2$CH$_2$CH$_3$), 29.0 (CH$_2$(CH$_2$)$_2$CH$_3$), 29.1 (CH$_2$(CH$_2$)$_3$CH$_3$), 30.3 (CH$_2$(CH$_2$)$_4$CH$_3$), 31.8 (CH$_2$(CH$_2$)$_5$CH$_3$), 50.5 (CH$_2$(CH$_2$)$_6$CH$_3$), 58.3 (d, $^2J_{CP}$ = 5.3, CH$_2$O), 123.7 (CH=), 128.7 (d, $^3J_{CP}$ = 13.2, C$_3$), 131.4 (d, $^1J_{CP}$ = 136.6, C$_1$), 131.8 (d, $^2J_{CP}$ = 10.3, C$_2$), 132.4 (d, J_{CP} = 2.8, C$_4$), 143.6 (d, $^3J_{CP}$ = 6.9, C=); [M+H]$^+$$_{found}$ = 412.2154, C$_{23}$H$_{31}$N$_3$O$_2$P requires 412.2153.

(1-Iso-octyl-1H-1,2,3-triazol-4-yl)methyl diphenylphosphinate (**13h**): Yield: 77% (0.49 g), yellow oil; ^{31}P NMR (CDCl$_3$) δ 33.3; ^{1}H NMR (CDCl$_3$) δ 0.89 (t, $^3J_{HH}$ = 6.7, 6H, CH$_3$), 1.02–1.41 (m, 8H, CH(CH$_2$)$_3$CH$_3$, CHCH$_2$CH$_3$), 1.74–1.97 (m, 1H, NCH$_2$CH), 4.21 (d, 2H, $^3J_{HH}$ = 6.9, NCH$_2$CH), 5.22 (d, $^3J_{HP}$ = 9.0, 2H, CH$_2$O), 7.34–7.58 (m, 6H, C$_3$H, C$_4$H), 7.59 (s, 1H, CH), 7.82 (dd, $^3J_{HP}$ = 12.5, $^3J_{HH}$ = 7.4, 4H, C$_2$H); ^{13}C NMR (CDCl$_3$) δ 10.5 (NCH$_2$CHCH$_2$CH$_3$), 14.0 (CH$_3$), 22.9 (CH$_2$CH$_3$), 23.7 (NCH$_2$CHCH$_2$CH$_3$), 28.5 (CH$_2$CH$_2$CH$_3$), 30.4 (CH$_2$(CH$_2$)$_2$CH$_3$), 40.3 (NCH$_2$CHCH$_2$CH$_3$), 53.6 (NCH$_2$CHCH$_2$CH$_3$), 58.2 (d, $^2J_{CP}$ = 5.3, CH$_2$O), 124.2 (CH=), 128.6 (d, $^3J_{CP}$ = 13.3, C$_3$), 131.3 (d, $^1J_{CP}$ = 136.7, C$_1$), 131.7 (d, $^2J_{CP}$ = 10.3, C$_2$), 132.4 (d, J_{CP} = 2.9, C$_4$), 143.4 (d, $^3J_{CP}$ = 6.2, C=); [M+H]$^+$$_{found}$ = 412.2154, C$_{23}$H$_{31}$N$_3$O$_2$P requires 412.2153.

(1-Cyclohexyl-1H-1,2,3-triazol-4-yl)methyl diphenylphosphinate (**13i**): Yield: 63% (0.24 g), white crystals; Mp: 122-124 °C; ^{31}P NMR (CDCl$_3$) δ 33.2; ^{1}H NMR (CDCl$_3$) δ 1.16–1.52 (m, 4H, C$_3$H$_{ax}$, C$_4$H$_{ax}$, C$_4$H$_{eq}$), 1.58–1.80 (m, 2H, C$_3$H$_{eq}$), 1.81–1.95 (m, 2H, C$_2$H$_{ax}$), 2.01–2.22 (m, 2H, C$_2$H$_{eq}$), 4.29–4.45 (m, 1H, C$_1$H), 5.21 (d, $^3J_{HP}$ = 9.0, 2H, CH$_2$O), 7.32–7.56 (m, 6H, C$_{3'}$H, C$_{4'}$H), 7.59 (s, 1H, CH), 7.82 (dd, $^3J_{HP}$ = 12.4, $^3J_{HH}$ = 7.3, 4H, C$_{2'}$H); ^{13}C NMR (CDCl$_3$) δ 25.2 (C$_4$), 25.3 (C$_3$), 33.6 (C$_2$), 58.4 (d, $^2J_{CP}$ = 5.3, CH$_2$O), 60.3 (C$_1$) 121.6 (CH=), 128.7 (d, $^3J_{CP}$ = 13.2, C$_{3'}$), 131.4 (d, $^1J_{CP}$ = 136.9, C$_{1'}$), 131.8 (d, $^2J_{CP}$ = 10.3, C$_{2'}$), 132.4 (d, J_{CP} = 2.8, C$_{4'}$), 143.1 (d, $^3J_{CP}$ = 6.5, C=); [M+H]$^+$$_{found}$ = 382.1677, C$_{21}$H$_{25}$N$_3$O$_2$P requires 382.1684.

(1-Phenyl-1H-1,2,3-triazol-4-yl)methyl diphenylphosphinate (**13j**): Yield: 82% (0.31 g), white crystals; Mp: 121-122 °C; ^{31}P NMR (CDCl$_3$) δ 33.5; ^{1}H NMR (CDCl$_3$) δ 5.30 (d, $^3J_{HP}$ = 9.1, 2H, CH$_2$O), 7.34–7.57 (m, 9H, C$_3$H, C$_4$H, C$_{3'}$H, C$_{4'}$H), 7.68 (d, $^3J_{HH}$ = 7.7, 2H, C$_2$H), 7.85 (dd, $^3J_{HP}$ = 12.4, $^3J_{HH}$ = 7.4, 4H, C$_{2'}$H), 8.05 (s, 1H, CH); ^{13}C NMR (CDCl$_3$) δ 58.2 (d, $^2J_{CP}$ = 5.3, CH$_2$O), 120.7 (C$_2$), 122.2 (CH=), 128.8 (d, $^3J_{CP}$ = 13.2, C$_{3'}$), 129.0 (C$_4$), 129.9 (C$_3$), 131.2 (d, $^1J_{CP}$ = 136.6, C$_{1'}$), 131.8 (d, $^2J_{CP}$ = 10.4, C$_{2'}$), 132.5 (d, J_{CP} = 2.9, C$_{4'}$), 137.0 (C$_1$), 144.4 (d, $^3J_{CP}$ = 6.0, C=); [M+H]$^+$$_{found}$ = 376.1210, C$_{21}$H$_{19}$N$_3$O$_2$P requires 376.1214.

(1-Benzyl-1H-1,2,3-triazol-4-yl)methyl diethyl phosphate (**14a**): Yield: 75% (0.24 g), pale yellow oil; ^{31}P NMR (CDCl$_3$) δ −1.0; ^{1}H NMR (CDCl$_3$) δ 1.27 (t, $^3J_{HH}$ = 6.9, 6H, OCH$_2$CH$_3$), 4.01–4.09 (m, 4H, OCH$_2$CH$_3$), 5.16 (d, $^3J_{HP}$ = 9.4, 2H, CH$_2$O), 5.53 (s, 2H, CH$_2$Ph), 7.25–7.30 (m, 2H, C$_2$H), 7.32–7.40 (m, 3H, C$_3$H, C$_4$H), 7.62 (s, 1H, CH); ^{13}C NMR (CDCl$_3$) δ 16.1 (d, $^3J_{CP}$ = 6.3, OCH$_2$CH$_3$), 54.4 (CH$_2$Ph), 60.6 (d, $^2J_{CP}$ = 6.2, CH$_2$O), 64.1 (d, $^2J_{CP}$ = 6.4, OCH$_2$CH$_3$), 123.5 (CH=), 128.2 (C$_2$), 129.0 (C$_4$), 129.2 (C$_3$), 134.5 (C$_1$), 143.8 (d, $^3J_{CP}$ = 8.2, C=); [M+H]$^+$$_{found}$ = 326.1259, C$_{14}$H$_{21}$N$_3$O$_4$P requires 326.1269.

[1-(4-Methylbenzyl)-1H-1,2,3-triazol-4-yl]methyl diethyl phosphate (**14b**): Yield: 69% (0.23 g), pale yellow oil; ^{31}P NMR (CDCl$_3$) δ −1.1; ^{1}H NMR (CDCl$_3$) δ 1.17 (t, $^3J_{HH}$ = 6.9, 6H, OCH$_2$CH$_3$), 2.27 (s, 3H, CH$_3$Ph), 3.87–4.05 (m, 4H, OCH$_2$CH$_3$), 5.07 (d, $^3J_{HP}$ = 9.2, 2H, CH$_2$O), 5.40 (s, 2H, CH$_2$Ph), 7.10 (s, 4H, C$_2$H, C$_3$H), 7.52 (s, 1H, CH); ^{13}C NMR (CDCl$_3$) δ 16.1 (d, $^3J_{CP}$ = 6.8, OCH$_2$CH$_3$), 21.2 (CH$_3$Ph), 54.1 (CH$_2$Ph), 60.6 (d, $^2J_{CP}$ = 5.2, CH$_2$O), 64.0 (d, $^2J_{CP}$ = 5.9, OCH$_2$CH$_3$), 123.3 (CH=), 128.3 (C$_2$), 129.9 (C$_3$), 131.5 (C$_1$), 138.9 (C$_4$), 143.7 (d, $^3J_{CP}$ = 6.9, C=); [M+H]$^+$$_{found}$ = 340.1414, C$_{15}$H$_{23}$N$_3$O$_4$P requires 340.1426.

[1-(2-Fluorobenzyl)-1H-1,2,3-triazol-4-yl]methyl diethyl phosphate (**14c**): Yield: 54% (0.19 g), pale yellow oil; ^{31}P NMR (CDCl$_3$) δ −1.0; ^{1}H NMR (CDCl$_3$) δ 1.28 (t, $^3J_{HH}$ = 7.1, 6H, OCH$_2$CH$_3$), 3.97–4.17 (m, 4H, OCH$_2$CH$_3$), 5.16 (d, $^3J_{HP}$ = 9.3, 2H, CH$_2$O), 5.59 (s, 2H, CH$_2$Ph), 7.07–7.20 (m, 2H, C$_3$H, C$_5$H), 7.24–7.31 (m, 1H, C$_6$H), 7.32–7.41 (m, 1H, C$_4$H), 7.70 (s, 1H, CH); ^{13}C NMR (CDCl$_3$) δ 16.2 (d, $^3J_{CP}$ = 6.8, OCH$_2$CH$_3$), 47.9 (d, $^3J_{CF}$ = 4.3, CH$_2$Ph), 60.6 (d, $^2J_{CP}$ = 5.3, CH$_2$O), 64.1 (d, $^2J_{CP}$ = 5.9, OCH$_2$CH$_3$), 116.0 (d, $^2J_{CF}$ = 21.1, C$_3$), 121.8 (d, $^2J_{CF}$ = 14.6, C$_1$), 123.6 (CH=), 125.0 (d, J_{CF} = 3.7, C$_5$), 130.8 (d, $^3J_{CF}$ = 3.2, C$_4$), 131.2 (d, $^3J_{CF}$ = 8.3, C$_6$), 143.9 (d, $^3J_{CP}$ = 7.0, C=), 160.7 (d, $^1J_{CF}$ = 248.0, C$_2$); [M+H]$^+$$_{found}$ = 344.1164, C$_{14}$H$_{20}$N$_3$O$_4$FP requires 344.1175.

[1-(3-Fluorobenzyl)-1H-1,2,3-triazol-4-yl]methyl diethyl phosphate (**14d**): Yield: 56% (0.20 g), pale yellow oil; ^{31}P NMR (CDCl$_3$) δ −0.9; ^{1}H NMR (CDCl$_3$) δ 1.28 (t, $^3J_{HH}$ = 7.1, 6H, OCH$_2$CH$_3$), 3.95–4.15 (m, 4H, OCH$_2$CH$_3$), 5.17 (d, $^3J_{HP}$ = 9.5, 2H, CH$_2$O), 5.53 (s, 2H, CH$_2$Ph), 6.89–7.00 (m, 1H, C$_4$H), 7.00–7.12 (m, 2H, C$_2$H, C$_6$H), 7.26–7.42 (m, 1H, C$_5$H), 7.66 (s, 1H, CH); ^{13}C NMR (CDCl$_3$) δ 16.2 (d, $^3J_{CP}$ = 6.8, OCH$_2$CH$_3$), 53.7 (d, J_{CF} = 1.9, CH$_2$Ph), 60.6 (d, $^2J_{CP}$ = 5.2, CH$_2$O), 64.1 (d, $^2J_{CP}$ = 6.0,

OCH$_2$CH$_3$), 115.2 (d, $^2J_{CF}$ = 22.3, C$_4$), 116.1 (d, $^2J_{CF}$ = 21.0, C$_2$), 123.6 (CH=), 123.7 (d, J_{CF} = 3.1, C$_6$), 131.0 (d, $^3J_{CF}$ = 8.3, C$_5$), 136.9 (d, $^3J_{CF}$ = 7.4, C$_1$), 144.2 (d, $^3J_{CP}$ = 6.8, C=), 163.2 (d, $^1J_{CF}$ = 248.1, C$_3$); [M+H]$^+$$_{found}$ = 344.1164, C$_{14}$H$_{20}$N$_3$O$_4$FP requires 344.1175.

[1-(4-Fluorobenzyl)-1H-1,2,3-triazol-4-yl]methyl diethyl phosphate (**14e**): Yield: 61% (0.22 g), pale yellow oil; ^{31}P NMR (CDCl$_3$) δ −1.0; ^{1}H NMR (CDCl$_3$) δ 1.28 (t, $^3J_{HH}$ = 7.1, 6H, OCH$_2$CH$_3$), 3.88–4.15 (m, 4H, OCH$_2$CH$_3$), 5.16 (d, $^3J_{HP}$ = 9.5, 2H, CH$_2$O), 5.50 (s, 2H, CH$_2$Ph), 7.06 (t, J_{HF} = $^3J_{HH}$ = 8.6, 2H, C$_2$H), 7.28 (dd, $^3J_{HF}$ = 8.0, $^3J_{HH}$ = 5.8, 2H, C$_3$H), 7.63 (s, 1H, CH); ^{13}C NMR (CDCl$_3$) δ 16.2 (d, $^3J_{CP}$ = 6.8, OCH$_2$CH$_3$), 53.6 (CH$_2$Ph), 60.6 (d, $^2J_{CP}$ = 5.2, CH$_2$O), 64.1 (d, $^2J_{CP}$ = 6.0, OCH$_2$CH$_3$), 116.3 (d, $^2J_{CF}$ = 21.8, C$_3$), 123.4 (CH=), 130.2 (d, $^3J_{CF}$ = 8.4, C$_2$), 130.4 (d, J_{CF} = 3.3, C$_1$), 144.0 (d, $^3J_{CP}$ = 6.9, C=), 163.1 (d, $^1J_{CF}$ = 248.3, C$_4$); [M+H]$^+$$_{found}$ = 344.1163, C$_{14}$H$_{20}$N$_3$O$_4$FP requires 344.1175.

[1-(4-Trifluoromethyl)benzyl)-1H-1,2,3-triazol-4-yl]methyl diethyl phosphate (**14f**): Yield: 68% (0.27 g), pale yellow oil; ^{31}P NMR (CDCl$_3$) δ −0.9; ^{1}H NMR (CDCl$_3$) δ 1.23 (t, $^3J_{HH}$ = 7.1, 6H, OCH$_2$CH$_3$), 3.92–4.12 (m, 4H, OCH$_2$CH$_3$), 5.13 (d, $^3J_{HP}$ = 9.4, 2H, CH$_2$O), 5.57 (s, 2H, CH$_2$Ph), 7.36 (d, $^3J_{HH}$ = 8.0, 2H, C$_2$H), 7.59 (d, $^3J_{HH}$ = 8.0, 2H, C$_3$H), 7.68 (s, 1H, CH); ^{13}C NMR (CDCl$_3$) δ 16.1 (d, $^3J_{CP}$ = 6.7, OCH$_2$CH$_3$), 53.6 (CH$_2$Ph), 60.5 (d, $^2J_{CP}$ = 5.2, CH$_2$O), 64.0 (d, $^2J_{CP}$ = 5.9, OCH$_2$CH$_3$), 123.7 (CH=), 123.8 (q, $^1J_{CF}$ = 272.4, CF$_3$), 126.1 (q, $^3J_{CF}$ = 3.8, C$_3$), 128.4 (C$_2$), 131.1 (q, $^2J_{CF}$ =32.8, C$_4$), 138.6 (C$_1$), 144.1 (d, $^3J_{CP}$ = 6.8, C=); [M+H]$^+$$_{found}$ = 394.1157, C$_{15}$H$_{20}$N$_3$O$_4$F$_3$P requires 394.1143.

(1-Octyl-1H-1,2,3-triazol-4-yl)methyl diethyl phosphate (**14g**): Yield: 60% (0.21 g), pale yellow oil; ^{31}P NMR (CDCl$_3$) δ −1.3; ^{1}H NMR (CDCl$_3$) δ 0.88 (t, $^3J_{HH}$ = 6.7, 3H, CH$_3$), 1.16–1.43 (m, 16H, CH$_2$CH$_2$(CH$_2$)$_5$CH$_3$, OCH$_2$CH$_3$), 1.81–2.01 (m, 2H, CH$_2$CH$_2$(CH$_2$)$_5$CH$_3$), 4.01–4.22 (m, 4H, OCH$_2$CH$_3$), 4.35 (t, 2H, $^3J_{HH}$ = 7.3, CH$_2$CH$_2$(CH$_2$)$_5$CH$_3$), 5.19 (d, $^3J_{HP}$ = 9.4, 2H, CH$_2$O), 7.68 (s, 1H, CH); ^{13}C NMR (CDCl$_3$) δ 14.2 (CH$_3$), 16.2 (d, $^3J_{CP}$ = 6.8, OCH$_2$CH$_3$), 22.7 (CH$_2$CH$_3$), 26.6 (CH$_2$CH$_2$CH$_3$), 29.1 (CH$_2$(CH$_2$)$_2$CH$_3$), 29.2 (CH$_2$(CH$_2$)$_3$CH$_3$), 30.4 (CH$_2$(CH$_2$)$_4$CH$_3$), 31.8 (CH$_2$(CH$_2$)$_5$CH$_3$), 50.6 (CH$_2$(CH$_2$)$_6$CH$_3$), 60.7 (d, $^2J_{CP}$ = 5.3, CH$_2$O), 64.1 (d, $^2J_{CP}$ = 6.0, OCH$_2$CH$_3$), 123.4 (CH=), 143.4 (d, $^3J_{CP}$ = 6.7, C=); [M+H]$^+$$_{found}$ = 348.2043, C$_{15}$H$_{31}$N$_3$O$_4$P requires 348.2052.

(1-Iso-octyl-1H-1,2,3-triazol-4-yl)methyl diethyl phosphate (**14h**): Yield: 54% (0.19 g), pale yellow oil; ^{31}P NMR (CDCl$_3$) δ −1.2; ^{1}H NMR (CDCl$_3$) δ 0.91 (t, $^3J_{HH}$ = 7.4, 6H, CH$_3$), 1.12–1.42 (m, 14H, CH(CH$_2$)$_3$CH$_3$, CHCH$_2$CH$_3$, OCH$_2$CH$_3$), 1.80–1.95 (m, 1H, NCH$_2$CH), 3.98–4.15 (m, 4H, OCH$_2$CH$_3$), 4.27 (d, 2H, $^3J_{HH}$ = 6.8, NCH$_2$CH), 5.19 (d, $^3J_{HP}$ = 9.4, 2H, CH$_2$O), 7.66 (s, 1H, CH); ^{13}C NMR (CDCl$_3$) δ 10.6 (NCH$_2$CHCH$_2$CH$_3$), 14.1 (CH$_3$), 16.2 (d, $^3J_{CP}$ = 6.7, OCH$_2$CH$_3$), 23.0 (CH$_2$CH$_3$), 23.8 (NCH$_2$CHCH$_2$CH$_3$), 28.6 (CH$_2$CH$_2$CH$_3$), 30.5 (CH$_2$(CH$_2$)$_2$CH$_3$), 40.5 (NCH$_2$CHCH$_2$CH$_3$), 53.7 (NCH$_2$CHCH$_2$CH$_3$), 60.7 (d, $^2J_{CP}$ = 5.3, CH$_2$O), 64.1 (d, $^2J_{CP}$ = 6.0, OCH$_2$CH$_3$), 124.0 (CH=), 143.3 (d, $^3J_{CP}$ = 6.6, C=); [M+H]$^+$$_{found}$ = 348.2043, C$_{15}$H$_{31}$N$_3$O$_4$P requires 348.2052.

(1-Cyclohexyl-1H-1,2,3-triazol-4-yl)methyl diethyl phosphate (**14i**): Yield: 51% (0.16 g), pale yellow oil; ^{31}P NMR (CDCl$_3$) δ −1.3; ^{1}H NMR (CDCl$_3$) δ 1.31 (t, $^3J_{HH}$ = 6.9, 6H, OCH$_2$CH$_3$), 1.17–1.55 (m, 4H, C$_3$H$_{ax}$, C$_4$H$_{ax}$, C$_4$H$_{eq}$), 1.67–1.82 (m, 2H, C$_3$H$_{eq}$), 1.88–1.98 (m, 2H, C$_2$H$_{ax}$), 2.15–2.26 (m, 2H, C$_2$H$_{eq}$), 4.02–4.16 (m, 4H, OCH$_2$CH$_3$), 4.38–4.52 (m, 1H, C$_1$H), 5.18 (d, $^3J_{HP}$ = 9.2, 2H, CH$_2$O), 7.70 (s, 1H, CH); ^{13}C NMR (CDCl$_3$) δ 16.2 (d, $^3J_{CP}$ = 6.8, OCH$_2$CH$_3$), 25.2 (C$_4$), 25.3 (C$_3$), 33.7 (C$_2$), 60.3 (C$_1$), 60.8 (d, $^2J_{CP}$ = 5.2, CH$_2$O), 64.0 (d, $^2J_{CP}$ = 5.9, OCH$_2$CH$_3$), 121.3 (CH=), 142.9 (d, $^3J_{CP}$ = 6.8, C=); [M+H]$^+$$_{found}$ = 318.1575, C$_{13}$H$_{25}$N$_3$O$_4$P requires 318.1583.

(1-Phenyl-1H-1,2,3-triazol-4-yl)methyl diethyl phosphate (**14j**): Yield: 73% (0.23 g), pale yellow oil; ^{31}P NMR (CDCl$_3$) δ −1.0; ^{1}H NMR (CDCl$_3$) δ 1.33 (t, $^3J_{HH}$ = 7.1, 6H, OCH$_2$CH$_3$), 3.99–4.20 (m, 4H, OCH$_2$CH$_3$), 5.28 (d, $^3J_{HP}$ = 9.2, 2H, CH$_2$O), 7.37–7.61 (m, 9H, C$_3$H, C$_4$H), 7.74 (d, $^3J_{HH}$ = 7.8, 2H, C$_2$H), 8.16 (s, 1H, CH); ^{13}C NMR (CDCl$_3$) δ 16.2 (d, $^3J_{CP}$ = 6.8, OCH$_2$CH$_3$), 60.5 (d, $^2J_{CP}$ = 5.1, CH$_2$O), 64.2 (d, $^2J_{CP}$ = 6.0, OCH$_2$CH$_3$), 120.7 (C$_2$), 121.8 (CH=), 129.1 (C$_4$), 129.9 (C$_3$), 137.0 (C$_1$), 144.1 (d, $^3J_{CP}$ = 7.0, C=); [M+H]$^+$$_{found}$ = 312.1104, C$_{13}$H$_{19}$N$_3$O$_4$P requires 312.1113.

4. Conclusions

In summary, we have developed a facile, efficient method for the synthesis of new (1-alkyl/aryl-1H-1,2,3-triazol-4-yl)methyl phosphinates or (1-alkyl/aryl-1H-1,2,3-triazol-4-yl)methyl diethyl phosphates by the copper(I)-catalyzed azide-alkyne cycloaddition of organic azides and prop-2-ynyl phosphinate or diethyl prop-2-ynyl phosphate. This method, which has the advantages of simple operation and mild reaction conditions, is a novel approach for the synthesis of the target products. Altogether, 20 new derivatives were synthesized and fully characterized.

Supplementary Materials: Supplementary data associated with this article are available online. Copies of ^{31}P, ^{1}H, and ^{13}C NMR spectra for all compounds synthesized are presented.

Author Contributions: Conceptualization, E.B.; methodology, E.B. and A.T.; validation, K.N. and P.T.S.; formal analysis, K.N. and P.T.S.; investigation, A.T.; writing—original draft preparation, E.B.; writing—review and editing, A.T.; visualization, E.B. and A.T.; supervision, E.B.; project administration, A.T.; funding acquisition, E.B.

Funding: The project was supported by the Hungarian Research Development and Innovation Office (FK123961). E.B. was supported by the János Bolyai Research Scholarship of the Hungarian Academy of Sciences (BO/00278/17/7), and by the ÚNKP-18-4-BME-131 New National Excellence Program of the Ministry of Human Capacities. A.T. was supported by the ÚNKP-18-3-I-BME-119 New National Excellence Program of the Ministry of Human Capacities.

Conflicts of Interest: The authors declare no conflict of interest.

References

1. Katritzky, A.R.; Rees, C.W.; Scriven, C.W.V. (Eds.) *Comprehensive Heterocyclic Chemistry*; Elsevier Science: Amsterdam, The Netherlands, 1996; Volume 4, pp. 1–126.
2. Thirumurugan, P.; Matosiuk, D.; Jozwiak, K. Click chemistry for drug development and diverse chemical-biology applications. *Chem. Rev.* **2013**, *113*, 4905–4979. [CrossRef] [PubMed]
3. Lauria, A.; Delisi, R.; Mingoia, F.; Terenzi, A.; Martorana, A.; Barone, G.; Almerico, A.M. 1,2,3-Triazole in heterocyclic compounds, endowed with biological activity, through 1,3-dipolar cycloadditions. *Eur. J. Org. Chem.* **2014**, 3289–3306. [CrossRef]
4. Dheer, D.; Singh, V.; Shankar, R. Medicinal attributes of 1,2,3-triazoles: Current developments. *Bioorg. Chem.* **2017**, *71*, 30–54. [CrossRef] [PubMed]
5. Williams, A. Opportunities for chiral agrochemicals. *Pestic. Sci.* **1996**, *46*, 3–9. [CrossRef]
6. Ali, G.Q.; El-Hiti, G.A.; Tomi, I.H.; Haddad, R.; Al-Qaisi, A.J.; Yousif, E. Photostability and performance of polystyrene films containing 1,2,4-triazole-3-thiol ring system schiff bases. *Molecules* **2016**, *21*, 1699–1711. [CrossRef] [PubMed]
7. Bouchemella, K.; Fauché, K.; Anak, B.; Jouffret, L.; Bencharif, M.; Cisnetti, F. Click 1,2,3-triazole derived fluorescent scaffold by mesoionic carbene–nitrene cyclization: an experimental and theoretical study. *New J. Chem.* **2018**, *42*, 18969–18978. [CrossRef]
8. Srividhya, D.; Manjunathan, S.; Thirumaran, S.; Saravanan, C.; Senthil, S. Synthesis and characterization of [1,2,3]-triazole containing liquid crystals through click reaction. *J. Mol. Struct.* **2009**, *927*, 7–13. [CrossRef]
9. Tajti, Á.; Keglevich, G. The importance of organophosphorus compounds as biologically active agents. In *Organophosphorus Chemistry*; Keglevich, G., Ed.; Walter de Gruyter GmbH: Berlin, Germany, 2018; pp. 53–65. ISBN 978-3-11-053453-5.
10. Li, L.; Hao, G.; Zhu, A.; Fan, X.; Zhang, G.; Zhang, L. A copper(I)-catalyzed three-component domino process: assembly of complex 1,2,3-triazolyl-5-phosphonates from azides, alkynes, and H-phosphates. *Chem. Eur. J.* **2013**, *19*, 14403–14406. [CrossRef]
11. Kee, J.M.; Villani, B.; Carpenter, L.R.; Muir, T.W. Development of stable phosphohistidine analogues. *J. Am. Chem. Soc.* **2010**, *132*, 14327–14329. [CrossRef]
12. Mukai, S.; Flematti, G.R.; Byrne, L.T.; Besant, P.G.; Attwood, P.V.; Piggott, M.J. Stable triazolylphosphonate analogues of phosphohistidine. *Amino Acids* **2012**, *43*, 857–874. [CrossRef]
13. Radi, S.; Lazrek, H.B. Synthesis and biological activity of new 1,2,3-triazole acyclonucleosides analogues of ACV. *J. Chem. Res. Synop.* **2002**, 264–266. [CrossRef]
14. Rostovtsev, V.V.; Green, L.G.; Fokin, V.V.; Sharpless, K.B. A stepwise huisgen cycloaddition process: Copper(I)-catalyzed regioselective "ligation" of azides and terminal alkynes. *Angew. Chem. Int. Ed.* **2002**, *41*, 2596–2599. [CrossRef]

15. Tornøe, C.W.; Christensen, C.; Meldal, M. Peptidotriazoles on solid phase: [1,2,3]-Triazoles by regiospecific copper(I)-catalyzed 1,3-dipolar cycloadditions of terminal alkynes to azides. *J. Org. Chem.* **2002**, *67*, 3057–3064. [CrossRef] [PubMed]
16. Thiery, E.; You, V.; Mora, A.-S.; Abarbri, M. Synthesis of 5-substituted 1,2,3-triazolyl-4-phosphonate through cross-coupling reactions of 5-iodo-1,2,3-triazolyl-4-phosphonate. *Eur. J. Org. Chem.* **2016**, 529–534. [CrossRef]
17. Skarpos, H.; Osipov, S.N.; Vorob'eva, D.V.; Odinets, I.L.; Lork, E.; Roschenthaler, G.V. Synthesis of functionalized bisphosphonates via click chemistry. *Org. Biomol. Chem.* **2007**, *5*, 2361–2367. [CrossRef] [PubMed]
18. Zhou, X.; Hartman, S.V.; Born, E.J.; Smits, J.P.; Holstein, S.A.; Wiemer, D.F. Triazole-based inhibitors of geranylgeranyltransferase II. *Bioorg. Med. Chem. Lett.* **2013**, *23*, 764–766. [CrossRef] [PubMed]
19. Vorobyeva, D.V.; Karimova, N.M.; Vasilyeva, T.P.; Osipov, S.N.; Shchetnikov, G.T.; Odinets, I.L.; Röschenthaler, G.-V. Synthesis of functionalized α-CF3-α-aminophosphonates via Cu(I)-catalyzed 1,3-dipolar cycloaddition. *J. Fluorine Chem.* **2010**, *131*, 378–383. [CrossRef]
20. Sykam, K.; Meka, K.K.R.; Donempudi, S. Intumescent phosphorus and triazole-based flame-retardant polyurethane foams from castor oil. *ACS Omega* **2019**, *4*, 1086–1094. [CrossRef]
21. Artyushin, O.I.; Vorob'eva, D.V.; Vasil'eva, T.P.; Osipov, S.N.; Röschenthaler, G.-V.; Odinets, I.L. Facile synthesis of phosphorylated azides in ionic liquids and their use in the preparation of 1,2,3-triazoles. *Heteroat. Chem.* **2008**, *19*, 293–300. [CrossRef]
22. Artyushin, O.I.; Matveeva, E.V.; Bushmarinov, I.S.; Odinets, I.L. Water as a promoting media for 1,3-dipolar cycloaddition of phosphorylated azides to internal alkynes. *Arkivoc* **2012**, *iv*, 252–263. [CrossRef]
23. Glowacka, I.E.; Balzarini, J.; Wroblewski, A.E. Design, synthesis, antiviral, and cytotoxic evaluation of novel phosphonylated 1,2,3-triazoles as acyclic nucleotide analogues. *Nucleosides Nucleotides Nucleic Acids* **2012**, *31*, 293–318. [CrossRef] [PubMed]
24. Veliscek-Carolan, J.; Rawal, A. Zirconium bistriazolylpyridine phosphonate materials for efficient, selective An(iii)/Ln(iii) separations. *Chem. Commun.* **2019**, *55*, 1168–1171. [CrossRef] [PubMed]
25. Erzunov, D.A.; Latyshev, G.V.; Averin, A.D.; Beletskaya, I.P.; Lukashev, N.V. CuAAC synthesis and anion binding properties of bile acid derived tripodal ligands. *Eur. J. Org. Chem.* **2015**, 6289–6297. [CrossRef]
26. Gann, A.W.; Amoroso, J.W.; Einck, V.J.; Rice, W.P.; Chambers, J.J.; Schnarr, N.A. A photoinduced, benzyne click reaction. *Org. Lett.* **2014**, *16*, 2003–2005. [CrossRef] [PubMed]
27. Womble, C.T.; Coates, G.W.; Matyjaszewski, K.; Noonan, K.J.T. Tetrakis(dialkylamino)phosphonium polyelectrolytes prepared by reversible addition–fragmentation chain transfer polymerization. *ACS Macro Lett.* **2016**, *5*, 253–257. [CrossRef]
28. Bao, M.; Lu, W.; Su, H.; Qiu, L.; Xu, X. A convergent formal [4 + 2] cycloaddition of 1,6-diynes and benzyl azides: construction of spiro-polyheterocycles. *Org. Biomol. Chem.* **2018**, *16*, 3258–3265. [CrossRef]
29. Wang, X.; Mei, T.-S.; Yu, J.-Q. Versatile Pd(OTf)$_2$ · 2H$_2$O-catalyzed ortho-fluorination using NMP as a promoter. *J. Am. Chem. Soc.* **2009**, *131*, 7520–7521. [CrossRef]
30. Colombano, G.; Albani, C.; Ottonello, G.; Ribeiro, A.; Scarpelli, R.; Tarozzo, G.; Daglian, J.; Jung, K.-M.; Piomelli, D.; Bandiera, T. O-(Triazolyl)methyl carbamates as a novel and potent class of fatty acid amide gydrolase (FAAH) inhibitors. *Chem. Med. Chem.* **2014**, *10*, 380–395. [CrossRef]
31. Tesch, M.; Kudruk, S.; Letzel, M.; Studer, A. Orthogonal click postfunctionalization of alternating copolymers prepared by nitroxide-mediated polymerization. *Chem. Eur. J.* **2017**, *23*, 5915–5919. [CrossRef]
32. Colombano, G.; Travelli, C.; Galli, U.; Caldarelli, A.; Chini, M.G.; Canonico, P.L.; Sorba, G.; Bifulco, G.; Tron, G.C.; Genazzani, A.A. A novel potent nicotinamide phosphoribosyltransferase inhibitor synthesized via click chemistry. *J. Med. Chem.* **2010**, *53*, 616–623. [CrossRef]
33. Swetha, M.; Ramana, P.V.; Shirodkar, S.G. Simple and efficient method for the synthesis of azides in water-THF solvent system. *Org. Prep. Proced. Int.* **2011**, *43*, 348–353. [CrossRef]
34. Jones, S.; Smanmoo, C. Phosphorylation of alcohols with N-phosphoryl oxazolidinones employing copper(II) triflate catalysis. *Org. Lett.* **2005**, *7*, 3271–3274. [CrossRef] [PubMed]

© 2019 by the authors. Licensee MDPI, Basel, Switzerland. This article is an open access article distributed under the terms and conditions of the Creative Commons Attribution (CC BY) license (http://creativecommons.org/licenses/by/4.0/).

Article

Synthesis of Terpyridines: Simple Reactions—What Could Possibly Go Wrong?

Dalila Rocco, Catherine E. Housecroft and Edwin C. Constable *

University of Basel, Department of Chemistry, BPR 1096, Mattenstrasse 24a, CH-4058 Basel, Switzerland; dalila.rocco@unibas.ch (D.R.); catherine.housecroft@unibas.ch (C.E.H.)
* Correspondence: edwin.constable@unibas.ch; Tel.: +41-61-207-1001

Academic Editor: Gianfranco Favi
Received: 15 April 2019; Accepted: 8 May 2019; Published: 9 May 2019

Abstract: The preparation of 2^4-functionalized $1^2,2^2{:}2^6,3^2$-terpyridines (4′-functionalized 3,2′:6′,3″-terpyridines) by the reaction of three 4-alkoxybenzaldehydes with 3-acetylpyridine and ammonia was investigated; under identical reaction conditions, two (R =^{n}C$_4$H$_9$, C$_2$H$_5$) gave the expected products whereas a third (R = ^{n}C$_3$H$_7$) gave only a cyclohexanol derivative derived from the condensation of three molecules of 3-acetylpyridine with two of 4-(n-propoxy)benzaldehyde. A comprehensive survey of "unexpected" products from reactions of ArCOCH$_3$ derivatives with aromatic aldehydes is presented. Three different types of alternative product are identified.

Keywords: terpyridines; 3,2′:6′,3″-terpyridine; cyclohexanol derivative; condensation; heterocyclic

1. Introduction

The 48 isomeric structures obtained by the linking together of three pyridine rings by three single bonds are known collectively as the terpyridines [1]. One of the commonest metal-binding scaffolds encountered in supramolecular and materials chemistry is 2,2′:6′,2″-terpyridine (**1**, 2,2′:6′,2″-tpy) [2–7]: Note that in this article, we use the nomenclature that is established in the community for the terpyridines, rather than the preferred IUPAC names (PINs) of $1^2,2^2{:}2^6,3^2$-terpyridine, $1^3,2^2{:}2^6,3^3$-terpyridine and $1^4,2^2{:}2^6,3^4$-terpyridine for 2,2′:6′,2″-terpyridine, 3,2′:6′,3″-terpyridine and 4,2′:6′,4″-terpyridine respectively [8]. Although the high kinetic and thermodynamic stability of 2,2′:6′,2″-tpy complexes confers unique properties, the main reason why this metal-binding domain is so commonly observed is the ease of synthesis of derivatives with a wide variety of substituents, in particular at the 4′-position of the central ring.

Although a vast array of synthetic strategies have been developed for the preparation of 2,2′:6′,2″-tpy derivatives, the majority are based upon the synthesis of an intermediate 1,5-bis(pyridin-2-yl)pentane-1,5-dione or 1,5-bis(pyridin-2-yl)pent-2-ene-1,5-dione (or their synthetic equivalents) which are subsequently cyclized to generate the central pyridine ring by reaction with a source of ammonia. Although the 1,5-bis(pyridin-2-yl)pent-2-ene-1,5-dione is at the correct oxidation state for the formation of the pyridine upon cyclization, the 1,5-bis(pyridin-2-yl)pentane-1,5-dione yields a 1′,2′- or 1′,4′-dihydro-2,2′:6′,2″-terpyridine which formally requires oxidation to the desired 2,2′:6′,2″-tpy (Scheme 1).

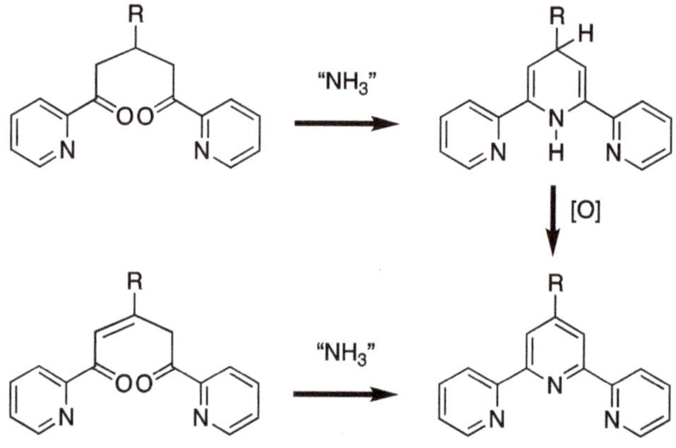

Scheme 1. The typical synthetic strategy for a 2,2':6',2''-tpy involves the cyclization of 1,5-bis(pyridin-2-yl)pentane-1,5-dione or 1,5-bis(pyridin-2-yl)pent-2-ene-1,5-dione or equivalent with ammonia. In the case of the 1,5-bis(pyridin-2-yl)pentane-1,5-dione, the intermediate dihydro-compound (the 1',4'-dihydro-2,2':6',2''-terpyridine is shown) requires an oxidation to generate the 2,2':6',2''-tpy.

The most common implementations of this strategy are based on the approach of Kröhnke (Scheme 2) [9]. In the classical Kröhnke synthesis, an intermediate enone (with the trivial name *chalcone* when two aromatic substituents are present [10]) obtained from the Claisen-Schmidt condensation [11,12] of 2-acetylpyridine with an aromatic aldehyde is reacted with an ammonia source and a "*2-pyridacylpyridinium*" salt; the latter is conveniently obtained by the Ortoleva-King reaction of 2-acetylpyridine with pyridine and iodine [13–15]. A convenient alternative is the direct reaction of two equivalents of 2-acetylpyridine with the aromatic aldehyde to give the 4-aryl-1,5-bis(pyridin-2-yl)pentane-1,5-dione directly or sequentially via the chalcone. Particularly attractive variations are solvent-free (or benign solvent) reactions and one-pot syntheses in which the reactants and an ammonia source, often ammonia and/or ammonium acetate are reacted directly or sequentially with the aromatic aldehyde and an acetylpyridine [16–20]. By replacing the 2-acetylpyridine by 3-acetylpyridine or 4-acetylpyridine, the synthetic approaches are readily adapted to the preparation of the symmetrical terpyridine isomers 3,2':6',3''-terpyridine (**2**, 3,2':6',3''-tpy, $1^3,2^2{:}2^6,3^3$-terpyridine) and 4,2':6',4''-terpyridine (**3**, 4,2':6',4''-tpy, $1^4,2^2{:}2^6,3^4$-terpyridine) (Figure 1). Although 2,2':6',2''-tpy is the most common motif encountered in coordination chemistry, 3,2':6',3''-tpy and 4,2':6',4''-tpy are increasingly finding application in the metal-directed formation of coordination networks [21–24]. In coordination networks incorporating 3,2':6',3''-tpy and 4,2':6',4''-tpy, the nitrogen atom of the central pyridine ring (ring 2 in the IUPAC recommendation), is never coordinated to a metal ion.

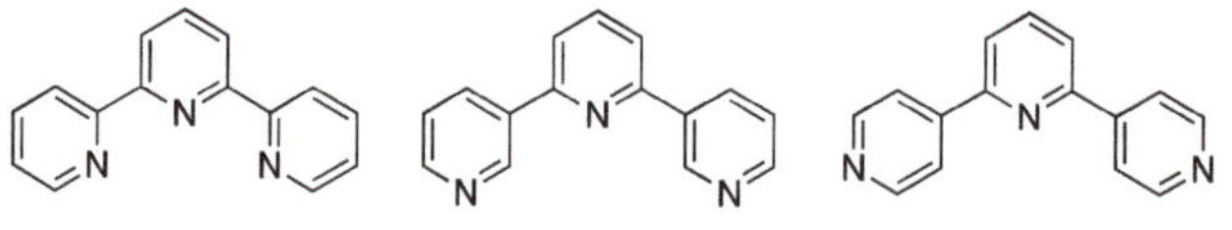

Figure 1. The structure of the most commonly encountered terpyridine, 2,2':6',2''-terpyridine (**1**), together with the closely related compounds 3,2':6',3''-terpyridine (**2**) and 4,2':6',4''-terpyridine (**3**).

In this paper, we discuss the synthesis of a series of 3,2':6',3''-tpy ligands (Figure 2) and describe the formation of an unexpected product in one case. The paper also provides a comprehensive review of alternative products which have been obtained from syntheses of terpyridines using the synthetic approaches above. We take this opportunity to issue a *caveat*: when utilizing these synthetic methods, the terpyridine products should be purified and fully characterized.

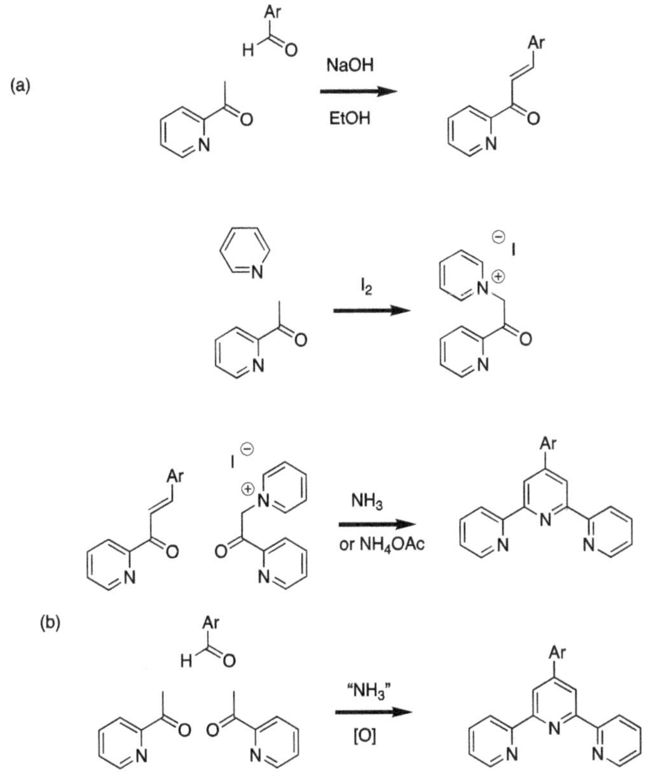

Scheme 2. Variations on the Kröhnke reaction for the preparation of 4′-aryl-2,2′:6′,2″-tpy ligands. (a) The classical Kröhnke route in which an enone (chalcone) is prepared from the aldol condensation of 2-acetylpyridine with an aromatic aldehyde followed by reaction with a "2-*pyridacylpyridinium*" salt (obtained from an Ortoleva-King reaction) in the presence of an ammonia source (often ammonium acetate) to give the 4′-aryl-2,2′:6′,2″-tpy. (b) The simpler "one-pot" approach in which two equivalents of 2-acetylpyridine and the aromatic aldehyde are reacted directly or sequentially with a source of ammonia. This latter reaction may be in solvent free-conditions by grinding the reactants, use benign solvents such as PEG or conventional solvents. Replacement of the 2-acetylpyridine by 3-acetylpyridine or 4-acetylpyridine allows the synthesis of 4′-aryl-3,2′:6′,3″-tpy and 4′-aryl-4,2′:6′,4″-tpy ligands.

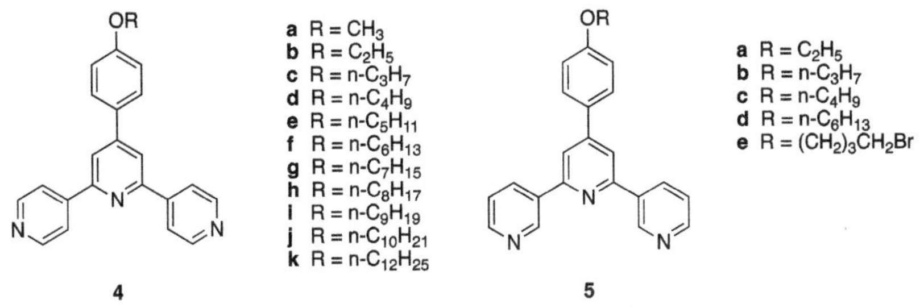

4
a R = CH$_3$
b R = C$_2$H$_5$
c R = n-C$_3$H$_7$
d R = n-C$_4$H$_9$
e R = n-C$_5$H$_{11}$
f R = n-C$_6$H$_{13}$
g R = n-C$_7$H$_{15}$
h R = n-C$_8$H$_{17}$
i R = n-C$_9$H$_{19}$
j R = n-C$_{10}$H$_{21}$
k R = n-C$_{12}$H$_{25}$

5
a R = C$_2$H$_5$
b R = n-C$_3$H$_7$
c R = n-C$_4$H$_9$
d R = n-C$_6$H$_{13}$
e R = (CH$_2$)$_3$CH$_2$Br

Figure 2. Structures of 4′-(4-alkyloxyphenyl)-4,2′:6′,4″-terpyridines and 4′-(4-alkyloxyphenyl)-3,2′:6′,3″-terpyridines. Compounds 5a–c are the subject of this publication.

2. Results and Discussion

2.1. Strategy and Ligand Design

We have shown that in the 4,2′:6′,4″-tpy compounds **4a–k** (R = n-C_nH_{2n+1}, n = 1–10, 12, Figure 2), the length of the alkyl chain on the alkyloxy substituent can influence the packing and the topology of coordination networks generated upon reaction with metal salts [22,25–30]. We are now beginning a systematic investigation of the effect of substituents on the assembly of coordination networks based on 3,2′:6′,3″-tpy ligands and commenced with the synthesis of derivatives **5a–c**. Compound **5d** has previously been reported [31]. The synthetic approach was a standard one-pot synthesis, involving the reaction of 3-acetylpyridine with the appropriate 4-alkyloxybenzaldehyde in basic solution, followed by the addition of aqueous ammonia.

2.2. Results

2.2.1. 4′-(4-Ethoxyphenyl)-3,2′:6′,3″-terpyridine and 4′-(4-butoxyphenyl)-3,2′:6′,3″-terpyridine

4′-(4-Ethoxyphenyl)-3,2′:6′,3″-terpyridine (**5a**) and 4′-(4-butoxyphenyl)-3,2′:6′,3″-terpyridine (**5c**) (Figure 3) were prepared following the one-pot method of Wang and Hanan [18]. 4-Ethoxybenzaldehyde or 4-butoxybenzaldehyde were reacted with 3-acetylpyridine in EtOH in the presence of KOH. After addition of aqueous NH$_3$, **5a** and **5c** precipitated over a period of about 16 h as white solids in 16.7 and 31.1% yields, respectively.

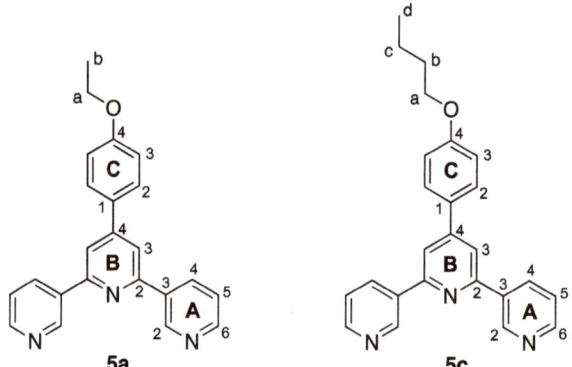

Figure 3. Structures of compounds **5a** and **5c** with atom labelling for NMR spectroscopic assignments.

The electrospray (ESI) mass spectra of compounds **5a** and **5c** showed base peaks at m/z 354.08 and 382.16, respectively, arising from the [M + H]$^+$ ions (Figures S1 and S2, see Supporting Information). The solution ^{1}H and ^{13}C NMR spectra were consistent with the structures shown in Figure 3 and were assigned using COSY, NOESY, HMQC and HMBC methods. Figure 4 displays a comparison of the ^{1}H NMR spectra, and confirms that the introduction of the different alkyloxy substituents has no significant influence on the spectroscopic signature of the 4′-phenyl-3,2′:6′,3″-tpy unit. ^{13}C{^{1}H} NMR spectra are compared in Figure S3. As expected, the solid-state IR spectra of **5a** and **5c** are very similar (Figures S4 and S5) and the solution absorption spectra (Figure 5) show intense absorptions in the UV region arising from spin-allowed $\pi^* \leftarrow \pi$ and $\pi^* \leftarrow n$ transitions.

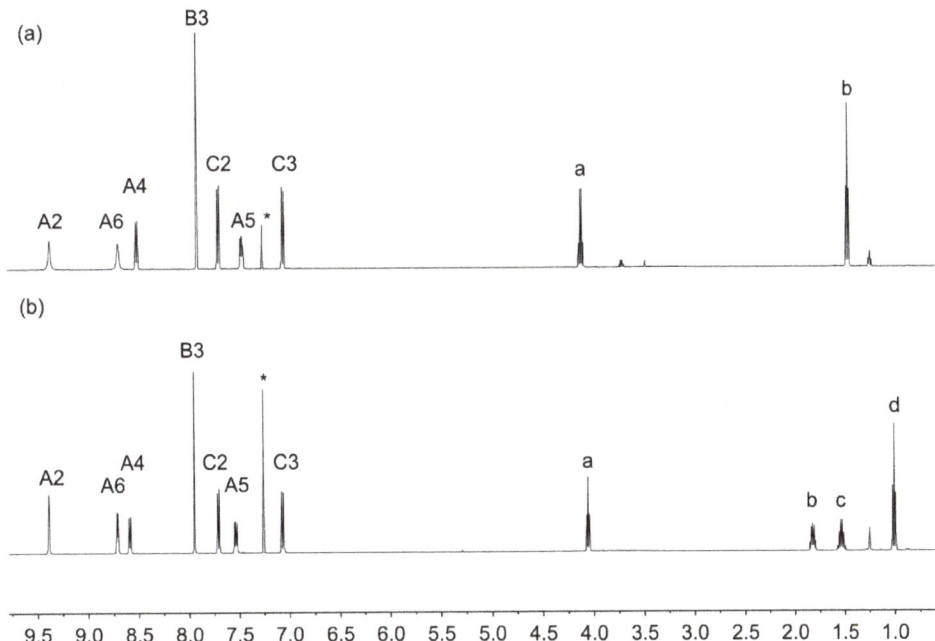

Figure 4. ^{1}H NMR spectra (500 MHz, CDCl$_3$, 298 K) of (**a**) **5a** and (**b**) **5c**. See Figure 3 for the atom labelling scheme. * = residual CHCl$_3$.

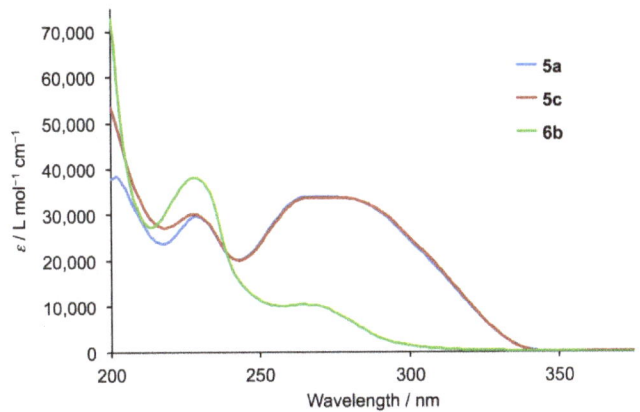

Figure 5. Solution absorption spectra of **5a**, **5c** and **6b** (MeCN, 3.3 × 10^{-5} mol dm^{-3}).

2.2.2. A Reaction that Works with Ethoxy and Butoxy Homologues, Fails for 4′-(4-propoxyphenyl)-3,2′:6′,3″-terpyridine

During attempts to prepare 4′-(4-propoxyphenyl)-3,2′:6′,3″-terpyridine (**5b**, Figure 2) by the reaction of 4-propoxybenzaldehyde with 3-acetylpyridine in the presence of KOH in ethanol, under identical conditions to those for the successful preparation of ligands **5a** and **5c**, followed by addition of aqueous NH$_3$, we noted that precipitation of a product began before ammonia was added. The white solid that was isolated exhibited IR (Figure S6) and absorption spectra (Figure 5) with different profiles from those of **5a** and **5c**. In particular, the IR spectrum of the product exhibited a sharp absorption at 3495 cm^{-1} attributable to an OH group. These observations were unexpected since we have previously reported the successful synthesis of the analogous series of 4,2':6',4''-tpy derivatives **4a–4k** (Figure 2) by the Hanan one-pot strategy [25]. The ESI mass spectrum of the product showed a base peak at *m/z* 656.31 (Figure S7) suggesting the formation of the cyclic product **6** shown in Figure 6.

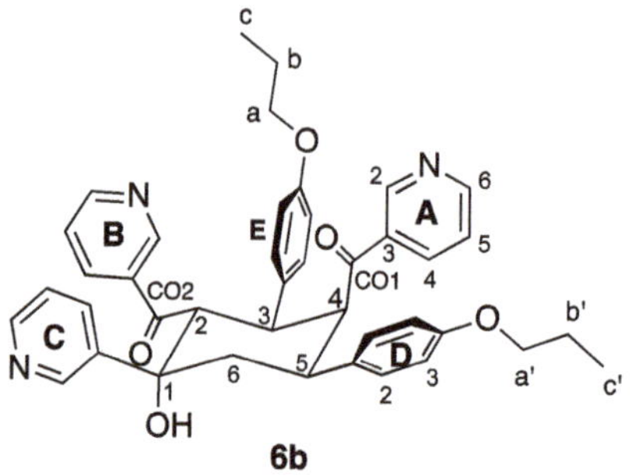

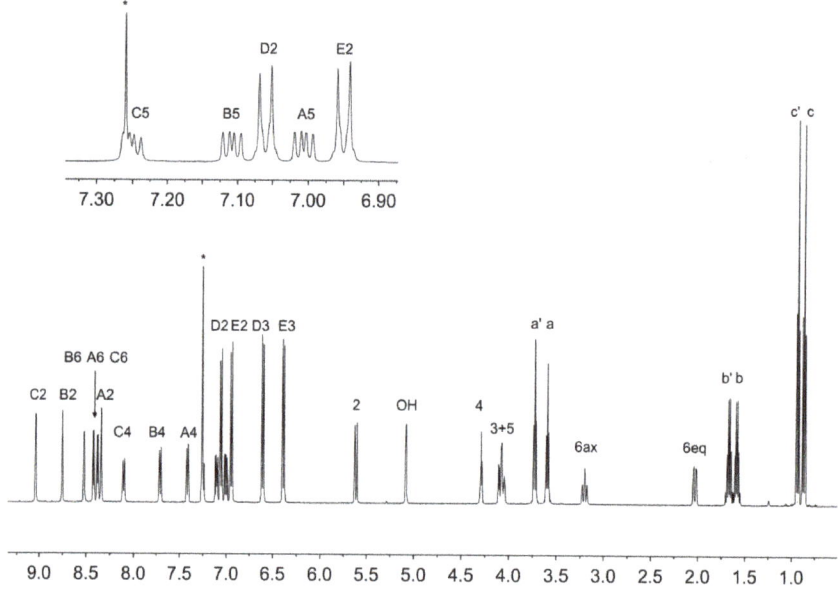

Figure 6. Structure of compound **6b** with atom labelling for NMR spectroscopic assignments.

The solution ^{1}H NMR spectrum of **6b** is shown in Figure 7 and is consistent with the presence of three pyridine environments (rings A, B and C) and two 4-propoxyphenyl environments (rings D and E). The spectrum was assigned using COSY, NOESY (Figure S8), HMQC (Figure S9) and HMBC (Figure S10) methods with critical NOESY crosspeaks being between H^{OH}/H^{C2}, $H^{D3}/H^{a'}$, H^{E3}/H^a, H^{D2}/H^5, H^{D2}/H^{6ax}, H^{E2}/H^3 and H^{E2}/H^2.

Figure 7. ^{1}H NMR spectrum (500 MHz, CDCl$_3$, 298 K) of compound **6b**. See Figure 6 for atom labelling. * = residual CHCl$_3$.

The spectroscopic characterization confirms the formation of **6b** although it offers no indication as to why this product precipitates from solution only in the case of the propoxy substituent. We noted that the reaction only involves the aldehyde and the 3-acetylpyridine and were prompted to perform the reactions in the absence of ammonia. We repeated the reaction of 4-propoxybenzaldehyde with 3-acetylpyridine at a 1:2 molar ratio in ethanol with KOH but without the addition of NH$_3$. This led to the formation of **6b** (confirmed by NMR spectroscopy, Figure 8b) in 42.0% yield. We then performed the analogous reactions of 4-ethoxybenzaldehyde or 4-butoxybenzaldehyde with 3-acetylpyridine in the presence of KOH in ethanol, but without NH$_3$. In both cases, white precipitates formed within 5 min. The ESI mass spectra of the products were consistent with their being analogues of **6b**. Base peaks at m/z = 628.29 and 684.35 were assigned to [**6a** + H]$^+$ and [**6c** + H]$^+$, respectively (Figures S11 and S12). The ^{1}H NMR spectra of **6a** and **6c** are shown in Figure 8a,c, and the similarity to that of **6b** (Figure 8b) is immediately apparent. The ^{1}H and ^{13}C NMR spectra of **6a** and **6c** (see Experimental Section) were assigned using routine 2D methods.

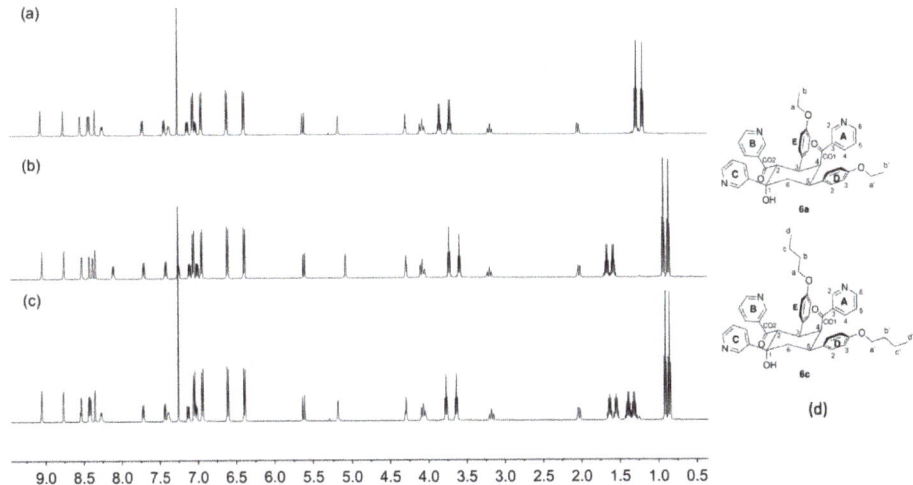

Figure 8. A comparison of the ^{1}H NMR (500 MHz, CDCl$_3$, 298 K) spectra of the cyclic products (**a**) **6a**, (**b**) **6b** and (**c**) **6c**. The singlet at δ 7.26 ppm in each spectrum is residual CHCl$_3$. (**d**) The structures of **6a** and **6c** with atom labelling for NMR spectroscopic assignments (see Sections 4.5 and 4.6).

3. Review of Literature

3.1. Introduction

Although the Claisen-Schmidt condensation is presented as a simple reaction with a hydroxyketone intermediate and a single enone product in many chemistry text books, the reality is often far more complex. Although this paper is concerned with the use of acetylpyridines in the Claisen-Schmidt reaction, it is instructive to review the broader literature regarding the condensation of aromatic aldehydes with aromatic ketones.

The enone products of the Claisen-Schmidt reaction are electrophilic and can react with nucleophilic enols or enolates at either the carbonyl carbon or by conjugate addition. The formation of the enone 1:1 products (1:1 ratio of ketone to aldehyde) involves the nucleophilic attack of an enol or enolate derived from the aromatic ketone upon the aromatic aldehyde. A number of publications have provided overviews of the reaction space and the types of products that are to be expected. If the rates of the subsequent reactions are significantly faster than the initial ones leading to the enone, domino [32–34] or tandem [35] reactions are expected, leading to the selective formation of one (or more) of a range of potential reaction products. In this section, we present a comprehensive overview of the products other than terpyridines that have been obtained from Kröhnke and related syntheses. We have already considered products of this type implicitly—the pentane-1,5-diones (**7**) arise from the addition of a second equivalent enol or enolate to the enone, giving products of 2:1 (ketone to aldehyde) stoichiometry (Figure 9).

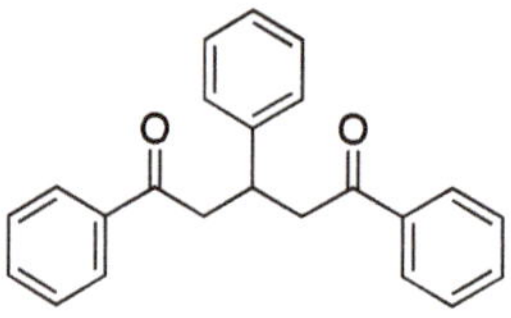

Figure 9. The structure of the 2:1 (ketone:aldehyde) pentane-1,5-dione that can arise from the reaction of acetophenone with benzaldehyde and the proposed structure of Kostanecki's triketone the 3:2 adduct that formally arises from the nucleophilic addition of the enol of enolate of the pentane-1,5-dione on the 1:1 enone.

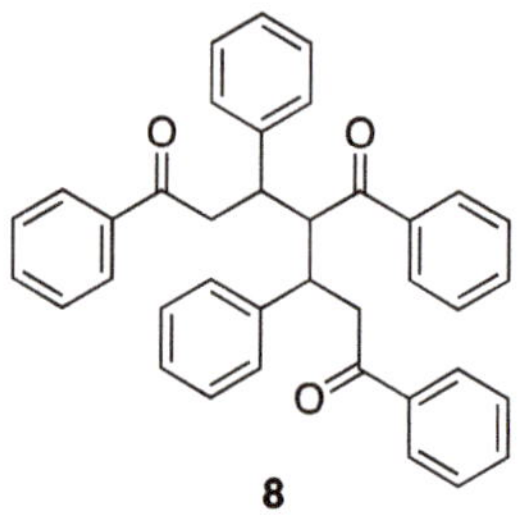

3.2. The 3:2 Products

The most common of the products that have been isolated from attempted terpyridine syntheses have a reactant stoichiometry of 3:2 ketone-aldehyde. In one respect, this is an old story dating back to 1892 when the compound claimed to be Kostanecki's triketone (**8**) was isolated as a product from the reaction of benzaldehyde and acetophenone [36]. Over the years, a number of materials purporting to be **8**, but possessing different physical properties have been reported [37,38]. A definitive report of the preparation and structural characterization of **8** describing the history of this compound together with an analysis of the various products isolated from the reaction of acetophenone with benzaldehyde appeared recently [39]. A large number of 3:2 condensation products have now been fully characterized by structural and spectroscopic means and shown to be cyclohexanols **9** arising from formally from the internal condensation of intermediate triketones. Naming the compounds as cyclohexanols for consistency (the IUPAC PIN is based upon 4-hydroxycyclohexane-1,3-diyl)bis(methanone)), the relative stereochemistry is 1*R**, 2*S**, 3*R**, 5*R**: the substituent in the 4-position may be either axial or equatorial giving both 4*R** and 4*S** (4-*R** **9** or 4-*S** **9**, Figure 10). The cyclic product is obtained in particularly high yield under phase transfer conditions from the reaction of benzaldehyde and acetophenone [40]. The reaction of enones with NaO^{t}Bu and N-heterocyclic carbenes, yielded **9** and aldehyde [41]. A significant range of compounds with various Ar and Ar$_1$ groups have been characterized, from one pot reactions of ArCHO and Ar$_1$COCH$_3$ or from sequential reactions of isolated enones. In addition to the parent compound (Ar = Ar$_1$ = C$_6$H$_5$) [39–46], a large number of derivatives with carbocyclic Ar and Ar$_1$ substituents have been obtained from reactions of substituted benzaldehydes and acetophenones [34,41,43,46–53]. Of most relevance to the chemistry reported in this manuscript, are compounds with heterocyclic substituents and, to date, examples with thiophen-2-yl [34,41,54–56], furan-2-yl [54,57], benzofuran-2-yl [33], benzofuran-3-yl [33], benzothiophen-2-yl [33], benzothiophen-3-yl [33], pyridin-2-yl [33,35,45,58–66] pyridin-3-yl [33,45,56] and pyridin-4-yl [16,20,33,45,56,67] groups have been reported. Cave and Raston have commented that reactions of 4-acetylpyridine with 4-alkoxybenzaldehydes in solution only yield the 3:2 products rather than the desired 4,2':6',4''-terpyridines [16,20]. We note, however, that in our hands and under the Hanan one-pot procedure [18], the 4'-(4-alkoxy)-4,2':6',4''-terpyridines shown in Figure 2 can be readily prepared [25].

To summarize, the formation of 3:2 condensation products is well-established with the stereochemistry at four of the five stereogenic centres defined. Examples of diastereoisomers with the 4-substituent in the axial or equatorial positions have been described and the less stable axial *R* diastereoisomer may be converted to the more stable equatorial *S* form with base [58].

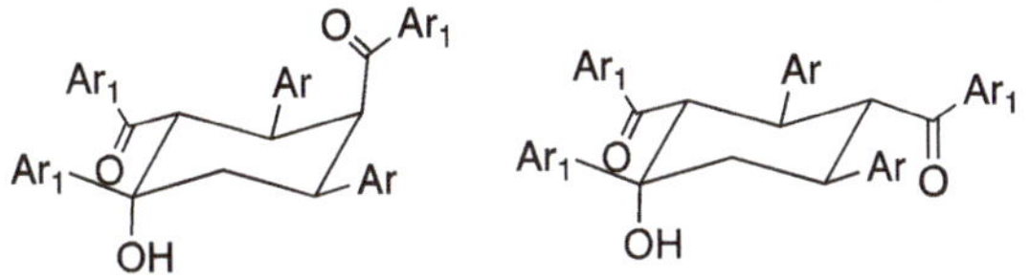

Figure 10. The structures of the 3:2 cyclohexanol derivatives arising from the cyclization of the triketones.

3.3. The 3:1 Products

A second type of cyclohexane derivative **10** (Figure 11) with a 3:1 (ketone-aldehyde) constitution is also known. We first described the formation of cyclohexane-1,3-diols from the condensation of 2-acetylpyridine with benzaldehyde derivatives in 1995 [58]. Although less commonly observed than the 3:2 condensation products, compounds of this type, arising from the aldol condensation of an intermediate pentane-1,5-dione with a third equivalent of ketone, have been isolated with a variety of substituents [32,60,68–73]. The relative stereochemistry in **10**, originally proposed on the basis of NMR studies [58], has been crystallographically confirmed [32,69,70,72]. Sequential reaction of the benzaldehyde derivative with two different ketones, Ar_1COCH_3 and Ar_2COCH_3 allows the synthesis of compounds **11** via the addition of Ar_2COCH_3 to the intermediate enone $Ar_1COCH=CHAr$.

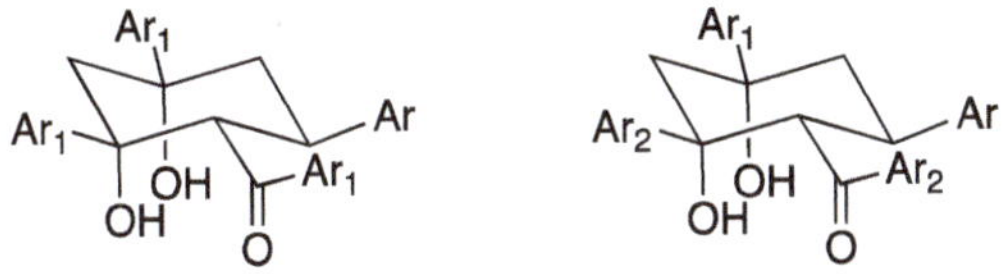

Figure 11. The structures of the 3:1 cyclohexane-1,3-diol derivatives arising from the condensation of the pentane-1,5-dione with Ar_1COCH_3.

3.4. An Unexpected Terpyridine Isomer

A third type of unexpected product obtained from attempted 4′-aryl-2,2′:6′,2″-terpyridine syntheses is the isomeric 6′-aryl-2,2′:4′,2″-terpyridine. This was first observed when 6′-(4-methylphenyl)-2,2′:4′,2″-terpyridine (**12**) was isolated as a side-product from the condensation of 2-acetylpyridine with 4-methylbenzaldehyde and ammonia [74–80]. This product presumably arises by the 1,2-attack of the enol or enolate of 2-acetylpyridine at the carbonyl (rather than the more usual 1,4-conjugate addition) of the intermediate enone, 3-(4-methylphenyl)-1-(pyridin-2-yl)prop-2-en-1-one (Scheme 3). Similarly, the preparation of 4′-phenyl-4,2′:6′,4″-terpyridine from 4-acetylpyridine and benzaldehyde gave appreciable amounts of 6′-(4-methylphenyl)-4,2′:4′,4″-terpyridine (**12**) [81].

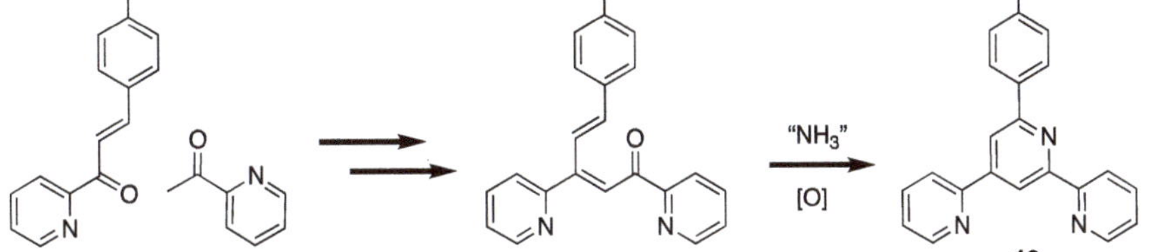

Scheme 3. The origin of the isomeric 2,2′:4′,2″-terpyridine in 2,2′:6′,2″-terpyridine synthesis.

4. Materials and Methods

^{1}H and ^{13}C and NMR spectra were recorded on a Bruker Avance III-500 spectrometer (Bruker BioSpin AG, Fällanden, Switzerland) at 298 K. The ^{1}H and ^{13}C NMR chemical shifts were referenced with respect to residual solvent peaks (δ TMS = 0) and all quoted coupling constants J are J_{HH} between protons. A Shimadzu LCMS-2020 instrument (Shimadzu Schweiz GmbH, Roemerstr., Switzerland) was used to record electrospray ionization (ESI) mass spectra; samples were introduced as 200–800 μM solutions in MeCN with the addition of formic acid. PerkinElmer UATR Two (Perkin Elmer, Bahnstrasse 8, 8603 Schwerzenbach, Switzerland) and Cary-5000 (Agilent Technologies Inc., Santa Clara, CA, United States) spectrometers were used to record FT-infrared (IR) and absorption spectra, respectively. Melting points were measured using a Bibby Melting Point Apparatus SMP30.

3-Acetylpyridine was purchased from Acros Organics (Chemie Brunschwig AG, Basel, Switzerland), 4-ethoxybenzaldehyde and 4-butoxybenzaldehyde from Sigma Aldrich (Riedstr. 2, 89555 Steinheim, Germany), and 4-propoxybenzaldehyde from Fluorochem and were used as received.

4.1. Compound 5a

4-Ethoxybenzaldehyde (1.50 g, 1.39 mL, 10.0 mmol) was dissolved in EtOH (50 mL), then 3-acetylpyridine (2.42 g, 2.20 mL, 20.0 mmol) and crushed KOH (1.12 g, 20.0 mmol) were added to the solution. Aqueous NH$_3$ (32%, 38.5 mL) was slowly added to the reaction mixture. This was stirred at room temperature overnight. The solid that formed was collected by filtration, washed with water (3 × 10 mL) followed by EtOH (3 × 10 mL), recrystallized from EtOH and dried *in vacuo*. Compound **5a** was isolated as a white powder (0.591 g, 1.67 mmol, 16.7%). M.p. = 120 °C. ^{1}H NMR (500 MHz, CDCl$_3$): δ/ppm = 9.37 (m, 2H, H^{A2}), 8.70 (m, 2H, H^{A6}), 8.51 (8.51 (dt, J = 8.0, 1.9 Hz, 2H, H^{A4}), 7.91 (s, 2H, H^{B3}), 7.70 (m, 2H, H^{C2}), 7.47 (m, 2H, H^{A5}), 7.05 (m, 2H, H^{C3}), 4.12 (q, J = 7.0 Hz, 2H, H^a), 1.47 (t, J = 7.0 Hz, 3H, H^b). ^{13}C{^{1}H} NMR (500 MHz, CDCl$_3$): δ/ppm = 160.1 (C^{C4}), 155.0 (C^{A3}), 150.3 (C^{B4}), 149.7 (C^{A6}), 148.0 (C^{A2}), 134.8 (C^{A4}), 134.6 (C^{B2}), 130.0 (C^{C1}), 128.2 (C^{C2}), 123.6 (C^{A5}), 117.1 (C^{B3}), 115.1 (C^{C3}), 63.6 (C^a), 14.7 (C^b). UV-VIS (CH$_3$CN, 3.3 × 10^{-5} mol dm^{-3}) λ/nm 229 (ε/dm^3 mol^{-1} cm^{-1} 29,700), 272 (33,800). ESI-MS m/z 354.08 [M + H]$^+$ (calc. 354.16). Found C 77.38, H 5.30, N 11.91; required for C$_{23}$H$_{19}$N$_3$O: C 78.16, H 5.42, N 11.89.

4.2. Compound 5c

4-Butoxybenzaldehyde (1.78 g, 1.73 mL, 10.0 mmol) was dissolved in EtOH (50 mL), then 3-acetylpyridine (2.42 g, 2.20 mL, 20.0 mmol) and crushed KOH (1.12 g, 20.0 mmol) were added to the solution. Aqueous NH$_3$ (32%, 38.5 mL) was slowly added to the reaction mixture. This was stirred at room temperature overnight. The solid that formed was collected by filtration, washed with water (3 × 10 mL) and EtOH (3 × 10 mL), then recrystallized from EtOH and dried *in vacuo*. Compound **5c** was isolated as a white powder (1.19 g, 3.11 mmol, 31.1%). M.p. = 113 °C. ^{1}H NMR (500 MHz, CDCl$_3$): δ/ppm = 9.39 ((d, J = 2.2 Hz, 2H, H^{A2}), 8.72 ((dd, J = 4.9, 1.6 Hz, 2H, H^{A6}), 8.59 (dt, J = 8.1, 2.0 Hz, 2H, H^{A4}), 7.95 (s, 2H, H^{B3}), 7.71 (m, 2H, H^{C2}), 7.53 (m, 2H, H^{A5}), 7.07 (m, 2H, H^{C3}), 4.06 (t, J = 6.5 Hz, 2H, H^a), 1.83 (m, 2H, H^b), 1.54 (m, 2H, H^c), 1.01 (t, J = 7.4 Hz, 3H, H^d). ^{13}C{^{1}H} NMR (500 MHz, CDCl$_3$): δ/ppm = 160.8 (C^{C4}), 154.9 (C^{A3}), 150.9 (C^{B4}), 149.0 (C^{A6}), 147.4 (C^{A2}), 135.8 (C^{A4}), 135.4 (C^{B2}), 129.9 (C^{C1}), 128.5 (C^{C2}), 124.2 (C^{A5}), 117.6 (C^{B3}), 115.5 (C^{C3}), 68.1 (C^a), 31.4 (C^b), 19.4 (C^c), 14.0 (C^d). UV-VIS (CH$_3$CN, 3.3 × 10^{-5} mol dm^{-3}) λ/nm 228 (ε/dm^3 mol^{-1} cm^{-1} 30,100), 274 (33,700). ESI-MS m/z 382.16 [M + H]$^+$ (calc. 382.19). Found C 78.37, H 5.93, N 11.05; required for C$_{25}$H$_{23}$N$_3$O: C 78.71, H 6.08, N 11.02.

4.3. Compound 6b: Method 1

4-Propoxybenzaldehyde (1.64 g, 1.58 mL, 10.0 mmol) was dissolved in EtOH (50 mL), then 3-acetylpyridine (2.42 g, 2.20 mL, 20.0 mmol) and crushed KOH (1.12 g, 20.0 mmol) were added to the solution. At this stage, a white precipitate began to form. Aqueous NH$_3$ (32%, 38.5 mL) was slowly added to the reaction mixture which was then stirred at room temperature overnight. The solid that formed was collected by filtration, washed with water (3 × 10 mL) then with EtOH (3 × 10 mL), and was recrystallized from EtOH and dried *in vacuo*. Compound **6b** was isolated as a white powder (0.495 g, 0.755 mmol, 15.1%). M.p. = 208 °C. ^{1}H NMR (500 MHz, CDCl$_3$): δ/ppm = 9.04 (d, J = 1.9 Hz, 1H, H^{C2}), 8.75 (d, J = 1.8 Hz, 1H, H^{B2}), 8.52 (dd, J = 4.8, 1.6 Hz, 1H, H^{B6}), 8.42 (dd, J = 4.8, 1.6 Hz, 1H, H^{A6}), 8.38 (dd, J = 4.8, 1.5 Hz, 1H, H^{C6}), 8.34 (d, J = 1.7 Hz, 1H, H^{A2}), 8.11 (dt, J = 8.1, 1.8 Hz, 1H, H^{C4}), 7.71 (dt, J = 8.1, 2.0 Hz, 1H, H^{B4}), 7.42 (dt, J = 8.1, 2.0 Hz, 1H, H^{A4}), 7.25 (m, 1H, H^{C5}), 7.11 (m, 1H, H^{B5}), 7.06 (m, 2H, H^{D2}), 7.01 (m, 1H, H^{A5}), 6.95 (m, 2H, H^{E2}), 6.61 (m, 2H, H^{D3}), 6.39 (m, 2H, H^{E3}), 5.62 (d, J = 11.9 Hz, 1H, H^2), 5.08 (d, J = 2.1 Hz, 1H, H^{OH}), 4.29 (dd, J = 4.7, 4.7 Hz, 1H, H^4), 4.08 (m, 2H, H^{3+5}), 3.73 (m, 2H, H$^{a'}$), 3.60 (m, 2H, H^a), 3.20 (ddd, J = 13.0, 13.0, 2.5 Hz, 1H, H^{6ax}), 2.03 (dd, J = 13.7, 3.4 Hz, 1H, H^{6eq}), 1.67 (m, 2H, H$^{b'}$), 1.59 (m, 2H, H^b), 0.94 (t, J = 7.4 Hz, 3H, H$^{c'}$), 0.87 (t, J = 7.4 Hz, 3H, H^c). ^{13}C{^{1}H} NMR (500 MHz, CDCl$_3$): δ/ppm = 206.8 (C^{CO2}), 206.1 (C^{CO1}), 158.3 (C^{E4}), 158.1 (C^{D4}), 153.7 (C^{B6}), 152.5 (C^{A6}), 149.6 (C^{B2}), 149.1 (C^{A2}), 147.9 (C^{C6}), 146.4 (C^{C2}), 142.6 (C^{C3}), 135.2 (C^{B4}), 135.0 (C^{A3}), 134.8 (C^{A4}), 133.8 (C^{C4}), 133.0 (C^{B3}), 132.7 (C^{D1}), 130.2 (C^{E1}), 129.7 (C^{E2}), 128.6 (C^{D2}), 123.5 (C^{C5}), 123.2 (C^{B5}), 122.9 (C^{A5}), 114.7 (C^{D3}), 114.6 (C^{E3}), 75.0 (C^1), 69.6 (C$^{a'}$), 69.4 (C^a), 53.5 (C^4), 51.0 (C^2), 46.8 (C^3), 41.2 (C^5), 38.8 (C^6), 22.6 (C$^{b'}$), 22.5 (C^b), 10.6 (C$^{c'}$), 10.5 (C^c). UV-VIS (CH$_3$CN, 3.3 × 10^{-5} mol dm^{-3}) λ/nm 228 (ε/dm^3 mol^{-1} cm^{-1} 38,100), 264 (10,400). ESI-MS *m/z* 656.31 [M + H]$^+$ (calc. 656.31). IR spectrum: see Figure S6. Found C 74.67, H 6.77, N 6.73; required for C$_{41}$H$_{41}$N$_3$O$_5$ C 75.09, H 6.30, N 6.41.

*4.4. Compound **6b**: Method 2*

The reaction was carried out as in method 1 without the addition of NH$_3$. Reagents and solvent: 4-propoxybenzaldehyde (0.82 g, 0.79 mL, 5.0 mmol), EtOH (25 mL), 3-acetylpyridine (1.21 g, 1.10 mL, 10.0 mmol) and crushed KOH (0.56 g, 10.0 mmol). **6b** was isolated as a white powder (0.690 g, 1.05 mmol, 42.0%). Characterization data matched those for the product reported in Section 4.3.

*4.5. Compound **6a***

4-Ethoxybenzaldehyde (0.75 g, 0.70 mL, 5.0 mmol) was dissolved in EtOH (25 mL), then 3-acetylpyridine (1.21 g, 1.10 mL, 10.0 mmol) and crushed KOH (0.56 g, 10.0 mmol) were added to the solution. After five minutes a white precipitate began to form. The reaction mixture was stirred at room temperature overnight. The solid that formed was collected by filtration, washed with water (3 × 10 mL) then EtOH (3 × 10 mL), recrystallized from EtOH and dried in vacuo. **6a** was isolated as a white powder (0.268 g, 0.427 mmol, 17.1%). M.p. = 210 °C. ^{1}H NMR (500 MHz, CDCl$_3$): δ/ppm = 9.06 (d, J = 1.9 Hz, 1H, H^{C2}), 8.76 (d, J = 1.8 Hz, 1H, H^{B2}), 8.54 (dd, J = 4.8, 1.6 Hz, 1H, H^{B6}), 8.44 (dd, J = 4.8, 1.6 Hz, 1H, H^{A6}), 8.41 (dd, J = 4.9, 1.3 Hz, 1H, H^{C6}), 8.34 (d, J = 1.8 Hz, 1H, H^{A2}), 8.25 (m, 1H, H^{C4}), 7.72 (dt, J = 8.0, 1.9 Hz, 1H, H^{B4}), 7.43 (dt, J = 8.0, 1.9 Hz, 1H, H^{A4}), 7.37 (m, 1H, H^{C5}), 7.13 (m, 1H, H^{B5}), 7.05 (m, 2H, H^{D2}), 7.02 (m, 1H, H^{A5}), 6.95 (m, 2H, H^{E2}), 6.61 (m, 2H, H^{D3}), 6.39 (m, 2H, H^{E3}), 5.62 (d, J = 11.9 Hz, 1H, H^2), 5.18 (d, J = 2.1 Hz, 1H, H^{OH}), 4.29 (dd, J = 4.6, 4.6 Hz, 1H, H^4), 4.08 (m, 2H, H^{3+5}), 3.85 (m, 2H, H$^{a'}$), 3.71 (m, 2H, H^a), 3.19 (ddd, J = 13.6, 13.6, 2.1 Hz, 1H, H^{6ax}), 2.04 (dd, J = 13.7, 3.4 Hz, 1H, H^{6eq}), 1.29 (t, J = 7.0 Hz, 3H, H$^{b'}$), 1.20 (t, J = 7.0 Hz, 3H, H^b). ^{13}C{^{1}H} NMR (500 MHz, CDCl$_3$): δ/ppm = 206.8 (C^{CO2}), 206.1 (C^{CO1}), 158.2 (C^{E4}), 158.0 (C^{D4}), 153.9 (C^{B6}), 152.6 (C^{A6}), 149.6 (C^{B2}), 149.1 (C^{A2}), 146.2 (C^{C6}), 144.8 (C^{C2}), 143.2 (C^{C3}), 135.3 (C^{C4}), 134.8 (C^{A3}), 134.9 (C^{B4}), 134.8 (C^{A4}), 132.8 (C^{B3}), 132.6 (C^{D1}), 130.1 (C^{E1}), 129.7 (C^{E2}), 128.6 (C^{D2}), 123.5 (C^{C5}), 123.4 (C^{B5}), 123.0 (C^{A5}), 114.7 (C^{D3}), 114.6 (C^{E3}), 75.0 (C^1), 63.5 (C$^{a'}$), 63.2 (C^a), 53.4 (C^4), 50.8 (C^2), 46.9 (C^3), 41.2 (C^5), 38.9 (C^6), 14.8 (C$^{b'}$), 14.7 (C^b). ESI-MS m/z 628.29 [M + H]$^+$ (calc. 628.28). IR spectrum: see Figure S13. Found C 74.26, H 5.88, N 6.53; required for C$_{39}$H$_{37}$N$_3$O$_5$ C 74.62, H 5.94, N 6.69.

*4.6. Compound **6c***

4-Butoxybenzaldehyde (0.89 g, 0.87 mL, 5.0 mmol) was dissolved in EtOH (25 mL), then 3-acetylpyridine (1.21 g, 1.10 mL, 10.0 mmol) and crushed KOH (0.56 g, 10.0 mmol) were added to the solution. After five minutes a white precipitate began to form. The reaction mixture was stirred at room temperature overnight. The precipitate was collected by filtration, washed with water (3 × 10 mL) and EtOH (3 × 10 mL), then recrystallized from EtOH and dried in vacuo. **6c** was isolated as a white powder (0.399 g, 0.585 mmol, 11.7%). M.p. = 185 °C. ^{1}H NMR (500 MHz, CDCl$_3$): δ/ppm = 9.06 (d, J = 1.8 Hz, 1H, H^{C2}), 8.77 (d, J = 1.8 Hz, 1H, H^{B2}), 8.54 (dd, J = 4.8, 1.6 Hz, 1H, H^{B6}), 8.43 (dd, J = 4.8, 1.7 Hz, 1H, H^{A6}), 8.42 (dd, J = 5.0, 1.4 Hz, 1H, H^{C6}), 8.35 (d, J = 1.7 Hz, 1H, H^{A2}), 8.27 (m, 1H, H^{C4}), 7.72 (dt, J = 8.0, 1.9 Hz, 1H, H^{B4}), 7.44 (dt, J = 8.0, 2.0 Hz, 1H, H^{A4}), 7.39 (m, 1H, H^{C5}), 7.13 (m, 1H, H^{B5}), 7.05 (m, 2H, H^{D2}), 7.02 (m, 1H, H^{A5}), 6.95 (m, 2H, H^{E2}), 6.61 (m, 2H, H^{D3}), 6.39 (m, 2H, H^{E3}), 5.63 (d, J = 11.9 Hz, 1H, H^2), 5.18 (d, J = 2.1 Hz, 1H, H^{OH}), 4.29 (dd, J = 4.6, 4.6 Hz, 1H, H^4), 4.08 (m, 2H, H^{3+5}), 3.77 (m, 2H, H$^{a'}$), 3.64 (m, 2H, H^a), 3.18 (ddd, J = 13.7, 13.7, 2.0 Hz, 1H, H^{6ax}), 2.04 (dd, J = 13.7, 3.4 Hz, 1H, H^{6eq}), 1.63 (m, 2H, H$^{b'}$), 1.55 (m, 2H, H^b), 1.39 (m, 2H, H$^{c'}$), 1.32 (m, 2H, H^c), 0.91 (t, J = 7.4 Hz, 3H, H$^{d'}$), 0.86 (t, J = 7.4 Hz, 3H, H^d). ^{13}C{^{1}H} NMR (500 MHz, CDCl$_3$): δ/ppm = 207.4 (C^{CO2}), 206.8 (C^{CO1}), 158.3 (C^{D4}), 158.1 (C^{E4}), 153.9 (C^{B6}), 152.4 (C^{A6}), 149.4 (C^{B2}), 149.0 (C^{A2}), 145.9 (C^{C6}), 144.5 (C^{C2}), 143.8 (C^{C3}), 135.6 (C^{C4}), 134.9 (C^{A3}), 135.2 (C^{B4}), 134.8 (C^{A4}), 132.9 (C^{B3}), 132.3 (C^{D1}), 130.0 (C^{E1}), 129.7 (C^{E2}), 128.5 (C^{D2}), 124.1 (C^{C5}), 123.3 (C^{B5}), 122.7 (C^{A5}), 114.8 (C^{D3}), 114.6 (C^{E3}), 75.4 (C^1), 67.7 (C$^{a'}$), 67.5 (C^a), 53.3 (C^4), 50.6 (C^2), 46.7 (C^3), 41.2 (C^5), 38.8 (C^6), 31.3 (C$^{b'}$), 31.2 (C^b), 19.3 (C$^{c'}$), 19.2 (C^c), 13.8 (C$^{d'}$), 13.7 (C^d). ESI-MS m/z 684.35 [M + H]$^+$ (calc. 684.34). IR spectrum: see Figure S14. Found C 75.31, H 6.67, N 6.11; required for C$_{43}$H$_{45}$N$_3$O$_5$ C 75.52, H 6.63, N 6.14.

5. Conclusions

The preparation of terpyridines from the reactions of acetylpyridines with aromatic aldehydes and ammonia is an established synthetic method. Nevertheless, the reaction can give a variety of products. This paper provides another example of an "unexpected" product and a systematic survey of the products of such reactions. Although the one-pot synthesis of terpyridines is presented in the literature as an infallible synthetic method, there is ample precedent for the formation of a variety of alternative products. In particular, the assumption that the material precipitating from the reaction mixture is the desired terpyridine is not always correct. As these reactions are commonly used in the coordination chemistry community, this paper serves as a useful caveat. We are currently further investigating these reactions with varying length alkyloxy chains and will report on the constitution of the full reaction space in the future.

Supplementary Materials: The following are available online. Figures S1 and S2: Mass spectra of **5a** and **5c**; Figure S3: ^{13}C{^{1}H} NMR spectra of **5a** and **5c**; Figures S4–S6: IR spectra of **5a**, **5c** and **6b**; Figure S7: Mass spectrum of **6b**; Figures S8–S10: NOESY, HMQC and HMBC spectra of **6b**; Figures S11 and S12: Mass spectra of **6a** and **6c**; Figures S3 and S4: IR spectra of **6a** and **6c**.

Author Contributions: Project conceptualization, administration, supervision and funding acquisition, E.C.C. and C.E.H.; investigation, D.R.; writing, E.C.C., C.E.H., D.R.

Funding: This research was partially funded by the Swiss National Science Foundation, grant number 200020_182000.

Acknowledgments: We gratefully acknowledge the support of the University of Basel. We thank Dr Alessandro Prescimone for time invested in attempts to obtain X-ray diffraction data on compound **6b**.

Conflicts of Interest: The authors declare no conflict of interest.

References

1. Favre, H.A.; Powell, W.H. *Nomenclature of Organic Chemistry. IUPAC Recommendations and Preferred Names 2013*; Royal Society of Chemistry: Cambridge, UK, 2013; Rule P-16.4.
2. Schubert, U.S.; Hofmeier, H.; Newkome, G.R. *Modern Terpyridine Chemistry*; Wiley-VCH Verlag & Co.: Weinheim, Germany, 2006.
3. Schubert, U.S.; Winter, A.; Newkome, G.R. *Terpyridine-Based Materials: For Catalytic, Optoelectronic and Life Science Applications*; Wiley-VCH Verlag & Co.: Weinheim, Germany, 2011.

4. Constable, E.C. 2,2′:6′,2″-Terpyridines: From chemical obscurity to common supramolecular motifs. *Chem. Soc. Rev.* **2007**, *36*, 246–253. [CrossRef]
5. Wei, C.; He, Y.; Shi, X.; Song, Z. Terpyridine-metal complexes: Applications in catalysis and supramolecular chemistry. *Coord. Chem. Rev.* **2019**, *385*, 1–19. [CrossRef] [PubMed]
6. Constable, E.C. Higher Oligopyridines as a Structural Motif in Metallosupramolecular Chemistry. *Progr. Inorg. Chem.* **1994**, *42*, 67–138. [CrossRef]
7. Constable, E.C. The Coordination Chemistry of 2,2′:6′,2″-Terpyridine and Higher Oligopyridines. *Adv. Inorg. Chem.* **1986**, *30*, 69–121. [CrossRef]
8. Favre, H.A.; Powell, W.H. *Nomenclature of Organic Chemistry. IUPAC Recommendations and Preferred Names 2013*; Royal Society of Chemistry: Cambridge, UK, 2013; Rule P-31.2.3.3.5.1.
9. Kröhnke, F. The Specific Synthesis of Pyridines and Oligopyridines. *Synthesis* **1976**, 1–24. [CrossRef]
10. Kostanecki, St. von; Tambor, J. Über die sechs isomeren Monooxybenzalacetophenone (Monooxychalkone). *Chem. Ber.* **1899**, *32*, 1921–1926. [CrossRef]
11. Wang, Z. *Comprehensive Organic Name Reactions and Reagents*; Wiley: New York, NY, USA, 2010; pp. 660–665.
12. Heathcock, C.H. *The Aldol Reaction: Group I and Group II Enolates in Comprehensive Organic Synthesis II*, 2nd ed.; Knochel, P., Molander, G.A., Eds.; Elsevier: Amsterdam, The Netherlands, 2014; Volume 2, pp. 340–395.
13. Ortoleva, G. Azione del jodio sull'acido malonico in soluzione piridica. *Gazz. Chim. Ital.* **1900**, *30*, 509–514.
14. King, L.C. The reaction of iodine with some ketones in the presence of pyridine. *J. Am. Chem. Soc.* **1944**, *66*, 894–895. [CrossRef]
15. Kröhnke, F. Syntheses Using Pyridinium Salts. *Angew. Chem. Int. Ed.* **1963**, *2*, 225. [CrossRef]
16. Cave, G.W.V.; Raston, C.L. Toward benign syntheses of pyridines involving sequential solvent free aldol and Michael addition reactions. *Chem. Commun.* **2000**, 2199–2200. [CrossRef]
17. Smith, N.M.; Raston, C.L.; Smith, C.B.; Sobolev, A.N. PEG mediated synthesis of amino-functionalised 2,4,6-triarylpyridines. *Green Chem.* **2007**, *9*, 1185–1190. [CrossRef]
18. Wang, J.; Hanan, G.S. A facile route to sterically hindered and non-hindered 4′-aryl-2,2′:6′,2″-terpyridines. *Synlett* **2005**, 1251–1254. [CrossRef]
19. Cooke, M.W.; Wang, J.; Theobald, I.; Hanan, G.S. Convenient One-Pot Procedures for the Synthesis of 2,2′:6′,2″-Terpyridine. *Synth. Commun.* **2006**, *36*, 1721–1726. [CrossRef]
20. Cave, G.W.V.; Raston, C.L. Efficient synthesis of pyridines via a sequential solventless aldol condensation and Michael addition. *J. Chem. Soc. Perkin Trans. 1* **2001**, 3258–3264. [CrossRef]
21. Housecroft, C.E. 4,2′:6′,4″-Terpyridines: Diverging and diverse building blocks in coordination polymers and metallomacrocycles. *Dalton Trans.* **2014**, *43*, 6594–6604. [CrossRef]
22. Housecroft, C.E. Divergent 4,2′:6′,4″- and 3,2′:6′,3″-terpyridines as linkers in 2- and 3-dimensional architectures. *CrystEngComm* **2015**, *17*, 7461–7468. [CrossRef]
23. Constable, E.C.; Housecroft, C.E. Tetratopic bis(4,2′:6′,4″-terpyridine) and bis(3,2′:6′,3″-terpyridine) ligands as 4-connecting nodes in 2D-coordination networks and 3D-frameworks. *J. Inorg. Organomet. Polym. Mater.* **2018**, *28*, 414–427. [CrossRef]
24. Housecroft, C.E.; Constable, E.C. Ditopic and tetratopic 4,2′:6′,4″-Terpyridines as Structural Motifs in 2D- and 3D-Coordination Assemblies. *Chimia* **2019**, in press.
25. Klein, Y.M.; Constable, E.C.; Housecroft, C.E.; Zampese, J.A.; Crochet, A. Greasy tails switch 1D-coordination [{Zn$_2$(OAc)(4′-(4-ROC$_6$H$_4$)-4,2′:6′,4″-tpy)}$_n$] polymers to discrete[Zn$_2$(OAc)(4′-(4-ROC$_6$H$_4$)-4,2′:6′,4″-tpy)$_2$] complexes. *CrystEngComm* **2014**, *16*, 9915–9929. [CrossRef]
26. Klein, Y.M.; Constable, E.C.; Housecroft, C.E.; Prescimone, A. Assembling coordination ladders with 4′-(4-methoxyphenyl)- 4,2′:6′,4″-terpyridine as rails and rungs. *Inorg. Chem. Commun.* **2014**, *49*, 41–43. [CrossRef]
27. Constable, E.C.; Zhang, G.; Housecroft, C.E.; Zampese, J.A. A matter of greasy tails: Interdigitation of alkyl chains in free and coordinated 4′-(4-dodecyloxyphenyl)-4,2′:6′,4″-terpyridines. *Inorg. Chem. Commun.* **2012**, *15*, 113–116. [CrossRef]
28. Klein, Y.M.; Prescimone, A.; Constable, E.C.; Housecroft, C.E. Manipulating connecting nodes through remote alkoxy chain variation in coordination networks with 4′-alkoxy-4,2′:6′,4″-terpyridine linkers. *CrystEngComm* **2015**, *17*, 6483–6492. [CrossRef]
29. Klein, Y.M.; Prescimone, A.; Constable, E.C.; Housecroft, C.E. 2-Dimensional networks assembled using 4′-functionalized 4,2′:6′,4″-terpyridines and Co(NCS)$_2$. *Polyhedron* **2016**, *103*, 58–65. [CrossRef]

30. Klein, Y.M.; Prescimone, A.; Pitak, M.B.; Coles, S.J.; Constable, E.C.; Housecroft, C.E. Constructing chiral MOFs by functionalizing 4,2′:6′,4″-terpyridine with long-chain alkoxy domains: Rare examples of *neb* nets. *CrystEngComm* **2016**, *18*, 4704–4707. [CrossRef]
31. Li, L.; Zhang, Y.Z.; Yang, C.; Liu, E.; Golen, J.A.; Zhang, G. One-dimensional copper(II) coordination polymers built on 4′-substituted 4,2′:6′,4″ and 3,2′:6′,3″-terpyridines: Syntheses, structures and catalytic properties. *Polyhedron* **2016**, *105*, 115–122. [CrossRef]
32. Chang, M.-Y.; Wu, M.-H. Domino cyclocondensation of arylaldehydes with 2-acetylpyridine. *Tetrahedron* **2012**, *68*, 9616–9623. [CrossRef]
33. Patel, P.N.; Chadha, A. A simple metal free highly diastereoselective synthesis of heteroaryl substituted (±) cyclohexanols by a branched domino reaction. *Tetrahedron* **2018**, *74*, 204–216. [CrossRef]
34. Gezegen, H.; Ceylan, M. Alternate Method for the Synthesis of Six-Membered Carbocycles with Five Stereocenters:1,2,3,4,6-Pentasubstituted-4-hydroxy-cyclohexanes. *Synth. Commun.* **2015**, *45*, 2344–2349. [CrossRef]
35. Li, C.-W.; Shen, T.-H.; Shih, T.L. Reinvestigation of synthesis of halo-substituted 3-phenyl-1-(2-pyridyl)-2-propen-1-ones (azachalcones). A tandem reaction for formation of penta-substituted cyclohexanols. *Tetrahedron* **2017**, *73*, 4644–4652. [CrossRef]
36. Kostanecki, S.V.; Rossbach, G. Ueber die Einwirkung von Benzaldehyd auf Acetophenon. *Ber. Dtsch. Chem. Ges.* **1896**, *29*, 1488. [CrossRef]
37. Hodnett, E.M.; Ross, W.W. Condensations of Aldehydes and Ketones Catalyzed by Potassium Cyanide. *Proc. Oklahoma Acad. Sci.* **1951**, *32*, 69–71.
38. Georgi, R.; Schwyzer, A. Versuche, d-Fenchon oder Campher an Benzalacetophenon oder an andere α,β-ungesättigte Ketone zu addieren. *J. Prakt. Chem.* **1913**, *86*, 273–276. [CrossRef]
39. Shan, Z.; Hu, X.; Hu, L.; Peng, X. First Authentication of Kostanecki's Triketone and Multimolecular Reaction of Aromatic Aldehydes with Acetophenone. *Helv. Chim. Acta* **2009**, *92*, 1102–1111. [CrossRef]
40. Inoue, K.; Noguchi, H.; Hidai, M.; Uchida, Y. Synthesis of a novel carbon ring compound, 2,4-dibenzoyl-1,3,5-triphenylcyclohexanol, from acetophenone and benzaldehyde under phase transfer conditions. *Yukagaku (J. Japan Oil Chem. Soc.)* **1983**, *32*, 219–226. [CrossRef]
41. Zhang, Y.; Wu, X.; Hao, L.; Wong, Z.R.; Lauw, S.J.L.; Yang, S.; Richard, D.; Webster, R.D.; Chi, Y.R. Trimerization of enones under air enabled by NHC/NaOtBu via a SET radical pathway. *Org. Chem. Front.* **2017**, *4*, 467–471. [CrossRef]
42. Vasileyev, B.K.; Bagrina, N.P.; Vysotskii, V.I.; Lindeman, S.V.; Struchkov, Y.T. Structure of Kostanecki's Triketone. *Acta Crystallogr. Sect. C Cryst. Struct. Commun.* **1990**, *46*, 2265–2267. [CrossRef]
43. Minyaev, M.E.; Roitershtein, D.M.; Nifant'ev, I.E.; Ananyev, I.V.; Minyaeva, T.V.; Mikhaylyev, T.A. A structural study of (1RS,2SR,3RS,4SR,5RS)-2,4-dibenzoyl-1,3,5-tri phenyl cyclo hexan-1-ol chloro form hemisolvate and (1RS,2SR,3RS,4SR,5RS)-2,4-dibenzoyl-1-phenyl-3,5-bis(2-methoxyphenyl)cyclohexan-1-ol. *Acta Crystallogr. Sect. C: Struct. Chem.* **2015**, *71*, 491–498. [CrossRef]
44. Zhang, J.-H.; He, Q.-P.; Wang, Y.; Wang, D.-Q. 2,4-Di benzoyl-1,3,5-tri phenyl cyclo hexan-1-ol di chloromethane hemisolvate. *Acta Crystallogr. Sect. E Struct. Rep. Online* **2007**, *63*, o4652. [CrossRef]
45. Yin, Y.G.; Cheung, K.K.; Wong, W.T. One-pot Synthesis of Substituted Cyclohexanes from Condensation of Aldehyde and Methylketone. *Chin. Chem. Lett.* **1998**, *9*, 329–332.
46. Rong, L.; Wei, X.; Lu, Y.; Zong, Z. A Facile and Efficient Synthesis of Polysubstituted Cyclohexanol Derivatives under Solvent-Free Conditions. *Chin. J. Org. Chem.* **2012**, *32*, 1999–2002. [CrossRef]
47. Mamedov, I.; Abbasoglu, R.; Bayramov, M.; Maharramov, A. Synthesis of a new 1,2,3,4,5-pentasubstituted cyclohexanol and determining its stereochemistry by NMR spectroscopy and quantum-chemical calculations. *Magn. Reson. Chem.* **2016**, *54*, 315–319. [CrossRef]
48. Luo, X.; Shan, Y. 2,4-Dibenzoyl-3,5-bis (4-methoxylphenyl)-1-phenyl cyclohexanol. *Acta Crystallogr. Sect. E Struct. Rep. Online* **2006**, *62*, o1631–o1632. [CrossRef]
49. Mukhtar, S.; Alsharif, M.A.; Alahmdi, M.I.; Parveen, H. Synthesis, Characterization, Stereochemistry and Biological Evaluation of Novel Cyclohexanol Derivatives. *Asian J. Chem.* **2018**, *30*, 1102–1108. [CrossRef]
50. Chen, W.-Y.; Peng, Y.-K. Microwave irradiated synthesis of substituted cyclohexanes. *Hecheng Huaxue* **2000**, *8*, 544–546.

51. Hussain, H.T.; Osama, M.; Hussain, W. Stereostructure, Antimicrobial and Cytotoxic Activity of Cyclohexene, Cyclohexanol and Pyridine Derivatives Synthesized from Chalcones. *Int. J. Pharm. Sci. Res.* **2014**, *5*, 2084–2094. [CrossRef]
52. Shan, Z.; Luo, X.; Hu, L.; Hu, X.-Y. New observation on a class of old reactions: Chemoselectivity for the solvent-free reaction of aromatic aldehydes with alkylketones catalyzed by a double-component inorganic base system. *Sci. China Chem.* **2010**, *53*, 1095–1101. [CrossRef]
53. Luo, X.; Shan, Z. Highly chemoselective synthesis of 1,2,3,4,5-pentasubstituted cyclohexanols under solvent-free condition. *Tetrahedron Lett.* **2006**, *47*, 5623–5627. [CrossRef]
54. Çelik, I.; Ersanl, C.C.; Akkurt, M.; Gezegen, H.; Köseoğlu, R. Crystal structure of racemic [(1R,2S,3R,4S,6S)-2,6-bis-(furan-2-yl)-4-hydroxy-4-(thiophen-2-yl)cyclohexane-1,3-diyl]bis(thiophen-2-yl methanone). *Acta Crystallogr. Sect. C Struct. Chem.* **2016**, *72*, 976–979. [CrossRef]
55. Wang, X.-F.; Huang, X.-Q. 2,4-Bis(4-chlorobenzoyl)-1-(4-chlorophenyl)-3,5-di-2-thienylcyclohexanol methanol hemisolvate. *Acta Crystallogr. Sect. E Struct. Rep. Online.* **2008**, *64*, o777. [CrossRef]
56. Vatsadze, S.Z.; Nuriev, V.N.; Leshcheva, I.F.; Zyk, N.V. New aspects of the aldol condensation of acetylpyridines with aromatic aldehydes. *Russ. Chem. Bull. Int. Ed.* **2004**, *53*, 911–915. [CrossRef]
57. Chen, X.-M.; Yin, Y.-G.; Chen, H.-R.; Ding, J. One-pot Synthesis and Crystal Structure of (1S,2R,3S,4R,5S)-1-Phenyl-2,4-dibenzoyl-3,5-difurylhexanol. *Chin. J. Struct. Chem.* **2006**, *25*, 699–703.
58. Thompson, A.M.W.C.; Constable, E.C.; Harverson, P.; Phillips, D.; Raithby, P.R.; Powell, H.R.; Ward, M.D. Condensation Reactions of 2-Acetylpyridine with Benzaldehydes—The Synthesis and Characterization of Some Cyclohexanols and Cyclohexanediols. *J. Chem. Res. (S)* **1995**, 122–123.
59. Constable, E.C.; Zhang, G.; Housecroft, C.E.; Neuburger, M.; Schaffner, S. Phase-separated hydrogen-bonded chloride ion–water–oxonium ion sheets and protonated 4′-(4-bromophenyl)-2,2′:6′,2″-terpyridine stacks, and condensation products of 2-acetylpyridine and benzaldehydes revisited. *CrystEngComm* **2009**, *11*, 1014–1021. [CrossRef]
60. Korall, P.; Börje, A.; Norrby, P.-O.; Åkermark, B. High Yield Preparation of 4′-(4-Bromophenyl)-2,2′:6′,2″-terpyridine by a Condensation Reaction. Determination of the Stereochemistry of Two Complex By-products by a Combination of Molecular Mechanics and NMR Spectroscopy. *Acta Chem. Scand.* **1997**, *51*, 760–766. [CrossRef]
61. Samshuddin, S.; Jasinski, J.P.; Butcher, R.J.; Neuhardt, E.A.; Narayana, B.; Yathirajan, H.S.; Glidewell, C. Three closely-related cyclo hexanols ($C_{35}H_{27}X_2N_3O_3$; X = F, Cl or Br): Similar molecular structures but different crystal structures. *Acta Crystallogr. Sect. C Struct. Chem.* **2014**, *70*, 953–959. [CrossRef]
62. Krishnapriya, K.R.; Sampath, N.; Aravindhan, S.; Ponnuswamy, M.N.; Kandaswamy, M. 2,4-Bis(pyridine-2-carbonyl)-1-(2-pyridyl)-3,5-di-ptolylcyclohex-1-ol. *Acta Crystallogr. Sect. E Struct. Rep. Online.* **2004**, *60*, 2353–2355. [CrossRef]
63. Fun, H.-K.; Ooi, C.W.; Samshuddin, S.; Narayanan, B.; Sarojini, B.K. [2,6-Bis(biphenyl-4-yl)-4-hydroxy-4-(pyridin-2-yl)cyclohexane-1,3-diyl]bis[(pyridin-2-yl)methanone]–butan-2-one (1/1). *Acta Crystallogr. Sect. E Struct. Rep. Online* **2012**, *68*, o1633–o1634. [CrossRef]
64. Downs, L.E.; Wolfe, D.M.; Schreiner, P.R. Organic Base-Mediated Condensation of Pyridinecarboxaldehydes to Azachalcones. *Adv. Synth. Catal.* **2005**, *347*, 235–238. [CrossRef]
65. Fry, D.; Huang, K.S.; Di Lello, P.; Mohr, P.; Müller, K.; So, S.-S.; Harada, T.; Stahl, M.; Vu, B.; Mauser, H. Design of Libraries Targeting Protein–Protein Interfaces. *ChemMedChem* **2013**, *8*, 726–732. [CrossRef]
66. Chamchoumis, C.; Potvin, P. Condensation Reactions of 2-Acetylpyridine and Benzaldehydes: New Cyclohexanol Products and an Improved Synthesis of 4′-p-Tolyl-2,2′:6′,2″-terpyridine. *J. Chem. Res. (S)* **1998**, 180–181. [CrossRef]
67. Kessler, H.; Mronga, S.; Kutscher, B.; Miiller, A.; Sheldrick, W.S. Synthesis of the 4-Pyridine Analog of Kostanecki's Triketone. Determination of Constitution and Stereochemistry by 2 D-NMR Spectroscopy and X-ray Structural Analysis. *Liebigs Ann. Chem.* **1991**, 1337–1341. [CrossRef]
68. Wang, H.; Chen, Y.; Ye, W.; Xu, J.; Liu, D.; Yang, J.; Kong, L.; Zhou, H.; Tian, Y.; Tao, Y. A facile and highly efficient green synthesis of carbazole derivatives containing a six-membered ring. *Dyes Pigm.* **2013**, *96*, 738–747. [CrossRef]
69. Ge, X.; Hao, F.; Li, S.; Jin, F.; Zhou, H. Synthesis and Crystal Structure of New Pyridyl Derivative. *Asian J. Chem.* **2014**, *26*, 1494–1496. [CrossRef]

70. Zhang, Y.; Pan, J.; Zhang, C.; Wang, H.; Zhang, G.; Kong, L.; Tian, Y.; Yang, J. High quantum yield both in solution and solid state based on cyclohexyl modified triphenylamine derivatives for picric acid detection. *Dyes Pigm.* **2015**, *123*, 257–266. [CrossRef]
71. Constable, E.C.; Harverson, P.; Smith, D.R.; Whall, L. The coordination chemistry of 4′-(4′-tertbutylphenyl)-2,2′:6′,2″-terpyridine-a solubilising oligopyridine. *Polyhedron* **1997**, *16*, 3615–3623. [CrossRef]
72. Constable, E.C.; Handel, R.; Housecroft, C.E.; Neuburger, M.; Schofield, E.R.; Zehnder, M. Efficient syntheses of 4′-(2-thienyl)- and 4′-(3-thienyl)-2,2′:6′,2″-terpyridine: Preparation and characterization of Fe(II), Ru(II), Os(II) and Co(II) complexes. *Polyhedron* **2004**, *23*, 135–143. [CrossRef]
73. Li, L.; Yang, J.-X.; Tian, Y.-P.; Liu, P.; Jin, B.-K.; Tao, X.-T.; Min-Hua Jiang, M.-H. Synthesis, characterization and crystal structure of 6-ferrocenyl-2,4-dihydroxy-2,4-di(pyridine-2-yl)cyclohexanecarbonyl ferrocene. *Transit. Met. Chem.* **2008**, *33*, 85–89. [CrossRef]
74. Bray, D.J.; Clegg, J.K.; Jolliffe, K.A.; Lindoy, L.F.; Wei, G. Synthesis and co-crystallisation behaviour of copper(II) complexes of two isomeric p-tolyl-terpyridines. *J. Coord. Chem.* **2008**, *61*, 3–13. [CrossRef]
75. Collin, J.-P.; Guillerez, S.; Sauvage, J.-P.; Barigelletti, F.; De Cola, L.; Flamigni, L.; Balzani, V. Photoinduced processes in dyads and triads containing a ruthenium (II)-bis (terpyridine) photosensitizer covalently linked to electron donor and acceptor groups. *Inorg. Chem.* **1991**, *30*, 4230–4238. [CrossRef]
76. Turonek, M.L.; Moore, P.; Errington, W. Synthesis of the Terpyridyl Pendant-arm Azamacrocycle 4′-(p-1,4,7-triazacyclonon-1-ylmethylphenyl)-2,2′:6′,2″-terpyridine (L) and Complexes of L with Copper(II) and Nickel(II). Crystal Structure of [Cu(HL)(H$_2$O)$_2$][PF$_6$]$_3$. *J. Chem. Soc. Dalton Trans.* **2000**, 441–444. [CrossRef]
77. Jing, B.-W.; Wu, T.; Zhang, M.-H.; Shen, T. Synthesis of the Polypyridine Ligands with Functional Groups. *Gaodeng Xuexiao Huaxue Xuebao (Chem. J. Higher Educational Inst.)* **2000**, *21*, 395–400.
78. Shirai, H.; Hanabusa, K.; Takahashi, M.F.; Hanada, K. Neurodegenerative disease therapeutics containing dipyridyltolylpyridine. *Jpn. Kokai Tokkyo Koho* **1996**. JP 08217676A 19960827 Heisei.
79. Rajalakshmanan, E.; Alexander, V. Synthesis, Luminescence, and Electrochemical Studies of Tris(homoleptic) Ruthenium(II) and Osmium(II) Complexes of 6′-Tolyl-2,2′:4′,2″-terpyridine. *Inorg. Chem.* **2007**, *46*, 6252–6260. [CrossRef]
80. Zhang, D.-Y.; Nie, Y.; Sang, H.; Suo, J.-J.; Li, Z.-J.; Gu, W.; Tian, J.-L.; Liu, X.; Yan, S.-P. Three structurally related Copper complexes with two isomers: DNA/BSA binding ability, DNA cleavage activity and excellent cytotoxicity. *Inorg. Chim. Acta* **2017**, *457*, 7–18. [CrossRef]
81. Anderson, H.L.; Anderson, S.; Sanders, J.K.M. Ligand binding by butadiyne-linked porphyrin dimers, trimers and tetramers. *J. Chem. Soc. Perkin Trans. 1* **1995**, 2231–2245. [CrossRef]

Sample Availability: Samples of the compounds are available from the authors.

© 2019 by the authors. Licensee MDPI, Basel, Switzerland. This article is an open access article distributed under the terms and conditions of the Creative Commons Attribution (CC BY) license (http://creativecommons.org/licenses/by/4.0/).

Article

Novel 6a,12b-Dihydro-6*H*,7*H*-chromeno[3,4-c]chromen-6-ones: Synthesis, Structure and Antifungal Activity

Jin-ping Bao [1], Cui-lian Xu [2,*], Guo-yu Yang [2,*], Cai-xia Wang [2], Xin Zheng [2] and Xin-xin Yuan [1]

[1] College of Resource and Environment; Henan Agricultural University, Zhengzhou 450002, China; baojinping666@126.com (J.-p.B.); 18339992125@163.com (X.-x.Y.)
[2] School of Science, Henan Agricultural University, Zhengzhou 450002, China; wcx670815@163.com (C.-x.W.); zhengxin@henau.edu.cn (X.Z.)
* Correspondence: xucuilian666@henau.edu.cn (C.-l.X.); yangguoyulxy@henau.edu.cn (G.-y.Y.); Tel.: +86-371-6355-8661 (C.-l.X.); Fax: +86-371-6355-8881 (C.-l.X.)

Academic Editor: Gianfranco Favi
Received: 19 March 2019; Accepted: 3 May 2019; Published: 5 May 2019

Abstract: A new series of coumarin derivatives, 7-hydroxy-7-(trifluoromethyl)-6a,12b-dihydro-6*H*,7*H*-chromeno[3,4-c]chromen-6-ones **3a–p**, were synthesized via Michael addition, transesterification and nucleophilic addition from the reaction of 3-trifluoroacetyl coumarins and phenols in the presence of an organic base. The products were characterized by infrared spectroscopy (IR), hydrogen nuclear magnetic resonance spectroscopy (^{1}H-NMR), carbon nuclear magnetic resonance spectroscopy (^{13}C-NMR) and high-resolution mass spectrometer (HRMS). Single crystal X-ray analysis of compounds **3a** and **3n** clearly confirmed their assigned chemical structures and their twisted conformations. Compound **3a** crystallized in the orthorhombic system, Pbca, in which a = 8.6244(2) Å, b = 17.4245(4) Å, c = 22.5188(6) Å, α = 90°, β = 90°, γ = 90°, v = 3384.02(14) Å^3, and z = 8. In addition, the mycelial growth rate method was used to examine the in vitro antifungal activities of the title compounds **3a–p** against *Fusarium graminearum* and *Fusarium monitiforme* at 500 μg/mL. The results showed that compound **3l** exhibited significant anti-*Fusarium monitiforme* activity with inhibitory index of 84.6%.

Keywords: dihydrocoumarins; synthesis; 3-trifluoroacetyl coumarins; phenols; antifungal activities

1. Introduction

Many natural compounds, such as coumarins, have commonly been used as meaningful lead compounds in the founding of some newer pharmaceuticals [1,2]. Coumarins belong to a class of notable heterocycle compounds containing oxygen, which exhibit various biological activities, including antioxidant [3], antitubercular [4], antitumor [5], antifungal [6], antibacterial [7], antiviral [8], antileishmanial [9], and anticancer [10] activities.

4-Arylcoumarins (neoflavones), present in many natural compounds, are considered important privileged structures due to their diverse biological activities, such as anticancer [11], antioxidant [12], antimicrobial [13], antiprotozoal [14] and antifungal activities [15]. Many studies have focused on the synthetic methods and structure-activity relationship research of 4-arylcoumarins, expecting to discover novel lead compounds [16].

Recently, heterocyclic fused coumarin derivatives from natural or non-natural products have attracted great interest from chemists and pharmaceutical scientists. For instance, pyrano-fused coumarin derivatives have shown much higher antifungal activities [17], and compounds containing thieno[3,2-c]coumarin and pyrazolo[4,3-c]coumarin frameworks demonstrated considerable antifungal and antibacterial activities in the in vitro test systems [18].

The attachment of a fluorine atom or fluorine-containing functional groups to heterocycle molecules often results in an increase in their lipophilicity, metabolic stability and bioactivity [19]. Trifluoromethyl ketones can be used as a typical structural framework to incorporate CF_3 in the target compounds [20–23]. Keeping these aspects in mind, we adopted 3-trifluoroacetyl coumarins as a building block to synthesize novel coumarins containing fluorine [24].

Here, we report the unexpected synthesis of a series of novel dihydrocoumarins or benzopyran fused dihydrocoumarins, chromeno [3,4-c]chromen-6-ones and the preliminary evaluation of their antifungal activities.

2. Results and Discussion

2.1. Chemistry

Fan et al. reported a tandem reaction of α,β-unsaturated trifluoromethyl ketones with 2-naphthol to obtain the corresponding benzo[f]chromene derivatives in the presence of catalysts with a one-pot reaction (Scheme 1) [25]. According to the literature, an 1,4-addition reaction should take place between α,β-unsaturated trifluoromethyl ketones and 2-naphthols under the catalysis of a weak base.

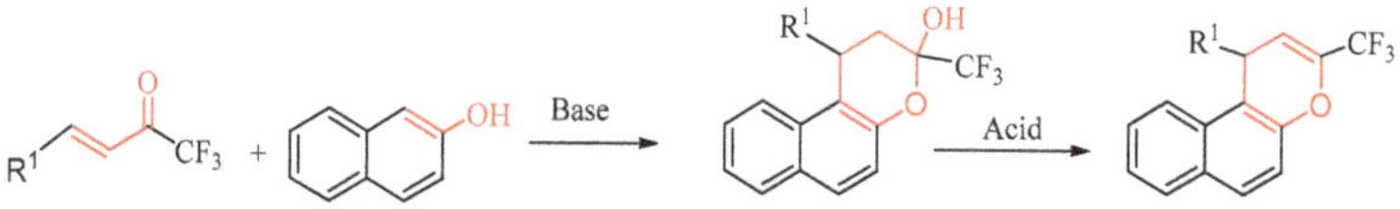

Scheme 1. A base-catalyzed cycloaddition reaction of unsaturated ketones with phenols.

Inspired by Fan's work, we anticipated synthesizing 3-hydroxy-3-(trifluoromethyl)-2a,10c-dihydro-2H,3H-benzo[f]chromeno[3,4-c]chromen-2-one (**A**) first and then 3-(trifluoromethyl)-2H,10cH-benzo[f]chromeno[3,4-c]chromen-2-one (**B**) by using coumarins containing CF_3 (**1a**) and 2-naphthol as starting materials in a preliminary experiment (Scheme 2). However, we did not obtain the expected products (**A** and **B**). Two new dihydrocoumarin derivatives, 1-(2-hydroxyphenyl)-3-oxo-2-(2,2,2-trifluoroacetyl)-2,3-dihydro-1H-benzo[f]chromen-2-ide (**C**) and 2-hydroxy-2-(trifluoromethyl)-2a,10c-dihydro-2H,3H-benzo[f]chromeno[3,4-c]chromen-3-one (**3a**), were unexpectedly synthesized instead.

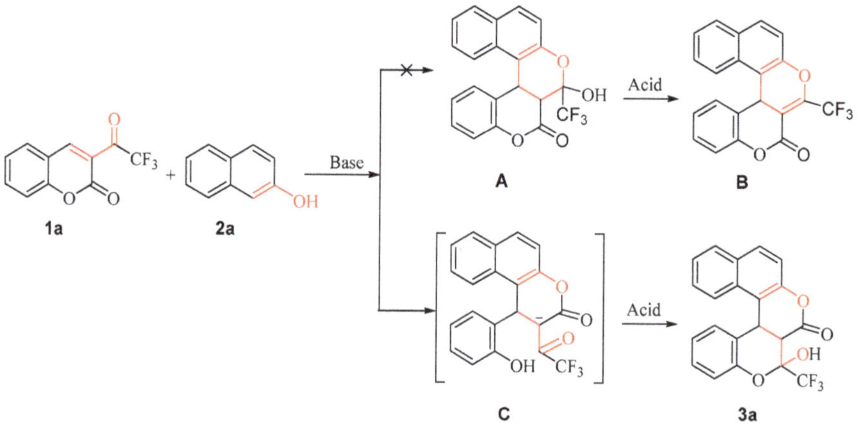

Scheme 2. Synthesis of 3-(trifluoroacetyl)coumarin with 2-naphthol.

Dihydrocoumarin is an important core structure of many bioactive compounds. For example, the derivatives of dihydrocoumarin have dual physiological activities in plants and various pharmacological effects on the human body, such as antioxidant, antiviral and antibacterial effects [26,27]. Therefore, it is necessary to explore whether this reaction can be developed into a new method for the preparation of dihydrocoumarin derivatives.

We initiated the model reaction of 3-(trifluoroacetyl)coumarin (1a) with 2-naphthol (2a) to establish optimal reaction conditions. The results of the optimization were presented in Table 1.

Table 1. The screening of catalysts and acidifiers [a].

Entry	Base	Acid	t (h)	Yield [b] (%)
1	i-Pr$_2$NEt	HCl	2	85
2	NEt$_3$	HCl	2	85
3	DBU	HCl	2	80
4	DMAP	HCl	2	4
5	NMI	HCl	2	3
6	Py	HCl	2	trace
7	NMM	HCl	2	62
8	TMEDA	HCl	2	29
9	DABCO	HCl	2	13
10	K$_2$CO$_3$	HCl	2	trace
11	K$_2$CO$_3$ + TEBA	HCl	2	24
12	KF	HCl	2	8
13	NEt$_3$	Conc. H$_2$SO$_4$	2	85
14	NEt$_3$	TFA	2	74
15	NEt$_3$	p-TsOH	2	20
16	NEt$_3$	Citric acid	2	38
17	NEt$_3$	Conc. H$_2$SO$_4$	DCE	78
18	NEt$_3$	Conc. H$_2$SO$_4$	THF	24
19	NEt$_3$	Conc. H$_2$SO$_4$	CHCl$_3$	7
20	NEt$_3$	Conc. H$_2$SO$_4$	dioxane	3
21 [c]	NEt$_3$	Conc. H$_2$SO$_4$	CH$_3$CN	85
22 [d]	NEt$_3$	Conc. H$_2$SO$_4$	CH$_3$CN	65
23 [e]	NEt$_3$	Conc. H$_2$SO$_4$	CH$_3$CN	71
24 [c,f]	NEt$_3$	Conc. H$_2$SO$_4$	CH$_3$CN	86
25 [c,g]	NEt$_3$	Conc. H$_2$SO$_4$	CH$_3$CN	87
26 [c,h]	NEt$_3$	Conc. H$_2$SO$_4$	CH$_3$CN	87
27 [c,i]	NEt$_3$	Conc. H$_2$SO$_4$	CH$_3$CN	89
28 [c,j]	NEt$_3$	Conc. H$_2$SO$_4$	CH$_3$CN	93
29 [c,k]	NEt$_3$	Conc. H$_2$SO$_4$	CH$_3$CN	92

[a] Reaction conditions: 1a (0.50 mmol), 2a (0.50 mmol), catalyst (0.50 mmol), acidifier (0.60 mmol) and dichloromethane (5 mL), 45 °C, 2 h, sealed tube. [b] Isolated yields. [c,d,e] Run at 45 °C, 25 °C, 65 °C [f] n(2a):n(1a):n(NEt$_3$) = 1:1.1:1. [g] n(2a):n(1a):n(NEt$_3$) = 1:1.2:1. [h] n(2a):n(1a):n(NEt$_3$) = 1:1.3:1. [i] n(2a):n(1a):n(NEt$_3$) = 1:1.2:1.3. [j] n(2a):n(1a):n(NEt$_3$) = 1:1.2:1.4. [k] n(2a):n(1a):n(NEt$_3$) = 1:1.2:1.5.

The reaction proceeded satisfactorily at 45 °C in dichloromethane (DCM) using N,N-diisopropylethylamine (i-Pr$_2$NEt) or triethylamine (NEt$_3$) as a catalyst, resulting in a yield of 85% (Table 1, Entries 1 and 2). If the base was changed to a strong base, such as 1,8-diazabicyclo[5.4.0]undec-7-ene (DBU), the yield slightly decreased (80%) (Table 1, Entry 3). None of the other bases, such as dimethylaminopyridine (DMAP), N-methyl imidazole (NMI), Py, N-methylmorpholine (NMM), tetramethylethylenediamine (TMEDA), 1,4-diazabicyclooctane triethylenediamine (DABCO) and inorganic bases, could produce good yields (Table 1, Entries 4–12). For acidifiers, the results showed that the best yield of 85% was obtained in the presence of hydrochloric acid or concentrated sulfuric acid (Table 1, Entries 2 and 13).

Further experiment indicated that this reaction was sensitive to solvents. The yield was compared to that in DCM when being carried out in CH$_3$CN, but dropped significantly when in 1,2-dichloroethane (1,2-DCE), tetrahydrofuran (THF), CHCl$_3$ and dioxane (Table 1, Entries 17–20). Moreover, increasing or lowering the reaction temperature resulted in a decreased yield (Table 1, Entries 21–23) and 45 °C was identified as the ideal reaction temperature for this reaction. Gratifyingly, when the mole ratio of 2a,

1a and NEt$_3$ was changed from 1:1:1 to 1:1.2:1.4, the reaction gave the best yield (93%) of **3a** (Table 1, Entries 24–29). Further increasing the ratio of **2a**, **1a** and NEt$_3$ failed to provide a better result.

Sixteen coumarin derivatives, 6a,12b-dihydro-6H,7H-chromeno[3,4-c] chromen-6-ones, were prepared from 3-(trifluoroacetyl)coumarin (**1**) and phenols or 2-naphthol (**2**), in the presence of NEt$_3$ as a catalyst and hydrochloric acid as an acidifier (Scheme 3 and Table 2).

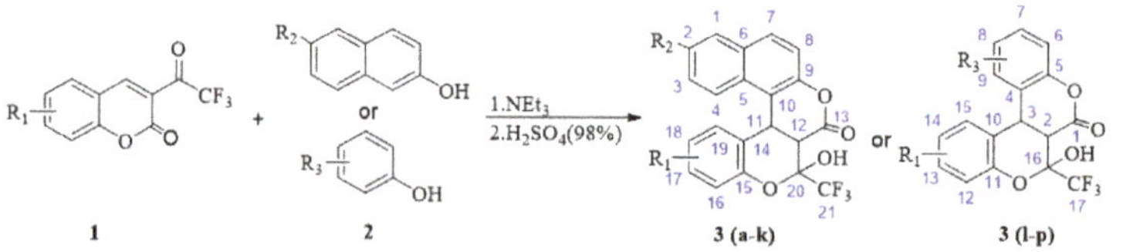

Scheme 3. Synthesis of compounds through reaction of **1** and **2**.

Table 2. Synthesis of compounds **3a–p**.

Entry	R$_1$	R$_2$	R$_3$	T (°C)	t/h	Product [a]	Yield [b] (%)
1	H	H	-	45	1.5	3a	93
2	8-OCH$_3$	H	-	45	0.3	3b	90
3	7-OCH$_3$	H	-	45	1.5	3c	94
4	6-CH$_3$	H	-	45	1.5	3d	91
5	6-Br	H	-	45	0.5	3e	92
6	H	Br	-	45	1.5	3f	94
7	8-OCH$_3$	Br	-	45	1	3g	89
8	7-OCH$_3$	Br	-	45	1	3h	90
9	6-Cl	Br	-	45	1	3i	91
10	H	CN	-	45	2	3j	88
11	7-OCH$_3$	CN	-	45	2	3k	89
12 [c]	H	-	H	65	5	3l	65
13 [c]	H	-	3-OH	65	3	3m	74
14 [c]	8-OCH$_3$	-	4-OH	65	6	3n	80
15 [c]	6-Cl	-	4-OH	65	4	3o	82
16 [c]	H	-	3-OH, 5-CH$_3$	65	6	3p	87

[a] Reaction conditions: **1a** (0.64 mmol), **2a** (0.50 mmol) and NEt$_3$ (0.71 mmol) carried out in a pressure vial at 45 °C by high-performance liquid chromatography (HPLC) tracking. [b] Yield of isolated product. [c] Run at 65 °C.

As seen from Table 2, a variety of coumarins bearing either electron-withdrawing or electron-donating groups on the aryl ring could be transformed into the corresponding novel dihydrocoumarin compounds with CF$_3$, **3a–p**, with good to excellent yields (65–94%) by this simple procedure (Scheme 3). The reactions between coumarins **1** and 2-naphthols proceeded smoothly in most cases to give the products **3a–k** at a relatively lower temperature (45 °C) in a short time (0.3–2 h) (Entries 1–11). However, the reactions of coumarins **1** with phenols required a higher temperature (65 °C) and longer time (3–6 h) for a lower yield, below 90% (Entries 13–16). From this point of view, the electronic density of 2-naphthols was crucial for this reaction, and the electron-donating group on the naphthalene ring was favorable to the reaction rate and yield. Indeed, the yields decreased slightly in the case of strong electron-withdrawing groups, such as CN, on the naphthalene ring, and a slightly longer reaction time of 2 h was needed (Entries 10 and 11). The results were similar for the reactions between coumarins **1** and phenols. It was not until there were strong electron-donating groups, such as OH, on the phenol ring that the reactions could run smoothly at a higher temperature (65 °C) and longer times of 3–6 h (Entries 12–16).

2.2. Structural Characterization of Chromenes 3a–p

The structures of the synthesized compounds were confirmed by ^{1}H-NMR, ^{13}C-NMR, IR, and HRMS. For example, the ^{1}H-NMR spectrum of **3a** showed two doublet signals for CH at 3.72 ppm and

5.26 ppm, respectively, a broad singlet signal for OH at 7.48 ppm and a multiplet signal at 6.59–8.06 ppm for aromatic protons. The downfield shift in the signal of two protons at the 3- and 4-positions, which appeared at 3.0 ppm and 4.2 ppm, respectively, in reported 3,4-dihydrocoumarins [13], was caused by the inductive effects of neighboring OH and CF$_3$ groups. The decoupled ^{13}C-NMR spectrum of **3a** showed 17 distinct resonance structures, in agreement with the proposed structure. Among them, the two quartet signals for C-CF$_3$ at 95.71 ppm and CF$_3$ at 122.13 ppm for which $^2J_{C,F}$ = 33 Hz and $^1J_{C,F}$ = 286 Hz, respectively, were observed. The IR spectrum of **3a** displayed characteristic OH, C=O and C-F vibrations at 3419, 1728 and 822 cm^{-1}, respectively. The HRMS data for **3a** showed the molecular ion peak at m/z [M − H]$^+$ was 385.0688, consistent with the calculated value of 385.0688.

By cooling the solutions of **3a** and **3n** in ethyl acetate, single crystals suitable for X-ray crystallographic analysis were obtained, and their crystal structures are shown in Figure 1. Compound **3a** has a twisted conformation. Its benzene ring and naphthalene ring are connected by the sp3 carbon atom of C11 with a C10–C11–C14 angle of 115.01(16)°, and a dihedral angle of 72.776° between the benzene ring and the naphthalene ring is observed. Interestingly, the rotation of the chiral carbon atoms of C11 and C12 are restricted by intramolecular hydrogen bonding between H on C12 and C11 and F in trifluoromethyl groups, with CH···F distances of 2.519 and 2.344 Å, respectively. Moreover, the adjacent molecules are connected by intermolecular hydrogen bonding with a distance of 2.610 Å, which is almost equal to the van der Waals radius (2.6 Å) [28]. In comparison to **3a**, **3n** also has a twisted conformation. However, two benzene rings are almost perpendicular to each other with a dihedral angle of 89.045°, and the C4–C3–C10 angle of 113.38(16)° is slightly narrowed in comparison to that in **3a**. In addition, **3n** shows weak intermolecular interactions with a centroid distance of 4.126 Å between adjacent methoxy-substituted benzene rings (Figure 1d). Both compounds **3a** and **3n** exhibit three chiral carbon atoms, C11, C12, and C21 and C2, C3, and C16, respectively. The refinement information of compounds **3a** (C$_{21}$H$_{13}$F$_3$O$_4$) and **3n** (C$_{18}$H$_{13}$F$_3$O$_6$) is summarized in the Supplementary Materials.

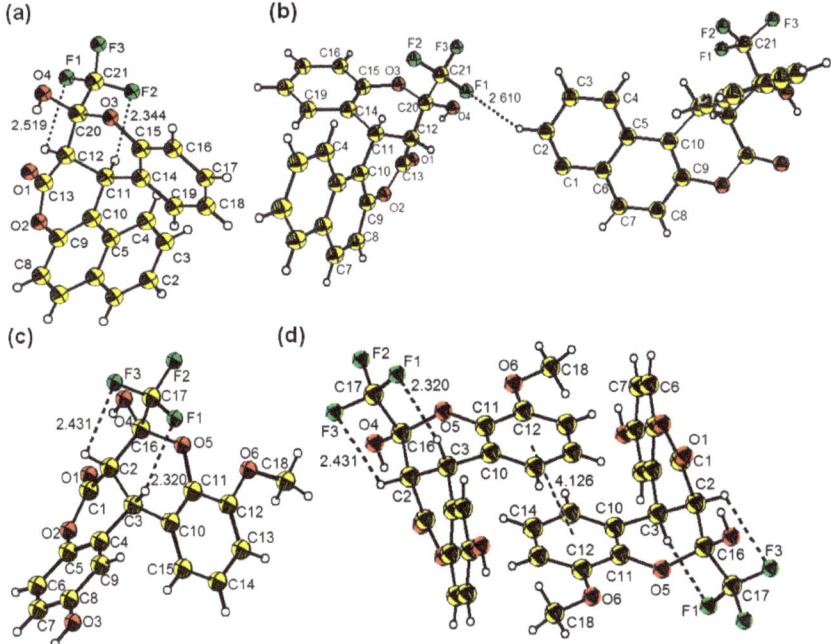

Figure 1. Thermal ellipsoid (30%) drawings of compound **3a** (a); packing of the crystal structure of **3a** (b); thermal ellipsoid (30%) drawings of compound **3n** (c); and intermolecular hydrogen bonding of compound **3n** (d). Note: yellow, carbon; white, hydrogen; red, oxygen; green, fluorine.

2.3. Reaction Mechanism

To speculate on the mechanism of the reaction, we obtained the crystals of compound **3a** before acidification, and it was ammonium salt **3a¹** of compound **3a**, being formed from the reaction between Et₃N and DCM (Figure 2). Compound **3a** exists in an open form with a free carbonyl group attached to -CF₃ in the presence of base and easily forms a semiacetal structure with a stable six-membered ring by internal nucleophilic attack of the OH group to an active C=O group after acidification.

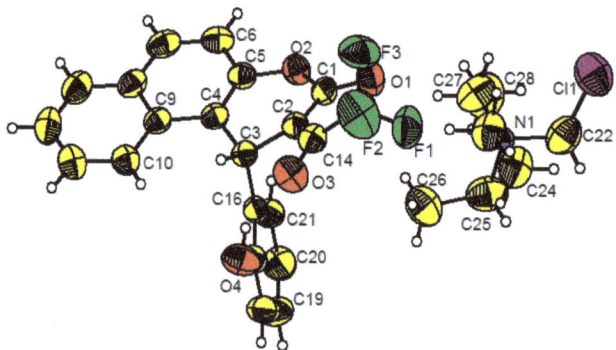

Figure 2. The ammonium salt **3a¹** of compound **3a** (CCDC 1900315). Note: yellow, carbon; white, hydrogen; red, oxygen; green, fluorine; blue, nitrogen; purple, chlorine.

Scheme 4 shows plausible pathways for the reaction. First, the 1,4-nucleophilic addition of 2-naphthol (**2**) to 3-trifluoroacetyl coumarin **1** with an α,β-unsaturated ketone structure generates the carbon anion intermediate **I**. It then forms enolate anion **II** by H-transformation. Intermediate **II** is next transformed into a new coumarin with phenoxy anion **III** through nucleophilic attack by a naphthalenyl oxyanion at the initial ester group of coumarin with slightly stronger basicity as the driving force. The subsequent addition of a phenoxy anion to the electrophilic carbon of -COCF₃ group happened to furnish annulation product **IV**, forming the final title compound **3a** with a semiacetal structure after acidification.

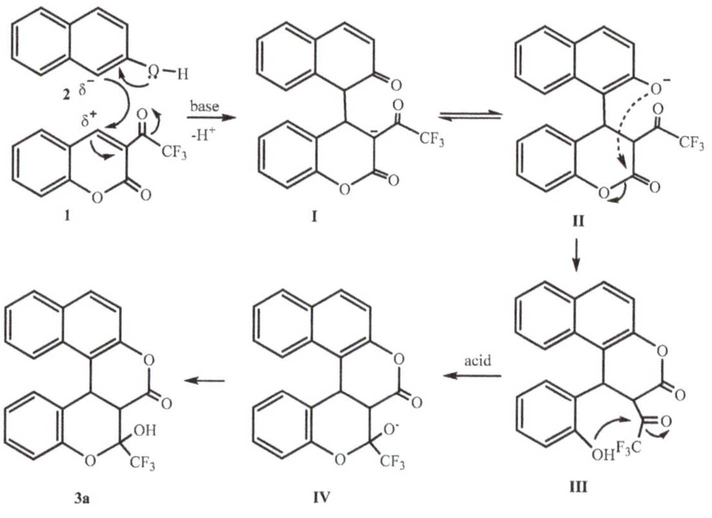

Scheme 4. Plausible mechanism for compound **3a**.

2.4. In Vitro Antifungal Assay

The antifungal activities of the target compounds **3a–p** were evaluated against *Fusarium graminearum* and *Fusarium monitiforme* by the mycelial growth rate method [29] at 500 μg/mL, using triazolone (100 μg/mL) as the positive control. The results of the antifungal activities were shown in Figure 3.

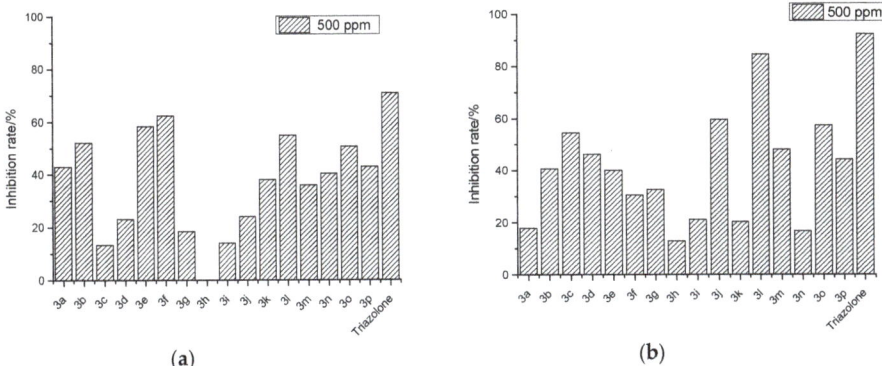

Figure 3. Antifungal activity of the target compounds **3a–p**, against—*Fusarium graminearum* (**a**) and *Fusarium monitiforme* (**b**).

The bioassay results for *Fusarium graminearum* indicated that, except for compound **3h**, all the target compounds showed antifungal activities to some extent, with an inhibitory index in the range of 13.4–62.4%. The inhibitory index of compounds **3a** and **3l** were 42.9% and 55%, respectively. From the activity results for the derivatives (**3b–k**) of compound **3a** obtained by structure modification, compared with compound **3a**, compounds **3b**, **3e** and **3f** showed greater inhibition of *F. graminearum*, with inhibitory indexes of 52.2%, 58.4% and 62.4%, respectively.

The obtained findings suggested that a functional group, such as bromine, introduced into compound **3a** could substantially contribute to the antifungal efficacy of its derivatives. However, the derivatives (**3m–p**) of compound **3l**, whether bearing electron-donating or electron-withdrawing groups, all showed lower antifungal activities against *F. graminearum* than that of compound **3l**.

Moreover, the inhibitory index of compound **3l** against *F. monitiforme* (84.6%) was much higher than that of compound **3a** (17.9%), which indicated that compared with the benzene ring, the naphthalene ring at the 4-position of dihydrocoumarin is unfavorable for activity against *F. monitiforme*. However, the compounds **3b–k** modified from compound **3a**, except compound **3h**, all had higher antifungal activities against *F. monitiforme* (inhibitory index: 21.2–59.6%) than compound **3a**. The reason may be that the introduction of a functional group can increase the antifungal activity. Similar to the antifungal results against *F. graminearum*, the compounds (**3m–p**) modified from compound **3l** all showed lower antifungal activities against *F. monitiforme* (inhibitory index: 16.8%–57.4%) than compound **3l**. Further research about structure and activity is in progress.

3. Materials and Methods

3.1. General Information

HPLC analyses were performed with Thermo Fisher U-3000 (Chromatographic column: XDB-C_{18} column with MeOH-H_2O as the eluent, (Thermo Fisher scientific Co., Waltham, MA, USA) equipped with an UltiMate 3000UV detector (Thermo Fisher scientific Co., Waltham, MA, USA). Melting points were determined on an X-5 digital microscopic melting-point apparatus (Beijing Tech Instruments Co., Beijing, China) and were uncorrected. High resolution mass spectra were obtained using a Waters

Q-Tof MicroTM instrument (Waters Co., MA, USA). NMR spectra were recorded on a Bruker DPX-400 spectrometer (Bruker Optics Co., Ettlingen, baden-wuerttemberg, Germany) in dimethyl sulfoxide-d_6 (DMSO-d_6) or chloroform-D (CDCl$_3$) using tetramethylsilane as the internal standard. Chemical shifts (δ) were reported in ppm, and J values were reported in Hertz. The IR spectra were recorded on a Thermo IS10 FT-IR spectrometer (Thermo scientific Co., Shenzhen, Guangdong, China), and the frequencies were reported in cm^{-1}. X-ray images were obtained with a Rigaku RAXIAS-IV type diffractometer (Rigaku Corporation Co., Tokyo, Japan).

The starting materials (**1**) were obtained according to the reported procedure [30]. All other reagents were acquired from commercial sources and utilized without further purification.

3.2. General Synthetic Procedures for Compounds 3a–p

Typical procedure for the synthesis of compound **3a**: compound **1** (0.64 mmol), naphthol (0.50 mmol), NEt$_3$ (0.71 mmol) and 5 mL CH$_3$CN were added to a pressure vial. The reaction mixture was stirred at 45 °C for 1.5 h (HPLC tracking reaction), and then cooled down to room temperature. After the addition of concentrated H$_2$SO$_4$ (0.71 mmol), the resulting mixture was stirred at room temperature for 2 h. The crude reaction mixture was concentrated by rotary evaporator, and dissolved in DCM 30 mL. Then it was washed with water (15 mL × 3) and the organic layers were dried with Na$_2$SO$_4$, filtered, and concentrated in vacuo. A white crystal was obtained after crystallization from ethyl acetate-petroleum ether (0.183 g, 93%).

3.3. Biological Assays

The antifungal activity of synthesized compounds **3a–p** was tested against two pathogenic fungi, namely, *Fusarium graminearum* and *Fusarium monitiforme*, by the poison plate technique at a concentration of 500 μg/mL. The two species of fungi were incubated in potato dextrose agar medium at 25 ± 1 °C for five days to obtain new mycelia for the antifungal assay, and then mycelia as disks of approximately 0.70 cm diameter cut from the culture medium were picked up with a sterilized inoculation needle and inoculated into the center of a PDA plate. The test compounds were dissolved in dimethyl sulfoxide (1 mL) then diluted to a 7500 μg/mL drug solution with 4% tween-80 emulsifier (1 mL) and added to the PDA medium (30 mL). The final concentration of compounds in the medium was adjusted to 500 μg/mL. The inoculated plates were incubated at 25 ± 1 °C for 3 days. Dimethyl sulfoxide was diluted with sterilized distilled water (4% tween-80) and used as the control, while the commercial fungicide triazolone (100 μg/mL) was used as the standard control. Three replicates of the experiments were performed. The radial growth of the fungal colonies was measured on the fourth day.

2-Hydroxy-2-(trifluoromethyl)-2a,10c-dihydro-2H,3H-benzo[f]chromeno[3,4-c]chromen-3-one (**3a**): Crystallization from ethyl acetate-petroleum ether, white crystal; M.p. 172.6~173.7 °C; ^{1}H-NMR (CDCl$_3$, 400 MHz) δ: 3.72 (d, J = 4 Hz, 1H, CH), 5.26 (d, J = 4 Hz, 1H, CH), 6.59 (d, J = 4 Hz, 1H, Ar-H), 6.70 (td, J_1 = 8 Hz, J_2 = 1.2 Hz, Ar-H), 7.12 (q, J = 4 Hz, 1H, Ar-H), 7.23~7.33 (m, 2H, Ar-H), 7.48 (s, 1H, -OH), 7.64 (td, J_1 = 8 Hz, J_2 = 0.8 Hz, 1H, Ar-H), 7.74 (td, J_1 = 8 Hz, J_2 = 1.2 Hz, 2H, Ar-H), 7.98~8.06 (m, 3H, Ar-H); ^{13}C-NMR (100 MHz, CDCl$_3$) δ: 31.12 (q, J = 3 Hz, C1), 39.46 (C1), 95.71 (q, J = 33 Hz, C20), 116.24 (C16), 116.94 (C18), 117.60 (C8), 118.29 (C10), 122.13 (q, J = 286 Hz, C21), 122.64 (C4), 122.90 (C2), 126.20 (C3), 127.61 (C17), 128.61 (C7), 129.13 (C1), 129.92 (C19), 130.88 (C6), 131.33 (C14), 131.82 (C5), 147.44 (C9), 152.64 (C15), 168.35 (C13); IR (KBr) v_{max} (cm^{-1}): 3419 (OH), 1728 (C=O), 1585 (Ar), 822 (CF$_3$); HRMS (ESI): m/z calcd for C$_{21}$H$_{12}$F$_3$O$_4$ [M − H]$^+$: 385.0688; found: 385.0688.

2-Hydroxy-14-methoxy-2-(trifluoromethyl)-2a,10c-dihydro-2H,3H-benzo[f]chromeno[3,4-c]chromen-3-one (**3b**): Crystallization from ethyl acetate-petroleum ether, white crystal; M.p. 173.7~174.5 °C; ^{1}H-NMR (CDCl$_3$, 400 MHz) δ: 3.67 (d, J = 4 Hz, 1H, CH), 3.88 (s, 3H, CH$_3$), 5.22 (d, J = 4 Hz, 1H, CH), 6.13 (d, J = 4 Hz, 1H, Ar-H), 6.68 (t, J = 8 Hz, 1H, Ar-H), 6.81 (d, J = 8 Hz, 1H, Ar-H), 7.27 (d, J = 4 Hz, 1H, Ar-H), 7.47 (s, 1H, -OH), 7.58 (td, J_1 = 8 Hz, J_2 = 0.8 Hz, 1H, Ar-H), 7.68 (td, J_1 = 8 Hz, J_2 = 0.8 Hz, 1H, Ar-H),

7.92~8.00 (m, 3H, Ar-H); ^{13}C-NMR (100 MHz, CDCl$_3$) δ: 31.16 (q, J = 2 Hz, C11), 39.38 (C12), 56.30 (-OCH$_3$), 95.97 (q, J = 26 Hz, C20), 112.34 (C17), 116.29 (C19), 116.87 (C18), 118.92 (C8), 119.40 (C10), 122.08 (q, J = 229 Hz, C21), 122.20 (C4), 122.90 (C2), 126.10 (C3), 128.49 (C7), 129.04 (C1), 130.79 (C6),, 131.27 (C14), 131.73 (C5), 142.31 (C16), 147.41 (C15), 148.65 (C9), 168.26 (C13); IR (KBr) ν$_{max}$ (cm^{-1}): 3391 (OH), 1736 (C=O), 1582, 1478 (Ar), 817 (CF$_3$); HRMS (ESI): m/z calcd for C$_{22}$H$_{15}$F$_3$O$_5$ [M − H]$^+$: 415.0793; found: 415.0795.

2-Hydroxy-13-methoxy-2-(trifluoromethyl)-2a,10c-dihydro-2H,3H-benzo[f]chromeno[3,4-c]chromen-3-one (**3c**): Crystallization from ethyl acetate-petroleum ether, white crystal; M.p. 173.6~174.3 °C; ^{1}H-NMR (CDCl$_3$, 400 MHz) δ: 3.64 (d, J = 4 Hz, 1H, CH), 3.72 (s, 3H, CH$_3$), 5.14 (d, J = 4 Hz, 1H, CH), 6.31 (dd, J_1 = 8 Hz, J_2 = 2 Hz, 1H, Ar-H), 6.42 (dd, J_1 = 4 Hz, J_2 = 0.8 Hz, 1H, Ar-H), 6.62 (d, J = 4 Hz, 1H, Ar-H), 7.69 (td, J_1 = 8 Hz, J_2 = 0.8 Hz, 1H, Ar-H), 7.92~8.00 (m, 3H, Ar-H); ^{13}C-NMR (100 MHz, CDCl$_3$) δ: 30.59 (q, J = 3 Hz, C11), 39.64 (C12), 55.39 (-OCH$_3$), 95.79 (q, J = 26 Hz, C20), 102.11 (C16), 109.75 (C18), 110.00 (C8), 116.38 (C14), 116.90 (C10), 122.05 (q, J = 229 Hz, C21), 122.87 (C4), 126.11 (C2), 128.25 (C3), 128.51 (C7), 129.06 (C1), 130.64 (C19), 131.29 (C6), 131.76 (C5), 147.28 (C9), 153.55 (C15), 160.87 (C17), 168.41 (C13); IR (KBr) ν$_{max}$ (cm^{-1}): 3402 (OH), 1736 (C=O), 1628, 1501 (Ar), 819 (CF$_3$); HRMS (ESI): m/z calcd for C$_{22}$H$_{15}$F$_3$O$_5$ [M − H]$^+$: 415.0793; found: 415.0792.

2-Hydroxy-12-methyl-2-(trifluoromethyl)-2a,10c-dihydro-2H,3H-benzo[f]chromeno[3,4-c]chromen-3-one (**3d**): Crystallization from ethyl acetate-petroleum ether, white crystal; M.p. 160.2~161.1 °C; ^{1}H-NMR (DMSO-d_6, 400 MHz) δ: 1.97 (s, 3H, CH$_3$), 3.61 (dd, J_1 = 6 Hz, J_2 = 1.2 Hz, 1H, CH), 5.60 (d, J = 6 Hz, 1H, CH), 6.17 (s, 1H, Ar-H), 7.02 (q, J = 8 Hz, 2H, Ar-H, OH), 7.25 (d, J = 8 Hz, 1H, Ar-H), 7.52 (t, J = 8 Hz, 1H, Ar-H), 7.62 (t, J = 8 Hz, 1H, Ar-H), 7.92 (d, J = 8 Hz, 1H, Ar-H), 8.00 (q, J = 4 Hz, 2H, Ar-H), 8.73 (d, J = 4 Hz, 1H, Ar-H); ^{13}C-NMR (100 MHz, DMSO-d_6) δ: 20.89 (-CH$_3$), 29.71 (C11), 41.54 (C12), 94.85 (q, J = 32 Hz, C20), 112.86 (C16), 115.82 (C8), 118.63 (C10), 122.66 (q, J = 288 Hz, C21), 123.31 (C4), 123.43 (C2), 125.11 (C3), 126.95 (C17), 128.00 (C7), 128.85 (C1), 129.10 (C18), 129.83 (C19), 130.61 (C14), 132.80 (C6), 133.52 (C5), 148.59 (C9), 150.30 (C15) (C13), 164.50 (C=O); IR (KBr) ν$_{max}$ (cm^{-1}): 3287 (OH), 1765 (C=O), 1628, 1600 (Ar), 812 (CF$_3$); HRMS (ESI): m/z calcd for C$_{22}$H$_{15}$F$_3$O$_4$ [M − H]$^+$: 399.0844; found: 399.0805.

12-Bromo-2-hydroxy-2-(trifluoromethyl)-2a,10c-dihydro-2H,3H-benzo[f]chromeno[3,4-c]chromen-3-one (**3e**): Crystallization from ethyl acetate-petroleum ether, white crystal; M.p. 158.7~159.9 °C; ^{1}H-NMR (CDCl$_3$, 400 MHz) δ: 3.65 (d, J = 4 Hz, 1H, CH), 5.18 (d, J = 4 Hz, 1H, CH), 7.64 (d, J = 4 Hz, 1H, Ar-H), 6.97 (d, J = 4 Hz, 1H, Ar-H), 7.28~7.32 (m, 2H, Ar-H), 7.42 (s, 1H, -OH), 7.61 (t, J = 4 Hz, 1H, Ar-H), 7.72 (t, J = 4 Hz, 1H, Ar-H), 7.95~7.99 (m, 3H, Ar-H); ^{13}C NMR (100 MHz, CDCl$_3$) δ: 30.96 (C11), 39.14 (C12), 95.84 (q, J = 26 Hz, C20), 115.03 (C16), 115.32 (C18), 116.88 (C8), 119.44 (C10), 120.78 (C4), 121.93 (q, J = 229 Hz, C21), 122.47 (C2), 126.36 (C3), 128.90 (C7), 129.26 (C1), 130.11 (C17), 131.28 (C6), 131.39 (C5), 131.49 (C14), 132.99 (C19), 147.43 (C9), 151.76 (C15), 168.00 (C13); IR (KBr) ν$_{max}$ (cm^{-1}): 3408 (OH), 1740 (C=O), 1628, 1471 (Ar), 809 (CF$_3$); HRMS (ESI): m/z calcd for C$_{21}$H$_{11}$BrF$_3$O$_4$ [M − H]$^+$: 462.9793; found: 462.9793.

7-Bromo-2-hydroxy-2-(trifluoromethyl)-2a,10c-dihydro-2H,3H-benzo[f]chromeno[3,4-c]chromen-3-one (**3f**): Eluent: EA:PE:AcOH = 2:8:0.1; white crystal; M.p.185.0~186.9 °C; ^{1}H-NMR (CDCl$_3$, 400 MHz) δ: 3.12 (d, J = 6 Hz, 1H, CH), 5.19 (d, J = 6 Hz, 1H, CH), 6.53 (d, J = 8 Hz, 1H, Ar-H), 6.80 (t, J = 8 Hz, 1H, Ar-H), 7.11 (d, J = 8 Hz, 1H, Ar-H), 7.26 (t, J = 8 Hz, 1H, Ar-H), 7.34 (d, J = 2 Hz, 1H, Ar-H), 7.41 (s, 1H, -OH), 7.80 (d, J = 8 Hz, 1H, Ar-H), 7.88~7.92 (t, J = 8 Hz, 2H, Ar-H), 8.17 (s, 1H, Ar-H); ^{13}C NMR (100 MHz, CDCl$_3$) δ: 31.19 (d, J = 3 Hz, C11), 39.37 (C12), 95.65 (q, J = 33 Hz, C20), 116.57 (C16), 117.74 (C18), 117.93 (C1), 118.21 (C8), 120.26 (C10), 122.09 (q, J = 287 Hz, C21), 122.73 (C4), 124.67 (C3), 127.37 (C7), 129.93 (C17), 130.10 (C19), 130.41 (C6), 131.11 (C2), 131.94 (C14), 132.39 (C5), 147.58 (C9), 152.61 (C15), 167.96 (C13); IR (KBr) ν$_{max}$ (cm^{-1}): 3360 (OH), 1723 (C=O), 1583, 1479 (Ar), 758 (CF$_3$); HRMS (ESI): m/z calcd for C$_{21}$H$_{11}$BrF$_3$O$_4$ [M − H]$^+$: 462.9793; found: 462.9791.

7-Bromo-2-hydroxy-14-methoxy-2-(trifluoromethyl)-2a,10c-dihydro-2H,3H-benzo[f]chromeno[3,4-c]chromen-3-one (**3g**): Crystallization from ethyl acetate-petroleum ether, white crystal; M.p. 204.3~204.8 °C; ^{1}H-NMR (CDCl$_3$, 400 MHz) δ: 3.72 (d, J = 4 Hz, 1H, CH), 5.19 (d, J = 4 Hz, 1H, CH), 6.09 (d, J = 8 Hz, 1H, Ar-H), 7.33 (d, J = 8 Hz, 1H, Ar-H), 7.45 (s, 1H, -OH), 7.79 (d, J = 8 Hz, 1H, Ar-H), 7.89 (t, J = 8 Hz, 2H, Ar-H), 8.16 (s, 1H, Ar-H); ^{13}C-NMR (100 MHz, CDCl$_3$) δ: 31.28 (q, J = 3 Hz, C11), 39.33 (C12), 56.35 (-OCH$_3$), 95.95 (q, J = 33 Hz, C20), 112.51 (C17), 116.67 (C19), 118.18 (C18), 118.69 (C1), 119.07 (C8), 120.20 (C10), 122.07 (q, J = 286 Hz, C21), 122.34 (C4), 124.72 (C3), 129.88 (C7), 130.37 (C6), 131.05 (C2), 131.87 (C14), 132.37 (C5), 142.33 (C16), 147.62 (C15), 148.77 (C9), 167.90 (C13); IR (KBr) ν$_{max}$ (cm^{-1}): 3416 (OH), 1762 (C=O), 1580 (Ar), 865 (CF$_3$); HRMS (ESI): *m/z* calcd for C$_{22}$H$_{14}$BrF$_3$O$_5$ [M − H]$^+$: 492.9898; found: 492.9850.

7-Bromo-2-hydroxy-13-methoxy-2-(trifluoromethyl)-2a,10c-dihydro-2H,3H-benzo[f]chromeno[3,4-c]chromen-3-one (**3h**): Eluent: EA:PE:AcOH = 1:8:0.25; white acicular crystal; M.p. 162.1~163.0 °C; ^{1}H-NMR (CDCl$_3$, 400 MHz) δ: 3.65 (d, J = 8 Hz, 1H, CH), 3.74 (s, 3H, CH$_3$), 5.08 (d, J = 8 Hz, 1H, CH), 6.32~6.38 (m, 2H, Ar-H), 6.63 (d, J = 4 Hz, 1H, Ar-H), 7.30 (d, J = 8 Hz, 1H, Ar-H), 7.37 (s, 1H, -OH), 7.76 (dd, J_1 = 8 Hz, J_2 = 2.4 Hz, 1H, Ar-H), 7.86 (t, J = 8 Hz, 1H, Ar-H), 8.13 (d, J = 1.6 Hz, 1H, Ar-H); ^{13}C NMR (100 MHz, CDCl$_3$) δ: 30.68 (q, J = 3 Hz, C11), 39.57 (C12), 55.45 (-OCH$_3$), 95.42 (q, J = 33 Hz, C20), 102.23 (C16), 109.62 (C18), 109.89 (C1), 116.74 (C8), 118.19 (C14), 120.20 (C10), 122.03 (q, J = 286 Hz, C21), 124.67 (C4), 128.08 (C3), 129.79 (C7), 130.37 (C6), 131.07 (C2), 131.88 (C19), 132.37 (C5), 147.48 (C9), 153.57 (C15), 161.02 (C17), 168.08 (C13); IR (KBr) ν$_{max}$ (cm^{-1}): 3419 (OH), 1737 (C=O), 1580, 1499 (Ar), 812 (CF$_3$); HRMS (ESI): *m/z* calcd for C$_{22}$H$_{14}$BrF$_3$O$_5$ [M − H]$^+$: 492.9898; found: 492.9510.

7-Bromo-12-chloro-2-hydroxy-2-(trifluoromethyl)-2a,10c-dihydro-2H,3H-benzo[f]chromeno[3,4-c]chromen-3-one (**3i**): Eluent: EA:PE:AcOH = 1:8:0.25; white crystal; M.p. 175.3~176.1 °C; ^{1}H-NMR (CDCl$_3$, 400 MHz) δ: 3.70 (d, J = 6 Hz, 1H, CH), 5.15 (d, J = 6 Hz, 1H, CH), 6.47 (s, 1H, Ar-H), 7.06 (d, J = 8 Hz, 1H, Ar-H), 7.22 (d, J = 4 Hz, 1H, Ar-H), 7.35 (d, J = 8 Hz, 1H, Ar-H), 7.40 (s, 1H, -OH), 7.80~7.93 (m, 3H, Ar-H), 8.18 (s, 1H, Ar-H); ^{13}C-NMR (100 MHz, CDCl$_3$) δ: 31.09 (d, J = 2 Hz, C11), 39.11 (C12), 95.83 (q, J = 33 Hz, C20), 115.72 (C16), 118.20 (C1), 119.21 (C8), 119.69 (C10), 120.48 (C4), 121.94 (q, J = 286 Hz, C21), 124.26 (C3), 127.07 (C18), 127.86 (C7), 130.12 (C17), 130.27 (C19), 130.36 (C6), 131.30 (C2), 132.26 (C14), 132.47 (C5), 147.63 (C9), 151.22 (C15), 167.64 (C13); IR (KBr) ν$_{max}$ (cm^{-1}): 3430 (OH), 1748 (C=O), 1583, 1504 (Ar), 812 (CF$_3$); HRMS (ESI): *m/z* calcd for C$_{21}$H$_{11}$BrClF$_3$O$_4$ [M + H]$^+$: 498.9560; found: 498.9331.

2-Hydroxy-3-oxo-2-(trifluoromethyl)-2a,10c-dihydro-2H,3H-benzo[f]chromeno[3,4-c]chromene-8-carbonitrile (**3j**): Eluent: EA:PE:AcOH = 1:4:0.25; white crystal; M.p. 174.7~175.5 °C; ^{1}H-NMR (CDCl$_3$, 400 MHz) δ: 3.32 (dd, J_1 = 16 Hz, J_2 = 2 Hz, 1H, CH), 5.34 (d, J = 8 Hz, 1H, CH), 5.70 (s, 1H, -OH), 6.55 (dd, J_1 = 8 Hz, J_2 = 0.8 Hz, 1H, Ar-H), 6.70 (t, J = 8 Hz, 1H, Ar-H), 6.80 (d, J = 4 Hz, 1H, Ar-H), 7.09 (td, J_1 = 8 Hz, J_2 = 1.6 Hz, 1H, Ar-H), 7.48 (d, J = 8 Hz, 1H, Ar-H), 7.59 (dd, J_1 = 8 Hz, J_2 = 1.6 Hz, 1H, Ar-H), 7.87 (d, J = 12 Hz, 1H, Ar-H), 7.94 (d, J = 12 Hz, 1H, Ar-H), 8.24 (d, J = 0.8 Hz, 1H, Ar-H); ^{13}C-NMR (100 MHz, CDCl$_3$) δ: 31.41 (C11), 35.04 (C12), 95.53 (q, J = 33 Hz, C20), 108.76 (C1), 115.60 (C16), 118.15 (-CN), 118.87 (C18), 119.51 (C7), 122.42 (q, J = 286 Hz, C21), 121.47 (C8), 124.67 (C10), 125.98 (C3), 127.84 (C17), 128.04 (C19), 129.11 (C4), 130.00 (C5), 130.35 (C6), 132.86 (C14)), 134.54 (C2), 152.40 (C15, C9), 167.22 (C13); IR (KBr) ν$_{max}$ (cm^{-1}): 3419 (OH), 2228 (CN), 1748 (C=O), 1628, 1585 (Ar), 753 (CF$_3$); HRMS (ESI): *m/z* calcd for C$_{22}$H$_{12}$F$_3$NO$_4$ [M − H]$^+$: 410.0640; found: 410.0600.

2-Hydroxy-13-methoxy-3-oxo-2-(trifluoromethyl)-2a,10c-dihydro-2H,3H-benzo[f]chromeno[3,4-c]chromene-8-carbonitrile (**3k**): Eluent: EA:PE:AcOH = 1:7:0.25; white acicular crystal; M.p. 169.7~170.4 °C; ^{1}H-NMR (CDCl$_3$, 400 MHz) δ: 3.68 (d, J = 6 Hz, 1H, CH), 3.74 (s, 3H, CH$_3$), 5.13 (d, J = 6 Hz, 1H, CH), 5.30 (s, 1H, -OH), 6.34 (d, J = 4 Hz, 1H, Ar-H), 6.64 (d, J = 4 Hz, 1H, Ar-H), 7.25 (d, J = 4 Hz, 1H, Ar-H), 7.43 (d, J = 12 Hz, 1H, Ar-H), 7.85 (dd, J_1 = 12 Hz, J_2 = 2 Hz, 1H, Ar-H), 8.03 (d, J = 8 Hz, 1H, Ar-H), 8.11 (d, J = 4 Hz, 1H, Ar-H), 8.36 (s, 1H, Ar-H); ^{13}C NMR (100 MHz, CDCl$_3$) δ: 30.69 (q, J = 3 Hz, C11), 39.48 (C12), 55.47 (-OCH$_3$), 95.63 (q, J = 30 Hz, C20), 102.34 (C16), 109.10 (C18), 109.93 (C1), 109.99 (-CN), 117.02 (C7), 118.43 (C14), 119.06 (C8), 121.96 (q, J = 286 Hz, C21), 124.39 (C10), 127.86

(C3), 129.08 (C4), 130.17 (C19), 131.36 (C5), 133.48 (C6), 134.92 (C2), 149.44 (C9), 153.56 (C15), 161.14 (C17), 167.55 (C13); IR (KBr) v_{max} (cm^{-1}): 3430 (OH), 2228 (CN), 1742 (C=O), 1619, 1580 (Ar), 817 (CF$_3$); HRMS (ESI): m/z calcd for C$_{23}$H$_{13}$F$_3$NO$_4$ [M − H]$^+$: 440.0746; found: 440.0743.

7-Hydroxy-7-(trifluoromethyl)-6a,12b-dihydro-6H,7H-chromeno[3,4-c]chromen-6-one (**3l**): Eluent: EA:PE:AcOH = 2:8:0.25; white acicular crystal; M.p. 151.8~152.6 °C; ^{1}H-NMR (CDCl$_3$, 400 MHz) δ: 3.67 (d, J = 6 Hz, 1H, CH), 4.59 (d, J = 6 Hz, 1H, CH), 6.69 (d, J = 8 Hz, 1H, Ar-H), 6.91 (t, J = 8 Hz, 1H, Ar-H), 7.07 (d, J = 8 Hz, 1H, Ar-H), 7.16 (d, J = 8 Hz, 1H, Ar-H), 7.23~7.28 (m, 1H, Ar-H), 7.38 (t, J = 8 Hz, 1H, Ar-H), 7.46~7.52 (m, 2H, Ar-H); ^{13}C-NMR (100 MHz, CDCl$_3$) δ: 34.98 (C3), 39.91 (C2), 95.50 (q, J = 32 Hz, C16), 117.47 (C12), 117.58 (C14), 118.41 (C6), 121.96 (C8), 122.01 (q, J = 287 Hz, C17), 122.54 (C13), 125.65 (C7), 127.45 (C15), 129.91 (C9), 129.99 (C10), 130.09 (C4), 149.42 (C5), 152.36 (C11), 168.11 (C1); IR (KBr) v_{max} (cm^{-1}): 3414 (OH), 1726 (C=O), 1588, 1487 (Ar), 756 (CF$_3$); HRMS (ESI): m/z calcd for C$_{17}$H$_{10}$F$_3$O$_4$ [M − H]$^+$: 335.0531; found: 335.0298.

1,7-Dihydroxy-7-(trifluoromethyl)-6a,12b-dihydro-6H,7H-chromeno[3,4-c]chromen-6-one (**3m**): Eluent: EA:PE:AcOH = 1:5:0.1; white solid; M.p. 138.5~139.3 °C; ^{1}H-NMR (CDCl$_3$, 400 MHz) δ: 3.60 (d, J = 4 Hz, 1H, CH), 4.47 (d, J = 4 Hz, 1H, CH), 5.12 (s, 1H, -OH), 6.61 (d, J = 2 Hz, 1H, Ar-H), 6.75~6.80 (m, 2H, Ar-H), 6.87 (t, J = 4 Hz, 1H, Ar-H), 7.01 (d, J = 8 Hz, 1H, Ar-H), 7.20 (t, J = 4 Hz, 1H, Ar-H), 7.31 (d, J = 8 Hz, 1H, Ar-H); ^{13}C NMR (100 MHz, CDCl$_3$) δ: 34.33 (q, J = 2 Hz, C3), 40.11 (C2), 95.47 (q, J = 26 Hz, C16), 104.79 (C8), 112.71 (C12), 114.01 (C6), 117.52 (C4), 118.82 (C14), 121.98 (q, J = 229 Hz, C17), 122.49 (C13), 127.41 (C7), 129.84 (C15), 130.75 (C10), 150.13 (C5), 152.30 (C11), 156.88 (C9), 168.02 (C1); IR (KBr) v_{max} (cm^{-1}): 3405 (OH), 1723 (C=O), 1625, 1597 (Ar), 758 (CF$_3$); HRMS (ESI): m/z calcd for C$_{17}$H$_{10}$F$_3$O$_5$ [M − H]$^+$: 351.0480; found: 351.0236.

2,7-Dihydroxy-9-methoxy-7-(trifluoromethyl)-6a,12b-dihydro-6H,7H-chromeno[3,4-c]chromen-6-one (**3n**): Eluent: DCM:PE:AcOH = 4:8:1; yellowish crystal; M.p. 187.8~188.5 °C; ^{1}H-NMR (DMSO-d_6, 400 MHz) δ: 3.51 (d, J = 4 Hz, 1H, CH), 3.79 (s, 3H, CH$_3$), 4.82 (d, J = 4 Hz, 1H, CH), 6.23 (s, 1H, -OH), 6.60 (dd, J$_1$ = 8 Hz, J$_2$ = 1.6 Hz, 1H, Ar-H), 6.87 (d, J = 8 Hz, 1H, Ar-H), 6.98~7.16 (m, 4H, Ar-H), 8.62 (s, 1H, Ar-H), 9.24 (s, 1H, -OH); ^{13}C NMR (100 MHz, DMSO-d^6) δ: 33.73 (C3), 42.18 (C2), 56.21 (-OCH$_3$), 95.23 (q, J = 26 Hz, C16), 112.63 (C13), 112.98 (C7), 114.68 (C9), 116.71 (C15), 118.36 (C14), 121.27 (C6), 122.65 (C10), 122.69 (q, J = 289 Hz, C17), 124.47 (C4), 140.17 (C5), 144.60 (C12), 148.90 (C11), 154.33 (C8), 164.44 (C1); IR (KBr) v_{max} (cm^{-1}): 3453 (OH), 1728 (C=O), 1605 (Ar), 730 (CF$_3$); HRMS (ESI): m/z calcd for C$_{18}$H$_{12}$F$_3$O$_6$ [M − H]$^+$: 381.0586; found: 381.0317.

11-Chloro-2,7-dihydroxy-7-(trifluoromethyl)-6a,12b-dihydro-6H,7H-chromeno[3,4-c]chromen-6-one (**3o**): Eluent: DCM:PE:AcOH = 4:8.5:0.5; white solid; M.p. 192.5~193.3 °C; ^{1}H-NMR (DMSO-d_6, 400 MHz) δ: 3.61 (d, J = 4 Hz, 1H, CH), 4.86 (d, J = 4 Hz, 1H, CH), 6.25 (s, 1H, -OH), 6.62 (d, J = 8 Hz, 1H, Ar-H), 6.90 (d, J = 8 Hz, 1H, Ar-H), 7.07 (d, J = 12 Hz, 1H, Ar-H), 7.43 (dd, J$_1$ = 8 Hz, J$_2$ = 2.4 Hz, 1H, Ar-H), 7.59 (s, 1H, Ar-H), 9.32 (s, 1H, -OH); ^{13}C-NMR (100 MHz, DMSO-d_6) δ: 33.46 (C3), 41.48 (C2), 95.35 (q, J = 32 Hz, C16), 112.77 (C12), 114.82 (C7), 116.88 (C9), 119.54 (C6), 122.25 (C14), 122.49 (q, J = 288 Hz, C17), 123.93 (C13), 126.56 (C15), 129.84 (C10), 130.79 (C4), 144.51 (C5), 149.64 (C11), 154.41 (C8), 164.13 (C1); IR (KBr) v_{max} (cm^{-1}): 3273 (OH), 1731 (C=O), 1611, 1482 (Ar), 825 (CF$_3$); HRMS (ESI): m/z calcd for C$_{17}$H$_9$ClF$_3$O$_5$ [M − H]$^+$: 385.0091; found: 384.9819.

1,7-Dihydroxy-3-methyl-7-(trifluoromethyl)-6a,12b-dihydro-6H,7H-chromeno[3,4-c]chromen-6-one (**3p**): Crystallization from ethyl acetate-petroleum ether, white solid; M.p. 170.8~171.5 °C; 1H-NMR (CDCl$_3$, 400 MHz) δ: 2.42 (s, 3H, CH$_3$), 3.53 (d, J = 4 Hz, 1H, CH), 4.58 (d, J = 4 Hz, 1H, CH), 5.02 (s, 1H, -OH), 6.46 (s, 1H, -OH), 6.59 (dt, J$_1$ = 8 Hz, J$_2$ = 0.8 Hz, 1H, Ar-H), 6.67 (d, J = 2.4 Hz, 1H, Ar-H), 6.86 (td, J$_1$ = 4 Hz, J$_2$ = 0.8 Hz, 1H, Ar-H), 7.03 (dd, J$_1$ = 8 Hz, J$_2$ = 0.8 Hz, 1H, Ar-H), 7.21 (d, J = 4 Hz, 1H, Ar-H), 7.43 (d, J = 0.8 Hz, 1H, Ar-H); ^{13}C-NMR (100 MHz, CDCl$_3$) δ: 18.95 (-CH$_3$), 31.20 (C3), 39.83 (C2), 95.82 (q, J = 26 Hz, C16), 102.53 (C12), 113.25 (C8), 114.46 (C6), 117.66 (C4), 118.43 (C14), 122.02 (q, J = 229 Hz, C17), 122.66 (C13), 126.75 (C15), 129.86 (C10), 138.93 (C7), 150.35 (C5), 152.57 (C11), 156.21 (C9), 168.39

(C1); IR (KBr) ν_{max} (cm^{-1}): 3461 (OH), 1706 (C=O), 1633, 1594 (Ar), 753 (CF$_3$); HRMS (ESI): m/z calcd for C$_{18}$H$_{12}$F$_3$O$_5$ [M − H]$^+$: 365.0637; found: 365.0378.

4. Conclusions

In conclusion, we have developed a novel strategy to obtain chromeno[3,4-c]chromen-6-ones 3a–p, a new kind of dihydrocoumarins, from the reactions of 3-trifluoroacetylcoumarin with phenols in one step. The chemical structures of the title compounds have been verified with the aid of several spectroscopic methods. The X-ray crystal structures of compounds 3a and 3n showed the twisted conformation in their cyclic structures. The antifungal activity of the compounds 3a–p was assessed against two fungal strains with the mycelial growth rate method. The preliminary results indicated that the phenyl ring at the dihydrocoumarins favors activity against *Fusarium monitiforme*. Compound 3l displayed the highest antifungal activity of 84.6% against *Fusarium monitiforme*. We continue our efforts to obtain new potent, broad spectrum and safe fluorine-bearing coumarins as antifungal drug-like candidates.

Supplementary Materials: The following are available online: Figure S1: Molecular packing of compound 3a (CCDC 1900313) and 3n (CCDC1900314); Table S1: The refinement information and crystallographic data of 3a and 3n; and Figures S2–S17:^{1}H-NMR, ^{13}C-NMR, and HRMS spectra of compounds 3a–p.

Author Contributions: Conceptualization and Methodology, C.-l.X.; Experiments and Writing-Original Draft Preparation, J.-p.B.; Writing-Review & Editing, G.-y.Y.; Analysis-Bioactivity tests, C.-x.W.; Analysis-Crystal analysis, X.Z.; Verification, X.-x.Y.

Funding: This work was supported by the Science and Technology Agency of Henan Province (152102110070, 182102110016) and Guidance Program for Key Scientific Research Items of Higher Education of Henan Province (19B210004).

Conflicts of Interest: The authors declare no conflict of interest.

References

1. Ayati, A.; Bakhshaiesh, T.O.; Moghimi, S.; Esmaeili, R.; Majidzadeh-A, K.; Safavi, M.; Firoozpour, L.; Emami, S.; Foroumadi, A. Synthesis and biological evaluation of new coumarins bearing 2,4-diaminothiazole-5-carbonyl moiety. *Eur. J. Med. Chem.* **2018**, *155*, 483–491. [CrossRef] [PubMed]
2. Chauhan, N.B.; Patel, N.B.; Patel, V.M.; Mistry, B.M. Synthesis and biological evaluation of coumarin clubbed thiazines scaffolds as antimicrobial and antioxidant. *Med. Chem. Res.* **2018**, *27*, 2141–2149. [CrossRef]
3. Al-Amiery, A.A.; Al-Majedy, Y.K.; Kadhum, A.A.; Mohamad, A.B. Novel macromolecules derived from coumarin: Synthesis and antioxidant activity. *Sci. Rep.* **2015**, *5*, 11825. [CrossRef]
4. Ambekar, S.P.; Dhananjaya, M.C.; Arunkumar, S.; Kumar, M.K.; Shobith, R.; Surender, M.; Obelannavar, K.; Rangappa, K.S. Synthesis of coumarin-benzotriazole hybrids and evaluation of their anti-tubercular activity. *Lett. Org. Chem.* **2017**, *15*, 23–31. [CrossRef]
5. Amin, K.M.; Taha, A.M.; George, R.F.; Mohamed, N.M.; Elsenduny, F.F. Synthesis, antitumor activity evaluation, and DNA-binding study of coumarin-based agents. *Arch. Pharm.* **2018**, *351*, e1700199. [CrossRef]
6. Zhang, R.R.; Liu, J.; Zhang, Y.; Hou, M.Q.; Zhang, M.Z.; Zhou, F.G.; Zhang, W.H. Microwave-assisted synthesis and antifungal activity of novel coumarin derivatives: Pyrano[3,2-c]chromene-2,5-diones. *Eur. J. Med. Chem.* **2016**, *116*, 76–83. [CrossRef] [PubMed]
7. Shi, X.; Lv, C.W.; Li, J.; Hou, Z.; Xiao Hui, Y.; Zhang, Z.D.; Luo, X.X.; Yuan, Z.; Li, M.K. Synthesis, photoluminescent, antibacterial activities and theoretical studies of three novel coumarin and dihydropyran derivatives containing a triphenylamine group. *Res. Chem. Intermed.* **2015**, *41*, 8965–8974. [CrossRef]
8. Shen, Y.F.; Liu, L.; Feng, C.Z.; Yang, H.; Chen, C.; Wang, G.X.; Zhu, B. Synthesis and antiviral activity of a new coumarin derivative against spring viraemia of carp virus. *Fish Shellfish Immun.* **2018**, *81*, 57–66. [CrossRef] [PubMed]
9. Sangshetti, J.N.; Khan, F.A.K.; Kulkarni, A.A.; Patil, R.H.; Pachpinde, A.M.; Lohar, K.S.; Shinde, D.B. Antileishmanial activity of novel indolyl–coumarin hybrids: Design, synthesis, biological evaluation, molecular docking study and in silico ADME prediction. *Bioorg. Med. Chem. Lett.* **2016**, *26*, 829–835. [CrossRef]

10. Morsy, S.A.; Farahat, A.A.; Nasr, M.N.A.; Tantawy, A.S. Synthesis, molecular modeling and anticancer activity of new coumarin containing compounds. *Saudi. Pharm. J.* **2017**, *25*, 873–883. [CrossRef]
11. Wu, J.; Peng, T.; Fang, C.; Leng, Y.; Tong, L.; Li, M.; Rong, Q.; Hua, X.; Jian, D.; Duan, W. Synthesis and Anticancer Activity of 7,8-dihydroxy-4-arylcoumarins. *Lett. Drug. Des. Discov.* **2015**, *12*, 366–373. [CrossRef]
12. Sun, J.; Ding, W.X.; Hong, X.P.; Zhang, K.Y. Antioxidant and antitumor activities of 4-arylcoumarins and 4-aryl-3,4-dihydrocoumarins. *Biochimie* **2014**, *107*, 203–210.
13. Sun, J.; Ding, W.X.; Hong, X.P.; Zhang, K.Y. Synthesis and antimicrobial activities of 4-aryl-3,4-dihydrocoumarins and 4-arylcoumarins. *Chem. Nat. Compd.* **2012**, *48*, 16–22. [CrossRef]
14. Pierson, J.T.; Dumètre, A.; Hutter, S.; Delmas, F.; Laget, M.; Finet, J.P.; Azas, N.; Combes, S. Synthesis and antiprotozoal activity of 4-arylcoumarins. *J. Cheminf.* **2010**, *41*, 864–869. [CrossRef] [PubMed]
15. Taechowisan, T.; Lu, C.h.; Shen, Y.m.; Lumyong, A. Secondary metabolites from endophytic Streptomyces aureofaciens CMUAc130 and their antifungal activity. *Microbiology* **2005**, *151*, 1691–1695. [CrossRef] [PubMed]
16. Jung, J.W.; Kim, N.J.; Yun, H.Y.; Han, Y. Recent advances in synthesis of 4-arylcoumarins. *Molecules* **2018**, *23*, 2417. [CrossRef]
17. Pourshojaei, Y.; Jadidi, M.H.; Eskandari, K.; Foroumadi, A.; Asadipour, A. An eco-friendly synthesis of 4-aryl-substituted pyrano-fuzed coumarins as potential pharmacological active heterocycles using molybdenum oxide nanoparticles as an effective and recyclable catalyst. *Res. Chem. Intermed.* **2018**, *44*, 4195–4212. [CrossRef]
18. El-Dean, A.M.K.; Zaki, R.M.; Geies, A.A.; Radwan, S.M.; Tolba, M.S. Synthesis and antimicrobial activity of new heterocyclic compounds containing thieno[3,2-]coumarin and pyrazolo[4,3-]coumarin frameworks. *Russ. J. Bioorg. Chem.* **2013**, *39*, 553–564.
19. Yin, S.; Yang, X.Y.; Zhang, M.; Zhou, Z.Q.; XU, H.H. Synthesis and fungicidal activities of novel fluorine-containing pyrimithamine derivatives. *Chin. J. Synth. Chem.* **2013**, *21*, 317–321.
20. Iaroshenko, V.O.; Ali, S.; Babar, T.M.; Dudkin, S.; Mkrtchyan, S.; Rama, N.H.; Villinger, A.; Langer, P. 4-Chloro-3-(trifluoroacetyl)coumarin as a novel building block for the synthesis of 7-(trifluoromethyl)-6H-chromeno[4,3-b]quinolin-6-ones. *Tetrahedron Lett.* **2011**, *52*, 373–376. [CrossRef]
21. Vdovenko, S.I.; Gerus, I.I.; Kukhar, V.P. Steric effects on the mechanism of reaction of nucleophilic substitution of β-substituted alkoxyvinyl trifluoromethyl ketones with four secondary amines. *J. Phys. Org. Chem.* **2007**, *20*, 190–200. [CrossRef]
22. Huang, D.F.; Zhang, X.H.; Wang, K.H.; Su, Y.P.; Hu, Y.L. Synthesis of 2-trifluoromethyl-2,3-dihydro-1,3,4-oxadiazole derivatives. *J. Northwest Normal Univ.* **2018**, *54*, 70–76.
23. Xu, C.L.; Yang, G.Y.; Zhao, M.Q.; Wang, C.X.; Fan, S.F.; Xie, P.H.; Li, X. Microwave assisted one-pot synthesis of novel trifluoromethyl coumarin thiosemicarbazones and their antifungal activities. *Curr. Microw. Chem.* **2016**, *3*, 60–67.
24. Yang, G.Y.; Yang, J.T.; Wang, C.X.; Fan, S.F.; Xie, P.H.; Xu, C.L. Microwave-assisted TsOH/SiO$_2$-catalyzed one-pot synthesis of novel fluoro-substituted coumarin hydrazones under solvent-free conditions. *J. Fluor. Chem.* **2014**, *168*, 1–8. [CrossRef]
25. Fan, H.F.; Wang, X.W.; Zhao, J.W.; Li, X.J.; Gao, J.M.; Zhu, S.Z. Synthesis of 3-trifluoromethyl-substituted benzo[f]chromene derivatives in a one-pot reaction. *Synth. Commun.* **2013**, *43*, 2883–2891. [CrossRef]
26. Xian, D.W. *Biological Activities Study on 4-arylcoumarins and 4-aryl-3,4 dihydrocoumarins*; Nanjing Agricultural University: Nanjing, China, 2011.
27. Zhang, J.; Yang, Y.; Liu, X.H.; Zhu, H.T. Progress in synthesis of chiral dihydrocoumarin derivatives. In Proceedings of the 3rd Symposium on Molecular Chirality in China, Guiyang, China, 30 July–1 August 2010; pp. 83–87.
28. Wallace, K.J.; Belcher, W.J.; Turner, D.R.; Syed, K.F.; Steed, J.W. Slow anion exchange, conformational equilibria, and fluorescent sensing in venus flytrap aminopyridinium-based anion hosts. *J. Am. Chem. Soc.* **2003**, *125*, 9699. [CrossRef] [PubMed]
29. Huang, Z.X. *Plant Chemical Protection Experiment Instruction*, 1st ed.; Agriculture Press: Beijing, China, 1993; pp. 51–52.

30. Yang, G.Y.; Wang, C.X.; Fan, S.F.; Xie, P.H.; Jin, Q.; Xu, C.L. Microwave assisted solvent-free synthesis of 3-(trifluoroacetyl)coumarins. *Chin. J. Org. Chem.* **2015**, *35*, 1173–1178. [CrossRef]

Sample Availability: Samples of the compounds are not available from the authors.

© 2019 by the authors. Licensee MDPI, Basel, Switzerland. This article is an open access article distributed under the terms and conditions of the Creative Commons Attribution (CC BY) license (http://creativecommons.org/licenses/by/4.0/).

Article

Traceless Solid-Phase Synthesis of Ketones via Acid-Labile Enol Ethers: Application in the Synthesis of Natural Products and Derivatives

Eva Schütznerová [1], Anna Krchňáková [2] and Viktor Krchňák [1,2,*]

1. Department of Organic Chemistry, Faculty of Science, Palacký University, 17. listopadu 12, 771 46 Olomouc, Czech Republic; eva.schutznerova@upol.cz
2. Department of Chemistry and Biochemistry, 251 Nieuwland Science Center, University of Notre Dame, Notre Dame, IN 46556, USA; viktor@torviq.com
* Correspondence: vkrchnak@nd.edu

Academic Editor: Gianfranco Favi
Received: 13 March 2019; Accepted: 8 April 2019; Published: 10 April 2019

Abstract: In solid-phase organic synthesis, Wang resin is traditionally used for the immobilization of acids, alcohols, phenols, and amines. We report the use of Wang resin for the traceless synthesis of ketones via acid-labile enol ethers. We demonstrate the practicality of this synthetic strategy on the solid-phase synthesis of pyrrolidine-2,4-diones, which represent the core structure of several natural products, including tetramic acid. Base-triggered condensation of pyrrolidine-2,4-diones yielded 4-hydroxy-1,1′,2′,5-tetrahydro-2H,5′H-[3,3′-bipyrrole]-2,5′-diones.

Keywords: solid-phase synthesis; ketone; traceless synthesis; natural products; enol ethers

---

1. Introduction

Solid-phase synthesis is a very attractive methodology for the time-efficient synthesis of diverse organic molecules [1–4]. The initial step in the entire synthetic sequence is the selection of the appropriate linker for the immobilization of the first building block. The Wang linker [5] is the most commonly used acid-labile linker, and it has been used to immobilize carboxylic acids, alcohols, phenols, and amines [6,7]. Typically, after finishing the synthesis, the product is released from the resin and the functional group that was initially used for immobilization will remain attached to the product. This functional group, referred to as the trace of the linker, may be an inherent part of the target molecules (peptides are the best examples), but for the synthesis of organic molecules that do not share a common functional group, the trace of the linker is undesirable. Therefore, numerous synthetic routes have been devised that enabled the synthesis to be performed in a traceless manner; heterocycles are undoubtedly the highly representative examples [8]. Here, we expand the application of Wang resin to the novel, traceless synthesis of ketones from acid-labile enol ethers, which were prepared via the Wittig reaction of resin-bound esters.

Enol ethers represent valuable synthons in organic synthesis, and numerous methods for the synthesis of enol ethers have been reported; however, the Wittig olefination of esters is used rarely [9]. This 'nonclassical' Wittig reaction of carboxylate esters suffers from sluggish reactivity due to the low electrophilicity of the carbonyl carbon towards phosphoranes compared to the electrophilicity of aldehydes and ketones [10]. The reactions typically require the use of microwave irradiation (conventional heating reportedly did not yield any product) [11] or reactive phosphoranes such as cyanomethylenetrimethylphosphorane [12]. An alternative approach to enol ethers is the alkylidenation of ester carbonyls with metal carbene complexes [13] used, for example, in traceless solid-phase synthesis of indoles [14,15].

On the other hand, the intramolecular Wittig cyclization of phosphonium salt proceeded smoothly, and this reaction was successfully used by Hercouet and Le Corre in 1979 for the synthesis of dihydrofurans and dihydropyrans [16,17], and later, this technique was used for the preparation of carbocycles [18–20] and heterocycles such as 2-alkylthiobenzimidazoles [21], indoles [22], 2-quinolones [23], and, recently, 4-alkoxy-1,5-dihydro-2H-pyrrol-2-ones [24,25].

Not surprisingly, although only a limited number of reports have described the application of enol ethers in the synthesis of ketones, the preparation of ketones from vinyl ethers using a Grignard reagent was reported in 1955 [26]. The hydrolysis of vinyl ethers has been studied on numerous occasions [27–29]; however, this technique has not been applied for general preparative use. Among the few recent reports in this area, silyl enol ethers were enantioselectively converted to ketones by Cheon and coworkers [30,31]. Ketones were prepared by the palladium-catalyzed regioselective arylation of vinyl ethers [32] and by the hydrolysis (MeOH/aq HCl, reflux) of vinyl ethers [33].

To summarize the prior work, neither the synthesis of enol ethers via a Wittig olefination nor the use of enol ethers in the synthesis of ketones is a method of choice for ketone synthesis. Here, we report a simple and practical synthesis of acid-labile Wang resin-bound enol ethers via the Wittig olefination of carboxylate esters and subsequent acid-mediated traceless release of the ketones from the resin.

2. Results and Discussion

To demonstrate the practical use of a Wang linker for the traceless synthesis of pharmacologically relevant ketones, we report the synthesis of pyrrolidine-2,4-diones. Pyrrolidine-2,4-dione, the core structure of tetramic acid, was selected as a pharmacologically relevant structure found in natural products [34,35]. Numerous chemical routes for the preparation of tetramic acid and its derivatives have been developed, and the reported syntheses have been reviewed [35–37].

2.1. Synthesis

The assembly of the acyclic precursor was efficiently carried out on a solid phase using optimized protocols for the individual transformations. Wang resin **1** was acylated with the N-[(9H-Fluoren-9-ylmethoxy)carbonyl] (Fmoc) N-alkyl amino acids (sarcosine, 4-OBzl-proline (Hyp(Bzl)), 2-indolinecarboxylic acid (Idc), and methyltyrosine (OtBu)), and the Fmoc protecting group was cleaved to yield resin **4** (Scheme 1, route I). Because of the limited number of commercially available N-alkyl amino acids, we also evaluated an alternative route using resin N-alkylation. Thus, Wang resin **1** was esterified with Fmoc–amino acids, the Fmoc group was cleaved, and amine **2** was reacted with 4-nitrobenzenesulfonyl chloride (Ns-Cl) (resin **3**) to facilitate Mitsunobu alkylation with alcohols [38], which introduced the N-substituent (Scheme 1, route II). This reaction sequence was designed to enhance the diversity of compounds available with this method. Removal of the Ns group yielded secondary amine **4**. Resin-bound amine **4** was then acylated with bromoacetic acid (resin **5**) and reacted with PPh$_3$ to form phosphonium salt **6**. The resin-bound phosphonium salt was not isolated, and the LC/MS analysis of trifluoroacetic acid (TFA)-cleaved sample revealed the expected molecular ion in all prepared compounds. The addition of trimethylamine (TEA) or 1,8-diazabicyclo[5.4.0]undec-7-ene (DBU) in N-methyl-2-pyrrolidone (NMP) triggered the Wittig olefination. The olefination of the ester proceeded smoothly at ambient temperature. TFA exposure released products **8**. Crude products were isolated and then purified by reversed-phase (RP) HPLC in an acidic mobile phase (aqueous formic acid or TFA), and tetramic acids **8** were fully characterized (Table 1). The structures of the building blocks are listed in Figure 1.

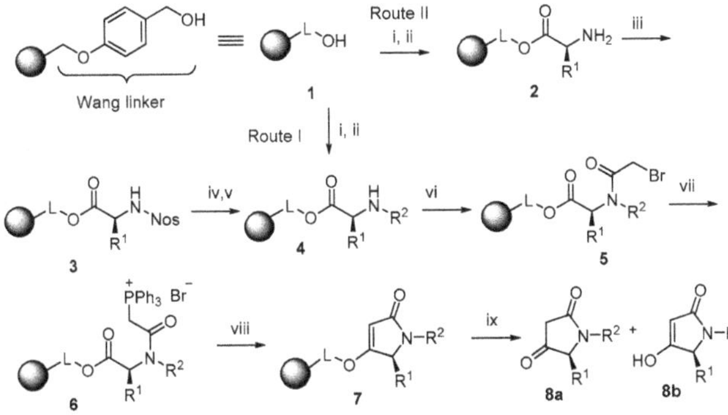

Scheme 1. Traceless solid-phase synthesis of pyrrolidine-2,4-dione. Reagents and conditions: (i) Fmoc-amino acid-OH (Fmoc–AA–OH), N,N′-diisopropylcarbodiimide (DIC), hydroxybenzotriazole (HOBt), 4-(dimethylamino)pyridine (DMAP), dichloromethane (DCM)/ dimethylformamid (DMF) (1:1), room temperature (rt), overnight; (ii) piperidine/DMF (1:1), rt, 15 min; (iii) Ns-Cl, 2,6-lutidine, DCM, rt, 2 h; (iv) alcohol, PPh$_3$, diisopropyl azodicarboxylate (DIAD), anhydrous THF, rt, 2 h; (v) mercaptoethanol, DBU, DMF, rt, 5 min; (vi) bromoacetic acid, DIC, DCM, 5 min, then the precipitated diisopropyl urea (DIU) was removed by filtration, N,N-diisopropylethylamine (DIEA) was added, and the solution was transferred to the reaction vessel with the resin, rt, 1 h; (vii) PPh$_3$, anhydrous NMP, rt, overnight; (viii) TEA, anhydrous NMP, rt, 2–48 h, or DBU, anhydrous NMP, 60 °C, overnight (see the Supplementary Materials Table S1 for details); (ix) TFA/DCM 1:1, rt, 1 h. Abbreviations:; AA: amino acid; DIC: N,N′-diisopropylcarbodiimide; HOBt = hydroxybenzotriazole; DMAP: 4-(dimethylamino)pyridine; DCM: dichloromethane; DMF: dimethylformamide; rt: room temperature; DIAD: diisopropyl azodicarboxylate; DIU: diisopropyl urea; DIEA: N,N-diisopropylethylamine.

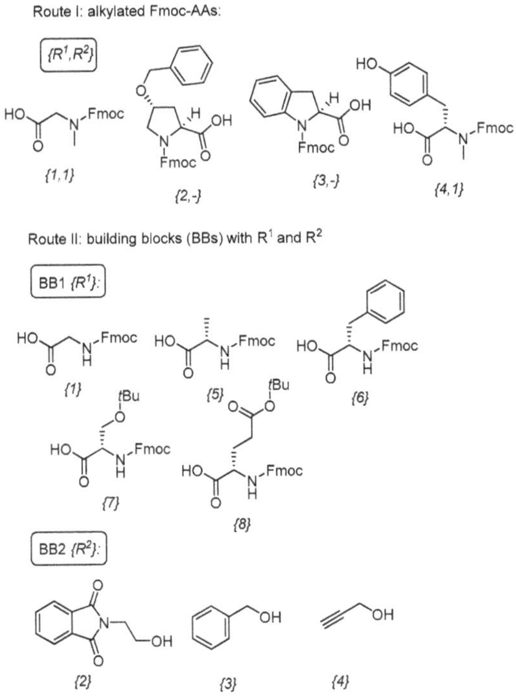

Figure 1. List of building blocks (BBs) used in synthesis.

To address the presence of tautomers **8a** and **8b**, we collected their NMR spectra in CDCl$_3$ and DMSO-d_6. As expected and in agreement with previously reported data [39,40], we observed the presence of one isomer (**8a**) in CDCl$_3$ based on the diagnostic proton resonances corresponding to the two methylene protons of isomer **8a**{1,2} (δ = 4.06 and 2.92 ppm) and the ketone carbon of isomer **8a**{1,2} (δ = 203.0 ppm). However, the NMR spectrum acquired in DMSO-d_6 revealed the presence of a mixture of tautomers **8a** and **8b** (in an approximately 1:1 ratio) based on the diagnostic proton resonances corresponding to the methylene protons of isomer **8a**{1,2} (δ = 3.89 and 2.91 ppm) and the olefinic proton of isomer **8b**{1,2} (δ = 4.62 ppm). Moreover, the ^{13}C NMR spectra unambiguously showed the presence of the ketone carbon of isomer **8a**{1,2} (δ = 205.0 ppm) as well as the olefin carbon of isomer **8b**{1,2} (δ = 93.4 ppm). The NMR spectra of all other compounds are presented in the Supplementary Materials.

Table 1. List of prepared compounds **8** and **9** (purity/yield).

Entry	Compound	Route	R^1	R^2	Purity of 8 (%)	Yield of 8 (%) [b]	Yield of 9 (%) [c]
1	{1,1}	I	H	Me	85	NI [d]	47
2	{1,2}	II	H	PhthN(CH$_2$)$_2$	81	70	53
3	{2,-}	I	Hyp(Bzl) [a]		81	NI [d]	37
4	{3,-}	I	Idc [a]		52	31	NT [e]
5	{4,1}	I	CH$_2$C$_6$H$_4$OH	Me	77	25	NT [e]
6	{5,3}	II	Me	Bn	56	NI [d]	23
7	{6,2}	II	Bn	PhthN(CH$_2$)$_2$	72	30	49
8	{6,4}	II	Bn	CH$_2$CCH	88	21	NT [e]
9	{7,2}	II	CH$_2$OH	PhthN(CH$_2$)$_2$	68	52	NT [e]
10	{8,2}	II	(CH$_2$)$_2$CO$_2$H	PhthN(CH$_2$)$_2$	52	21	NT [e]

Note: In the compound labeling, the first digit refers to the R^1 substituent and the second digit refers to the R^2 substituent; [a] compounds were prepared using cyclic amino acids Hyp(Bzl) for {2,-} and Idc for {3,-}; [b] HPLC purification in acetonitrile/aqueous TFA or formic acid; [c] purification in acetonitrile/aqueous ammonium acetate; [d] NI, not isolated, crude compounds **8** were converted to **9**; [e] NT, not tested.

2.2. Self-Condensation

Today's search for new drugs is focused on the design and synthesis of compounds structurally resembling natural products, referred to as biology-oriented synthesis [41–45]. Because our synthetic route provides traceless access to ketones, we investigated their potential for self-condensation that would convert pyrrolidine-2,4-diones **8** to the natural product derivatives 4-hydroxy-1,1′,2′,5-tetrahydro-2H,5′H-[3,3′-bipyrrole]-2,5′-diones **9** (Scheme 2). This condensed product has not been exploited as a potential pharmacologically relevant structure, although an analogous self-condensation was reported as a side-reaction in 1985 [39]. It is important to emphasize that the 2-substituted pyrrolidine-2,4-diones do not form condensed products.

In a search for reaction conditions for condensation, we found that pyrrolidine-2,4-diones **8** undergo self-condensation in basic solution at ambient temperature and form bisheterocycles **9**. Exposure of purified pyrrolidine-2,4-dione **8**{1,2} to 10 mM aqueous ammonium acetate buffer in acetonitrile triggered quantitative conversion within 24 h, and compound **9**{1,2} was isolated in 80% yield. We also purified crude compound **8**{1,2} by reversed-phase HPLC in an acetonitrile/aqueous ammonium acetate buffer and isolated clean **9**{1,2} in an overall 53% yield (without purification of the **8**{1,2}). The ^{1}H and ^{13}C NMR spectra and HRMS analysis confirmed the bisheterocyclic structure of **9**{1,2}, formed by the self-condensation of two molecules of tetramic acid.

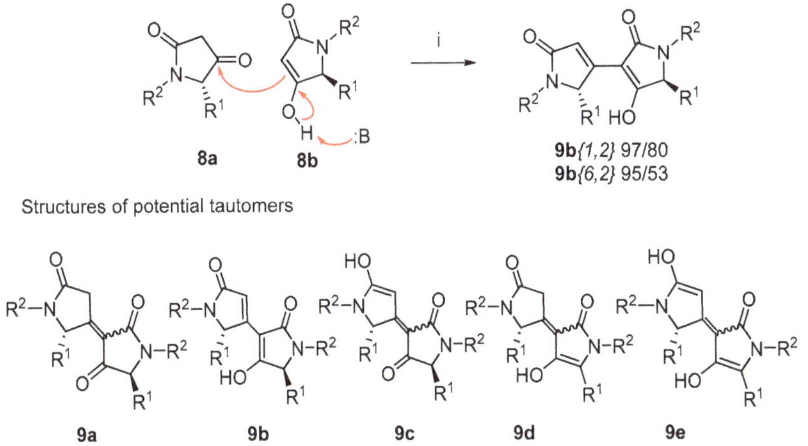

Scheme 2. Self-condensed pyrrolidine-2,4-diones and possible tautomers. Reagents and conditions: (i) aqueous ammonium acetate buffer, MeCN, rt, 24 h.

Based on the results described above, we also tested the stability of purified pyrrolidine-2,4-dione 8{1,2}. The compound was stable and no spontaneous conversion into 9{1,2} was observed after storage of the HPLC-purified sample in DMSO at 4 °C (in a refrigerator) for one month.

2.3. Structure Determination

Base-triggered condensed product 9 can exist as several tautomers (Scheme 2). To determine the structure of the tautomer present in solution, NMR spectra were measured both in DMSO-d_6 and CDCl$_3$, and they exhibited analogous patterns. The proton NMR spectrum of 9{1,2} showed two methylene singlets (δ = 4.43 and 3.99) and one singlet corresponding to an olefinic proton (δ = 6.15). These resonances indicated the presence of tautomer 9b or 9c. The ^{13}C NMR spectra did not show the presence of a ketone carbon (δ = 205.0 in the case of compound 8), eliminating tautomer 9c. Analogous findings were observed with compound 9{5,3}, which exhibited one diagnostic olefin singlet and two quartets coupled to two methyl groups, suggesting the same type of tautomer 9b. The LC/MS analysis and NMR spectra of 9 also did not indicate the presence of any diastereomer, confirming that the optical integrity of the amino acid chiral carbon was preserved.

3. Conclusion

We demonstrated a general and novel application of a Wang linker for the traceless solid-phase synthesis of ketones from acid-labile enol ethers. The synthesis of tetramic acid derivatives, including self-condensed 4-hydroxy-1,1',2',5-tetrahydro-2H,5'H-[3,3'-bipyrrole]-2,5'-diones, highlighted the practical use of this protocol for the synthesis of pharmacologically relevant natural products. The advantages of solid-phase synthesis enabled time-efficient synthesis and preparation of products with any combination of building blocks. An extension of this synthetic route to the synthesis of other natural products and its application in self-condensation is in progress and will be reported in due course.

4. Experimental Procedures

4.1. General Information

Solvents were used without further purification. The Wang linker (100–200 mesh, 1% DVB, 0.9 mmol/g) was used. Synthesis was carried out on Domino Blocks [46] in disposable polypropylene reaction vessels. The volume of wash solvent was 10 mL per 1 g of resin. For washing, resin slurry was shaken with the fresh solvent for at least 1 min before changing the solvent. After adding a reagent solution, the resin slurry was manually vigorously shaken to break any potential resin clumps. Resin-bound intermediates were dried by a stream of nitrogen for prolonged storage and/or quantitative analysis.

4.2. Esterification with Fmoc–AA (Resins 2 and 4)

Resin **1** (1 g) was washed with DCM (3 × 10 mL) and treated with a solution of Fmoc–amino acid (2 mmol) and HOBt·H_2O (306 mg, 2 mmol), DMAP (61 mg, 0.5 mmol), and DIC (312 µL, 2 mmol) in 10 mL of DMF/DCM (1:1), and the reaction slurry was shaken overnight at room temperature. The resin was washed with DMF (3 × 10 mL) and DCM (5 × 10 mL).

Quantification of Resin Loading: A sample of resin was washed 5 times with DCM, 3 times with MeOH, and then dried with nitrogen. A 10-mg portion of resin was cleaved with 50% TFA in DCM for 30 min. The cleavage cocktail was evaporated by a stream of nitrogen, and the cleaved compound was extracted into 1 mL of MeOH. This sample of Fmoc derivate was analyzed by LC/MS, and the quantity was determined by comparison with a standard (Fmoc–Ala–OH; concentration: 1 mg/mL). The loading of the resin was determined by external standard method by integration of the UV response at 300 nm.

Fmoc deprotection: The resin (1 g) was washed with DCM (3 × 10 mL) and DMF (3 × 10 mL) and treated with a solution of 50% piperidine in DMF (10 mL) for 15 min at room temperature. The resin was thoroughly washed with DMF (5 × 10 mL) and DCM (3 × 10 mL).

4.3. Reaction with Ns-Cl (Resin 3)

Resin **2** (1 g) was swollen with DCM, washed with DCM (3 times), and a solution of Ns-Cl (3 mmol, 663 mg) and 2,6-lutidine (1 mmol, 382 µL) in DCM (10 mL) was added. The reaction slurry was shaken at rt for 2 h, and then the resin was washed with DCM (3 times).

4.4. Fukuyama N-Alkylation (Resin 4)

The Ns resin **3** (1 g) was swollen in anhydrous THF, a solution of alcohol (2 mmol) with PPh_3 (2 mmol, 524 mg) in anhydrous THF (8 mL) was added to the resin, and the slurry was left in the freezer for 30 min. Subsequently, cooled DIAD (2 mmol, 393 µL) in anhydrous THF (2 mL) was added to the resin, and the reaction mixture was shaken at rt for 2 h. The resin was then washed with THF (3 times) and DCM (3 times).

Resin was swollen in DCM and washed with DMF (3 times), and the Ns group was cleaved with 2-mercaptoethanol (6 mmol, 420 µL) and DBU (2 mmol, 300 µL) in DMF (10 mL) for 5 min. The resin was washed with DMF (3 times) and DCM (3 times).

4.5. Acylation with Bromoacetic Acid (Resin 5)

A solution of bromoacetic acid (700 mg, 5 mmol) was prepared in another syringe with a frit, and DIC (386 µL, 2.5 mmol) was added. After 5 min, DIU was removed by filtration, lutidine (292 µL, 2.5 mmol) was added, and the solution was transferred to the syringe with resin **4**. The slurry was shaken for 1 h at room temperature. The resin was washed with DCM (3 × 10 mL).

4.6. Preparation of the Triphenylphosphonium Salt (Resin 6)

Resin **5** (1 g) was washed with DCM (3 × 10 mL) and anhydrous NMP (3 × 10 mL). A solution of triphenylphosphine (1.05 g, 4 mmol) in anhydrous NMP (10 mL) was added to the resin, and the slurry was shaken overnight at room temperature. The resin was washed with NMP (3 × 10 mL) and DCM (3 × 10 mL).

4.7. Wittig Olefination (Resins 7)

Resin-bound triphenylphosphonium salt **6** (500 mg), prepared according to a recently published procedure [24,25] was washed with DCM (3 × 10 mL) and anhydrous NMP (3 × 10 mL). A solution of TEA (70 µL, 0.5 mmol) or DBU (71 µL, 0.5 mmol) in anhydrous NMP (5 mL) was added to the resin, and the slurry was shaken at room temperature (see Table S1 in the Supplementary Materials). The resin was washed with NMP (3 × 10 mL) and DCM (3 × 10 mL).

4.8. Cleavage from the Resin (Compounds 8)

The cyclized resin **7** (250 mg) was washed with DCM (3 × 10 mL). The resin was treated with 3 mL of a solution of 50% TFA in DCM for 1 h at room temperature. The TFA solution was collected, and then the resin was washed with 10% TFA in DCM (5 mL) and DCM (5 mL), and the combined extracts were concentrated under a stream of nitrogen. The crude product was dissolved in 3 mL of MeOH and purified by semipreparative RP HPLC in MeCN/aqueous TFA or formic acid.

4.9. Self-Condensation (Compounds 9)

Two compounds, **8**{1,2} and **8**{6,2}, purified by RP HPLC in MeCN/aqueous 0.1% TFA or formic acid, were dissolved in 600 µL of DMSO and 5 mL of 10 mM aqueous ammonium acetate was added. The solution was left at rt overnight, and then condensed compounds **9** were purified in MeCN/10 mM aqueous ammonium acetate (Table S2). The remaining compounds were subjected to self-condensation without purification; however, the self-condensed products **9** were purified.

Supplementary Materials: The following are available online: ^{1}H and ^{13}C NMR spectral data and figures of all compounds.

Author Contributions: V.K. conceived and designed the experiments and wrote the manuscript; E.S. and A.K. performed the experiments.

Funding: This research was supported by the Department of Chemistry and Biochemistry, University of Notre Dame, and by research grant 16-06446S from the Grant Agency of the Czech Republic (GACR).

Acknowledgments: We thank Adam Přibylka, for the solid-phase synthesis of several model compounds.

Conflicts of Interest: The authors declare no conflict of interest.

References

1. Balkenhohl, F.; von dem Bussche-Hünnefeld, C.; Lansky, A.; Zechel, C. Combinatorial Synthesis of Small Organic Molecules. *Angew. Chem. Int. Ed.* **1996**, *35*, 2288–2337.
2. Orzaez, M.; Mora, P.; Mondragon, L.; Perez-Paya, E.; Vicent, M.J. Solid-Phase Chemistry: A Useful Tool to Discover Modulators of Protein Interactions. *Int. J. Peptide Res. Therap.* **2007**, *13*, 281–293. [CrossRef]
3. Nielsen, T.E.; Meldal, M. Solid-Phase Synthesis of Complex and Pharmacologically Interesting Heterocycles. *Curr. Opin. Drug Discovery Dev.* **2009**, *12*, 798–810.
4. Krchnak, V. *Solid-Phase Synthesis of Nitrogenous Heterocycles*; Springer: Cham, Switzerland, 2017.
5. Wang, S.-S. P-Alkoxybenzyl Alcohol Resin and P-Alkoxybenzyloxycarbonylhydrazide Resin for Solid Phase Synthesis of Protected Peptide Fragments. *J. Am. Chem. Soc.* **1973**, *95*, 1328–1333. [PubMed]
6. James, I.W. Linkers for Solid Phase Organic Synthesis. *Tetrahedron* **1999**, *55*, 4855–4946. [CrossRef]
7. Soural, M.; Hlavac, J.; Krchnak, V. Linkers for Solid-Phase Peptide Synthesis. In *Solid-Phase Peptide Synthesis*; Hughes, A.B., Ed.; Willey-VCH: Weinheim, Germany, 2011.
8. Krchnak, V.; Holladay, M.W. Solid Phase Heterocyclic Chemistry. *Chem. Rev.* **2002**, *102*, 61–91. [CrossRef]
9. Winternheimer, D.J.; Shade, R.E.; Merlic, C.A. Methods for Vinyl Ether Synthesis. *Synthesis* **2010**, *2010*, 2497–2511.
10. Murphy, P.J.; Lee, S.E. Recent Synthetic Applications of the Non-Classical Wittig Reaction. *J. Chem. Soc. Perkin Trans. 1* **1999**, *21*, 3049–3066.
11. Sabitha, G.; Reddy, M.M.; Srinivas, D.; Yadov, J.S. Microwave Irradiation: Wittig Olefination of Lactones and Amides. *Tetrahedron Lett.* **1999**, *40*, 165–166. [CrossRef]

12. Tsunoda, T.; Takagi, H.; Takaba, D.; Kaku, H.; Ito, S. Cyanomethylenetrimethylphosphorane, a Powerful Reagent for the Wittig Olefination of Esters, Lactones and Imides. *Tetrahedron Lett.* **2000**, *41*, 235–237. [CrossRef]
13. Mortimore, M.; Kocienski, P. A New Synthesis of Spiroacetals Via Alkylidenation of Ester Carbonyls With Metal Carbene Complexes. *Tetrahedron Lett.* **1988**, *29*, 3357–3360.
14. Macleod, C.; Hartley, R.C.; Hamprecht, D.W. Novel Functionalized Titanium(IV) Benzylidenes for the Traceless Solid- Phase Synthesis of Indoles. *Org. Lett.* **2002**, *4*, 75–78.
15. Macleod, C.; McKiernan, G.J.; Guthrie, E.J.; Farrugia, L.J.; Hamprecht, D.W.; Macritchie, J.; Hartley, R.C. Synthesis of 2-Substituted Benzofurans and Indoles Using Functionalized Titanium Benzylidene Reagents on Solid Phase. *J. Org. Chem.* **2003**, *68*, 387–401. [PubMed]
16. Hercouet, A.; Le Corre, M. Une Voie D'Acces Simple Aux Dihydro-2,3 Furannes Et Aux Dihydro-2,3 Pyrannes. *Tetrahedron Lett.* **1979**, *20*, 5–6. [CrossRef]
17. Zhu, J.; Kayser, M.M. Synthesis of Enol Lactones Under a Solid/Liquid Phase Transfer Witting Reaction. *Synth. Commun.* **1994**, *24*, 1179–1186.
18. Sakhautdinov, I.M.; Khalikov, I.G.; Galin, F.Z.; Egorov, V.A.; Lakeev, S.N.; Maidanova, I.O. The Comparative Study of Intramolecular Cyclization of Phthalimide Containing Sulfur and Phosphonium γ-Ylides. *Bashk. Khim. Zh.* **2007**, *14*, 96–99.
19. Zhao, G.; Zhang, Q.; Zhou, H. Propargyl-Allenyl Isomerizations and Electrocyclizations for the Functionalization of Phosphonium Salts: One-Pot Synthesis of Polysubstituted Vinylbenzenes and Naphthalenes. *Adv. Synth. Catal.* **2013**, *355*, 3492–3496.
20. Rahim, M.A.; Sasaki, H.; Saito, J.; Fujiwara, T.; Takeda, T. Intramolecular Carbonyl Olefination of Esters. Regioselective Preparation of Enol Ethers of Cyclic Ketones by the Titanocene(II)-Promoted Reaction of Alkyl ω,ω-Bis(Phenylthio)Alkanoates. *Chem. Commun. (Cambridge, U.K.)* **2001**, *7*, 625–626.
21. Slade, R.M.; Phillips, M.A.; Berger, J.G. Application of an Almost Traceless Linker in the Synthesis of 2-Alkylthiobenzimidazole Combinatorial Libraries. *Mol. Diversity* **1998**, *4*, 215–219. [CrossRef]
22. Hughes, I. Application of Polymer-Bound Phosphonium Salts As Traceless Supports for Solid Phase Synthesis. *Tetrahedron Lett.* **1996**, *37*, 7595–7598. [CrossRef]
23. Desai, V.G.; Shet, J.B.; Tilve, S.G.; Mali, R.S. Intramolecular Wittig Reactions. A New Synthesis of Coumarins and 2-Quinolones. *J. Chem. Res. Synop.* **2003**, *10*, 628–629. [CrossRef]
24. Schütznerova, E.; Pribylka, A.; Krchnak, V. N^a-Amino Acid Containing Privileged Structures: Design, Synthesis and Use in Solid-Phase Peptide Synthesis. *Org. Biomol. Chem.* **2018**, *16*, 5359–5362. [CrossRef] [PubMed]
25. Schütznerova, E.; Oliver, A.G.; Pribylka, A.; Krchnak, V. Solid-Phase Synthesis of Tetramic Acid Via Resin-Bound Enol Ethers As a Privileged Scaffold in Drug Discovery. *Adv. Synth. Catal.* **2018**, *360*, 3693–3699.
26. Hill, C.M.; Prigmore, R.M.; Moore, G.J. Grignard Reagents and Unsaturated Ethers. IV. The Synthesis and Reaction of Several Vinyl Ethers With Grignard Reagents. *J. Am. Chem. Soc.* **1955**, *77*, 352–354.
27. Fife, T.H. Vinyl Ether Hydrolysis. The Facile General Acid Catalyzed Conversion of 2-Ethoxy-1-Cyclopentene-1-Carboxylic Acid to Cyclopentanone. *J. Am. Chem. Soc.* **1965**, *87*, 1084–1089. [PubMed]
28. Fedor, L.R.; McLaughlin, J. Vinyl Ether Hydrolysis. Specific Acid Catalyzed Hydrolysis of 4-Methoxy-3-Buten-2-One. *J. Am. Chem. Soc.* **1969**, *91*, 3594–3597. [CrossRef] [PubMed]
29. Bergman, N.A.; Halvarsson, T. Hydrolysis of the Vinyl Ether Functional Group in a Model for Prostacyclin in Which the Carboxyl Group Has Been Replaced by a Pyridine Ring. *J. Org. Chem.* **1989**, *54*, 2137–2142. [CrossRef]
30. Cheon, C.H.; Yamamoto, H. A Brönsted Acid Catalyst for the Enantioselective Protonation Reaction. *J. Am. Chem. Soc.* **2008**, *130*, 9246–9247. [PubMed]
31. Cheon, C.H.; Kanno, O.; Toste, F.D. Chiral Brönsted Acid From a Cationic Gold(I) Complex: Catalytic Enantioselective Protonation of Silyl Enol Ethers of Ketones. *J. Am. Chem. Soc.* **2011**, *133*, 13248–13251. [CrossRef]
32. Liu, M.; Hyder, Z.; Sun, Y.; Tang, W.; Xu, L.; Xiao, J. Efficient Synthesis of Alkyl Aryl Ketones & Ketals Via Palladium-Catalyzed Regioselective Arylation of Vinyl Ethers. *Org. Biomol. Chem.* **2010**, *8*, 2012–2015.

33. Huang, W.; Rong, H.Y.; Xu, J. Cyclic a-Alkoxyphosphonium Salts From (2-(Diphenylphosphino)Phenyl)Methanol and Aldehydes and Their Application in Synthesis of Vinyl Ethers and Ketones Via Wittig Olefination. *J. Org. Chem.* **2015**, *80*, 6628–6638. [PubMed]
34. Mo, X.; Li, Q.; Ju, J. Naturally Occurring Tetramic Acid Products: Isolation, Structure Elucidation and Biological Activity. *RSC Adv.* **2014**, *4*, 50566–50593.
35. Bai, W.-J.; Lu, C.; Wang, X. Recent Advances in the Total Synthesis of Tetramic Acid-Containing Natural Products. *J. Chem.* **2016**. [CrossRef]
36. Royles, B.J.L. Naturally Occurring Tetramic Acids: Structure, Isolation, and Synthesis. *Chem. Rev.* **1995**, *95*, 1981–2001. [CrossRef]
37. Athanasellis, G.; Igglessi-Markopoulou, O.; Markopoulos, J. Tetramic and Tetronic Acids As Scaffolds in Bioinorganic and Bioorganic Chemistry. *Bioinorg. Chem. Appl.* **2010**. [CrossRef]
38. Fukuyama, T.; Jow, C.-K.; Cheung, M. 2- and 4-Nitrobenzenesulfonamides: Exceptionally Versatile Means for Preparation of Secondary Amines and Protection of Amines. *Tetrahedron Lett.* **1995**, *36*, 6373–6374. [CrossRef]
39. Heinicke, G.W.; Morella, A.M.; Orban, J.; Prager, R.H.; Ward, A.D. Central-Nervous-System Active Compounds .XVI. Some Chemistry of 6-Oxo Caprolactams Derived From an Enamine Ring-Expansion Synthesis. *Aust. J. Chem.* **1985**, *38*, 1847–1856. [CrossRef]
40. Jeong, Y.-C.; Moloney, M.G. Tetramic Acids as Scaffolds: Synthesis, Tautomeric and Antibacterial Behaviour. *Synlett* **2009**, *15*, 2487–2491.
41. Wilk, W.; Zimmermann, T.J.; Kaiser, M.; Waldmann, H. Principles, Implementation, and Application of Biology-Oriented Synthesis (BIOS). *Biol. Chem.* **2010**, *391*, 491–497.
42. Kaiser, M.; Wetzel, S.; Kumar, K.; Waldmann, H. Biology-Inspired Synthesis of Compound Libraries. *Cell. Mol. Life Sci.* **2008**, *65*, 1186–1201. [CrossRef]
43. Wetzel, S.; Klein, K.; Renner, S.; Rauh, D.; Oprea, T.I.; Mutzel, P.; Waldmann, H. Interactive Exploration of Chemical Space With Scaffold Hunter. *Nat. Chem. Biol.* **2009**, *5*, 581–583. [CrossRef] [PubMed]
44. Wetzel, S.; Bon, R.S.; Kumar, K.; Waldmann, H. Biology-Oriented Synthesis. *Angew. Chem. Int. Ed.* **2011**, *50*, 10800–10826. [CrossRef] [PubMed]
45. Van Hattum, H.; Waldmann, H. Biology-Oriented Synthesis: Harnessing the Power of Evolution. *J. Am. Chem. Soc.* **2014**, *136*, 11853–11859. [CrossRef] [PubMed]
46. Krchnak, V.; Padera, V. The Domino Blocks: A Simple Solution for Parallel Solid Phase Organic Synthesis. *Bioorg. Med. Chem. Lett.* **1998**, *22*, 3261–3264. [CrossRef]

Sample Availability: Samples of the compounds are not available from the authors.

© 2019 by the authors. Licensee MDPI, Basel, Switzerland. This article is an open access article distributed under the terms and conditions of the Creative Commons Attribution (CC BY) license (http://creativecommons.org/licenses/by/4.0/).

Article

Green and Facile Assembly of Diverse Fused N-Heterocycles Using Gold-Catalyzed Cascade Reactions in Water

Xiuwen Jia [1], Pinyi Li [1], Xiaoyan Liu [1], Jiafu Lin [1,*], Yiwen Chu [1], Jinhai Yu [2], Jiang Wang [3,4], Hong Liu [3,4,*] and Fei Zhao [1,*]

[1] Antibiotics Research and Re-evaluation Key Laboratory of Sichuan Province, Sichuan Industrial Institute of Antibiotics, Chengdu University, Chengdu 610052, China; jiaxiuwen2018@126.com (X.J.); pinyiLi19950206@126.com (P.L.); 19940826097@163.com (X.L.); siiakyb@139.com (Y.C.)
[2] School of Biological Science and Technology, University of Jinan, Jinan 250022, China; bio_yujh@ujn.edu.cn
[3] State Key Laboratory of Drug Research and CAS Key Laboratory of Receptor Research, Shanghai Institute of Materia Medical, Chinese Academy of Sciences, Shanghai 201203, China; jwang@simm.ac.cn
[4] University of Chinese Academy of Sciences, Beijing 100049, China

Academic Editor: Gianfranco Favi
Received: 4 February 2019; Accepted: 4 March 2019; Published: 11 March 2019

Abstract: The present study describes an AuPPh$_3$Cl/AgSbF$_6$-catalyzed cascade reaction between amine nucleophiles and alkynoic acids in water. This process proceeds in high step economy with water as the sole coproduct, and leads to the generation of two rings, together with the formation of three new bonds in a single operation. This green cascade process exhibits valuable features such as low catalyst loading, good to excellent yields, high efficiency in bond formation, excellent selectivity, great tolerance of functional groups, and extraordinarily broad substrate scope. In addition, this is the first example of the generation of an indole/thiophene/pyrrole/pyridine/naphthalene/benzene-fused N-heterocycle library through gold catalysis in water from readily available materials. Notably, the discovery of antibacterial molecules from this library demonstrates its high quality and potential for the identification of active pharmaceutical ingredients.

Keywords: amine nucleophiles; alkynoic acids; cascade reaction; gold catalysis; fused N-heterocycles

---

1. Introduction

Rapid advances in genomics and proteomics have resulted in the identification of an increasing number of novel therapeutic targets [1–5], and the existing compound libraries can no longer well meet the needs of drug screening. Therefore, it is highly demanding to develop robust synthetic methods to construct new compound libraries for drug discovery aimed at these targets [6–8]. Considering the structurally diverse targets in a wide "biological space", high-throughput screening (HTS) of skeletally diverse compounds, which occupy a broader "chemical space", can apparently enhance the hit rates [9,10]. In addition to skeletal diversity, drug-like properties of the compounds are equally important for the generation of high-quality compound libraries [11–14], which can increase the possibility of identifying drug-like hit compounds. As a result, privileged structures have received wide attention in drug discovery because they are widely found in natural and pharmaceutical products [15–19]. Although privileged substructure-based diversity-oriented synthesis (pDOS) provides a useful access to assemble compound libraries with high-quality [20–26], it is still challenging to develop efficient and practical approaches to generate a variety of molecular frameworks embedded with privileged structures, especially in a green and sustainable manner. With our interests to develop green and efficient protocols to synthesize valuable N-heterocycles [27–31], we herein construct a library of privileged substructure-based N-heterocycles with diverse scaffolds using gold-catalyzed cascade reactions in water. To the best of our knowledge, this is the first example of the generation of pDOS compound library encompassing skeletal diversity, molecular complexity, and drug-like properties through gold catalysis in water.

Alkynoic acids are extensively used to react with amine nucleophiles to furnish heterocyclic compounds because of the efficient cascade reaction developed by Dixon's group [32], in which, an activated cyclic enol lactone species, which derives from alkynoic acids, is involved as the key intermediate (Scheme 1a). Dixon and co-workers disclosed linear aliphatic terminal alkynoic acids reacted smoothly with amine nucleophiles bearing a nucleophilic carbon atom in toluene or xylene to produce pyrrole- or indole-based heterocyclic frameworks catalyzed by $AuPPh_3Cl/AgOTf$. It should be noted that, based on Dixon's pioneering work, the reactions of alkynoic acids with amine nucleophiles in aprotic solvents such as toluene, xylene, 1,2-dichloroethane, dichloromethane, etc., have been well studied by Patil's group as well as our group [33–43]. However, the reactivity of alkynoic acids and amine nucleophiles in the environmentally friendly, abundant, and cheap solvent—water—was seldom explored [44,45], mainly because the ring opening reaction of the activated enol lactone intermediate in water at elevated temperature may lead to the failure of the cascade reaction. This reason prevented researchers' steps from investigating the nature of the cascade reaction in water. To date, only two successful examples in water were reported by our group [44,45], although the alkynoic acids were only limited to linear aliphatic terminal alkynoic acids, amine nucleophiles were only limited to amine nucleophiles carrying a nucleophilic carbon atom, and complicated Au catalysts were required (Scheme 1b). However, considering water often displays unique reactivity and selectivity which can't be obtained in common organic solvents [46–51] and it is an environmentally benign solvent, we aim to develop a more general cascade process between various alkynoic acids and amine nucleophiles in water with great interest. Despite the possibility of hydrolysis of enol lactone intermediate in water at high temperature which may result in the failure of the cascade reaction, we hypothesize that a more general cascade process could also be achieved in water but with a proper catalytic system. In the present study, we develop a greener, and more general and efficient catalytic system ($AuPPh_3Cl/AgSbF_6/CF_3CO_2H$) in water, which tolerates a broader substrate scope of alkynoic acids as well as amine nucleophiles. As shown in Scheme 1c, not only linear aliphatic terminal and internal alkynoic acids but also cyclic aromatic terminal and internal alkynoic acids are well tolerated. In addition, extraordinarily broad amine nucleophiles bearing a nucleophilic carbon/nitrogen/oxygen atom turned out to be suitable substrates. Besides, the reaction mechanism in water was carefully checked and studied for the first time. Interestingly, when D_2O was used as the reaction solvent, we observed the highly deuterated products generated from the H–D exchange between the reaction substrates/intermediates and D_2O. Herein, we also present the library construction of skeletally diverse N-heterocycles embedded with privileged structures employing different alkynoic acids and various amine nucleophiles as the building blocks. To our delight, five antimicrobial compounds were identified from this library after biological evaluation. The production of N-heterocycles with diverse scaffolds and the discovery of active pharmaceutical ingredients (APIs) demonstrate the power of this method in both organic synthesis and medicinal chemistry.

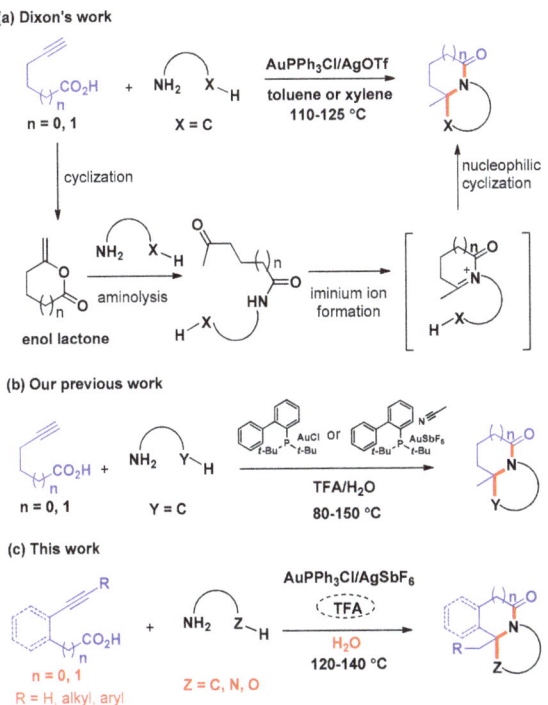

Scheme 1. Gold-catalyzed tandem reactions between alkynoic acids and amine nucleophiles. (a). Dixon's work. (b). Our previous work. (c). This work.

2. Results and Discussion

4-pentynoic acid (**1a**) and tryptamine (**2a**) were employed as the model substrates to optimize the cascade reaction conditions (Table 1). Pleasingly, treatment of starting materials **1a** and **2a** in water without any catalyst at 100 °C for 24 h gave the desired product **SF1a**, albeit with a low yield (entry 1). Then various metal catalysts were screened to improve the reaction yield. A screening of Pd, Cu, Ni, and Mn complexes (entries 2–11), disclosed, at first, that Pd(II) sources, such as $PdCl_2(CH_3CN)_2$ and $Pd(PPh_3)_2Cl_2$, NiOAc and $Mn(OAc)_2$, had little enhancement on the yield; while Pd(0) sources such as $Pd(PPh_3)_4$ and $Pd_2(dba)_3$; copper catalysts, such as $Cu(OAc)_2$ and $Cu(OH)_2$, and CuCl; and $Ni(PPh_3)_2Cl$ inhibited this transformation. A subsequent survey of Ag, Ru, and Co catalysts (entries 12–16) revealed that Ag_2CO_3, AgOAc, $[RuCl_2(p\text{-cym})]_2$, $CoCl_2$, and $Co(acac)_2$ could obviously increase the yield, affording the product **SF1a** in moderate yields (38–57%). Notably, a good yield (70%) was achieved with $AuPPh_3Cl$ (entry 17), and increasing the temperature to 120 °C was much more efficient in improving the yield as compared with extending the reaction time to 48 h (entries 18 and 19). Encouraged by this result, an investigation of other Au catalysts at 120 °C was carried out (entries 20–24). Among them, none displayed higher catalytic reactivity than $AuPPh_3Cl$. To further improve the yield of the product, a series of silver salts, which were proved to be able to increase the catalytic reactivity of gold catalysts [52–57], were screened as the additives (entries 25–28). To our delight, $AgSbF_6$ was found to be the best choice, with which product **SF1a** was obtained in 91% yield (entry 28). In addition, other typical organic solvents were also tested as the reaction solvents, and toluene, xylene, and DCE also turned out to be suitable solvents, in which similar high isolated yields were observed (See Supplementary Materials for details). However, considering water is more environmentally benign, we then decided to use water as the solvent to explore the substrate scope of this method.

In this way, the optimal reaction conditions were identified using a catalytic system consisting of AuPPh$_3$Cl/AgSbF$_6$ in water at 120 °C for 24 h.

Table 1. Reaction condition optimization for the tandem synthesis of compound SF1a [a].

Entry	Catalyst/Additive	Solvent	T (°C)	Yield (%) [b]
1	-	H$_2$O	100	17
2	PdCl$_2$(CH$_3$CN)$_2$	H$_2$O	100	26
3	Pd(PPh$_3$)$_2$Cl$_2$	H$_2$O	100	31
4	Pd(PPh$_3$)$_4$	H$_2$O	100	0
5	Pd$_2$(dba)$_3$	H$_2$O	100	0
6	Cu(OAc)$_2$	H$_2$O	100	trace
7	Cu(OH)$_2$	H$_2$O	100	5
8	CuCl	H$_2$O	100	trace
9	NiOAc	H$_2$O	100	33
10	Ni(PPh$_3$)$_2$Cl	H$_2$O	100	11
11	Mn(OAc)$_2$	H$_2$O	100	25
12	Ag$_2$CO$_3$	H$_2$O	100	51
13	AgOAc	H$_2$O	100	57
14	[RuCl$_2$(p-cym)]$_2$	H$_2$O	100	48
15	CoCl$_2$	H$_2$O	100	38
16	Co(acac)$_2$	H$_2$O	100	50
17	AuPPh$_3$Cl	H$_2$O	100	70
18 [c]	AuPPh$_3$Cl	H$_2$O	100	76
19	AuPPh$_3$Cl	H$_2$O	120	83
20	AuBr$_3$	H$_2$O	120	52
21	AuI	H$_2$O	120	46
22	Au^1 catalyst [d]	H$_2$O	120	81
23	Au^2 catalyst [e]	H$_2$O	120	78
24	Au^3 catalyst [f]	H$_2$O	120	80
25	AuPPh$_3$Cl/Ag$_2$CO$_3$	H$_2$O	120	85
26	AuPPh$_3$Cl/AgOAc	H$_2$O	120	86
27	AuPPh$_3$Cl/AgOTf	H$_2$O	120	89
28	AuPPh$_3$Cl/AgSbF$_6$	H$_2$O	120	91

[a] Reaction conditions: 4-pentynoic acid 1a (0.6 mmol), tryptamine 2a (0.5 mmol), catalyst/additive (0.005 mmol), and solvent (4.0 mL). [b] Yield refers to isolated yield. [c] The reaction was performed for 48 h. [d] Au^1 catalyst = Chloro[(1,1′-biphenyl-2-yl)di-tert-butylphosphine]gold(I). [e] Au^2 catalyst = Chloro[1,3-bis(2,6-diisopropylphenyl)imidazol-2-ylidene]gold(I). [f] Au^3 catalyst = (Acetonitrile)[(2-biphenyl)di-tert-butylphosphine]gold(I) hexafluoroantimonate.

After determining the optimal reaction conditions, we then began to construct a high-quality library of privileged substructure-based N-heterocycles with diverse scaffolds. We first examined the generality of the process with various amine nucleophiles 2 containing a nucleophilic carbon on a heteroaromatic or aromatic ring. In general, this process tolerated a variety of amine nucleophiles 2 and alkynoic acids 1, and 26 scaffolds embedded with privileged structures were furnished in good to high yields in water under optimal or modified reaction conditions (Scheme 2). For example, tryptamines reacted smoothly with terminal alkynoic acids such as 4-pentynoic acid under standard conditions to give products SF1a and SF1b in high yields. The reactions of tryptamines with other terminal alkynoic acids, such as 5-hexynoic acid, 2-ethynylbenzoic acid, and 2-(2-ethynylphenyl)acetic acid, also afforded the corresponding products (SF2a, SF2b, SF3a, SF3b, and SF4a) in good to high yields under a modified two-step one-pot process, in which CF$_3$CO$_2$H (TFA) was added to promote the iminium ion formation. In addition, internal alkynoic acids, such as 5-phenylpent-4-ynoic acid, 6-phenylhex-5-ynoic acid, 2-(phenylethynyl)benzoic acid, and 2-(2-(phenylethynyl)phenyl)acetic acid, were also tested as the substrates. Unfortunately, they all failed to react with tryptamines to give the desired products (**SF1c**, **SF2c**, **SF3c**, and **SF4b**) even under further improved conditions. Interestingly, 2-(1*H*-indol-2-yl)ethylamines could undergo this transformation with terminal alkynoic acids as well as internal alkynoic acids to yield the desired products **SF5–SF8** in 35–96% yields. It should be noted that no N1 ring closure products were observed when 2-(1*H*-indol-2-yl)ethylamines were used as the amine nucleophiles, suggesting excellent selectivity of this cascade process. This may be because the C3 nucleophilicity is stronger than N1 nucleophilicity [58]. Likewise, the reactivity of 2-(1*H*-indol-1-yl)ethanamines in this cascade reaction was very similar to that of tryptamines. They reacted well with terminal alkynoic acids while their reactions with internal alkynoic acids failed to give the desired products (**SF9–SF12**). Interestingly, the protocol was also compatible with 3-(1*H*-indol-1-yl)propan-1-amines, which furnished products **SF13** carrying a seven-membered ring in 54–80% yields. Similarly, products **SF14–SF16** were obtained in 59–86% yields when 2-(1*H*-indol-1-yl)anilines and alkynoic acids were subjected to the modified conditions. Notably, the indole-containing polycyclic frameworks, represented by compounds **SF1–SF16,** are regarded as valuable N-heterocycles considering their ubiquitous presence in biologically active molecules [59–63]. Subsequently, 2-(1*H*-pyrrol-2-yl)ethanamines, 2-(1*H*-pyrrol-1-yl)ethanamines, 2-(thiophen-2-yl)ethanamines, and 2-(thiophen-3-yl)ethanamines were employed as amine nucleophiles. Their reactions with diverse alkynoic acids took place successfully to provide pyrrole- or thiophene-fused compounds **SF17–SF23** in moderate to high yields, despite the fact that stronger conditions were required. In particular, 2-phenylethanamines with electron-donating substituents on the benzene ring were also well tolerated, leading to the formation of benzene-fused heterocyclic products **SF24–SF26** in yields ranging from 47 to 88%. It is also worth noting that excellent selectivity was achieved in the reactions of 2-(1*H*-pyrrol-2-yl)ethanamine, 2-(thiophen-3-yl)ethanamine or 2-phenylethanamines, even though two potential cyclization sites existed in the final step. It should be noted that compound **SF26** is the analog of tetrahydroberbines, which were extracted from the Chinese herb *Corydalis ambigua* and exhibited a broad range of biological activities [64,65].

To further broaden the substrate scope of this approach, amine nucleophiles **3** containing a nucleophilic heteroatom (Z = amide/aniline N, acid/alcohol O) were tested as substrates. Overall, this protocol was also applicable to diverse amine nucleophiles **3**, and 17 scaffolds were constructed with high efficiency (Scheme 3). For instance, the reactions of 2-aminobenzamides with various alkynoic acids worked successfully, affording benzene-fused polycyclic products **SF27–SF30** in moderate to high yields. Gratifyingly, this approach was compatible with 2-(aminomethyl)anilines and benzene-1,2-diamines, although the desired benzene-based heterocyclic products **SF31–SF34** were obtained in lower yields. Remarkably, in the case of substrates such as 2-aminobenzamides and 2-(aminomethyl)anilines, which contain two nitrogen atoms as the nucleophiles, the nitrogen atom with stronger nucleophilicity tended to attack the enol lactone intermediate and therefore the other nitrogen atom with weaker nucleophilicity attacked the iminium ion intermediate to selectively provide the corresponding products, while not in the reverse way. Besides, 2-aminobenzoic acids, 3-amino-2-naphthoic acids, or 2-aminonicotinic acids reacted with various alkynoic acids smoothly, producing the corresponding benzene-, naphthalene-, or pyridine-fused heterocyclic products **SF35–SF41** in yields of 50–96%. Surprisingly, 2-aminobenzyl alcohols were also found to be suitable substrates, which could undergo the cascade reaction with alkynoic acids to give the desired benzene-fused polycyclic products **SF42** and **SF43**, albeit with lower yields. In addition, we also tried to synthesize the indole-fused compounds **SF44–SF46** embedded with an eight- or nine-membered ring; unfortunately, we failed. This may be attributed to the low reactivities of the amine nucleophiles and the instability of the large rings in energetics.

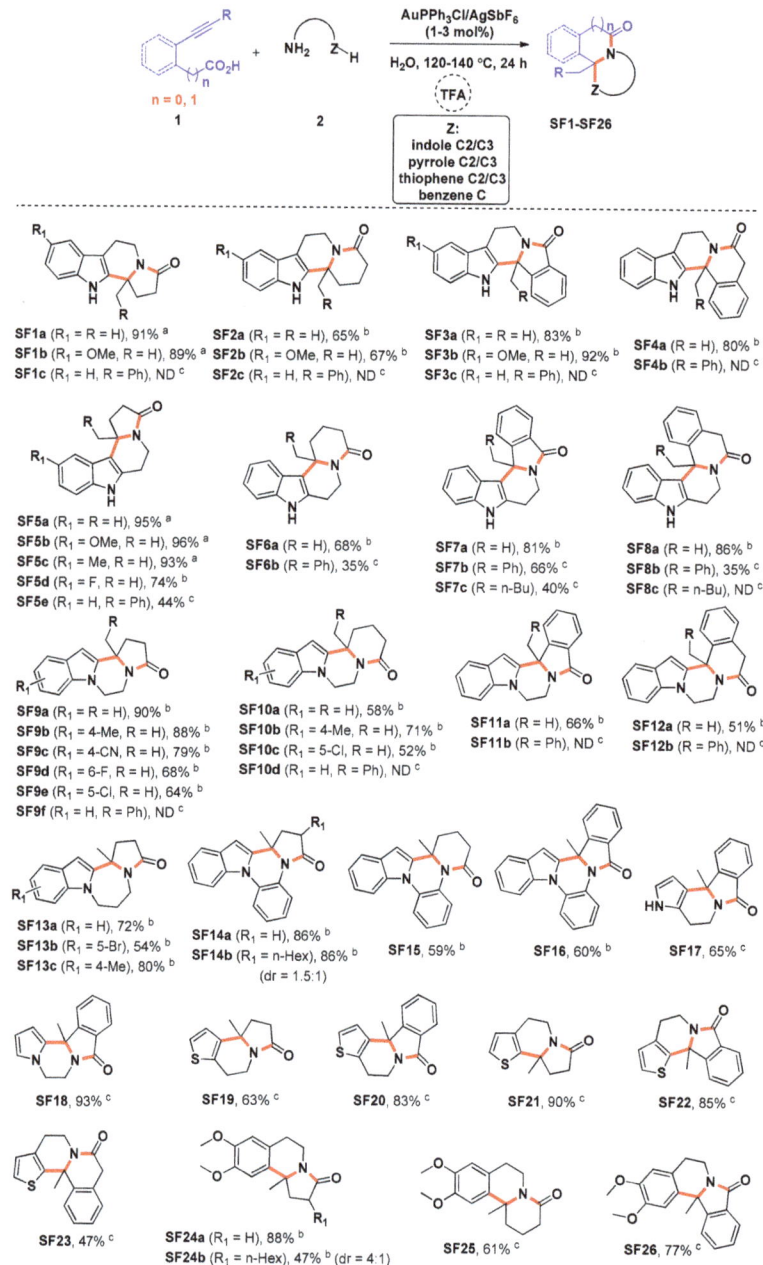

Scheme 2. Generation of scaffold diversity. [a] Reaction conditions: alkynoic acids **1** (0.6 mmol), amine nucleophiles **2** (0.5 mmol), AuPPh$_3$Cl/AgSbF$_6$ (0.005 mmol), H$_2$O (4.0 mL), and 120 °C, 24 h. [b] Reaction conditions: (i) alkynoic acids **1** (0.6 mmol), amine nucleophiles **2** (0.5 mmol), AuPPh$_3$Cl/AgSbF$_6$ (0.01 mmol), H$_2$O (4.0 mL), 120 °C, 20 h; (ii) TFA (0.5 mmol) was added, and then 120 °C, 4 h. [c] Reaction conditions: (i) alkynoic acids **1** (0.6 mmol), amine nucleophiles **2** (0.5 mmol), AuPPh$_3$Cl/AgSbF$_6$ (0.015 mmol), H$_2$O (4.0 mL), 140 °C, 20 h; (ii) TFA (0.5 mmol) was added, and then 140 °C, 4 h. ND = Not detected.

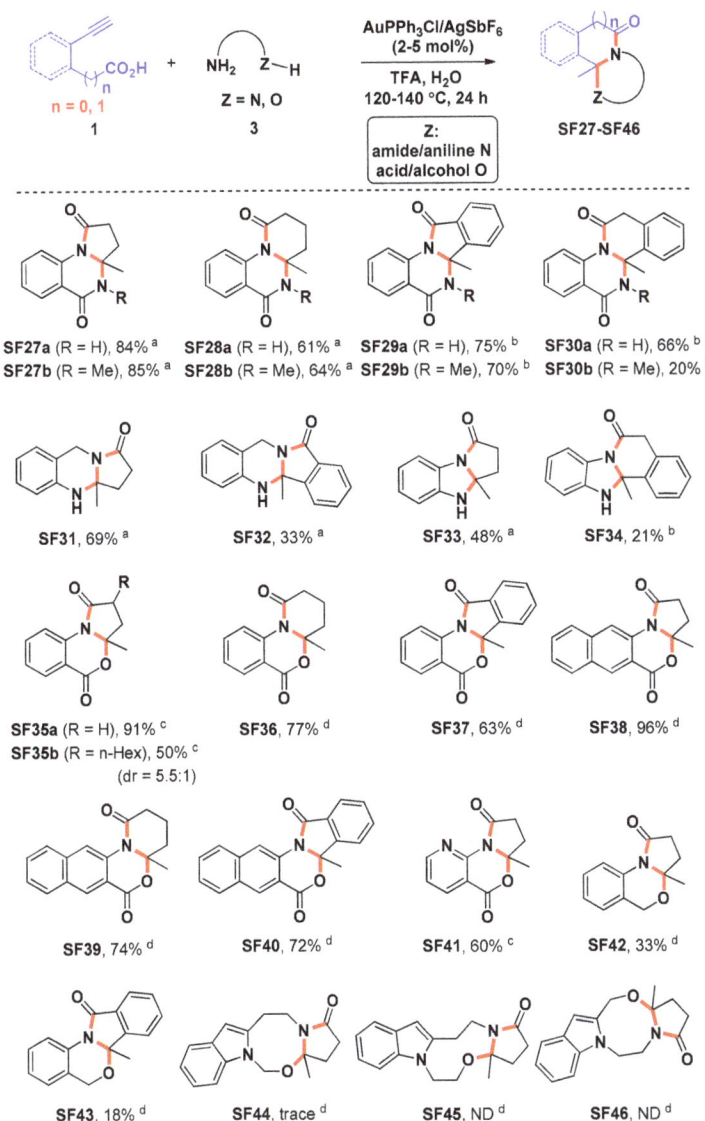

Scheme 3. Generation of scaffold diversity. [a] Reactions conditions: (i) Alkynoic acids **1** (0.6 mmol), amine nucleophiles **3** (0.5 mmol), AuPPh$_3$Cl/AgSbF$_6$ (0.01 mmol), H$_2$O (4.0 mL), 120 °C, 20 h; (ii) TFA (0.5 mmol) was added, and then 120 °C, 4 h. [b] The reaction was carried out at 140 °C. [c] The reaction was carried out with 5 mol% AuPPh$_3$Cl/AgSbF$_6$. [d] The reaction was performed with 5 mol% AuPPh$_3$Cl/AgSbF$_6$ at 140 °C. ND = Not detected.

The diversity of the library can be further expanded by the derivatization of the target compounds. We herein introduce the derivatization of the target compounds based on the simple reduction of the carbonyl group, and the selected results are shown in Scheme 4. Nine different scaffolds represented by indole-, pyrrole-, or thiophene-based polycyclic compounds **SF47–SF55** containing a tertiary amine were produced conveniently through an easy reduction of the corresponding precursors with LiAlH$_4$/AlCl$_3$. Notably, these scaffolds are very similar to those found in natural and pharmaceutical agents [65]. We expect the screening of these compounds towards specific biological targets might lead to the identification of bioactive molecules.

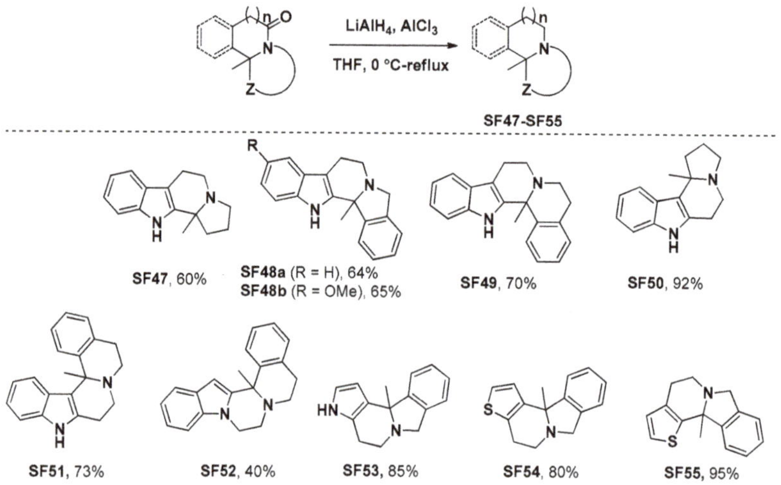

Scheme 4. Derivatization of the target compounds in the library.

Thus, using various amine nucleophiles and alkynoic acids as the building blocks, a library of privileged substructure-based N-heterocycles with diverse scaffolds was constructed through gold catalysis in a green and efficient manner. It is worth noting this cascade process constructs three new bonds together with two rings in one chemical process, suggesting the high efficiency of this cascade reaction in synthesizing nitrogen-containing heterocyclic compounds. Regarding the large occurrence of nitrogen-containing heterocyclic compounds in APIs [66–68], the method presented in this paper is prospective since it could provide an environmentally benign and useful platform for the preparation of diverse nitrogen-containing heterocyclic compounds.

To verify our expectation that this approach could provide useful scaffolds with attractive bioactivities, a bioactivity study of this library was carried out. An initial pharmacological study of these nitrogen-containing heterocyclic compounds led to the discovery of five antimicrobial compounds—**SF9d**, **SF29b**, **SF33**, **SF36**, and **SF41**. The minimal inhibitory concentration (MIC) results revealed that compound **SF36** displayed the most potent antibacterial activity against *S. aureus* strain, with a MIC$_{90}$ value of 10–25 μg/mL (Table 2) (Time-kill assays and colony-forming unit studies of compounds **SF9d**, **SF29b**, **SF33**, **SF36** and **SF41** could be found in Supplementary Materials).

Table 2. MIC$_{90}$ of compounds **SF9d**, **SF29b**, **SF33**, **SF36**, and **SF41** against the *S. aureus* strain.

Compound	MIC$_{90}$ (μg/mL)
SF9d	100–200
SF29b	50
SF33	100–200
SF36	10–25
SF41	100

Mechanistic studies were carried out with deuterium-labeling experiments, and the hydrogens of the products were assigned at first by the analysis of the ^{1}H-NMR, ^{13}C-NMR, HSQC, HMBC, and ^{1}H-^{1}H COSY spectrum to confirm the deuterated positions (See Supplementary Materials for details). Interestingly, the reaction of 4-pentynoic acid **1a** with tryptamine **2a** in D$_2$O under the standard conditions afforded the deuterated product [D]$_n$-**SF1a**, not only at the methyl position but also at the β-position of the carbonyl (Scheme 5a). Specifically, a 96% deuteration at the methyl position and a similar deuteration (90%) of the two unequal hydrogens at the β-position of the carbonyl were observed. The reactions of 2-(1H-indol-2-yl)ethylamines with 4-pentynoic acid **1a** in D_2O gave the similar results (Scheme 5b,c).

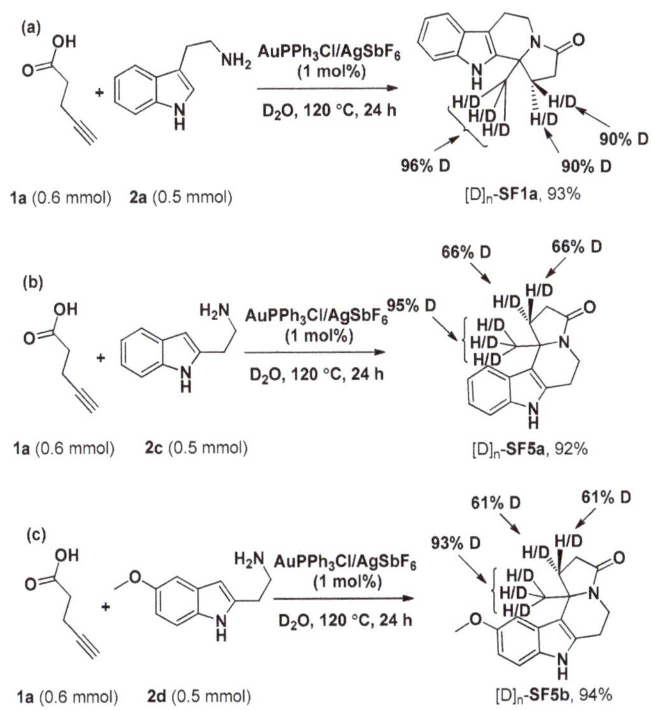

Scheme 5. Deuterium-labeling experiments.; (a) the deuterium-labeling experiment of substrates **1a** and **2a** in D_2O; (b) the deuterium-labeling experiment of substrates **1a** and **2c** in D_2O; (c) the deuterium-labeling experiment of substrates **1a** and **2d** in D_2O.

According to the results of deuterium-labeling experiments, we hypothesize two possible reaction pathways (Schemes 6 and 7). The reaction of with 4-pentynoic acid **1a** with tryptamine **2a** in the presence of Au catalyst in D_2O is taken as the example to illustrate the reaction pathway. The first hypothetic reaction pathway (Scheme 6) may involve the gold-catalyzed hydration of carbon–carbon triple bond, which was observed in our previous work [69]. The H–D exchange between the carboxyl group of 4-pentynoic acid **1a** and D_2O leads to the formation of intermediate **A1**. Gold-catalyzed addition of D_2O to the carbon–carbon triple bond of intermediate **A1** produces intermediate **A2**, which undergoes two keto–enol tautomerizations to give intermediate **A4**. The subsequent H–D exchange between the hydroxyl group of intermediate **A4** and D_2O affords intermediate **A5**, which undergoes keto–enol tautomerization again to give intermediate **A6**. The two acidic protons at the α-position of the carbonyl in intermediate **A6** undergo H–D exchange with D_2O via keto–enol tautomerizations to yield intermediate **A7**. The following condensation between intermediate **A7** and tryptamine **2a**, the subsequent iminium ion formation, and the final cyclization achieve the product [D]$_n$-**SF1a**. The second hypothetic reaction pathway is shown in Scheme 7; the H–D exchange between the carboxyl group of 4-pentynoic acid **1a** and D_2O produces intermediate **A1**. Gold-catalyzed intramolecular cyclization of intermediate **A1** produces enol lactone intermediate **B1**, which is attacked by tryptamine **2a** to give intermediate **B2**. The subsequent H–D exchange between the hydroxyl group of intermediate **B2** and D_2O yields intermediate **B3**, which undergoes enol–keto tautomerization to provide intermediate **B4**. Similarly, the three acidic protons at the α-position of the carbonyl in intermediate **B4** undergo H–D exchange with D$_2$O via keto–enol tautomerizations to yield intermediate **C1**, which is converted into the product [D]$_n$-**SF1a** via an iminium ion formation/cyclization sequence.

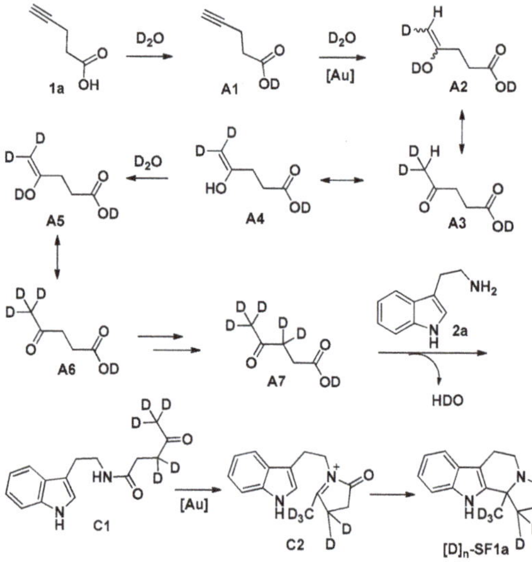

Scheme 6. Hypothetic reaction pathway 1.

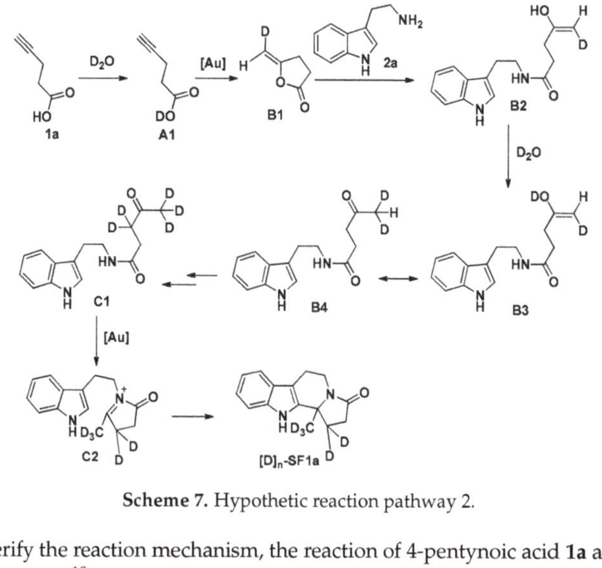

Scheme 7. Hypothetic reaction pathway 2.

To further verify the reaction mechanism, the reaction of 4-pentynoic acid **1a** and tryptamine **2a** under gold catalysis in O^{18}-labeled water was carried out first. This reaction in H$_2$O^{18} was stopped after 0.5 h to track the reaction intermediates. As shown in Scheme 8a, apart from the remaining starting materials, intermediate **2a″** and the product **SF1a** were obtained in 21% and 15% yield, respectively. While the O^{18}-labeled intermediate **2a′** was not observed. This result clearly indicates the hydration of alkyne moiety, which is proposed in Scheme 6, is not involved. By contrast, the commercially purchased enol lactone **D** reacted smoothly with tryptamine **2a** without the gold catalyst (Scheme 8b).

This result shows that the enol lactone species **B1**, which is proposed in Scheme 7, is likely to be the key intermediate.

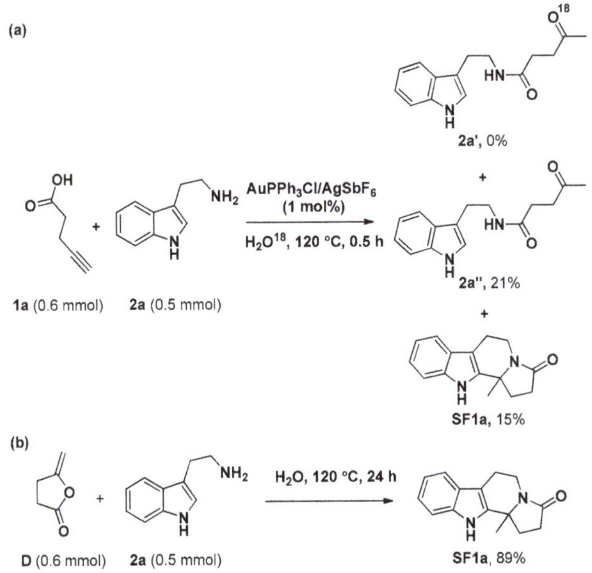

Scheme 8. Mechanistic study experiments. (**a**) the mechanistic experiment of substrates **1a** and **2a** in H_2O^{18}; (**b**) the mechanistic experiment of substrates **D** and **2a** without catalyst/additive.

On the basis of the above results of mechanistic experiments, a final proposed reaction mechanism is outlined in Scheme 9. The proposed mechanism commences with the coordination of the gold catalyst to the carbon–carbon triple bond of alkynoic acids to produce intermediate **I1**. The subsequent intramolecular *exo* cyclization of **I1** yields intermediate **I2**. The following protodemetalation of intermediate **I2** takes place to produce the enol lactone species **I3** with the regeneration of the catalyst. Then intermediate **I3** undergoes aminolysis by amine nucleophiles to give intermediate **I4**, which tautomerizes to produce intermediate **I5**. Intermediate **I5** is converted into the iminium ion **I8** under the catalysis of the gold catalyst. The final nucleophilic cyclization of intermediate **I8** affords the desired products with the release of the gold catalyst. The stronger nucleophilicity of amine nucleophiles compared to that of H_2O results in the aminolysis of enol lactone **I3** by amine nucleophiles instead of hydrolysis by H_2O. It should be noted that the reaction solvent H_2O participates in this cascade reaction via the H–H exchange with the carboxyl group of alkynoic acids, the hydroxyl group of intermediate **I4**, and the α hydrogen atoms of the carbonyl group of intermediate **I5**, as demonstrated by deuterium-labeling experiments. Besides, TFA could promote this reaction by accelerating the formation of iminium ion **I8**.

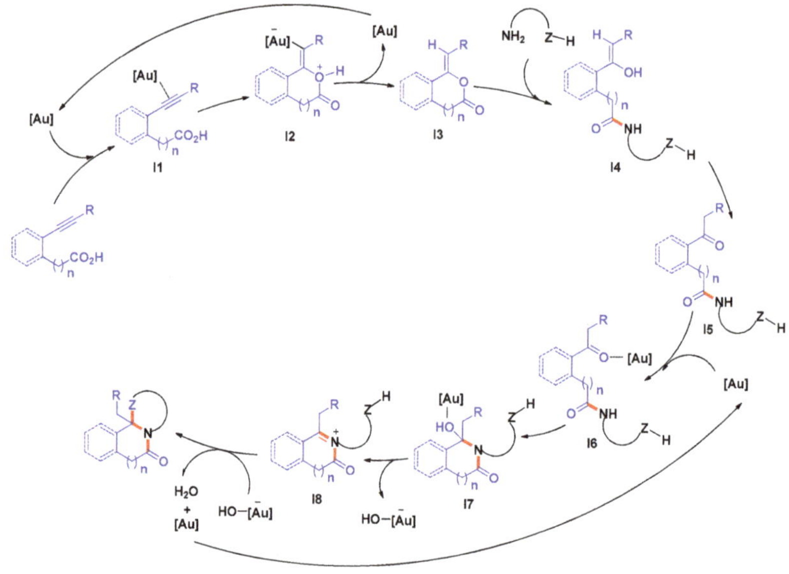

Scheme 9. A final proposed reaction mechanism.

3. Materials and Methods

3.1. General Information

If not otherwise specified, the starting materials were obtained from commercial sources and used directly without purification. Analytical thin-layer chromatography (TLC): HSGF 254 (0.15–0.2 mm thickness). Detection under UV light at 254 nm. Column chromatography: Separations were carried out on silica gel FCP 200–300. Yields refer to isolated compounds. Melting point apparatus: a micro melting point apparatus, values are uncorrected. Nuclear magnetic resonance (NMR) apparatus: a Brucker instrument. Chemical shifts (δ) are given in ppm. Proton coupling patterns were recorded as singlet (s), doublet (d), triplet (t), quartet (q), and multiplet (m). LRMS (low-resolution mass) and HRMS (high-resolution mass) were measured on a spectrometer with an electrospray ionization (ESI) source.

3.2. General Procedure for the Preparation of Compounds SF1a, SF1b, SF5a, SF5b, and SF5c

A suspension of alkynoic acids 1 (0.6 mmol), amine nucleophiles 2 (0.5 mmol), and AuPPh$_3$Cl/AgSbF$_6$ (0.005 mmol) in H$_2$O (4.0 mL) was stirred at 120 °C for 24 h. At ambient temperature, saturated Na$_2$CO$_3$ solution (25.0 mL) was added to the reaction mixture. The resulting mixture was then extracted with ethyl acetate (3 × 15 mL). The combined organic layers were washed with brine, and dried over Na$_2$SO$_4$. After filtration and removal of the solvents in vacuo, the crude product was purified by flash chromatography on silica gel to provide the desired product.

11b-Methyl-5,6,11,11b-tetrahydro-1H-indolizino[8,7-b]indol-3(2H)-one (**SF1a**): white solid (109.7 mg, yield 91%), mp 260–261 °C. ^{1}H-NMR (500 MHz, DMSO-d_6) δ 1.54 (s, 3H), 2.07–1.99 (m, 1H), 2.32–2.20 (m, 2H), 2.66–2.55 (m, 2H), 2.75–2.67 (m, 1H), 3.11–3.00 (m, 1H), 4.25–4.16 (m, 1H), 7.01–6.94 (m, 1H), 7.10-7.02 (m, 1H), 7.32 (d, J = 8.0 Hz, 1H), 7.39 (d, J = 7.8 Hz, 1H), 11.06 (s, 1H); ^{13}C-NMR (100 MHz, DMSO-d_6) δ 171.9 (CO), 139.0 (C, Ar), 135.9 (C, Ar), 126.3 (C, Ar), 121.0 (CH, Ar), 118.6 (CH, Ar), 118.0 (CH, Ar), 111.1 (CH, Ar), 104.7 (C, Ar), 58.9 (C), 34.3 (CH$_2$), 32.6 (CH$_2$), 30.1 (CH$_2$), 25.0 (CH$_3$), 20.9 (CH$_2$); ESI-LRMS *m/z*: 241 [M + H]$^+$; ESI-HRMS *m/z* calcd for M + H$^+$ 241.1335, found: 241.1331. The characterization data is in accordance with that reported in [32].

8-Methoxy-11b-methyl-5,6,11,11b-tetrahydro-1H-indolizino[8,7-b]indol-3(2H)-one (**SF1b**): yellow oil (120.2 mg, yield 89%). ^{1}H-NMR (500 MHz, DMSO-d_6) δ 1.53 (s, 3H), 2.07–1.97 (m, 1H), 2.31–2.20 (m, 2H), 2.64–2.52 (m, 2H), 2.73–2.64 (m, 1H), 3.10–2.98 (m, 1H), 3.74 (s, 3H), 4.24–4.15 (m, 1H), 6.70 (dd, J = 8.7, 2.4 Hz, 1H), 6.89 (d, J = 2.3 Hz, 1H), 7.20 (d, J = 8.7 Hz, 1H), 10.88 (s, 1H); ^{13}C-NMR (125 MHz, DMSO-d_6) δ 171.9 (CO), 153.2 (C, Ar), 139.7 (C, Ar), 130.9 (C, Ar), 126.6 (C, Ar), 111.7 (CH, Ar), 110.8 (CH, Ar), 104.6 (C, Ar), 100.2 (CH, Ar), 58.9 (C), 55.4 (OCH$_3$), 34.3 (CH$_2$), 32.6 (CH$_2$), 30.1 (CH$_2$), 25.0 (CH$_3$), 21.0 (CH$_2$); ESI-LRMS m/z: 271 [M + H]$^+$; ESI-HRMS m/z calcd for M + H$^+$ 271.1441, found: 271.1437. The characterization data is in accordance with that reported in [43].

11c-Methyl-5,6,7,11c-tetrahydro-1H-indolizino[7,8-b]indol-3(2H)-one (**SF5a**): yellow solid (114.2 mg, yield 95%), mp 96–97 °C. ^{1}H-NMR (500 MHz, DMSO-d_6) δ 1.54 (s, 3H), 2.02–1.92 (m, 1H), 2.27–2.17 (m, 1H), 2.54–2.47 (m, 1H), 2.65–2.54 (m, 1H), 2.81–2.67 (m, 2H), 3.16–3.04 (m, 1H), 4.27–4.18 (m, 1H), 7.00–6.92 (m, 1H), 7.08–7.00 (m, 1H), 7.30 (d, J = 8.0 Hz, 1H), 7.47 (d, J = 7.8 Hz, 1H), 10.90 (s, 1H); ^{13}C-NMR (100 MHz, DMSO-d_6) δ 171.5 (CO), 135.9 (C, Ar), 130.5 (C, Ar), 123.9 (C, Ar), 120.5 (CH, Ar), 118.6 (CH, Ar), 117.9 (CH, Ar), 115.3 (C, Ar), 111.1 (CH, Ar), 59.2 (C), 33.2 (CH$_2$), 33.2 (CH$_2$), 30.2 (CH$_2$), 25.0 (CH$_3$), 22.8 (CH$_2$); ESI-LRMS m/z: 241 [M + H]$^+$; ESI-HRMS m/z calcd for M + H$^+$ 241.1335, found: 241.1332. The characterization data is in accordance with that reported in [43].

10-Methoxy-11c-methyl-5,6,7,11c-tetrahydro-1H-indolizino[7,8-b]indol-3(2H)-one (**SF5b**): pale yellow solid (129.5 mg, yield 96%), mp 193–194 °C. ^{1}H-NMR (500 MHz, DMSO-d_6) δ 1.54 (s, 3H), 2.01–1.91 (m, 1H), 2.27–2.18 (m, 1H), 2.64–2.52 (m, 2H), 2.79–2.65 (m, 2H), 3.14–3.03 (m, 1H), 3.77 (s, 3H), 4.25–4.15 (m, 1H), 6.69 (dd, J = 8.7, 2.4 Hz, 1H), 6.93 (d, J = 2.3 Hz, 1H), 7.19 (d, J = 8.7 Hz, 1H), 10.72 (s, 1H); ^{13}C-NMR (100 MHz, DMSO-d_6) δ 171.5 (CO), 153.1 (C, Ar), 131.3 (C, Ar), 131.0 (C, Ar), 124.2 (C, Ar), 115.1 (C, Ar), 111.7 (CH, Ar), 109.8 (CH, Ar), 100.6 (CH, Ar), 59.2 (C), 55.5 (OCH$_3$), 33.2 (CH$_2$), 33.0 (CH$_2$), 30.1 (CH$_2$), 24.8 (CH$_3$), 22.9 (CH$_2$); ESI-LRMS m/z: 271 [M + H]$^+$; ESI-HRMS m/z calcd for M + H$^+$ 271.1441, found: 271.1437. The characterization data is in accordance with that reported in [43].

10,11c-Dimethyl-5,6,7,11c-tetrahydro-1H-indolizino[7,8-b]indol-3(2H)-one (**SF5c**): pale yellow oil (118.4 mg, yield 93%). ^{1}H-NMR (500 MHz, DMSO-d_6) δ 1.53 (s, 3H), 2.02–1.91 (m, 1H), 2.29–2.17 (m, 1H), 2.38 (s, 3H), 2.50–2.46 (m, 1H), 2.64–2.54 (m, 1H), 2.80–2.65 (m, 2H), 3.15–3.03 (m, 1H), 4.26–4.15 (m, 1H), 6.86 (dd, J = 8.2, 1.0 Hz, 1H), 7.18 (d, J = 8.2 Hz, 1H), 7.25 (s, 1H), 10.75 (s, 1H); ^{13}C-NMR (125 MHz, DMSO-d_6) δ 171.5 (CO), 134.2 (C, Ar), 130.5 (C, Ar), 127.1 (C, Ar), 124.1 (C, Ar), 122.0 (CH, Ar), 117.6 (CH, Ar), 114.8 (C, Ar), 110.8 (CH, Ar), 59.2 (C), 33.3 (CH$_2$), 33.2 (CH$_2$), 30.2 (CH$_2$), 25.0 (CH$_3$), 22.8 (CH$_2$), 21.3 (CH$_3$); ESI-LRMS m/z: 255 [M + H]$^+$; ESI-HRMS m/z calcd for M + H$^+$ 255.1492, found: 255.1489. The characterization data is in accordance with that reported in [43].

3.3. General Procedure for the Preparation of Compounds SF1c, SF2–SF4, SF5d, SF5e and SF6–SF46

A suspension of alkynoic acids **1** (0.6 mmol), amine nucleophiles **2** or **3** (0.5 mmol), and AuPPh$_3$Cl/AgSbF$_6$ (with the amount indicated) in H$_2$O (4.0 mL) was stirred at the temperature indicated for 20 h. Then the reaction mixture was cooled to room temperature, and CF$_3$COOH (0.5 mmol) was added, and the resulting mixture was stirred for another 4 h at the temperature indicated. At ambient temperature, saturated Na$_2$CO$_3$ solution (25.0 mL) was added to the reaction mixture. The resulting mixture was then extracted with ethyl acetate (3 × 15 mL). The combined organic layers were washed with brine, and dried over Na$_2$SO$_4$. After filtration and removal of the solvents in vacuo, the crude product was purified by flash chromatography on silica gel to yield the desired product.

12b-Methyl-1,2,3,6,7,12b-hexahydroindolo[2,3-a]quinolizin-4(12H)-one (**SF2a**): white solid (82.6 mg, yield 65%), mp 255–256 °C. ^{1}H-NMR (500 MHz, DMSO-d_6) δ 1.60 (s, 3H), 1.80–1.68 (m, 2H), 1.97–1.86 (m, 1H), 2.32–2.21 (m, 1H), 2.45–2.32 (m, 2H), 2.61–2.53 (m, 1H), 2.69–2.62 (m, 1H), 2.99–2.87 (m, 1H), 4.90–4.81 (m, 1H), 7.00–6.93 (m, 1H), 7.10–7.02 (m, 1H), 7.32 (d, J = 8.0 Hz, 1H), 7.39 (d, J = 7.8 Hz, 1H), 10.92 (s, 1H); ^{13}C-NMR (125 MHz, DMSO-d_6) δ 167.9 (CO), 139.7 (C, Ar), 136.0 (C, Ar), 126.2 (C,

Ar), 120.9 (CH, Ar), 118.5 (CH, Ar), 117.9 (CH, Ar), 111.1 (CH, Ar), 105.8 (C, Ar), 56.4 (C), 35.6 (CH$_2$), 34.8 (CH$_2$), 31.8 (CH$_2$), 25.3 (CH$_3$), 21.0 (CH$_2$), 16.3 (CH$_2$); ESI-LRMS *m/z*: 255 [M + H]$^+$; ESI-HRMS *m/z* calcd for M + H$^+$ 255.1492, found: 255.1488. The characterization data is in accordance with that reported in [43].

9-Methoxy-12b-methyl-1,2,3,6,7,12b-hexahydroindolo[2,3-a]quinolizin-4(12H)-one (**SF2b**): pale yellow solid (95.4 mg, yield 67%), mp 190–191 °C. ^{1}H-NMR (400 MHz, DMSO-d_6) δ 1.59 (s, 3H), 1.79–1.66 (m, 2H), 1.97–1.84 (m, 1H), 2.42–2.21 (m, 3H), 2.66–2.51 (m, 2H), 2.99–2.86 (m, 1H), 3.74 (s, 3H), 4.90–4.79 (m, 1H), 6.70 (dd, *J* = 8.7, 2.4 Hz, 1H), 6.89 (d, *J* = 2.4 Hz, 1H), 7.20 (d, *J* = 8.7 Hz, 1H), 10.73 (s, 1H); ^{13}C-NMR (125 MHz, DMSO-d_6) δ 167.9 (CO), 153.2 (C, Ar), 140.4 (C, Ar), 131.0 (C, Ar), 126.5 (C, Ar), 111.7 (CH, Ar), 110.7 (CH, Ar), 105.7 (C, Ar), 100.1 (CH, Ar), 56.4 (C), 55.4 (OCH$_3$), 35.6 (CH$_2$), 34.8 (CH$_2$), 31.8 (CH$_2$), 25.4 (CH$_3$), 21.1 (CH$_2$), 16.3 (CH$_2$); ESI-LRMS *m/z*: 285 [M + H]$^+$; ESI-HRMS *m/z* calcd for M + H$^+$ 285.1598, found: 285.1593. The characterization data is in accordance with that reported in [43].

13b-Methyl-7,8,13,13b-tetrahydro-5H-benzo[1,2]indolizino[8,7-b]indol-5-one (**SF3a**): white solid (119.4 mg, yield 83%), mp 283–284 °C. ^{1}H-NMR (400 MHz, DMSO-d_6) δ 1.86 (s, 3H), 2.75–2.63 (m, 1H), 2.85–2.75 (m, 1H), 3.47–3.36 (m, 1H), 4.59–4.47 (m, 1H), 7.03–6.93 (m, 1H), 7.15–7.05 (m, 1H), 7.44–7.34 (m, 2H), 7.58–7.49 (m, 1H), 7.79–7.68 (m, 2H), 8.32 (d, *J* = 7.9 Hz, 1H), 11.35 (s, 1H); ^{13}C-NMR (100 MHz, DMSO-d_6) δ 167.2 (CO), 149.3 (C, Ar), 136.2 (C, Ar), 135.2 (C, Ar), 132.2 (CH, Ar), 130.3 (C, Ar), 128.6 (CH, Ar), 126.0 (C, Ar), 123.2 (CH, Ar), 122.8 (CH, Ar), 121.6 (CH, Ar), 118.9 (CH, Ar), 118.3 (CH, Ar), 111.2 (CH, Ar), 106.3 (C, Ar), 62.0 (C), 35.4 (CH$_2$), 25.9 (CH$_3$), 21.4 (CH$_2$); ESI-LRMS *m/z*: 289 [M + H]$^+$; ESI-HRMS *m/z* calcd for M + H$^+$ 289.1335, found: 289.1330. The characterization data is in accordance with that reported in [43].

10-Methoxy-13b-methyl-7,8,13,13b-tetrahydro-5H-benzo[1,2]indolizino[8,7-b]indol-5-one (**SF3b**): white solid (146.9 mg, yield 92%), mp 164–165 °C. ^{1}H-NMR (400 MHz, DMSO-d_6) δ 1.84 (s, 3H), 2.72–2.60 (m, 1H), 2.81–2.73 (m, 1H), 3.45–3.35 (m, 1H), 3.73 (s, 3H), 4.57–4.45 (m, 1H), 6.73 (dd, *J* = 8.7, 2.1 Hz, 1H), 6.89 (d, *J* = 2.1 Hz, 1H), 7.26 (d, *J* = 8.7 Hz, 1H), 7.58–7.47 (m, 1H), 7.77–7.67 (m, 2H), 8.29 (d, *J* = 7.9 Hz, 1H), 11.18 (s, 1H); ^{13}C-NMR (125 MHz, DMSO-d_6) δ 167.2 (CO), 153.3 (C, Ar), 149.4 (C, Ar), 135.8 (C, Ar), 132.2 (CH, Ar), 131.2 (C, Ar), 130.2 (C, Ar), 128.6 (CH, Ar), 126.3 (C, Ar), 123.2 (CH, Ar), 122.8 (CH, Ar), 111.9 (CH, Ar), 111.5 (CH, Ar), 106.2 (C, Ar), 100.3 (CH, Ar), 62.0 (C), 55.4 (OCH$_3$), 35.5 (CH$_2$), 26.0 (CH$_3$), 21.5 (CH$_2$); ESI-LRMS *m/z*: 319 [M + H]$^+$; ESI-HRMS *m/z* calcd for M + H$^+$ 319.1441, found: 319.1435. The characterization data is in accordance with that reported in [43].

14b-Methyl-8,9,14,14b-tetrahydroindolo[2′,3′:3,4]pyrido[2,1-a]isoquinolin-6(5H)-one (**SF4a**): pale yellow solid (120.7 mg, yield 80%), mp 137–138 °C. ^{1}H-NMR (400 MHz, DMSO-d_6) δ 1.84 (s, 3H), 2.50–2.41 (m, 1H), 2.86–2.75 (m, 1H), 2.96–2.86 (m, 1H), 3.63 (d, *J* = 19.4 Hz, 1H), 4.08 (d, *J* = 19.2 Hz, 1H), 4.96–4.85 (m, 1H), 7.08–6.98 (m, 1H), 7.18–7.12 (m, 1H), 7.24–7.18 (m, 1H), 7.31–7.24 (m, 2H), 7.52–7.43 (m, 3H), 11.56 (s, 1H); ^{13}C-NMR (100 MHz, DMSO-d_6) δ169.0 (CO), 139.9 (C, Ar), 136.1 (C, Ar), 135.1 (C, Ar), 132.3 (C, Ar), 127.8 (CH, Ar), 127.5 (CH, Ar), 126.4 (CH, Ar), 126.0 (CH, Ar), 124.1 (CH, Ar), 121.4 (CH, Ar), 118.8 (CH, Ar), 118.2 (CH, Ar), 111.3 (CH, Ar), 109.2 (C, Ar), 60.9 (C), 38.0 (CH$_2$), 37.9 (CH$_2$), 26.2 (CH$_3$), 21.0 (CH$_2$); ESI-LRMS *m/z*: 303 [M + H]$^+$; ESI-HRMS *m/z* calcd for M + H$^+$ 303.1492, found: 303.1487. The characterization data is in accordance with that reported in [43].

10-Fluoro-11c-methyl-5,6,7,11c-tetrahydro-1H-indolizino[7,8-b]indol-3(2H)-one (**SF5d**): pale yellow solid (95.1 mg, yield 74%), mp 104–105 °C. ^{1}H-NMR (400 MHz, DMSO-d_6) δ 1.52 (s, 3H), 1.99–1.87 (m, 1H), 2.28–2.16 (m, 1H), 2.64–2.51 (m, 2H), 2.83–2.67 (m, 2H), 3.16–3.01 (m, 1H), 4.28–4.15 (m, 1H), 6.93–6.82 (m, 1H), 7.34–7.21 (m, 2H), 11.00 (s, 1H); ^{13}C-NMR (100 MHz, DMSO-d_6) δ 171.5 (CO), 156.7 (d, J_{C-F} = 231.1 Hz, CF, Ar), 132.9 (C, Ar), 132.5 (C, Ar), 124.0 (d, J_{C-F} = 10.1 Hz, C, Ar), 115.7 (d, J_{C-F} = 4.5 Hz, C, Ar), 111.9 (d, J_{C-F} = 9.9 Hz, CH, Ar), 108.3 (d, J_{C-F} = 25.8 Hz, CH, Ar), 102.9 (d, J_{C-F} = 23.4 Hz, CH, Ar), 59.0 (C), 33.1 (CH$_2$), 32.9 (CH$_2$), 30.1 (CH$_2$), 24.8 (CH$_3$), 22.9 (CH$_2$); ESI-LRMS *m/z*: 259 [M + H]$^+$; ESI-HRMS *m/z* calcd for M + H$^+$ 259.1241, found: 259.1238. The characterization data is in accordance with that reported in [43].

11c-Benzyl-5,6,7,11c-tetrahydro-1H-indolizino[7,8-b]indol-3(2H)-one (**SF5e**): pale yellow oil (69.4 mg, yield 44%). ^{1}H-NMR (500 MHz, CDCl$_3$) δ 1.78–1.65 (m, 1H), 2.22–2.09 (m, 2H), 2.72–2.60 (m, 2H), 2.97–2.85 (m, 1H), 3.08–2.98 (m, 1H), 3.21 (d, *J* = 13.9 Hz, 1H), 3.30 (d, *J* = 13.9 Hz, 1H), 4.50 (dd, *J* = 12.9, 6.4 Hz, 1H), 7.12–7.06 (m, 2H), 7.22–7.13 (m, 2H), 7.26–7.23 (m, 3H), 7.35 (d, *J* = 8.0 Hz, 1H), 7.52 (d, *J* = 7.6 Hz, 1H), 8.33 (s, 1H); ^{13}C-NMR (125 MHz, CDCl$_3$) δ 174.3 (CO), 136.7 (C, Ar), 136.2 (C, Ar), 130.6 (C, Ar), 130.2 (2 × CH, Ar), 128.5 (2 × CH, Ar), 127.0 (CH, Ar), 124.4 (C, Ar), 121.8 (CH, Ar), 119.9 (CH, Ar), 118.5 (CH, Ar), 115.9 (C, Ar), 111.3 (CH, Ar), 63.8 (C), 44.6 (CH$_2$), 34.3 (CH$_2$), 31.6 (CH$_2$), 31.0 (CH$_2$), 23.1 (CH$_2$); ESI-LRMS *m/z*: 317 [M + H]$^+$; ESI-HRMS *m/z* calcd for M + H$^+$ 317.1648, found: 317.1651. The characterization data is in accordance with that reported in [43].

12c-Methyl-1,2,3,6,7,12c-hexahydroindolo[3,2-a]quinolizin-4(8H)-one (**SF6a**): white solid (86.8 mg, yield 68%), mp 132–133 °C. ^{1}H-NMR (400 MHz, DMSO-*d*$_6$) δ 1.63 (s, 3H), 1.77–1.65 (m, 2H), 2.01–1.85 (m, 1H), 2.32–2.19 (m, 1H), 2.43–2.32 (m, 1H), 2.78–2.60 (m, 3H), 3.01–2.89 (m, 1H), 4.92–4.80 (m, 1H), 6.99–6.90 (m, 1H), 7.07–6.99 (m, 1H), 7.29 (d, *J* = 8.0 Hz, 1H), 7.51 (d, *J* = 7.8 Hz, 1H), 10.91 (s, 1H); ^{13}C-NMR (125 MHz, DMSO-*d*$_6$) δ 168.1 (CO), 136.1 (C, Ar), 131.9 (C, Ar), 124.0 (C, Ar), 120.3 (CH, Ar), 118.6 (CH, Ar), 118.5 (CH, Ar), 115.7 (C, Ar), 111.1 (CH, Ar), 57.0 (C), 35.6 (CH$_2$), 34.7 (CH$_2$), 31.9 (CH$_2$), 25.1 (CH$_3$), 23.3 (CH$_2$), 16.5 (CH$_2$); ESI-LRMS *m/z*: 255 [M + H]$^+$; ESI-HRMS *m/z* calcd for M + H$^+$ 255.1492, found: 255.1488. The characterization data is in accordance with that reported in [43].

12c-Benzyl-1,2,3,6,7,12c-hexahydroindolo[3,2-a]quinolizin-4(8H)-one (**SF6b**): pale yellow oil (57.9 mg, yield 35%). ^{1}H-NMR (400 MHz, DMSO-*d*$_6$) δ 1.52–1.41 (m, 1H), 1.68–1.54 (m, 1H), 1.90–1.78 (m, 1H), 2.24–2.05 (m, 2H), 2.61–2.53 (m, 1H), 2.81–2.64 (m, 3H), 3.24 (d, *J* = 13.6 Hz, 1H), 3.41 (d, *J* = 13.5 Hz, 1H), 4.85–4.74 (m, 1H), 6.94–6.86 (m, 1H), 7.06–6.96 (m, 3H), 7.23–7.14 (m, 3H), 7.29 (d, *J* = 8.0 Hz, 1H), 7.33 (d, *J* = 7.9 Hz, 1H), 10.95 (s, 1H); ^{13}C-NMR (150 MHz, DMSO-*d*$_6$) δ 169.2 (CO), 137.8 (C, Ar), 136.0 (C, Ar), 133.0 (C, Ar), 130.4 (2 × CH, Ar), 127.9 (2 × CH, Ar), 126.4 (CH, Ar), 124.3 (C, Ar), 120.2 (CH, Ar), 119.0 (CH, Ar), 118.5 (CH, Ar), 114.2 (C, Ar), 111.1 (CH, Ar), 60.9 (C), 44.3 (CH$_2$), 35.0 (CH$_2$), 33.9 (CH$_2$), 31.7 (CH$_2$), 23.0 (CH$_2$), 16.3 (CH$_2$); ESI-LRMS *m/z*: 331 [M + H]$^+$; ESI-HRMS *m/z* calcd for M + H$^+$ 331.1805, found: 331.1812. The characterization data is in accordance with that reported in [43].

13b-Methyl-6,7-dihydro-5H-benzo[1,2]indolizino[7,8-b]indol-9(13bH)-one (**SF7a**): pale yellow solid (116.4 mg, yield 81%), mp 235–236 °C. ^{1}H-NMR (500 MHz, DMSO-*d*$_6$) δ 1.90 (s, 3H), 2.80–2.70 (m, 1H), 2.91–2.80 (m, 1H), 3.45–3.37 (m, 1H), 4.56–4.46 (m, 1H), 7.12–7.03 (m, 2H), 7.34–7.26 (m, 1H), 7.52–7.44 (m, 1H), 7.69–7.64 (m, 1H), 7.71 (d, *J* = 7.5 Hz, 1H), 8.12–8.03 (m, 1H), 8.27 (d, *J* = 7.8 Hz, 1H), 11.10 (s, 1H); ^{13}C-NMR (125 MHz, DMSO-*d*$_6$) δ 167.3 (CO), 151.4 (C, Ar), 135.9 (C, Ar), 132.3 (C, Ar), 132.0 (CH, Ar), 130.2 (C, Ar), 128.1 (CH, Ar), 124.3 (C, Ar), 123.2 (CH, Ar), 123.1 (CH, Ar), 120.7 (CH, Ar), 119.0 (CH, Ar), 119.0 (CH, Ar), 111.6 (C, Ar), 111.3 (CH, Ar), 63.6 (C), 34.6 (CH$_2$), 25.7 (CH$_3$), 23.4 (CH$_2$); ESI-LRMS *m/z*: 289 [M + H]$^+$; ESI-HRMS *m/z* calcd for M + H$^+$ 289.1335, found: 289.1330. The characterization data is in accordance with that reported in [43].

13b-Benzyl-6,7-dihydro-5H-benzo[1,2]indolizino[7,8-b]indol-9(13bH)-one (**SF7b**): white solid (121.1 mg, yield 66%), mp 281–282 °C. ^{1}H-NMR (500 MHz, DMSO-*d*$_6$) δ 2.91–2.72 (m, 2H), 3.49–3.38 (m, 1H), 3.57 (d, *J* = 13.8 Hz, 1H), 3.85 (d, *J* = 13.9 Hz, 1H), 4.53 (dd, *J* = 13.0, 5.6 Hz, 1H), 6.95–6.87 (m, 2H), 7.08–6.98 (m, 3H), 7.20–7.08 (m, 2H), 7.39–7.30 (m, 2H), 7.46 (d, *J* = 7.4 Hz, 1H), 7.66–7.58 (m, 1H), 8.32 (d, *J* = 7.8 Hz, 1H), 8.43 (d, *J* = 7.8 Hz, 1H), 11.14 (s, 1H); ^{13}C-NMR (125 MHz, DMSO-*d*$_6$) δ 167.7 (CO), 148.8 (C, Ar), 135.9 (C, Ar), 135.7 (C, Ar), 132.8 (C, Ar), 131.4 (CH, Ar), 131.0 (C, Ar), 130.0 (2 × CH, Ar), 127.8 (CH, Ar), 127.3 (2 × CH, Ar), 126.2 (CH, Ar), 124.4 (C, Ar), 123.9 (CH, Ar), 122.6 (CH, Ar), 120.7 (CH, Ar), 119.4 (CH, Ar), 119.1 (CH, Ar), 111.3 (CH, Ar), 111.1 (C, Ar), 67.1 (C), 42.7 (CH$_2$), 34.8 (CH$_2$), 23.4 (CH$_2$); ESI-LRMS *m/z*: 365 [M + H]$^+$; ESI-HRMS *m/z* calcd for M + H$^+$ 365.1648, found: 365.1649. The characterization data is in accordance with that reported in [43].

13b-Pentyl-6,7-dihydro-5H-benzo[1,2]indolizino[7,8-b]indol-9(13bH)-one (**SF7c**): pale yellow oil (68.7 mg, yield 40%). ^{1}H-NMR (400 MHz, CDCl$_3$) δ 0.77 (t, *J* = 6.7 Hz, 3H), 0.89–0.83 (m, 1H), 1.20–1.02 (m, 5H),

2.24–2.10 (m, 1H), 2.76–2.58 (m, 2H), 3.13–2.98 (m, 1H), 3.38–3.24 (m, 1H), 4.77 (dd, J = 13.1, 6.1 Hz, 1H), 7.23–7.12 (m, 2H), 7.31 (d, J = 7.8 Hz, 1H), 7.45–7.38 (m, 1H), 7.62–7.54 (m, 1H), 7.85 (d, J = 7.5 Hz, 1H), 7.99 (d, J = 7.7 Hz, 1H), 8.06 (d, J = 7.7 Hz, 1H), 8.23 (s, 1H); ^{13}C-NMR (150 MHz, CDCl$_3$) δ 169.4 (CO), 149.6 (C, Ar), 136.1 (C, Ar), 132.0 (CH, Ar), 131.9 (C, Ar), 131.8 (C, Ar), 128.1 (CH, Ar), 124.9 (C, Ar), 123.9 (CH, Ar), 122.9 (CH, Ar), 121.7 (CH, Ar), 120.0 (CH, Ar), 119.4 (CH, Ar), 113.5 (C, Ar), 111.4 (CH, Ar), 67.4 (C), 38.2 (CH$_2$), 35.3 (CH$_2$), 31.8 (CH$_2$), 24.1 (CH$_2$), 23.1 (CH$_2$), 22.5 (CH$_2$), 14.1 (CH$_3$); ESI-LRMS m/z: 345 [M + H]$^+$; ESI-HRMS m/z calcd for M + H$^+$ 345.1961, found: 345.1966. The characterization data is in accordance with that reported in [43].

14c-Methyl-8,9,10,14c-tetrahydroindolo[3',2':3,4]pyrido[2,1-a]isoquinolin-6(5H)-one (**SF8a**): yellow solid (129.8 mg, yield 86%), mp 250–251 °C. ^{1}H-NMR (400 MHz, DMSO-d_6) δ 1.89 (s, 3H), 2.72–2.60 (m, 1H), 2.84–2.73 (m, 1H), 3.02–2.89 (m, 1H), 3.62 (d, J = 19.4 Hz, 1H), 4.10 (d, J = 19.2 Hz, 1H), 4.95–4.86 (m, 1H), 7.15–7.04 (m, 3H), 7.30–7.19 (m, 2H), 7.44–7.38 (m, 2H), 7.59 (d, J = 7.6 Hz, 1H), 11.26 (s, 1H); ^{13}C-NMR (100 MHz, DMSO-d_6) δ 169.3 (CO), 141.4 (C, Ar), 136.0 (C, Ar), 134.5 (C, Ar), 132.4 (C, Ar), 127.7 (CH, Ar), 127.1 (CH, Ar), 126.2 (C, Ar), 126.1 (CH, Ar), 124.7 (CH, Ar), 120.5 (CH, Ar), 120.2 (CH, Ar), 119.2 (CH, Ar), 111.5 (CH, Ar), 111.2 (C, Ar), 61.5 (C), 38.2 (CH$_2$), 36.9 (CH$_2$), 25.7 (CH$_3$), 23.3 (CH$_2$); ESI-LRMS m/z: 303 [M + H]$^+$; ESI-HRMS m/z calcd for M + H$^+$ 303.1492, found: 303.1486. The characterization data is in accordance with that reported in [43].

14c-Benzyl-8,9,10,14c-tetrahydroindolo[3',2':3,4]pyrido[2,1-a]isoquinolin-6(5H)-one (**SF8b**): pale yellow oil (66.5 mg, yield 35%). ^{1}H-NMR (500 MHz, CDCl$_3$) δ 1.84–1.73 (m, 1H), 2.48–2.38 (m, 1H), 2.67–2.56 (m, 1H), 3.40 (d, J = 13.8 Hz, 1H), 3.96–3.76 (m, 3H), 4.79 (dd, J = 12.7, 4.3 Hz, 1H), 6.80–6.71 (m, 2H), 7.08–7.00 (m, 2H), 7.16–7.08 (m, 2H), 7.28–7.20 (m, 2H), 7.33–7.28 (m, 2H), 7.50–7.42 (m, 1H), 7.66 (d, J = 7.9 Hz, 1H), 7.96–7.88 (m, 1H), 8.36 (s, 1H); ^{13}C-NMR (125 MHz, CDCl$_3$) δ 170.6 (CO), 141.0 (C, Ar), 136.4 (C, Ar), 136.1 (C, Ar), 135.7 (C, Ar), 132.3 (C, Ar), 130.4 (2 × CH, Ar), 128.2 (2 × CH, Ar), 128.0 (CH, Ar), 127.7 (CH, Ar), 127.3 (C, Ar), 126.9 (CH, Ar), 126.8 (CH, Ar), 126.0 (CH, Ar), 121.6 (CH, Ar), 121.4 (CH, Ar), 120.5 (CH, Ar), 111.7 (CH, Ar), 109.9 (C, Ar), 66.5 (C), 45.0 (CH$_2$), 39.1 (CH$_2$), 38.7 (CH$_2$), 23.3 (CH$_2$); ESI-LRMS m/z: 379 [M + H]$^+$; ESI-HRMS m/z calcd for M + H$^+$ 379.1805, found: 379.1815. The characterization data is in accordance with that reported in [43].

12b-Methyl-1,5,6,12b-tetrahydropyrrolo[2',1':3,4]pyrazino[1,2-a]indol-3(2H)-one (**SF9a**): white solid (107.8 mg, yield 90%), mp 121–122 °C. ^{1}H-NMR (400 MHz, DMSO-d_6) δ 1.59 (s, 3H), 2.33–2.17 (m, 2H), 2.47–2.36 (m, 1H), 2.65–2.53 (m, 1H), 3.49–3.38 (m, 1H), 3.81–3.68 (m, 1H), 4.34–4.22 (m, 2H), 6.34 (s, 1H), 7.08–7.01 (m, 1H), 7.15–7.09 (m, 1H), 7.38 (d, J = 8.1 Hz, 1H), 7.51 (d, J = 7.7 Hz, 1H); ^{13}C-NMR (125 MHz, DMSO-d_6) δ 172.1 (CO), 142.1 (C, Ar), 135.2 (C, Ar), 127.6 (C, Ar), 120.7 (CH, Ar), 119.9 (CH, Ar), 119.8 (CH, Ar), 109.6 (CH, Ar), 95.5 (CH, Ar), 58.6 (C), 40.9 (CH$_2$), 34.3 (CH$_2$), 33.6 (CH$_2$), 29.7 (CH$_2$), 27.2 (CH$_3$); ESI-LRMS m/z: 241 [M + H]$^+$; ESI-HRMS m/z calcd for M + H$^+$ 241.1335, found: 241.1330. The characterization data is in accordance with that reported in [43].

11,12b-Dimethyl-1,5,6,12b-tetrahydropyrrolo[2',1':3,4]pyrazino[1,2-a]indol-3(2H)-one (**SF9b**): pale yellow oil (111.3 mg, yield 88%). ^{1}H-NMR (300 MHz, CDCl$_3$) δ 1.65 (s, 3H), 2.49–2.37 (m, 3H), 2.54 (s, 3H), 2.73–2.57 (m, 1H), 3.47–3.32 (m, 1H), 3.96–3.81 (m, 1H), 4.20 (dd, J = 11.7, 4.7 Hz, 1H), 4.52 (dd, J = 13.6, 5.1 Hz, 1H), 6.30 (s, 1H), 6.98–6.89 (m, 1H), 7.15–7.09 (m, 2H); ^{13}C-NMR (100 MHz, CDCl$_3$) δ 173.2 (CO), 141.1 (C, Ar), 135.4 (C, Ar), 130.0 (C, Ar), 127.9 (C, Ar), 121.8 (CH, Ar), 120.7 (CH, Ar), 106.8 (CH, Ar), 94.8 (CH, Ar), 59.5 (C), 41.3 (CH$_2$), 35.0 (CH$_2$), 34.3 (CH$_2$), 30.4 (CH$_2$), 27.8 (CH$_3$), 18.8 (CH$_3$); ESI-LRMS m/z: 255 [M + H]$^+$; ESI-HRMS m/z calcd for M + H$^+$ 255.1492, found: 255.1488. The characterization data is in accordance with that reported in [45].

12b-Methyl-3-oxo-1,2,3,5,6,12b-hexahydropyrrolo[2',1':3,4]pyrazino[1,2-a]indole-11-carbonitrile (**SF9c**): yellow oil (104.3 mg, yield 79%). ^{1}H-NMR (300 MHz, CDCl$_3$) δ 1.66 (s, 3H), 2.54–2.32 (m, 3H), 2.73–2.59 (m, 1H), 3.49–3.33 (m, 1H), 4.02–3.88 (m, 1H), 4.26 (dd, J = 11.7, 4.8 Hz, 1H), 4.56 (dd, J = 13.7, 5.2 Hz, 1H), 6.52 (s, 1H), 7.26–7.19 (m, 1H), 7.52–7.44 (m, 2H); ^{13}C-NMR (125 MHz, CDCl$_3$) δ 173.1 (CO), 144.6 (C, Ar), 135.4 (C, Ar), 129.7 (C, Ar), 125.7 (CH, Ar), 121.2 (CH, Ar), 118.7 (CN), 113.9 (CH, Ar), 102.8 (C,

Ar), 95.6 (CH, Ar), 59.3 (C), 41.5 (CH$_2$), 34.8 (CH$_2$), 33.9 (CH$_2$), 30.2 (CH$_2$), 27.6 (CH$_3$); ESI-LRMS *m/z*: 266 [M + H]$^+$; ESI-HRMS *m/z* calcd for M + H$^+$ 266.1288, found: 266.1282. The characterization data is in accordance with that reported in [45].

9-Fluoro-12b-methyl-1,5,6,12b-tetrahydropyrrolo[2',1':3,4]pyrazino[1,2-a]indol-3(2H)-one (**SF9d**): pale yellow solid (43.6 mg, yield 68%), mp 71–72 °C. ^{1}H-NMR (500 MHz, DMSO-d_6) δ 1.58 (s, 3H), 2.30–2.18 (m, 2H), 2.44–2.36 (m, 1H), 2.63–2.52 (m, 1H), 3.49–3.38 (m, 1H), 3.77–3.67 (m, 1H), 4.32–4.22 (m, 2H), 6.36 (s, 1H), 6.94–6.86 (m, 1H), 7.26 (dd, *J* = 10.2, 2.2 Hz, 1H), 7.50 (dd, *J* = 8.6, 5.4 Hz, 1H); ^{13}C-NMR (125 MHz, DMSO-d_6) δ 172.1 (CO), 158.5 (d, J_{C-F} = 234.3 Hz, CF, Ar), 142.9 (d, J_{C-F} = 3.7 Hz, C, Ar), 135.2 (d, J_{C-F} = 12.5 Hz, C, Ar), 124.2 (C, Ar), 120.9 (d, J_{C-F} = 10.0 Hz, CH, Ar), 108.1 (d, J_{C-F} = 24.3 Hz, CH, Ar), 96.3 (d, J_{C-F} = 26.3 Hz, CH, Ar), 95.6 (CH, Ar), 58.6 (C), 41.1 (CH$_2$), 34.3 (CH$_2$), 33.5 (CH$_2$), 29.6 (CH$_2$), 27.3 (CH$_3$); ESI-LRMS *m/z*: 259 [M + H]$^+$; ESI-HRMS *m/z* calcd for M + H$^+$ 259.1241, found: 259.1238.

10-Chloro-12b-methyl-1,5,6,12b-tetrahydropyrrolo[2',1':3,4]pyrazino[1,2-a]indol-3(2H)-one (**SF9e**): pale yellow oil (87.5 mg, yield 64%). ^{1}H-NMR (300 MHz, CDCl$_3$) δ 1.63 (s, 3H), 2.52–2.32 (m, 3H), 2.71–2.56 (m, 1H), 3.45–3.32 (m, 1H), 3.94–3.81 (m, 1H), 4.17 (dd, *J* = 11.6, 4.6 Hz, 1H), 4.52 (dd, *J* = 13.7, 5.0 Hz, 1H), 6.23 (s, 1H), 7.20–7.11 (m, 2H), 7.52 (s, 1H); ^{13}C-NMR (100 MHz, CDCl$_3$) δ 173.1 (CO), 143.0 (C, Ar), 134.1 (C, Ar), 129.1 (C, Ar), 126.1 (C, Ar), 121.8 (CH, Ar), 119.9 (CH, Ar), 110.1 (CH, Ar), 96.0 (CH, Ar), 59.4 (C), 41.3 (CH$_2$), 34.8 (CH$_2$), 34.0 (CH$_2$), 30.2 (CH$_2$), 27.7 (CH$_3$); ESI-LRMS *m/z*: 277 ([M + H]$^+$, Cl^{37}), 275 ([M + H]$^+$, Cl^{35}); ESI-HRMS *m/z* calcd for M + H$^+$ 275.0946, found: 275.0943. The characterization data is in accordance with that reported in [45].

13b-Methyl-2,3,6,7-tetrahydro-1H-pyrido[2',1':3,4]pyrazino[1,2-a]indol-4(13bH)-one (**SF10a**): white solid (73.5 mg, yield 58%), mp 113–114 °C. ^{1}H-NMR (500 MHz, DMSO-d_6) δ 1.67 (s, 3H), 1.78–1.69 (m, 1H), 1.98–1.88 (m, 1H), 2.07–1.99 (m, 1H), 2.42–2.26 (m, 3H), 3.40–3.25 (m, 1H), 3.81–3.70 (m, 1H), 4.33–4.24 (m, 1H), 4.95–4.85 (m, 1H), 6.34 (s, 1H), 7.06–7.01 (m, 1H), 7.14–7.07 (m, 1H), 7.37 (d, *J* = 8.1 Hz, 1H), 7.50 (d, *J* = 7.8 Hz, 1H); ^{13}C-NMR (125 MHz, DMSO-d_6) δ 168.1 (CO), 142.4 (C, Ar), 135.1 (C, Ar), 127.7 (C, Ar), 120.6 (CH, Ar), 119.8 (CH, Ar), 119.7 (CH, Ar), 109.4 (CH, Ar), 95.2 (CH, Ar), 56.7 (C), 41.2 (CH$_2$), 36.6 (CH$_2$), 34.5 (CH$_2$), 31.7 (CH$_2$), 28.2 (CH$_3$), 16.9 (CH$_2$); ESI-LRMS *m/z*: 255 [M + H]$^+$; ESI-HRMS *m/z* calcd for M + H$^+$ 255.1492, found: 255.1488. The characterization data is in accordance with that reported in [43].

12,13b-Dimethyl-2,3,6,7-tetrahydro-1H-pyrido[2',1':3,4]pyrazino[1,2-a]indol-4(13bH)-one (**SF10b**): colorless oil (95.4 mg, yield 71%). ^{1}H-NMR (300 MHz, CDCl$_3$) δ 1.73 (s, 3H), 2.06–1.86 (m, 2H), 2.26–2.11 (m, 1H), 2.50–2.32 (m, 2H), 2.65–2.51 (m, 4H), 3.39–3.23 (m, 1H), 4.01–3.87 (m, 1H), 4.22–4.10 (m, 1H), 5.16 (dd, *J* = 13.7, 4.5 Hz, 1H), 6.26 (s, 1H), 6.99–6.88 (m, 1H), 7.18–7.07 (m, 2H); ^{13}C-NMR (100 MHz, CDCl$_3$) δ 169.5 (CO), 141.4 (C, Ar), 135.2 (C, Ar), 129.9 (C, Ar), 127.9 (C, Ar), 121.6 (CH, Ar), 120.5 (CH, Ar), 106.6 (CH, Ar), 94.3 (CH, Ar), 57.4 (C), 41.7 (CH$_2$), 37.5 (CH$_2$), 35.3 (CH$_2$), 32.2 (CH$_2$), 28.9 (CH$_3$), 18.8 (CH$_3$), 17.4 (CH$_2$); ESI-LRMS *m/z*: 269 [M + H]$^+$; ESI-HRMS *m/z* calcd for M + H$^+$ 269.1648, found: 269.1645. The characterization data is in accordance with that reported in [45].

11-Chloro-13b-methyl-2,3,6,7-tetrahydro-1H-pyrido[2',1':3,4]pyrazino[1,2-a]indol-4(13bH)-one (**SF10c**): pale yellow oil (75.5 mg, yield 52%). ^{1}H-NMR (300 MHz, CDCl$_3$) δ 1.70 (s, 3H), 2.20–1.83 (m, 3H), 2.63–2.27 (m, 3H), 3.37–3.22 (m, 1H), 3.98–3.86 (m, 1H), 4.14 (dd, *J* = 11.6, 4.2 Hz, 1H), 5.16 (dd, *J* = 13.8, 4.2 Hz, 1H), 6.18 (s, 1H), 7.22–7.09 (m, 2H), 7.55–7.48 (m, 1H); ^{13}C-NMR (100 MHz, CDCl$_3$) δ 169.4 (CO), 143.4 (C, Ar), 134.0 (C, Ar), 129.1 (C, Ar), 126.0 (C, Ar), 121.6 (CH, Ar), 119.8 (CH, Ar), 110.0 (CH, Ar), 95.5 (CH, Ar), 57.3 (C), 41.7 (CH$_2$), 37.4 (CH$_2$), 35.1 (CH$_2$), 32.1 (CH$_2$), 28.7 (CH$_3$), 17.3 (CH$_2$); ESI-LRMS *m/z*: 291 ([M + H]$^+$, Cl^{37}), 289 ([M + H]$^+$, Cl^{35}); ESI-HRMS *m/z* calcd for M + H$^+$ 289.1102, found: 289.1099. The characterization data is in accordance with that reported in [45].

13b-Methyl-6,7-dihydroisoindolo[1',2':3,4]pyrazino[1,2-a]indol-9(13bH)-one (**SF11a**): pale yellow solid (95.6 mg, yield 66%), mp 149–150 °C. ^{1}H-NMR (400 MHz, DMSO-d_6) δ 1.91 (s, 3H), 3.90–3.75 (m, 2H), 4.45–4.31 (m, 1H), 4.68–4.57 (m, 1H), 6.84 (s, 1H), 7.08–7.00 (m, 1H), 7.17–7.09 (m, 1H), 7.38 (d, *J* = 8.1 Hz,

1H), 7.59–7.48 (m, 2H), 7.80–7.69 (m, 2H), 8.24 (d, J = 7.6 Hz, 1H); ^{13}C-NMR (100 MHz, DMSO-d_6) δ 166.6 (CO), 150.3 (C, Ar), 137.8 (C, Ar), 135.2 (C, Ar), 132.8 (CH, Ar), 129.9 (C, Ar), 128.7 (CH, Ar), 127.3 (C, Ar), 123.1 (2 × CH, Ar), 121.2 (CH, Ar), 120.0 (2 × CH, Ar), 109.9 (CH, Ar), 97.8 (CH, Ar), 61.4 (C), 41.1 (CH$_2$), 34.2 (CH$_2$), 28.2 (CH$_3$); ESI-LRMS m/z: 289 [M + H]$^+$; ESI-HRMS m/z calcd for M + H$^+$ 289.1335, found: 289.1332. The characterization data is in accordance with that reported in [43].

15b-Methyl-8,9-dihydro-5H-indolo[2',1':3,4]pyrazino[2,1-a]isoquinolin-6(15bH)-one (**SF12a**): yellow solid (77.6 mg, yield 51%), mp 235–236 °C. ^{1}H-NMR (400 MHz, DMSO-d_6) δ 1.89 (s, 3H), 3.20–3.08 (m, 1H), 3.68 (d, J = 19.5 Hz, 1H), 3.81–3.72 (m, 1H), 4.10 (d, J = 19.4 Hz, 1H), 4.40–4.31 (m, 1H), 4.99–4.90 (m, 1H), 6.83 (s, 1H), 7.15–7.08 (m, 1H), 7.23–7.15 (m, 2H), 7.33–7.25 (m, 2H), 7.44–7.38 (m, 2H), 7.67 (d, J = 7.7 Hz, 1H); ^{13}C-NMR (125 MHz, DMSO-d_6) δ 168.3 (CO), 138.6 (C, Ar), 136.9 (C, Ar), 136.5 (C, Ar), 132.1 (C, Ar), 128.0 (CH, Ar), 127.6 (CH, Ar), 127.2 (C, Ar), 126.5 (CH, Ar), 124.2 (CH, Ar), 121.4 (CH, Ar), 120.3 (CH, Ar), 119.9 (CH, Ar), 109.7 (CH, Ar), 101.4 (CH, Ar), 61.8 (C), 42.8 (CH$_2$), 37.7 (CH$_2$), 37.4 (CH$_2$), 28.6 (CH$_3$); ESI-LRMS m/z: 303 [M + H]$^+$; ESI-HRMS m/z calcd for M + H$^+$ 303.1492, found: 303.1487. The characterization data is in accordance with that reported in [43].

13b-Methyl-5,6,7,13b-tetrahydro-1H-pyrrolo[2',1':3,4][1,4]diazepino[1,2-a]indol-3(2H)-one (**SF13a**): colorless oil (91.6 mg, yield 72%). ^{1}H-NMR (400 MHz, CDCl$_3$) δ 1.72 (s, 3H), 2.14–1.88 (m, 3H), 2.53–2.42 (m, 2H), 2.80–2.70 (m, 1H), 3.17–3.04 (m, 1H), 4.20 (ddd, J = 14.8, 9.1, 2.5 Hz, 1H), 4.56–4.33 (m, 2H), 6.45 (s, 1H), 7.13–7.06 (m, 1H), 7.24–7.19 (m, 1H), 7.30 (d, J = 8.3 Hz, 1H), 7.57 (d, J = 7.8 Hz, 1H); ^{13}C-NMR (100 MHz, CDCl$_3$) δ 174.2 (CO), 142.5 (C, Ar), 137.7 (C, Ar), 126.9 (C, Ar), 122.0 (CH, Ar), 120.7 (CH, Ar), 119.9 (CH, Ar), 109.0 (CH, Ar), 99.6 (CH, Ar), 62.4 (C), 43.4 (CH$_2$), 38.5 (CH$_2$), 35.0 (CH$_2$), 30.5 (CH$_2$), 28.0 (CH$_2$), 26.4 (CH$_3$); ESI-LRMS m/z: 255 [M + H]$^+$; ESI-HRMS m/z calcd for M + H$^+$ 255.1492, found: 255.1489. The characterization data is in accordance with that reported in [45].

11-Bromo-13b-methyl-5,6,7,13b-tetrahydro-1H-pyrrolo[2',1':3,4][1,4]diazepino[1,2-a]indol-3(2H)-one (**SF13b**): pale yellow oil (89.3 mg, yield 54%). ^{1}H-NMR (300 MHz, CDCl$_3$) δ 1.71 (s, 3H), 2.15–1.91 (m, 3H), 2.51–2.40 (m, 2H), 2.77–2.66 (m, 1H), 3.16–3.03 (m, 1H), 4.18 (ddd, J = 14.8, 8.8, 2.7 Hz, 1H), 4.46–4.32 (m, 2H), 6.37 (s, 1H), 7.14 (d, J = 8.8 Hz, 1H), 7.30–7.26 (m, 1H), 7.66 (d, J = 1.9 Hz, 1H); ^{13}C-NMR (100 MHz, CDCl$_3$) δ 174.1 (CO), 143.7 (C, Ar), 136.4 (C, Ar), 128.5 (C, Ar), 124.8 (CH, Ar), 123.1 (CH, Ar), 112.9 (C, Ar), 110.6 (CH, Ar), 99.1 (CH, Ar), 62.3 (C), 43.7 (CH$_2$), 38.5 (CH$_2$), 34.9 (CH$_2$), 30.4 (CH$_2$), 27.9 (CH$_2$), 26.3 (CH$_3$); ESI-LRMS m/z: 335 ([M + H]$^+$, Br^{81}), 333 ([M + H]$^+$, Br^{79}); ESI-HRMS m/z calcd for M + H$^+$ 333.0597, found: 333.0595. The characterization data is in accordance with that reported in [45].

12,13b-Dimethyl-5,6,7,13b-tetrahydro-1H-pyrrolo[2',1':3,4][1,4]diazepino[1,2-a]indol-3(2H)-one (**SF13c**): colorless oil (107.1 mg, yield 80%). ^{1}H-NMR (300 MHz, CDCl$_3$) δ 1.73 (s, 3H), 2.03–1.82 (m, 2H), 2.17–2.07 (m, 1H), 2.54–2.44 (m, 2H), 2.55 (s, 3H), 2.85–2.70 (m, 1H), 3.17–3.04 (m, 1H), 4.25–4.13 (m, 1H), 4.52–4.33 (m, 2H), 6.46 (s, 1H), 6.95–6.88 (m, 1H), 7.18–7.12 (m, 2H); ^{13}C-NMR (100 MHz, CDCl$_3$) δ 174.0 (CO), 141.8 (C, Ar), 137.3 (C, Ar), 130.1 (C, Ar), 126.5 (C, Ar), 122.1 (CH, Ar), 120.0 (CH, Ar), 106.6 (CH, Ar), 97.9 (CH, Ar), 62.3 (C), 43.5 (CH$_2$), 38.4 (CH$_2$), 34.8 (CH$_2$), 30.4 (CH$_2$), 28.0 (CH$_2$), 26.3 (CH$_3$), 18.6 (CH$_3$); ESI-LRMS m/z: 269 [M + H]$^+$; ESI-HRMS m/z calcd for M + H$^+$ 269.1648, found: 269.1644. The characterization data is in accordance with that reported in [45].

14b-Methyl-1,14b-dihydroindolo[1,2-a]pyrrolo[2,1-c]quinoxalin-3(2H)-one (**SF14a**): pale yellow solid (123.5 mg, yield 86%), mp 179–180 °C. ^{1}H-NMR (400 MHz, DMSO-d_6) δ 1.39 (s, 3H), 2.69–2.45 (m, 3H), 2.95–2.81 (m, 1H), 6.64 (s, 1H), 7.25–7.17 (m, 1H), 7.35–7.26 (m, 2H), 7.46–7.39 (m, 1H), 7.67 (d, J = 7.7 Hz, 1H), 8.13–8.04 (m, 2H), 8.16 (d, J = 8.1 Hz, 1H); ^{13}C-NMR (100 MHz, DMSO-d_6) δ 172.2 (CO), 142.7 (C, Ar), 133.1 (C, Ar), 129.3 (C, Ar), 128.8 (C, Ar), 125.8 (CH, Ar), 125.7 (C, Ar), 124.1 (CH, Ar), 123.0 (CH, Ar), 122.7 (CH, Ar), 121.4 (CH, Ar), 121.2 (CH, Ar), 117.2 (CH, Ar), 111.8 (CH, Ar), 97.6 (CH, Ar), 59.4 (C), 31.2 (CH$_2$), 30.0 (CH$_2$), 25.9 (CH$_3$); ESI-LRMS m/z: 289 [M + H]$^+$; ESI-HRMS m/z calcd for M + H$^+$ 289.1335, found: 289.1331. The characterization data is in accordance with that reported in [43].

2-Hexyl-14b-methyl-1,14b-dihydroindolo[1,2-a]pyrrolo[2,1-c]quinoxalin-3(2H)-one (**SF14b**): pale yellow oil (159.8 mg, yield 86% (dr = 1.5:1)), and the two diastereomers were inseparable by chromatography. ^{1}H-NMR (600 MHz, CDCl$_3$) δ 1.00–0.80 (m, 3.02H), 1.53–1.22 (m, 11.93H), 2.09–1.96 (m, 1.03H), 2.14–2.08 (m, 0.34H), 2.35–2.27 (m, 0.53H), 2.62–2.52 (m, 0.31H), 2.81–2.73 (m, 0.53H), 2.91–2.82 (m, 0.53H), 3.03–2.94 (m, 0.32H), 6.44 (s, 0.51H), 6.47 (s, 0.34H), 7.25–7.15 (m, 1.61H), 7.36–7.27 (m, 1.91H), 7.43–7.37 (m, 0.41H), 7.68–7.61 (m, 0.92H), 7.86 (dd, J = 7.9, 1.3 Hz, 0.34H), 8.06–7.95 (m, 1.91H), 8.28 (dd, J = 8.1, 1.4 Hz, 0.53H); ^{13}C-NMR (150 MHz, CDCl$_3$) δ 175.3 (CO), 174.1 (CO), 143.4 (C, Ar), 141.9 (C, Ar), 134.4 (C, Ar), 133.6 (C, Ar), 130.6 (C, Ar), 129.8 (C, Ar), 129.5 (C, Ar), 129.4 (C, Ar), 126.4 (CH, Ar), 126.2 (C, Ar), 125.6 (CH, Ar), 125.4 (CH, Ar), 124.3 (2 × CH, Ar), 123.1 (2 × CH, Ar), 122.8 (CH, Ar), 121.6 (CH, Ar), 121.3 (2 × CH, Ar), 117.2 (CH, Ar), 117.1 (CH, Ar), 111.9 (CH, Ar), 111.7 (CH, Ar), 97.4 (CH, Ar), 97.1 (CH, Ar), 58.3 (C), 41.9 (CH), 41.5 (CH), 39.5 (CH$_2$), 37.7 (CH$_2$), 32.1 (CH$_2$), 31.9 (CH$_2$), 31.8 (CH$_2$), 30.9 (CH$_2$), 29.3 (CH$_2$), 29.3 (CH$_2$), 28.9 (CH$_3$), 27.6 (CH$_2$), 27.2 (CH$_2$), 26.2 (CH$_3$), 22.8 (CH$_2$), 22.7 (CH$_2$), 14.2 (2 × CH$_3$); ESI-LRMS m/z: 373 [M + H]$^+$; ESI-HRMS m/z calcd for M + H$^+$ 373.2274, found: 373.2273.

15b-Methyl-2,3-dihydro-1H-indolo[1,2-a]pyrido[2,1-c]quinoxalin-4(15bH)-one (**SF15**): pale yellow oil (89.4 mg, yield 59%). ^{1}H-NMR (400 MHz, DMSO-d_6) δ 1.29 (s, 3H), 1.97–1.75 (m, 2H), 2.42–2.28 (m, 2H), 2.71–2.56 (m, 2H), 6.60 (s, 1H), 7.23–7.17 (m, 1H), 7.33–7.24 (m, 2H), 7.47–7.40 (m, 1H), 7.71–7.64 (m, 2H), 8.08–8.01 (m, 2H); ^{13}C-NMR (100 MHz, DMSO-d_6) δ 168.2 (CO), 142.5 (C, Ar), 133.2 (C, Ar), 130.5 (C, Ar), 129.1 (C, Ar), 128.4 (CH, Ar), 127.8 (C, Ar), 126.6 (CH, Ar), 123.4 (CH, Ar), 122.8 (CH, Ar), 121.2 (2 × CH, Ar), 116.9 (CH, Ar), 111.5 (CH, Ar), 96.8 (CH, Ar), 57.3 (C), 33.3 (CH$_2$), 32.9 (CH$_2$), 27.7 (CH$_3$), 17.1 (CH$_2$); ESI-LRMS m/z: 303 [M + H]$^+$; ESI-HRMS m/z calcd for M + H$^+$ 303.1492, found: 303.1486. The characterization data is in accordance with that reported in [43].

15b-Mmethylindolo[1,2-a]isoindolo[1,2-c]quinoxalin-11(15bH)-one (**SF16**): white solid (100.5 mg, yield 60%), mp 149–150 °C. ^{1}H-NMR (400 MHz, DMSO-d_6) δ 1.65 (s, 3H), 6.77 (s, 1H), 7.24–7.18 (m, 1H), 7.35–7.29 (m, 1H), 7.47–7.40 (m, 1H), 7.57–7.51 (m, 1H), 7.65 (d, J = 7.5 Hz, 1H), 7.76–7.68 (m, 1H), 7.96–7.87 (m, 2H), 8.04 (dd, J = 7.9, 1.5 Hz, 1H), 8.15 (d, J = 8.4 Hz, 1H), 8.31–8.26 (m, 1H), 8.35 (d, J = 7.6 Hz, 1H); ^{13}C-NMR (100 MHz, DMSO-d_6) δ 164.1 (CO), 146.4 (C, Ar), 137.1 (C, Ar), 133.9 (C, Ar), 133.5 (CH, Ar), 129.7 (C, Ar), 129.5 (CH, Ar), 129.3 (C, Ar), 128.9 (C, Ar), 126.5 (CH, Ar), 124.6 (C, Ar), 124.4 (CH, Ar), 124.0 (CH, Ar), 123.9 (CH, Ar), 123.7 (CH, Ar), 123.5 (CH, Ar), 121.6 (CH, Ar), 121.3 (CH, Ar), 117.5 (CH, Ar), 111.9 (CH, Ar), 99.3 (CH, Ar), 61.2 (C), 26.3 (CH$_3$); ESI-LRMS m/z: 337 [M + H]$^+$; ESI-HRMS m/z calcd for M + H$^+$ 337.1335, found: 337.1329. The characterization data is in accordance with that reported in [43].

11b-Methyl-4,5-dihydro-3H-pyrrolo[3',2':3,4]pyrido[2,1-a]isoindol-7(11bH)-one (**SF17**): pale yellow solid (77.1 mg, yield 65%), mp 237–238 °C. ^{1}H-NMR (500 MHz, DMSO-d_6) δ 1.66 (s, 3H), 2.66–2.54 (m, 2H), 3.41–3.27 (m, 1H), 4.48–4.37 (m, 1H), 6.31–6.26 (m, 1H), 6.62–6.55 (m, 1H), 7.49–7.42 (m, 1H), 7.68–7.60 (m, 2H), 7.97 (d, J = 7.6 Hz, 1H), 10.56 (s, 1H); ^{13}C-NMR (100 MHz, DMSO-d_6) δ 166.8 (CO), 152.2 (C, Ar), 131.9 (CH, Ar), 130.0 (C, Ar), 127.8 (CH, Ar), 122.7 (CH, Ar), 122.7 (C, Ar), 122.3 (CH, Ar), 119.9 (C, Ar), 116.8 (CH, Ar), 104.0 (CH, Ar), 62.3 (C), 34.5 (CH$_2$), 27.7 (CH$_3$), 22.6 (CH$_2$); ESI-LRMS m/z: 239 [M + H]$^+$; ESI-HRMS m/z calcd for M + H$^+$ 239.1179, found: 239.1175. The characterization data is in accordance with that reported in [43].

12b-Methyl-5,6-dihydropyrrolo[2',1':3,4]pyrazino[2,1-a]isoindol-8(12bH)-one (**SF18**): pale yellow solid (110.9 mg, yield 93%), mp 156–157 °C. ^{1}H-NMR (500 MHz, DMSO-d_6) δ 1.76 (s, 3H), 3.69–3.60 (m, 1H), 3.78–3.70 (m, 1H), 4.08 (dd, J = 12.0, 3.6 Hz, 1H), 4.45 (dd, J = 13.3, 4.2 Hz, 1H), 6.06–5.98 (m, 1H), 6.39–6.31 (m, 1H), 6.67–6.60 (m, 1H), 7.55–7.47 (m, 1H), 7.75–7.65 (m, 2H), 8.06 (d, J = 7.9 Hz, 1H); ^{13}C-NMR (100 MHz, DMSO-d_6) δ 166.8 (CO), 151.1 (C, Ar), 132.6 (CH, Ar), 130.0 (C, Ar), 129.6 (C, Ar), 128.4 (CH, Ar), 122.9 (CH, Ar), 122.8 (CH, Ar), 119.4 (CH, Ar), 107.8 (CH, Ar), 104.2 (CH, Ar), 61.3 (C), 43.8 (CH$_2$), 34.9 (CH$_2$), 28.5 (CH$_3$); ESI-LRMS m/z: 239 [M + H]$^+$; ESI-HRMS m/z calcd for M + H$^+$ 239.1179, found: 239.1175. The characterization data is in accordance with that reported in [43].

9a-Methyl-4,5,9,9a-tetrahydrothieno[2,3-g]indolizin-7(8H)-one (**SF19**): pale yellow solid (65.1 mg, yield 63%), mp 128–129 °C. ^{1}H-NMR (400 MHz, DMSO-d_6) δ 1.42 (s, 3H), 1.90–1.79 (m, 1H), 2.31–2.15 (m, 2H), 2.61–2.52 (m, 1H), 2.74–2.62 (m, 1H), 2.85–2.76 (m, 1H), 3.09–2.98 (m, 1H), 4.15 (dd, J = 13.2, 5.6 Hz, 1H), 6.99 (d, J = 5.2 Hz, 1H), 7.37 (d, J = 5.1 Hz, 1H); ^{13}C-NMR (100 MHz, DMSO-d_6) δ 171.6 (CO), 141.8 (C, Ar), 131.3 (C, Ar), 124.2 (CH, Ar), 124.1 (CH, Ar), 60.4 (C), 33.7 (CH$_2$), 33.4 (CH$_2$), 30.1 (CH$_2$), 25.8 (CH$_3$), 24.3 (CH$_2$); ESI-LRMS m/z: 208 [M + H]$^+$; ESI-HRMS m/z calcd for M + H$^+$ 208.0791, found: 208.0788. The characterization data is in accordance with that reported in [43].

11b-Methyl-4,5-dihydrothieno[3',2':3,4]pyrido[2,1-a]isoindol-7(11bH)-one (**SF20**): yellow solid (105.6 mg, yield 83%), mp 199–200 °C. ^{1}H-NMR (400 MHz, DMSO-d_6) δ 1.75 (s, 3H), 2.82–2.71 (m, 1H), 2.93–2.84 (m, 1H), 3.46–3.38 (m, 1H), 4.47 (dd, J = 13.4, 5.7 Hz, 1H), 7.39 (d, J = 5.3 Hz, 1H), 7.56–7.48 (m, 2H), 7.74–7.63 (m, 2H), 8.16 (d, J = 7.6 Hz, 1H); ^{13}C-NMR (100 MHz, DMSO-d_6) δ 166.8 (CO), 150.2 (C, Ar), 137.8 (C, Ar), 132.8 (C, Ar), 132.3 (CH, Ar), 130.1 (C, Ar), 128.4 (CH, Ar), 125.3 (CH, Ar), 124.0 (CH, Ar), 123.0 (2 × CH, Ar), 63.4 (C), 34.7 (CH$_2$), 27.0 (CH$_3$), 24.8 (CH$_2$); ESI-LRMS m/z: 256 [M + H]$^+$; ESI-HRMS m/z calcd for M + H$^+$ 256.0791, found: 256.0785. The characterization data is in accordance with that reported in [43].

9a-Methyl-4,5,9,9a-tetrahydrothieno[3,2-g]indolizin-7(8H)-one (**SF21**): colorless oil (93.3 mg, yield 90%). ^{1}H-NMR (500 MHz, DMSO-d_6) δ 1.50 (s, 3H), 2.03–1.92 (m, 1H), 2.32–2.19 (m, 2H), 2.61–2.51 (m, 2H), 2.72–2.63 (m, 1H), 3.07–2.96 (m, 1H), 4.11 (dd, J = 13.3, 6.2 Hz, 1H), 6.80 (d, J = 5.1 Hz, 1H), 7.40 (d, J = 5.0 Hz, 1H); ^{13}C-NMR (100 MHz, DMSO-d_6) δ 171.7 (CO), 141.8 (C, Ar), 132.3 (C, Ar), 127.1 (CH, Ar), 123.5 (CH, Ar), 60.4 (C), 35.1 (CH$_2$), 33.6 (CH$_2$), 30.2 (CH$_2$), 27.9 (CH$_3$), 25.1 (CH$_2$); ESI-LRMS m/z: 208 [M + H]$^+$; ESI-HRMS m/z calcd for M + H$^+$ 208.0791, found: 208.0788. The characterization data is in accordance with that reported in [43].

11b-Methyl-4,5-dihydrothieno[2',3':3,4]pyrido[2,1-a]isoindol-7(11bH)-one (**SF22**): pale yellow solid (108.6 mg, yield 85%), mp 137–138 °C. ^{1}H-NMR (500 MHz, DMSO-d_6) δ 1.82 (s, 3H), 2.62 (ddd, J = 16.2, 11.7, 6.5 Hz, 1H), 2.76 (dd, J = 16.2, 4.5 Hz, 1H), 3.40 (ddd, J = 13.4, 12.0, 4.7 Hz, 1H), 4.43 (dd, J = 13.5, 6.2 Hz, 1H), 6.82 (d, J = 5.1 Hz, 1H), 7.45 (d, J = 5.1 Hz, 1H), 7.57–7.49 (m, 1H), 7.76–7.67 (m, 2H), 7.97–7.90 (m, 1H); ^{13}C-NMR (100 MHz, DMSO-d_6) δ 166.5 (CO), 150.2 (C, Ar), 138.0 (C, Ar), 133.4 (C, Ar), 132.5 (CH, Ar), 130.0 (C, Ar), 128.7 (CH, Ar), 127.1 (CH, Ar), 124.4 (CH, Ar), 123.1 (CH, Ar), 122.4 (CH, Ar), 63.1 (C), 34.3 (CH$_2$), 28.8 (CH$_3$), 25.6 (CH$_2$); ESI-LRMS m/z: 256 [M + H]$^+$; ESI-HRMS m/z calcd for M + H$^+$ 256.0791, found: 256.0789. The characterization data is in accordance with that reported in [43].

12b-Methyl-8,12b-dihydro-4H-thieno[2',3':3,4]pyrido[2,1-a]isoquinolin-7(5H)-one (**SF23**): pale yellow solid (63.4 mg, yield 47%), mp 155–156 °C. ^{1}H-NMR (500 MHz, DMSO-d_6) δ 1.87 (s, 3H), 2.50–2.44 (m, 1H), 2.76–2.68 (m, 1H), 3.10–2.99 (m, 1H), 3.64 (d, J = 20.3 Hz, 1H), 3.97 (d, J = 20.2 Hz, 1H), 4.90–4.79 (m, 1H), 6.91 (d, J = 5.1 Hz, 1H), 7.32–7.22 (m, 3H), 7.55 (d, J = 5.1 Hz, 1H), 7.77–7.71 (m, 1H); ^{13}C-NMR (100 MHz, DMSO-d_6) δ 167.9 (CO), 139.0 (C, Ar), 138.8 (C, Ar), 136.4 (C, Ar), 131.3 (C, Ar), 128.0 (CH, Ar), 127.6 (CH, Ar), 127.4 (CH, Ar), 126.5 (CH, Ar), 124.8 (CH, Ar), 124.5 (CH, Ar), 62.6 (C), 37.0 (CH$_2$), 36.7 (CH$_2$), 31.7 (CH$_3$), 25.2 (CH$_2$); ESI-LRMS m/z: 270 [M + H]$^+$; ESI-HRMS m/z calcd for M + H$^+$ 270.0947, found: 270.0942. The characterization data is in accordance with that reported in [43].

8,9-Dimethoxy-10b-methyl-1,5,6,10b-tetrahydropyrrolo[2,1-a]isoquinolin-3(2H)-one (**SF24a**): white solid (114.5 mg, yield 88%), mp 53–54 °C. ^{1}H-NMR (500 MHz, DMSO-d_6) δ 1.45 (s, 3H), 1.93–1.82 (m, 1H), 2.25–2.15 (m, 1H), 2.45–2.37 (m, 1H), 2.59–2.47 (m, 1H), 2.70–2.62 (m, 2H), 3.05–2.96 (m, 1H), 3.71 (s, 3H), 3.75 (s, 3H), 4.08–4.00 (m, 1H), 6.66 (s, 1H), 6.78 (s, 1H); ^{13}C-NMR (100 MHz, DMSO-d_6) δ 171.3 (CO), 147.7 (C, Ar), 147.4 (C, Ar), 134.9 (C, Ar), 124.1 (C, Ar), 111.9 (CH, Ar), 108.7 (CH, Ar), 60.3 (C), 55.8 (OCH$_3$), 55.5 (OCH$_3$), 34.3 (CH$_2$), 33.4 (CH$_2$), 30.2 (CH$_2$), 27.7 (CH$_2$), 26.8 (CH$_3$); ESI-LRMS m/z: 262 [M + H]$^+$; ESI-HRMS m/z calcd for M + H$^+$ 262.1438, found: 262.1434. The characterization data is in accordance with that reported in [43].

2-Hexyl-8,9-dimethoxy-10b-methyl-1,5,6,10b-tetrahydropyrrolo[2,1-a]isoquinolin-3(2H)-one (**SF24b**): white solid (81.2 mg, yield 47% (dr = 4:1)), mp 65–67 °C, and the two diastereomers were inseparable by chromatography. ^{1}H-NMR (400 MHz, DMSO-d_6) δ 0.85–0.72 (m, 3H), 1.37–0.99 (m, 10H), 1.47–1.35 (m, 3H), 1.73–1.57 (m, 1H), 2.43–2.35 (m, 1H), 2.52–2.44 (m, 1H), 2.70–2.52 (m, 2H), 3.12–2.88 (m, 1H), 3.65–3.59 (m, 3H), 3.68 (s, 3H), 4.00–3.89 (m, 1H), 6.61–6.51 (m, 1H), 6.77–6.69 (m, 1H); ^{13}C-NMR (100 MHz, DMSO-d_6) δ 174.9 (CO), 147.5 (C, Ar), 147.4 (C, Ar), 134.7 (C, Ar), 124.4 (C, Ar), 112.0 (CH, Ar), 108.8 (CH, Ar), 59.5 (C, Ar), 55.7 (OCH$_3$), 55.4 (OCH$_3$), 41.2 (CH), 39.5 (CH$_2$), 34.2 (CH$_2$), 31.8 (CH$_2$), 31.2 (CH$_2$), 30.8 (CH$_3$), 28.6 (CH$_2$), 27.0 (CH$_2$), 26.9 (CH$_2$), 22.0 (CH$_2$), 14.0 (CH$_3$); ESI-LRMS m/z: 346 [M + H]$^+$; ESI-HRMS m/z calcd for M + H$^+$ 346.2377, found: 346.2375.

9,10-Dimethoxy-11b-methyl-2,3,6,7-tetrahydro-1H-pyrido[2,1-a]isoquinolin-4(11bH)-one (**SF25**): white solid (84.4 mg, yield 61%), mp 75–76 °C. ^{1}H-NMR (400 MHz, DMSO-d_6) δ 1.55 (s, 3H), 1.64–1.57 (m, 1H), 1.75–1.64 (m, 1H), 1.96–1.81 (m, 1H), 2.37–2.16 (m, 2H), 2.49–2.41 (m, 1H), 2.68–2.53 (m, 2H), 2.87–2.75 (m, 1H), 3.71 (s, 3H), 3.74 (s, 3H), 4.75–4.63 (m, 1H), 6.65 (s, 1H), 6.84 (s, 1H); ^{13}C-NMR (100 MHz, DMSO-d_6) δ 167.7 (CO), 147.4 (C, Ar), 147.2 (C, Ar), 135.2 (C, Ar), 125.5 (C, Ar), 111.7 (CH, Ar), 109.3 (CH, Ar), 58.2 (C), 55.8 (OCH$_3$), 55.4 (OCH$_3$), 36.7 (CH$_2$), 34.8 (CH$_2$), 31.5 (CH$_2$), 28.5 (CH$_2$), 27.3 (CH$_3$), 16.6 (CH$_2$); ESI-LRMS m/z: 276 [M + H]$^+$; ESI-HRMS m/z calcd for M + H$^+$ 276.1594, found: 276.1589. The characterization data is in accordance with that reported in [43].

2,3-Dimethoxy-12b-methyl-5,6-dihydroisoindolo[1,2-a]isoquinolin-8(12bH)-one (**SF26**): white solid (118.8 mg, yield 77%), mp 187–188 °C. ^{1}H-NMR (500 MHz, DMSO-d_6) δ 1.82 (s, 3H), 2.83–2.67 (m, 2H), 3.42–3.34 (m, 1H), 3.69 (s, 3H), 3.83 (s, 3H), 4.39–4.31 (m, 1H), 6.69 (s, 1H), 7.39 (s, 1H), 7.53–7.47 (m, 1H), 7.72–7.65 (m, 2H), 8.30 (d, J = 7.7 Hz, 1H); ^{13}C-NMR (125 MHz, DMSO-d_6) δ 166.5 (CO), 151.1 (C, Ar), 147.8 (C, Ar), 147.5 (C, Ar), 132.2 (CH, Ar), 130.8 (C, Ar), 130.3 (C, Ar), 128.3 (CH, Ar), 125.3 (C, Ar), 123.4 (CH, Ar), 122.8 (CH, Ar), 112.2 (CH, Ar), 110.2 (CH, Ar), 63.4 (C), 56.1 (OCH$_3$), 55.5 (OCH$_3$), 34.6 (CH$_2$), 28.7 (CH$_2$), 28.2 (CH$_3$); ESI-LRMS m/z: 310 [M + H]$^+$; ESI-HRMS m/z calcd for M + H$^+$ 310.1438, found: 310.1432. The characterization data is in accordance with that reported in [43].

3a-Methyl-2,3,3a,4-tetrahydropyrrolo[1,2-a]quinazoline-1,5-dione (**SF27a**): white solid (91.2 mg, yield 84%), mp 177–178 °C. ^{1}H-NMR (500 MHz, DMSO-d_6) δ 1.42 (s, 3H), 2.30–2.19 (m, 2H), 2.56–2.49 (m, 1H), 2.78–2.68 (m, 1H), 7.32–7.25 (m, 1H), 7.65–7.57 (m, 1H), 7.91 (dd, J = 7.7, 1.4 Hz, 1H), 8.07 (d, J = 8.0 Hz, 1H), 8.94 (s, 1H); ^{13}C-NMR (100 MHz, DMSO-d_6) δ 171.9 (CO), 161.2 (CO), 135.7 (C, Ar), 133.2 (CH, Ar), 127.7 (CH, Ar), 124.5 (CH, Ar), 119.9 (CH, Ar), 119.8 (C, Ar), 74.0 (C), 32.4 (CH$_2$), 29.6 (CH$_2$), 26.4 (CH$_3$); ESI-LRMS m/z: 217 [M + H]$^+$; ESI-HRMS m/z calcd for M + H$^+$ 217.0972, found: 217.0969. The characterization data is in accordance with that reported in [43].

3a,4-Dimethyl-2,3,3a,4-tetrahydropyrrolo[1,2-a]quinazoline-1,5-dione (**SF27b**): white solid (97.5 mg, yield 85%), mp 105–107 °C. ^{1}H-NMR (600 MHz, CDCl$_3$) δ 1.46 (s, 3H), 2.30 (ddd, J = 12.0, 6.1, 4.0 Hz, 1H), 2.48–2.39 (m, 1H), 2.70–2.63 (m, 2H), 3.07 (s, 3H), 7.27–7.24 (m, 1H), 7.56–7.51 (m, 1H), 8.08 (dd, J = 7.8, 1.6 Hz, 1H), 8.26 (dd, J = 8.2, 0.8 Hz, 1H); ^{13}C-NMR (150 MHz, CDCl$_3$) δ 171.3 (CO), 162.1 (CO), 135.2 (C, Ar), 133.4 (CH, Ar), 128.6 (CH, Ar), 125.0 (CH, Ar), 119.7 (CH, Ar), 119.4 (C, Ar), 78.5 (C), 32.4 (CH$_2$), 30.3 (CH$_2$), 27.8 (CH$_3$), 21.8 (CH$_3$); ESI-LRMS m/z: 231 [M + H]$^+$; ESI-HRMS m/z calcd for M + H$^+$ 231.1128, found: 231.1127. The characterization data is in accordance with that reported in [43].

4a-Methyl-3,4,4a,5-tetrahydro-1H-pyrido[1,2-a]quinazoline-1,6(2H)-dione (**SF28a**): white solid (70.7 mg, yield 61%), mp 198–199 °C. ^{1}H-NMR (500 MHz, DMSO-d_6) δ 1.35 (s, 3H), 1.90–1.74 (m, 2H), 2.09–2.02 (m, 1H), 2.17–2.09 (m, 1H), 2.48–2.39 (m, 1H), 2.62–2.54 (m, 1H), 7.36–7.28 (m, 1H), 7.60–7.52 (m, 1H), 7.66 (d, J = 8.0 Hz, 1H), 7.86 (dd, J = 7.7, 1.4 Hz, 1H), 8.85 (s, 1H); ^{13}C-NMR (100 MHz, DMSO-d_6) δ 168.8 (CO), 162.3 (CO), 138.1 (C, Ar), 131.8 (CH, Ar), 126.5 (CH, Ar), 126.2 (CH, Ar), 125.2 (CH, Ar), 123.7 (C, Ar), 71.1 (C), 34.7 (CH$_2$), 33.1 (CH$_2$), 28.3 (CH$_3$), 16.6 (CH$_2$); ESI-LRMS m/z: 231 [M + H]$^+$; ESI-HRMS m/z calcd for M + H$^+$ 231.1128, found: 231.1126. The characterization data is in accordance with that reported in [43].

4a,5-Dimethyl-3,4,4a,5-tetrahydro-1H-pyrido[1,2-a]quinazoline-1,6(2H)-dione (**SF28b**): white solid (78.3 mg, yield 64%), mp 138–139 °C. ^{1}H-NMR (600 MHz, CDCl$_3$) δ 1.39 (s, 3H), 1.98–1.86 (m, 2H), 2.17–2.10 (m, 1H), 2.44–2.36 (m, 1H), 2.63–2.55 (m, 1H), 2.74–2.65 (m, 1H), 3.12 (s, 3H), 7.32–7.28 (m, 1H), 7.54–7.49

(m, 1H), 7.62 (dd, J = 8.1, 0.9 Hz, 1H), 8.02 (dd, J = 7.8, 1.6 Hz, 1H); ^{13}C-NMR (150 MHz, CDCl$_3$) δ 169.3 (CO), 163.6 (CO), 137.4 (C, Ar), 132.0 (CH, Ar), 127.8 (CH, Ar), 126.2 (CH, Ar), 126.0 (CH, Ar), 123.9 (C, Ar), 75.5 (C), 34.4 (CH$_2$), 33.7 (CH$_2$), 27.5 (CH$_3$), 24.6 (CH$_3$), 16.9 (CH$_2$); ESI-LRMS m/z: 245 [M + H]$^+$; ESI-HRMS m/z calcd for M + H$^+$ 245.1285, found: 245.1284. The characterization data is in accordance with that reported in [43].

6a-Methyl-6,6a-dihydroisoindolo[2,1-a]quinazoline-5,11-dione (**SF29a**): white solid (99.3 mg, yield 75%), mp 219–220 °C. ^{1}H-NMR (400 MHz, DMSO-d_6) δ 1.72 (s, 3H), 7.42–7.36 (m, 1H), 7.70–7.64 (m, 1H), 7.77–7.71 (m, 1H), 7.85–7.78 (m, 1H), 7.89 (d, J = 7.5 Hz, 1H), 7.95 (d, J = 7.6 Hz, 1H), 8.04–7.98 (m, 2H), 9.48 (s, 1H); ^{13}C-NMR (100 MHz, DMSO-d_6) δ 164.4 (CO), 162.5 (CO), 146.0 (C, Ar), 135.2 (C, Ar), 133.7 (CH, Ar), 133.6 (CH, Ar), 130.1 (CH, Ar), 129.4 (C, Ar), 128.0 (CH, Ar), 125.1 (CH, Ar), 124.0 (CH, Ar), 122.9 (CH, Ar), 121.2 (CH, Ar), 119.8 (C, Ar), 74.2 (C), 27.3 (CH$_3$); ESI-LRMS m/z: 265 [M + H]$^+$; ESI-HRMS m/z calcd for M + H$^+$ 265.0972, found: 265.0966. The characterization data is in accordance with that reported in [43].

6,6a-Dimethyl-6,6a-dihydroisoindolo[2,1-a]quinazoline-5,11-dione (**SF29b**): white solid (97.1 mg, yield 70%), mp 182–184 °C. ^{1}H-NMR (600 MHz, CDCl$_3$) δ 1.72 (s, 3H), 3.25 (s, 3H), 7.37–7.32 (m, 1H), 7.67–7.63 (m, 2H), 7.76–7.70 (m, 2H), 8.05–7.99 (m, 2H), 8.15 (dd, J = 8.1, 1.0 Hz, 1H); ^{13}C-NMR (150 MHz, CDCl$_3$) δ 165.0 (CO), 163.0 (CO), 143.3 (C, Ar), 134.8 (C, Ar), 133.5 (CH, Ar), 132.8 (CH, Ar), 131.4 (C, Ar), 130.5 (CH, Ar), 128.9 (CH, Ar), 125.5 (CH, Ar), 125.2 (CH, Ar), 124.6 (CH, Ar), 121.8 (CH, Ar), 120.6 (C, Ar), 78.1 (C), 29.6 (CH$_3$), 23.3 (CH$_3$); ESI-LRMS m/z: 279 [M + H]$^+$; ESI-HRMS m/z calcd for M + H$^+$ 279.1128, found: 279.1127. The characterization data is in accordance with that reported in [43].

4b-Methyl-4bH-isoquinolino[2,1-a]quinazoline-6,12(5H,13H)-dione (**SF30a**): yellow solid (91.5 mg, yield 66%), mp 225–226 °C. ^{1}H-NMR (400 MHz, DMSO-d_6) δ 1.71 (s, 3H), 3.84 (d, J = 21.2 Hz, 1H), 4.16 (d, J = 21.1 Hz, 1H), 7.30–7.23 (m, 1H), 7.45–7.36 (m, 3H), 7.68–7.61 (m, 1H), 7.79–7.71 (m, 1H), 7.82 (d, J = 7.8 Hz, 1H), 7.93 (dd, J = 7.7, 1.5 Hz, 1H), 8.89 (s, 1H); ^{13}C-NMR (100 MHz, DMSO-d_6) δ 165.5 (CO), 161.9 (CO), 137.6 (C, Ar), 133.1 (C, Ar), 132.1 (CH, Ar), 128.8 (CH, Ar), 128.6 (C, Ar), 127.5 (CH, Ar), 127.2 (CH, Ar), 126.6 (CH, Ar), 126.2 (CH, Ar), 126.1 (CH, Ar), 125.7 (CH, Ar), 123.3 (C, Ar), 73.9 (C), 35.2 (CH$_2$), 30.5 (CH$_3$); ESI-LRMS m/z: 279 [M + H]$^+$; ESI-HRMS m/z calcd for M + H$^+$ 279.1128, found: 279.1122. The characterization data is in accordance with that reported in [43].

4b,5-Dimethyl-4bH-isoquinolino[2,1-a]quinazoline-6,12(5H,13H)-dione (**SF30b**): yellow oil (29.1 mg, yield 20%). ^{1}H-NMR (600 MHz, CDCl$_3$) δ 1.86 (s, 3H), 2.77 (s, 3H), 3.91 (s, 2H), 7.28 (d, J = 7.5 Hz, 1H), 7.45–7.39 (m, 2H), 7.49–7.45 (m, 1H), 7.61–7.55 (m, 2H), 7.63 (d, J = 8.0 Hz, 1H), 8.08 (dd, J = 7.9, 1.3 Hz, 1H); ^{13}C-NMR (150 MHz, DMSO-d_6) δ 169.2 (CO), 162.1 (CO), 138.6 (C, Ar), 131.9 (CH, Ar), 131.8 (C, Ar), 130.6 (C, Ar), 129.9 (CH, Ar), 128.9 (CH, Ar), 128.0 (CH, Ar), 127.7 (CH, Ar), 127.0 (CH, Ar), 126.9 (CH, Ar), 126.7 (CH, Ar), 125.2 (C, Ar), 76.7 (C), 36.2 (CH$_2$), 29.9 (CH$_3$), 25.7 (CH$_3$); ESI-LRMS m/z: 293 [M + H]$^+$; ESI-HRMS m/z calcd for M + H$^+$ 293.1285, found: 293.1284.

3a-Methyl-2,3,3a,4-tetrahydropyrrolo[2,1-b]quinazolin-1(9H)-one (**SF31**): white solid (69.4 mg, yield 69%), mp 141–143 °C. ^{1}H-NMR (600 MHz, CDCl$_3$) δ 1.54 (s, 3H), 2.15–2.05 (m, 2H), 2.52–2.45 (m, 1H), 2.60–2.52 (m, 1H), 4.17 (d, J = 16.7 Hz, 1H), 5.02 (d, J = 16.8 Hz, 1H), 6.58 (dd, J = 8.0, 0.8 Hz, 1H), 6.81–6.76 (m, 1H), 7.07–6.99 (m, 2H); ^{13}C-NMR (150 MHz, CDCl$_3$) δ 174.3 (CO), 141.9 (C, Ar), 127.6 (CH, Ar), 127.0 (CH, Ar), 119.3 (CH, Ar), 117.4 (C, Ar), 116.5 (CH, Ar), 71.9 (C), 38.6 (CH$_2$), 33.0 (CH$_2$), 29.6 (CH$_2$), 25.6 (CH$_3$); ESI-LRMS m/z: 203 [M + H]$^+$; ESI-HRMS m/z calcd for M+H$^+$ 203.1179, found: 203.1178. The characterization data is in accordance with that reported in [43].

4b-Methyl-4b,5-dihydroisoindolo[1,2-b]quinazolin-12(10H)-one (**SF32**): pale yellow solid (41.1 mg, yield 33%), mp 222–223 °C. ^{1}H-NMR (600 MHz, CDCl$_3$) δ 1.71 (s, 3H), 4.24 (s, 1H), 4.45 (d, J = 16.9 Hz, 1H), 5.32 (d, J = 17.0 Hz, 1H), 6.69 (d, J = 8.3 Hz, 1H), 6.90–6.85 (m, 1H), 7.12–7.08 (m, 1H), 7.14 (d, J = 7.5 Hz, 1H), 7.56–7.52 (m, 1H), 7.63 (d, J = 3.9 Hz, 2H), 7.88 (d, J = 7.6 Hz, 1H); ^{13}C-NMR (150 MHz, CDCl$_3$) δ 165.8 (CO), 147.8 (C, Ar), 140.2 (C, Ar), 132.3 (CH, Ar), 131.5 (C, Ar), 129.6 (CH, Ar), 127.9 (CH, Ar), 127.2 (CH, Ar), 124.4 (CH, Ar), 120.7 (CH, Ar), 120.5 (CH, Ar), 118.7 (C, Ar), 118.1 (CH, Ar), 71.5

(C), 38.0 (CH$_2$), 23.9 (CH$_3$); ESI-LRMS m/z: 251 [M + H]$^+$; ESI-HRMS m/z calcd for M + H$^+$ 251.1179, found: 251.1178. The characterization data is in accordance with that reported in [43].

3a-Methyl-2,3,3a,4-tetrahydro-1H-benzo[d]pyrrolo[1,2-a]imidazol-1-one (**SF33**): colorless oil (45.1 mg, yield 48%). ^{1}H-NMR (600 MHz, CDCl$_3$) δ 1.51 (s, 3H), 2.44–2.33 (m, 2H), 2.54 (ddd, J = 16.8, 8.5, 1.6 Hz, 1H), 2.78 (ddd, J = 16.8, 11.7, 8.5 Hz, 1H), 6.68 (dd, J = 7.7, 0.7 Hz, 1H), 6.84–6.79 (m, 1H), 6.98–6.93 (m, 1H), 7.43 (dd, J = 7.6, 1.1 Hz, 1H); ^{13}C-NMR (150 MHz, CDCl$_3$) δ 173.9 (CO), 142.8 (C, Ar), 128.7 (C, Ar), 125.4 (CH, Ar), 120.3 (CH, Ar), 115.5 (CH, Ar), 110.7 (CH, Ar), 85.7 (C), 37.8 (CH$_2$), 33.7 (CH$_2$), 26.3 (CH$_3$); ESI-LRMS m/z: 189 [M + H]$^+$; ESI-HRMS m/z calcd for M + H$^+$ 189.1022, found: 189.1023. The characterization data is in accordance with that reported in [34].

12a-Methyl-12,12a-dihydrobenzo[4,5]imidazo[2,1-a]isoquinolin-6(5H)-one (**SF34**): colorless oil (26.1 mg, yield 21%). ^{1}H-NMR (600 MHz, CDCl$_3$) δ 1.69 (s, 3H), 3.71 (d, J = 19.0 Hz, 1H), 3.88 (d, J = 19.0 Hz, 1H), 4.48 (s, 1H), 6.85 (dd, J = 7.6, 0.5 Hz, 1H), 6.93–6.89 (m, 1H), 7.02–6.97 (m, 1H), 7.24 (d, J = 7.5 Hz, 1H), 7.33–7.30 (m, 1H), 7.37–7.33 (m, 1H), 7.41 (dd, J = 7.6, 0.9 Hz, 1H), 8.02–7.98 (m, 1H);^{13}C-NMR (150 MHz, CDCl$_3$) δ 165.5 (CO), 139.3 (C, Ar), 139.1 (C, Ar), 131.3 (C, Ar), 129.8 (C, Ar), 128.4 (CH, Ar), 128.0 (CH, Ar), 127.6 (CH, Ar), 125.0 (CH, Ar), 123.2 (CH, Ar), 121.6 (CH, Ar), 116.6 (CH, Ar), 112.0 (CH, Ar), 82.3 (C), 38.7 (CH$_2$), 29.5 (CH$_3$); ESI-LRMS m/z: 251 [M + H]$^+$; ESI-HRMS m/z calcd for M + H$^+$ 251.1179, found: 251.1178. The characterization data is in accordance with that reported in [43].

3a-Methyl-3,3a-dihydro-1H-benzo[d]pyrrolo[2,1-b][1,3]oxazine-1,5(2H)-dione (**SF35a**): white solid (98.9 mg, yield 91%), mp 114–115 °C. ^{1}H-NMR (400 MHz, DMSO-d_6) δ 1.63 (s, 3H), 2.48–2.41 (m, 2H), 2.68–2.59 (m, 1H), 2.82–2.71 (m, 1H), 7.44–7.36 (m, 1H), 7.83–7.76 (m, 1H), 8.08–7.92 (m, 2H); ^{13}C-NMR (125 MHz, DMSO-d_6) δ 171.7 (CO), 161.1 (CO), 136.1 (C, Ar), 135.7 (CH, Ar), 129.9 (CH, Ar), 125.5 (CH, Ar), 120.5 (CH, Ar), 115.7 (C, Ar), 95.4 (C), 31.7 (CH$_2$), 29.1 (CH$_2$), 24.0 (CH$_3$); ESI-LRMS m/z: 218 [M + H]$^+$; ESI-HRMS m/z calcd for M + H$^+$ 218.0812, found: 218.0809. The characterization data is in accordance with that reported in [43].

2-Hexyl-3a-methyl-3,3a-dihydro-1H-benzo[d]pyrrolo[2,1-b][1,3]oxazine-1,5(2H)-dione (**SF35b**): colorless oil (75.4 mg, yield 50% (dr = 5.5:1)), and the two diastereomers were inseparable by chromatography. ^{1}H-NMR (600 MHz, DMSO-d_6) for the major isomer: δ 0.91–0.84 (m, 3H), 1.42–1.21 (m, 9H), 1.65–1.61 (m, 3H), 1.85–1.78 (m, 1H), 2.13–2.07 (m, 1H), 2.71–2.62 (m, 1H), 2.91–2.83 (m, 1H), 7.41–7.37 (m, 1H), 7.81–7.78 (m, 1H), 8.00 (dd, J = 7.8, 1.5 Hz, 1H), 8.10–8.07 (m, 1H); ^{13}C-NMR (150 MHz, DMSO-d_6) for major isomer: δ 173.0 (CO), 161.0 (CO), 136.0 (CH, Ar), 135.8 (CH, Ar), 129.9 (CH, Ar), 125.3 (CH, Ar), 119.8 (CH, Ar), 115.1 (C, Ar), 93.5 (C, Ar), 39.8 (CH), 38.5 (CH$_2$), 31.2 (CH$_2$), 29.6 (CH$_2$), 28.6 (CH$_2$), 26.3 (CH$_2$), 23.5 (CH$_3$), 22.1 (CH$_2$), 14.0 (CH$_3$); ESI-LRMS m/z: 302 [M + H]$^+$; ESI-HRMS m/z calcd for M + H$^+$ 302.1751, found: 302.1750.

4a-Methyl-2,3,4,4a-tetrahydrobenzo[d]pyrido[2,1-b][1,3]oxazine-1,6-dione (**SF36**): colorless oil (88.5 mg, yield 77%). ^{1}H-NMR (500 MHz, DMSO-d_6) δ 1.55 (s, 3H), 1.86–1.77 (m, 1H), 1.95–1.86 (m, 1H), 2.31–2.21 (m, 2H), 2.55–2.48 (m, 1H), 2.68–2.58 (m, 1H), 7.46–7.39 (m, 1H), 7.77–7.70 (m, 2H), 7.98–7.93 (m, 1H); ^{13}C-NMR (125 MHz, DMSO-d_6) δ 169.0 (CO), 161.8 (CO), 138.6 (C, Ar), 134.1 (CH, Ar), 128.7 (CH, Ar), 126.1 (2 × CH, Ar), 119.9 (C, Ar), 92.7 (C), 34.9 (CH$_2$), 32.9 (CH$_2$), 25.9 (CH$_3$), 16.1 (CH$_2$); ESI-LRMS m/z: 232 [M + H]$^+$; ESI-HRMS m/z calcd for M + H$^+$ 232.0968, found: 232.0965. The characterization data is in accordance with that reported in ref.43. The characterization data is in accordance with that reported in [43].

6a-Methyl-5H-benzo[4,5][1,3]oxazino[2,3-a]isoindole-5,11(6aH)-dione (**SF37**): pale yellow solid (83.6 mg, yield 63%), mp 138–139 °C. ^{1}H-NMR (400 MHz, DMSO-d_6) δ 1.94 (s, 3H), 7.51–7.45 (m, 1H), 7.79–7.73 (m, 1H), 7.93–7.85 (m, 2H), 7.95 (d, J = 7.5 Hz, 1H), 8.06–8.00 (m, 2H), 8.08 (dd, J = 7.8, 1.2 Hz, 1H); ^{13}C-NMR (125 MHz, DMSO-d_6) δ 164.1 (CO), 161.3 (CO), 143.8 (C, Ar), 136.0 (CH, Ar), 135.7 (C, Ar), 134.4 (CH, Ar), 131.3 (CH, Ar), 130.2 (CH, Ar), 129.4 (C, Ar), 125.7 (CH, Ar), 124.2 (CH, Ar), 123.2 (CH, Ar), 121.2 (CH, Ar), 115.3 (C, Ar), 92.5 (C), 23.4 (CH$_3$); ESI-LRMS m/z: 266 [M + H]$^+$; ESI-HRMS

m/z calcd for M + H$^+$ 266.0812, found: 266.0807. The characterization data is in accordance with that reported in [43].

3a-Methyl-3,3a-dihydro-1H-naphtho[2,3-d]pyrrolo[2,1-b][1,3]oxazine-1,5(2H)-dione (**SF38**): yellow solid (128.1 mg, yield 96%), mp 228–230 °C. ^{1}H-NMR (600 MHz, CDCl$_3$) δ 1.72 (s, 3H), 2.50–2.42 (m, 1H), 2.68–2.61 (m, 1H), 2.76–2.68 (m, 1H), 2.84–2.77 (m, 1H), 7.56–7.51 (m, 1H), 7.67–7.62 (m, 1H), 7.91 (d, J = 8.3 Hz, 1H), 7.95 (d, J = 8.2 Hz, 1H), 8.44 (s, 1H), 8.72 (s, 1H); ^{13}C-NMR (150 MHz, CDCl$_3$) δ 171.7 (CO), 162.3 (CO), 136.7 (C, Ar), 133.2 (CH, Ar), 131.1 (C, Ar), 130.5 (C, Ar), 129.9 (CH, Ar), 129.7 (CH, Ar), 128.0 (CH, Ar), 126.9 (CH, Ar), 119.4 (CH, Ar), 115.7 (C, Ar), 95.7 (C), 32.3 (CH$_2$), 29.6 (CH$_2$), 25.7 (CH$_3$); ESI-LRMS m/z: 268 [M + H]$^+$; ESI-HRMS m/z calcd for M + H$^+$ 268.0968, found: 268.0962. The characterization data is in accordance with that reported in [35].

4a-Methyl-2,3,4,4a-tetrahydronaphtho[2,3-d]pyrido[2,1-b][1,3]oxazine-1,6-dione (**SF39**): pale yellow oil (103.7 mg, yield 74%). ^{1}H-NMR (600 MHz, CDCl$_3$) δ 1.64 (s, 3H), 1.94–1.87 (m, 1H), 2.22–2.12 (m, 2H), 2.52–2.46 (m, 1H), 2.74–2.61 (m, 2H), 7.57–7.53 (m, 1H), 7.66–7.61 (m, 1H), 7.90 (d, J = 8.3 Hz, 1H), 7.96 (d, J = 8.2 Hz, 1H), 8.22 (s, 1H), 8.67 (s, 1H); ^{13}C-NMR (150 MHz, CDCl$_3$) δ 169.3 (CO), 163.3 (CO), 135.9 (C, Ar), 133.0 (C, Ar), 131.8 (CH, Ar), 130.8 (C, Ar), 129.6 (CH, Ar), 129.5 (CH, Ar), 128.2 (CH, Ar), 127.1 (CH, Ar), 124.9 (CH, Ar), 119.2 (C, Ar), 92.5 (C), 36.3 (CH$_2$), 33.7 (CH$_2$), 27.7 (CH$_3$), 16.7 (CH$_2$); ESI-LRMS m/z: 282 [M + H]$^+$; ESI-HRMS m/z calcd for M + H$^+$ 282.1125, found: 282.1117. The characterization data is in accordance with that reported in [43].

4b-Methyl-4bH-naphtho[2′,3′:4,5][1,3]oxazino[2,3-a]isoindole-6,14-dione (**SF40**): white solid (113.2 mg, yield 72%), mp 210–212 °C. ^{1}H-NMR (600 MHz, CDCl$_3$) δ 1.98 (s, 3H), 7.58–7.54 (m, 1H), 7.69–7.64 (m, 2H), 7.78–7.74 (m, 1H), 7.81–7.79 (m, 1H), 8.02–7.95 (m, 3H), 8.49 (s, 1H), 8.79 (s, 1H); ^{13}C-NMR (150 MHz, CDCl$_3$) δ 164.7 (CO), 162.4 (CO), 144.0 (C, Ar), 136.9 (C, Ar), 133.9 (CH, Ar), 133.5 (CH, Ar), 131.1 (CH, Ar), 130.9 (C, Ar), 130.4 (C, Ar), 130.3 (C, Ar), 130.0 (CH, Ar), 129.9 (CH, Ar), 128.0 (CH, Ar), 126.8 (CH, Ar), 124.8 (CH, Ar), 122.6 (CH, Ar), 119.4 (CH, Ar), 115.0 (C, Ar), 92.7 (C), 25.0 (CH$_3$); ESI-LRMS m/z: 316 [M + H]$^+$; ESI-HRMS m/z calcd for M + H$^+$ 316.0968, found: 316.0959. The characterization data is in accordance with that reported in [43].

6a-Methyl-7,8-dihydro-5H-pyrido[2,3-d]pyrrolo[2,1-b][1,3]oxazine-5,9(6aH)-dione (**SF41**): colorless oil (65.1 mg, yield 60%). ^{1}H-NMR (600 MHz, CDCl$_3$) δ 1.72 (s, 3H), 2.52–2.41 (m, 1H), 2.68–2.60 (m, 1H), 2.86–2.69 (m, 2H), 7.36 (dd, J = 7.7, 4.9 Hz, 1H), 8.42 (dd, J = 7.7, 1.9 Hz, 1H), 8.81 (dd, J = 4.9, 1.9 Hz, 1H); ^{13}C-NMR (150 MHz, CDCl$_3$) δ 171.6 (CO), 161.3 (CO), 155.1 (CH, Ar), 149.1 (C, Ar), 139.4 (CH, Ar), 122.0 (CH, Ar), 113.0 (C, Ar), 96.1 (C), 32.3 (CH$_2$), 29.8 (CH$_2$), 25.4 (CH$_3$); ESI-LRMS m/z: 219 [M + H]$^+$; ESI-HRMS m/z calcd for M + H$^+$ 219.0764, found: 219.0763. The characterization data is in accordance with that reported in [35].

3a-Methyl-2,3,3a,5-tetrahydro-1H-benzo[d]pyrrolo[2,1-b][1,3]oxazin-1-one (**SF42**): pale yellow oil (33.8 mg, yield 33%). ^{1}H-NMR (500 MHz, DMSO-d_6) δ 1.45 (s, 3H), 2.07–1.97 (m, 1H), 2.29–2.19 (m, 1H), 2.43 (ddd, J = 17.1, 10.0, 2.1 Hz, 1H), 2.76–2.65 (m, 1H), 4.88 (d, J = 16.0 Hz, 1H), 4.99 (d, J = 16.0 Hz, 1H), 7.16–7.10 (m, 1H), 7.21–7.16 (m, 1H), 7.32–7.25 (m, 1H), 8.20 (d, J = 8.1 Hz, 1H); ^{13}C-NMR (125 MHz, DMSO-d_6) δ 170.9 (CO), 132.6 (C, Ar), 127.1 (CH, Ar), 124.7 (CH, Ar), 123.7 (CH, Ar), 123.2 (C, Ar), 119.2 (CH, Ar), 89.6 (C), 62.0 (CH$_2$), 32.5 (CH$_2$), 29.7 (CH$_2$), 20.7 (CH$_3$); ESI-LRMS m/z: 204 [M + H]$^+$; ESI-HRMS m/z calcd for M + H$^+$ 204.1019, found: 204.1017. The characterization data is in accordance with that reported in [43].

6a-Methyl-5H-benzo[4,5][1,3]oxazino[2,3-a]isoindol-11(6aH)-one (**SF43**): colorless oil (22.4 mg, yield 18%). ^{1}H-NMR (600 MHz, CDCl$_3$) δ 1.74 (s, 3H), 4.98 (d, J = 15.3 Hz, 1H), 5.20 (d, J = 15.3 Hz, 1H), 7.14–7.10 (m, 1H), 7.19–7.15 (m, 1H), 7.40–7.35 (m, 1H), 7.59–7.54 (m, 1H), 7.68–7.60 (m, 2H), 7.91 (d, J = 7.5 Hz, 1H), 8.25 (d, J = 8.2 Hz, 1H); ^{13}C-NMR (150 MHz, CDCl$_3$) δ 165.2 (CO), 146.1 (C, Ar), 133.1 (CH, Ar), 132.6 (C, Ar), 131.2 (C, Ar), 130.2 (CH, Ar), 127.9 (CH, Ar), 124.3 (2 × CH, Ar), 124.2 (CH, Ar), 123.0 (C, Ar), 121.8 (CH, Ar), 121.5 (CH, Ar), 88.1 (C), 63.3 (CH$_2$), 20.8 (CH$_3$); ESI-LRMS m/z: 252 [M + H]$^+$;

ESI-HRMS m/z calcd for M + H$^+$ 252.1019, found: 252.1018. The characterization data is in accordance with that reported in [43].

3.4. General Procedure of the Reductive Preparation of Compounds SF47–SF55

To a solution of substrates (0.3 mmol) in dry THF was added AlCl$_3$ (0.6 mmol), then LiAlH$_4$ (0.6 mmol) was added portionwise at 0 °C. After that, the mixture was heated to reflux for 4 h. After the reaction was cooled, the reaction mixture was diluted with dichloromethane (120.0 mL), and then water was added dropwise at 0 °C to quench the reaction under vigorous stirring conditions. The solid which precipitated out was removed by filtration, and the organic layer obtained was dried over Na$_2$SO$_4$. After the removal of the solvents in vacuo, the residue was purified to give **SF47–SF55**.

11b-Methyl-2,3,5,6,11,11b-hexahydro-1H-indolizino[8,7-b]indole (**SF47**): pale yellow oil (40.5 mg, yield 60%). ^{1}H-NMR (400 MHz, DMSO-d_6) δ 1.67–1.54 (m, 1H), 1.80 (s, 3H), 2.04–1.93 (m, 1H), 2.20–2.08 (m, 1H), 2.94–2.85 (m, 1H), 3.07–2.96 (m, 1H), 3.30–3.22 (m, 2H), 3.56–3.47 (m, 3H), 7.05–6.99 (m, 1H), 7.15–7.09 (m, 1H), 7.35 (d, J = 8.1 Hz, 1H), 7.47 (d, J = 7.8 Hz, 1H), 11.28 (s, 1H); ^{13}C-NMR (125 MHz, DMSO-d_6) δ 136.3 (C, Ar), 125.7 (C, Ar), 121.8 (CH, Ar), 119.0 (CH, Ar), 118.3 (CH, Ar), 111.3 (CH, Ar), 104.5 (C, Ar), 48.9 (CH$_2$), 42.5 (CH$_2$), 36.4 (CH$_2$), 24.1 (CH$_3$), 21.0 (CH$_2$), 15.3 (CH$_2$); ESI-LRMS m/z: 227 [M + H]$^+$; ESI-HRMS m/z calcd for M + H$^+$ 227.1543, found: 227.1540.

13b-Methyl-7,8,13,13b-tetrahydro-5H-benzo[1,2]indolizino[8,7-b]indole (**SF48a**): pale yellow solid (52.3 mg, yield 64%), mp 204–206 °C. ^{1}H-NMR (400 MHz, CDCl$_3$) δ 1.95 (s, 3H), 2.62–2.53 (m, 1H), 3.28–3.13 (m, 1H), 3.54–3.38 (m, 2H), 4.27–4.18 (m, 2H), 7.13–7.03 (m, 2H), 7.21–7.13 (m, 2H), 7.31–7.22 (m, 2H), 7.85–7.74 (m, 2H), 7.93 (s, 1H); ^{13}C-NMR (125 MHz, DMSO-d_6) δ 135.5 (C, Ar), 131.8 (C, Ar), 126.5 (CH, Ar), 126.4 (CH, Ar), 124.9 (C, Ar), 122.8 (CH, Ar), 122.4 (CH, Ar), 120.2 (CH, Ar), 119.0 (CH, Ar), 118.3 (CH, Ar), 110.8 (CH, Ar), 52.7 (CH$_2$), 41.0 (CH$_2$), 26.2 (CH$_3$), 17.1 (CH$_2$); ESI-LRMS m/z: 275 [M + H]$^+$; ESI-HRMS m/z calcd for M + H$^+$ 275.1543, found: 275.1541. The characterization data is in accordance with that reported in [43].

10-Methoxy-13b-methyl-7,8,13,13b-tetrahydro-5H-benzo[1,2]indolizino[8,7-b]indole (**SF48b**): pale yellow oil (59.2 mg, yield 65%). ^{1}H-NMR (400 MHz, CDCl$_3$) δ 1.86 (s, 3H), 2.60 (ddd, J = 15.8, 4.3, 1.4 Hz, 1H), 3.13 (ddd, J = 16.2, 11.1, 6.4 Hz, 1H), 3.54–3.43 (m, 2H), 3.84 (s, 3H), 4.26–4.15 (m, 2H), 6.78 (dd, J = 8.7, 2.5 Hz, 1H), 6.93 (d, J = 2.4 Hz, 1H), 7.15 (d, J = 8.7 Hz, 1H), 7.25–7.18 (m, 2H), 7.33–7.27 (m, 1H), 7.45 (d, J = 7.4 Hz, 1H), 7.60 (s, 1H); ^{13}C-NMR (100 MHz, CDCl$_3$) δ 154.2 (C, Ar), 144.9 (C, Ar), 139.0 (C, Ar), 137.8 (C, Ar), 131.3 (C, Ar), 127.7 (CH, Ar), 127.7 (C, Ar), 127.4 (CH, Ar), 123.4 (CH, Ar), 121.1 (CH, Ar), 111.7 (CH, Ar), 111.7 (CH, Ar), 106.9 (C, Ar), 100.9 (CH, Ar), 65.0 (C), 56.1 (OCH$_3$), 53.2 (CH$_2$), 41.7 (CH$_2$), 26.0 (CH$_3$), 16.4 (CH$_2$); ESI-LRMS m/z: 305 [M + H]$^+$; ESI-HRMS m/z calcd for M + H$^+$ 305.1648, found: 305.1645. The characterization data is in accordance with that reported in [43].

14b-Methyl-5,6,8,9,14,14b-hexahydroindolo[2',3':3,4]pyrido[2,1-a]isoquinoline (**SF49**): pale yellow oil (60.2 mg, yield 70%). ^{1}H-NMR (400 MHz, CDCl$_3$) δ 1.95 (s, 3H), 2.85–2.66 (m, 2H), 3.07–2.91 (m, 2H), 3.14–3.07 (m, 1H), 3.19–3.15 (m, 1H), 3.26 (ddd, J = 13.8, 6.1, 2.6 Hz, 1H), 3.76 (ddd, J = 13.9, 10.0, 5.8 Hz, 1H), 7.19–7.05 (m, 3H), 7.25–7.20 (m, 1H), 7.31–7.27 (m, 1H), 7.37–7.31 (m, 1H), 7.49 (d, J = 7.6 Hz, 1H), 7.65 (d, J = 7.8 Hz, 1H), 7.87 (s, 1H); ^{13}C-NMR (100 MHz, CDCl$_3$) δ140.0 (C, Ar), 137.5 (C, Ar), 135.8 (C, Ar), 134.5 (C, Ar), 130.1 (CH, Ar), 127.7 (CH, Ar), 126.8 (CH, Ar), 126.6 (CH, Ar), 126.0 (CH, Ar), 121.8 (CH, Ar), 119.6 (CH, Ar), 118.5 (CH, Ar), 111.0 (CH, Ar), 105.9 (C, Ar), 58.3 (C), 46.8 (CH$_2$), 45.5 (CH$_2$), 30.4 (CH$_3$), 28.8 (CH$_2$), 17.9 (CH$_2$); ESI-LRMS m/z: 289 [M + H]$^+$; ESI-HRMS m/z calcd for M + H$^+$ 289.1699, found: 289.1695. The characterization data is in accordance with that reported in [43].

11c-Methyl-2,3,5,6,7,11c-hexahydro-1H-indolizino[7,8-b]indole (**SF50**): pale yellow solid (62.2 mg, yield 92%), mp 249–250 °C. ^{1}H-NMR (400 MHz, DMSO-d_6) δ 1.66–1.51 (m, 1H), 1.82 (s, 3H), 2.04–1.92 (m, 1H), 2.27–2.15 (m, 1H), 2.61–2.52 (m, 1H), 3.02–2.87 (m, 1H), 3.24–3.13 (m, 1H), 3.31–3.25 (m, 1H), 3.60–3.44 (m, 3H), 7.04–6.97 (m, 1H), 7.13–7.07 (m, 1H), 7.35 (d, J = 8.0 Hz, 1H), 7.51 (d, J = 7.9 Hz, 1H), 11.24 (s, 1H); ^{13}C-NMR (125 MHz, DMSO-d_6) δ 136.2 (C, Ar), 129.5 (C, Ar), 123.3 (C, Ar), 121.3 (CH,

Ar), 119.0 (CH, Ar), 118.0 (CH, Ar), 111.4 (CH, Ar), 48.7 (CH$_2$), 41.8 (CH$_2$), 36.1 (CH$_2$), 24.3 (CH$_3$), 21.2 (CH$_2$), 16.8 (CH$_2$); ESI-LRMS m/z: 227 [M + H]$^+$; ESI-HRMS m/z calcd for M + H$^+$ 227.1543, found: 227.1541.

14c-Methyl-5,6,8,9,10,14c-hexahydroindolo[3',2':3,4]pyrido[2,1-a]isoquinoline (**SF51**): pale yellow oil (63.5 mg, yield 73%). ^{1}H-NMR (600 MHz, CDCl$_3$) δ 2.21 (s, 3H), 2.82–2.75 (m, 1H), 2.97 (dd, J = 17.9, 6.5 Hz, 1H), 3.23–3.17 (m, 1H), 3.32–3.27 (m, 2H), 3.42–3.38 (m, 1H), 3.50–3.43 (m, 1H), 4.15–4.07 (m, 1H), 7.10–7.05 (m, 2H), 7.14–7.10 (m, 2H), 7.20–7.16 (m, 1H), 7.31 (d, J = 8.0 Hz, 1H), 7.65 (d, J = 7.9 Hz, 1H), 7.79 (d, J = 7.9 Hz, 1H), 9.64 (s, 1H); ^{13}C-NMR (150 MHz, CDCl$_3$) δ 142.9 (C, Ar), 136.6 (C, Ar), 132.7 (C, Ar), 131.2 (C, Ar), 129.2 (CH, Ar), 128.1 (CH, Ar), 127.3 (C, Ar), 126.3 (2 × CH, Ar), 120.8 (CH, Ar), 120.7 (CH, Ar), 119.5 (CH, Ar), 115.3 (C, Ar), 111.2 (CH, Ar), 59.8 (C), 45.8 (CH$_2$), 45.3 (CH$_2$), 29.9 (CH$_3$), 23.6 (CH$_2$), 23.1 (CH$_2$); ESI-LRMS m/z: 289 [M + H]$^+$; ESI-HRMS m/z calcd for M + H$^+$ 289.1699, found: 289.1698.

15b-Methyl-6,8,9,15b-tetrahydro-5H-indolo[2',1':3,4]pyrazino[2,1-a]isoquinoline (**SF52**): pale yellow oil (34.3 mg, yield 40%). ^{1}H-NMR (600 MHz, CDCl$_3$) δ 1.94 (s, 3H), 2.92–2.83 (m, 1H), 3.07–2.99 (m, 1H), 3.17–3.09 (m, 1H), 3.40–3.32 (m, 1H), 3.55–3.47 (m, 1H), 3.63–3.56 (m, 1H), 4.08–4.01 (m, 1H), 4.15–4.10 (m, 1H), 6.55 (s, 1H), 7.15–7.10 (m, 2H), 7.20–7.15 (m, 3H), 7.27 (d, J = 8.0 Hz, 1H), 7.63–7.57 (m, 2H); ^{13}C-NMR (150 MHz, CDCl$_3$) δ 140.9 (C, Ar), 139.8 (C, Ar), 136.7 (C, Ar), 132.8 (C, Ar), 129.5 (CH, Ar), 128.0 (C, Ar), 127.8 (CH, Ar), 126.6 (CH, Ar), 126.1 (CH, Ar), 121.2 (CH, Ar), 120.3 (CH, Ar), 120.1 (CH, Ar), 109.1 (CH, Ar), 100.6 (CH, Ar), 59.2 (C), 45.7 (CH$_2$), 45.6 (CH$_2$), 40.4 (CH$_2$), 31.5 (CH$_3$), 25.3 (CH$_2$); ESI-LRMS m/z: 289 [M + H]$^+$; ESI-HRMS m/z calcd for M + H$^+$ 289.1699, found: 289.1696.

11b-Methyl-4,5,7,11b-tetrahydro-3H-pyrrolo[3',2':3,4]pyrido[2,1-a]isoindole (**SF53**): pale yellow oil (57.2 mg, yield 85%). ^{1}H-NMR (600 MHz, CDCl$_3$) δ 1.78 (s, 3H), 3.07–2.97 (m, 1H), 3.49–3.36 (m, 2H), 3.72–3.65 (m, 1H), 4.18–4.11 (m, 1H), 4.28–4.20 (m, 1H), 6.08–6.04 (m, 1H), 6.62–6.58 (m, 1H), 7.19–7.14 (m, 2H), 7.29–7.23 (m, 1H), 7.45–7.40 (m, 1H), 7.83 (s, 1H); ^{13}C-NMR (150 MHz, CDCl$_3$) δ 148.1 (C, Ar), 137.8 (C, Ar), 127.2 (CH, Ar), 126.8 (CH, Ar), 122.9 (C, Ar), 122.7 (CH, Ar), 121.9 (1CH + 1C, Ar), 116.6 (CH, Ar), 105.2 (CH, Ar), 65.8 (C), 53.5 (CH$_2$), 42.1 (CH$_2$), 28.2 (CH$_3$), 17.2 (CH$_2$); ESI-LRMS m/z: 225 [M + H]$^+$; ESI-HRMS m/z calcd for M + H$^+$ 225.1386, found: 225.1385.

11b-Methyl-4,5,7,11b-tetrahydrothieno[3',2':3,4]pyrido[2,1-a]isoindole (**SF54**): pale yellow solid (57.9 mg, yield 80%), mp 73–74 °C. ^{1}H-NMR (600 MHz, CDCl$_3$) δ 1.76 (s, 3H), 2.60 (ddd, J = 16.5, 4.4, 1.5 Hz, 1H), 3.22–3.13 (m, 1H), 3.49–3.38 (m, 2H), 4.20–4.12 (m, 2H), 6.94 (d, J = 5.2 Hz, 1H), 7.00 (dd, J = 5.3, 0.7 Hz, 1H), 7.21–7.17 (m, 2H), 7.28–7.25 (m, 1H), 7.45 (d, J = 7.6 Hz, 1H); ^{13}C-NMR (150 MHz, CDCl$_3$) δ 146.7 (C, Ar), 139.5 (C, Ar), 138.6 (C, Ar), 132.4 (C, Ar), 127.1 (2 × CH, Ar), 125.6 (CH, Ar), 122.8 (CH, Ar), 122.3 (CH, Ar), 122.1 (CH, Ar), 67.0 (C, Ar), 53.6 (CH$_2$), 42.1 (CH$_2$), 27.9 (CH$_3$), 19.2 (CH$_2$); ESI-LRMS m/z: 242 [M + H]$^+$; ESI-HRMS m/z calcd for M + H$^+$ 242.0998, found: 242.0997.

11b-Methyl-4,5,7,11b-tetrahydrothieno[2',3':3,4]pyrido[2,1-a]isoindole (**SF55**): yellow oil (68.6 mg, yield 95%). ^{1}H-NMR (600 MHz, CDCl$_3$) δ 1.85 (s, 3H), 2.47 (ddd, J = 16.4, 4.5, 1.4 Hz, 1H), 3.07–2.97 (m, 1H), 3.45–3.30 (m, 2H), 4.22–4.12 (m, 2H), 6.69 (d, J = 5.1 Hz, 1H), 7.11 (d, J = 5.0 Hz, 1H), 7.18–7.16 (m, 1H), 7.21–7.18 (m, 1H), 7.30–7.26 (m, 1H), 7.46 (d, J = 7.6 Hz, 1H); ^{13}C-NMR (150 MHz, CDCl$_3$) δ 147.2 (C, Ar), 142.4 (C, Ar), 138.4 (C, Ar), 132.5 (C, Ar), 127.3 (2 × CH, Ar), 126.9 (CH, Ar), 123.1 (CH, Ar), 122.8 (CH, Ar), 122.1 (CH, Ar), 67.0 (C), 53.7 (CH$_2$), 41.8 (CH$_2$), 29.8 (CH$_3$), 20.2 (CH$_2$); ESI-LRMS m/z: 242 [M + H]$^+$; ESI-HRMS m/z calcd for M + H$^+$ 242.0998, found: 242.0997.

4. Conclusions

In conclusion, a green and general tandem reaction between alkynoic acids and amine nucleophiles through gold catalysis in water has been developed. This process proceeds with high efficiency leading to the formation of two rings and three new bonds in a single operation. This approach features low catalyst loading, good to excellent yields, high efficiency in bond formation, high step economy, excellent selectivity, great functional group tolerance, and extraordinarily broad substrate scope, and has been successfully employed to construct a high-quality library of indole/thiophene/pyrrole/pyridine/naphthalene/benzene-fused N-heterocycles. In addition, five antimicrobial compounds were discovered from the library, suggesting the value of our strategy to identify APIs. This is the first example of the generation of pDOS compound library encompassing skeletal diversity, molecular complexity, and drug-like properties from readily available materials through gold catalysis in water. We anticipate that these valuable N-heterocycles will find more pharmaceutical applications after our further investigations.

Supplementary Materials: The following are available online. Table S1: Survey of the solvents on the yield of product **SF1a**, Figures S1–S12: NMR (^{1}H-NMR, ^{13}C-NMR, HSQC, HMBC, and ^{1}H-^{1}H COSY) and ESI(+)MS spectrum of **SF5a**, **[D]$_n$-SF5a**, **SF5b**, **[D]$_n$-SF5b**, **SF1a**, and **[D]$_n$-SF1a**. Figure S13: preliminary screening of antibacterial activities of compounds at 100 µg/mL. Figures S14–S19: time-kill results of **SF9d**, **SF29b**, **SF33**, **SF36**, and **SF41** again*st S. aureus* strain. Figures S20–S24: CFU results of compounds **SF9d**, **SF29b**, **SF33**, **SF36** and **SF41**; copies of ^{1}H and ^{13}C-NMR spectra of new compounds.

Author Contributions: Conceptualization, H.L. and F.Z.; Methodology, X.J. and X.L.; Formal Analysis, J.Y., J.L., and F.Z.; Investigation, X.J., P.L., and J.L.; Resources, Y.C. and X.L.; Writing—Original Draft Preparation, J.L. and F.Z.; Writing—Review and Editing, F.Z., J.W., and H.L.; Visualization, F.Z.; Supervision, H.L. and F.Z.; Project Administration, F.Z.; Funding Acquisition, J.L., H.L., and F.Z.

Funding: This research was funded by the National Natural Science Foundation of China (grants 21602022, 81620108027, and 21632008), the Major Project of Chinese National Programs for Fundamental Research and Development (grant 2015CB910304), Sichuan Science and Technology Program (grants 2018JY0345 and 2018HH007), and Chengdu Municipal Government Program of Science and Technology (grant 2016-XT00-00023-GX). The APC was funded by Chengdu University New Faculty Start-up Funding (grant 2081915037).

Acknowledgments: F.Z. gratefully acknowledges the support from the 1000 Talents Program of Sichuan Province and Chengdu Talents Program.

Conflicts of Interest: The authors declare no conflict of interest.

References

1. Yarmush, M.L.; Jayaraman, A. Advances in proteomic technologies. *Annu. Rev. Biomed. Eng.* **2002**, *4*, 349–373. [CrossRef] [PubMed]
2. Aebersold, R.; Cravatt, B.F. Proteomics–advances, applications and the challenges that remain. *Trends Biotechnol.* **2002**, *20*, S1–S2. [CrossRef]
3. Yee, A.; Pardee, K.; Christendat, D.; Savchenko, A.; Edwards, A.M.; Arrowsmith, C.H. Structural proteomics: Toward high-throughput structural biology as a tool in functional genomics. *Acc. Chem. Res.* **2003**, *36*, 183–189. [CrossRef] [PubMed]
4. Cottingham, K. Government & society: New cancer center will focus on proteomics and genomics. *J. Proteome Res.* **2007**, *6*, 3369.
5. Yates, J.R.; Osterman, A.L. Introduction: Advances in genomics and proteomics. *Chem. Rev.* **2007**, *107*, 3363–3366.
6. Bleicher, K.H.; Böhm, H.J.; Müller, K.; Alanine, A.I. A guide to drug discovery: Hit and lead generation: Beyond high-throughput screening. *Nat. Rev. Drug Discov.* **2003**, *2*, 369–378. [CrossRef] [PubMed]
7. Drewry, D.H.; Macarron, R. Enhancements of screening collections to address areas of unmet medical need: An industry perspective. *Curr. Opin. Chem. Biol.* **2010**, *14*, 289–298. [CrossRef] [PubMed]
8. Dandapani, S.; Marcaurelle, L.A. Grand challenge commentary: Accessing new chemical space for 'undruggable' targets. *Nat. Chem. Biol.* **2010**, *6*, 861–863. [CrossRef] [PubMed]
9. Sauer, W.H.B.; Schwarz, M.K. Molecular shape diversity of combinatorial libraries: A prerequisite for broad bioactivity. *J. Chem. Inf. Comput. Sci.* **2003**, *43*, 987–1003. [CrossRef] [PubMed]
10. Haggarty, S.J. The principle of complementarity: Chemical *versus* biological space. *Curr. Opin. Chem. Biol.* **2005**, *9*, 296–303. [CrossRef] [PubMed]
11. Quinn, R.J.; Carroll, A.R.; Pham, N.B.; Baron, P.; Palframan, M.E.; Suraweera, L.; Pierens, G.K.; Muresan, S. Developing a drug-like natural product library. *J. Nat. Prod.* **2008**, *71*, 464–468. [CrossRef] [PubMed]
12. Camp, D.; Davis, R.A.; Campitelli, M.; Ebdon, J.; Quinn, R.J. Drug-like properties: Guiding principles for the design of natural product libraries. *J. Nat. Prod.* **2012**, *75*, 72–81. [CrossRef] [PubMed]

13. Dandapani, S.; Rosse, G.; Southall, N.; Salvino, J.M.; Thomas, C.J. Selecting, acquiring, and using small molecule libraries for high-throughput screening. *Curr. Protoc. Chem. Biol.* **2012**, *4*, 177–191. [PubMed]
14. Rzuczek, S.G.; Southern, M.R.; Disney, M.D. Studying a drug-like, RNA-focused small molecule library identifies compounds that inhibit RNA toxicity in myotonic dystrophy. *ACS Chem. Biol.* **2015**, *10*, 2706–2715. [CrossRef] [PubMed]
15. Evans, B.E.; Rittle, K.E.; Bock, M.G.; DiPardo, R.M.; Freidinger, R.M.; Whitter, W.L.; Lundell, G.F.; Veber, D.F.; Anderson, P.S.; Chang, R.S.L.; et al. Methods for drug discovery: Development of potent, selective, orally effective cholecystokinin antagonists. *J. Med. Chem.* **1988**, *31*, 2235–2246. [CrossRef] [PubMed]
16. Horton, D.A.; Bourne, G.T.; Smythe, M.L. The combinatorial synthesis of bicyclic privileged structures or privileged substructures. *Chem. Rev.* **2003**, *103*, 893–930. [CrossRef] [PubMed]
17. DeSimone, R.W.; Currie, K.S.; Mitchell, S.A.; Darrow, J.W.; Pippin, D.A. Privileged structures: Applications in drug discovery. *Comb. Chem. High Throughput Screen.* **2004**, *7*, 473–494. [CrossRef] [PubMed]
18. Welsch, M.E.; Snyder, S.A.; Stockwell, B.R. Privileged scaffolds for library design and drug discovery. *Curr. Opin. Chem. Biol.* **2010**, *14*, 347–361. [CrossRef] [PubMed]
19. González, J.F.; Ortín, I.; de la Cuesta, E.; Menéndez, J.C. Privileged scaffolds in synthesis: 2,5-piperazinediones as templates for the preparation of structurally diverse heterocycles. *Chem. Soc. Rev.* **2012**, *41*, 6902–6915. [CrossRef] [PubMed]
20. Schreiber, S.L. Target-oriented and diversity-oriented organic synthesis in drug discovery. *Science* **2000**, *287*, 1964–1969. [CrossRef] [PubMed]
21. Reayi, A.; Arya, P. Natural product-like chemical space: Search for chemical dissectors of macromolecular interactions. *Curr. Opin. Chem. Biol.* **2005**, *9*, 240–247. [CrossRef] [PubMed]
22. Spandl, R.J.; Bender, A.; Spring, D.R. Diversity-oriented synthesis; a spectrum of approaches and results. *Org. Biomol. Chem.* **2008**, *6*, 1149–1158. [CrossRef] [PubMed]
23. Oh, S.; Park, S.B. A design strategy for drug-like polyheterocycles with privileged substructures for discovery of specific small-molecule modulators. *Chem. Commun.* **2011**, *47*, 12754–12761. [CrossRef] [PubMed]
24. Liu, H. Construction of biologically potential library by diversity-oriented synthesis. *Org. Chem. Curr. Res.* **2013**, *2*, 1000e123. [CrossRef]
25. Kim, H.; Tung, T.T.; Park, S.B. Privileged substructure-based diversity-oriented synthesis pathway for diverse pyrimidine-embedded polyheterocycles. *Org. Lett.* **2013**, *15*, 5814–5817. [CrossRef] [PubMed]
26. Kim, J.; Jung, J.; Koo, J.; Cho, W.; Lee, W.S.; Kim, C.; Park, W.; Park, S.B. Diversity-oriented synthetic strategy for developing a chemical modulator of protein–protein interaction. *Nat. Commun.* **2016**, *7*, 13196. [CrossRef] [PubMed]
27. Zhang, X.; Ye, D.; Sun, H.; Guo, D.; Wang, J.; Huang, H.; Zhang, X.; Jiang, H.; Liu, H. Microwave-assisted synthesis of quinazolinone derivatives by efficient and rapid iron-catalyzed cyclization in water. *Green Chem.* **2009**, *11*, 1881–1888. [CrossRef]
28. Ye, D.; Wang, J.; Zhang, X.; Zhou, Y.; Ding, X.; Feng, E.; Sun, H.; Liu, G.; Jiang, H.; Liu, H. Gold-catalyzed intramolecular hydroamination of terminal alkynes in aqueous media: Efficient and regioselective synthesis of indole-1-carboxamides. *Green Chem.* **2009**, *11*, 1201–1208. [CrossRef]
29. Zhou, Y.; Zhai, Y.; Li, J.; Ye, D.; Jiang, H.; Liu, H. Metal-free tandem reaction in water: An efficient and regioselective synthesis of 3-hydroxyisoindolin-1-ones. *Green Chem.* **2010**, *12*, 1397–1404. [CrossRef]
30. Zhang, X.; Zhou, Y.; Wang, H.; Guo, D.; Ye, D.; Xu, Y.; Jiang, H.; Liu, H. Silver-catalyzed intramolecular hydroamination of alkynes in aqueous media: Efficient and regioselective synthesis for fused benzimidazoles. *Green Chem.* **2011**, *13*, 397–405. [CrossRef]
31. Xu, S.; Zhou, Y.; Xu, J.; Jiang, H.; Liu, H. Gold-catalyzed Michael addition/intramolecular annulation cascade: An effective pathway for the chemoselective- and regioselective synthesis of tetracyclic indole derivatives in water. *Green Chem.* **2013**, *15*, 718–726. [CrossRef]
32. Yang, T.; Campbell, L.; Dixon, D.J. A Au(I)-catalyzed N-acyl iminium ion cyclization cascade. *J. Am. Chem. Soc.* **2007**, *129*, 12070–12071. [CrossRef] [PubMed]
33. Muratore, M.E.; Holloway, C.A.; Pilling, A.W.; Storer, R.I.; Trevitt, G.; Dixon, D.J. Enantioselective Brønsted acid-catalyzed N-acyliminium cyclization cascades. *J. Am. Chem. Soc.* **2009**, *131*, 10796–10797. [CrossRef] [PubMed]

34. Patil, N.T.; Mutyala, A.K.; Lakshmi, P.G.V.V.; Gajula, B.; Sridhar, B.; Pottireddygari, G.R.; Rao, T.P. Au(I)-catalyzed cascade reaction involving formal double hydroamination of alkynes bearing tethered carboxylic groups: An easy access to fused dihydrobenzimidazoles and tetrahydroquinazolines. *J. Org. Chem.* **2010**, *75*, 5963–5975. [CrossRef] [PubMed]
35. Feng, E.; Zhou, Y.; Zhang, D.; Zhang, L.; Sun, H.; Jiang, H.; Liu, H. Gold(I)-catalyzed tandem transformation: A simple approach for the synthesis of pyrrolo/pyrido[2,1-a][1,3]benzoxazinones and pyrrolo/pyrido[2,1-a]quinazolinones. *J. Org. Chem.* **2010**, *75*, 3274–3282. [CrossRef] [PubMed]
36. Zhou, Y.; Zhai, Y.; Ji, X.; Liu, G.; Feng, E.; Ye, D.; Zhao, L.; Jiang, H.; Liu, H. Gold(I)-catalyzed one-pot tandem coupling/cyclization: An efficient synthesis of pyrrolo-/pyrido[2,1-b]benzo[d][1,3]oxazin-1-ones. *Adv. Synth. Catal.* **2010**, *352*, 373–378. [CrossRef]
37. Zhou, Y.; Li, J.; Ji, X.; Zhou, W.; Zhang, X.; Qian, W.; Jiang, H.; Liu, H. Silver- and gold-mediated domino transformation: A strategy for synthesizing benzo[e]indolo[1,2-a]pyrrolo/pyrido[2,1-c][1,4]diazepine-3,9-diones. *J. Org. Chem.* **2011**, *76*, 1239–1249. [CrossRef] [PubMed]
38. Patil, N.T.; Lakshmi, P.G.V.V.; Sridhar, B.; Patra, S.; Bhadra, M.P.; Patra, C.R. New linearly and angularly fused quinazolinones: Synthesis through gold(I)-catalyzed cascade reactions and anticancer activities. *Eur. J. Org. Chem.* **2012**, 1790–1799. [CrossRef]
39. Li, Z.; Li, J.; Yang, N.; Chen, Y.; Zhou, Y.; Ji, X.; Zhang, L.; Wang, J.; Xie, X.; Liu, H. Gold(I)-catalyzed cascade approach for the synthesis of tryptamine-based polycyclic privileged scaffolds as α_1-adrenergic receptor antagonists. *J. Org. Chem.* **2013**, *78*, 10802–10811. [CrossRef] [PubMed]
40. Ji, X.; Zhou, Y.; Wang, J.; Zhao, L.; Jiang, H.; Liu, H. Au(I)/Ag(I)-catalyzed cascade approach for the synthesis of benzo[4,5]imidazo[1,2-c]pyrrolo[1,2-a]quinazolinones. *J. Org. Chem.* **2013**, *78*, 4312–4318. [CrossRef] [PubMed]
41. Patil, N.T.; Shinde, V.S.; Sridhar, B. Relay catalytic branching cascade: A technique to access diverse molecular scaffolds. *Angew. Chem. Int. Ed.* **2013**, *52*, 2251–2255. [CrossRef] [PubMed]
42. Naidu, S.; Reddy, S.R. Copper-catalyzed tandem reaction in ionic liquid: An efficient reusable catalyst and solvent media for the synthesis of fused poly hetero cyclic compounds. *RSC Adv.* **2016**, *6*, 62742–62746. [CrossRef]
43. Qiao, J.; Jia, X.; Li, P.; Liu, X.; Zhao, J.; Zhou, Y.; Wang, J.; Liu, H.; Zhao, F. Gold-catalyzed rapid construction of nitrogen-containing heterocyclic compound library with scaffold diversity and molecular complexity. *Adv. Synth. Catal.* **2019**, *361*. [CrossRef]
44. Zhou, Y.; Ji, X.; Liu, G.; Zhang, D.; Zhao, L.; Jiang, H.; Liu, H. Gold(I)-catalyzed cascade for synthesis of pyrrolo[1,2-a:2′,1′-c]-/Pyrido[2,1-c]pyrrolo[1,2-a]quinoxalinones. *Adv. Synth. Catal.* **2010**, *352*, 1711–1717. [CrossRef]
45. Feng, E.; Zhou, Y.; Zhao, F.; Chen, X.; Zhang, L.; Jiang, H.; Liu, H. Gold-catalyzed tandem reaction in water: An efficient and convenient synthesis of fused polycyclic indoles. *Green Chem.* **2012**, *14*, 1888–1895. [CrossRef]
46. Aplander, K.; Hidestal, O.; Katebzadeh, K.; Lindstrom, U.M. A green and facile route to γ- and δ-lactones via efficient Pinner-cyclization of hydroxynitriles in water. *Green Chem.* **2006**, *8*, 22–24. [CrossRef]
47. Li, C.J.; Chen, L. Organic chemistry in water. *Chem. Soc. Rev.* **2006**, *35*, 68–82. [CrossRef] [PubMed]
48. Chanda, A.; Fokin, V.V. Organic synthesis "on water". *Chem. Rev.* **2009**, *109*, 725–748. [CrossRef] [PubMed]
49. Kumaravel, K.; Vasuki, G. Multi-component reactions in water. *Curr. Org. Chem.* **2009**, *13*, 1820–1841. [CrossRef]
50. Gupta, M.; Paul, S.; Gupta, R. General aspects of 12 basic principles of green chemistry with applications. *Curr. Sci.* **2010**, *99*, 1341–1360.
51. Butler, R.N.; Coyne, A.G. Water: Nature's reaction enforcer—Comparative effects for organic synthesis "in-water" and "on-water". *Chem. Rev.* **2010**, *110*, 6302–6337. [CrossRef] [PubMed]
52. Hashmi, A.S.K. Gold-catalyzed organic reactions. *Chem. Rev.* **2007**, *107*, 3180–3211. [CrossRef] [PubMed]
53. Corma, A.; Leyva-Pérez, A.; Sabater, M.J. Gold-catalyzed carbon−heteroatom bond-forming reactions. *Chem. Rev.* **2011**, *111*, 1657–1712. [CrossRef] [PubMed]
54. Huang, H.; Zhou, Y.; Liu, H. Recent advances in the gold-catalyzed additions to C–C multiple bonds. *Beilstein J. Org. Chem.* **2011**, *7*, 897–936. [CrossRef] [PubMed]
55. Krause, N.; Winter, C. Gold-catalyzed nucleophilic cyclization of functionalized allenes: A powerful access to carbo- and heterocycles. *Chem. Rev.* **2011**, *111*, 1994–2009. [CrossRef] [PubMed]

56. Dorel, R.; Echavarren, A.M. Gold(I)-catalyzed activation of alkynes for the construction of molecular complexity. *Chem. Rev.* **2015**, *115*, 9028–9072. [CrossRef] [PubMed]
57. Gao, F.; Zhou, Y.; Liu, H. Recent advances in the synthesis of heterocycles via gold-catalyzed cascade reactions: A review. *Curr. Org. Chem.* **2017**, *21*, 1530–1566. [CrossRef]
58. Dalpozzo, R. Strategies for the asymmetric functionalization of indoles: An update. *Chem. Soc. Rev.* **2015**, *44*, 742–778. [CrossRef] [PubMed]
59. Lindner, A.; Claassen, V.; Hendriksen, T.W.J.; Kralt, T. Reserpine analogs; phenethylamine derivatives[1]. *J. Med. Chem.* **1963**, *6*, 97–101. [CrossRef] [PubMed]
60. Daugan, A.; Grondin, P.; Ruault, C.; de Gouville, A.C.L.M.; Coste, H.; Kirilovsky, J.; Hyafil, F.; Labaudinière, R. The discovery of Tadalafil: A novel and highly selective PDE5 inhibitor. 1: 5,6,11,11a-tetrahydro-1H-imidazo[1′,5′:1,6]pyrido[3,4-b]indole-1,3(2H)-dione analogues. *J. Med. Chem.* **2003**, *46*, 4525–4532. [CrossRef] [PubMed]
61. Daugan, A.; Grondin, P.; Ruault, C.; de Gouville, A.C.L.M.; Coste, H.; Linget, J.M.; Kirilovsky, J.; Hyafil, F.; Labaudinière, R. The discovery of Tadalafil: A novel and highly selective PDE5 inhibitor. 2: 2,3,6,7,12,12a-hexahydropyrazino[1′,2′:1,6]pyrido[3,4-b]indole-1,4-dione analogues. *J. Med. Chem.* **2003**, *46*, 4533–4542. [CrossRef] [PubMed]
62. Kochanowska-Karamyan, A.J.; Hamann, M.T. Marine indole alkaloids: Potential new drug leads for the control of depression and anxiety. *Chem. Rev.* **2010**, *110*, 4489–4497. [CrossRef] [PubMed]
63. Silvestri, R. New prospects for vinblastine analogues as anticancer agents. *J. Med. Chem.* **2013**, *56*, 625–627. [CrossRef] [PubMed]
64. Sun, H.; Zhu, L.; Yang, H.; Qian, W.; Guo, L.; Zhou, S.; Gao, B.; Li, Z.; Zhou, Y.; Jiang, H.; et al. Asymmetric total synthesis and identification of tetrahydroprotoberberine derivatives as new antipsychotic agents possessing a dopamine D_1, D_2 and serotonin 5-HT_{1A} multi-action profile. *Bioorg. Med. Chem.* **2013**, *21*, 856–868. [CrossRef] [PubMed]
65. Guo, D.; Li, J.; Lin, H.; Zhou, Y.; Chen, Y.; Zhao, F.; Sun, H.; Zhang, D.; Li, H.; Shoichet, B.K.; et al. Design, synthesis, and biological evaluation of novel tetrahydroprotoberberine derivatives (THPBs) as selective α_{1A}-adrenoceptor antagonists. *J. Med. Chem.* **2016**, *59*, 9489–9502. [CrossRef] [PubMed]
66. Stempel, E.; Gaich, T. Cyclohepta[b]indoles: A privileged structure motif in natural products and drug design. *Acc. Chem. Res.* **2016**, *49*, 2390–2402. [CrossRef] [PubMed]
67. Homer, J.A.; Sperry, J. Mushroom-derived indole alkaloids. *J. Nat. Prod.* **2017**, *80*, 2178–2187. [CrossRef] [PubMed]
68. Klas, K.R.; Kato, H.; Frisvad, J.C.; Yu, F.; Newmister, S.A.; Fraley, A.E.; Sherman, D.H.; Tsukamoto, S.; Williams, R.M. Structural and stereochemical diversity in prenylated indole alkaloids containing the bicyclo[2.2.2]diazaoctane ring system from marine and terrestrial fungi. *Nat. Prod. Rep.* **2018**, *35*, 532–558. [CrossRef] [PubMed]
69. Sun, H.; Huang, H.; Zhang, D.; Feng, E.; Qian, W.; Zhang, L.; Chen, K.; Liu, H. Synthesis of 4-aryl-2(5H)-furanones by gold(I)-catalyzed intramolecular annulation. *Adv. Synth. Catal.* **2011**, *353*, 1413–1419. [CrossRef]

Sample Availability: Samples of the compounds **SF1–SF43, SF47–SF55** are available from the authors.

© 2019 by the authors. Licensee MDPI, Basel, Switzerland. This article is an open access article distributed under the terms and conditions of the Creative Commons Attribution (CC BY) license (http://creativecommons.org/licenses/by/4.0/).

Review

Last Decade of Unconventional Methodologies for the Synthesis of Substituted Benzofurans

Lucia Chiummiento *, Rosarita D'Orsi, Maria Funicello and Paolo Lupattelli

Department of Science, Via dell'Ateneo Lucano, 10, 85100 Potenza, Italy; rosarita.dorsi@gmail.com (R.D.); maria.funicello@unibas.it (M.F.); paolo.lupattelli@unibas.it (P.L.)
* Correspondence: lucia.chiummiento@unibas.it

Academic Editor: Gianfranco Favi
Received: 22 April 2020; Accepted: 13 May 2020; Published: 16 May 2020

Abstract: This review describes the progress of the last decade on the synthesis of substituted benzofurans, which are useful scaffolds for the synthesis of numerous natural products and pharmaceuticals. In particular, new intramolecular and intermolecular C–C and/or C–O bond-forming processes, with transition-metal catalysis or metal-free are summarized. (1) Introduction. (2) Ring generation via intramolecular cyclization. (2.1) C7a–O bond formation: (route a). (2.2) O–C2 bond formation: (route b). (2.3) C2–C3 bond formation: (route c). (2.4) C3–C3a bond formation: (route d). (3) Ring generation via intermolecular cyclization. (3.1) C7a-O and C3–C3a bond formation (route a + d). (3.2) O–C2 and C2–C3 bond formation: (route b + c). (3.3) O–C2 and C3–C3a bond formation: (route b + d). (4) Benzannulation. (5) Conclusion.

Keywords: synthesis of benzofurans; intra-molecular approach; inter-molecular approach

---

1. Introduction

Benzofuran (**BF**) and 2,3-dihydrobenzofuran (**2,3-DBF**) are key structural units in a variety of biologically active natural products (Figure 1) and represent the core structures of many approved drugs, as well as lead-design developments from natural products [1–4].

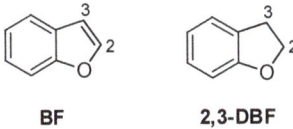

Figure 1. Structures of benzofuran (**BF**) and 2,3-dihydrobenzofuran (**2,3-DBF**).

BF is a completely aromatic flat structure, while 2,3-DBF bears two prochiral sp^3 carbons on the heterocycle, placing the substituents out of the benzofuryl plane. Naturally occurring compounds bearing BF and their derivatives show a broad range of pharmacological activities. Among them, amurensin H (or viniferifuran) **1** (Figure 2) displayed anti-inflammatory effect on an asthma-like reaction induced in mice [5], while Anigopreissin A **2** showed low antimicrobial activity against *Staphylococcus. aureus* and *S. pyogene* [6] and was also discovered as an inhibitor of HIV-1 reverse transcriptase (IC50 = 8 mM), including two mutant enzymes resistant to the clinical drug nevirapine [7].

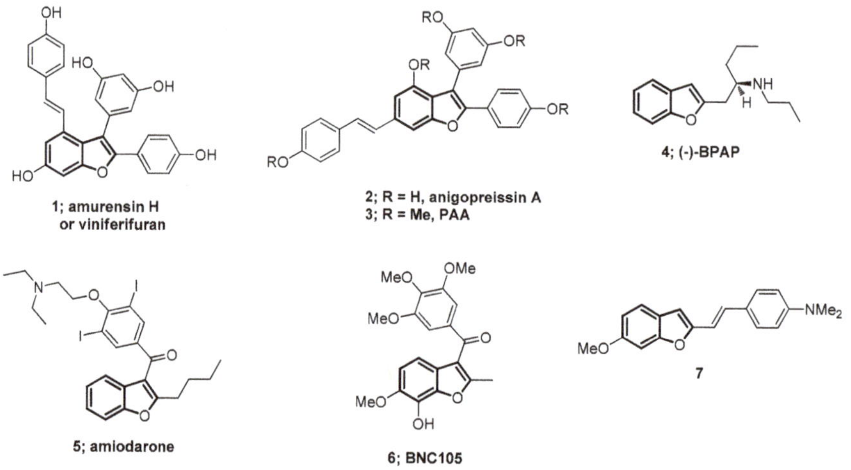

Figure 2. Representative pharmacologically active BF derivatives.

Permethylated anigopreissin A (PAA) **3** showed inhibitory activity for human hepatoma cell proliferation [8,9], while different benzofuran derivatives have shown pharmacological properties such as anticancer [10,11], antiviral [12,13], anti-Alzheimer's disease [14,15], together with antiparasitic [16], antitubercular [17], and antibacterial [18,19] activities. Prescribed agents featuring the benzofuran scaffold include the antidepressant (−)-BPAP **4**, the antiarrythmic amiodarone **5** [20], the clinical candidate drug for renal and ovarian cancers BNC105 **6** [21], and the inhibitor of Aβ fibril formation **7** [2]. Such a variety of biological and pharmacological activities make BF an important pharmacophore for the development of new drugs.

Thus, synthetic access to benzofurans is of considerable interest, and numerous approaches to this scaffold have been disclosed in the literature. Herein, we deal with the most recent literature, which is not included in the appeared reviews of the last decade (2009–2020) [22].

The review of De Luca et al. [23] dealt with the synthesis of 2-substituted-benzofurans up to 2009, while the one of Abu-Hashem et al. [24] was an overview about the different approaches to benzofurans. Very recently, from the same authors, a chapter in Advances in Heterocyclic Chemistry [25] and three reviews have been published in which the full perspective of reactivity of benzofurans [26], advances in the synthesis of biologically potent compounds bearing at least one benzofuran moiety in their structures [27] and the recent reports on the total synthesis of natural products containing at least one benzofuran moiety in their complex structures [28] have been discussed. A review on synthetic routes for synthesis of benzofuran-based compounds appeared in 2017 [29] and at the least one summarizes the recent studies on the various aspects of benzofurans derivatives, including their natural sources, biological activities and drug prospects, and chemical synthesis, as well as the relationship between the bioactivities and structures [30].

In this plethora of methodologies, different classifications were used, subdividing the data into synthesis of 2- or 3-substituted or 2,3-disubstituted benzofurans [31,32], or into transition-metal-catalyzed [33–37] vs. metal-free approaches, or pointing out the most recent applications of metal catalyzed C–H insertion [38]. We chose to compare the intra-molecular and inter-molecular methodologies used to build selected bonds.

In this review, the synthetic approaches are classified according to the method by which the core BF structure is constructed. We divided the methods into intra-molecular and inter-molecular approaches. Both are classified according to which bond is formed in the key reaction (Scheme 1).

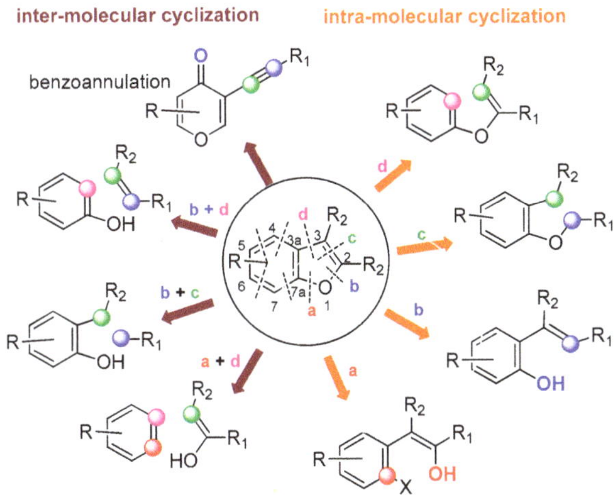

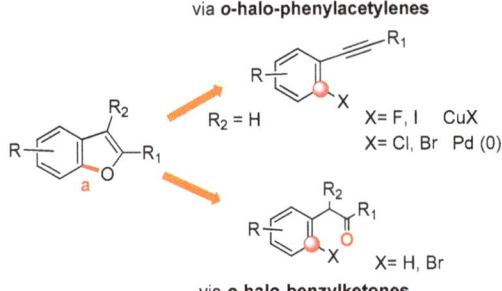

Scheme 1. Intra-molecular and inter-molecular retrosynthetic approaches to benzofuran scaffold.

Intra-molecular approaches are the most dated and commonly used, so only newly developed catalytic systems and reaction conditions are introduced herein. The real novelty of this decade is represented by the advent of the inter-molecular strategies often employing transition metal as catalysts (Rh, Fe, and Pd), using one-pot protocols, [3 + 2] cycloaddition reactions, or sigmatropic rearrangements as Claisen's one.

2. Ring Generation via Intra-Molecular Cyclization

2.1. C7a–O Bond Formation: (Route a)

The main approaches for the formation of C7a-O bond are collected below (Scheme 2).

Scheme 2. Approaches for C7a-O bond formation.

2.1.1. From o-Halophenylacetylenes

An interesting alternative for in situ generating of 2-alkynylphenols is represented by the hydroxylation coupling of 2-haloarylalkynes. This strategy has been successfully employed in the domino hydroxylation–cyclization in which the conversion of 2-haloarylalkyne into a 2-phenylbenzo[b]furan happened in the presence of hydroxide anions.

An efficient copper-promoted hydration/annulation reaction and its application in the synthesis of benzofuran and benzothiophene derivatives has been presented, starting from readily available 2-fluorophenylacetylene derivatives (Scheme 3). This strategy involved a domino hydration of the C–F bond of 2-fluorophenylacetylene derivatives, using CuI, KOH, H$_2$O, and KI in DMSO at 80 °C, followed by an intramolecular annulation to afford benzo[b]furan and benzo[b]thiophene derivatives [39]. A very similar copper-mediated hydroxylation of aryl iodide with hydroxide salts has been performed, as reported in Reference [36].

An extension of the hydroxylation on o-halide phenylacetylenes (Br and Cl) has been performed using a mixture of tris(dibenzylideneacetone)dipalladium(0) (Pd$_2$dba$_3$) and 5-(di-tert-butylphosphino)-1′,3′,5′-triphenyl-1′H-[1,4′]bipyrazole (Bippyphos), showing to be a robust and efficient catalyst system under mild conditions and with broad substrate scope (Scheme 3). Notably, a significant number of the reported reactions proceeded at room temperature, on the benchtop under air using unpurified solvents with negligible loss in reactivity vs. related transformations conducted under inert atmosphere conditions [40].

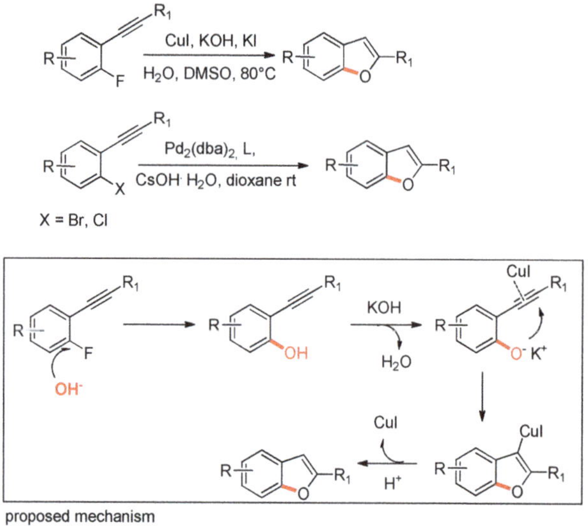

Scheme 3. Synthesis of benzofurans via hydroxylation of o-halophenylacetylene.

2.1.2. From o-Halo-Benzylketones

It was known that copper catalysts [36] were successfully applied to the synthesis of benzofurans by ring closure of aryl o-bromobenzyl ketones. The analogous palladium-catalyzed ring closure provided a straightforward route to 2-arylbenzofurans (Scheme 4). The best results were obtained when 1,3-bis(2,6-diisopropylphenyl) imidazolium tetrafluoroborate (IPr) was used as ligand of Pd$_2$(dba)$_3$, Cs$_2$CO$_3$ as base in o-xylene at 100 °C, giving nearly full conversionconversion of 2-bromobenzyl phenylketone [41]. Different aryl 2-bromobenzylketones gave rise to benzofurans if 10% mol of FeCl$_3$ (of 98% or of 99.995% purity) or sub-mol % quantities of CuCl$_2$ (of 99.995% purity) was used as catalyst and Cs$_2$CO$_3$ as base in DMF [42].

A one-pot approach for the preparation of highly substituted benzofurans was proposed, starting from simple 1-aryl- or 1-alkylketones, involving regioselective iron(III)-catalyzed halogenation, followed by metal-mediated O-arylation, as well as demonstrating the use of parts per million (ppm) copper loading to perform C–O cyclization [43].

A variety of benzofurans were achieved by way of a FeCl$_3$-mediated intramolecular cyclization of electron-rich-aryl ketones. This method allowed the construction of benzofuran rings by linking the O-atom on the side chain to the benzene ring via direct oxidative aromatic C–O bond formation. The alkoxy substituent on the benzene ring in the substrates was essential for an efficient cyclization to occur [44].

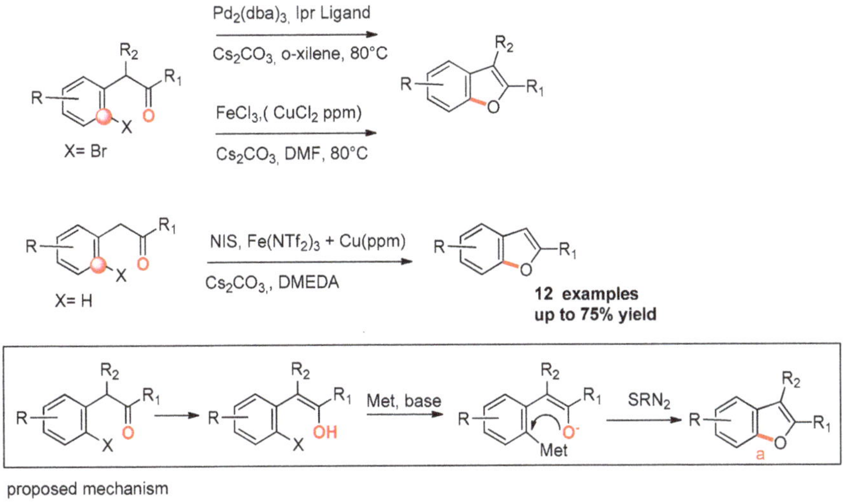

Scheme 4. Synthesis of benzofurans from *o*-halo-benzylketones.

2.2. O–C2 Bond Formation: (Route b)

The main approaches for the formation of O-C2 bond dealt in this section are summarized below (Scheme 5).

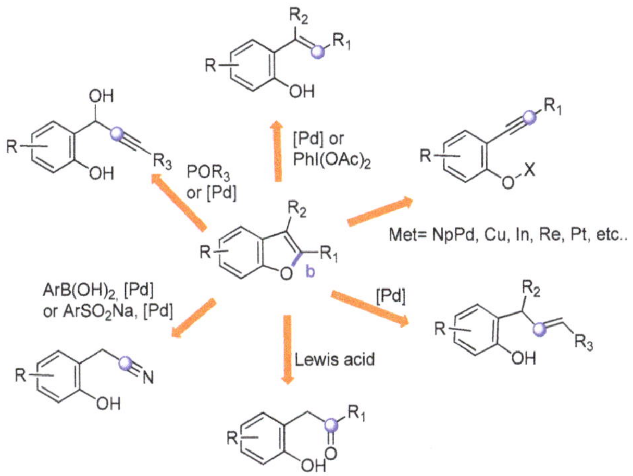

Scheme 5. Approaches for O–C2 bond formation.

The formation of O–C2 bond as the last bond represents the most popular between the intramolecular approaches. Several methods have been collected for this type of disconnection in the reviews of Abu-Hashem et al. [24] and Cacchi [35,36].

2.2.1. Via C-H Activation of *o*-Alkenylphenols

The C-H functionalization represents markedly a different approach from traditional ones, which exploit functional group transformations [45,46]. In 2014, a new, unprecedent palladium-catalyzed method for the synthesis of benzofurans was discovered [47] in which 2-hydroxystyrenes and iodobenzenes were involved in a C–H tandem activation/oxidation reaction. After careful analysis of

243

the whole process, it was clear that the formation of benzofurans resulted by tandem Pd-catalyzed Heck reaction/oxidative cyclization sequence, although the detailed mechanism is still unknown. Following this method, the overall efficiency of the synthesis of decursivine and its analogues was improved (Scheme 6).

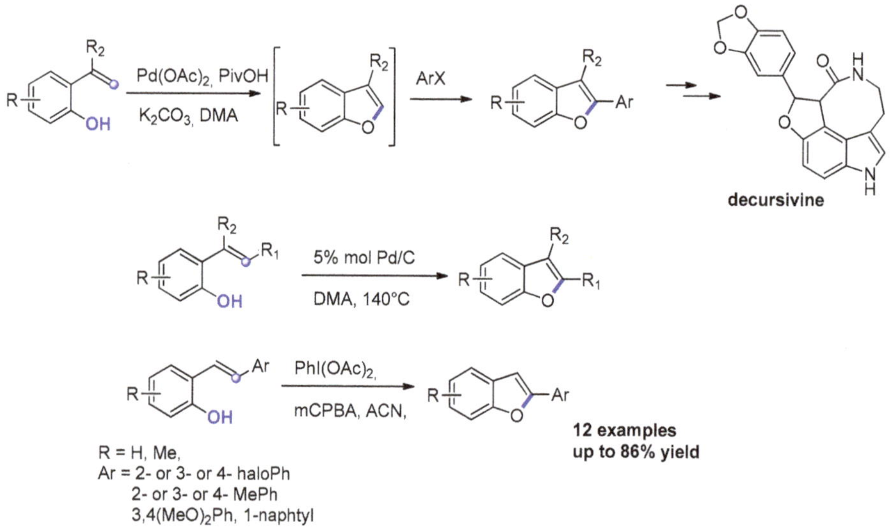

Scheme 6. Synthesis of benzofurans via oxidative cyclization of o-alkenylphenols.

Although significant and indicative progress has been made in the realm of oxidative C-H functionalization with stoichiometric oxidants, the C-H oxygenation involved in the one-step conversion of o-alkenylphenols to benzofurans without oxidants and sacrificing acceptors is scarcely reported. Liu and co-workers reported [48] this cyclization reaction catalyzed by palladium on carbon (Pd/C) without any oxidants and presented the perspectives of the method in the utility of ubiquitous C(sp2)-H bonds as latent functional groups for the construction of C(sp2)-O bonds

An alternative route for the synthesis of 2-arylbenzofurans without the use of Pd catalyst is described by iodine(III)-catalyzed oxidative cyclization of 2-hydroxystilbenes, using 10 mol% (diacetoxyiodo)benzene [PhI(OAc)$_2$] as catalyst in the presence of m-chloroperbenzoic acid. The 2-arylbenzofurans were isolated in good-to-excellent yields [49].

2.2.2. From o-Alkynylphenols

Transition metal-catalyzed hydroalkoxylation of alkynes provides a reliable method for synthesizing C2-substituted benzofurans from readily available o-alkynylphenols [22–30]. Into this large sea of strategies, we report some new examples that overcome some limitations of the previous ones or represent extensions of applicability of methodologies.

For example, a highly active heterogeneous Pd-nanoparticle catalyst for the intramolecular addition of phenols to alkynes was developed and employed in a continuous flow reaction system [50].

Indium(III) halides catalyzed the hydroalkoxylation reaction of alkynylphenols to afford benzofurans in good yields. The reaction proceeded with 5-endo-dig regioselectivity, with a variety of phenols functionalized at the arene and alkyne moieties in high yields, using InI$_3$ (5 mol %) in DCE. Experimental and computational studies supported a mechanism based on the indium(III) π-Lewis acid activation of the alkyne, followed by nucleophilic addition of the phenol and final protodemetalation to afford the corresponding benzofuran [51]. Such cyclization was also reported to be efficient with

Cu(I) [52], Rh(I) [53,54], Zeolite [55], Au [56], and pTsOH-MW [57], in aqueous conditions [58], or basic conditions in the last case, to afford 2-trifluoromethyl benzofurans [59] (Scheme 7).

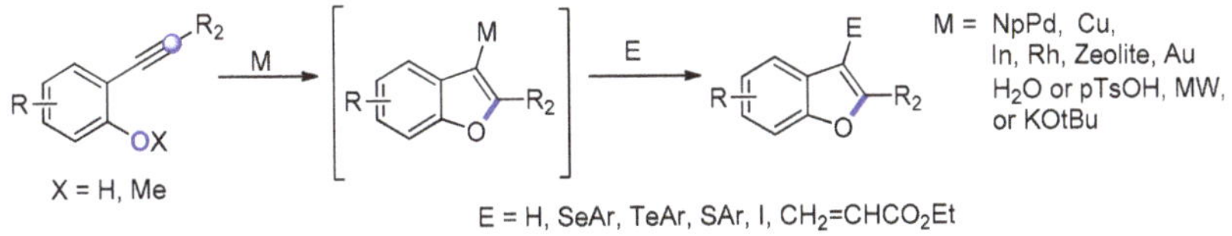

Scheme 7. Synthesis of benzofurans via cyclization of o-alkynylphenols.

Organoboron compounds and heterocycles are powerful building blocks and precursors for organic synthesis, including for drug discovery and agrochemical and material synthesis. Blum and co-workers first studied direct oxyboration toward the formation of borylated benzofurans, where a preformed boron–oxygen σ bond was added across an alkyne activated by a carbophilic gold catalyst. Detailed mechanistic and kinetic studies of this class of reactions were reported [60,61] (Scheme 8).

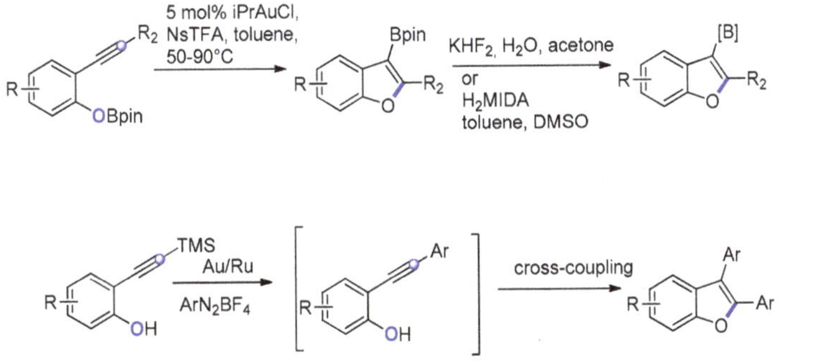

Scheme 8. Synthesis of 2,3-disubstituted benzofurans via cyclization of o-alkynylphenols.

A comparative study has been carried out of the catalytic activities of five-, six-, and seven membered carbene complexes [(NHC)AuX], [(Ph$_3$P)AuX], and [(Me$_2$S)AuX], and inorganic compounds of gold in model reactions of indole and benzofuran synthesis. A selective and convenient synthesis of 2,3-diarylbenzofurans has been developed with the aid of light and taking advantage of a cooperative gold/photoredox-catalyzed two-fold arylation reaction of TMS-terminated alkynols. The photoredox sequence involving 2-[(trimethylsilyl)ethynyl]phenol exclusively afforded 2,3-diarylbenzofurans. The reaction of terminal alkynes proceeded in poor yields, while the use of bulkier silyl groups, such as TIPS, resulted unproductive. Apparently, the C(sp) arylation reaction was the first event on the domino bis-arylative sequence. These results could be explained through the intermediation of arylgold(III) species and several single electron transfer processes [62] (Scheme 8).

A rhenium-catalyzed carboalkoxylation and carboamination of alkyne was reported by Zi and co-workers, providing an efficient route to synthesize de novo C3-substituted benzofurans and indoles under mild conditions in moderate-to-good yields [63]. Mechanistic studies revealed that the rhenium played the role of a π acid catalyst to activate the alkynes, followed by a charge-accelerated [3,3]-sigmatropic rearrangement. An analogue activation of the alkyne was proved by the electrophilic Pt-species, which enables nucleophilic attack by the heteroelement, resulting in *trans* alkoxyplatination [64]. This process formally generates an allyl cation that reacts to the most nucleophilic position of the ring to give the product and regenerate the catalyst. This notion suggests that other entities R$_1$, which are able to stabilize positive charge, might transfer in a similar fashion. Moreover, Ferreira's group investigated on the use of platinum catalysis to generate α,β-unsaturated carbene intermediates via an intramolecular nucleophilic addition into alkynes bearing propargylic ethers. These carbenes have been demonstrated to undergo cycloadditions, hydrogen migrations, and vinylogous nucleophilic additions [65], where β-diketones, ketoesters, and ketoamides all successfully added onto the platinum carbene intermediate (Scheme 9).

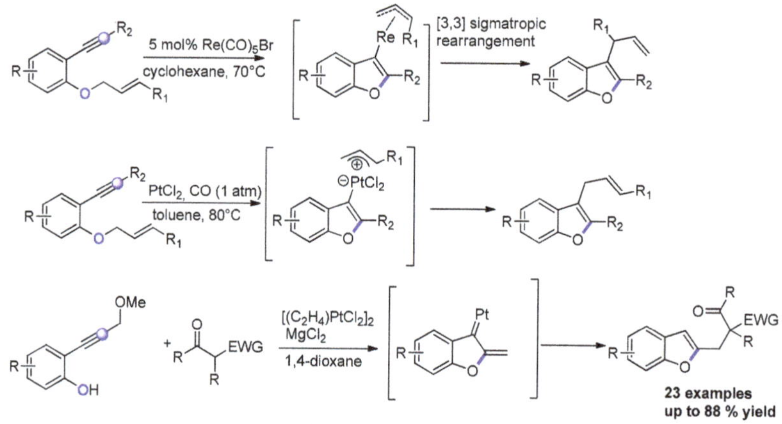

Scheme 9. Synthesis of benzofurans via cyclization of o-alkynylphenyl allyl ethers.

2.2.3. From o-Gem-Dibromoalkenyl Phenols

The popular strategy involving a Sonogashira coupling/cyclization protocol, using ethoxycarbonylethenyl-derived 2-halophenols and alkynes, worked well for certain compounds but failed, for example, for egonol and other related compounds. This failure was attributed to the electronic nature of both the 2-halophenols and the alkynes, and the related complications when applied in a multistep synthesis. Thus, this protocol did not facilitate a rapid and concise synthesis of benzofuran natural products. A seminal strategy was reported by Lautens et al., in 2009, in which 2-bromobenzofurans were generated by a Cu-catalyzed cyclization of 2-(2,2-dibromovinyl)-phenols [66]. The same approach was used by Kim's group, to synthesize different natural products [67,68], while recently, Rao et al. explored the application of a Pd-catalysed domino cyclization/coupling protocol in a pooled approach for the synthesis of benzofuran natural products [69,70] (Scheme 10).

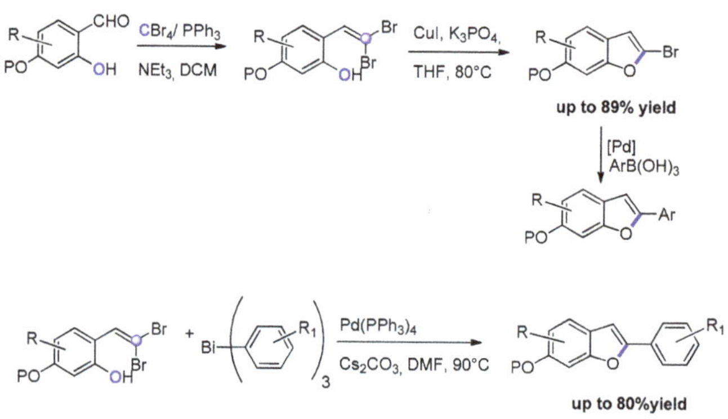

Scheme 10. Synthesis of benzofurans via o-gem-dibromoalkenyl phenols.

2.2.4. From o-Allylphenols

A special position in palladium catalysis constitutes palladium on carbon (Pd/C). Palladium on carbon (Pd/C) has some unique advantages, like stability in air, easy removal by simple filtration, sustainability, and commercial availability at a relatively low cost [71]. Kokotos and co-workers described a cheap and easy-to-execute strategy for the synthesis of benzofurans, employing Pd/C as the promoter. A variety of substituted allyl-phenols were converted into the desired products in good to excellent yields. Recycling of Pd/C was possible up to five cycles, keeping similar levels of reactivity [72] (Scheme 11). From the natural product honokiol, which contains a o-allylphenol fragment, a benzofuran scaffold was produced through a Wacker-type intramolecular cyclization, using $PdCl_2$, NaOAc, and O_2, in DMA/H_2O [73]. Different substituted o-allylphenol derivatives, prepared via a Friedel–Crafts alkylation of cinnamyl alcohol with phenols, using Re_2O_7 catalyst in acetonitrile as solvent, underwent oxidative cyclization, using $PdCl_2(C_2H_4)_2$ as catalyst and BQ as oxidant [74]. A simple sequential reaction protocol has been developed for the synthesis of functionalized 2-benzyl benzofurans via Friedel–Crafts alkylation of phenols with cinnamyl alcohols in the presence of Re_2O_7 catalyst, followed by Pd(II)-catalyzed oxidative annulation of in situ generated o-cinnamyl phenols. Synthesis of 2-benzyl benzofurans was achieved in good yields (disconnection **b+d**). In the work of Li's group, the o-allylphenol derivatives were generated from aryne, using Kobayashi benzyne precursor, and aryl allyl sulfoxide [75] (Scheme 11).

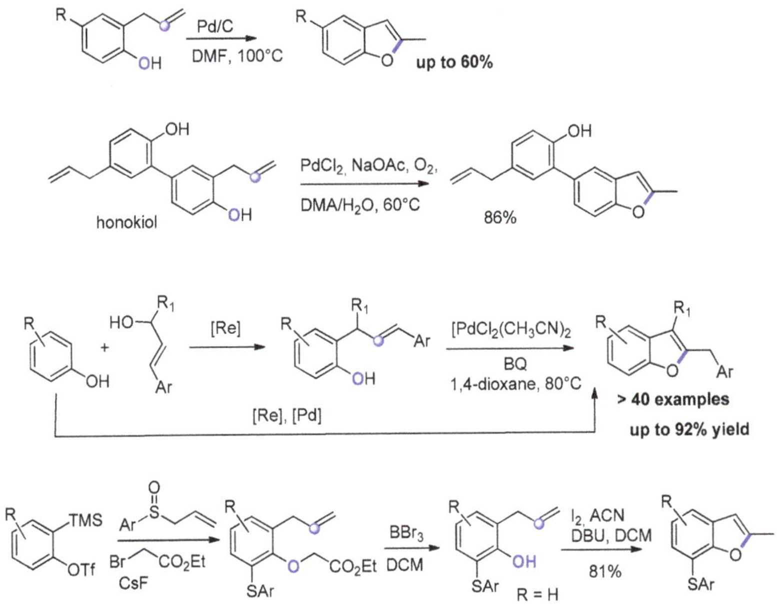

Scheme 11. Synthesis of bezofurans via cyclization of o-allyl phenols.

2.2.5. From o-Hydroxybenzyl Ketones

A new method to generate o-hydroxybenzyl ketones was proposed recently by Patil and co-workers, using an o-oxygenative 1,2-difunctionalization of diarylalkynes. This procedure allowed under merged gold/organophotoredox catalysis to access highly functionalized 2-(2-hydroxyaryl)-2-alkoxy-1-arylethan-1-ones [76]. Detailed mechanistic studies suggested a relay process, initiating with gold-catalyzed hydroalkoxylation of alkynes, to generate enol-ether followed by a key formal [4 + 2]-cycloaddition reaction. This type of oxidation of alkynes depended strongly on the nature of substituents on the aryl. Symmetrical diarylalkynes with electron withdrawing groups gave rise to the corresponding α-methoxyketones with yield up to 68%, while with asymmetrical alkyne lower yields were obtained. Not all the functionalities were well tolerated. The successful application of the present methodology was also shown for the synthesis of benzofurans (Scheme 12).

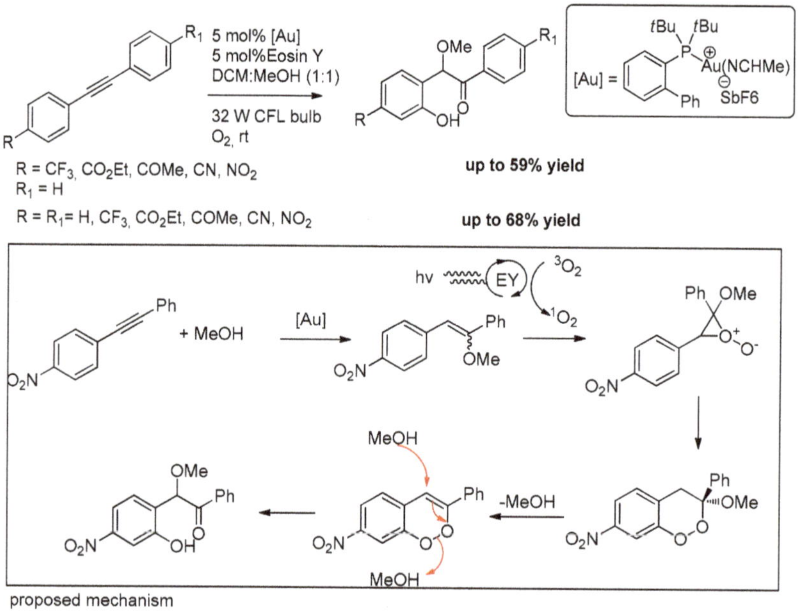

Scheme 12. Synthesis of benzofurans via o-hydroxybenzyl ketones.

An efficient and practical method for construction of 2-arylbenzofurans from 2-methoxychalcone epoxides has been reported. Catalyzed by 2 mol % of BF$_3$·Et$_2$O, 2-methoxychalcone epoxides went through the Meerwein rearrangement, followed by deformylation in one-pot to successfully afforded 2-methoxydeoxybenzoins. Afterward, 2-arylbenzofurans were obtained in high yields (87–100%) via inter-molecular cyclodehydration of 2-methoxydeoxybenzoins with 48% HBr [77] (Scheme 13).

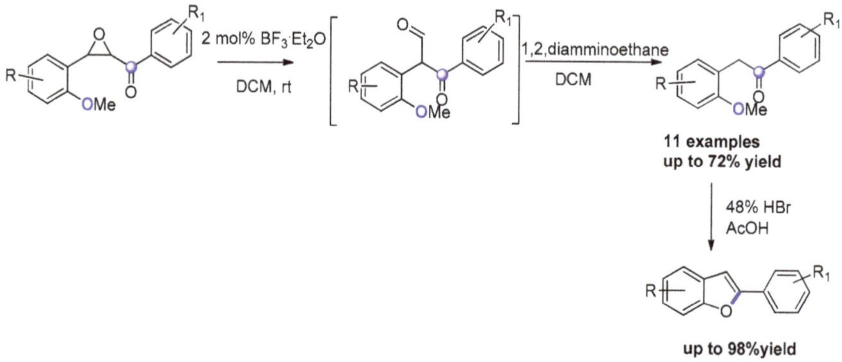

Scheme 13. Synthesis of benzofurans via o-methoxychalcone epoxides.

2.2.6. From o-(Cyanomethyl) Phenols

A seminal study for exploring the potential of palladium-catalyzed addition of potassium organotrifluoroborates to a nitrile group, which is usually more challenging than an aldehyde or acyl chloride group, has been performed by Whu's group [78,79]. This work has provided a new method for the synthesis of alkyl aryl ketones (or dicarbonyl compounds) by Pd-catalyzed addition of arylboronic acid or potassium aryltrifluoroborates to aliphatic nitriles (or dinitriles) and the one-step synthesis of 2-arylbenzofuran derivatives (Scheme 14).

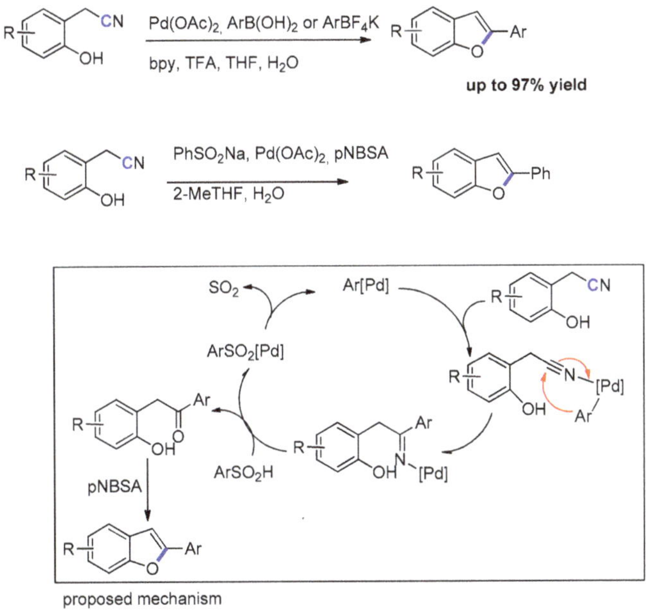

Scheme 14. Synthesis of benzofurans via o-(cyanomethyl) phenols.

The first example of the palladium-catalyzed one-pot synthesis of 2-arylbenzofurans in moderate to excellent yields via a tandem reaction of 2-hydroxyarylacetonitriles with sodium sulfinates was reported in 2014. A plausible mechanism for the formation of 2-arylbenzofurans involving desulfinative addition and intramolecular annulation reactions is proposed. Moreover, the present synthetic route to benzofurans could be readily scaled up to the gram quantity, without any difficulty. Thus, the method represents a convenient and practical strategy for synthesis of benzofuran derivatives [80,81].

2.2.7. From 1-(2-hydroxyphenyl) Propargyl Alcohol Derivatives

A novel metal-free one-pot protocol for the effective and efficient synthesis of 3-phosphinoylbenzofurans via a phospha-Michael addition/cyclization of H-phosphine oxides and in situ generated o-quinone methides was described. Based on the expeditious construction of C(sp2)–P bonds, asymmetric synthesis of optically pure 3-phosphinoylbenzofurans containing chiral P-stereogenic center has also been probed by using chiral RP-(−)-menthyl phenylphosphine oxide [82]. A metal-free procedure used BF_3-Et_2O to catalyze the cyclization and the substitution with a nucleophle [83]. Moreover, Pd-catalyzed cyclization was reported on these substrates [84,85] (Scheme 15).

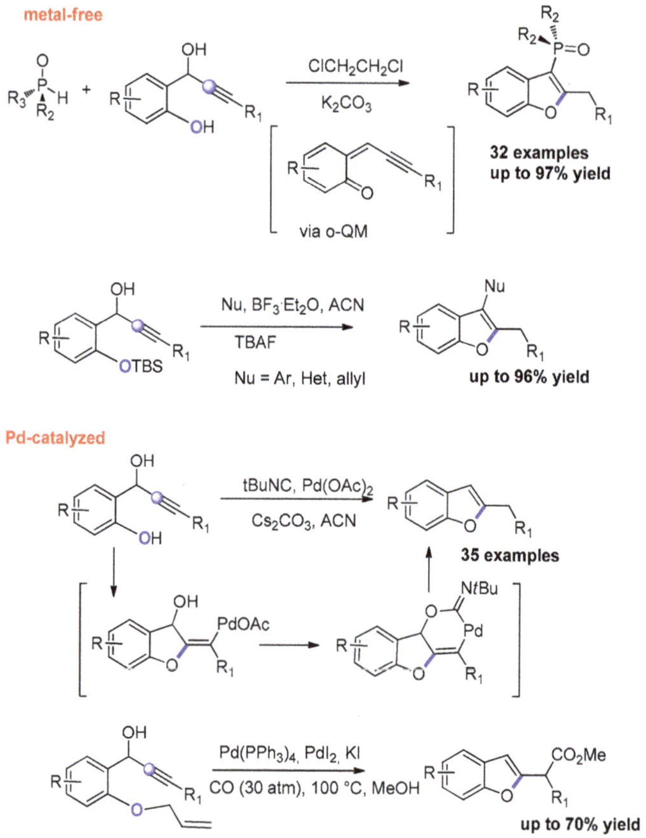

Scheme 15. Synthesis of benzofurans via 1-(2-hydroxyphenyl)propargyl alcohol derivatives.

2.3. C2–C3 Bond Formation: (Route c)

The main approaches for the formation of C3-C2 bond dealt in this section are summarized below (Scheme 16).

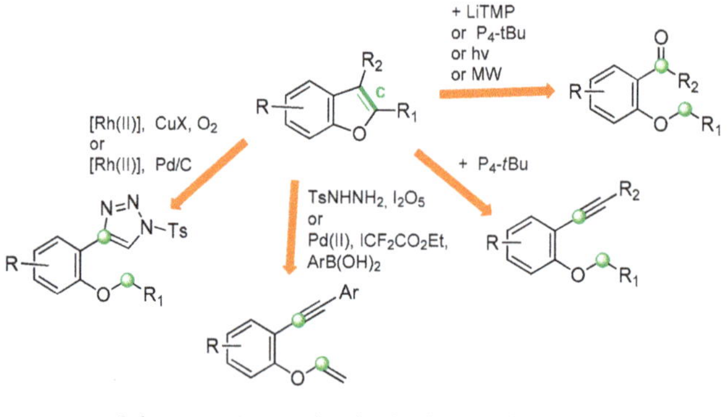

Scheme 16. Approaches for C2–C3 bond formation.

2.3.1. From o-(Alkoxy)Phenyl Arylketones

Dehydrative decarboxylation of o-acylphenoxyacetic acids or esters on treatment with a base represents an old methodology to prepare benzofurans. A more recent approach used benzylic deprotonation, using LiTMP, followed by an intramolecular cyclization between the carbanion and carbonyl group of the corresponding arylketone and subsequent dehydration acid catalyzed (p-TsOH·H$_2$O), to deliver benzofuran, the core structure for the synthesis of the natural and biologically relevant products, Malibatol A and Shoreaphenol (Scheme 17) [86].

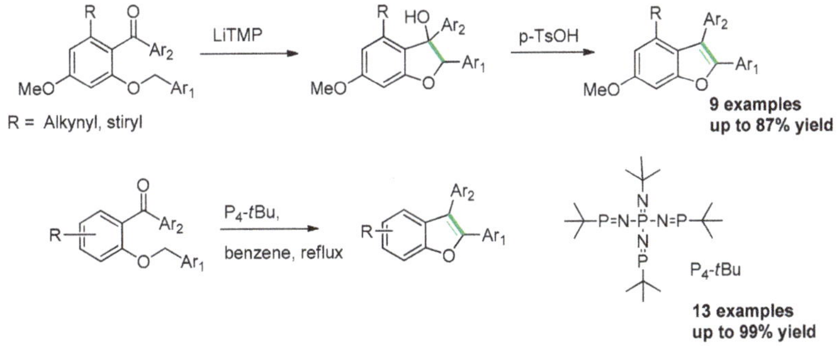

Scheme 17. Synthesis of benzofurans via o-benzyloxyphenyl arylketones.

Moreover, 2,3-Diarylbenzofurans were also efficiently generated by the cyclization of o-benzyloxybenzophenones, using the hindered phosphazene base P$_4$-t-Bu [87].

Condensation of carbonyls with non-acidic methylenes such as those adjacent to heteroatoms and allylic types to generate C=C bonds is challenging but highly desirable. An advanced method overcame this limitation. Li and co-workers reported a simple, clean, and high-yielding protocol promoted by UV-light, to achieve condensation of non-acidic methylenes with carbonyls. As examples to demonstrate the power of this methodology, benzofurans were synthesized with broad functional group compatibility [88] (Scheme 18).

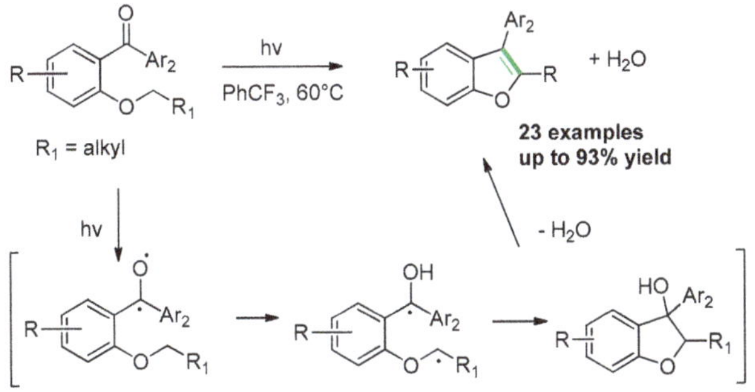

Scheme 18. Synthesis of benzofurans via o-alkoxyphenyl arylketones.

2.3.2. From o-Alkynylphenyl benzyl (or Allyl)Ethers

Terada group demonstrated that the organic superbase phosphazene P$_4$-tBu worked as an active catalyst for intramolecular cyclization of o-alkynylphenylethers, reporting the carbon–carbon bond formation under mild reaction conditions without the need for a metal catalyst, providing an efficient synthetic method for 2,3-disubstituted benzofurans derivatives [89] (Scheme 19).

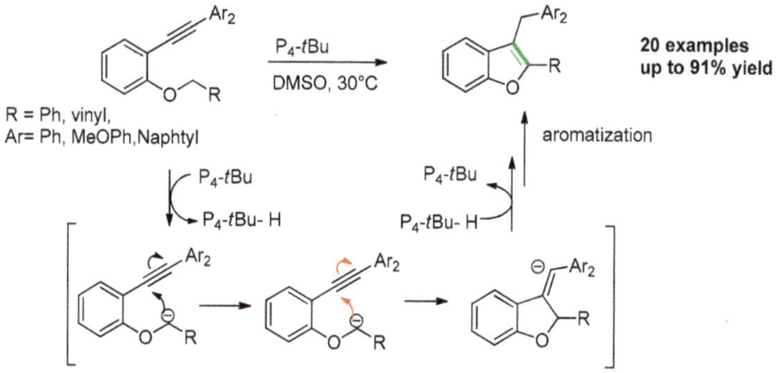

Scheme 19. Synthesis of benzofurans via *o*-alkynylphenyl ethers.

2.3.3. From *o*-Alkynylphenyl Vinylethers

A simple I_2O_5-mediated method has been developed under metal-free conditions for the construction of sulfonylated benzofurans. The present reaction was efficiently achieved through the oxidative cyclization of 1,6-enynes and arylsulfonylhydrazides, which provided an attractive approach to a series of sulfonylated benzofurans in moderate to good yields [90] (Scheme 20).

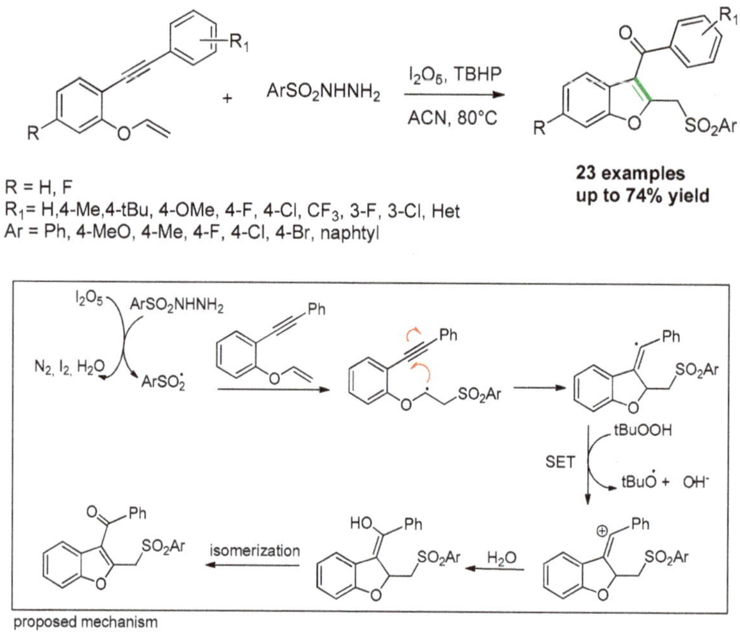

Scheme 20. Synthesis of benzofurans via *o*-alkynylphenyl vinylethers.

Very recently, the same substrates were used by Zhang's group to introduce difluoroalkylated substituent on benzofurans, according to their expertise in the transition-metal catalyzed cross-coupling of difluoroalkylhalide and boronic acids. Difluoroalkylated benzofuran derivatives were constructed via palladium-catalyzed cascade difluoroalkylation-arylation of 1,6-enyne. Moreover, final difluoroalkylated benzofurans were obtained through an isomerization process catalyzed by $Fe(OTf)_3$ [91] (Scheme 21).

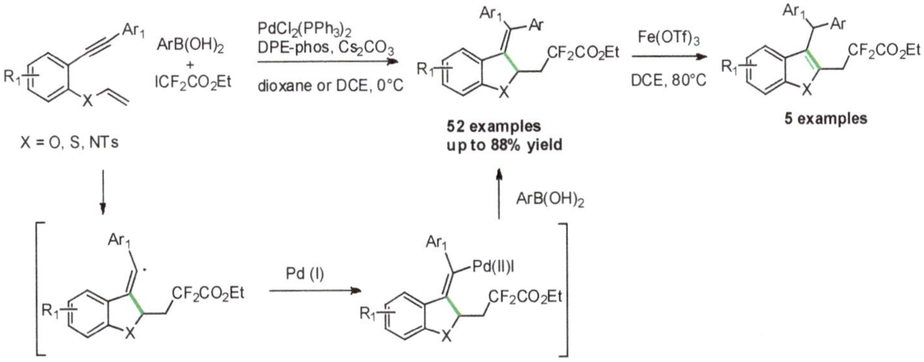

Scheme 21. Synthesis of benzofurans via *o*-alkynylphenyl vinylethers.

Within the broad field of C-H bond functionalization, the insertion of carbenes into C-H bonds is arguably the best approach for directly transforming a C-H bond directly into a C-C bond. Dong strategy [92] for C-H bond functionalization was inspired by Adrian Brook's discovery of the unique ability of acylsilanes to undergo thermal and photochemically induced 1,2 silicon-to-oxygen migration [93]. This Brook rearrangement of acylsilanes could be considered an umpolung process, where the acylsilanes acted as a carbonyl anion equivalents. Thus, a thermally induced Brook rearrangement generated a transient siloxycarbene that underwent to a rapid insertion into a neighboring C-H bond (Scheme 22). Thus, this new approach furnished 2,3-dihydrobenzofuran and benzofuran derivatives under microwave irradiation, in which the solvent played an important role to determine the generated species.

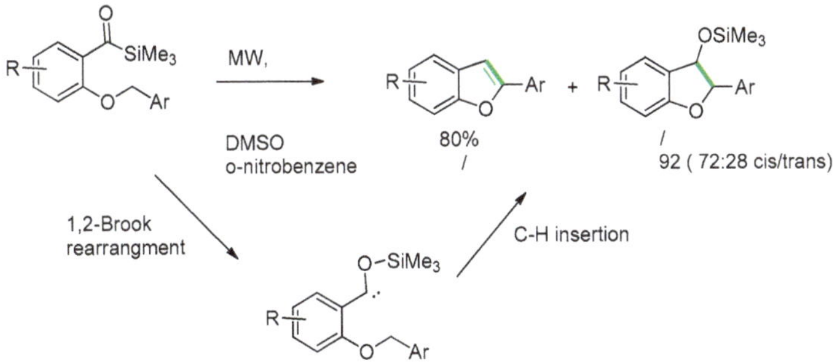

Scheme 22. Synthesis of benzofurans via *o*-acylsylanesphenyl benzyl ethers.

2.3.4. From *o*-Triazole-Phenyl Benzylethers

Recently, N-sulfonyl-1,2,3-triazoles have emerged as alternative precursors for the formation of metallocarbenes [94]. Independent studies from Kang [95] and Chen [96] reported intramolecular sp3 C–H insertion reaction of α-imino rhodium carbene generated from N-sulfonyl-1,2,3-triazoles. The first one reported the use of O_2 as the oxidant to obtain benzofurans, whereas the second one used Pd/C, to allow the isomerization of the allylic portion of enamine, which, in presence of H_2, afforded to amine derivatives. Both methodologies furnished a number of benzofuran derivatives in good to excellent yields (Scheme 23).

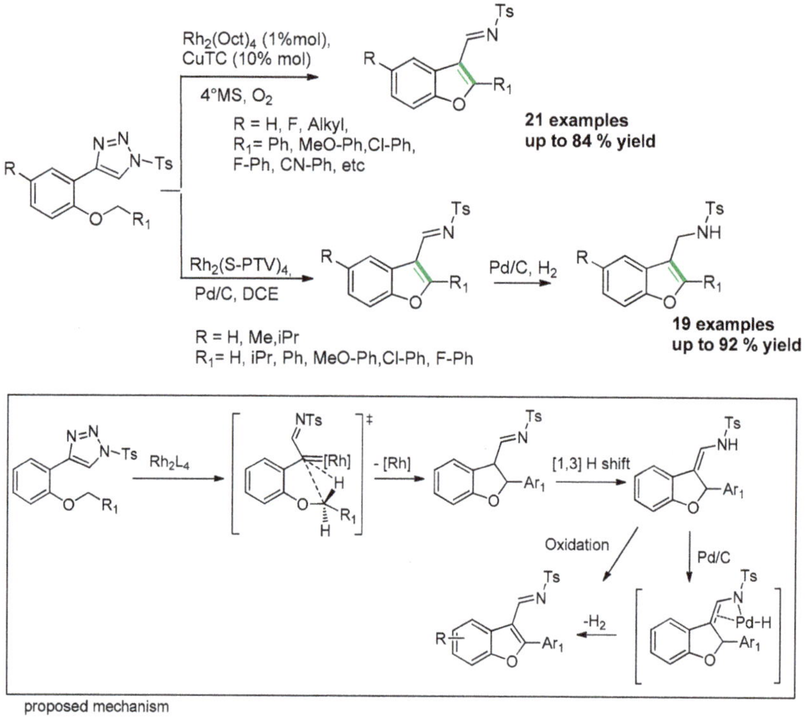

Scheme 23. Synthesis of benzofurans via *o*-N-sulfonyl1,2,3-triazolphenyl bezylethers.

2.4. C3–C3a Bond Formation: (Route d)

The main approaches for the formation of C3-C3a bond dealt in this section are summarized below (Scheme 24).

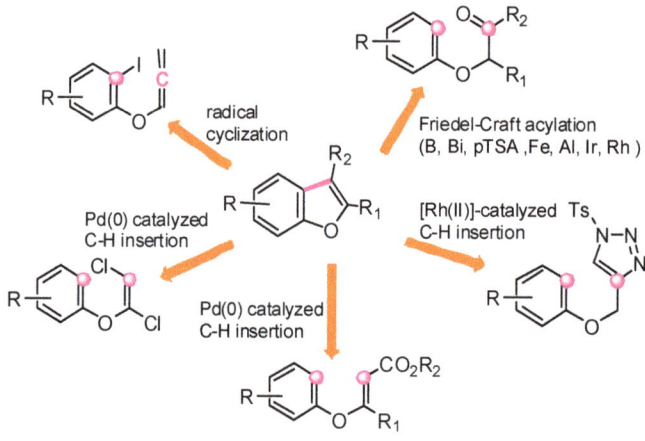

Scheme 24. Approaches for the C3–C3a bond formation.

2.4.1. Via Friedel–Crafts Acylation

Outstanding total synthesis of multisubstituted benzofurans were achieved by intra-molecular Friedel–Crafts acylation of α-aryloxyaryl ketones, which were prepared from an inter-molecular O-alkylation of α-haloarylketones with phenoxide. Many examples of Lewis acid and transition-metal-catalyzed direct intramolecular cyclodehydration of the resulting α-aryloxyaryl ketones have been developed by several research groups. Kim's group used BBl$_3$ [97], pTSA [98], or Bi(OTf)$_3$ [99] in the total synthesis of natural stilbenoids, while Chang's group used Ga(OTf)$_3$ [100] for the first time in such cyclization. Arava's group was concerned with AlCl$_3$ or FeCl$_3$ [101], while Tang investigated on TiCl$_4$ [102]. Shibata and co-workers used In(III) species generated from [CpIrCl$_2$]$_2$ and AgSbF$_6$, in which the presence of the acetyl group on the aryl group allowed the Ir insertion in C-H bond [103]. Recently, Xu, employing 10 mol% [Rh(cod)(MeCN)$_2$]BF$_4$ and 12 mol% DPPF in THF, developed a cascade transformation initiated by regioselective activation of benzocyclobutenone, followed by insertion into C=O and spontaneous aromatization, which generated 2,3-disubstituted benzofurans [104] (Scheme 25).

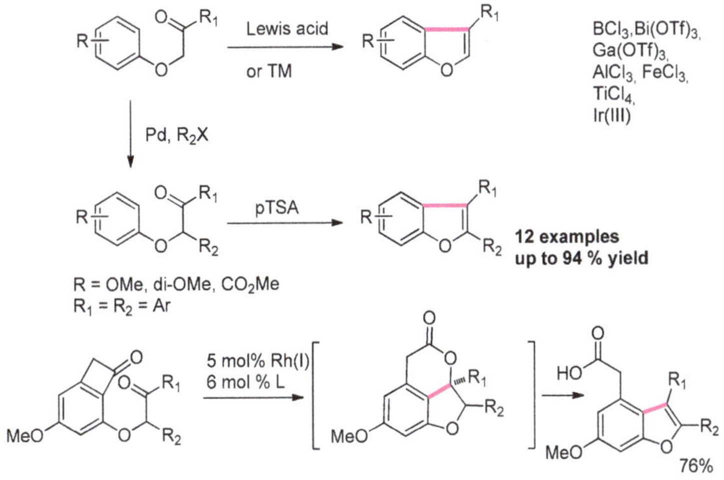

Scheme 25. Synthesis of benzofurans via Friedel–Crafts acylation.

2.4.2. Via [Ru(II)] C-H Insertion of N-Sulfonyl-1,2,3-Triazole Derivatives

The selective synthesis of substituted 3-methylene-2,3-dihydrobenzofurans and 3-methylbenzofurans was developed in the seminal work of Shi through Rh(II)-catalyzed denitrogenative annulation of N-sulfonyl-1,2,3-triazole at ambient to mild heating condition, respectively [105] (Scheme 26). Further, a one-pot strategy was also developed by using substituted O-propargylphenols through Cu/Rh-catalyzed method, to rapidly construct 3-methylene-2,3-dihydrobenzofuran derivatives [106].

2.4.3. Via [Pd(0)] C–H Insertion

In 2008, Hultin's group reported the synthesis of 2-substituted benzofurans from simple phenols, boronic acids or other organoboron reagents, and trichloroethylene. The overall process required only two synthetic steps, with the key step being a one-pot sequential Pd-catalyzed Suzuki cross-coupling/direct arylation reaction. The method tolerated many useful functional groups and did not require the installation of any other activating functionality [107,108].

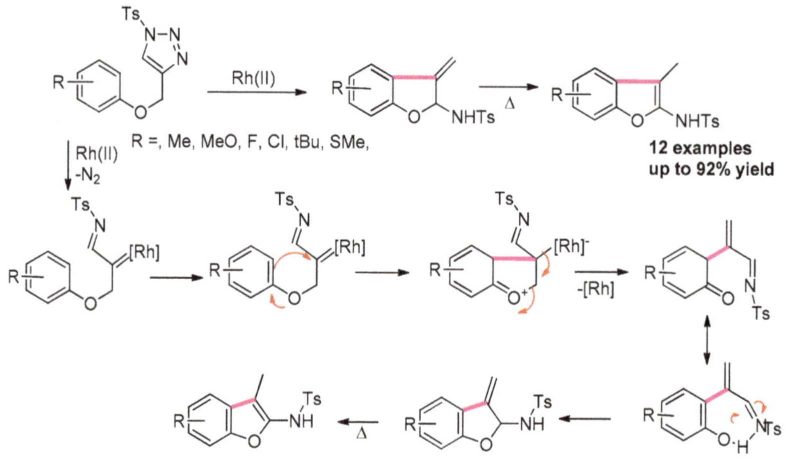

Scheme 26. Synthesis of benzofurans via [Rh(II)] C–H insertion of N-sulfonyl-1,2,3-triazol derivatives.

In 2011, Wang and co-workers developed a synthesis of benzofurans from commercially available phenols and propiolate through the direct oxidative cyclization. In the presence of Pd(OAc)$_2$/PPh$_3$ and CF$_3$CO$_2$Ag, (E)-type 3-phenoxyacrylates underwent reaction smoothly to generate the corresponding benzofurans in good yields in benzene at 110 °C, with no need for an inert atmosphere. This transformation of phenols into benzofurans was also carried out in one-pot, in a simple and efficient way [109] (Scheme 27).

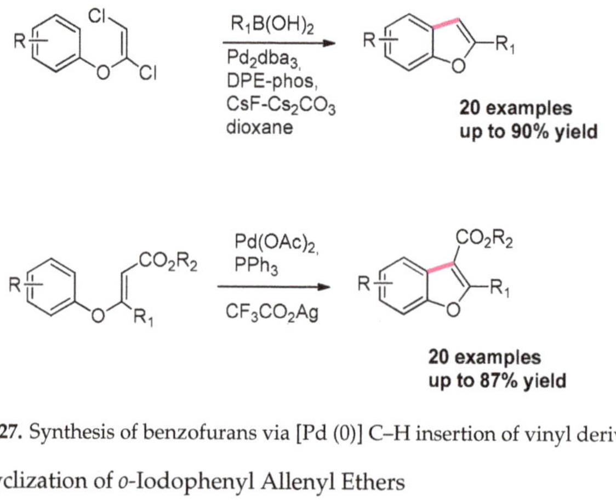

Scheme 27. Synthesis of benzofurans via [Pd (0)] C–H insertion of vinyl derivatives.

2.4.4. Via Radical Cyclization of o-Iodophenyl Allenyl Ethers

Recently, a mild and broadly applicable methodology to prepare complex benzofurylethylamine derivatives through a unique radical cyclization cascade mechanism was reported [110]. Single-electron transfer (SET) from 2-azaallyl anions to 2-iodo aryl allenyl ethers initiated a radical cyclization that was followed by intermolecular radical–radical coupling. A diverse series of benzofurylethylamine derivatives was prepared in good-to-excellent yields, in three steps, from 2-iodophenols (Scheme 28). This methodology could also be extended to build larger heterocycles.

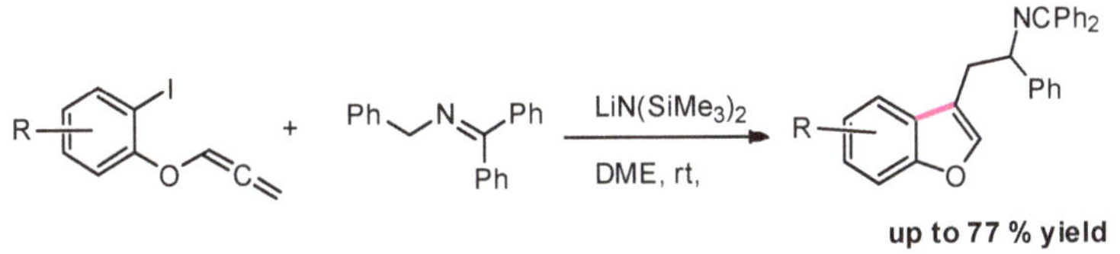

Scheme 28. Synthesis of benzofurans via *o*-iodophenyl allenylethers.

3. Ring Generation via Intermolecular Cyclization

3.1. C7a–O and C3–C3a Bond Formation: (Route a + d)

The main approaches for the formation of C7a-O and C3-C3a bond dealt in this section are summarized below (Scheme 29).

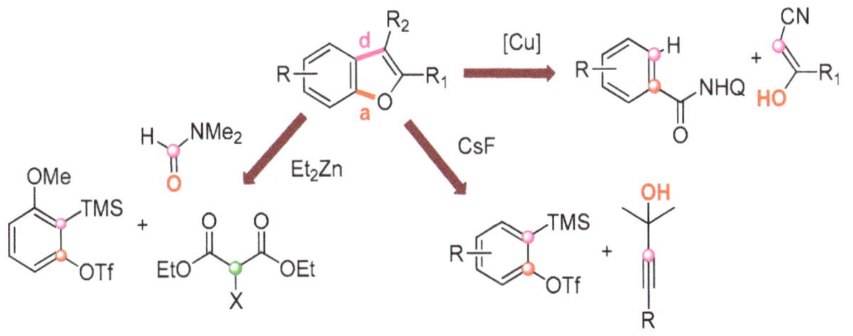

Scheme 29. Approaches for C7a-O and C3–C3a bond formation.

3.1.1. Via *o*-C-H alkylation/Decarboxylation

Transition-metal-catalyzed directing-group-assisted C-H bond functionalization has proven to be a powerful strategy for the construction of carbon–carbon and carbon–heteroatom bonds because of its great potential for step-economy and environmental sustainability. A range of carboxyl directed *o*-C-H alkylation/decarboxylation reactions [111–115] has been exploited. In all transformations, the directing group plays the key role for reaction efficiency and regioselectivity. However, the installation and disconnection of a directing group require extra and tricky steps, which determine severe limitations on application. A very recent copper-mediated synthesis of 2,3-disubstituted benzofurans from readily available benzamides and benzoylacetonitriles was described, in which the assistance of an 8-aminoquinolyl auxiliary was shown [116]. To overcome the above shortcomings, carboxyl group was successfully used as a traceless directing group, which could introduce a target functional group into a specific position of the substrate and then be completely removed by decarboxylation. In this strategy, the C3–C3a bond was successfully constructed via C–H activation, and C7a–O bond was subsequently formed at the original position of the amide group in a one-pot manner. The amide directing group was removed simultaneously under the reaction conditions through C–C bond cleavage (Scheme 30).

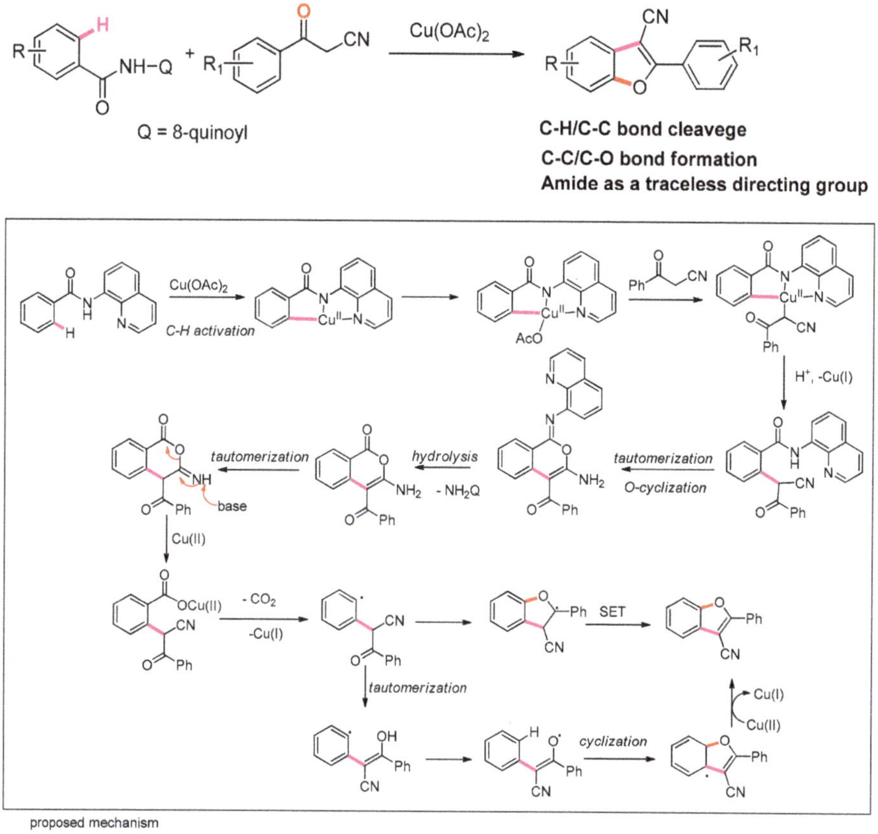

Scheme 30. Synthesis of benzofurans via C-H alkylation/decarboxylation.

3.1.2. Via Propargyl Claisen Rearrangement/Cycloaddition

Synthetic methodologies for the preparation of benzofuran derivatives through transition-metal/ noble-metal catalysts, Lewis/Brønsted acids, and base-promoted cyclizations reported propargylic alcohols or their derivatives as starting materials. However, accessing such preferred scaffolds using aryne chemistry is less known. Palakodety and co-workers [117] reported an unprecedented base-mediated cyclization of propargylic alcohols with aryne, providing a novel method for the synthesis of 3-benzofuryl-2-oxindole and 3-spirooxindole benzofuran scaffolds via a propargyl Claisen rearrangement/cycloaddition pathway (Scheme 31). The nature of the substituent on acetylene group of propargylic alcohol influenced the outcome of the reaction. The protocol offered a transition-metal-free and operationally simple methodology with broad substrate scope as a ready access to complex oxindole-linked heterocyclic compounds.

3.1.3. Via Addition of Zinc-Enolate to Methines

Miyabe's group developed an efficient insertion of arynes, which were generated in situ from o-(trimethylsilyl)aryl triflates and the fluoride ion, into the C=O π-bond of formamides. The subsequent addition of zinc enolates of α-chlorinated methines gave rise to benzofurans, via the addition of an ethyl anion to the dihydrobenzofurans bearing a ketone group and the retro-aldolic process inducted by Et_2Zn [118] (Scheme 32).

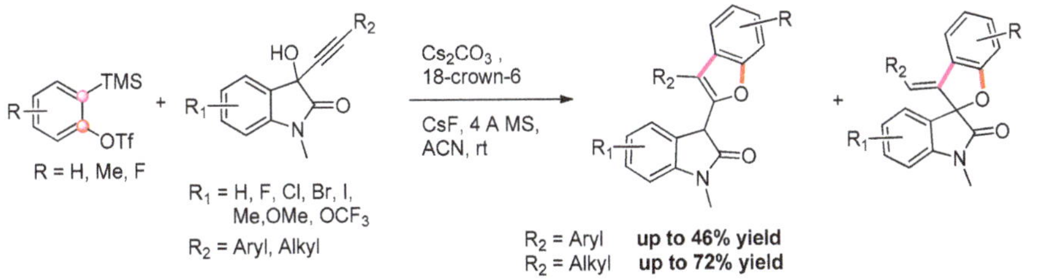

Scheme 31. Synthesis of benzofurans via propargyl Claisen rearrangement/cycladdition.

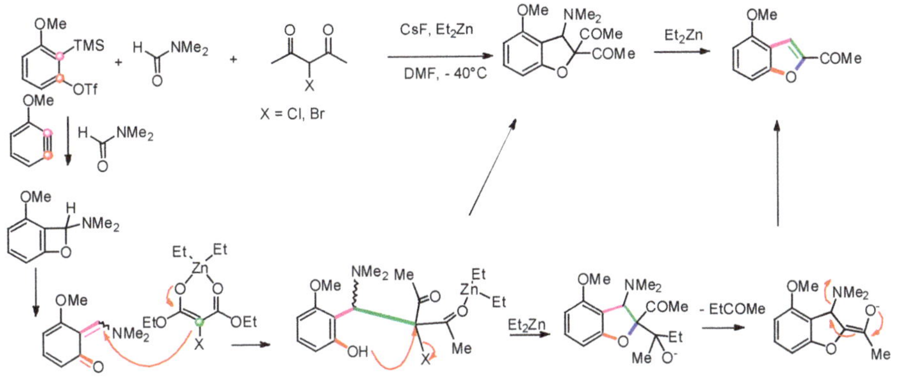

Scheme 32. Synthesis of benzofurans via addition of zinc-enolate to methynes.

3.2. O–C2 and C2–C3 Bond Formation: (Route b + c)

The main approaches for the formation of O–C2 and C2–C3 bond dealt in this section are summarized below (Scheme 33).

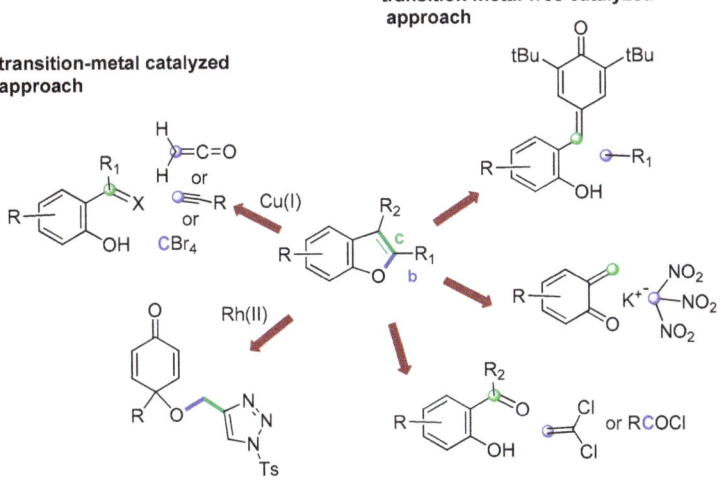

Scheme 33. Approaches for the O–C2 and C2–C3 bond formation.

3.2.1. Via Transition-Metal-Free Catalyzed Approaches: *p*-Quinone Methides

Recently, *p*-quinone methides (*p*-QMs) were subjected to extensive investigation for their interesting chemical properties. A typical reaction of *p*-QMs involves rearomatization via nucleophilic addition by a variety of carbon nucleophiles.

A one-pot protocol for the synthesis of 2,3-diarylbenzo[b]furan derivatives through an *N*-heterocyclic carbine catalyzed 1,6-conjugate addition of aromatic aldehydes to 2-hydroxyphenyl-substituted *p*-quinone methides followed by acid-mediated dehydrative annulation has been developed. This protocol allows access to a wide range of 2,3-diarylbenzofuran derivatives in moderate-to-good yields [119] (Scheme 34).

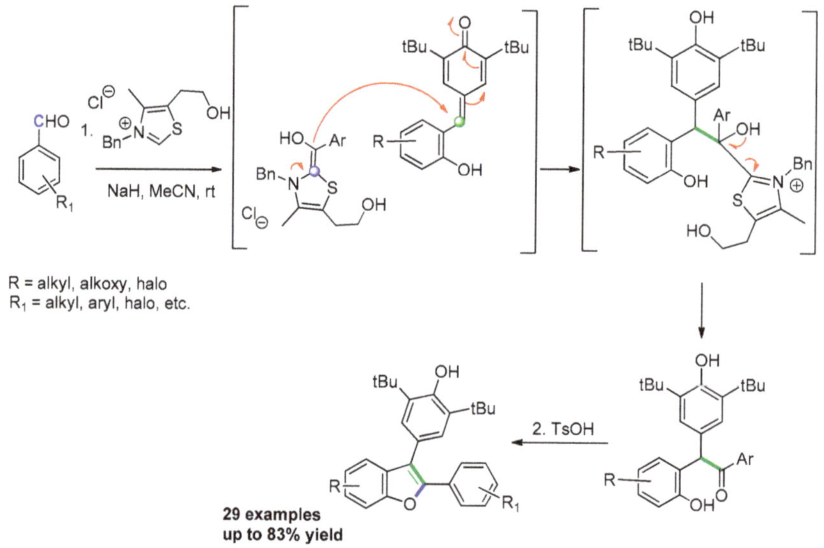

Scheme 34. Synthesis of benzofurans via addition of aromatic aldehyde to *p*-quinone methide.

In this context, an efficient synthesis of functionalized benzofurans was obtained under mild and metal-free conditions from the *p*-QMs bearing an *o*-hydroxy group, treated with phosphine, acyl chloride, and base. Through a 1,6-phospha-Michael addition, O-acylation, and subsequent Wittig pathway, this protocol was demonstrated to be useful for the synthesis of benzofurans [120] (Scheme 35).

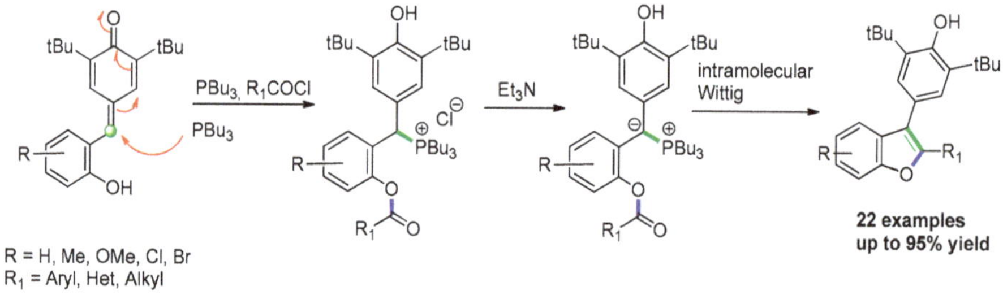

Scheme 35. Synthesis of benzofurans via addition of phosphines to *p*-quinone methides.

3.2.2. Via transition-Metal-Free Catalyzed Approaches: *o*-Quinone Methides

o-Quinone methides (*o*-QMs) are highly reactive and useful species that have been implicated in the reaction with nucleophiles as 1,4-Michael acceptors [121,122]. Moreover, ambiphilic synthons, which contain both electrophilic and nucleophilic centers in the same molecule, are widely used in organic synthesis as useful building blocks. The development of new synthetic methods using ambiphiles has great potential in the elaboration of new high-step-economy reactions. The first example of the use of potassium trinitromethanide as a 1,1-ambiphilic synthon equivalent for the construction of a benzofuran moiety, mediated by triethylamine, has been developed. The method tolerates a variety of functional groups on the starting quaternary ammonium salt and has been successfully extended to polysubstituted benzofurans. The formation of an *o*-quinone methide intermediate is postulated as a key step in this cascade process [123] (Scheme 36).

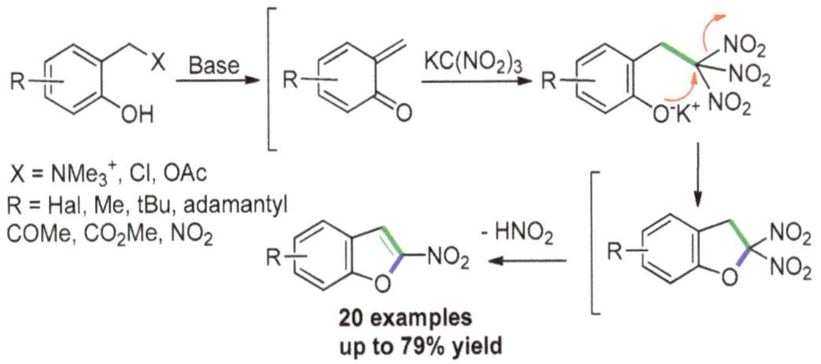

Scheme 36. Synthesis of benzofurans via addition of trinitromethanide to *o*-quinone methides.

3.2.3. Via Transition-Metal-Free Catalyzed Approaches: *o*-Hydroxyphenone or Salicylaldehydes

The transition metal-free preparation of highly functionalized benzofurans by a unique and connective transformation has been reported. Base-catalyzed condensation of *o*-hydroxyphenones with 1,1-dichloroethylene generated the corresponding chloromethide benzofurans. These labile intermediates underwent a facile rearrangement into benzofuran carbaldehydes, under mild acidic conditions [124] (Scheme 37).

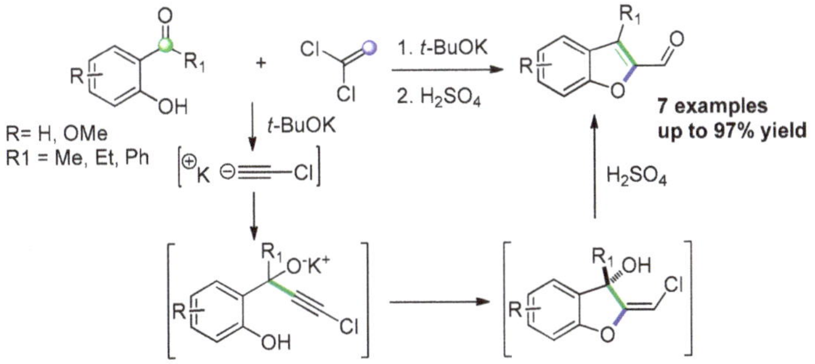

Scheme 37. Synthesis of benzofurans via addition of 1,1-dichloroethylene to *o*-hydroxyphenones.

The preparation of new types of highly functional benzofurans was realized via intramolecular Wittig reactions with the corresponding ester functionality. The key phosphorus ylide intermediate presumably resulted from the addition of Bu$_3$P toward salicylaldehydes followed by acylation and deprotonation. The umpolung reactivity of carbonyl carbon of the aldehyde allowed the synthesis of functional benzofurans [125] (Scheme 38).

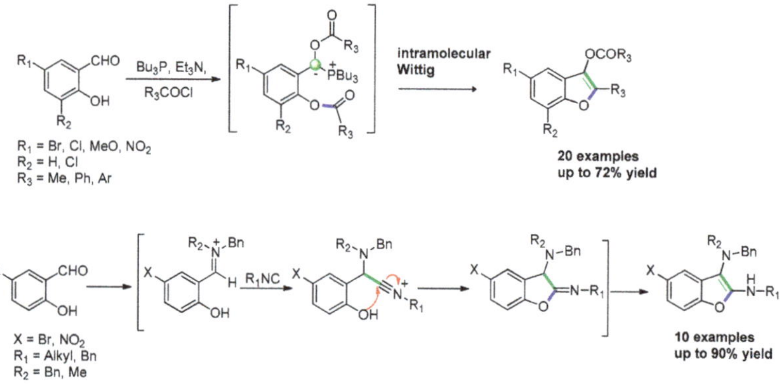

Scheme 38. Synthesis of benzofurans via reactions with salicyldehydes.

In an alternative approach, the addition of an isocyanide on an iminium ion intermediate, formed from an electronpoor salicylaldehyde derivative and a secondary amine in the presence of silica gel, proceeded smoothly at room temperature and afforded benzofuran derivatives in high yields [126] (Scheme 38).

3.2.4. Transition-Metal-Catalyzed Approaches: [Rh(II)] Catalyzed Addition of N-Sulfonyl-1,2,3-Triazole

A rhodium-catalyzed intramolecular denitrogenative transannulation of N-sulfonyl-1,2,3-triazole-tethered cyclohexadienones has been described for the synthesis of benzofurans and cyclopropa[cd]indole-carbaldehydes in an operationally simple procedure. Remarkably, the reaction pathway is fully dependent on heteroatom (O or N) in the linker between the cyclohexadienone unit and triazole moiety. In the case of O-linked triazoles, a cascade sequence consisting of intramolecular cyclopropanation and rearrangement took place, leading to the formation of benzofurans [127] (Scheme 39).

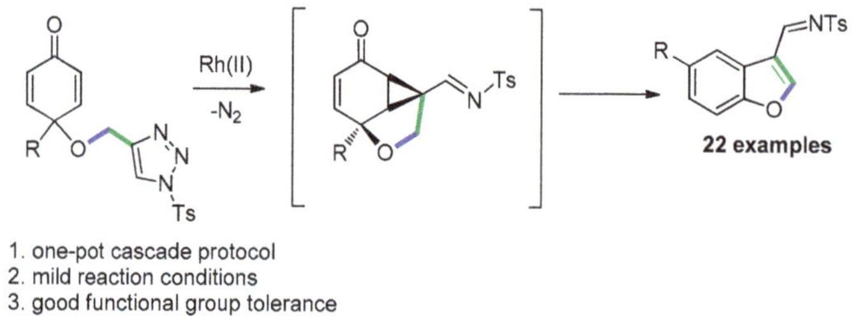

Scheme 39. Synthesis of benzofurans via [Rh]-catalyzed addition of N-sulfonyl-1,2,3-triazole.

3.2.5. Transition–Metal Catalyzed Approaches: [Cu(I)]-Catalyzed Addition to o-Hydroxybenzophenones/Salicylaldehydes

Dominguez and co-workers have described the synthesis of a series of 3-arylbenzofurans [128], using o-hydroxy-benzophenone, CuOAc (50%mol), 8-HQ (8-hydroxyquinoline) (50%mol), and K_2CO_3 (1 equiv), in DMA (N,N-dimethoxyacetamide), at 140 °C, under O_2 atmosphere. The optimized conditions were extended to different diarylketone derivatives affording benzofurans in good yields, in which halogen, alkyl, and alkoxy functional groups were well tolerated under these oxidative conditions. It was demonstrated that DMA took part to reaction furnishing the additional carbon which was involved through ketene intermediate to the formation of 2-hydroxy-α-phenylstyrene or ester α,β-unsaturated. In the last step, the Cu-catalyzed oxidation of the double bond or Cu-catalyzed Wacker cyclization gave rise to benzofurans (Scheme 40).

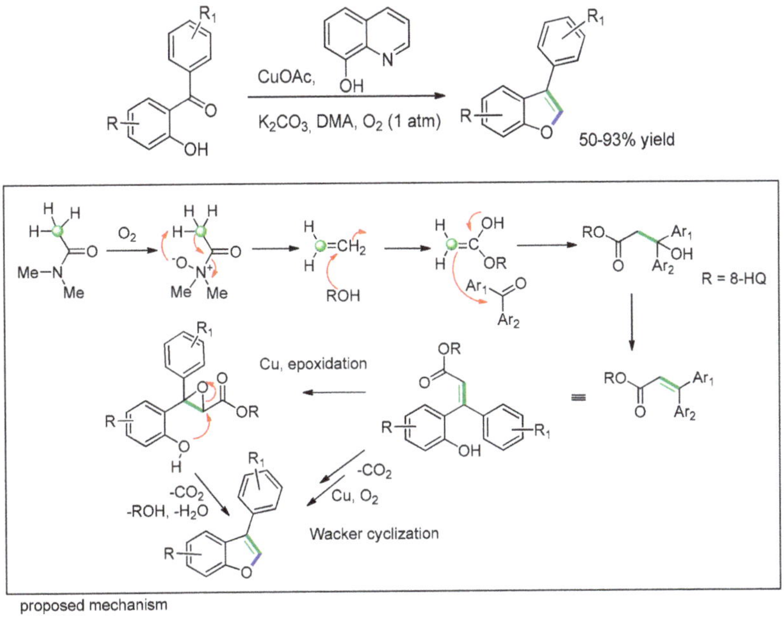

Scheme 40. [Cu(I)]-catalyzed synthesis of benzofurans via addition of ketene to *o*-hydroxybenzophenone.

As it was said before, the catalytic functionalization of unactivated C-H bonds is an increasingly viable method for organic synthesis. In particular, C-H activations that lead to the formation of C–O bonds have recently provided step-economical access to substituted phenols. A versatile ruthenium(II) complex, [{RuCl$_2$(p-cymene)}$_2$], in presence of PhI(OTFA)$_2$ as the terminal oxidant in DME allowed the preparation of different salicylaldehydes by a site selective C-H oxygenations with weakly-coordinating aldehydes. The challenging C–H functionalizations proceeded with high chemoselectivity by rate-determining C–H metalation.

The new method featured an ample substrate scope, which set the stage for the step-economical preparation of various heterocycles, among these benzofurans [129]. Wang J. and coworkers [130] developed a new method to prepare benzofurans by using an economically convenient ligand-free CuBr which catalyzed coupling/cyclization of terminal alkynes with N-tosyl-hydrazones, derived from o-hydroxybenzaldehydes (Scheme 41). N-tosylhydrazones were involved in the synthesis of substituted allenes via Cu(I)-catalyzed coupling of with terminal alkynes [131]. A wide range of functional groups on the aryls and alkynes was found to tolerate the reaction conditions.

Previously, we reported the synthesis of substituted 2-bromobenzofuran compounds from the intramolecular cyclization of gem-dibromoalkenes (prepared via Ramirez olefination) and the subsequent Suzuki cross-coupling for the synthesis of poly-substituted benzofurans (see Section 2.2.3). The aforementioned methods require additional protection–deprotection techniques and are less divergent. An overcoming advanced procedure was envisioned by Lee and co-workers proposing a divergent-pooled route for benzofuran analogues, using 2-bromo-6-hydroxybenzofurans, which were prepared in a one-pot sequence of reactions, using a modified Ramirez olefination and the intramolecular cyclization of the derived *gem*-dibromoalkenes. The best results were obtained when using Cs$_2$CO$_3$ (3.5 equiv.) and CuI (5 mol %) at 85 °C, giving the desired cyclized compounds with complete selectivity in 65% yield [132] (Scheme 42).

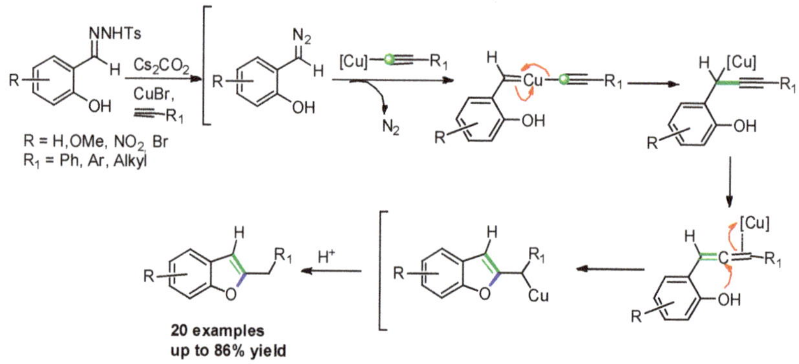

Scheme 41. [Cu(I)]-catalyzed synthesis of benzofurans via addition of acetylene to N-tosylhydrazones.

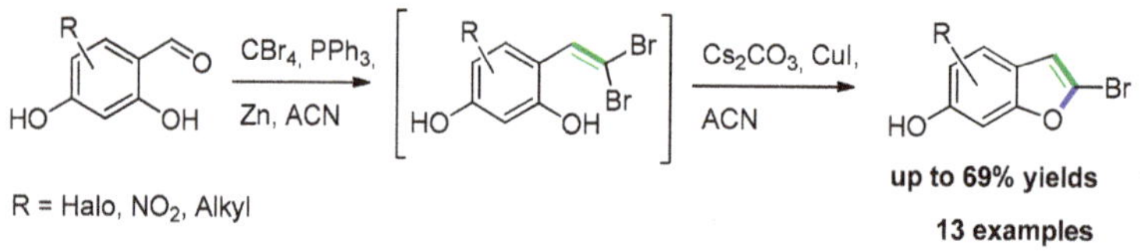

Scheme 42. [Cu (I)] catalyzed synthesis of benzofurans via Ramirez olefinations.

3.2.6. Miscellaneous

The Brönsted acid-catalyzed cascade synthesis of densely substituted benzofurans from easily available salicyl alcohols and biomass-derived furans has been performed. The disclosed sequence included the formation of 2-(2-hydroxybenzyl)furans that quickly rearranged into functionalized benzofurans. The established protocol was applied for the total synthesis of sugikurojinol B [133] (Scheme 43).

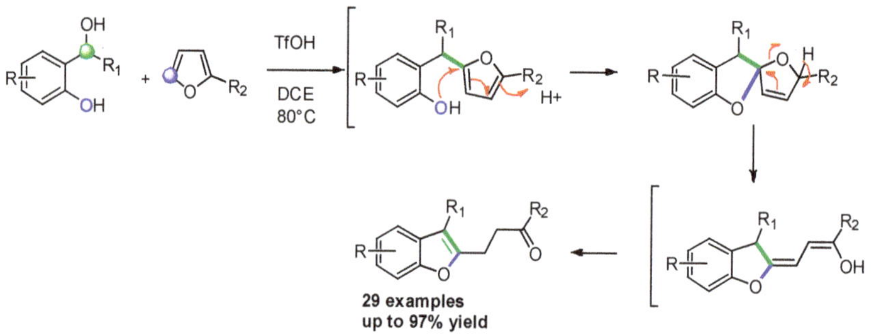

Scheme 43. Acid-catalyzed synthesis of benzofurans via salicyl alcohols and furans.

Chi and co-workers, in 2012, described a convenient method of synthesizing C2-substituted benzofurans from carbamate of 2-hydroxyphenylacetonitrile. In situ two-step reactions using t-BuOK in the absence of oxygen and microwave/silica gel treatment provided several C2-derivatized benzofurans in 52–89% yields. Furthermore, straightforward purification of final product by filtration from silica gel avoided the need for column chromatography. This method is quite convenient, because various starting compounds could be easily prepared from commercially available carbonyl chlorides, such as carbamoyl chloride, thiocarbamoyl chloride, chloroformate, and acid chloride, and because further derivatization of benzofurans at the C3 position could be used to find biologically active benzofurans [134] (Scheme 44).

A similar mechanism was reported in a procedure of a Pd-catalyzed three-component coupling reaction of o-(cyanomethyl)phenol, aryl halide, and carbon monoxide [135].

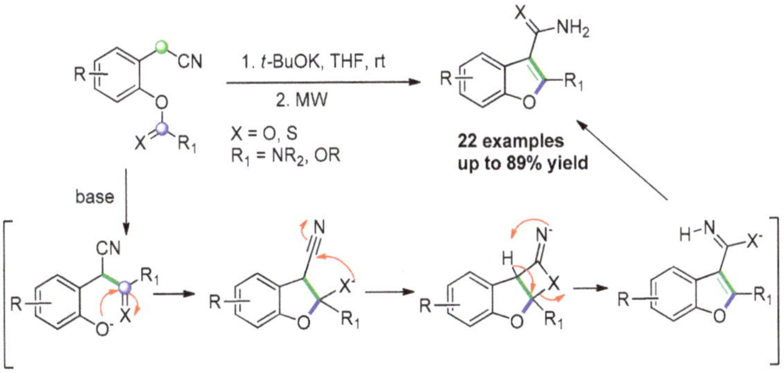

Scheme 44. Synthesis of benzofurans via o-(cyanomethyl)phenol derivatives.

3.3. O–C2 and C3–C3a Bond Formation: (Route b + d)

The main approaches for the formation of O–C2 and C3–C3a bond dealt in this section are summarized below (Scheme 45).

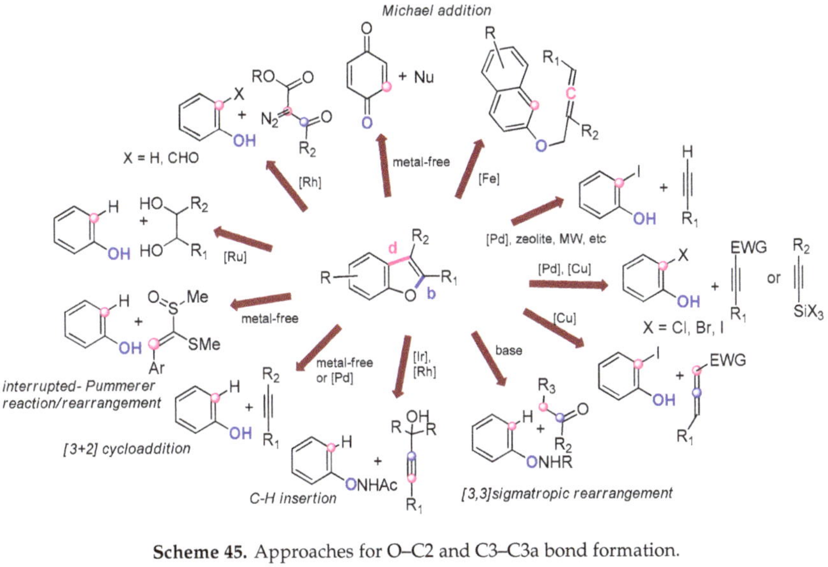

Scheme 45. Approaches for O–C2 and C3–C3a bond formation.

3.3.1. From o-Halophenols and Terminal Alkynes

Pd-catalyzed one-pot synthesis from 2-halophenols and terminal alkynes by a Sonogashira coupling cyclization sequence is a useful and reliable way to construct 2-substituted benzo[b]furans [136,137]. Furthermore, 2-Iodo- and 2-bromophenols have been widely used as 2-halophenols.

A catalyst composed of Pd and hydroxyterphenylphosphine was found to be effective for one-pot benzo[b]-furan synthesis from 2-chlorophenols and alkynes [138]. Moreover, 2,3-Disubstituted benzofurans possessing 2-hydroxyphenyl moiety at the C-3 position were synthesized from readily available 2-chlorophenols and terminal alkynes by hydroxy-directed o-Sonogashira coupling and subsequent oxypalladation/reductive elimination, using Pd-dihydroxyterphenylphosphine as the catalyst. The catalyst accelerated not only the Sonogashira coupling but also the introduction of 2-hydroxyphenyl group at the C-3 position of benzofuran [139] (Scheme 46).

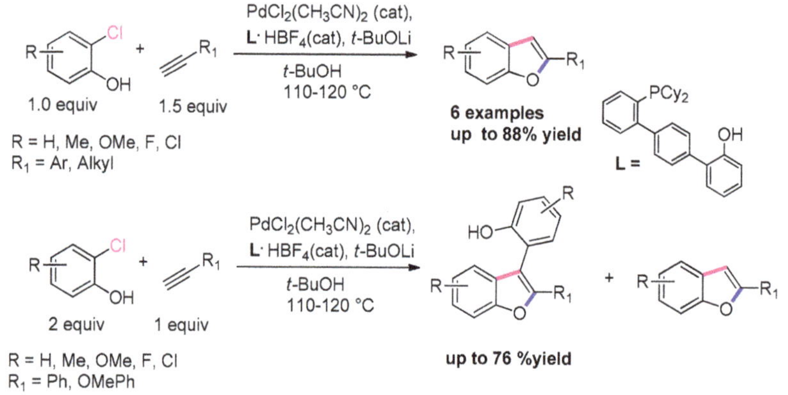

Scheme 46. Synthesis of benzofurans via o-halophenol and terminal alkynes.

The development of a multicatalytic one-pot synthesis of 2-arylbenzofurans starting from aryl halides and 2-halophenols (bromide or frequently iodide) has been described. The protocol involved two Sonogashira coupling reactions, followed by 2-ethynylphenol cyclization, leading to 2-arylbenzofuran derivatives (Scheme 47). The process occured smoothly under mild conditions, giving products in good yields, and was applied to many 2-arylbenzofurans substituted both at 2-aryl position and in the benzodifuran moiety. Substituents such as halogens, hydroxyl, cyano, nitro, and amino groups were tolerated, enabling further functionalization of the system [140,141].

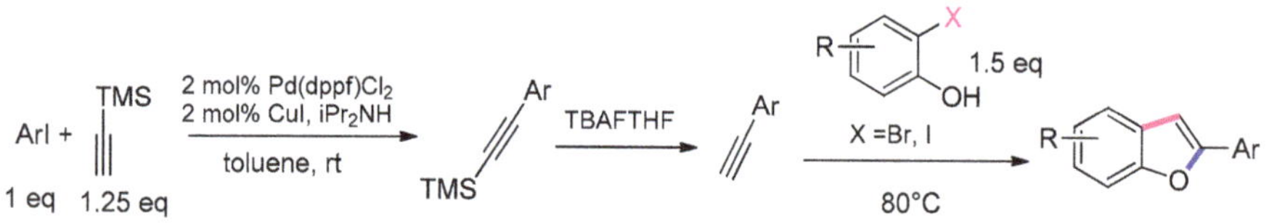

Scheme 47. Synthesis of benzofurans via multicatalytic system of o-halophenol and terminal alkynes.

A multicatalytic system was also used in a cascade transformation of polyenynes into a polyaromatic structure [142]. Nanoparticles of Pd doped by carbon [143] or supported by N,O-dual-doped hierarchical porous carbon [144], as well as NpPd in water copper- and ligand-free [145], have all been new catalytic systems used to generate benzofurans.

In 2013, Larock proposed a one-pot three component MW assisted protocol to generate 2,3-disubstituted benzofurans [146]. MWs were used by Elofsson to assist the synthesis of benzofuran core of some natural products [147]. Furthermore, 2-TMS-benzofuran was used in the asymmetric synthesis of the natural product (+)(R)-concentricolide [148]. Moreover, syntheses of benzofurans were proposed in good yield and tolerance of functional groups, using CuI, diaminecyclohexane, and KOtBu, in 1,4-dioxane condition [149] or Cu scorpionate complex and P450 mediated oxidation [150], to generate a methylene-bridged bis-benzofuran system.

3.3.2. From o-Halophenols and Internal Alkynes

Among all methods reported to obtain selectively 2,3-subsituted benzofurans, the Larock procedure, starting from 2-iodophenols and internal alkynes, appeared the most versatile procedure.

However, these procedures rely on the use of soluble palladium catalysts; thus, they involve significant difficulties, including the high contamination of the products by palladium and ligand, which is not tolerable in the context of biological applications. Obviously, an analogous catalytic heterogeneous method would eliminate all of these drawbacks. The easily homemade [Pd(NH$_3$)$_4$]/NaY catalyst appeared to be the best choice for both indoles and benzofurans syntheses, even in reactions where the original Larock procedure failed and for which previous successes required the use of expensive ligand systems [151] (Scheme 48). Recently, Ghosh's group described a convenient one-pot tandem procedure, a Hiyama alkynylation/cyclization reaction of 2-iodophenol with a range of triethoxysilylalkyne compounds in the presence of palladium acyclic diaminocarbene triflate complexes, which produced 2-substituted benzofurans [152]. A novel approach was developed for the synthesis of 2-substituted-3-functionalized benzofurans, in which the first step was the conjugate addition of phenol to an ynone in the presence of a base (K$_3$PO$_4$ gave the highest yield). Subsequently, an intramolecular Heck reaction (Pd(OAc)$_2$, PPh$_3$, Ag$_2$CO$_3$ in ACN) gave rise to the benzofuran core in a good high yield (up to 97%). This strategy was further applied in the first enantioselective total synthesis of Daphnodorin B [153].

Recently, the same approach was used with the iodo-derivative of tyrosine and several propargyl aldehydes. The atmosphere applied to the reaction medium directly influenced the formation of the products. When an inert atmosphere of nitrogen was applied, a 2-aryl-3-formyl-5-alanylbenzofuran core was selectively obtained via a Heck intramolecular reaction, while under a carbon monoxide atmosphere, the reactions led exclusively to 6-alanyl-2-arylflavone derivatives via reductive intramolecular acylation [154] (Scheme 48).

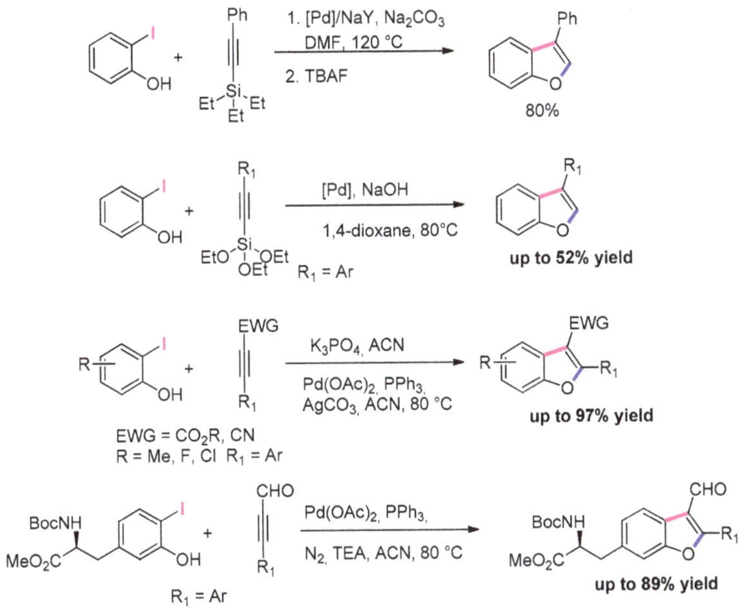

Scheme 48. Synthesis of benzofurans via o-halophenol and internal alkynes.

3.3.3. From o-Halophenols and Allenes

Overall, 2-vinylbenzofurans have been synthesized via the copper-catalyzed one-pot, three-component reactions of o-iodophenols, in situ generated allenes, and dichloromethane. Cascade transformation of oxa-Michael addition, C-arylation, and sp^3 C–H/sp^3 C–Cl conversion-based vinylation has been involved in realizing the construction of this 2-vinylbenzofuran framework [155] (Scheme 49).

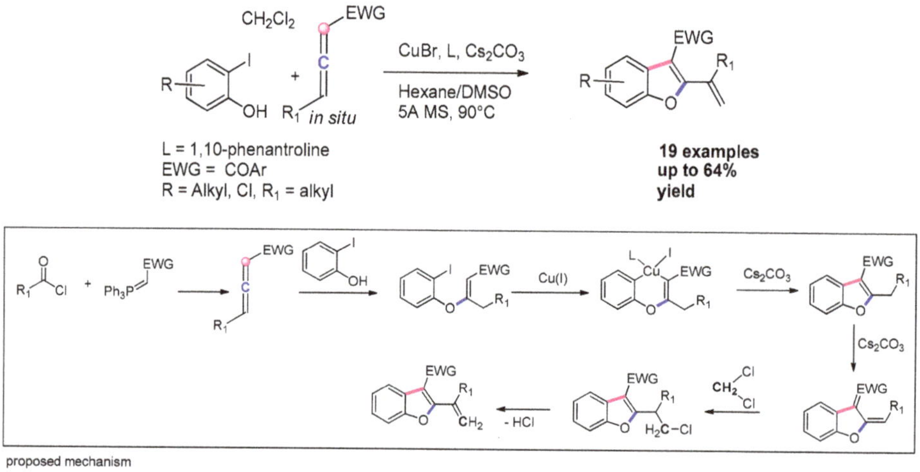

Scheme 49. Synthesis of benzofurans via o-halophenol and allenes.

3.3.4. From Phenols: O-aryloxime/[3,3]-Sigmatropic Rearrangement/Cyclization

One century after their discovery, [3,3]-sigmatropic rearrangements occupy an irreplaceable role in the synthesis of complex organic molecules and continue to be intensively investigated. Among the methods available to prepare benzofurans one of the most synthetically accessible involves [3,3]-sigmatropic rearrangement of preformed O-aryl oxime ethers promoted by Brønsted or Lewis acids [156]. Although high efficiency has been achieved in the synthesis of indoles, as well as benzofurans, under mild reaction conditions via cleavage of O–N bonds, a unified approach to access diverse oxa-heterocycles is highly desirable. The introduction of an O–N bond may encompass the elevated temperature required by the classical Claisen [3,3]-sigmatropic rearrangement. The other challenge is that the annulation/aromatization may not occur readily after the rearrangement step. O-aryl oxime ethers was synthesized by the Cu-catalyzed arylation of N-hydroxyphthalimide with arylboronic acids, followed by cleavage with hydrazine [157] (Scheme 50). Buchwald and co-workers recently reported a more general palladium-catalyzed arylation of ethyl acetohydroxamate with aryl halides in the presence of air-sensitive alkyl-arylphosphine ligands. Ethyl acetohydroxamate served as an efficient hydroxylamine equivalent for C–O cross-coupling, thereby allowing for the preparation of O-arylhydroxylamines from simple aryl halides. Short reaction times and broad substrate scope, including heteroaryl coupling partners, allowed access to O-arylhydroxylamines that would be difficult to prepare in a single step by traditional methods. Moreover, the O-arylated products so formed could be directly transformed into substituted benzofurans in a single operation [158]. Ethyl acetohydroxamate was efficiently arylated with diaryliodonium salts at room temperature under transition-metal-free conditions. The obtained O-arylated products were reacted in situ with ketones, under acidic conditions, to yield substituted benzo[b]furans through oxime formation, [3,3]-rearrangement, and cyclization, in a fast and operationally simple one-pot fashion, without using an excess of reagents. Alternatively, the O-arylated products could be isolated or transformed in situ to aryloxyamines or O-arylaldoximes. The methodology was applied to the synthesis of Stemofuran A and the formal syntheses of Coumestan, Eupomatenoid 6, and (+)-machaeriol B [159].

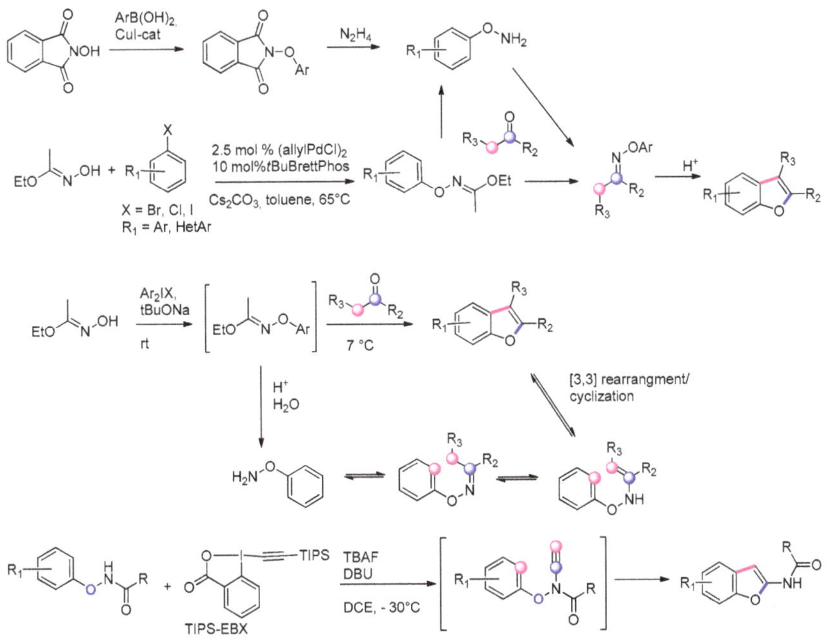

Scheme 50. Synthesis of benzofurans via oxime formation/[3,3]-sigmatropic rearrangement/cyclization.

An efficient method to selectively construct benzofuran and dihydrobenzofuro[2,3-d]oxazole derivatives has been successfully established by means of base-controlled cyclization of N-phenoxyamides with 1-[(triisopropylsilyl)ethynyl]-1,2-benziodoxol-3(1H)-one (TIPS-EBX). N-phenoxyamides, as multitasking reagents, have triggered two different cascade-reaction sequences. This was the first example of using TIPS-EBX for the transformation of C(sp) to either $C(sp^2)$ or $C(sp^3)$ under metal-free conditions [160,161].

3.3.5. Via Transition-Metal-Catalyzed Annulation of N-Aryloxyacetamides and Propargyl Alcohols

Propargylic alcohols are some of the most useful building blocks with two functional groups. These units have been involved in numerous cascade synthetic transformations in organic chemistry, providing an opportunity to discover novel cascade processes [162].

In 2018, Yi revealed an efficient and mild Ir(III)-catalyzed C–H annulation of N-aryloxyacetamides with tertiary propargyl alcohols to deliver benzofurans [163], in which the efficiency of protocol was influenced by the position and the nature of substituents on phenol. In the same year, Yi developed the Rh(III)-catalyzed and solvent-controlled C–H functionalization of N-aryloxyacetamides with secondary or primary propargyl alcohols for the divergent synthesis of chalcones and benzofurans [164]. By virtue of a synergically dual-directing-group (the O–NHAc part and the hydroxyl group)-assisted strategy, the efficient and practical Rh(III)-catalyzed regioselective redox neutral C–H functionalization of diverse N-phenoxyacetamides with propargyl alcohols has been realized, which led to the divergent synthesis of privileged benzofuran and chalcone frameworks in a solvent-controlled chemoselective manner (Scheme 51). Experimental and computational studies revealed that the formation of the hydrogen bonding between dual directing groups and the subsequent coordination interaction between the hydroxyl group and the Rh(III) catalyst play a decisive role in promoting the regioselective migratory insertion of the alkyne moiety. Thereafter, two solvent-controlled switchable reaction pathways occured to deliver the corresponding products with excellent chemoselectivity.

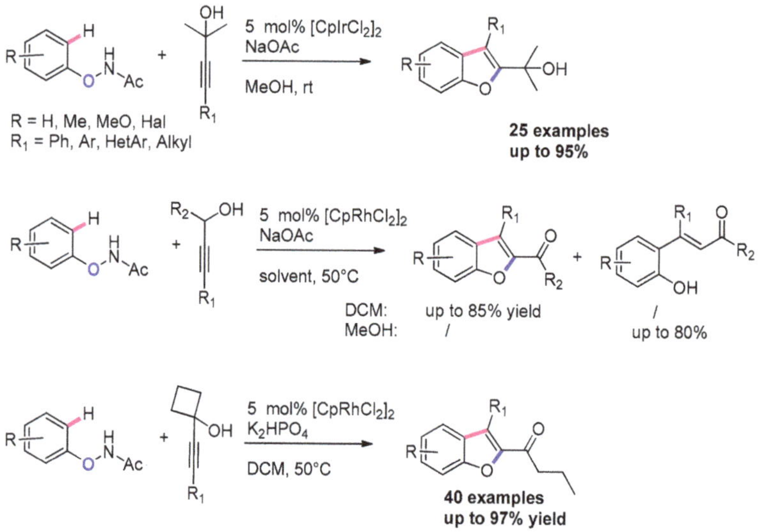

Scheme 51. Synthesis of benzofurans via TM-catalyzed annulation of *N*-aryloxyacetamides and propargyl alcohols.

A cascade [3 + 2] annulation of *N*-aryloxyacetamides with 1-alkynylcyclobutanols via Rh-(III)-catalyzed redox-neutral C–H/C–C activations, using internal oxidative O–NHAc and –OH as the dual directing groups, has been achieved, as well, with the subsequent ring-opening of cyclobutanol. This reaction, performed with [Rh]-complex and KH_2PO_4 in DCM, provided an efficient and regioselective approach to benzofuran derivatives, with good functional group compatibility and high yields [165].

3.3.6. Metal-Free [3 + 2] Annulation of Phenols with Acetylenes

The metal-free [3 + 2] annulation of phenols with propargylic alcohols generated benzofurans as well as naphthofurans, in a highly atom-economy manner. This reaction utilized C1 and C2 carbons of propargylic alcohols for the annulation in a two-step process involving (i) an acid catalyzed intermolecular C–C bond formation between C1 of the propargylic alcohol and the α-position of phenol or β-naphthol, i.e., α-propargylation; and (ii) a base-catalyzed intramolecular O–C bond formation between C2 of the propargylic alcohol and –OH of phenol or naphthol [166].

An unprecedented acid-mediated cascade [3 + 2] annulation process for the generation of complex naphthofurans (benzofurans) from propargylic alcohols and β-naphthols (resorcinols) was proposed by Baire and co-worker [167] (Scheme 52). They described the synthesis of naphthofurans (and benzofurans) in a cascade manner, via an annulation utilizing C2 and C3 carbons of propargylic alcohols with β-naphthols (resorcinols).

Densely substituted amino-functionalized benzofurans were concisely accessed via the first one-pot domino oxidation/[3 + 2] cyclization of a hydroquinone ester and easily accessible ynamides under mild conditions in short time. The complex benzofurans were able to be efficiently synthesized, all from simple and inexpensive starting materials, in two steps [168].

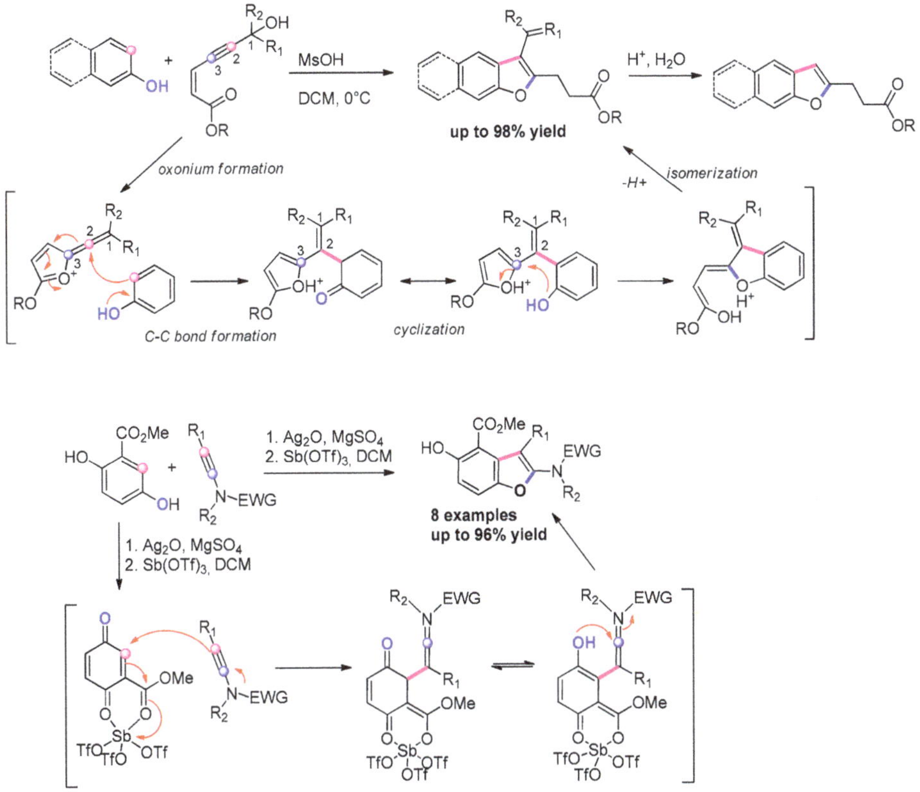

Scheme 52. Synthesis of benzofurans via metal-free [3 + 2] annulation of phenols and acetylenes.

3.3.7. [Pd]-Catalyzed [3 + 2] Annulation of Phenols with Internal Alkynes

While the transition metal (TM)-catalyzed annulation between aniline and unactivated alkynes provides indoles andtheir derivatives easily, a detailed survey of the literature reveals that the corresponding one step synthesis of benzofurans from readily available phenols and unactivated alkynes has remained elusive so far. The reactivity of phenol toward unactivated alkynes presents challenges as the participation of an unfavorable four-membered oxygen-containing metallacycle, the difficulties associated with the formation of the C-O bond through reductive elimination of the putative Pd(II) intermediates, and the sensitivity of phenols to strong oxidants like TMs.

Despite these cumulative challenges, Sahoo and co-workers [38,169] developed an unprecedented one-step synthesis of benzofurans by the Pd-catalyzed oxidative annulation of readily accessible phenols and unactivated internal alkynes.

Moreover, benzo furans were prepared in one-pot, based on the addition/palladium-catalyzed C-H bond functionalization of phenols with bromoalkynes. The addition reactions of phenols to bromoalkynes generated (Z)-2-bromovinyl phenyl ethers in high yields with excellent regio- and stereoselectivity. The obtained (Z)-2-bromovinyl phenyl ethers subsequently proceeded by cyclization, affording 2-substituted benzofurans in good yields. It is important to note that the transformation of phenols with bromoalkynes into benzofurans could be carried out in one-pot with a simple and efficient tandem procedure [170] (Scheme 53).

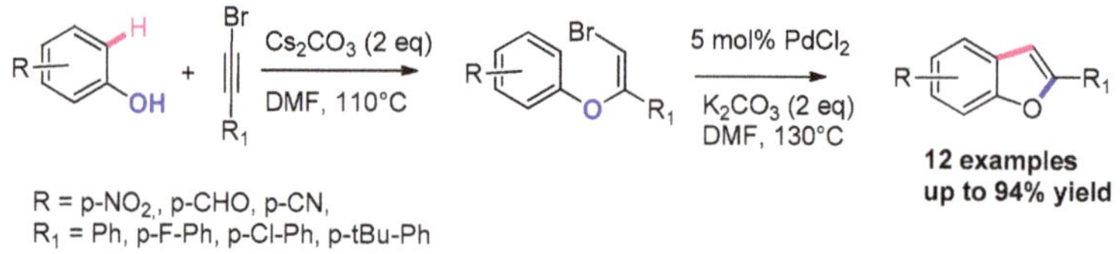

Scheme 53. Synthesis of benzofurans via [Pd] catalyzed [3 + 2] annulation of phenols and acetylenes.

Palladium-catalyzed oxidative annulations between phenols and alkenylcarboxylic acids produced a library of benzofuran compounds. Depending on the nature of the substitution of the phenol precursor, either 2,3-dialkylbenzofurans or 2-alkyl-3-methylene-2,3-dihydrobenzofurans were synthesized with excellent regioselectivity [38,171–173].

3.3.8. Via Interrupted Pummerer Reaction/[3,3] Sigmatropic Rearrangement/Cyclization

The Pummerer reaction is a reaction of an alkyl sulfoxide with a Lewis acidic activator (LA+), such as acid anhydride, to yield a α-functionalized alkyl sulfide. Interrupted Pummerer reactions are different from other Pummerer-type reactions in terms of the reaction mode: The cationic sulfur center is directly attacked, or interrupted, by a nucleophile [174]. In 2010, Yorimitsu and co-workers prepared 2-methylthio-3-trifluoromethyl-substituted benzofurans from phenol and ketene dithioacetal monoxides (KDM) [175]. Subsequently, the same group extended the methodology by using a wide range of KDMs activated by trifluoroacetic anhydride (TFFA), in order to avoid the fast decomposition of dicationic intermediate [176–178].

In 2018, they accomplished a facile synthesis of fluorinated benzofurans from polyfluorophenols by means of a sigmatropic dearomatization/defluorination strategy composed of three processes: (1) interrupted Pummerer reaction of ketene dithioacetal monoxides, activated by TFFA, with polyfluorophenols followed by [3,3] sigmatropic rearrangement; (2) Zn-mediated smooth reductive removal of fluoride from the dearomatized intermediate; and (3) acid-promoted cyclization/aromatization to lead to benzofuran in 90% overall yield. Some of the fluorinated benzofurans were transformed by utilizing the 2-methylsulfanyl moieties [179] (Scheme 54).

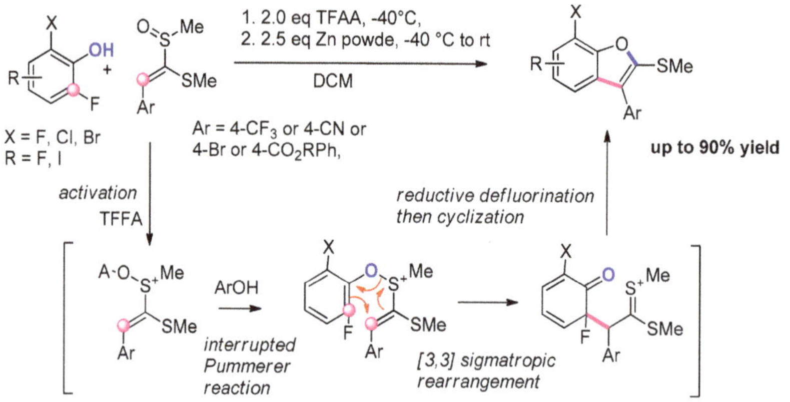

Scheme 54. Synthesis of benzofurans via interrupted Pummerer reaction/[3,3]-sigmatropic rearrangement/annulation of perfluorophenols.

Procter's group reported a transition-metal-free synthesis of benzofurans from benzothiophenes and phenols which exploited the unique reactivity of sulfoxides [180]. Through a sequence involving an interrupted Pummerer reaction and [3,3] sigmatropic rearrangement, phenols were combined with readily accessible, yet synthetically unexplored, benzothiophene S-oxides to provide 3-arylated benzofurans. The products from this approach underwent subsequent functionalization, to gain access to a range of important benzofuran derivatives (Scheme 55). Sulfinate salts are a class of versatile compounds that have recently found application as coupling partners in palladium-catalyzed cross-coupling reactions. In fact, they were subjected to the subsequent desulfinative cross-coupling of substituted aryl halides, known to be easily available [181]. This approach established sulfoxides as a traceless activating group for C–H functionalization in this method. Thus, the intermediate aryl sulfinates, formed from treatment of the sulfones with base, underwent desulfinative palladium-catalyzed cross-coupling in the same pot, to provide the desired biphenyl benzofurans. This procedure gave good-to-excellent yields for all substrates tested; o-, m-, and p-substituted substrates all gave similarly high yields. It is worth noting that no trace of the sulfoxide group was present in the unreacted starting material.

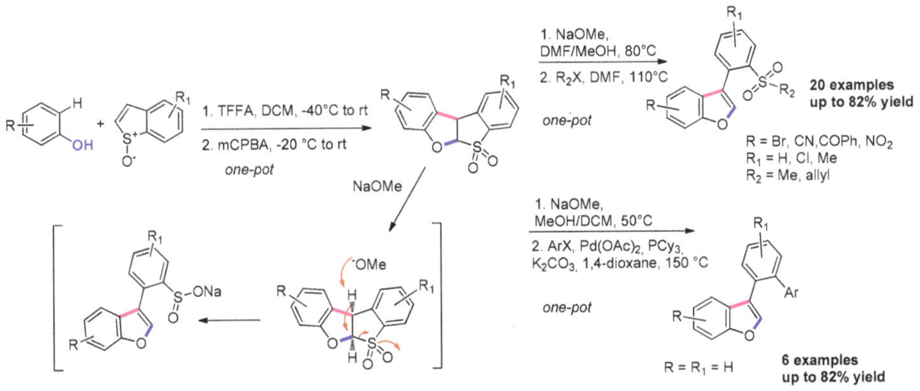

Scheme 55. Synthesis of benzofurans via interrupted Pummerer reaction/[3,3]-sigmatropic rearrangement/annulation/desulfinative reactions.

3.3.9. Via Fries-type O-C Rearrangement/Michael Addition of Phenols

Recently, the direct synthesis of naphthofurans and benzofurans was reported from readily available phenols and α-haloketones. It was promoted by titanium tetrachloride (TiCl$_4$) which combined Friedel–Crafts-like alkylation and intramolecular cyclodehydration into one step. High levels of regioselectivity, broad substrate scope, and moderate-to-excellent yields were obtained [182].

An unusual and facile approach for the synthesis of 2-benzofuranyl- 3-hydroxyacetones from 6-acetoxy-β-pyrones and phenols was described by Ramasastry [183] (Scheme 56). The synthetic sequence involved a cascade transacetalisation, Fries-type O–C rearrangement followed by Michael addition, and ring-opening aromatization. The unexpected cascade event also provided new possible considerations in the β-pyrone-involved organic synthesis.

Seggi and co-workers reported that 3-(2-bromoethyl)benzofurans were readily obtained from commercially available bis[(trimethylsilyl)oxy]cyclobutene and various phenols via a Brønsted acid-mediated nucleophilic addition–carbocyclic rearrangement cascade reaction; this is a one-pot, metal-free process that operates in mild conditions [184]. In the presence of a Brønsted acid, 2-hydroxycyclobutanone and its precursor bis[(trimethylsilyl)oxy]cyclobutene behaved as electrophilic acceptors for intermolecular nucleophilic addition, followed by a ring closure–ring fission process. This mild and facile strategy was applied for the synthesis of a series of 5-HT serotonin receptor agonists, underlining its potential for the syntheses of bioactive compounds and natural products.

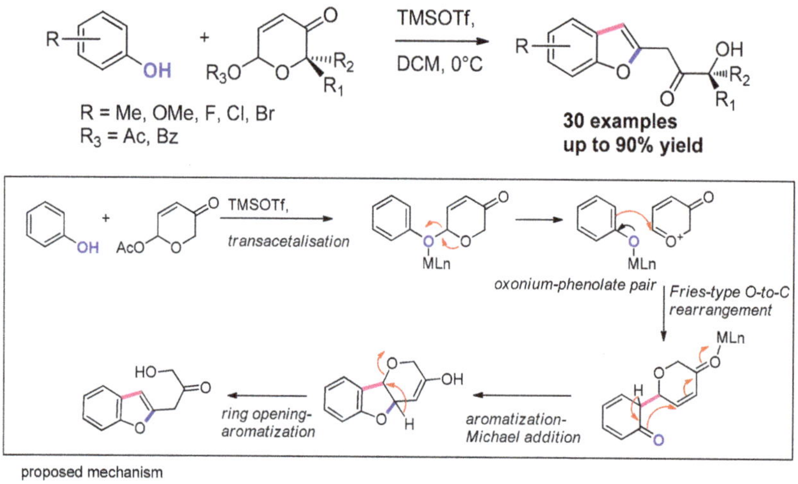

Scheme 56. Synthesis of benzofurans via Fries-type O-C rearrangement/Michael addition.

3.3.10. Via [Ru]-Catalyzed C–H Alkylation of Phenols with 1,2-Diols

Alcohols have been rarely employed as the substrate for the catalytic C–H coupling reactions, because of their tendency for undergoing energetically more favorable alkoxylation and oxidation reactions over the respective C–O bond cleavage reaction. Yi and co-workers discovered an exceptionally selective dehydrative C–H alkylation reaction of alkenes with alcohols that was catalyzed by a well-defined cationic ruthenium hydride complex [$(C_6H_6)(PCy_3)(CO)RuH$]$^+BF_4^-$. This cationic Ru–H complex also catalyzed the dehydrative C–H alkylation reaction of phenols with alcohols to form ortho-substituted phenol products so that benzofuran derivatives were efficiently synthesized from the dehydrative C–H alkenylation and annulation reaction of phenols with 1,2-diols [185]. The catalytic C–H coupling method employed cheaply available phenols and alcohols, exhibited a broad substrate scope, tolerated carbonyl and amine functional groups, and formed water as the only byproduct (Scheme 57).

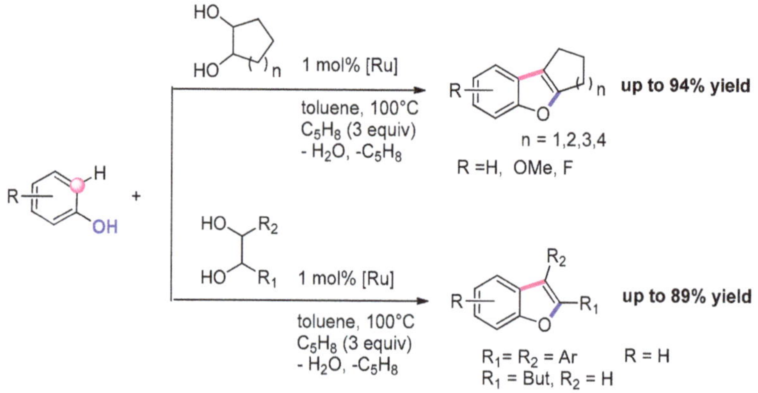

Scheme 57. *Cont.*

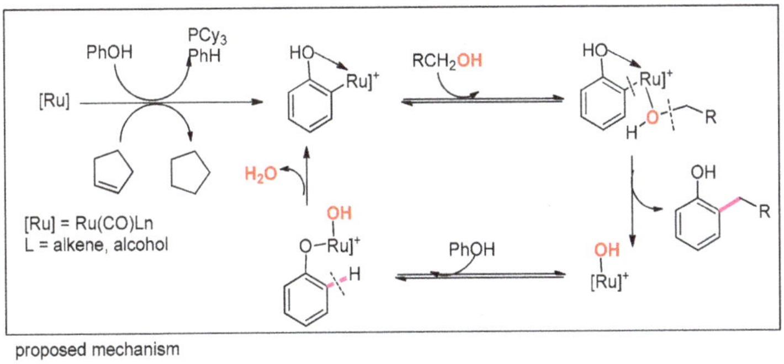

Scheme 57. Synthesis of benzofurans via [Ru]-catalyzed C–H alkylation of phenols with 1,2-diols.

3.3.11. Via [Rh]-Catalyzed Carbene Insertion with Phenols/Salicylaldehydes

Transition-metal carbene X–H insertion reactions (X=N or O) have been employed in the simple conversion of anilines and phenols into indoles and benzofurans, respectively. Thus, copper(II) catalyzed N–H insertion reactions of α-diazo-β-ketoesters with N-methylanilines, followed by treatment with acidic ion exchange resin gave indoles. In a similar manner, dirhodium(II) catalyzed O–H insertion reactions of α-diazo-β-ketoesters with phenols, followed by treatment with polyphosphoric acid (PPA) gave benzofurans [186] (Scheme 58).

A Rh(III)-catalyzed annulation between salicylaldehydes and diazo compounds with controllable chemoselectivity was described by Lin and Yao [187]. AgNTf$_2$ favored benzofurans via a tandem C–H activation/decarbonylation/annulation process, while AcOH led to a chromones through a C–H activation/annulation pathway. The reaction exhibited good functional group tolerance and scalability. Moreover, only a single regioisomer of benzofuran was obtained due to the in situ decarbonylation orientation effect. Reactions of salicylaldehyde and its cyclic acetals with diazocarbonyl compounds in the presence of copper and rhodium catalysts have been studied. The reaction pathway and product yields were determined by the nature of the initial reactants and catalyst [188] (Scheme 58).

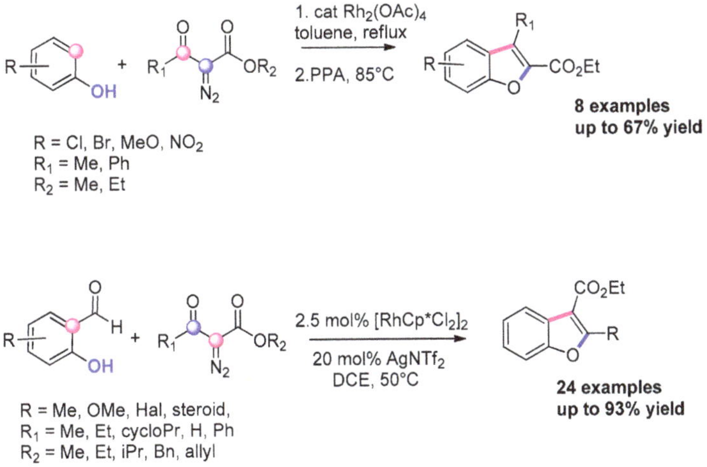

Scheme 58. Cont.

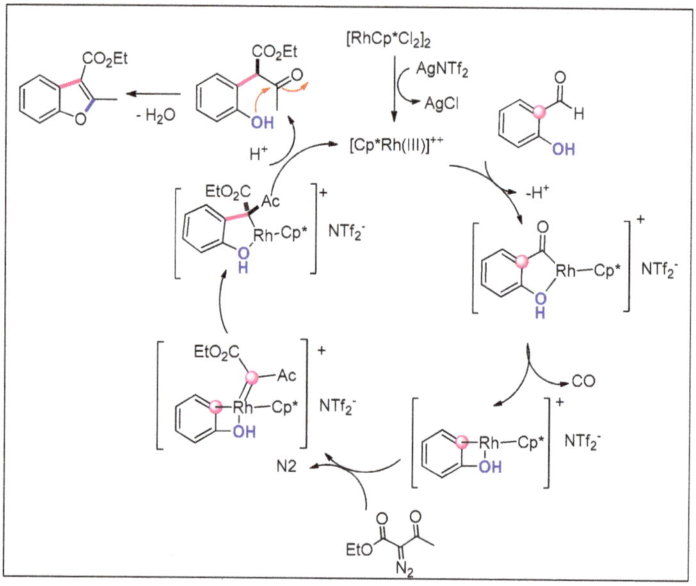

Scheme 58. Synthesis of benzofurans via [Ru]-catalyzed carbene insertion with phenols/salicylaldehydes.

3.3.12. Via Michael Addition/Cyclization of Nucleophiles on Benzoquinones

Benzofuran derivatives were synthesized through the sequential Michael addition and cyclization of 1,3-dicarbonyl compounds with 1,4-benzoquinones. However, ketones are rarely used in this reaction because of their low nucleophilicities. In this study, this problem was solved by utilizing triethyl orthoformate, which enabled the formation of a vinyl ethyl ether as an additive. As a result, the nucleophilicity of ketones increased. Many important 5-hydroxybenzofuran derivatives, not previously available by synthesis, were also prepared by these newly established reactions [189] (Scheme 59).

A convenient metal-free one-pot synthesis of benzofuran derivatives starting from simple ynones has been developed by Cui and co-workers [190]. Various functionalized benzofurans, closely related to bioactive molecules, were obtained in moderate-to-good yields (up to 90%) through aza-Michael/Michael/annulation sequence (a mechanism similar to the previously described method). Preparative scale synthesis of benzofurans was successfully achieved, as well. The application of the benzofuran products was shown by easy transformations to highly functionalized molecules, holding significant promise for medicinal chemistry and organic material chemistry.

An efficient synthesis of benzofuran derivatives via the cross-coupling of catechols and hydroxycoumarins in H_2O, using O_2 as an ideal oxidant, was reported by Maeno and co-workers (Scheme 60). The above reaction allowed the direct use of substrates without prefunctionalization, involved formation of C–C and C–O bonds in a cascade manner, and afforded H_2O as the sole by-product. This simple and clean reaction was achieved by the development of an $AlPO_4$-supported Rh nanoparticle catalyst. The catalyst was applicable to the synthesis of a wide range of benzofurans. This catalytic method was successfully utilized for total synthesis of flemichapparin C, one of the naturally occurring coumestans exhibiting bioactivity [191].

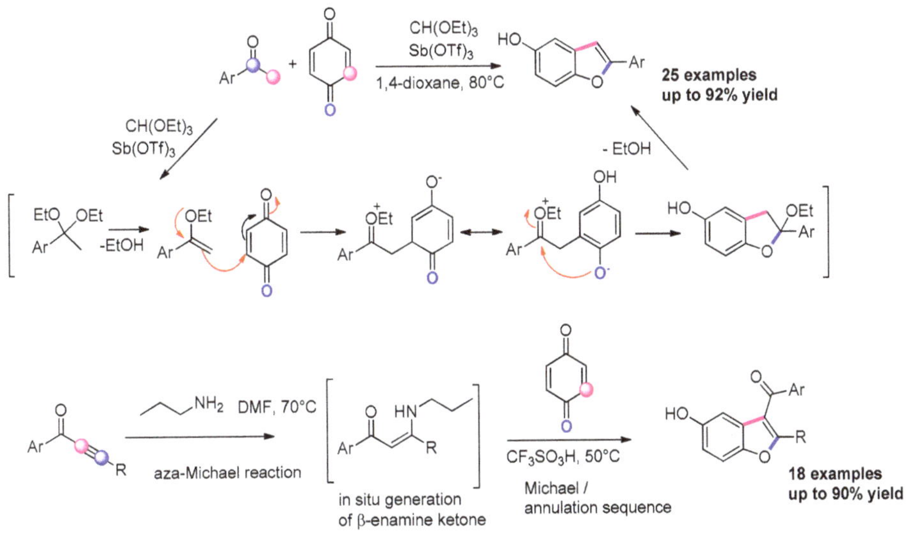

Scheme 59. Synthesis of benzofurans via Michael addition/cyclization of nucleophiles on p-benzoquinones.

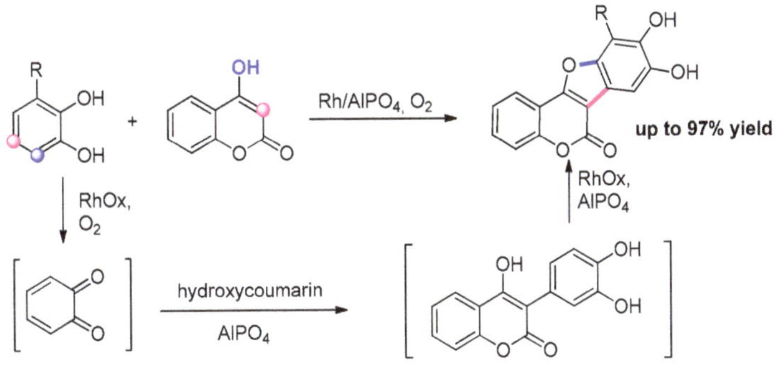

Scheme 60. Synthesis of benzofurans via Michael addition/cyclization of nucleophiles on o-benzoquinones.

3.3.13. Via FeCl₃-Catalyzed Allenic Claisen Rearrangement/ Dehydrogenative Cyclization

A FeCl$_3$-catalyzed allenic Claisen rearrangement/regio- and chemoselective aerobic dehydrogenative cyclization domino reaction is developed, providing a wide range of 2-aryl/alkyl, 3-(substituted-vinyl)naphtho[2,1-b]-furans in high yields at 95–130 °C in an atom- and step economic fashion. Mechanistic studies suggested that the FeCl$_3$ catalyst was responsible for the high regio- and chemoselectivity in reaction. A blue-emitting product showed a quantum yield of 0.95. The reaction proceeded readily on the gram scale, and synthetic applications of the products were also demonstrated [192] (Scheme 61).

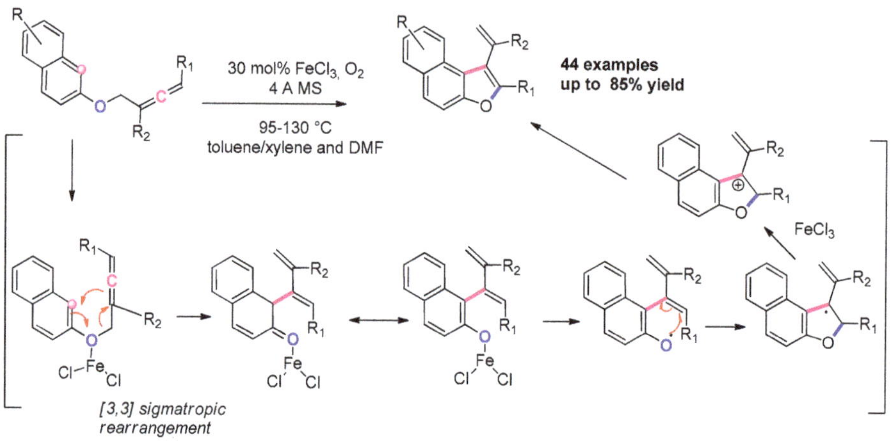

Scheme 61. Synthesis of benzofurans via FeCl$_3$-catalyzed allenic Claisen rearrangement/dehydrogenative cyclization.

4. Benzoannulation

This last section reports examples in which the benzofuran scaffold was built through benzannulation or aromatization. Few recent examples were already reported in References [28] and [30].

Recently, Mehta and co-workers, during an attempt to perform a Tanabe annulation [193] on 4-hydroxy-cyclohexanone with 1,1-dimethoxy acetone, obtained the corresponding 3-methybenzofuran as the only product in 78% yield [194] (Scheme 62). This method was extended to naphthofurans and other related frameworks. The utility of this adaptable methodology was applied to concise syntheses of natural products, stereumene B, paeoveitol D, and (±)-paeoveitol.

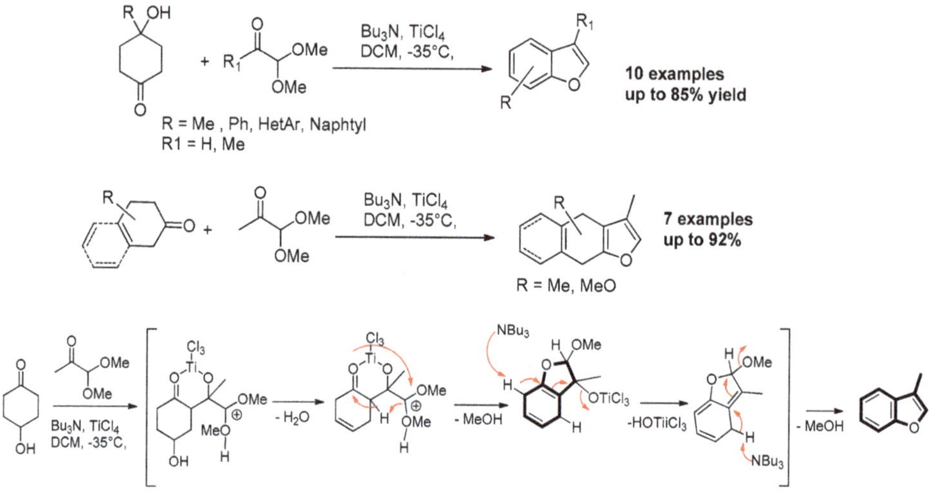

Scheme 62. Synthesis of benzofurans via Tanabe-type annulation with cyclohexanone and 1,1-dimethoxy acetone.

A novel cascade de novo reaction to access benzofurans from suitable 2-hydroxy-1,4-diones was reported by Sha [195]. The hydroxyl group played a crucial role in the reaction, allowing the dehydration reaction to form the key C-C bond during the cascade reaction process. This facile one-pot method for the preparation of benzofurans in moderate-to-good yields (R_1 = Me, Ph, 2-naphtyl, etc.; R_2 = Me, EtO, BnO, etc.; R = H, Et, MeO_2C, Ph, etc.) runs via cyclization/oxidative aromatization cascade reaction of 2-hydroxy-1,4-diones, using trifluoroacetic acid as a catalyst and N-bromosuccinimide as oxidant. Such 2-hydroxy-1,4-diones also showed as a supplement of the Paal–Knorr furan synthesis. A preliminary study was undertaken, as well, to support the proposed mechanism, during which a novel 1,6-conjugate addition reaction was revealed (Scheme 63).

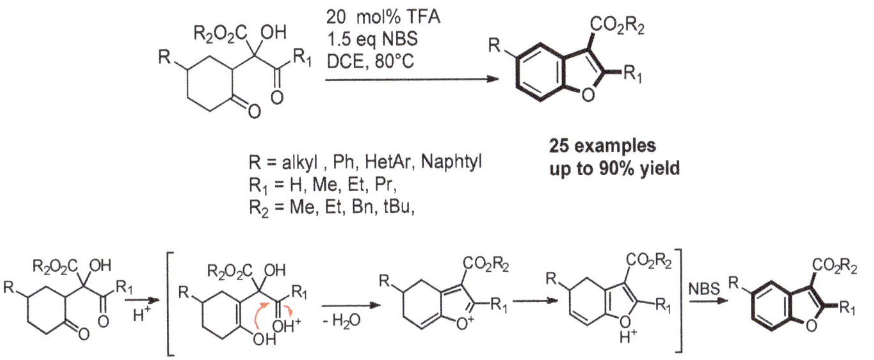

Scheme 63. Synthesis of benzofurans via cyclization/oxidative aromatization cascade reaction of 2-hydroxy-1,4-diones.

Recently, the coupe Lewis acid/NBS as catalyst and oxidant, respectively, was proposed as partners of reaction for the facile way to construct a six-and-five two-aromatic-ring fused heterocycle, namely benzofuran. Starting from easily available chemicals, acrolein dimer and 1,3-dicarbonyl compounds, 2,3-disubstituted benzofurans were synthesized in good yield (Scheme 64). The method succeded to synthesize two commercial drug molecules, benzbromarone and amiodarone [196].

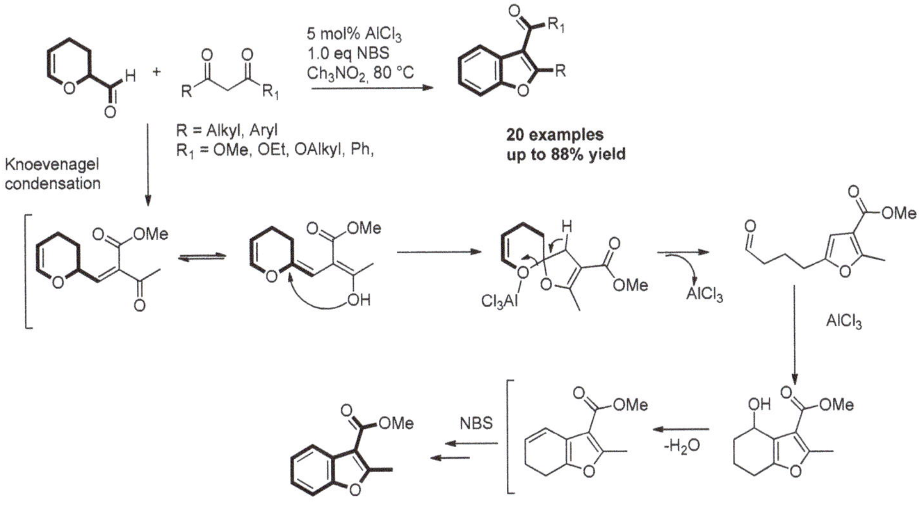

Scheme 64. Synthesis of benzofurans via Knoevenagel condensation with acrolein dimer and acetoacetates.

At last, a very interesting methodology was proposed by Zhu and co-worker in 2020. This is an unprecedented decostructive reorganization strategy for the preparation of hydroxylated benzofurans from either kojic acid or maltol-derived alkynes [197] (Scheme 65). With the aim to develop new dearomatic cascade rearrangement of pyrones, the authors reported a study in which both the benzene and furan rings were simultaneously estabilished via an arene cycloisomerization tandem reaction. A range of substitution patterns was achieved, and a large number of hydroxylated benzofurans were prepared in one-step, with 100% atom economy, enabling a collective total synthesis of different kinds of natural products.

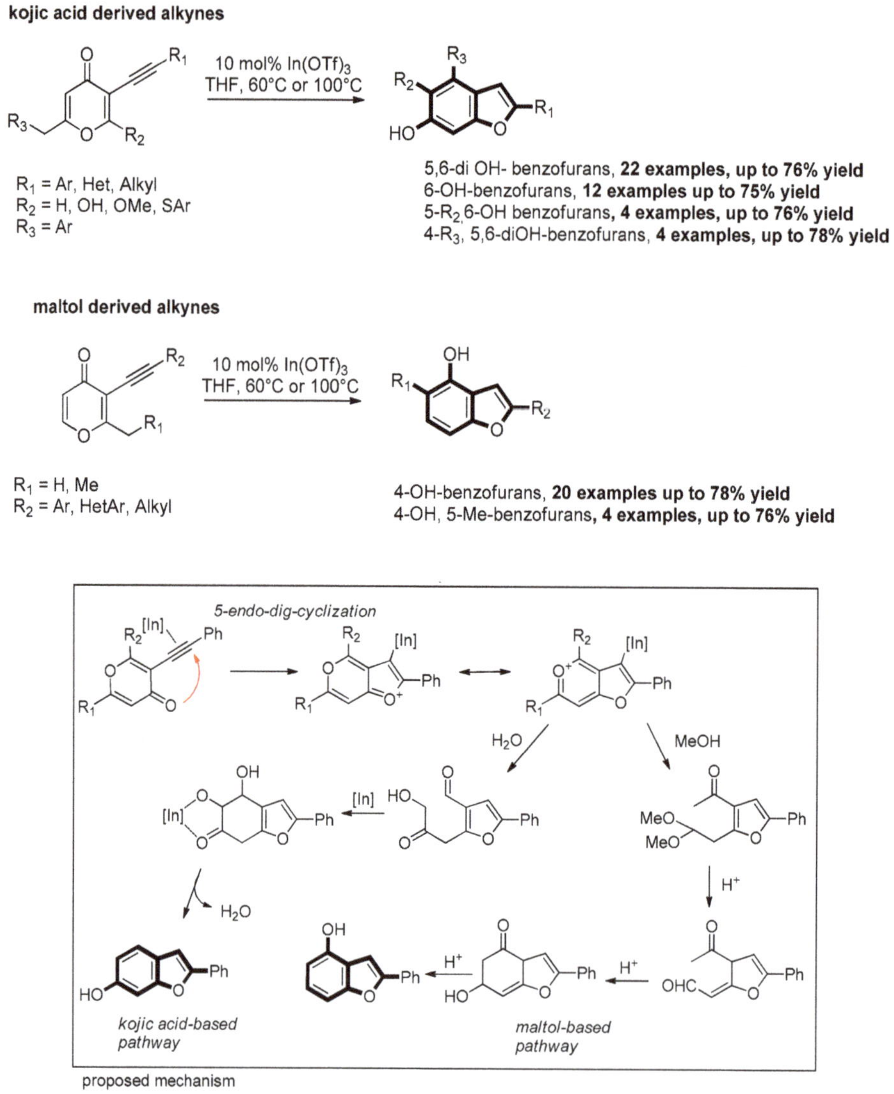

Scheme 65. Synthesis of benzofurans via [In]-catalyzed decostructive reorganization strategy.

5. Conclusions

This review has described recent progress in transition-metal-catalyzed and metal-free couplings for the synthesis of polysubstituted benzo[b]furans. Due to their high efficiency, economy and versatility, transition-metal-catalyzed one-pot processes, especially those involving multiple C–C/C–O bond-forming cascades in an inter-molecular approach, are powerful methods and thus have been extensively investigated. However, the development of more sustainable catalytic systems and more practical synthetic methods, starting from simple and readily available feedstocks, is still highly desirable. Due to the large amount of publications on this topic, a selection of the most relevant had to be done. Hopefully, this review could be a reference of new synthetic strategies which have never appeared in previous reviews.

Author Contributions: All authors contributed to the writing of this paper. All authors have read and agreed to the published version of the manuscript.

Funding: This research was funded by the Department of Science of University of Basilicata, Potenza, Italy.

Conflicts of Interest: The authors declare no conflict of interest.

References

1. Radadiya, A.; Shah, A. Bioactive benzofuran derivatives: An insight on lead developments, radioligands and advances of the last decade. *Eur. J. Med. Chem.* **2015**, *97*, 356–376. [CrossRef]
2. Shamsuzzaman, H.K. Bioactive Benzofuran derivatives: A review. *Eur. J. Med. Chem.* **2015**, *97*, 483–504. [CrossRef]
3. Nevagi, R.J.; Dighe, S.N.; Dighe, S.N. Biological and medicinal significance of benzofuran. *Eur. J. Med. Chem.* **2015**, *97*, 561–581. [CrossRef] [PubMed]
4. Laurita, T.; D'Orsi, R.; Chiummiento, L.; Funicello, M.; Lupattelli, P. Recent Advances in Synthetic Strategies to 2,3-Dihydrobenzofurans. *Synthesis* **2020**, *52*, 1452–1477. [CrossRef]
5. Li, Y.T.; Yao, C.S.; Bai, J.Y.; Lin, M.; Cheng, G.F. Anti-inflammatory effect of amurensin H on asthma-like reaction induced by allergen in sensitized mice. *Acta Pharmacol. Sin.* **2006**, *27*, 735–740. [CrossRef] [PubMed]
6. Brkljača, R.; White, J.M.; Urban, S. Phytochemical Investigation of the Constituents Derived from the Australian Plant Macropidia fuliginosa. *J. Nat. Prod.* **2015**, *78*, 1600–1608. [CrossRef]
7. Vo, D.D.; Elofsson, M. Total Synthesis of Viniferifuran, Resveratrol-Piceatannol Hybrid, Anigopreissin A and Analogues–Investigation of Demethylation Strategies. *Adv. Synth. Catal.* **2016**, *358*, 4085–4092. [CrossRef]
8. Chiummiento, L.; Funicello, M.; Lopardo, M.T.; Lupattelli, P.; Choppin, S.; Colobert, F. Concise total synthesis of permethylated anigopreissin A, a novel benzofuryl resveratrol dimer. *Eur. J. Org. Chem.* **2012**, 188–192. [CrossRef]
9. Convertini, P.; Tramutola, F.; Iacobazzi, V.; Lupattelli, P.; Chiummiento, L.; Infantino, V. Permethylated Anigopreissin A inhibits human hepatoma cell proliferation by mitochondria-induced apoptosis. *Chem. Biol. Interact.* **2015**, *237*, 1–8. [CrossRef]
10. Gaisina, I.N.; Gallier, F.; Ougolkov, A.V.; Kim, K.H.; Kurome, T.; Guo, S.; Holzle, D.; Luchini, D.N.; Blond, S.Y.; Billadeau, D.D.; et al. From a natural product lead to the identification of potent and selective benzofuran-3-yl-(indol-3-yl)maleimides as glycogen synthase kinase 3b inhibitors that suppress proliferation and survival of pancreatic cancer cells. *J. Med. Chem.* **2009**, *52*, 1853–1863. [CrossRef]
11. Naik, R.; Harmalkar, D.S.; Xu, X.Z.; Jang, K.; Lee, K. Bioactive benzofuran derivatives: Moracins A-Z in medicinal chemistry. *Eur. J. Med. Chem.* **2015**, *90*, 379–393. [CrossRef] [PubMed]
12. Zhong, M.; Peng, E.; Huang, N.; Huang, Q.; Quq, A.; Lau, M.; Colonno, R.; Li, L. Discovery of novel potent HCV NS5B polymerase non-nucleoside inhibitors bearing a fused benzofuran scaffold. *Bioorg. Med. Chem. Lett.* **2018**, *28*, 963–968. [CrossRef] [PubMed]
13. Galal, S.A.; El-All, A.S.A.; Abdallah, M.M.; El-Diwani, H.I. Synthesis of potent antitumor and antiviral benzofuran derivatives. *Bioorg. Med. Chem. Lett.* **2009**, *19*, 2420–2428. [CrossRef] [PubMed]
14. Hiremathad, A.; Chand, K.; Tolayan, L.; Rajeshwari; Keri, R.S.; Esteves, A.R.; Cardoso, S.M.; Chaves, S.; Santos, M.A. Hydroxypyridinone-benzofuran hybrids with potential protective roles for Alzheimer's disease therapy. *J. Inorg. Biochem.* **2018**, *179*, 82–96. [CrossRef]

15. Deepti, G.; Amandeep, K.; Bhupesh, G. Benzofuran and indole: Promising Scaffolds for drug development in Alzheimer's disease. *Chem. Med. Chem.* **2018**, *13*, 1275–1299. [CrossRef]
16. Marion, T.; Sylviane, T.; Philippe, G.; Joelle, D. Synthesis of polysubstituted benzofuran derivatives as novel inhibitors of parasitic growth. *Bioorg. Med. Chem.* **2013**, *21*, 4885–4892. [CrossRef]
17. Sangeeta, B.; Deepti, R. Synthetic routes and biological activities of benzofuran and its derivatives: A review. *Lett. Org. Chem.* **2017**, *14*, 381–402. [CrossRef]
18. Hiremathad, A.; Patil, M.R.; Chand, K.; Santos, M.A.; Keri, R.S. Benzofuran: An emerging scaffold for antimicrobial agents. *RSC Adv.* **2015**, *5*, 96809–96828. [CrossRef]
19. Simonetti, S.O.; Larghi, E.L.; Bracca, A.B.J.; Kaufman, T.S. Angular tricyclic benzofurans and related natural products of fungal origin. Isolation, biological activity and synthesis. *Nat. Prod. Rep.* **2013**, *30*, 941–969. [CrossRef]
20. Shimazu, S.; Takahata, K.; Katsuki, H.; Tsunekawa, H.; Tanigawa, A.; Yoneda, F.; Knoll, J.; Akaike, A. (−)-1-(Benzofuran-2-yl)-2-propylaminopentane enhances locomotor activity in rats due to its ability to induce dopamine release. *Eur. J. Pharmacol.* **2001**, *421*, 181–189. [CrossRef]
21. Lavranos, T.C.; Leske, A.F.; Inglis, D.J.; Brown, C.K.; Bibby, D.C.; Kremmidiotis, G. Abstract 2774: Anti-cancer activity of the tumor-selective, hypoxia-inducing, agent BNC105 in platinum resistant ovarian cancer. *Cancer Res.* **2012**, *72*, 2774. [CrossRef]
22. Joule, J.A.; Mills, K. *Heterocyclic Chemistry*, 5th ed.; Wiley-Blackwell: New York, NY, USA, 2010; pp. 437–443.
23. De Luca, L.; Nieddu, G.; Porcheddu, A.; Giacomelli, G. Some Recent Approaches to the Synthesis of 2-Substituted Benzofurans. *Curr. Med. Chem.* **2009**, *16*, 1–20. [CrossRef] [PubMed]
24. Abu-Hashem, A.A.; Hussein, H.A.R.; Aly, A.S.; Gouda, M.A. Synthesis of benzofuran derivatives via different methods. *Synth. Commun.* **2014**, *44*, 2285–2312. [CrossRef]
25. Heravi, M.M.; Zadsirjan, V. The recent advances in the synthesis of benzo[b]furans. *Adv. Heterocycl. Chem.* **2015**, *117*, 261–376. [CrossRef]
26. Heravi, M.M.; Zadsirjan, V.; Dehghani, M. Reactivity of Benzo[b]furans: A Full Perspective. *Curr. Org. Chem.* **2016**, *20*, 1069–1134. [CrossRef]
27. Heravi, M.M.; Zadsirjan, V. Recent Advances in the Synthesis of Biologically Active Compounds Containing Benzo[b]Furans as a Framework. *Curr. Org. Synth.* **2016**, *13*, 780–833. [CrossRef]
28. Heravi, M.M.; Zadsirjan, V.; Hamidi, H.; Hajiabbas, P.; Amiri, T. Total synthesis of natural products containing benzofuran rings. *RSC Adv.* **2017**, *7*, 24470–24521. [CrossRef]
29. More, K.R. Review on Synthetic Routes for Synthesis of Benzofuran-Based Compounds. *J. Chem. Pharm. Res.* **2017**, *9*, 210–220.
30. Miao, Y.-h.; Hu, Y.-h.; Yang, J.; Liu, T.; Sun, J.; Wang, X.-j. Natural source, bioactivity and synthesis of benzofuran derivatives. *RSC Adv.* **2019**, *9*, 27510–27540. [CrossRef]
31. Katritzky, A.R.; Ji, Y.; Fang, Y.; Prakash, I. Novel Syntheses of 2,3-Disubstituted Benzofurans. *J. Org. Chem.* **2001**, *66*, 5613–5615. [CrossRef]
32. Carrër, A.; Florent, J.-C.; Auvrouin, E.; Rousselle, P.; Bertounesque, E. Synthesis of 3-Aryl-2-arylamidobenzofurans Based on the Curtius Rearrangement. *J. Org. Chem.* **2011**, *76*, 2502–2520. [CrossRef] [PubMed]
33. Alonso, F.; Beletskaya, I.P.; Yus, M. Transition-Metal-Catalyzed Addition of Heteroatom–Hydrogen Bonds to Alkynes. *Chem. Rev.* **2004**, *104*, 3079–3160. [CrossRef] [PubMed]
34. Zeni, G.; Larock, R.C. Synthesis of Heterocycles via Palladium-Catalyzed Oxidative Addition. *Chem. Rev.* **2006**, *106*, 4644–4680. [CrossRef] [PubMed]
35. Cacchi, S.; Fabrizi, G.; Goggiamani, A. The palladium-catalyzed assembly and functionalization of benzo [b] furans. *Curr. Org. Chem.* **2006**, *10*, 1423–1455. [CrossRef]
36. Cacchi, S.; Fabrizi, G.; Goggiamani, A. Copper catalysis in the construction of indole and benzo[b]furan rings. *Org. Biomol. Chem.* **2011**, *9*, 641–652. [CrossRef]
37. Sadig, J.E.R.; Willis, M.C. Palladium- and Copper-Catalyzed Aryl Halide Amination, Etherification and Thioetherification Reactions in the Synthesis of Aromatic Heterocycles. *Synthesis* **2011**, 1–22. [CrossRef]
38. Agasti, S.; Dey, A.; Maiti, D. Palladium-catalyzed benzofuran and indole synthesis by multiple C–H functionalizations. *Chem. Commun.* **2017**, *53*, 6544–6566. [CrossRef]

39. Li, Y.; Cheng, L.; Liu, X.; Li, B.; Sun, N. Copper-promoted hydration and annulation of 2-fluorophenylacetylene derivatives: From alkynes to benzo[b]furans and benzo[b]thiophenes. *Beilstein J. Org. Chem.* **2014**, *10*, 2886–2891. [CrossRef]
40. Lavery, C.B.; Rotta-Loria, N.L.; McDonald, R.; Stradiotto, M. Pd$_2$dba$_3$/Bippyphos: A Robust Catalyst System for the Hydroxylation of Aryl Halides with Broad Substrate Scope. *Adv. Synth. Catal.* **2013**, *355*, 981–987. [CrossRef]
41. Faragó, J.; Kotschy, A. Synthesis of Benzo[b]furans by Palladium–NHC Catalyzed Ring Closure of o-Bromobenzyl Ketones. *Synthesis* **2009**, 85–90. [CrossRef]
42. Bonnamour, J.; Piedrafita, M.; Bolm, C. Iron and copper salts in the synthesis of benzo[b]furan. *Adv. Synth. Catal.* **2010**, *352*, 1577–1581. [CrossRef]
43. Henry, M.C.; Sutherland, A. Synthesis of Benzo[b]furans by Intramolecular C–O Bond Formation Using Iron and Copper Catalysis. *Org. Lett.* **2020**, *22*, 2766–2770. [CrossRef] [PubMed]
44. Liang, Z.; Hou, W.; Du, Y.; Zhang, Y.; Pan, Y.; Mao, D.; Zhao, K. Oxidative Aromatic C–O Bond Formation: Synthesis of 3-Functionalized Benzo[b]furans by FeCl$_3$-Mediated Ring Closure of -Aryl Ketones. *Org. Lett.* **2009**, *11*, 4978–4981. [CrossRef]
45. Lyons, T.W.; Sanford, M.S. Palladium-Catalyzed Ligand-Directed C–H Functionalization Reactions. *Chem. Rev.* **2010**, *110*, 1147–1169. [CrossRef] [PubMed]
46. Li, B.-J.; Shi, Z.-J. From C(sp2)–H to C(sp3)–H: Systematic studies on transition metal-catalyzed oxidative C–C formation. *Chem. Soc. Rev.* **2012**, *41*, 5588–5598. [CrossRef] [PubMed]
47. Guo, L.; Zhang, F.; Hu, W.; Li, L.; Jia, Y. Palladium-catalyzed synthesis of benzofurans via C–H activation/oxidation tandem reaction and its application to the synthesis of decursivine and serotobenine. *Chem. Commun.* **2014**, *50*, 3299–3302. [CrossRef] [PubMed]
48. Yang, D.; Zhu, Y.; Yang, N.; Jiang, Q.; Liua, R. One-Step Synthesis of Substituted Benzofurans from ortho-Alkenylphenols via Palladium-Catalyzed C–H functionalization. *Adv. Synth. Catal.* **2016**, *358*, 1731–1735. [CrossRef]
49. Singh, F.V.; Mangaonkar, S.R. Hypervalent iodine(III)-Catalyzed Synthesis of 2-Arylbenzofurans. *Synthesis* **2018**, *50*, 4940–4948. [CrossRef]
50. Huang, W.; Hung-Chang Liu, J.; Alayoglu, P.; Li, Y.; Witham, C.A.; Tsung, C.-K.; Toste, F.D.; Somorjai, G.A. Highly active heterogeneous palladium nanoparticle catalysts for homogeneous electrophilic reactions in solution and the utilization of a continuous flow reactor. *J. Am.Chem. Soc.* **2010**, *132*, 16771–16773. [CrossRef]
51. Maranon, L.A.; Martínez, M.M.; Sarandeses, L.A.; Gomez-Bengoa, E.; Sestel, J.P. Indium(III)-Catalyzed Synthesis of Benzo[b]furans by Intramolecular Hydroalkoxylation of ortho-Alkynylphenols: Scope and Mechanistic Insights. *J. Org. Chem.* **2018**, *83*, 7970–7980. [CrossRef]
52. Rong, Z.; Gao, K.; Zhou, L.; Lin, J.; Qian, G. Facile synthesis of 2-substituted benzo[b]furans and indoles by copper-catalyzed intramolecular cyclization of 2-alkynyl phenols and tosylanilines. *RSC Adv.* **2019**, *9*, 17975–17978. [CrossRef]
53. Boyer, A.; Isono, N.; Lackner, S.; Lautens, M. Domino rhodium(I)-catalysed reactions for the efficient synthesis of substituted benzofurans and indoles. *Tetrahedron* **2010**, *66*, 6468–6482. [CrossRef]
54. Sarbajna, A.; Pandey, P.; Rahaman, S.M.W.; Singh, K.; Tyagi, A.; Dixneuf, P.H.; Bera, J.K. A Triflamide-Tethered N-Heterocyclic Carbene–Rhodium(I)Catalyst for Hydroalkoxylation Reactions: Ligand-Promoted Nucleophilic Activation of Alcohols. *ChemCatChem* **2017**, *9*, 1397–1401. [CrossRef]
55. Rubio-Marqus, P.; Rivero-Crespo, M.A.; Leyva-Pérez, A.; Corma, A. Well-Defined Noble Metal Single Sites in Zeolites as an Alternative to Catalysis by Insoluble Metal Salts. *J. Am. Chem. Soc.* **2015**, *137*, 11832–11837. [CrossRef] [PubMed]
56. Morozov, O.S.; Lunchev, A.V.; Bush, A.A.; Tukov, A.A.; Asachenko, A.F.; Khrustalev, V.N.; Zalesskiy, S.S.; Ananikov, V.P.; Nechaev, M.S. Expanded-Ring N-Heterocyclic Carbenes Efficiently Stabilized Gold(I) Cations, Leading to High Activity in p-Acid-Catalyzed Cyclizations. *Chem. Eur. J.* **2014**, *20*, 6162–6170. [CrossRef]
57. Jacubert, M.; Hamze, A.; Provot, O.; Peyrat, J.-F.; Brion, J.-D.; Alami, M. p-Toluenesulfonic acid-mediated cyclization of o-(1-alkynyl)anisoles or thioanisoles: Synthesis of 2-arylsubstituted benzofurans and benzothiophenes. *Tetrahedron Lett.* **2009**, *50*, 3588–3592. [CrossRef]
58. Sun, S.-X.; Wang, J.-J.; Xu, Z.-J.; Cao, L.-Y.; Shi, Z.-F.; Zhang, H.-L. Highly efficient heterogeneous synthesis of benzofurans under aqueous condition. *Tetrahedron* **2014**, *70*, 3798–3806. [CrossRef]

59. Zhou, H.; Niu, J.-J.; Xu, J.-W.; Hu, S.-J. Novel Route to 2-Trifluoromethylated Benzofurans. *Synth. Commun.* **2009**, *39*, 716–732. [CrossRef]
60. Hirner, J.J.; Faizi, D.J.; Blum, S.A. Alkoxyboration: Ring-Closing Addition of B–O σ Bonds across Alkynes. *J. Am. Chem. Soc.* **2014**, *136*, 4740–4745. [CrossRef]
61. Issaian, A.; Tu, K.N.; Blum, S.A. Boron–Heteroatom Addition Reactions via Borylative Heterocyclization: Oxyboration, Aminoboration, and Thioboration. *Acc. Chem. Res.* **2017**, *50*, 2598–2609. [CrossRef]
62. Alcaide, B.; Almendros, P.; Busto, E.; Lázaro-Milla, C. Photoinduced Gold-Catalyzed Domino C(sp) Arylation/Oxyarylation of TMS-Terminated Alkynols with Arenediazonium Salts. *J. Org. Chem.* **2017**, *82*, 2177–2186. [CrossRef] [PubMed]
63. Rong, M.-G.; Qin, T.-Z.; Zi, W. Rhenium-Catalyzed Intramolecular Carboalkoxylation and Carboamination of Alkynes for the Synthesis of C3-Substituted Benzofurans and Indoles. *Org. Lett.* **2019**, *21*, 5421–5425. [CrossRef] [PubMed]
64. Furstner, A.; Davies, P.W. Heterocycles by PtCl$_2$-Catalyzed Intramolecular Carboalkoxylation or Carboamination of Alkynes. *J. Am. Chem. Soc.* **2005**, *127*, 15024–15025. [CrossRef] [PubMed]
65. Allegretti, P.A.; Huynh, K.; Ozumerzifon, T.J.; Ferreira, E.M. Lewis Acid Mediated Vinylogous Additions of Enol Nucleophiles into an α,β-Unsaturated Platinum Carbene. *Org. Lett.* **2016**, *18*, 64–67. [CrossRef]
66. Newman, S.G.; Aureggi, V.; Bryan, C.S.; Lautens, M. Intramolecular cross-coupling of *gem*-dibromoolefins: A mild approach to 2-bromo benzofused heterocycles. *Chem. Commun.* **2009**, 5236–5238. [CrossRef]
67. Kim, C.G.; Jun, J.G. An Efficient Total Synthesis of Mulberrofuran B and L. *Bull. Korean Chem. Soc.* **2015**, *36*, 2278–2283. [CrossRef]
68. Damodar, K.; Kim, J.K.; Jun, J.G. Unified syntheses of gramniphenols F and G, cicerfuran, morunigrol C and its derivative. *Tetrahedron Lett.* **2016**, *57*, 1183–1186. [CrossRef]
69. Rao, M.L.N.; Murty, V.N. Rapid Access to Benzofuran-Based Natural Products through a Concise Synthetic Strategy. *Eur. J. Org. Chem.* **2016**, 2177–2186. [CrossRef]
70. Rao, M.N.; Murty, V.N.; Nanda, S. Functional group manoeuvring for tuning stability and reactivity: Synthesis of cicerfuran, moracins (D, E, M) and chromene-fused benzofuran-based natural products. *Org. Biomol. Chem.* **2017**, *15*, 9415–9423. [CrossRef]
71. Liu, X.; Astruc, D. Development of the Applications of Palladium on Charcoal in Organic Synthesis. *Adv. Synth. Catal.* **2018**, *360*, 3426–3459. [CrossRef]
72. Savvidou, A.; Tzaras, D.I.; Koutoulogenis, G.S.; Theodorou, A.; Kokotos, C.G. Synthesis of Benzofuran and Indole Derivatives Catalyzed by Palladium on Carbon. *Eur. J. Org. Chem.* **2019**, 3890–3897. [CrossRef]
73. Lin, D.; Wang, L.; Yan, Z.Z.; Ye, J.; Hu, A.X.; Liao, H.D.; Liu, J.; Peng, J.M. Semi-synthesis, structural modification and biological evaluation of 5-arylbenzofuran neolignans. *RSC Adv.* **2018**, *8*, 34331–34342. [CrossRef]
74. Rehan, M.; Nallagonda, R.; Das, B.G.; Meena, T.; Ghorai, P. Synthesis of Functionalized Benzo[*b*]furans via Oxidative Cyclization of *o*-Cinnamyl Phenols. *J. Org. Chem.* **2017**, *82*, 3411–3424. [CrossRef] [PubMed]
75. Li, Y.; Qiu, D.; Gu, R.; Wang, J.; Shi, J.; Li, Y. Aryne 1,2,3-Trifunctionalization with Aryl Allyl Sulfoxides. *J. Am. Chem. Soc.* **2016**, *138*, 10814–10817. [CrossRef]
76. Sancheti, S.P.; Akram, M.O.; Roy, R.; Bedi, V.; Kundu, S.; Patil, N.T. ortho-Oxygenative 1,2-Difunctionalization of Diarylalkynes under Merged Gold/Organophotoredox Relay Catalysis. *Chem. Asian J.* **2019**, *14*, 4601–4606. [CrossRef]
77. Ruan, L.; Shi, M.; Mao, S.; Yu, L.; Yang, F.; Tang, J. An efficient approach to construct 2-arylbenzo[*b*]furans from 2-methoxychalcone epoxides. *Tetrahedron* **2014**, *70*, 1065–1070. [CrossRef]
78. Wang, X.; Liu, M.; Xu, L.; Wang, Q.; Chen, J.; Ding, J.; Wu, H. Palladium-Catalyzed Addition of Potassium Aryltrifluoroborates to Aliphatic Nitriles: Synthesis of Alkyl Aryl Ketones, Diketone Compounds, and 2-Arylbenzo[*b*]furans. *J. Org. Chem.* **2013**, *78*, 5273–5281. [CrossRef]
79. Wang, X.; Wang, X.; Liu, M.; Ding, J.; Chen, J.; Wu, H. Palladium-Catalyzed Reaction of Arylboronic Acids with Aliphatic Nitriles: Synthesis of Alkyl Aryl Ketones and 2-Arylbenzofurans. *Synthesis* **2013**, *45*, 2241–2244. [CrossRef]
80. Chen, J.; Li, J.; Su, W. Palladium-catalyzed tandem reaction of 2-hydroxyarylacetonitriles with sodium sulfinates: One-pot synthesis of 2-arylbenzofurans. *Org. Biomol. Chem.* **2014**, *12*, 4078–4083. [CrossRef]

81. Skillinghaug, B.; Skold, C.; Rydfjord, J.; Svensson, F.; Behrends, M.; Savmarker, J.; Sjoberg, P.J.; Larhed, M. Palladium(II)-Catalyzed Desulfitative Synthesis of Aryl Ketones from Sodium Arylsulfinates and Nitriles: Scope, Limitations, and Mechanistic Studies. *J. Org. Chem.* **2014**, *79*, 12018–12032. [CrossRef]
82. Du, J.-Y.; Ma, Y.-H.; Yuan, R.-Q.; Xin, N.; Nie, S.-Z.; Ma, C.-L.; Li, C.-Z.; Zhao, C.-Q. Metal-Free One-Pot Synthesis of 3-Phosphinoylbenzofurans via Phospha-Michael Addition/Cyclization of H-Phosphine Oxides and in Situ Generated ortho-Quinone Methides. *Org. Lett.* **2018**, *20*, 477–480. [CrossRef] [PubMed]
83. Reddy, C.R.; Krishna, G.; Kavitha, N.; Latha, B.; Shin, D.-S. Access to 2,3-Disubstituted Benzofurans through One-Pot Acid-Catalyzed Nucleophilic Substitution/TBAF-Mediated Oxacycloisomerization. *Eur. J. Org. Chem.* **2012**, 5381–5388. [CrossRef]
84. Mancuso, R.; Gabriele, B. The sequential homobimetallic catalysis concept applied to the synthesis of benzofuran. *Chem. Heterocycl. Compd.* **2014**, *50*, 160–170. [CrossRef]
85. Rajesh, M.; Thirupathi, N.; Reddy, T.J.; Kanojiya, S.; Reddy, M.S. Pd-Catalyzed Isocyanide Assisted Reductive Cyclization of 1-(2-Hydroxyphenyl)-propargyl Alcohols for 2-Alkyl/Benzyl Benzofurans and Their Useful Oxidative Derivatization. *J. Org. Chem.* **2015**, *80*, 12311–12320. [CrossRef]
86. Chen, D.Y.-K.; Kang, Q.; Wu, T.R. Modular Synthesis of Polyphenolic Benzofurans, and Application in the Total Synthesis of Malibatol A and Shoreaphenol. *Molecules* **2010**, *15*, 5909–5927. [CrossRef]
87. Kraus, G.A.; Gupta, V. A new synthetic strategy for the synthesis of bioactive stilbene dimers. A direct synthesis of amurensin H. *Tetrahedron Lett.* **2009**, *50*, 7180–7183. [CrossRef]
88. Liu, W.; Chen, N.; Yang, X.; Li, L.; Li, C.-J. Dehydrative condensation of carbonyls with non-acidic ethylenes enabled by light: Synthesis of benzofurans. *Chem. Commun.* **2016**, *52*, 13120–13123. [CrossRef]
89. Kanazawa, C.; Goto, K.; Terada, M. Phosphazene base-catalyzed intramolecular cyclization for efficient synthesis of benzofurans via carbon–carbon bond formation. *Chem. Commun.* **2009**, 5248–5250. [CrossRef]
90. Wang, L.; Zhang, Y.; Zhang, M.; Bao, P.; Lv, X.; Liu, H.-G.; Zhao, X.; Li, J.-S.; Luo, Z.; Wei, W. Metal-free I$_2$O$_5$-mediated oxidative synthesis of sulfonylated benzofurans through cyclization reaction of 1,6-enynes and arylsulfonylhydrazides. *Tetrahedron. Lett.* **2019**, *60*, 1845–1848. [CrossRef]
91. Zhang, P.; Wang, C.; Cui, M.; Du, M.; Li, W.; Jia, Z.; Zhao, Q. Synthesis of Difluoroalkylated Benzofuran, Benzothiophene, and Indole Derivatives via Palladium-Catalyzed Cascade Difluoroalkylation and Arylation of 1,6-Enynes. *Org. Lett.* **2020**, *22*, 1149–1154. [CrossRef]
92. Shen, Z.; Dong, V.M. Benzofurans Prepared by C-H Bond Functionalization with Acylsilanes. *Angew. Chem. Int. Ed.* **2009**, *48*, 784–786. [CrossRef] [PubMed]
93. Brook, A.G. Molecular rearrangements of organosilicon compounds. *Acc. Chem. Res.* **1974**, *7*, 77–84. [CrossRef]
94. Davies, H.M.L.; Alford, J.S. Reactions of metallocarbenes derived from N-sulfonyl-1, 2, 3-triazoles. *Chem. Soc. Rev.* **2014**, *41*, 5151–5162. [CrossRef] [PubMed]
95. Li, L.; Xia, X.-H.; Wang, Y.; Bora, P.P.; Kang, Q. Synthesis of Benzofurans via Tandem Rhodium-Catalyzed C(sp$_3$)-H Insertion and Copper-Catalyzed Dehydrogenation. *Adv. Synth. Catal.* **2015**, *357*, 2089–2097. [CrossRef]
96. Ma, X.; Wu, F.; Yi, X.; Wang, H.; Chen, W. One-pot synthesis of 2,3-disubstituted dihydrobenzofurans and benzofurans via rhodium-catalyzed intramolecular C–H insertion reaction. *Chem. Commun.* **2015**, *51*, 6862–6865. [CrossRef]
97. Kim, K.; Kim, I. Total Synthesis of Diptoindonesin G via a Highly Efficient Domino Cyclodehydration/Intramolecular Friedel-Crafts Acylation/Regioselective Demethylation Sequence. *Org. Lett.* **2010**, *12*, 5314–5317. [CrossRef]
98. Lee, J.H.; Kim, M.; Kim, I. Palladium-Catalyzed α-Arylation of Aryloxyketones for the Synthesis of 2,3-Disubstituted Benzofurans. *J. Org. Chem.* **2014**, *79*, 6153–6163. [CrossRef]
99. Kim, I.; Choi, J. versatile approach to oligostilbenoid natural products synthesis of permethylated analogues of viniferifuran, malibatol A, and shoreaphenol. *Org. Biomol. Chem.* **2009**, *7*, 2788–2795. [CrossRef]
100. Wang, H.-S.; Chan, C.-K.; Chang, M.-Y. Ga(OTf)3-mediated synthesis of substituted benzofurans. *Tetrahedron* **2016**, *72*, 5132–5141. [CrossRef]
101. Umareddy, P.; Arava, V.R. Facile synthesis of 3-aryl benzofurans, 3-aryl benzothiophenes, 2-aryl indoles and their dimers. *Synth. Commun.* **2019**, *49*, 2156–2167. [CrossRef]
102. Zhang, Q.; Luo, J.; Wang, B.; Xiao, X.; Gan, Z.; Tang, Q. Titanium tetrachloride promoted cyclodehydration of aryloxyketones: Facile synthesis of benzofurans and naphthofurans with high regioselectivity. *Tetrahedron Lett.* **2019**, *60*, 1337–1340. [CrossRef]

103. Shibata, T.; Hashimoto, Y.-K.; Otsuka, M.; Tsuchikama, K.; Endo, K. Ir(III)-Catalyzed Room-Temperature Synthesis of Multisubstituted Benzofurans Initiated by C–H Activation of α-Aryloxy Ketones. *Synlett* **2011**, *14*, 2075–2079. [CrossRef]
104. Qin, Y.; Zhan, J.-L.; Shan, T.-t.; Xu, T. Total synthesis of penta-Me amurensin H and diptoindonesin G featuring a Rh-catalyzed carboacylation/aromatization cascade enabled by C-C activation. *Tetrahedron Lett.* **2019**, *60*, 925–927. [CrossRef]
105. Chen, Y.-Y.; Chen, K.-L.; Tyan, Y.-C.; Liang, C.-F.; Lin, P.-C. Synthesis of substituted 3-methylene-2,3-dihydrobenzofurans and 3-methylbenzofurans by rhodium (II)-catalyzed annulation. *Tetrahedron* **2015**, *71*, 6210–6218. [CrossRef]
106. Tang, X.-Y.; Zhang, Y.-S.; He, L.; Wei, Y.; Shi, M. Intramolecular Annulation of Aromatic Rings with N-Sulfonyl 1,2,3-Triazoles: Divergent Synthesis of 3-Methylene-2,3-Dihydrobenzofurans and 3-Methylene-2,3-Dihydroindoles. *Chem. Commun.* **2015**, *51*, 133–136. [CrossRef] [PubMed]
107. Geary, L.M.; Hultin, P.G. Modular Construction of 2-Substituted Benzo[b]furans from 1,2-Dichlorovinyl Ethers. *Org. Lett.* **2009**, *11*, 5478–5481. [CrossRef]
108. Geary, L.M.; Hultin, P.G. 2-Substituted Benzo[b]furans from (E)-1,2-Dichlorovinyl Ethers and Organoboron Reagents: Scope and Mechanistic Investigations into the One-Pot Suzuki Coupling/Direct Arylation. *Eur. J. Org. Chem.* **2010**, 5563–5573. [CrossRef]
109. Li, C.; Zhang, Y.; Li, P.; Wang, L. Palladium-Catalyzed Oxidative Cyclization of 3-Phenoxyacrylates: An Approach To Construct Substituted Benzofurans from Phenols. *J. Org. Chem.* **2011**, *76*, 4692–4696. [CrossRef]
110. Deng, G.; Li, M.; Yu, K.; Liu, C.; Liu, Z.; Duan, S.; Chen, W.; Yang, X.; Zhang, H.; Walsh, P.J. Synthesis of Benzofuran Derivatives through Cascade Radical Cyclization/Intermolecular Coupling of 2-Azaallyls. *Angew. Chem. Int. Ed.* **2019**, *58*, 2826–2830. [CrossRef]
111. Quan, Y.; Xie, Z. Iridium Catalyzed Regioselective Cage Boron Alkenylation of o-Carboranes via Direct Cage B–H Activation. *J. Am. Chem. Soc.* **2014**, *136*, 15513–15516. [CrossRef]
112. Kumar, N.Y.P.; Bechtoldt, A.; Raghuvanshi, K.; Ackermann, L. Ruthenium(II)-Catalyzed Decarboxylative C–H Activation: Versatile Routes to *meta*-Alkenylated Arenes. *Angew. Chem. Int. Ed.* **2016**, *55*, 6929–6932. [CrossRef] [PubMed]
113. Zhang, J.; Shrestha, R.; Hartwig, J.F.; Zhao, P. A decarboxylative approach for regioselective hydroarylation of alkynes. *Nat. Chem.* **2016**, *8*, 1144–1151. [CrossRef] [PubMed]
114. Mandal, A.; Sahoo, H.; Dana, S.; Baidya, M. Ruthenium(II)-Catalyzed Hydroarylation of Maleimides Using Carboxylic Acids as a Traceless Directing Group. *Org. Lett.* **2017**, *19*, 4138–4141. [CrossRef] [PubMed]
115. Kim, K.; Vasu, D.; Im, H.; Hong, S. Palladium(II)-Catalyzed Tandem Synthesis of Acenes Using Carboxylic Acids as Traceless Directing Groups. *Angew. Chem. Int. Ed.* **2016**, *55*, 8652–8655. [CrossRef] [PubMed]
116. Yu, S.; Lv, N.; Liu, Z.; Zhang, Y. Cu(II)-Mediated C-C/C-O Bond Formation via C-H/C-C Bond Cleavage: Access to Benzofurans Using Amide as a Traceless Directing Group. *Adv. Synth. Catal.* **2020**, *362*, 118–125. [CrossRef]
117. Kalvacherla, B.; Batthula, S.; Balasubramanian, S.; Palakodety, R.K. Transition-Metal-Free Cyclization of Propargylic Alcohols with Aryne: Synthesis of 3-Benzofuranyl-2-oxindole and 3-Spirooxindole Benzofuran Derivatives. *Org. Lett.* **2018**, *20*, 3824–3828. [CrossRef]
118. Yoshioka, E.; Tanaka, H.; Kohtani, S.; Miyabe, H. Straightforward Synthesis of Dihydrobenzofurans and Benzofurans from Arynes. *Org. Lett.* **2013**, *15*, 3938–3941. [CrossRef]
119. Singh, G.; Goswami, P.; Sharma, S.; Anand, R.V. A One-Pot Approach to 2,3-Diarylbenzo[b]furans through N-Heterocyclic Carbene-Catalyzed 1,6-Conjugate Addition Followed by Acid Mediated Dehydrative Annulation. *J. Org. Chem.* **2018**, *83*, 10546–10554. [CrossRef]
120. Liou, Y.-C.; Karanam, P.; Jang, Y.-J.; Lin, W. Synthesis of Functionalized Benzofurans from para-Quinone Methides via Phospha-1,6-Addition/O-Acylation/Wittig Pathway. *Org. Lett.* **2019**, *21*, 8008–8012. [CrossRef]
121. Mattson, A.E.; Scheid, K.A. Nucleophilic Acylation of o-Quinone Methides: An Umpolung Strategy for the Synthesis of α-Aryl Ketones and Benzofurans. *J. Am. Chem. Soc.* **2007**, *129*, 4508–4509. [CrossRef]
122. Rokita, S.E. *Quinone Methides*; Wiley: Hoboken, NJ, USA, 2009; p. 464.
123. Osyanin, V.A.; Osipov, D.V.; Demidov, M.R.; Klimochkin, Y.N. Potassium Trinitromethanide as a 1,1-Ambiphilic Synthon Equivalent: Access to 2-Nitroarenofurans. *J. Org. Chem.* **2014**, *79*, 1192–1198. [CrossRef] [PubMed]

124. Schevenels, F.; Marko, I.E. Efficient and Connective Assembly of Highly Functionalized Benzofurans Using o-Hydroxyphenones and Dichloroethylene. *Org. Lett.* **2012**, *14*, 1298–1301. [CrossRef] [PubMed]
125. Syu, S.; Lee, Y.-T.; Jang, Y.-J.; Lin, W. Preparation of Functional Benzofurans, Benzothiophenes, and Indoles Using Ester, Thioester, and Amide via Intramolecular Wittig Reactions. *Org. Lett.* **2011**, *13*, 2971–2973. [CrossRef] [PubMed]
126. Ramazani, A.; Mahyari, A.T.; Rouhani, M.; Rezaei, A. A novel three-component reaction of a secondary amine and a 2-hydroxybenzaldehyde derivative with an isocyanide in the presence of silica gel: An efficient one-pot synthesis of benzo[b]furan derivatives. *Tetrahedron Lett.* **2009**, *50*, 5625–5627. [CrossRef]
127. Sontakke, G.S.; Pal, K.; Volla, C.M.R. Rh(II)-Catalyzed Denitrogenative Transannulation of N-Sulfonyl-1,2,3-triazolyl Cyclohexadienones for the Synthesis of Benzofurans and Cyclopropa[cd]indole-carbaldehydes. *J. Org. Chem.* **2019**, *84*, 12198–12208. [CrossRef] [PubMed]
128. Moure, M.; SanMartin, J.R.; Dominguez, E. Benzofurans from Benzophenones and Dimethylacetamide: Copper-Promoted Cascade Formation of Furan O1-C2 and C2–C3 Bonds Under Oxidative Conditions. *Angew. Chem. Int. Ed.* **2012**, *51*, 3220–3224. [CrossRef]
129. Yang, F.; Rauch, K.; Kettelhoit, K.; Ackermann, L. Aldehyde-Assisted Ruthenium(II)-Catalyzed C-H Oxygenations. *Angew. Chem. Int. Ed.* **2014**, *53*, 11285–11288. [CrossRef]
130. Zhou, L.; Shi, Y.; Xiao, Q.; Liu, Y.; Ye, F.; Zhang, Y.; Wang, J. CuBr-Catalyzed Coupling of N-Tosylhydrazones and Terminal Alkynes: Synthesis of Benzofurans and Indoles. *Org. Lett.* **2011**, *13*, 968–971. [CrossRef]
131. Xiao, Q.; Xia, Y.; Li, H.; Zhang, Y.; Wang, J. Coupling of N-Tosylhydrazones with Terminal Alkynes Catalyzed by Copper(I): Synthesis of Trisubstituted Allenes. *Angew. Chem. Int. Ed.* **2011**, *50*, 1114–1117. [CrossRef]
132. Sivaraman, A.; Harmalkar, D.S.; Kang, J.; Choi, Y.; Lee, K. A protecting group-free divergent synthesis of natural benzofurans via one-pot synthesis of 2-bromo-6-hydroxybenzofurans. *Org. Biomol. Chem.* **2019**, *17*, 2153–2161. [CrossRef]
133. Makarov, A.S.; Kekhvaeva, A.E.; Chalikidi, P.N.; Abaev, V.T.; Trushkov, I.V.; Uchuskin, M.G. A Simple Synthesis of Densely Substituted Benzofurans by Domino Reaction of 2-Hydroxybenzyl Alcohols with 2-Substituted Furans. *Synthesis* **2019**, *51*, 3747–3757. [CrossRef]
134. Kim, H.J.; Seo, J.W.; Lee, M.H.; Shin, D.S.; Kim, H.B.; Cho, H.; Lee, B.S.; Chi, D.Y. New synthetic method for benzofurans from 2-(cyanomethyl)phenyl derivatives. *Tetrahedron* **2012**, *68*, 3942–3947. [CrossRef]
135. Murai, M.; Okamoto, K.; Miki, K.; Ohe, K. Palladium-catalyzed three-component coupling reactions of 2-(cyanomethyl)phenol, aryl halides, and carbon monoxide. *Tetrahedron* **2015**, *71*, 4432–4437. [CrossRef]
136. Shea, K.M. *Palladium in Heterocyclic Chemistry*, 2nd ed.; Elsevier: Amsterdam, The Netherlands, 2007; p. 303.
137. Heravi, M.M.; Sadjadi, S. Recent advances in the application of the Sonogashira method in the synthesis of heterocyclic compounds. *Tetrahedron* **2009**, *65*, 7761–7775. [CrossRef]
138. Wang, J.-R.; Manabe, K. Hydroxyterphenylphoshine-Palladium Catalyst for Benzo[b]furan Synthesis from 2-Chlorophenols. Bifunctional Ligand Strategy for Cross-Coupling of Chloroarenes. *J. Org. Chem.* **2010**, *75*, 5340–5353. [CrossRef]
139. Yamaguchi, M.; Ozawa, H.; Katsumata, H.; Akiyama, T.; Manabe, K. One-pot synthesis of 2,3-disubstituted benzofurans from 2-chlorophenols using palladium–dihydroxyterphenylphosphine catalyst. *Tetrahedron Lett.* **2018**, *59*, 3175–3178. [CrossRef]
140. Bosiak, M.J. A Convenient Synthesis of 2-Arylbenzo[b]furans from Aryl Halides and 2-Halophenols by Catalytic One-Pot Cascade Method. *ACS Catal.* **2016**, *6*, 2429–2434. [CrossRef]
141. Tréguier, B.; Rasolofonjatovo, E.; Hamze, A.; Provot, O.; Wdzieczak-Bakala, J.; Dubois, J.; Brion, J.-D.; Alami, M. Synthesis of 2-(1-phenylvinyl)benzofurans and 2-(1-phenylvinyl)indoles as antimitotic agents by a tandem palladium-assisted coupling-cyclization reaction between 1-phenylvinyl iodides and ortho-substituted arylalkynes. *Eur. J. Org. Chem.* **2011**, 4868–4876. [CrossRef]
142. Byers, P.M.; Rashid, J.I.; Mohamed, R.K.; Alabugin, I.V. Polyaromatic Ribbon/Benzofuran Fusion via Consecutive Endo Cyclizations of Enediynes. *Org. Lett.* **2012**, *14*, 6032–6035. [CrossRef]
143. Song, T.; Yang, Y. Metal Nanoparticles Supported on Biomass-Derived Hierarchical Porous Heteroatom-Doped Carbon from Bamboo Shoots Design, Synthesis and Applications. *Chem. Rec.* **2019**, *19*, 1283–1301. [CrossRef]
144. Ji, G.; Duan, Y.; Zhang, S.; Yang, Y. Synthesis of benzofurans from terminal alkynes and iodophenols catalyzed by recyclable palladium nanoparticles supported on N,O-dual doped hierarchical porous carbon under copper- and ligand-free conditions. *Catalysis Today* **2019**, *330*, 101–108. [CrossRef]

145. Saha, D.; Dey, R.; Ranu, B.C. A Simple and Efficient One-Pot Synthesis of Substituted Benzo[b]furans by Sonogashira Coupling–5-*endo-dig* Cyclization Catalyzed by Palladium Nanoparticles in Water Under Ligand- and Copper-Free Aerobic Conditions. *Eur. J. Org. Chem.* **2010**, 6067–6071. [CrossRef]
146. Markina, N.A.; Chen, Y.; Larock, R.C. Efficient microwave-assisted one-pot three-component synthesis of 2,3-disubstituted benzofurans under Sonogashira conditions. *Tetrahedron* **2013**, *69*, 2701–2713. [CrossRef] [PubMed]
147. Vo, D.D.; Elofsson, M. Synthesis of 4-Formyl-2-arylbenzofuran Derivatives by PdCl(C_3H_5)dppb-Catalyzed Tandem Sonogashira Coupling-Cyclization under Microwave Irradiation- Application to the Synthesis of Viniferifuran Analogues. *Chem. Sel.* **2017**, *2*, 6245–6248. [CrossRef]
148. Chang, C.-W.; Chein, R.-J. Absolute Configuration of Anti-HIV-1 Agent (-)-Concentricolide: Total Synthesis of (+)-(R)-Concentricolide. *J. Org. Chem.* **2011**, *76*, 4154–4157. [CrossRef] [PubMed]
149. Thomas, A.M.; Asha, S.; Menon, R.; Anilkumar, G. One-Pot Synthesis of Benzofurans via Cu–Catalyzed Tandem Sonogashira coupling-Cyclization Reactions. *Chemistry Select* **2019**, *4*, 5544–5547. [CrossRef]
150. Mertens, M.A.S.; Thomas, F.; Noth, M.; Moegling, J.; El-Awaad, I.; Sauer, D.F.; Dhoke, G.V.; Xu, W.; Pich, A.; Herres-Pawlis, S.; et al. One-Pot Two-Step Chemoenzymatic Cascade for the Synthesis of a Bis-benzofuran Derivative. *Eur. J. Org. Chem.* **2019**, 6341–6346. [CrossRef]
151. Batail, N.; Bendjeriou, A.; Lomberget, T.; Barret, R.; Dufaud, V.; Djakovitch, L. First Heterogeneous Ligand-and Salt-Free Larock Indole Synthesis. *Adv. Synth. Catal.* **2009**, *351*, 2055–2062. [CrossRef]
152. Singh, C.; Prakasham, A.P.; Ghosh, P. Palladium Acyclic Diaminocarbene (ADC) Triflate Complexes as Effective Precatalysts for the Hiyama Alkynylation/Cyclization Reaction Yielding Benzofuran Compounds: Probing the Influence of the Triflate Co-Ligand in the One-Pot Tandem Reaction. *Chem. Sel.* **2019**, *4*, 329–336. [CrossRef]
153. Yuan, H.; Bi, K.-J.; Li, B.; Yue, R.-C.; Ye, J.; Shen, Y.-H.; Shan, L.; Jin, H.-Z.; Sun, Q.-Y.; Zhang, W.-D. Construction of 2-Substituted-3-Functionalized Benzofurans via Intramolecular Heck Coupling: Application to Enantioselective Total Synthesis of Daphnodorin B. *Org. Lett.* **2013**, 4742–4745. [CrossRef]
154. Vasconcelos, S.N.S.; de Oliveira, I.M.; Shamim, A.; Zukerman-Schpector, J.; Pimenta, D.C.; Stefani, H.A. Stereoselective Oxa-Michael Addition of Tyrosine to Propargyl Aldehyde/Esters: Formation of Benzofurans and Flavones. *Adv. Synth. Catal.* **2019**, *361*, 4243–4254. [CrossRef]
155. Wan, J.-P.; Wang, H.; Liu, Y.; Ding, H. Synthesis of 2-Vinylbenzofurans via the Copper-Catalyzed Multicomponent Reactions Involving an Oxa-Michael/Arylation/Vinylation Cascade. *Org. Lett.* **2014**, *16*, 5160–5163. [CrossRef] [PubMed]
156. Tabolin, A.A.; Ioffe, S.L. Rearrangement of N-Oxyenamines and Related Reactions. *Chem. Rev.* **2014**, *114*, 5426–5476. [CrossRef] [PubMed]
157. Contiero, F.; Jones, K.M.; Matts, E.A.; Porzelle, A.; Tomkinson, N.C.O. Direct Preparation of Benzofurans from O-Arylhydroxylamines. *Synlett* **2009**, 3003–3006. [CrossRef]
158. Maimone, T.J.; Buchwald, S.L. Pd-Catalyzed O-Arylation of Ethyl Acetohydroximate: Synthesis of O-Arylhydroxylamines and Substituted Benzofurans. *J. Am. Chem. Soc.* **2010**, *132*, 9990–9991. [CrossRef]
159. Ghosh, R.; Stridfeldt, E.; Olofsson, B. Metal-Free One-Pot Synthesis of Benzofurans. *Chem. Eur. J.* **2014**, *20*, 8888–8892. [CrossRef]
160. Li, M.; Wang, J.-H.; Li, W.; Lin, C.-D.; Zhang, L.-B.; Wen, L.-R. N-Phenoxyamides as Multitasking Reagents: Base-Controlled Selective Construction of Benzofurans or Dihydrobenzofuro[2,3-d]oxazoles. *J. Org. Chem.* **2019**, *84*, 8523–8530. [CrossRef]
161. Yan, D.; Jiang, H.; Sun, W.; Wei, W.; Zhao, J.; Zhang, X.; Wu, Y.-D. Synthesis of Benzofurans and Benzoxazoles through a [3,3]- Sigmatropic Rearrangement: O−NHAc as a Multitasking Functional Group. *Org. Process Res. Dev.* **2019**, *23*, 1646–1653. [CrossRef]
162. Alcaide, B.; Almendros, P.; Quirs, M.T.; Lopez, R.; Menndez, M.I.; Sochacka-C'wikła, A. Unveiling the Reactivity of Propargylic Hydroperoxides under Gold Catalysis. *J. Am. Chem. Soc.* **2013**, *135*, 898–905. [CrossRef]
163. Chen, W.; Liu, F.-X.; Gong, W.; Zhou, Z.; Gao, H.; Shi, J.; Wu, B.; Yi, W. Hydroxyl Group-Prompted and Iridium(III)-Catalyzed Regioselective C–H Annulation of N-phenoxyacetamides with Propargyl Alcohols. *Adv. Synth. Catal.* **2018**, *360*, 2470–2475. [CrossRef]

164. Yi, W.; Chen, W.; Liu, F.-X.; Zhong, Y.; Wu, D.; Zhou, Z.; Gao, H. Rh(III)-Catalyzed and Solvent-Controlled Chemoselective Synthesis of Chalcone and Benzofuran Frameworks via Synergistic Dual Directing Groups Enabled Regioselective C–H Functionalization: A Combined Experimental and Computational Study. *ACS Catal.* **2018**, *8*, 9508–9519. [CrossRef]
165. Pan, J.-L.; Liu, C.; Chen, C.; Liu, T.-Q.; Wang, M.; Sun, Z.; Zhang, S.-Y. Dual Directing-Groups-Assisted Redox-Neutral Annulation and Ring Opening of N-Aryloxyacetamides with 1-Alkynylcyclobutanols via Rhodium(III)-Catalyzed C–H/C–C Activations. *Org. Lett.* **2019**, *21*, 2823–2827. [CrossRef] [PubMed]
166. Li, W.-T.; Nan, W.-H.; Luo, Q.-L. Metal-free sequential reaction via a propargylation, annulation and isomerization sequence for the one-pot synthesis of 2,3-disubstituted benzofurans. *RSC Adv.* **2014**, *4*, 34774–34779. [CrossRef]
167. Baire, B.; Tharra, P. Regioselective, Cascade [3 + 2] Annulation of β-Naphthols (Resorcinols) with Z-enoate Propargylic alcohols: A Novel entry into Complex Naphtho(*benzo*)furans. *Chem. Comm.* **2016**, *52*, 14290–14293. [CrossRef]
168. Zhang, D.; Man, J.; Chen, Y.; Yin, L.; Zhong, J.; Zhang, Q.-F. Synthesis of poly-functionalized benzofurans via one-pot domino oxidation/[3 + 2] cyclization reactions of a hydroquinone ester and ynamides. *RSC Adv.* **2019**, *9*, 12567. [CrossRef]
169. Kuram, M.R.; Bhanuchandra, M.; Sahoo, A.K. Direct Access to Benzo[b]furans through Palladium-Catalyzed Oxidative Annulation of Phenols and Unactivated Internal Alkynes. *Angew. Chem. Int. Ed.* **2013**, *125*, 4705–4710. [CrossRef]
170. Wang, S.; Li, P.; Yu, L.; Wang, L. Sequential and one-pot reactions of phenols with bromoalkynes for the synthesis of (Z)-2-bromovinyl phenylethers and benzo[b]furans. *Org. Lett.* **2011**, *13*, 5968–5971. [CrossRef]
171. Maji, A.; Reddi, Y.; Sunoj, R.B.; Maiti, D. Mechanistic Insights on Orthogonal Selectivity in Heterocycle Synthesis. *ACS Catal.* **2018**, *8*, 10111–10118. [CrossRef]
172. Agasti, S.; Maity, S.; Szabo, K.J.; Maiti, D. Palladium-Catalyzed Synthesis of 2,3-Disubstituted Benzofurans: An Approach Towards the Synthesis of Deuterium Labeled Compounds. *Adv. Synth. Catal.* **2015**, *357*, 2331–2338. [CrossRef]
173. Agasti, S.; Sharma, U.; Naveen, T.; Maiti, D. Orthogonal selectivity with cinnamic acids in 3-substituted benzofuran synthesis through C–H olefination of phenols. *Chem. Commun.* **2015**, *51*, 5375–5378. [CrossRef]
174. Yorimitsu, H. Cascades of Interrupted Pummerer Reaction-Sigmatropic Rearrangement. *Chem. Rec.* **2017**, *17*, 1156–1167. [CrossRef] [PubMed]
175. Kobatake, T.; Fujino, D.; Yoshida, S.; Yorimitsu, H.; Oshima, K. Synthesis of 3-Trifluoromethylbenzo[b]furans from Phenols via Direct *Ortho* Functionalization by Extended Pummerer Reaction. *J. Am. Chem. Soc.* **2010**, *132*, 11838–11840. [CrossRef] [PubMed]
176. Ookubo, Y.; Wakamiya, A.; Yorimitsu, H.; Osuka, A. Synthesis of a Library of Fluorescent 2-Aryl-3-trifluoromethylnaphthofurans from Naphthols by Using a Sequential Pummerer-Annulation/Cross-Coupling Strategy and their Photophysical Properties. *Chem. Eur. J.* **2012**, *18*, 12690–12697. [CrossRef] [PubMed]
177. Murakami, K.; Yorimitsu, H.; Osuka, A. Two-Step, Practical, and Diversity-Oriented Synthesis of Multisubstituted Benzofurans from Phenols through Pummerer Annulation Followed by Cross-coupling. *Bull. Chem. Soc. Jpn.* **2014**, *87*, 1349–1366. [CrossRef]
178. Murakami, K.; Yorimitsu, H.; Osuka, A. Practical, Modular, and General Synthesis of Benzofurans through Extended Pummerer Annulation/Cross-Coupling Strategy. *Angew. Chem. Int. Ed.* **2014**, *53*, 7510–7513. [CrossRef] [PubMed]
179. Okamoto, K.; Hori, M.; Yanagi, T.; Murakami, K.; Nogi, K.; Yorimitsu, H. Sigmatropic Dearomatization/Defluorination Strategy for C-F Transformation: Synthesis of Fluorinated Benzofurans from Polyfluorophenols. *Angew. Chem. Int. Ed.* **2018**, *57*, 14230–14234. [CrossRef]
180. Yang, K.; Pulis, A.P.; Perry, G.J.P.; Procter, D.J. Transition-Metal-Free Synthesis of C3-Arylated Benzofurans from Benzothiophenes and Phenols. *Org. Lett.* **2018**, *20*, 7498–7503. [CrossRef]
181. Tramutola, F.; Chiummiento, L.; Funicello, M.; Lupattelli, P. Practical and efficient ipso-iodination of arylboronic acids via KF/I_2 System. *Tetrahedron Lett.* **2015**, *56*, 1122–1123. [CrossRef]
182. Wang, B.; Zhang, Q.; Luo, J.; Gan, Z.; Jiang, W.; Tang, Q. One-Step Regioselective Synthesis of Benzofurans from Phenols and α-Haloketones. *Molecules* **2019**, *24*, 2187. [CrossRef]

183. Bankar, S.K.; Mathew, J.; Ramasastry, S.S.V. Synthesis of benzofurans via an acid catalyzed transacetalisation/Fries-type O–C rearrangement/Michael addition/ring-opening aromatisation cascade of β-pyrones. *Chem. Commun.* **2016**, *52*, 5569–5572. [CrossRef]
184. Porcu, S.; Demuro, S.; Luridiana, A.; Cocco, A.; Frongia, A.; Aitken, D.J.; Charnay-Pouget, F.; Guillot, R.; Sarais, G.; Secci, F. Brønsted Acid Mediated Cascade Reaction To Access 3-(2-Bromoethyl)benzofurans. *Org. Lett.* **2018**, *20*, 7699–7702. [CrossRef] [PubMed]
185. Lee, D.-H.; Kwon, K.-H.; Yi, C.S. Dehydrative C–H Alkylation and Alkenylation of Phenols with Alcohols: Expedient Synthesis for Substituted Phenols and Benzofurans. *J. Am. Chem. Soc.* **2012**, *134*, 7325–7328. [CrossRef] [PubMed]
186. Honey, M.A.; Blake, A.J.; Campbell, I.B.; Judkins, B.D.; Moody, C.J. One-pot synthesis of N-methylindoles from N-methylanilines and of benzofurans from phenols using transition-metal carbene X–H insertion reactions. *Tetrahedron* **2009**, *65*, 8995–9001. [CrossRef]
187. Sun, P.; Gao, S.; Yang, C.; Guo, S.; Lin, A.; Yao, H. Controllable Rh(III)-Catalyzed Annulation between Salicylaldehydes and Diazo Compounds: Divergent Synthesis of Chromones and Benzofurans. *Org. Lett.* **2016**, *18*, 6464–6467. [CrossRef] [PubMed]
188. Sakhabutdinova, G.N.; Raskil'dina, G.Z.; Zlotskii, S.S.; Sultanova, R.M. Rhodium(II)-Catalyzed Reaction of Salicylaldehyde and Its Derivatives with Diazocarbonyl Compounds. *Russ. J. Organic Chem.* **2018**, *54*, 1772–1776. [CrossRef]
189. Wu, F.; Bai, R.; Gu, Y. Synthesis of Benzofurans from Ketones and 1,4-Benzoquinones. *Adv. Synth. Catal.* **2016**, *358*, 2307–2316. [CrossRef]
190. Cui, H.-L.; Deng, H.-Q.; Lei, J.-J. Metal-free one-pot synthesis of benzofurans with ynones and quinones through aza-Michael/Michael/annulation sequence. *Tetrahedron* **2017**, *73*, 7282–7290. [CrossRef]
191. Maeno, Z.; Yamamoto, M.; Mitsudome, T.; Mizugaki, T.; Jitsukawa, K. Efficient Synthesis of Benzofurans via Cross-Coupling of Catechols with Hydroxycoumarins Using O_2 as an Oxidant Catalyzed by $AlPO_4$-Supported Rh Nanoparticle. *Chem. Sel.* **2019**, *4*, 11394–11397. [CrossRef]
192. Zhang, C.; Zhen, L.; Yao, Z.; Jiang, L. Iron(III)-Catalyzed Domino Claisen Rearrangement/Regio- and Chemoselective Aerobic Dehydrogenative Cyclization of β-Naphthyl-Substituted-Allenylmethyl Ether. *Org. Lett.* **2019**, *21*, 955–959. [CrossRef]
193. Tanabe, Y.; Mitarai, K.; Higashi, T.; Misaki, T.; Nishii, Y. Efficient one-step synthesis of trialkylsubstituted 2(5H)-furanones utilizing direct Ti-crossed aldol condensation and its application to the straightforward synthesis of (R)-mintlactone and (R)-menthofuran. *Chem. Commun.* **2002**, 2542–2543. [CrossRef]
194. Rashid, S.; Bhat, B.A.; Mehta, G. A vicarious, one-pot synthesis of benzo- and naphthofurans: Applications to the syntheses of stereumene B and paeoveitols. *Tetrahedron Lett.* **2019**, *60*, 1122–1125. [CrossRef]
195. Qiang, S.; Haixuan, L. De novo synthesis of benzofurans via trifluoroacetic acid catalyzed cyclization/oxidative aromatization cascade reaction of 2-hydroxy-1,4-diones. *Org. Biomol. Chem.* **2019**, *17*, 7547–7551. [CrossRef]
196. Huang, W.; Xu, J.; Liu, C.; Chen, Z.; Gu, Y. Lewis Acid-Catalyzed Synthesis of Benzofurans and 4,5,6,7-Tetrahydrobenzofurans from Acrolein Dimer and 1,3-Dicarbonyl Compounds. *J. Org. Chem.* **2019**, *84*, 2941–2950. [CrossRef] [PubMed]
197. Zhang, L.; Cao, T.; Jiang, H.; Zhu, S. Deconstructive Reorganization: De Novo Synthesis of Hydroxylated Benzofuran. *Angew. Chem. Int. Ed.* **2020**, *59*, 4670–4677. [CrossRef]

© 2020 by the authors. Licensee MDPI, Basel, Switzerland. This article is an open access article distributed under the terms and conditions of the Creative Commons Attribution (CC BY) license (http://creativecommons.org/licenses/by/4.0/).

Review

An Overview of Saturated Cyclic Ethers: Biological Profiles and Synthetic Strategies

Qili Lu [1], Dipesh S. Harmalkar [1,2], Yongseok Choi [2] and Kyeong Lee [1,*]

[1] College of Pharmacy, Dongguk University-Seoul, Goyang 10326, Korea; luqili220@gmail.com (Q.L.); dsharmalkar@gmail.com (D.S.H.)
[2] College of Life Sciences and Biotechnology, Korea University, Seoul 02841, Korea; ychoi@korea.ac.kr
* Correspondence: kaylee@dongguk.edu

Received: 23 September 2019; Accepted: 19 October 2019; Published: 21 October 2019

Abstract: Saturated oxygen heterocycles are widely found in a broad array of natural products and other biologically active molecules. In medicinal chemistry, small and medium rings are also important synthetic intermediates since they can undergo ring-opening and -expansion reactions. These applications have driven numerous studies on the synthesis of oxygen-containing heterocycles and considerable effort has been devoted toward the development of methods for the construction of saturated oxygen heterocycles. This paper provides an overview of the biological roles and synthetic strategies of saturated cyclic ethers, covering some of the most studied and newly discovered related natural products in recent years. This paper also reports several promising and newly developed synthetic methods, emphasizing 3–7 membered rings.

Keywords: saturated oxygen heterocycles; cyclic ethers; total synthesis

---

1. Introduction

Constituting more than half of all the known organic compounds, heterocyclic compounds play an important role in organic chemistry. Among these, saturated cyclic ethers are abundant, appearing in a large number of biologically active natural products and pharmaceutically active compounds. The many FDA-approved cyclic ether rings containing therapeutic compounds (Figure 1) suggest and are evidence that cyclic ethers are significant motifs during the development of potential drug molecules. They have also been frequently found as key structural units in synthetic pharmaceuticals and agrochemicals. Additionally, a large number of natural products containing cyclic ethers have a wide range of interesting biological activities. For example, thousands of marine products that have oxacyclic moieties are isolated each year, providing rich sources for new drug candidates [1].

Over previous decades, considerable efforts have been devoted to the development of simple and efficient methods for constructing saturated oxygen heterocycles [2]. Given that most natural products occur as single enantiomers, and that chiral drugs on the market are regulated to be single enantiomers, special attention has been devoted to the asymmetric synthesis of heterocyclic compounds, as they play fundamental biological roles.

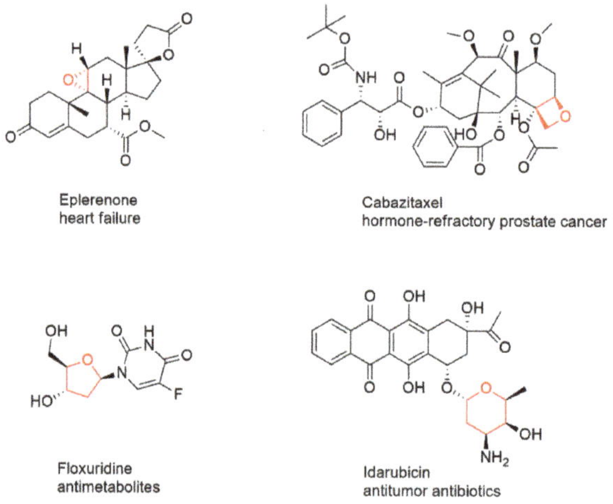

Figure 1. Structure and usage of FDA-approved drugs containing cyclic ether rings.

Here, we attempted to provide an overview of saturated oxygen heterocycles. It would be an impossible endeavor to compose a comprehensive review of all the great achievements that have been made in this field. Therefore, we have collated the activities of some of the most heavily studied and newly discovered related natural products with the aim of showing the biological profiles of cyclic ethers. Several total synthesis methods are given as examples to show the general synthetic strategies used to generate cyclic ethers. We then glance at recent advances in the synthesis of cyclic ethers within the last 5 years that may be applied widely in organic synthesis in the future.

2. Epoxides

Epoxides are common in natural products and present a wide range of biological activities [3,4]. Synthetically, epoxides are very versatile intermediates [5]. Synthetic organic chemists can take advantage of regio- and stereoselective ring openings to easily convert epoxides into diols, amino alcohols, ethers, etc. [6]. Therefore, the formation of enantiomerically pure epoxides is an essential step in the asymmetric synthesis of organic chiral compounds.

2.1. Natural Epoxides Containing Products and Biological Activities

Triptolide (TPL) **1** (Figure 2) is the major active component in an epoxy-diterpene structure; it is isolated from *Tripterygium wilfordii* Hook. f. (TWHF), a vine-like plant widely distributed throughout Eastern and Southern China [7–10]. In Chinese traditional herbal medicine, the crude root extracts of TWHF have been used for centuries to treat autoimmune and inflammatory diseases such as rheumatoid arthritis and lupus erythematosus. TPL has also been recognized as a potential drug for a variety of cancers [11–13]. Recent research on TPL has been focused on mechanisms of action. Hu et al.'s studies showed that TPL significantly inhibited the growth of COC1/DDP cells ($p < 0.05$) at a low concentration of 3 ng/mL [14]. Animal results indicated that TPL + DDP significantly enhanced the inflammatory factor-2 (IL-2) and tumor necrosis factor-α (TNF-α) in serum of mice [15]. Song et al. [16] observed that TPL suppresses the growth of lung cancer cells by targeting hyaluronan-CD44/RHAMM signaling. Gao et al. [17] reported that TPL induces the proliferation and apoptosis of MCF-7 breast cancer cells, potentially via autophagy and p38/Erk/mTOR phosphorylation. Minnelide, a more water-soluble synthetic analogue of TPL that is converted to TPL in vivo [18] has entered Phase II clinical trials for pancreatic cancer [19]. Triptolide is one of the most promising phytochemicals.

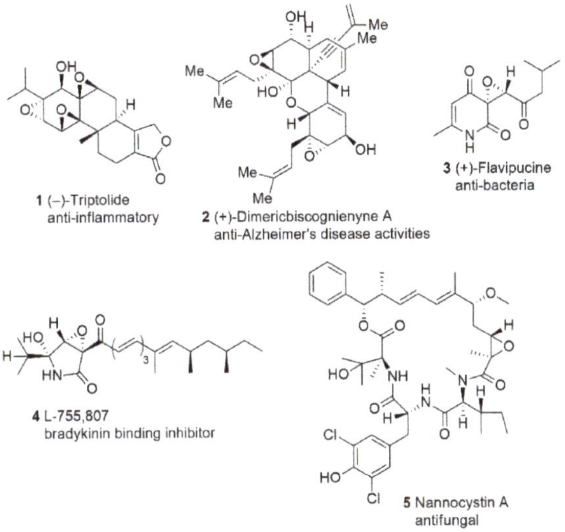

Figure 2. Epoxides containing natural products and biological activities.

In 2016, Zhao et al. [20] isolated five new compounds from secondary metabolites of *Biscogniauxia* sp., including the isolation of one new skeleton diisoprenyl-cyclohexene-type of meroterpenoid dimer—dimericbiscognienyne A **2** (Figure 2). In their anti-Alzheimer's disease (AD) fly assay study, dimericbiscognienyne A showed short-term memory enhancement activities in AD flies [20].

(+)-Flavipucine **3** (Figure 2) is a pyridione epoxide isolated from the culture extract of *Phoma* sp., and Loesgen et al. [21] determined the absolute configuration by comparing the experimental and calculated CD spectra. Since its enantiomer (−)-flavipucine had been previously reported to possess antibacterial and antifungal activity [22,23], Loesgen et al. evaluated the biological activities of (+)-flavipucine **3** as well. To their delight, (+)-flavipucine **3** also exhibited good antibacterial and antifungal activity [21]. In 2019, Kusakabe et al. [24] synthesized and conducted antibacterial and cytotoxic evaluations of flavipucine and its derivatives. The antibacterial activity of the analogues, racemic flavipucine and both its enantiomers against Gram-positive *Bacillus subtilis* (*B. subtilis*) and Gram-negative *Escherichia coli* (*E. coli*), were evaluated via broth microdilution assay. Flavipucine was the most potent among the tested compounds. Furthermore, the results indicate that the pyridione epoxide moiety is a pharmacophore for antibacterial activity against *B. subtilis*. The cytotoxicity assay against cancer cells revealed that flavipucine has strong cytotoxic activity against HL-60 cells (IC_{50} = 1.8 µM). Surprisingly, there were no significant differences observed in the biological activity of the racemates or enantiomers of flavipucine [24].

Several natural products containing an epoxy-γ-lactam ring have been found to induce neurite outgrowth and are regarded as potential therapeutic agents for AD [25]. Tanaka et al. [26] devoted their interest to an epoxy-γ-lactam ring natural product, L-755, 807 (Figure 2). Isolated from an endophytic *Microsphueropsis* sp., L-755,807 consists of an epoxy-γ-lactam moiety and was identified as a bradykinin binding inhibitor with an IC_{50} of 71 µM. Tanaka et al. [27] completed the first total synthesis of (−)-L-755,807 and its stereoisomers. Recently, they carried out the establishment of relative and absolute configurations of L-755,807 and accomplished the structure–activity relationship (SAR) study for the first time. The biological evaluations revealed that the L-755,807 and its stereoisomers display potent inhibitory activities against amyloid-β aggregation (IC_{50} = 5–21 µM) which indicates that L-755,807 and related compounds could be a promising lead as compounds developed as therapeutic agents against AD [28].

Nannocystin A **5** (Figure 2), an epoxide-carrying compound isolated from a myxobacterium *Nannocystis* sp., was reported by Hoffmann et al. [29] and Krastel et al. [30] in 2015. According to Hoffmann et al., nannocystin A has a strong antifungal effect against *C. albicans* and displays potent cell proliferation inhibitive properties by inducing apoptosis early in tested cell lines. Parallel to Hoffmann's research, Krastel et al. found that nannocystin A shows antiproliferative properties against 472 cancer cell lines in the nanomolar concentration range (IC_{50} values ranging from 0.5 µM to 5 nM). Moreover, combined genetic and proteomic approaches strongly suggest that the primary target protein of nannocystins is elongation factor 1-α (EF-1α). These studies indicate that nannocystin A may serve as a lead candidate for anticancer therapy. Due to the fact of its promising biological profiles, this novel 21 membered macrocycle immediately attracted chemists' attention in 2016 [31–34]. Further SAR study by Tian's group [35] demonstrated that the epoxide region does not interact directly with the bind site of the target eEF1a but is responsible for controlling the macrocyclic conformation.

2.2. Synthetic Strategies Used in Total Synthesis of Epoxides Containing Natural Products

Oxidation of alkenes is a general strategy to provide epoxides. Peroxy acids, such as hydrogen peroxide (H_2O_2) and *meta*-chloroperoxybenzoic acid (mCPBA), are commonly used oxidizing agents [36–39].

Yang et al. [40] completed the first enantioselective total synthesis of (−)-TPL in 2000. The fascinating structure and distinguished biological activity of TPL lead to considerable continued interest in their total synthesis and structure modification. In recent years, divergent total synthesis of TPL and its analogues have been reported [41–44]. However, these newly developed synthetic routes all adopted Yang et al.'s strategy to assemble the three successive epoxide groups (Scheme 1). Diol **6** was converted to monoepoxide **7** using the Adler reaction [45]. Epoxidation of **7** by in situ-generated methyl(trifluoromethyl)dioxirane (TFDO) and further epoxidation with alkaline hydrogen peroxide (H_2O_2/NaOH) successfully introduced the C9,C11 and C12,C13 epoxides, respectively, to give compound **9**. Reduction of **9** with $NaBH_4$ in MeOH in the presence of $Eu(fod)_3$ furnished (−)-TPL.

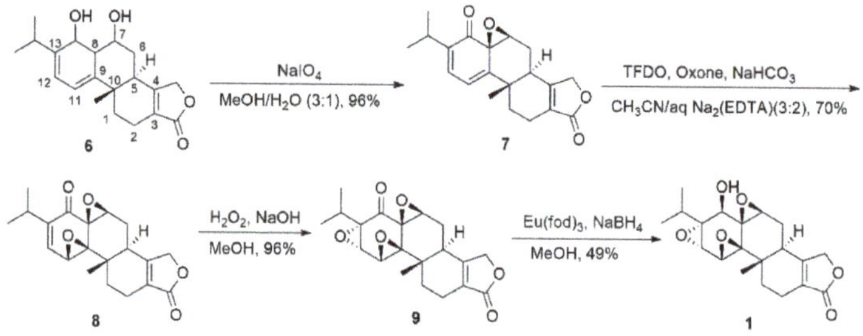

Scheme 1. The generation of epoxide groups in the total synthesis of triptolide (TPL).

The Darzens reaction is a non-oxidative method used to construct epoxycarbonyl from a halocarbonyl and an aldehyde in the presence of a base in organic solvents. The α- and β-Epoxy-γ-lactams are a highly valuable skeleton that can be generated from the Darzens reaction; they serve as attractive building blocks to access more complex molecular architectures.

Tanaka et al. [27] accomplished the first asymmetric total synthesis of L-755,807 via a diastereoselective Darzens reaction (Scheme 2). Alcohol **10** was converted to an intermediate aldehyde using Parikh–Doering oxidation. Without further purification, the aldehyde was treated with bromo di-tert-butyl malonate pre-treated with lithium bis(trimethylsilyl)amide (LHMDS); the reaction proceeded cleanly to give only the desired diastereomer **11** in high yield. This strategy decreased the number of reaction steps and avoided side reactions. Further study showed that this reaction can be applied on aldehydes bearing a branched or an unbranched alkyl side chain. Moreover, two aromatic aldehydes were evaluated but a low yield was observed in each case [46].

The essential factor for high diastereoselectivity and yield might be attributed to the formation of a metal-cation-mediated rigid structure during the reaction [28]. Because researchers are still developing new chiral organocatalysts, the substitute scope of the highly enantioselective asymmetric Darzens reaction is expanding [47]. Moreover, shorter reaction times were achieved by using aqueous media in the presence of a Li$^+$-containing base, a phase-transfer catalyst and granular polytetrafluoroethylene under mechanical stirring [48]. These developments make the Darzens reaction a promising method for the synthesis of natural products in an efficient and green way.

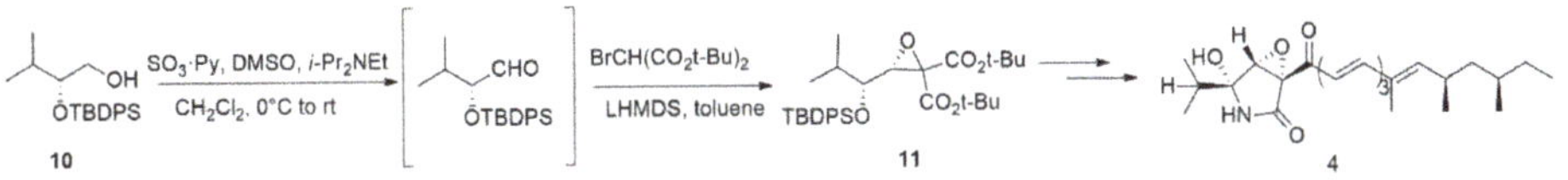

Scheme 2. The total synthesis of L-755,807 using the highly diastereoselective Darzens reaction.

2.3. Recent Advances in Epoxidation

Due to the high need for enantiomerically pure epoxides, numerous powerful and efficient catalytic asymmetric reactions have been introduced and developed to generate epoxides [49]. Among these processes, Sharpless asymmetric epoxidation (SAE), Jecobsen–Katsuki epoxidation, Shi epoxidation, etc., are classic, powerful, and still popular [50–52]. The products of asymmetric epoxidation (AE) often show enantiomeric excesses above 90%. Nowadays, the challenge for AE is to explore more sustainable and efficient catalyst systems that are environmentally friendly.

In an intriguing epoxidation of olefins with H_2O_2, Dai et al. [53] demonstrated that *cis*, *trans*, and terminal together with trisubstituted olefins can be converted to epoxides using an inexpensive and readily available in situ-formed manganese complex in excellent yields and enantioselectivities (Scheme 3). The additive adamantane carboxylic acid (aca) was found to be essential in improving the enantioselectivity. The supposed reason is that the sterically hindered aca could impart a highly rigid environment around the metal center.

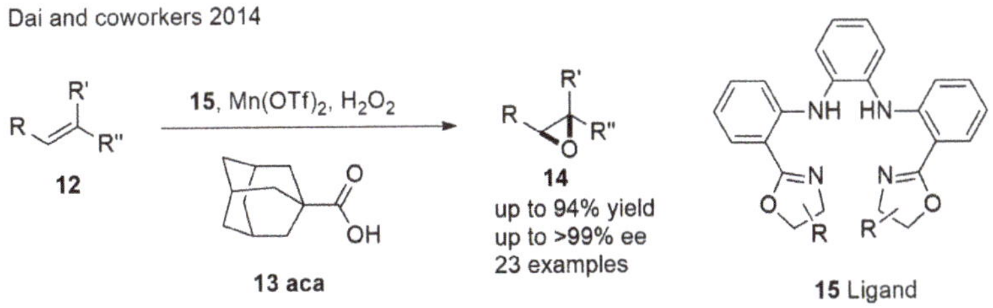

Scheme 3. Synthesis of epoxides from trisubstituted olefins.

Sharpless asymmetric epoxidation of allylic alcohols is a reliable and commonly used method of obtaining chiral epoxy alcohols; it gives high asymmetric induction for various types of allylic alcohols and provides predictable configuration of the products. However, it requires strict anhydrous conditions. A good solution to this problem may be the development of vanadium–chiral hydroxamic acid (V–HA) complex-catalyzed AE. Noji et al. [54] disclosed a highly potent approach to obtaining chiral epoxy alcohols of 2,3,3-trisubstituted allylic alcohols using the vanadium–binaphthylbishydroxamic acid (BBHA) complex (Scheme 4). This method allows a simple reaction procedure which can be conducted in aqueous TBHP solutions and offers good yields and *ee*. Noji et al. [55] recently reported a great achievement in this field. They developed an immobilized polymer-supported vanadium-binaphthylbishydroxamic acid (PS-VBHA) that can be easily recycled and reused over five consecutive runs without significant sacrifice of catalytic activity or enantioselectivity. It would not be unrealistic to say that the application of chiral hydroxamic-acid ligands as enantioselective catalysts and the development of PS-VBHA will contribute to the application of sustainable green processes for various asymmetric oxidations.

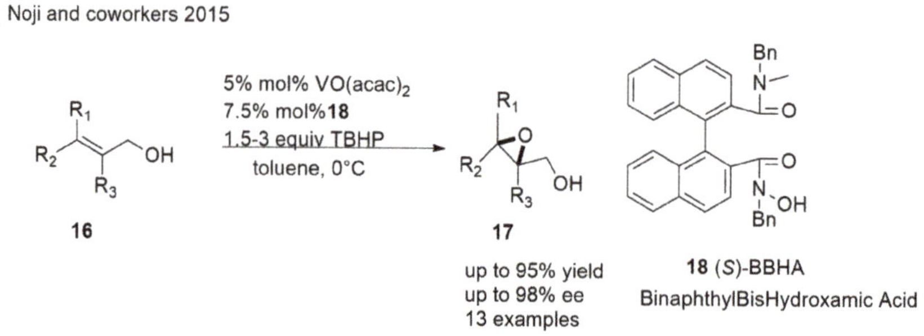

Noji and coworkers 2015

Scheme 4. Synthesis of epoxides from allylic alcohols.

Quinone epoxides are important synthetic intermediates for biologically active molecules. Nevertheless, it is difficult to apply AE on quinones because of their highly symmetric and planar structures with two carbonyl groups; therefore, differentiating the *si*- and *re*-faces of the olefins with chiral oxidants or catalysts is challenging. Kawaguchi et al. [56] developed an asymmetric epoxidation of 1,4-naphthoquinones catalyzed by guanidine–urea bifunctional organocatalysts with TBHP as an oxidant, resulting in the desired epoxides with 85:15–95:5 *er* in 71%–98% yields (Scheme 5).

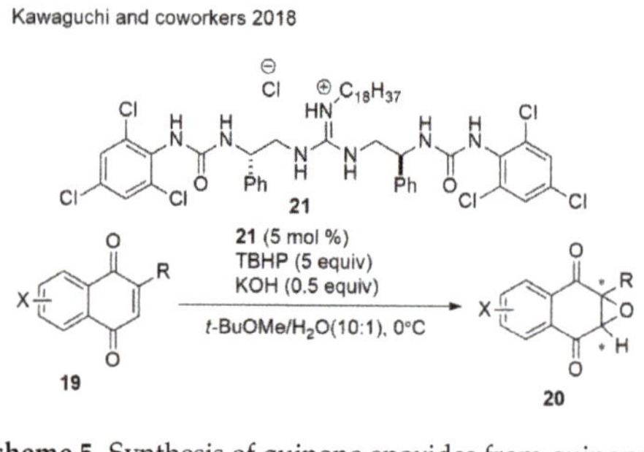

Kawaguchi and coworkers 2018

Scheme 5. Synthesis of quinone epoxides from quinones.

3. Oxetanes

Compared to epoxides, the oxetane ring appears in relatively few natural product structures, but when it is present, the natural products often display strong and intriguing biological activities. Oxetanes have been identified as efficient hydrogen-bond acceptors and have significant impacts on key physico- and biochemical properties [57]. Additionally, oxetanes are versatile building blocks of many other natural products and pharmaceutically active compounds [58]. Therefore, oxetanes have received a lot of interest as versatile precursors in synthetic chemistry [59,60].

3.1. Natural Oxetanes Containing Products and Biological Activities

The most famous example could be Taxol (Figure 3) which is widely used to treat many types of cancers, such as breast cancer [61–63], ovarian cancer [64–66], lung cancer [67–69], cervical cancer [70–72], etc. Taxol and its semi-synthetic derivative cabazitaxel furthered interest in the study of oxetanes. Despite anticancer activity, oxetanes containing natural products exhibit many more pharmacological properties. Merrilactone A **23** has a unique sesquiterpene bearing two γ-lactones; an oxetane ring was isolated from the pericarps of *Illicium merrillianum*. The presence of an oxetane ring was required for neurotrophic activity. However, the isolation yield was only 0.004% [73]. Hence, total synthesis is fundamental for further biological research of merrilactone A. Mitrephorone A **24** is a compound isolated from the Bornean shrub *Mitrephora glabra* and displays potent cytotoxicity against a panel of cancer cells as well as featuring excellent antimicrobial activity [74]. It contains a fully substituted oxetane embedded in a pentacyclic carbon skeleton with a rare 1,2-diketone. This combination makes this natural product a veritable challenge for synthetic chemistry. The biological activities of some oxetanes containing natural products remain unavailable for many years due to the rareness and difficulty of chemical preparation, namely, (+)-dictyoxetane **25** [75].

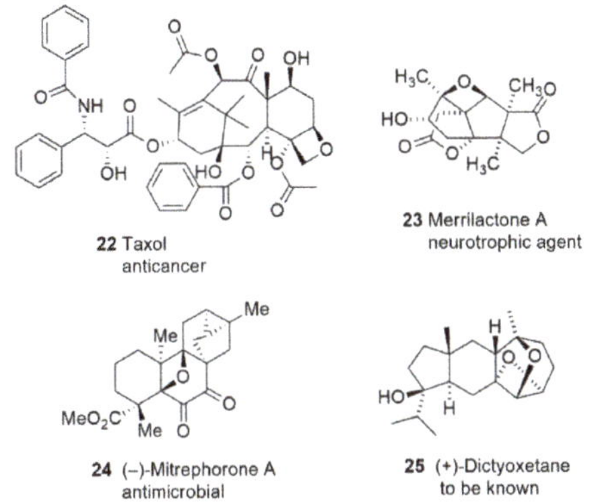

Figure 3. Oxetanes containing natural products and biological activities.

3.2. Synthesis Strategies Used in the Total Synthesis of Oxetanes Containing Natural Products

Unlike epoxides, there is a paucity of synthetic methods available for the construction of oxetanes. In general, oxetanes can be synthesized via (a) Paternò–Büchi [2 + 2] photocycloaddition [76]; (b) C–O bond-forming cyclisation; (c) ring expansion of epoxides; and (d) C–C bond-forming cyclisation.

A good example of using epoxide ring expansion to form oxetanes is the total synthesis of merrilactone A. Since its isolation in 2000, merrilactone A has been consistently appealing to researchers. Over the years, several different routes towards the total synthesis of merrilactone A have been reported [77–81]. Most recently, Liu and Wang [82] designed and achieved a concise synthesis of merrilactone A in a racemic form (Scheme 6). In their synthesis work, exploiting dimethyldioxirane (DMDO) on **27** allowed the epoxide intermediate that was subjected to acidic conditions to afford synthetic merrilactone A through the epoxide-opening oxetane formation.

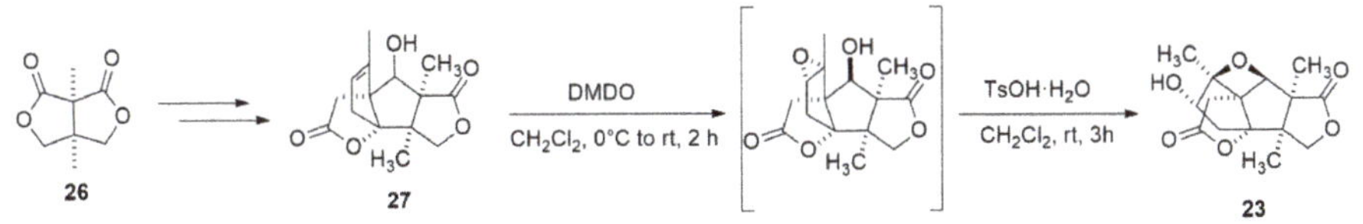

Scheme 6. The total synthesis of merrilactone A.

Richter et al. [83] reported the first and enantioselective synthesis of (−)-mitrephorone A in 2018 by using a novel late-stage oxidative cyclisation (Scheme 7). Synthesis commenced with methacrolein **28** and gave **29** in steps. Finally, the pivotal oxetane moiety was generated via one-pot deprotection of silyl ether **29** (TASF) and subsequent reaction with Koser's reagent (PhI(OH)OTs) to successfully complete the total synthesis of (−)-mitrephorone A.

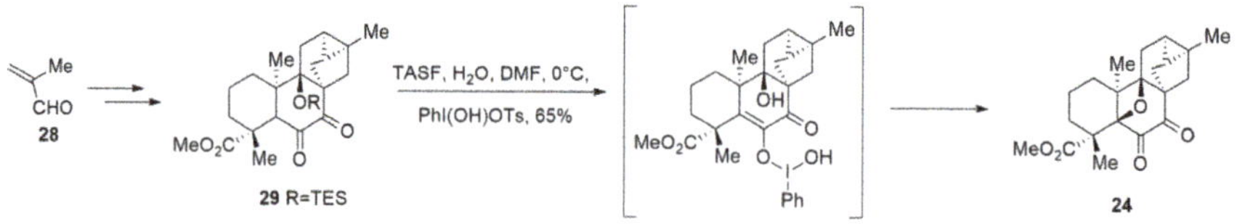

Scheme 7. The total synthesis of mitrephorone A.

3.3. Recent Advances in Oxetane Synthesis

Impressive synthetic approaches toward C–C bond-forming hydrogenations and transfer hydrogenations were developed. Guo et al. [84] achieved access to oxetanes bearing all-carbon quaternary stereocenters readily prepared through the iridium catalyzed anti-diastereo and enantioselective C–C coupling of primary alcohols and isoprene oxide (Scheme 8). A group of primary alcohols **30** were exposed to isoprene oxide (300 mol%) and potassium phosphate (5 mol%) in the presence of the chromatographically purified π-allyliridium C,O-benzoate complex modified by (S)-Tol-BINAP in a tetrahydrofuran solvent at 60 °C to give adducts **31**. Conversion of the diol-containing adducts to the corresponding oxetanes **32** were accomplished through highly chemoselective tosylation of the primary alcohol moiety followed by S_N2 cyclisation [84].

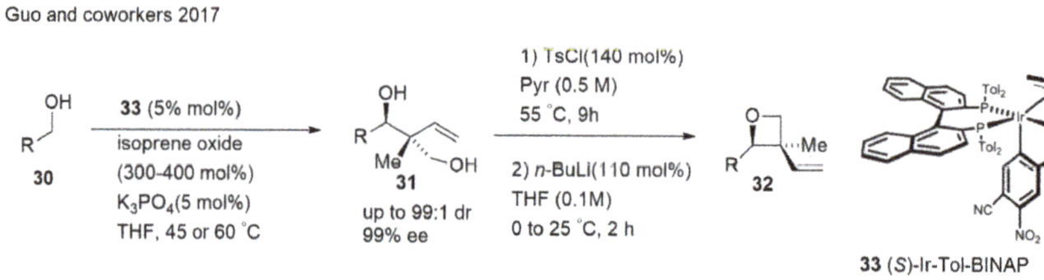

Scheme 8. Synthesis of oxetanes using an iridium catalyst.

4. Tetrahydrofurans

Tetrahydrofuran (THF) moieties occur in many natural products with a wide array of bioactivities [85]; it has encouraged the development of a variety of synthetic methods [86,87].

4.1. Natural THF-Containing Products and Biological Activities

Annonaceous acetogenins (AGEs) have been widely considered and extensively researched over the past three decades [88]. Structurally, AGEs are characterized by linear 32 or 34 carbon chains containing oxygenated functional groups. Most of them have one, two or three THF rings located along the hydrocarbon chain. Annonaceous acetogenins have been isolated from more than 50 different species of plants [89]. Extensive studies have indicated that members of this class of natural compounds possess a broad spectrum of bioactivity, featuring anticancer, antiparasitic, insecticidal, and immunosuppressive effects [90]. Recently, five new acetogenins were isolated from the roots of *Annona purpurea*, among which the most potent compound was annopurpuricins A **34** (Figure 4). Antiproliferative activity evaluation indicated that annopurpuricins A inhibited the growth of HeLa and HepG2 cells significantly with GI50 values of 0.06 nM and 0.45 mM, respectively. The THF rings may play an important role in these results [91]. The study of the antitumor mechanisms of acetogenins is also attractive to scientists [92].

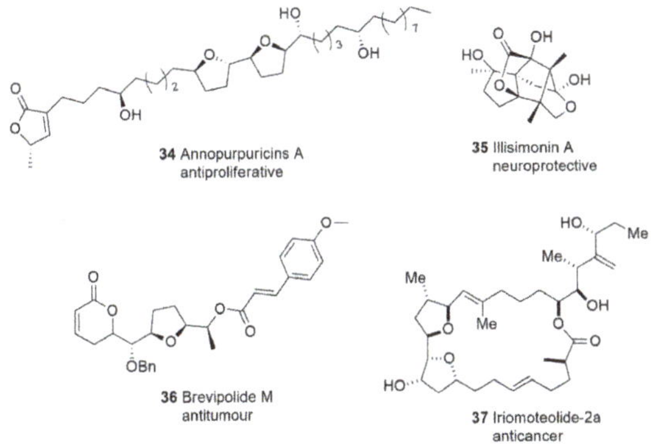

Figure 4. Tetrahydrofuran (THF)-containing natural products and biological activities.

The THF units are also found in many special skeletons. In 2017, Ma et al. [93] isolated illisimonin A **35** (Figure 4), which features a previously unreported tricyclic carbon framework from the fruits of *Illicium simonsii*. The structure and absolute configuration of illisimonin A were determined to be a caged 2-oxatricyclo[3,3,0,1^{4,7}]nonane ring system fused to a five-membered carbocyclic ring and a five-membered lactone ring. These results were determined by the extensive use of spectroscopic evidence and electronic circular dichroism (ECD) calculations. Illisimonin A displayed potent neuroprotective effects against oxygen and glucose deprivation (OGD)-induced cell injury in SH-SY5Y cells, and its unique structure has inspired research into 3-Oxabicyclo [3.2.0]heptan-2-one core building blocks [94] and total synthesis [95].

In 2016, Suárez-Ortiz et al. [96] isolated and described the absolute configuration of five new compounds named brevipolides K–O from *H. brevipes*. Taking brevipolide M **36** (Figure 4) as an example, it contains a distinctive THF ring in the structure. These compounds displayed cytotoxicity against a variety of tumor cell lines including nasopharyngeal (KB) and cervix (HeLa) cancer cells with IC_{50} values of 1.7–10 µM.

Iriomoteolide-2a **37** (Figure 4) is a new anticancer macrolide isolated from the cultured broth of the benthic dinoflagellate *Amphidinium* sp. (HYA024 strain) collected off Iriomote Island, Okinawa, by the Tsuda group [97]. Significantly, it possesses potent cytotoxic activities against human B lymphoma DG75 cells and human cervix adenocarcinoma HeLa cells with IC_{50} values of 6 and 30 ng/mL, respectively.

4.2. Synthesis Strategies Used in Total Synthesis of THFs Containing Natural Products

Nucleophilic substitution chemistry plays an important role in the synthesis of cyclic ethers. Many THF-containing natural products have been constructed by employing intramolecular S_N2 reactions between a hydroxyl group and a leaving group.

Raju et al. [98] reported the first stereoselective total synthesis of brevipolide M with the readily available (−)-DET (Scheme 9). In this synthesis, allylic alcohol **39** was treated with Sharpless asymmetric epoxidation to provide the chiral epoxy alcohol **40** followed by tosylation of the primary alcohol. Treating tosyl compound **41** with PTSA in MeOH/CH$_2$Cl$_2$ (1:1) at rt for 2 h resulted in the deprotection of the acetonide group and the spontaneous cyclisation of hydroxy epoxide, giving the desired syn-tetrahydrofuran **42** in an 85% yield. Other key steps involved Brown's allylation, the RCM reaction to install an α- and β-unsaturated lactone ring, and the inversion of the C-6′ stereogenic hydroxyl group using the Mitsunobu reaction furnished brevipolide M.

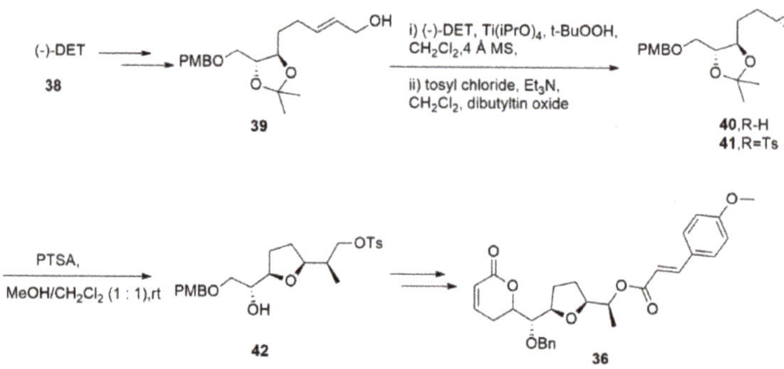

Scheme 9. The total synthesis of brevipolide M.

Another example of THF construction is via the intramolecular addition of alcohols to epoxide. A good example of using this strategy is the total synthesis of iriomotelide-2a by Sakamoto (Scheme 10) [99]. The gross structure of iriomotelide-2a consists of an unusual 23 membered macrocyclic backbone with a characteristic bis(tetrahydrofuran) substructure and a complex side chain containing four stereogenic centers. The efforts to achieve the bis(tetrahydrofuran) unit is a fundamental part of total synthesis. In Sakamoto's approach, Sharpless epoxidation of 44 followed by the first THF ring formation using the intramolecular addition of alcohols to epoxide gave alcohol 46. After mesylation, cleavage of the benzoyl group and concomitant cycloetherification in the basic condition furnished bis(tetrahydrofuran) 47 [100].

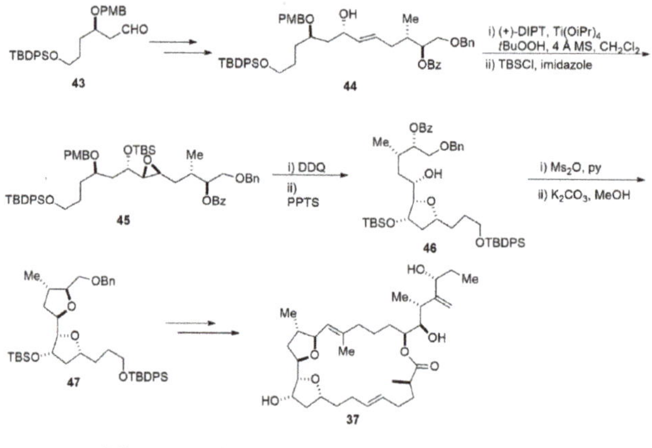

Scheme 10. The total synthesis of iriomoteolide-2a.

4.3. Recent Advances in THF Synthesis

A stereodivergent intramolecular Rh-catalyzed azavinyl carbenoid C(sp^3)–H insertion reaction was achieved by Lindsay et al. (Scheme 11), which allowed the formation of cis-2,3-disubstituted THFs. Good yields were observed and the resulting THF products were transformed to ring-fused THFs efficiently [101].

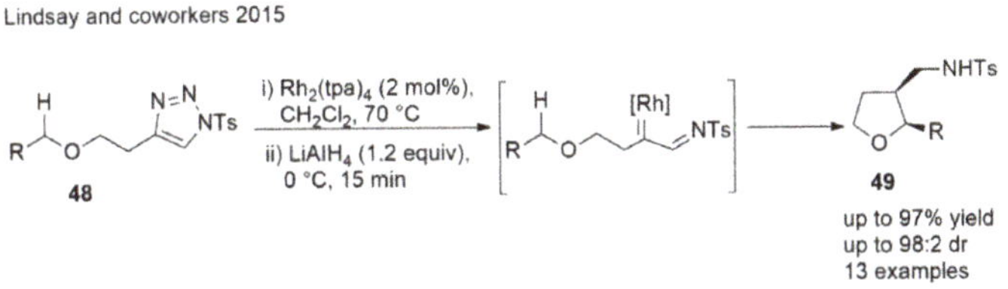

Lindsay and coworkers 2015

Scheme 11. Synthesis of THF using C(sp3)–H insertion.

Oxiranes have become interesting precursors in the last few years. Yuan et al. [102] demonstrated an efficient diastereo- and enantioselective [3 + 2] cycloaddition of heterosubstituted alkenes with oxiranes via selective C–C bond cleavage of epoxides to give chiral THFs (Scheme 12). The reaction was catalyzed by a chiral N,N'-dioxide/Ni(II) catalyst which was derived from L-ramipril (Ra) by complexing with Ni(BF$_4$)$_2$·6H$_2$O. The enantioselectivity was found to increase little by little as the steric hindrance at the *ortho* positions of the aniline of N,N'-dioxide ligands or on the heterosubstituted alkenes became larger.

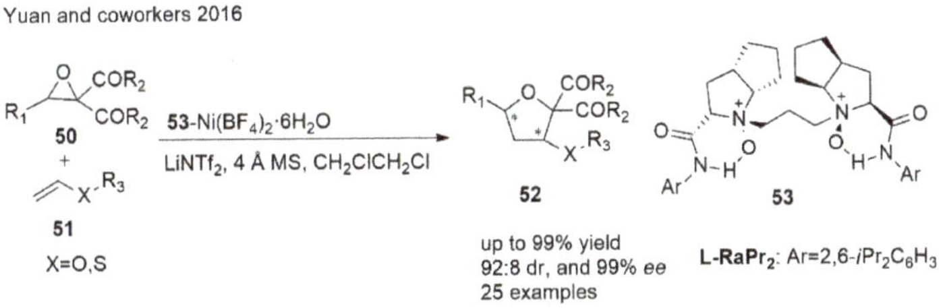

Yuan and coworkers 2016

Scheme 12. Synthesis of THF using [3 + 2] cycloaddition.

Asymmetric nucleophilic additions to keto aldehydes in the presence of enantiomerically pure imidophosporimidates (IDPis) interestingly shows that ketone reacts preferentially over the aldehyde. This method provides 2,2-disubstituted THF analogues with tetrasubstituted stereogenic centers starting from 1,4-dicarbonyl compounds [103] (Scheme 13). Moreover, 2,2,5,5-tetrasubstituted THFs can be readily prepared using the described method.

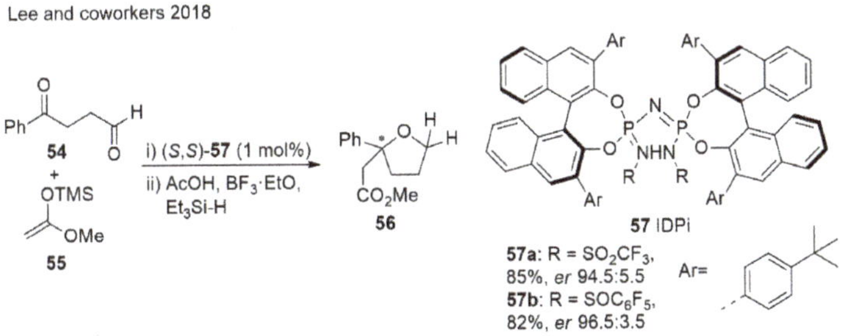

Lee and coworkers 2018

Scheme 13. Synthesis of THF using an IDPi catalyst.

5. Tetrahydropyrans

Tetrahydropyrans (THPs) have received a considerable amount of attention from many biologists and synthetic organic chemists due to the prevalence of these substructures in biologically interesting natural products [104,105].

5.1. *Natural THPs Containing Products and Biological Activities*

Salinomycin (SAL) **59** (Figure 5) has shown a broad spectrum of bioactivity, including antibacterial, antifungal, antiviral, antiparasitic, and anticancer activity, proving its significant therapeutic potential [106,107]. Many research groups around the world are currently performing intensive studies to discover novel aspects of the biological activity of SAL and its derivatives. Namely, Tyagi et al. [108] recently reported that a follow-up treatment of SAL may be a promising strategy against cisplatin (*cis*-diamminedichloro-platinum, CDDP)-resistant breast cancer cells and metastasis and help reduce CDDP-induced side effects as it reduces the growth, proliferation, and metastasis of cisplatin-resistant breast cancer cells via NF-kB deregulation. Another discovery of SAL's effects on breast cancer by Dewangan et al. indicated that SAL inhibits breast cancer progression via targeting HIF-1α/VEGF-mediated tumor angiogenesis [109].

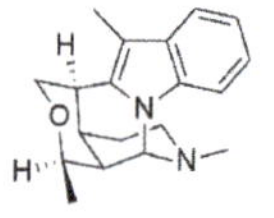

58 Alstoscholarisine A
neural stem cells activitor

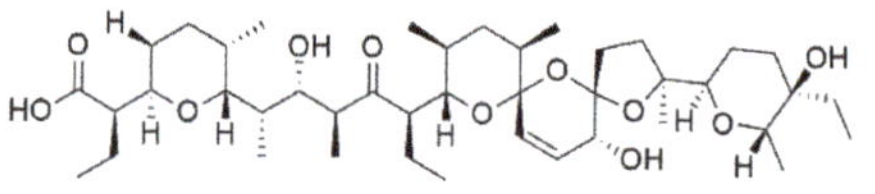

59 Salinomycin
anticancer

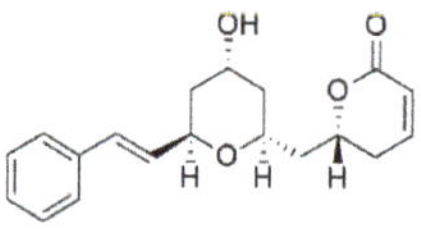

60 (+)-Cryptoconcatone H
NO inhibition

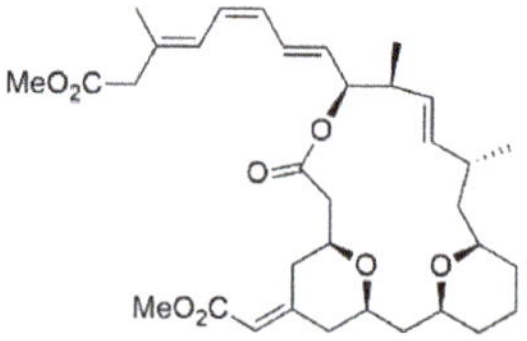

61 (−)-Exiguolide
anticancer

Figure 5. THPs containing natural products and biological activities.

In 2014, Yang et al. [110] isolated five unprecedented monoterpenoid indole alkaloids from *Alstonia scholaris*. The most potent compound, alstoscholarisine A **58** (Figure 5), promoted adult neural stem cell proliferation significantly with a concentration of 0.1 μg/mL in a dosage-dependent manner and did not affect the proliferation of neuroblastoma cells. This finding attracted interest from the synthetic community to achieve the total synthesis of this series of compounds with a new skeleton [111]. In 2016, Yang et al. [112] published the new compounds they isolated from the leaves and twigs of *C. concinna*. Among them, the structure of cryptoconcatone H **60** (Figure 5) was proposed as an *S* absolute configuration through the interplay of Mosher's ester methodology and ROESY experiments whereas Della-Felice et al. revised it as all *R* stereoisomer shown in Figure 5 [113]. Cryptoconcatone H displayed the inhibition of NO production induced by LPS in RAW 264.7 macrophages with an IC_{50} value of 4.2 μM [112].

5.2. *Synthesis Strategies Used in the Total Synthesis of THP-Containing Natural Products*

A large number of valuable and high-quality contributions have been made in the construction of THP rings [104,114]. The Prins cyclisation reaction and its variants are extremely powerful methods for constructing THF [115]/THP [116] rings and widely applied in total synthesis [117–119].

Li et al. [120] described a Prins cyclisation/homobromination process that involved dienyl alcohol with aldehyde to construct *cis*-THP containing an exocyclic *E*-alkene by using $TMSBr/InBr_3$ as a combined bromide source and a Lewis acid. This approach provides good-to-excellent *cis*-E stereochemical control in one step, and this reaction was soon employed in the total synthesis of (−)-exiguolide **61** (Scheme 14). The ester substituted **65** was delivered at 71% yield with a *cis/trans* ratio of ≥95:5 and a *Z/E* ratio of 95:5 [121].

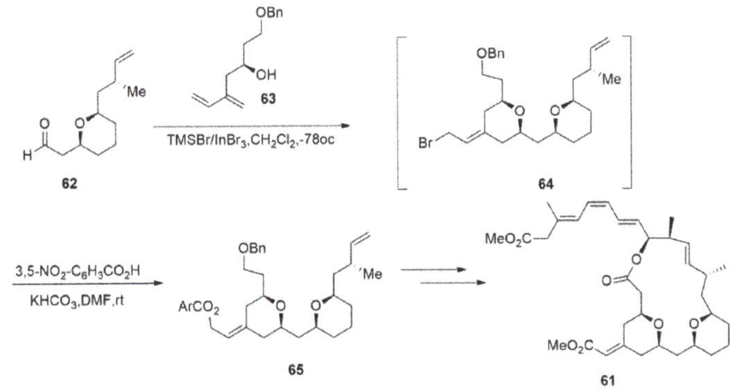

Scheme 14. Total synthesis of (−)-exiguolide.

The THP formation was observed during Jacobsen et al.'s study of the enantioselective intramolecular opening of oxetanes for obtaining enantioenriched THFs [122]. In 2014, Yadav et al. [123] disclosed that exposing oxetanes to acids in the presence of aprotic solvents can afford THPs smoothly. The best solvent condition was found to be a mixture of CH_2Cl_2 and i-PrOH (15:1), in which oxetane substrates were converted into THPs at high yields within two hours. This newly developed methodology was successfully applied to the synthesis of the THP motif of salinomycin (Scheme 15).

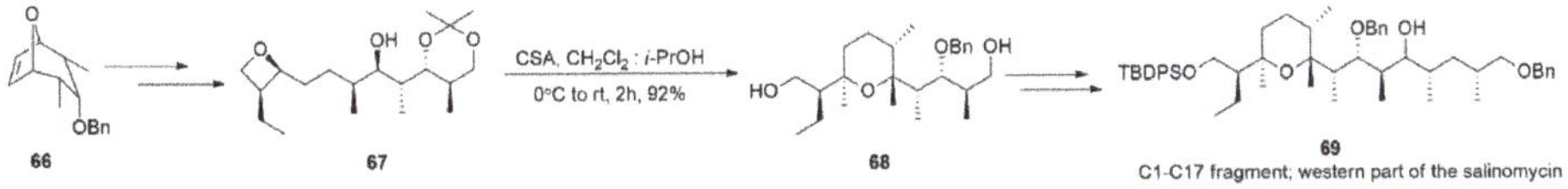

Scheme 15. Synthesis of the C1–C17 fragment of salinomycin.

5.3. Recent Advances in THP Synthesis

Srinivas et al. [124] demonstrated an efficient one-pot protocol to obtain THPs via the diastereoselective tandem dihydroxylation of ζ-mesyloxy α,β-unsaturated esters followed by S_N2 cyclisation (Scheme 16). The highlight of this method is that it allows both *cis*- and *trans*-THP rings to be synthesized starting from a common precursor. This protocol also demonstrated the formal synthesis of (+)-muconin in a concise and highly stereoselective manner.

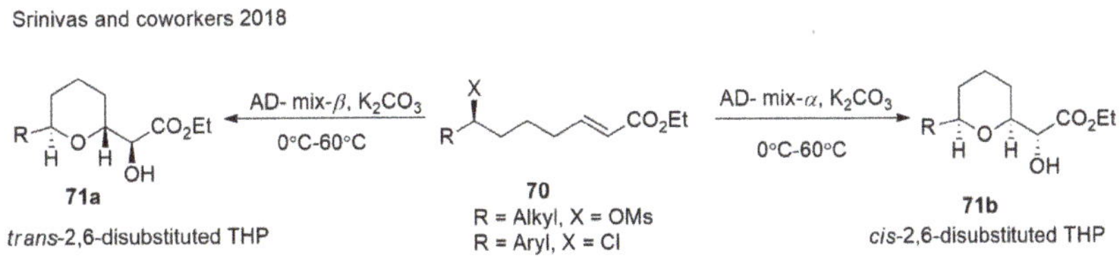

Scheme 16. Synthesis of both *cis*- and *trans*-THPs from a common precursor.

In 2018, Sergio [125] reported a direct and general method for the synthesis of naturally occurring 2,3,4,5,6-pentasubstituted THPs employing β,γ-unsaturated N-acyl oxazolidin-2-ones as key starting materials (Scheme 17). The combination of the Evans aldol addition and Prins cyclisation allowed the diastereoselective and efficient generation of the target highly substituted THPs.

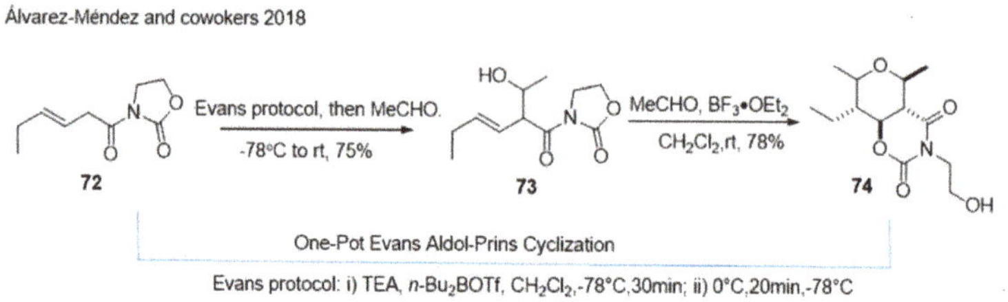

Álvarez-Méndez and cowokers 2018

Scheme 17. One-Pot Evans–Prins cyclisation to construct THPs.

6. Oxepanes

Natural products containing oxepane scaffolds are not rare, particularly among marine products. Their interesting biological properties can include anticancer, antibacterial, and antifungal activities. Oxepane motif in known natural products examples is often flanked by aromatic moieties or polycyclics [126,127], making the structures overwhelmingly complex and unusual. Simple oxepane compounds, such as isolaurepinnacin 77 (Figure 6), have been synthetic targets of considerable interest.

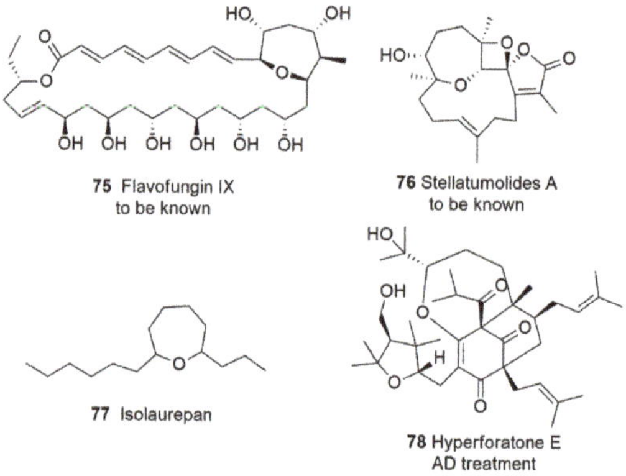

Figure 6. Oxepane-containing natural products and biological activities.

6.1. Natural Oxepane-Containing Products and Biological Activities

Only a few new oxepane-containing natural products have been isolated in recent years. In 2017, flavofungin IX 75 (Figure 6) was obtained by Wang et al. from mangrove-derived *streptomyces* sp. ZQ4BG [128]. In 2018, Ahmed et al. isolated stellatumolides A 76 (Figure 6) from soft coral *Sarcophyton stellatum* [129]. However, there have been no further reports of these new compounds. It would be interesting to discover their bioactivities or synthetic methods. Guo et al. isolated 15 new polycyclic polyprenylated acylphloroglucinols (PPAPs) from the stems and leaves of *Hypericum perforatum*, including six oxepane ring-containing products hyperforatones I-J [130]. Hyperforatone E 78 (Figure 6) was found to exhibit dual inhibitory activities against AChE and BACE1. Preliminary molecular docking studies have shown that it has strong interactions with the major active sites of BACE1 and AChE. These initial studies suggest that hyperforatone E may be further developed into a potential candidate or lead compound for AD treatment.

6.2. Synthesis Strategies Used in the Total Synthesis of Oxepane-Containing Natural Products

The oxepanyl rings in natural products are commonly bonded to a great diversity of rings and chains. Very few total syntheses of such complicated oxepane-containing compounds have been achieved in recent years due to the challenge of oxepane formation. The development of the construction of oxepane motifs would be beneficial for the synthesis of bioactive molecules and the total synthesis in the future.

In 2016, Hernàndez-Torres et al. [131] developed a synthesis method of 7–9 membered cyclic ethers by the reductive cyclisation of hydroxy ketones and successfully applied this to the total synthesis of isolaurepan (Scheme 18). Full hydrogenation and PMB cleavage of 80 under hydrogen atmosphere afforded hydroxy ketone 81. Then, reductive cyclisation by treatment with Et$_3$SiH and TMSOTf in CH$_2$Cl$_2$ gave the desired products. Even though many synthesis routes towards isolaurepan have been reported, this methodology stands out for its few steps and comparably high yield.

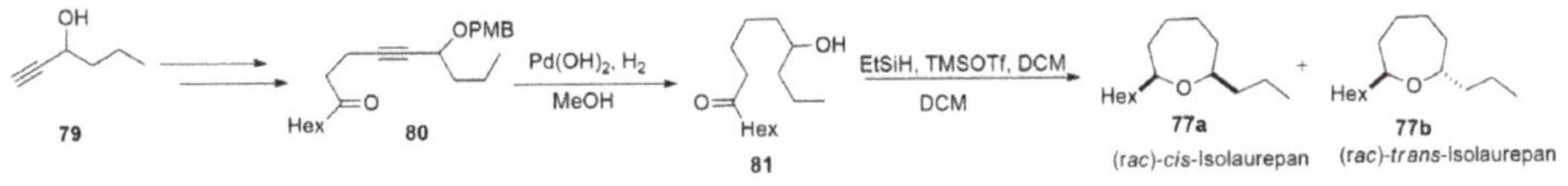

Scheme 18. Synthesis of isolaurepan.

6.3. Recent Advances in Oxepane Synthesis

Armbrust et al. [132] disclosed a successful approach to assembling 6 and 7 membered oxygen heterocycles (Scheme 19). In their method, di/trisubstituted epoxides were activated by Rhodium catalyzed via π-coordination and oxidative addition into the vinylic C–O bond of the epoxide.

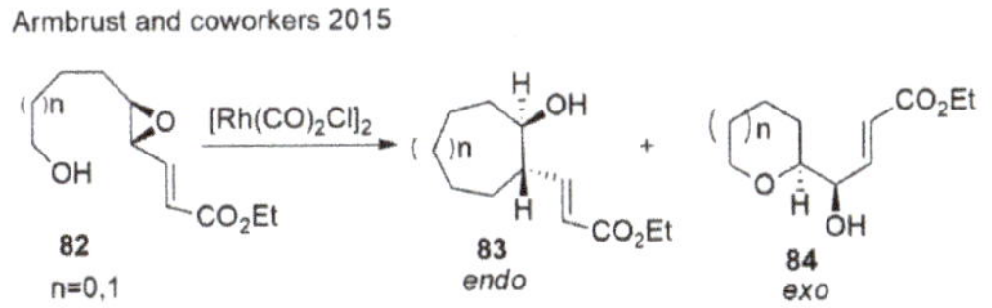

Scheme 19. Synthesis of oxepanes by cascade reaction.

The cyclodehydration of diols offers good access to 7 membered rings. In 2018, Sun et al. [133] achieved oxepane via the heteropoly acids (HPAs)-catalyzed cyclodehydration of hexane-1,6-diol in 80% yield (Scheme 20). There are no substituted oxepane examples given in the literature. It would be interesting to try making substituted oxepanes using this protocol.

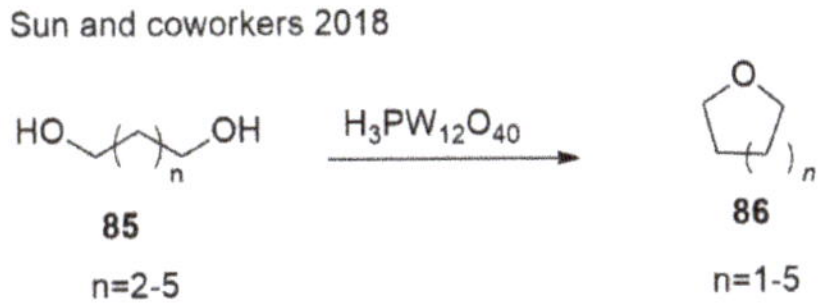

Scheme 20. Synthesis of oxepanes by HPA-catalyzed cyclodehydration.

7. Conclusions

This review, although covering only a small fraction of cyclic ether chemistry, demonstrates that small- and medium-ether rings appear in a large variety of natural resources and most of these compounds display promising biological and pharmacological properties, such as antibiotic, antibacterial, and antitumor activities. Even though cyclic ethers are not privileged heterocycles, they represent an undeniably important class of compounds. Many reliable synthetic methods have been efficiently applied to the total synthesis of related natural products, while numerous new developments of cycle ether synthesis have been achieved in organic chemistry. However, there is still a large developing space left to explore: discovering and creating new pharmacologically active molecules, extending the substitute scope of synthetic strategies, overcoming the difficulty of constructing large rings, and achieving new drugs.

Author Contributions: Conceptualization, K.L. and Q.L.; writing—original draft preparation, Q.L.; writing—review and editing, K.L.; supervision, Q.L., D.S.H., Y.C., K.L.

Funding: The authors acknowledge the National Research Foundation of Korea's (NRF) grant funded by the Korea government (MSIT) (NRF 2018R1A5A2023127) for the financial support of this work.

Conflicts of Interest: The authors declare no conflict of interest.

References

1. Blunt, J.W.; Copp, B.R.; Keyzers, R.A.; Munro, M.H.; Prinsep, M.R. Marine natural products. *Nat. Prod. Rep.* **2016**, *33*, 382–431. [CrossRef] [PubMed]
2. Martín, T.; Padron, J.I.; Martin, V.S. Strategies for the synthesis of cyclic ethers of marine natural products. *Synlett* **2014**, *25*, 12–32. [CrossRef]
3. Marco-Contelles, J.; Molina, M.T.; Anjum, S. Naturally occurring cyclohexane epoxides: Sources, biological activities, and synthesis. *Chem. Rev.* **2004**, *104*, 2857–2900. [CrossRef] [PubMed]
4. Hassan, H.M.; Rateb, M.E.; Hassan, M.H.; Sayed, A.M.; Shabana, S.; Raslan, M.; Amin, E.; Behery, F.A.; Ahmed, O.M.; Bin Muhsinah, A.; et al. New Antiproliferative Cembrane Diterpenes from the Red Sea Sarcophyton Species. *Mar. Drugs* **2019**, *17*, 411. [CrossRef]
5. Davis, R.L.; Stiller, J.; Naicker, T.; Jiang, H.; Jørgensen, K.A. Asymmetric organocatalytic epoxidations: Reactions, scope, mechanisms, and applications. *Angew. Chem. Int. Ed.* **2014**, *53*, 7406–7426. [CrossRef]
6. da Silva, A.R.; dos Santos, D.A.; Paixão, M.W.; Corrêa, A.G. Stereoselective Multicomponent Reactions in the Synthesis or Transformations of Epoxides and Aziridines. *Molecules* **2019**, *24*, 630. [CrossRef]
7. Shen, J.; He, C. Isolation and Purification of Triptolide from the Leaves of Tripterygium wilfordii Hook F. *Chin. J. Chem. Eng.* **2010**, *18*, 750–754. [CrossRef]
8. Gu, W.-Z.; Chen, R.; Brandwein, S.; McAlpine, J.; Burres, N. Isolation, purification, and characterization of immunosuppressive compounds from tripterygium: Triptolide and tripdiolide. *Int. J. Immunopharmacol.* **1995**, *17*, 351–356. [CrossRef]
9. Ye, H.; Ignatova, S.; Luo, H.; Li, Y.; Peng, A.; Chen, L.; Sutherland, I. Preparative separation of a terpenoid and alkaloids from Tripterygium wilfordii Hook. f. using high-performance counter-current chromatography: Comparison of various elution and operating strategies. *J. Chromatogr. A* **2008**, *1213*, 145–153. [CrossRef]
10. Hewitson, P.; Ignatova, S.; Ye, H.; Chen, L.; Sutherland, I. Intermittent counter-current extraction as an alternative approach to purification of Chinese herbal medicine. *J. Chromatogr. A* **2009**, *1216*, 4187–4192. [CrossRef]
11. Ziaei, S.; Halaby, R. Immunosuppressive, anti-inflammatory and anti-cancer properties of triptolide: A mini review. *Avicenna J. Phytomed.* **2016**, *6*, 149–164. [PubMed]
12. Wang, X.; Matta, R.; Shen, G.; Nelin, L.D.; Pei, D.; Liu, Y. Mechanism of triptolide-induced apoptosis: Effect on caspase activation and Bid cleavage and essentiality of the hydroxyl group of triptolide. *J. Mol. Med.* **2006**, *84*, 405. [CrossRef] [PubMed]
13. Noel, P.; Von Hoff, D.D.; Saluja, A.K.; Velagapudi, M.; Borazanci, E.; Han, H. Triptolide and Its Derivatives as Cancer Therapies. *Trends Pharmacol. Sci.* **2019**, *40*, 327–341. [CrossRef] [PubMed]
14. Hu, H.; Luo, L.; Liu, F.; Zou, D.; Zhu, S.; Tan, B.; Chen, T. Anti-cancer and Sensibilisation Effect of Triptolide on Human Epithelial Ovarian Cancer. *J. Cancer* **2016**, *7*, 2093–2099. [CrossRef] [PubMed]
15. Hu, H.; Huang, G.; Wang, H.; Li, X.; Wang, X.; Feng, Y.; Tan, B.; Chen, T. Inhibition effect of triptolide on human epithelial ovarian cancer via adjusting cellular immunity and angiogenesis. *Oncol. Rep.* **2018**, *39*, 1191–1196. [CrossRef] [PubMed]
16. Song, J.M.; Molla, K.; Anandharaj, A.; Cornax, I.; OSullivan, M.G.; Kirtane, A.R.; Panyam, J.; Kassie, F. Triptolide suppresses the in vitro and in vivo growth of lung cancer cells by targeting hyaluronan-CD44/RHAMM signaling. *Oncotarget* **2017**, *8*, 26927–26940. [CrossRef] [PubMed]
17. Gao, H.; Zhang, Y.; Dong, L.; Qu, X.Y.; Tao, L.N.; Zhang, Y.M.; Zhai, J.H.; Song, Y.Q. Triptolide induces autophagy and apoptosis through ERK activation in human breast cancer MCF-7 cells. *Exp Ther Med.* **2018**, *15*, 3413–3419. [CrossRef]
18. Chugh, R.; Sangwan, V.; Patil, S.P.; Dudeja, V.; Dawra, R.K.; Banerjee, S.; Schumacher, R.J.; Blazar, B.R.; Georg, G.I.; Vickers, S.M. A preclinical evaluation of Minnelide as a therapeutic agent against pancreatic cancer. *Sci. Transl. Med.* **2012**, *4*, 156ra139. [CrossRef]

19. Propper, D.; Han, H.; Hoff, D.V.; Borazanci, E.; Reya, T.; Ghergurovich, J.; Pshenichnaya, I.; Antal, C.; Condjella, R.; Sharma, S. Abstract CT165: Phase II open label trial of minnelide™ in patients with chemotherapy refractory metastatic pancreatic cancer. *Cancer Res.* **2019**, *79*, CT165. [CrossRef]
20. Zhao, H.; Chen, G.D.; Zou, J.; He, R.R.; Qin, S.Y.; Hu, D.; Li, G.Q.; Guo, L.D.; Yao, X.S.; Gao, H. Dimericbiscognienyne A: A Meroterpenoid Dimer from Biscogniauxia sp. with New Skeleton and Its Activity. *Org. Lett.* **2017**, *19*, 38–41. [CrossRef]
21. Loesgen, S.; Bruhn, T.; Meindl, K.; Dix, I.; Schulz, B.; Zeeck, A.; Bringmann, G. (+)-Flavipucine, the Missing Member of the Pyridone Epoxide Family of Fungal Antibiotics. *Eur. J. Org. Chem.* **2011**, *2011*, 5156–5162. [CrossRef]
22. Girotra, N.; Patchett, A.; Zimmerman, S.; Achimov, D.; Wendler, N. Synthesis and biological activity of flavipucine analogs. *J. Med. Chem.* **1980**, *23*, 209–213. [CrossRef] [PubMed]
23. Wagner, C.; Anke, H.; Besla, H.; Sterner, O. Flavipucine and brunnescin, two antibiotics from cultures of the mycophilic fungus Cladobotryum rubrobrunnescens. *Z. Naturforsch. C* **1995**, *50*, 358–364. [CrossRef] [PubMed]
24. Kusakabe, Y.; Mizutani, S.; Kamo, S.; Yoshimoto, T.; Tomoshige, S.; Kawasaki, T.; Takasawa, R.; Tsubaki, K.; Kuramochi, K. Synthesis, antibacterial and cytotoxic evaluation of flavipucine and its derivatives. *Bioorg. Med. Chem. Lett.* **2019**, *29*, 1390–1394. [CrossRef] [PubMed]
25. Mitsuhashi, S.; Shindo, C.; Shigetomi, K.; Miyamoto, T.; Ubukata, M. (+)-Epogymnolactam, a novel autophagy inducer from mycelial culture of Gymnopus sp. *Phytochemistry* **2015**, *114*, 163–167. [CrossRef] [PubMed]
26. Lam, Y.K.T.; Hensens, O.D.; Ransom, R.; Giacobbe, R.A.; Polishook, J.; Zink, D. L-755,807, A new non-peptide bradykinin binding inhibitor from an endophytic Microsphaeropsis sp. *Tetrahedron* **1996**, *52*, 1481–1486. [CrossRef]
27. Tanaka, K., 3rd; Kobayashi, K.; Kogen, H. Total Synthesis of (-)-L-755,807: Establishment of Relative and Absolute Configurations. *Org. Lett.* **2016**, *18*, 1920–1923. [CrossRef]
28. Tanaka, K.; Honma, Y.; Yamaguchi, C.; Aoki, L.; Saito, M.; Suzuki, M.; Arahata, K.; Kinoshita, K.; Koyama, K.; Kobayashi, K. Total synthesis, stereochemical assignment, and biological evaluation of L-755,807. *Tetrahedron* **2019**, *75*, 1085–1097. [CrossRef]
29. Hoffmann, H.; Kogler, H.; Heyse, W.; Matter, H.; Caspers, M.; Schummer, D.; Klemke-Jahn, C.; Bauer, A.; Penarier, G.; Debussche, L. Discovery, Structure Elucidation, and Biological Characterization of Nannocystin A, a Macrocyclic Myxobacterial Metabolite with Potent Antiproliferative Properties. *Angew. Chem.* **2015**, *54*, 10145–10148. [CrossRef]
30. Krastel, P.; Roggo, S.; Schirle, M.; Ross, N.T.; Perruccio, F.; Aspesi, P., Jr.; Aust, T.; Buntin, K.; Estoppey, D.; Liechty, B. Nannocystin A: An Elongation Factor 1 Inhibitor from Myxobacteria with Differential Anti-Cancer Properties. *Angew. Chem.* **2015**, *54*, 10149–10154. [CrossRef]
31. Liao, L.; Zhou, J.; Xu, Z.; Ye, T. Concise Total Synthesis of Nannocystin A. *Angew. Chem.* **2016**, *55*, 13263–13266. [CrossRef] [PubMed]
32. Huang, J.; Wang, Z. Total Syntheses of Nannocystins A and A0, Two Elongation Factor 1 Inhibitors. *Org. Lett.* **2016**, *18*, 4702–4705. [CrossRef] [PubMed]
33. Yang, Z.; Xu, X.; Yang, C.-H.; Tian, Y.; Chen, X.; Lian, L.; Pan, W.; Su, X.; Zhang, W.; Chen, Y. Total Synthesis of Nannocystin A. *Org. Lett.* **2016**, *18*, 5768–5770. [CrossRef] [PubMed]
34. Liu, Q.; Hu, P.; He, Y. Asymmetric Total Synthesis of Nannocystin A. *J. Org. Chem.* **2017**, *82*, 9217–9222. [CrossRef]
35. Tian, Y.; Xu, X.; Ding, Y.; Hao, X.; Bai, Y.; Tang, Y.; Zhang, X.; Li, Q.; Yang, Z.; Zhang, W. Synthesis and biological evaluation of nannocystin analogues toward understanding the binding role of the (2R,3S)-Epoxide in nannocystin A. *Eur. J. Med. Chem.* **2018**, *150*, 626–632. [CrossRef]
36. Limnios, D.; Kokotos, C. 2, 2, 2-Trifluoroacetophenone: An organocatalyst for an environmentally friendly epoxidation of alkenes. *J. Org. Chem.* **2014**, *79*, 4270–4276. [CrossRef]
37. Wójtowicz-Młochowska, H. Synthetic utility of metal catalyzed hydrogen peroxide oxidation of CH, CC and C=C bonds in alkanes, arenes and alkenes: Recent advances. *Arkivoc* **2017**, *2017*, 12–58. [CrossRef]
38. Wang, C.; Yamamoto, H. Asymmetric epoxidation using hydrogen peroxide as oxidant. *Chem. Asian J.* **2015**, *10*, 2056–2068. [CrossRef]

39. Quideau, S.; Lyvinec, G.; Marguerit, M.; Bathany, K.; Ozanne-Beaudenon, A.; Buffeteau, T.; Cavagnat, D.; Chénedé, A. Asymmetric Hydroxylative Phenol Dearomatization through In Situ Generation of Iodanes from Chiral Iodoarenes and m-CPBA. *Angew. Chem.* **2009**, *48*, 4605–4609. [CrossRef]
40. Yang, D.; Ye, X.-Y.; Xu, M. Enantioselective Total Synthesis of (−)-Triptolide, (−)-Triptonide, (+)-Triptophenolide, and (+)-Triptoquinonide. *J. Org. Chem.* **2000**, *65*, 2208–2217. [CrossRef]
41. Xu, H.; Tang, H.; Feng, H.; Li, Y. Divergent total synthesis of triptolide, triptonide, tripdiolide, 16-hydroxytriptolide, and their analogues. *J. Org. Chem.* **2014**, *79*, 10110–10122. [CrossRef] [PubMed]
42. Xu, H.; Tang, H.; Yang, Z.; Feng, H.; Li, Y. Synthesis and biological evaluation of 20-hydroxytriptonide and its analogues. *Tetrahedron* **2014**, *70*, 3107–3115. [CrossRef]
43. Goncalves, S.; Hellier, P.; Nicolas, M.; Wagner, A.; Baati, R. Diastereoselective formal total synthesis of (+/−)-triptolide via a novel cationic cyclization of 2-alkenyl-1,3-dithiolane. *ChemComm* **2010**, *46*, 5778–5780. [CrossRef] [PubMed]
44. Xu, W.D.; Li, L.Q.; Li, M.M.; Geng, H.C.; Qin, H.B. Catalytic Asymmetric Formal Total Synthesis of (−)-Triptophenolide and (+)-Triptolide. *Nat. Prod. Bioprospect* **2016**, *6*, 183–186. [CrossRef]
45. Yang, D.; Ye, X.-Y.; Xu, M.; Pang, K.-W.; Zou, N.; Letcher, R.M. A Concise Total Synthesis of Triptolide. *J. Org. Chem.* **1998**, *63*, 6446–6447. [CrossRef]
46. Tanaka, K.; Kobayashi, K.; Takatori, K.; Kogen, H. Efficient synthesis of syn -α-alkoxy epoxide via a diastereoselective Darzens reaction. *Tetrahedron* **2017**, *73*, 2062–2067. [CrossRef]
47. Ashokkumar, V.; Siva, A.; Ramaswamy Chidambaram, R. A highly enantioselective asymmetric Darzens reaction catalysed by proline based efficient organocatalysts for the synthesis of di- and tri-substituted epoxides. *ChemComm* **2017**, *53*, 10926–10929. [CrossRef]
48. Li, B.; Li, C. Darzens reaction rate enhancement using aqueous media leading to a high level of kinetically controlled diastereoselective synthesis of steroidal epoxyketones. *J. Org. Chem.* **2014**, *79*, 8271–8277. [CrossRef]
49. Day, D.P.; Sellars, P.B. Recent Advances in Iminium-Salt-Catalysed Asymmetric Epoxidation. *Eur. J. Org. Chem.* **2017**, *2017*, 1034–1044. [CrossRef]
50. Heravi, M.M.; Lashaki, T.B.; Poorahmad, N. Applications of Sharpless asymmetric epoxidation in total synthesis. *Tetrahedron: Asymmetry* **2015**, *26*, 405–495. [CrossRef]
51. Ramesh, P.; Reddy, Y.N. A three-step total synthesis of goniothalesdiol A using a one-pot Sharpless epoxidation/regioselective epoxide ring-opening. *Tetrahedron Lett.* **2017**, *58*, 1037–1039. [CrossRef]
52. Zhu, Y.; Wang, Q.; Cornwall, R.G.; Shi, Y. Organocatalytic asymmetric epoxidation and aziridination of olefins and their synthetic applications. *Chem. Rev.* **2014**, *114*, 8199–8256. [CrossRef] [PubMed]
53. Dai, W.; Shang, S.; Chen, B.; Li, G.; Wang, L.; Ren, L.; Gao, S. Asymmetric epoxidation of olefins with hydrogen peroxide by an in situ-formed manganese complex. *J. Org. Chem.* **2014**, *79*, 6688–6694. [CrossRef] [PubMed]
54. Noji, M.; Kobayashi, T.; Uechi, Y.; Kikuchi, A.; Kondo, H.; Sugiyama, S.; Ishii, K. Asymmetric Epoxidation of Allylic Alcohols Catalyzed by Vanadium–Binaphthylbishydroxamic Acid Complex. *J. Org. Chem.* **2015**, *80*, 3203–3210. [CrossRef]
55. Noji, M.; Kondo, H.; Yazaki, C.; Yamaguchi, H.; Ohkura, S.; Takanami, T. An immobilized vanadium-binaphthylbishydroxamic acid complex as a reusable catalyst for the asymmetric epoxidation of allylic alcohols. *Tetrahedron Lett.* **2019**, *60*, 1518–1521. [CrossRef]
56. Kawaguchi, M.; Nakano, K.; Hosoya, K.; Orihara, T.; Yamanaka, M.; Odagi, M.; Nagasawa, K. Asymmetric Epoxidation of 1, 4-Naphthoquinones Catalyzed by Guanidine–Urea Bifunctional Organocatalyst. *Org. Lett.* **2018**, *20*, 2811–2815. [CrossRef]
57. Wuitschik, G.; Carreira, E.M.; Wagner, B.; Fischer, H.; Parrilla, I.; Schuler, F.; Rogers-Evans, M.; Muller, K. Oxetanes in drug discovery: Structural and synthetic insights. *J. Med. Chem.* **2010**, *53*, 3227–3246. [CrossRef]
58. Burkhard, J.A.; Wuitschik, G.; Rogers-Evans, M.; Müller, K.; Carreira, E.M. Oxetanes as versatile elements in drug discovery and synthesis. *Angew. Chem.* **2010**, *49*, 9052–9067. [CrossRef]
59. Mahal, A. Oxetanes as versatile building blocks in the total synthesis of natural products: An overview. *Eur. J. Chem.* **2015**, *6*, 357–366. [CrossRef]
60. Bull, J.A.; Croft, R.A.; Davis, O.A.; Doran, R.; Morgan, K.F. Oxetanes: Recent advances in synthesis, reactivity, and medicinal chemistry. *Chem. Rev.* **2016**, *116*, 12150–12233. [CrossRef]

61. Fraguas-Sánchez, A.; Martín-Sabroso, C.; Fernández-Carballido, A.; Torres-Suárez, A. Current status of nanomedicine in the chemotherapy of breast cancer. *Cancer Chemother. Pharmacol.* **2019**, *84*, 289–706. [CrossRef] [PubMed]
62. Peng, J.; Chen, J.; Xie, F.; Bao, W.; Xu, H.; Wang, H.; Xu, Y.; Du, Z. Herceptin-conjugated paclitaxel loaded PCL-PEG worm-like nanocrystal micelles for the combinatorial treatment of HER2-positive breast cancer. *Biomaterials* **2019**, *222*, 119420. [CrossRef] [PubMed]
63. Shetti, D.; Zhang, B.; Fan, C.; Mo, C.; Lee, B.H.; Wei, K. Low Dose of Paclitaxel Combined with XAV939 Attenuates Metastasis, Angiogenesis and Growth in Breast Cancer by Suppressing Wnt Signaling. *Cells* **2019**, *8*, 892. [CrossRef] [PubMed]
64. Khalifa, A.; Elsheikh, M.A.; Khalifa, A.; Elnaggar, Y.S. Current strategies for different paclitaxel-loaded Nano-delivery Systems towards therapeutic applications for ovarian carcinoma: A review article. *J. Control. Release* **2019**, *311–312*, 125–137. [CrossRef]
65. Cong, J.; Liu, R.; Hou, J.; Wang, X.; Jiang, H.; Wang, J. Therapeutic effect of bevacizumab combined with paclitaxel and carboplatin on recurrent ovarian cancer. *JBUON* **2019**, *24*, 1003–1008.
66. Park, G.-B.; Jeong, J.-Y.; Kim, D. Gliotoxin Enhances Autophagic Cell Death via the DAPK1-TAp63 Signaling Pathway in Paclitaxel-Resistant Ovarian Cancer Cells. *Mar. Drugs* **2019**, *17*, 412. [CrossRef]
67. Morgensztern, D.; Karaseva, N.; Felip, E.; Delgado, I.; Burdaeva, O.; Dómine, M.; Lara, P.; Paik, P.K.; Lassen, U.; Orlov, S. An open-label phase IB study to evaluate GSK3052230 in combination with paclitaxel and carboplatin, or docetaxel, in FGFR1-amplified non-small cell lung cancer. *Lung Cancer* **2019**, *136*, 74–79. [CrossRef]
68. Villaruz, L.C.; Cobo, M.; Syrigos, K.; Mavroudis, D.; Zhang, W.; Kim, J.S.; Socinski, M.A. A phase II study of nab-paclitaxel and carboplatin chemotherapy plus necitumumab in the first-line treatment of patients with stage IV squamous non-small cell lung cancer. *Lung Cancer* **2019**, *136*, 52–56. [CrossRef]
69. Wang, J.P.; Yan, J.P.; Xu, J.; Yin, T.H.; Zheng, R.Q.; Wang, W. Paclitaxel-loaded nanobubble targeted to pro-gastrin-releasing peptide inhibits the growth of small cell lung cancer. *Cancer Manag. Res.* **2019**, *11*, 6637–6649. [CrossRef]
70. Liu, J.J.; Ho, J.Y.; Lee, H.W.; Baik, M.W.; Kim, O.; Choi, Y.J.; Hur, S.Y. Inhibition of Phosphatidylinositol 3-kinase (PI3K) Signaling Synergistically Potentiates Antitumor Efficacy of Paclitaxel and Overcomes Paclitaxel-Mediated Resistance in Cervical Cancer. *Int. J. Mol. Sci.* **2019**, *20*, 3383. [CrossRef]
71. Suzuki, K.; Nagao, S.; Shibutani, T.; Yamamoto, K.; Jimi, T.; Yano, H.; Kitai, M.; Shiozaki, T.; Matsuoka, K.; Yamaguchi, S. Phase II trial of paclitaxel, carboplatin, and bevacizumab for advanced or recurrent cervical cancer. *Gynecol. Oncol.* **2019**, *154*, 554–557. [CrossRef] [PubMed]
72. Flores-Villaseñor, S.E.; Peralta-Rodríguez, R.D.; Padilla-Vaca, F.; Meléndez-Ortiz, H.I.; Ramirez-Contreras, J.C.; Franco, B. Preparation of Peppermint Oil-Based Nanodevices Loaded with Paclitaxel: Cytotoxic and Apoptosis Studies in HeLa Cells. *AAPS PharmSciTech* **2019**, *20*, 198. [CrossRef] [PubMed]
73. Huang, J.M.; Yokoyama, R.; Yang, C.S.; Fukuyama, Y. Merrilactone A, a novel neurotrophic sesquiterpene dilactone from Illicium merrillianum. *Tetrahedron Lett.* **2000**, *41*, 6111–6114. [CrossRef]
74. Li, C.; Lee, D.; Graf, T.N.; Phifer, S.S.; Nakanishi, Y.; Burgess, J.P.; Riswan, S.; Setyowati, F.M.; Saribi, A.M.; Soejarto, D.D. A Hexacyclic ent-Trachylobane Diterpenoid Possessing an Oxetane Ring from Mitrephora glabra. *Org. Lett.* **2005**, *7*, 5709–5712. [CrossRef] [PubMed]
75. Hugelshofer, C.L.; Magauer, T. A Bioinspired Cyclization Sequence Enables the Asymmetric Total Synthesis of Dictyoxetane. *J. Am. Chem. Soc.* **2016**, *138*, 6420–6423. [CrossRef] [PubMed]
76. Thompson, M.P.; Agger, J.; Wong, L.S. Paternò–Büchi Reaction as a Demonstration of Chemical Kinetics and Synthetic Photochemistry Using a Light Emitting Diode Apparatus. *J. Chem. Educ.* **2015**, *92*, 1716–1720. [CrossRef]
77. Birman, V.B.; Danishefsky, S.J. The Total Synthesis of (±)-Merrilactone A. *J. Am. Chem. Soc.* **2002**, *124*, 2080–2081. [CrossRef]
78. Inoue, M.; Sato, T.; Hirama, M. Total synthesis of merrilactone A. *J. Am. Chem. Soc.* **2003**, *125*, 10772–10773. [CrossRef]
79. Inoue, M.; Lee, N.; Kasuya, S.; Sato, T.; Hirama, M.; Moriyama, M.; Fukuyama, Y. Total synthesis and bioactivity of an unnatural enantiomer of merrilactone a: Development of an enantioselective desymmetrization strategy. *J. Org. Chem.* **2007**, *72*, 3065–3075. [CrossRef]

80. Nazef, N.; Davies, R.D.; Greaney, M.F. Formal Synthesis of Merrilactone A Using a Domino Cyanide 1, 4-Addition–Aldol Cyclization. *Org. Lett.* **2012**, *14*, 3720–3723. [CrossRef]
81. Chen, J.; Gao, P.; Yu, F.; Yang, Y.; Zhu, S.; Zhai, H. Total Synthesis of (±)-Merrilactone A. *Angew. Chem.* **2012**, *51*, 5897–5899. [CrossRef] [PubMed]
82. Liu, W.; Wang, B. Synthesis of (+/−)-Merrilactone A by a Desymmetrization Strategy. *Chemistry* **2018**, *24*, 16511–16515. [CrossRef] [PubMed]
83. Richter, M.J.; Schneider, M.; Brandstatter, M.; Krautwald, S.; Carreira, E.M. Total Synthesis of (-)-Mitrephorone A. *J. Am. Chem. Soc.* **2018**, *140*, 16704–16710. [CrossRef] [PubMed]
84. Guo, Y.A.; Lee, W.; Krische, M.J. Enantioselective Synthesis of Oxetanes Bearing All-Carbon Quaternary Stereocenters via Iridium-Catalyzed C–C Bond-Forming Transfer Hydrogenation. *Chem. Eur. J.* **2017**, *23*, 2557–2559. [CrossRef] [PubMed]
85. Lorente, A.; Lamariano-Merketegi, J.; Albericio, F.; Alvarez, M. Tetrahydrofuran-containing macrolides: A fascinating gift from the deep sea. *Chem. Rev.* **2013**, *113*, 4567–4610. [CrossRef] [PubMed]
86. Torre, A.; Cuyamendous, C.; Bultel-Poncé, V.; Durand, T.; Galano, J.M.; Oger, C. Recent advances in the synthesis of tetrahydrofurans and applications in total synthesis. *Terahedorn* **2016**, *33*, 5003–5025. [CrossRef]
87. Chen, L.Y.; Chen, J.R.; Cheng, H.G.; Lu, L.Q.; Xiao, W.J. Enantioselective Synthesis of Tetrahydrofuran Derivatives by Sequential Henry Reaction and Iodocyclization of γ, δ-Unsaturated Alcohols. *Eur. J. Org. Chem.* **2014**, *2014*, 4714–4719. [CrossRef]
88. Li, H.; Li, Y.; Ao, H.; Bi, D.; Han, M.; Guo, Y.; Wang, X. Folate-targeting annonaceous acetogenins nanosuspensions: Significantly enhanced antitumor efficacy in HeLa tumor-bearing mice. *Drug Deliv.* **2018**, *25*, 880–887. [CrossRef]
89. Bermejo, A.; Figadère, B.; Zafra-Polo, M.-C.; Barrachina, I.; Estornell, E.; Cortes, D. Acetogenins from Annonaceae: Recent progress in isolation, synthesis and mechanisms of action. *Nat. Pro. Rep.* **2005**, *22*, 269–303. [CrossRef]
90. Liaw, C.C.; Wu, T.Y.; Chang, F.R.; Wu, Y.C. Historic perspectives on Annonaceous acetogenins from the chemical bench to preclinical trials. *Planta Medica* **2010**, *76*, 1390–1404. [CrossRef]
91. Hernández-Fuentes, G.A.; García-Argáez, A.N.; Peraza Campos, A.L.; Delgado-Enciso, I.; Muñiz-Valencia, R.; Martínez-Martínez, F.J.; Toninello, A.; Gómez-Sandoval, Z.; Mojica-Sánchez, J.P.; Dalla Via, L. Cytotoxic Acetogenins from the Roots of Annona purpurea. *Int. J. Mol. Sci.* **2019**, *20*, 1870. [CrossRef]
92. Juang, S.H.; Chiang, C.Y.; Liang, F.P.; Chan, H.H.; Yang, J.S.; Wang, S.H.; Lin, Y.C.; Kuo, P.C.; Shen, M.R.; Thang, T.D. Mechanistic Study of Tetrahydrofuran-acetogenins In Triggering Endoplasmic Reticulum Stress Response-apotoposis in Human Nasopharyngeal Carcinoma. *Sci. Rep.* **2016**, *6*, 39251. [CrossRef] [PubMed]
93. Ma, S.-G.; Li, M.; Lin, M.-B.; Li, L.; Liu, Y.-B.; Qu, J.; Li, Y.; Wang, X.-J.; Wang, R.-B.; Xu, S. Illisimonin A, a Caged Sesquiterpenoid with a Tricyclo[5.2.1.01,6]decane Skeleton from the Fruits of Illicium simonsii. *Org. Lett.* **2017**, *19*, 6160–6163. [CrossRef] [PubMed]
94. Liu, W.; Wang, B. Stereoselective Synthesis of a Common 3-Oxabicyclo[3.2.0]heptan-2-one Core Building Block Toward Illicium Sesquiterpenes via Desymmetrization. *Chem. Res. Chin. Univ.* **2018**, *34*, 867–870. [CrossRef]
95. Burns, A.S.; Rychnovsky, S.D. Total Synthesis and Structure Revision of (−)-Illisimonin A, a Neuroprotective Sesquiterpenoid from the Fruits of Illicium simonsii. *J. Am. Chem. Soc.* **2019**, *141*, 13295–13300. [CrossRef] [PubMed]
96. Suárez-Ortiz, G.A.; Cerda-García-Rojas, C.M.; Fragoso-Serrano, M.; Pereda-Miranda, R. Complementarity of DFT Calculations, NMR Anisotropy, and ECD for the Configurational Analysis of Brevipolides K–O from Hyptis brevipes. *J. Nat. Prod.* **2017**, *80*, 181–189. [CrossRef] [PubMed]
97. Kumagai, K.; Tsuda, M.; Masuda, A. Iriomoteolide-2a, a Cytotoxic 23-Membered Macrolide from Marine Benthic Dinoflagellate Amphidinium Species. *Heterocycles* **2015**, *91*, 265–274. [CrossRef]
98. Shiva Raju, K.; Sabitha, G. First stereoselective total synthesis of brevipolide M. *Org. Biomol. Chem.* **2017**, *15*, 6393–6400. [CrossRef]
99. Sakamoto, K.; Hakamata, A.; Tsuda, M.; Fuwa, H. Total Synthesis and Stereochemical Revision of Iriomoteolide-2a. *Angew. Chem. Int. Ed. Engl.* **2018**, *57*, 3801–3805. [CrossRef]
100. Sakamoto, K.; Hakamata, A.; Iwasaki, A.; Suenaga, K.; Tsuda, M.; Fuwa, H. Total Synthesis, Stereochemical Revision, and Biological Assessment of Iriomoteolide-2a. *Chemistry* **2019**, *25*, 8528–8542. [CrossRef]

101. Lindsay, V.N.; Viart, H.l.n.M.-F.; Sarpong, R. Stereodivergent intramolecular C (sp3)–H functionalization of azavinyl carbenes: Synthesis of saturated heterocycles and fused N-heterotricycles. *J. Am. Chem. Soc.* **2015**, *137*, 8368–8371. [CrossRef] [PubMed]
102. Yuan, X.; Lin, L.; Chen, W.; Wu, W.; Liu, X.; Feng, X. Synthesis of chiral tetrahydrofurans via catalytic asymmetric [3+ 2] cycloaddition of heterosubstituted alkenes with oxiranes. *J. Org. Chem.* **2016**, *81*, 1237–1243. [CrossRef] [PubMed]
103. Lee, S.; Bae, H.Y.; List, B. Can a ketone be more reactive than an aldehyde? Catalytic asymmetric synthesis of substituted tetrahydrofurans. *Angew. Chem. Int. Ed. Engl.* **2018**, *57*, 12162–12166. [CrossRef] [PubMed]
104. Nasir, N.M.; Ermanis, K.; Clarke, A.P. Strategies for the construction of tetrahydropyran rings in the synthesis of natural products. *Org. Biomol. Chem.* **2014**, *12*, 3323–3335. [CrossRef]
105. Ghosh, A.K.; Brindisi, M. Achmatowicz reaction and its application in the syntheses of bioactive molecules. *RSC Adv.* **2016**, *6*, 111564–111598. [CrossRef]
106. Antoszczak, M.; Huczyński, A. Salinomycin and its derivatives–A new class of multiple-targeted "magic bullets". *Eur. J. Med. Chem.* **2019**, *176*, 208–227. [CrossRef]
107. Markowska, A.; Kaysiewicz, J.; Markowska, J.; Huczyński, A. Doxycycline, salinomycin, monensin and ivermectin repositioned as cancer drugs. *Bioorg. Med. Chem. Lett.* **2019**, *29*, 1549–1554. [CrossRef]
108. Tyagi, M.; Patro, B.S. Salinomycin reduces growth, proliferation and metastasis of cisplatin resistant breast cancer cells via NF-kB deregulation. *Toxicol. In Vitro* **2019**, *60*, 125–133. [CrossRef]
109. Dewangan, J.; Srivastava, S.; Mishra, S.; Divakar, A.; Kumar, S.; Rath, S.K. Salinomycin inhibits breast cancer progression via targeting HIF-1α/VEGF mediated tumor angiogenesis in vitro and in vivo. *Biochem. Pharmacol.* **2019**, *164*, 326–335. [CrossRef]
110. Yang, X.W.; Yang, C.P.; Jiang, L.P.; Qin, X.J.; Liu, Y.P.; Shen, Q.S.; Chen, Y.B.; Luo, X.D. Indole Alkaloids with New Skeleton Activating Neural Stem Cells. *Org. Lett.* **2014**, *16*, 5808–5811. [CrossRef]
111. Mason, J.D.; Weinreb, S.M. Synthesis of Alstoscholarisines A–E, Monoterpene Indole Alkaloids with Modulating Effects on Neural Stem Cells. *J. Org. Chem.* **2018**, *83*, 5877–5896. [CrossRef] [PubMed]
112. Yang, B.Y.; Kong, L.Y.; Wang, X.B.; Zhang, Y.M.; Li, R.J.; Yang, M.H.; Luo, J.G. Nitric oxide inhibitory activity and absolute configurations of arylalkenyl α, β-unsaturated δ/γ-lactones from Cryptocarya concinna. *J. Nat. Prod.* **2016**, *79*, 196–203. [CrossRef] [PubMed]
113. Della-Felice, F.; Sarotti, A.M.; Pilli, R. Catalytic Asymmetric Synthesis and Stereochemical Revision of (+)-Cryptoconcatone H. *J. Org. Chem.* **2017**, *82*, 9191–9197. [CrossRef] [PubMed]
114. Zhang, Z.; Tong, R. Synthetic Approaches to 2, 6-trans-Tetrahydropyrans. *Synthesis* **2017**, *49*, 4899–4916. [CrossRef]
115. Zhao, L.M.; Dou, F.; Sun, R.; Zhang, A. Regioselective Synthesis of Substituted Tetrahydrofurans through Prins Cyclization. *Synlett* **2014**, *25*, 1431–1434. [CrossRef]
116. Reddy, B.S.; Swathi, V.; Bhadra, M.P.; Raju, M.K.; Kunwar, A. Tandem vinylcyclopropane ring opening/Prins cyclization for the synthesis of 2, 3-disubstituted tetrahydropyrans. *Tetrahedron Lett.* **2016**, *57*, 1889–1891. [CrossRef]
117. Millán, A.; Smith, J.R.; Chen, J.L.Y.; Aggarwal, V.K. Tandem Allylboration–Prins Reaction for the Rapid Construction of Substituted Tetrahydropyrans: Application to the Total Synthesis of (−)-Clavosolide A. *Angew. Chem.* **2016**, *55*, 2498–2502. [CrossRef]
118. Han, X.; Peh, G.; Floreancig, P.E. Prins-Type Cyclization Reactions in Natural Product Synthesis. *Eur. J. Org. Chem.* **2013**, *2013*, 1193–1208. [CrossRef]
119. Vetica, F.; Chauhan, P.; Dochain, S.; Enders, D. Asymmetric organocatalytic methods for the synthesis of tetrahydropyrans and their application in total synthesis. *Chem. Soc. Rev.* **2017**, *46*, 1661–1674. [CrossRef]
120. Li, L.; Sun, X.; He, Y.; Gao, L.; Song, Z. TMSBr/InBr3-promoted Prins cyclization/homobromination of dienyl alcohol with aldehyde to construct cis-THP containing an exocyclic E-alkene. *ChemComm* **2015**, *51*, 14925–14928. [CrossRef]
121. Zhang, Z.; Xie, H.; Li, H.; Gao, L.; Song, Z. Total Synthesis of (−)-Exiguolide. *Org. Lett.* **2015**, *17*, 4706–4709. [CrossRef] [PubMed]
122. Loy, R.N.; Jacobsen, E.N. Enantioselective Intramolecular Openings of Oxetanes Catalyzed by (salen)Co(III) Complexes: Access to Enantioenriched Tetrahydrofurans. *J. Am. Chem. Soc.* **2009**, *131*, 2786–2787. [CrossRef] [PubMed]

123. Yadav, J.S.; Singh, V.K.; Srihari, P. Formation of Substituted Tetrahydropyrans through Oxetane Ring Opening: Application to the Synthesis of C1–C17 Fragment of Salinomycin. *Org. Lett.* **2014**, *16*, 836–839. [CrossRef] [PubMed]
124. Srinivas, B.; Reddy, D.S.; Mallampudi, N.A.; Mohapatra, D.K.J.O.l. A General Diastereoselective Strategy for Both cis-and trans-2, 6-Disubstituted Tetrahydropyrans: Formal Total Synthesis of (+)-Muconin. *Org. Lett.* **2018**, *20*, 6910–6914. [CrossRef]
125. Alvarez-Mendez, S.J.; Farina-Ramos, M.; Villalba, M.L.; Perretti, M.D.; Garcia, C.; Moujir, L.M.; Ramirez, M.A.; Martin, V.S. Stereoselective Synthesis of Highly Substituted Tetrahydropyrans through an Evans Aldol-Prins Strategy. *J. Org. Chem.* **2018**, *83*, 9039–9066. [CrossRef]
126. Reekie, T.A.; Kavanagh, M.E.; Longworth, M.; Kassiou, M. Synthesis of Biologically Active Seven-Membered-Ring Heterocycles. *Synthesis* **2013**, *45*, 3211–3227. [CrossRef]
127. Barbero, H.; Diez-Poza, C.; Barbero, A. The Oxepane Motif in Marine Drugs. *Mar. Drugs* **2017**, *15*. [CrossRef]
128. Wang, W.; Song, T.; Chai, W.; Chen, L.; Chen, L.; Lian, X.Y.; Zhang, Z. Rare Polyene-polyol Macrolides from Mangrove-derived Streptomyces sp. ZQ4BG. *Sci. Rep.* **2017**, *7*, 1703. [CrossRef]
129. Ahmed, A.F.; Chen, Y.W.; Huang, C.Y.; Tseng, Y.J.; Lin, C.C.; Dai, C.F.; Wu, Y.C.; Sheu, J.H. Isolation and Structure Elucidation of Cembranoids from a Dongsha Atoll Soft Coral Sarcophyton stellatum. *Mar. Drugs* **2018**, *16*. [CrossRef]
130. Guo, Y.; Zhang, N.; Sun, W.; Duan, X.; Zhang, Q.; Zhou, Q.; Chen, C.; Zhu, H.; Luo, Z.; Liu, J.; et al. Bioactive polycyclic polyprenylated acylphloroglucinols from Hypericum perforatum. *Org. Biomol. Chem.* **2018**, *16*, 8130–8143. [CrossRef]
131. Hernández-Torres, G.; Mateo, J.; Colobert, F.; Urbano, A.; Carreño, M.C. Synthesis of Medium-Sized 2,ω-cis-Disubstituted Cyclic Ethers by Reductive Cyclization of Hydroxy Ketones. *ChemistrySelect* **2016**, *1*, 4101–4107. [CrossRef]
132. Armbrust, K.W.; Beaver, M.G.; Jamison, T.F. Rhodium-Catalyzed Endo-Selective Epoxide-Opening Cascades: Formal Synthesis of (−)-Brevisin. *J. Am. Chem. Soc.* **2015**, *137*, 6941–6946. [CrossRef] [PubMed]
133. Sun, Y.; Huang, Y.; Li, M.; Lu, J.; Jin, N.; Fan, B. Synthesis of cyclic ethers by cyclodehydration of 1, n-diols using heteropoly acids as catalysts. *Royal Soc. Open Sci.* **2018**, *5*, 180740. [CrossRef] [PubMed]

© 2019 by the authors. Licensee MDPI, Basel, Switzerland. This article is an open access article distributed under the terms and conditions of the Creative Commons Attribution (CC BY) license (http://creativecommons.org/licenses/by/4.0/).

Review

Recent Advances in the Synthesis of 2*H*-Pyrans

David Tejedor [1,*], Samuel Delgado-Hernández [1,2], Raquel Diana-Rivero [1,2], Abián Díaz-Díaz [1,2] and Fernando García-Tellado [1,*]

1. Instituto de Productos Naturales y Agrobiología, Consejo Superior de Investigaciones Científicas, Astrofísico Francisco Sánchez 3, 38206 La Laguna, Tenerife, Spain
2. Doctoral and Postgraduate School, Universidad de La Laguna, Astrofísico Francisco Sánchez, SN, 38200 La Laguna, Tenerife, Spain

* Correspondence: dtejedor@ipna.csic.es (D.T.); fgarcia@ipna.csic.es (F.G.-T.); Tel.: +34-922-256847 (D.T. & F.G.-T.)

Academic Editor: Gianfranco Favi
Received: 25 July 2019; Accepted: 9 August 2019; Published: 9 August 2019

Abstract: In this review, we discuss the nature of the different physicochemical factors affecting the valence isomerism between 2*H*-pyrans (2HPs) and 1-oxatrienes, and we describe the most versatile synthetic methods reported in recent literature to access to 2HPs, with the only exception of 2HPs fused to aromatic rings (i.e., 2*H*-chromenes), which are not included in this review.

Keywords: 2*H*-pyran; heterocycles; synthesis; valence isomerism; 1-oxa-triene; dienone; oxa-electrocyclization; Knoevenagel; propargyl Claisen; cycloisomerization

1. Introduction

The 2*H*-pyran (2HP) ring constitutes a structural motif present in many natural products (Figure 1) [1] and is a strategic key intermediate in the construction of many of these structures [2,3]. In spite of their importance, the literature of 2HPs is relatively scarce [4–9], mainly due to the instability associated with the heterocyclic ring, which makes these heterocycles establish an equilibrium with their opened isomeric forms (Scheme 1). Fusion of a 2HP to an aromatic ring confers stability to these heterocycles. Thus, while simple 2HPs are difficult to synthesize as pure and isolated compounds, many of their benzo derivatives (i.e., 2*H*-chromenes) constitute stable molecules, with a broad spectrum of biological activities and a widespread representation in the higher plants (Figure 1). Because the chemistry and reactivity of 2*H*-chromenes have been already previously revised [1,10–15], they will not be included in this review. Instead, we will focus on the recent advances on accessing 2HPs, either as simple and stable monocyclic structures or as part of fused polycyclic structures, excluding the 2*H*-chromene system.

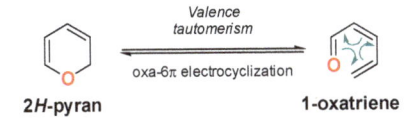

Scheme 1. Valence tautomerism of 2*H*-pyrans (2HPs).

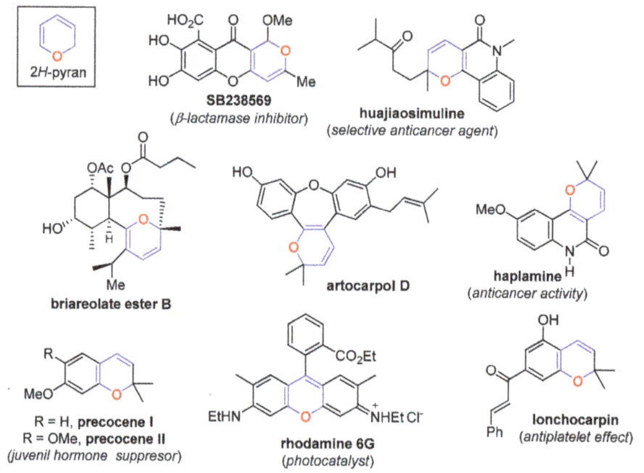

Figure 1. Examples of natural products containing the 2HP motif.

2. Dienone/2HP Equilibrium

2HPs are prone to undergo spontaneous valence isomerization [16] to the corresponding 1-oxatrienes through a reversible pericyclic oxa-6π-electrocyclization process (Scheme 1) [17]. This valence tautomerism determines the chemistry of these heterocycles, which commonly exist as a mixture of valence tautomers (isomers) [1]. In this sense, it is important to take into account that because this interconversion is fast in the majority of cases, the method of synthesis used to access these structures does not determine the valence tautomer obtained.

Although this valence isomerization was invoked to explain some enigmatic results found in earlier examples with these molecules, the first clear-cut example of it came from the irradiation of *trans*-β-ionone (**1**) (Scheme 2) [18]. Authors found that the irradiation afforded a mixture of *cis*-β-ionone (**2**) and 2HP **3**, with a value for the equilibrium constant K = 4.61 at 327 °K (K = 1.52 at 386 °K), and values for $k_1 = 1.4 \times 10^{-3} \cdot s^{-1}$ and $k_{-1} = 1.3 \times 10^{-4} \cdot s^{-1}$. In addition, measurements at different temperatures gave activation energies (E_a) of 20 Kcal/mol for the *cis*-dienone to 2HP reaction, and 27 Kcal/mol for the reverse process [19].

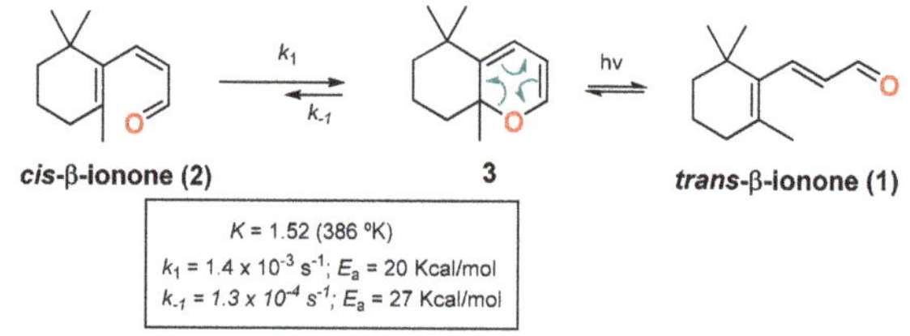

K = 1.52 (386 °K)
$k_1 = 1.4 \times 10^{-3}\ s^{-1}$; E_a = 20 Kcal/mol
$k_{-1} = 1.3 \times 10^{-4}\ s^{-1}$; E_a = 27 Kcal/mol

Scheme 2. Valence isomerism of *cis*-β-ionone (**2**) and 2HP **3**.

Further studies on the conformation of conjugated dienones allowed the establishment of some general patterns for these dienone/2HP equilibria (Scheme 3) [20]. It was observed that steric destabilization of the dienones shifted the equilibria toward the 2HPs. This was the case for tetrasubstituted dienones **4** and **5**, which fully isomerized to the corresponding 2HPs. On the other hand, simpler dienones **8–12**, which could adopt a stable planar conformation, existed in the opened form. Likewise, trisubstituted dienones **6–7**, which are representative examples of dienones featuring non-stable planar conformations, preferred their closed forms. Along with these results, the authors also observed that the substitution of a δ-alkyl substituent (R^5 or R^6) by a substituent able to extend the conjugation of the π-system (e.g, vinyl group) favored the dienone form (Scheme 3). A main conclusion from these and other studies [20,21] is that the existence of the 2HP form depends primarily on the steric destabilization of the dienone rather than on its specific substitution pattern. Thus, the design of stable 2HPs must include, among other structural/electronic criteria, enough steric destabilization on the dienone to penalize the valence isomerization.

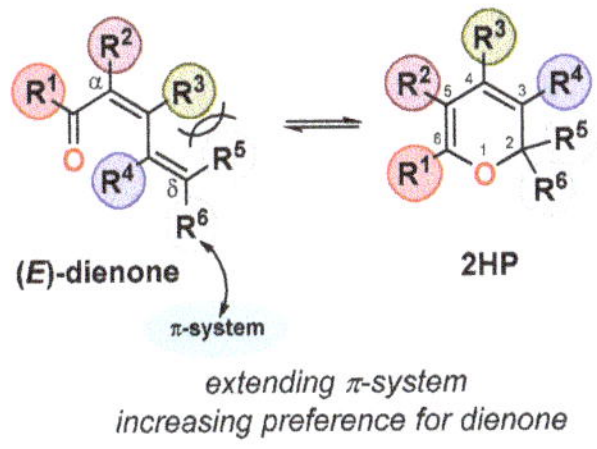
(E)-dienone
2HP
π-system
extending π-system
increasing preference for dienone

	Substituent	dienone	2HP
4	$R^1 = R^3 = R^4 = R^5 = Me$	-	+
5	$R^1 = R^3 = R^5 = R^6 = Me$	-	+
6	$R^1 = R^5 = R^6 = Me$	-	+
7	$R^1 = Me$; $R^5 = R^6 = {}^nBu$	-	+
8	$R^1 = R^6 = Me$	+	-
9	$R^1 = R^5 = R^6 = Me$	+	-
10	$R^1 = R^4 = R^6 = Me$	+	-
11	R^1-$R^2 = R^5$-$R^6 = -(CH_2)_4-$	+	-
12	$R^1 = {}^iBu$; $R^4 = Me$	+	-

Stable planar conformations

Scheme 3. Implications of the *cis*-dienone conformation on the valence isomerism.

More recently, Krasnaya et al. carried out a systematic investigation on the influence of substituents and solvents on the valence isomerization of trisubstituted α-acyl-dienones **13** (Scheme 4). In this study, the authors quantified the equilibrium compositions of 26 differently substituted α-acyl-dienones **13** (Table 1), and determined the thermodynamic and activation parameters for some of these equilibria (Scheme 5) [22].

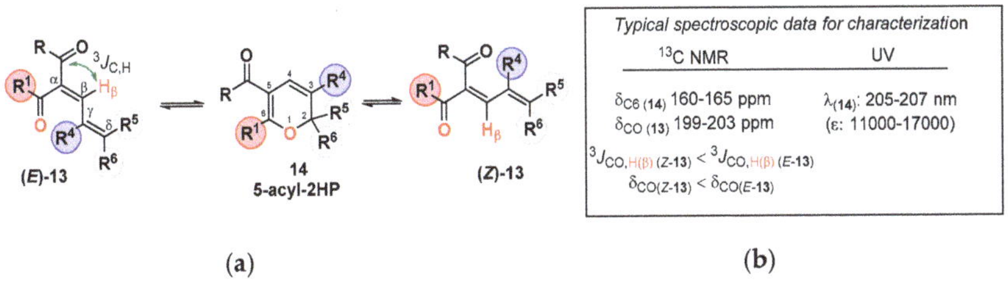
(E)-13
14
5-acyl-2HP
(Z)-13
Typical spectroscopic data for characterization
13C NMR
UV
δC6 (14) 160-165 ppm
δCO (13) 199-203 ppm
λ(14): 205-207 nm
(ε: 11000-17000)
3JCO,H(β) (Z-13) < 3JCO,H(β) (E-13)
δCO(Z-13) < δCO(E-13)
(a)
(b)

Scheme 4. (**a**) Valence tautomerism of 5-acyl-2HPs **14**. (**b**) Spectroscopic characteristic of tautomers.

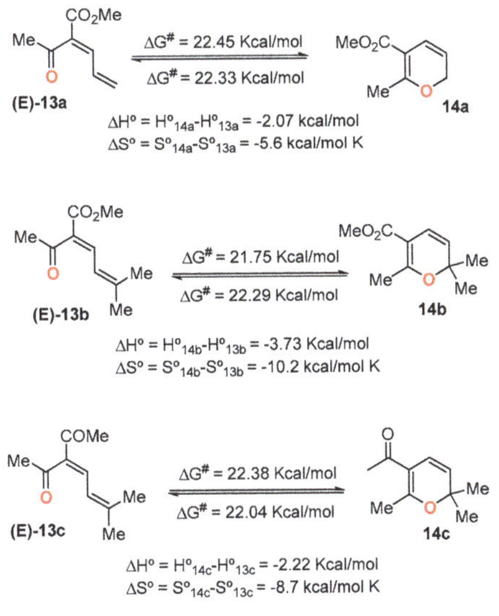
ΔG# = 22.45 Kcal/mol
ΔG# = 22.33 Kcal/mol
(E)-13a
14a
ΔH° = H°14a-H°13a = -2.07 kcal/mol
ΔS° = S°14a-S°13a = -5.6 kcal/mol K
ΔG# = 21.75 Kcal/mol
ΔG# = 22.29 Kcal/mol
(E)-13b
14b
ΔH° = H°14b-H°13b = -3.73 Kcal/mol
ΔS° = S°14b-S°13b = -10.2 kcal/mol K
ΔG# = 22.38 Kcal/mol
ΔG# = 22.04 Kcal/mol
(E)-13c
14c
ΔH° = H°14c-H°13c = -2.22 Kcal/mol
ΔS° = S°14c-S°13c = -8.7 kcal/mol K

Scheme 5. Thermodynamic data for **13a–c**/**14a–c** valence isomerization.

Table 1 summarizes the earlier observed importance of structural effects on the valence equilibrium, and it confirms some general patterns:

1. The successive substitution at position C_2 in the ring (C_δ on dienone) leads to an increase in the content of 2HP (steric strain on the dienone) (compare entries 1–3 and 7–8).
2. The elongation of the conjugated system results in an increase in the content of dienone (resonance delocalization) (compare entries 15, 25 and 17, 26).
3. Substitution at the C_2-position of the ring (C_δ on dienone) with two methyl groups strongly shifts the equilibrium toward the 2HP (entries 17, 19). In this case, it is possible to observe only the 2HP (compare entries 7, 17 with 11, 19).
4. Aprotic polar solvent shifts the equilibrium toward the formation of the dienone.
5. Although the electronic effect of the acyl group at the C_5-position of the ring (C_α in the dienone) is masked in Table 1, other studies have shown that the presence of an electron-withdrawing substituent(s) at the ring, preferentially at this C_5-position, favors the 2HP [23–25]. Table 1 shows that although this effect could be operative in α-acyl-dienones **13**, it can be completely surpassed by other structural/electronic effects (Table 1, entries 11–14).

Table 1. α-Acyl-dienones **13** and their equilibrium isomeric compositions.[a]

Entry	R	R^1	R^4	R^5	R^6	(E)-13	2HP 14	(Z)-13
1	EtO	Me	H	H	Me	30	30	40
2	EtO	Me	H	H	H	45	30	25
3	EtO	Me	H	Me	Me	17	68	15
4	MeO	Me	H	H	H	43	37	30
5	McO	Me	H	H	Me	40	40	20
6	MeO	Me	H	Me	Me	26	62	12
7	Me	Me	H	H	Me	72	28	-
8	Me	Me	H	Me	Me	64	36	-
9	t-BuO	Me	H	Me	Me	18	17	65
10	EtO	Ar[b]	H	Me	Me	84	9	7
11	EtO	Ph	H	H	Me	67	-	33
12	EtO	Ph	H	Me	Me	86	-	14
13	EtO	Me	H	H	Ph	60	-	40
14	Me	Me	H	H	Ph	100	-	-
15	MeO	Me	Me	Me	Me	-	83	17
16	MeO	Me	Me	H	Ph	-	100	-
17	Me	Me	Me	Me	Me	-	100	-
18	Me	Me	Me	H	Ph	-	100	-
19	EtO	Ph	Me	Me	Me	-	100	-
20	MeO	Me	H	H	c-C$_6$H$_{11}$	30	23	47
21	MeO	Me	Me	H	c-C$_6$H$_{11}$	-	100	-
22	MeO	Me	H	–(CH$_2$)$_5$–		16	67	17
23	MeO	Me	Me	–(CH$_2$)$_5$–		-	100	-
24	MeO	Me	H	–(CH$_2$)$_4$–		47	31	22
25	MeO	Me	Me	H	HC=CMe$_2$	75	-	25
26	Me	Me	Me	H	HC=CMe$_2$	100[c]	-	-

[a] The composition was determined by ^{1}H-NMR in CDCl$_3$ at 30 °C. [b] Ar = p-nitrophenyl. [c] The E and Z are topomers.

With regard to thermodynamic parameters of some of these equilibria (Scheme 5), Krasnaya et al. found that, in all cases, the enthalpies of the α-acyl-dienones **13** were appreciably higher than those of 5-acyl-2HPs **14**, which is in full agreement with the observed increase in the dienone content with the increase in temperature. As should be expected, the entropy contents were also higher for the closed structures. In all the investigated cases, $\Delta G^{\#}$ values were on the order of 21.88 Kcal/mol to 22.86 Kcal/mol.

Finally, other structural factors, such as annulation, also favor the 2HP form. It has been well established that annulation favors the closed form by restricting conformational freedom (entropic trap), and it has been used as a design element in the synthesis of stable 2HPs [26,27]. Scheme 6 graphically summarizes the main conclusions of these studies. Structures **3, 15, 16** represent prototypical examples of room temperature stable 2HPs.

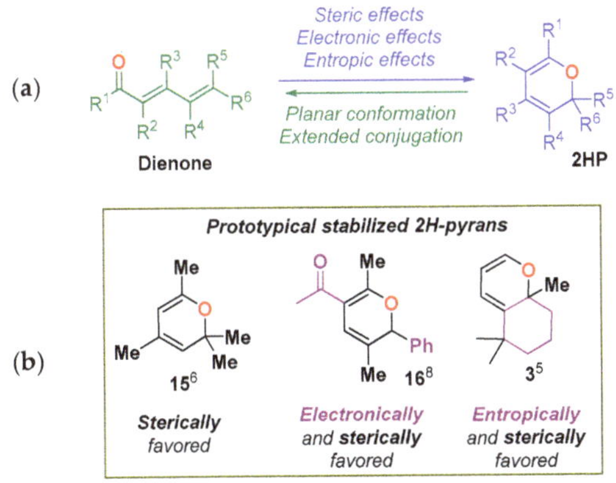

Scheme 6. (a) Summary of parameters affecting the valence isomerization. (b) Prototypical examples of stable 2HPs.

3. Synthesis of the 2HP Core

The most common route for synthesizing these heterocycles relies on the oxa-6π-electrocyclization of dienones, the so-called 1-oxatriene pathway [28]. As already discussed in the previous section, this methodology requires endowing the 1-oxatriene unit with structural or electronic information, or both, to shift the valence equilibrium toward the 2HP form (Scheme 7). Thus, different synthetic pathways to the 1-oxatriene core have been successfully explored, involving, among others, the classic Knoevenagel condensation between active methylene compounds and α, β-unsaturated aldehydes (enals), Claisen rearrangements of propargyl vinyl ethers, Stille coupling of vinyl stannanes and vinyl iodides, and cycloisomerization of dienols (Scheme 7).

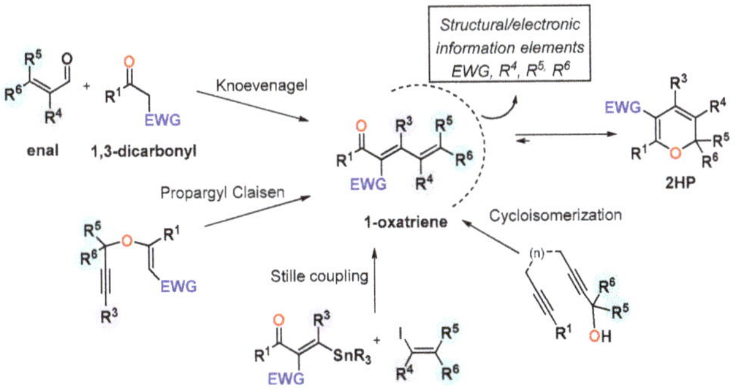

Scheme 7. Synthesis of 1-oxatriene core incorporating structural/electronic information.

3.1. The Knoevenagel/Electrocyclization Protocol

The Knoevenagel condensation constitutes the most common access to 1-oxatrienes, and most generally involves the reaction of an enal with a 1,3-dicarbonyl compound [29]. The sequential performance of the Knoevenagel condensation and the electrocyclization reaction generates 2HPs (Scheme 8). From a synthetic point of view, the whole tandem process can be considered a formal [3 + 3] cycloaddition [2,28]. There is a plethora of examples of this strategy in the literature, mainly in the field of total synthesis of natural products. In this review, we will pay attention exclusively to established synthetic methods that allow or have allowed general access to these heterocycles. Specific cases utilized to access a particular structure or a specific natural product will not be covered. We refer the reader to the excellent published reviews covering this issue [2,3].

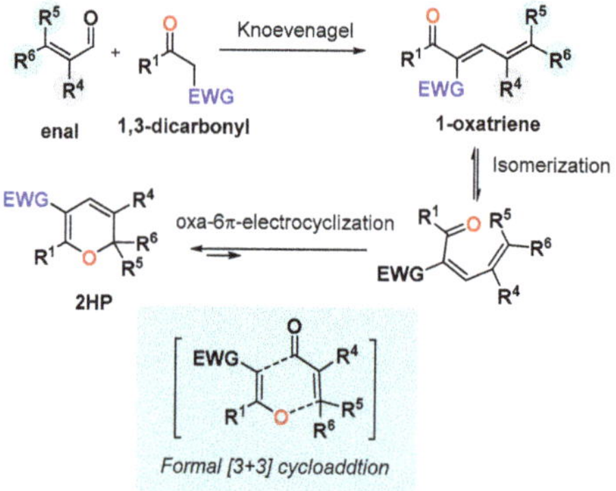

Scheme 8. Knoevenagel/electrocyclization strategy.

Fusion to a ring favors the electrocyclization of the 1-oxatriene intermediate, and it has been used as a steering element to synthesize stable 2HPs. As an earlier example, the pyridine-mediated condensation of different cyclic 1,3-dicarbonyl compounds **17** and functionalized enals **18** generated the stable bicyclic 2HPs **19** in good yields (Scheme 9a) [30]. Therefore, the double substitution at the terminal position of the enal also contributed to the global stability of 2HPs **19**. Using this methodology, the same authors synthesized the alkaloid flindersine (**21**) in 86% yield and in just one synthetic step (Scheme 9b).

a) General method.

17 **18** **19**

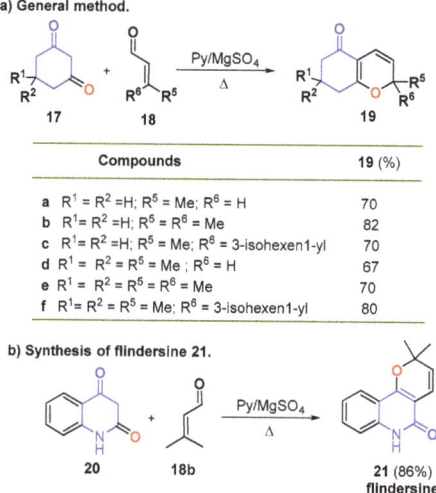

	Compounds	19 (%)
a	$R^1 = R^2 = H$; $R^5 = Me$; $R^6 = H$	70
b	$R^1 = R^2 = H$; $R^5 = R^6 = Me$	82
c	$R^1 = R^2 = H$; $R^5 = Me$; R^6 = 3-isohexen1-yl	70
d	$R^1 = R^2 = R^5 = Me$; $R^6 = H$	67
e	$R^1 = R^2 = R^5 = R^6 = Me$	70
f	$R^1 = R^2 = R^5 = Me$; R^6 = 3-isohexen1-yl	80

b) Synthesis of flindersine 21.

20 **18b** **21** (86%) **flindersine**

Scheme 9. Knoevenagel/electrocyclization: (**a**) synthesis of annellated 2HPs **19** and (**b**) synthesis of flindersine (**21**).

The iminium-mediated Knoevenagel condensation (IMKC) [31,32] has been currently used to condense 1,3-dicarbonyl (active methylene) compounds with 2-alkenyliminiums (activated enals), and it constitutes a very versatile route to 1-oxatrienes [2,33]. The chemical outcome of the reaction is that of a formal [3 + 3] cycloaddition between enols and enals (see Scheme 8). The reaction is productive when functionalized cyclohexane-1,3-diones (e.g., **21**) (Scheme 10) or 4-hydroxypyrones **25** (Scheme 11) are used as the active methylene compounds in the process. In this manner, 2HPs **23a–g** were synthesized from the functionalized cyclohexa-1,3-dione **21** and different functionalized enals **22** (Scheme 10) [34]. These 2HPs were used as key intermediates in the total synthesis of (−)-daurichromenic acid and analogues. The use of Lewis [35] or Brønsted [36] acids, In^{3+} [37], or iodine [38] as catalysts resulted complementary to the iminium formation and afforded similar reaction outcomes.

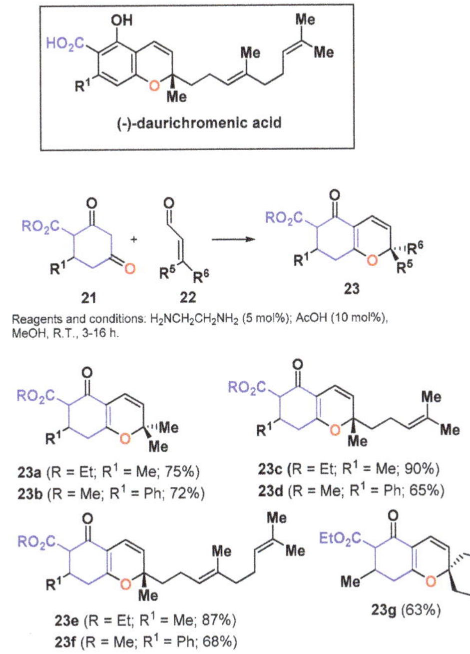

Scheme 10. Synthesis of 2HPs **23** by iminium-mediated Knoevenagel condensation (IMKC) of cyclohexa-1,3-dione **21** and enals **22**.

This methodology is well suited for use in diversity oriented synthesis programs [39], as long as the structural control elements are incorporated into the library design. Thus, a small and structurally varied library of 2HPs **26** was constructed using the β-alanine-mediated IMKC between 4-hydroxypyranone **25** and different enals **24** (Scheme 11) [40]. In vitro studies of antiproliferative/cytotoxic activity with human SH-SY5Y neuroblastoma cells showed IC_{50} values ranging from 6.7 to >200 μM. 2HP **26a** exhibited the highest cytotoxicity to the neuroblastome cells and necrotic effects on the human IPC melanoma cells.

Although the use of cyclic 1,3-dienones has been beneficial for the synthesis and stability of the resulting 2HPs, it is not a mandatory requirement, and acyclic active methylene compounds, such as methyl acetoacetate **27**, have been successfully condensed with 2-alkyl-2-enals **28** to deliver the corresponding stable 2,3,6-trisubstituted 2HPs **29** (Scheme 12) [41].

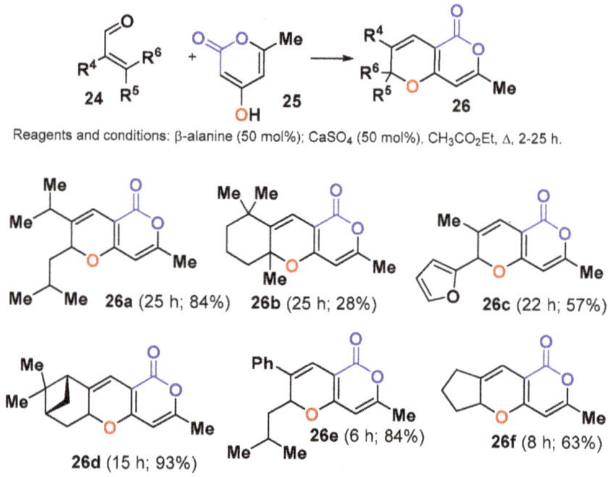

Scheme 11. Construction of a small library of annulated 2HPs **26** by the IMKC of 4-hydroxypyranone **25** and functionalized enals **24**. A selection of library members is shown.

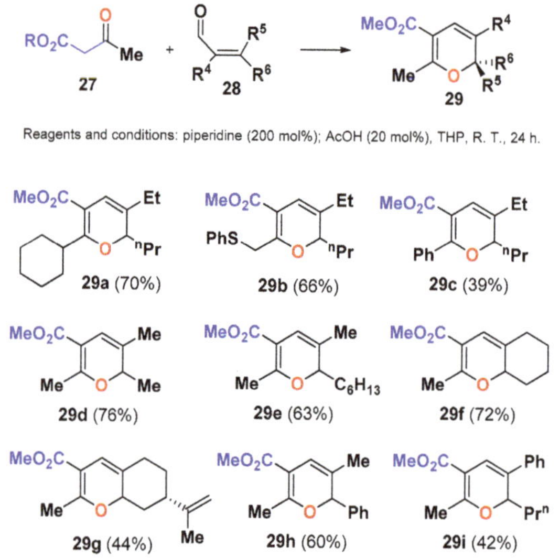

Reagents and conditions: β-alanine (50 mol%); CaSO₄ (50 mol%), CH₃CO₂Et, Δ, 2-25 h.

26a (25 h; 84%), 26b (25 h; 28%), 26c (22 h; 57%), 26d (15 h; 93%), 26e (6 h; 84%), 26f (8 h; 63%)

Reagents and conditions: piperidine (200 mol%); AcOH (20 mol%), THF, R. T., 24 h.

29a (70%), 29b (66%), 29c (39%), 29d (76%), 29e (63%), 29f (72%), 29g (44%), 29h (60%), 29i (42%)

Scheme 12. Synthesis of 2,3,6-trisubstituted 2HPs **29** by IMKC of methyl acetoacetate **27** and enals **28**.

Pyrano[3,2-c]quinolone is a core structural motif in alkaloids and is endowed with important pharmacological and therapeutic activities. As part of a research program aimed at developing efficient synthesis of natural product-like small molecules, a small 23-membered library focused on carbohydrate fused pyrano[3,2-c]quinolone structures **32** was synthesized and subjected to antiproliferative activity studies (Scheme 13) [42]. The library was synthesized using the microwave assisted pyrrolidine-AcOH catalyzed IMKC of formyl galactal (30-Gal) and formyl glucal (30-Glu) with 4-hydroxyquinolones **31**, and although the electron donating or electron withdrawing character of groups R^1, R^2, R^3, and R^4 of 4-hydroxyquinolones significantly affected neither the yield nor the reaction completion time, the best yields were obtained when unsubstituted **31** was used (70% with 30-Gal and 71% with 30-Glu). The other combinations gave yields ranging from 62 to 69%.

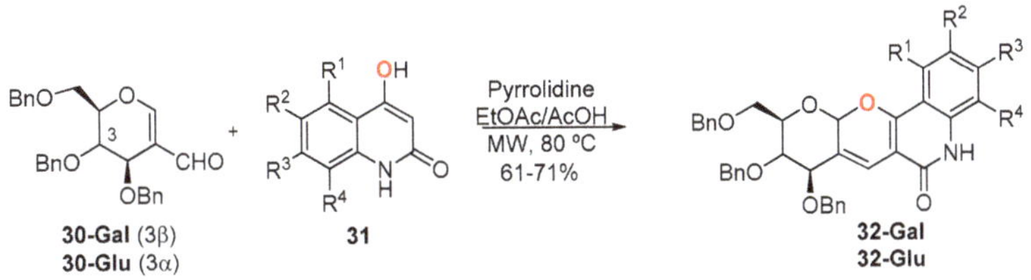

Scheme 13. Carbohydrate fused pyrano[3,2-c]quinolone library.

The Knoevenagel/electrocyclization strategy is suitable to be performed in water (Scheme 14) [43]. This methodology was applied to the synthesis of biologically interesting 2HPs of general structure 39, comprising pyranocoumarin, pyranoquinolinone, and pyranonaphthoquinone derivatives along with selected natural and non-natural products (X = CH_2, O, NH). The reactions were performed by mixing the 1,3-dicarbonyl compound 33–37 with enal 38 in water at 80 °C for 4–6 h. Although authors did not specify the physical conditions of these reactions, the high hydrophobicity of the reactants suggested that these reactions were carried out as aqueous suspensions (the so-called "on water" conditions [44], rather than as homogeneous solutions. Solvent-free protocols have been also described for the IMKC reaction [45,46].

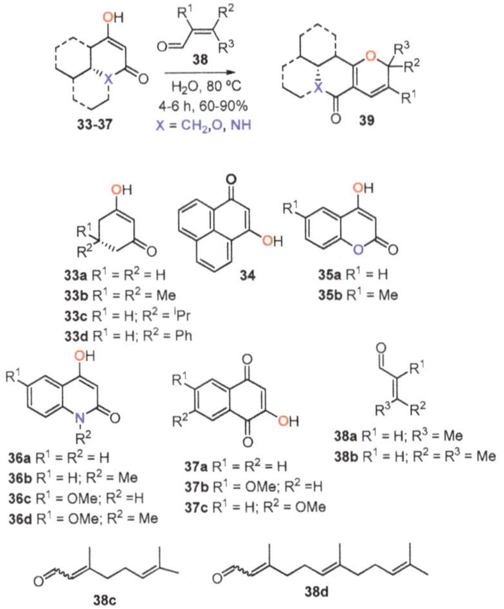

Scheme 14. Knoevenagel/electrocyclization in water.

3.2. From Other Heterocycles

The condensation of methyl coumalate (40) with a wide range of active methylene compounds 41 has been implemented to access an extensive series of 2,3,5,6-tetrasubstituted 2HPs 42 (Scheme 15) [47]. The reaction involved a domino 1,6-Michael/6π-electrocyclic ring opening/[1,5]-H transfer/(decarboxylation)/6π-electrocyclization reaction. The methyl substituent allocated at C_2 position of the 2HP ring corresponds to the α-methine group alpha to the lactone in the coumalate ring (highlighted as CH in Scheme 15).

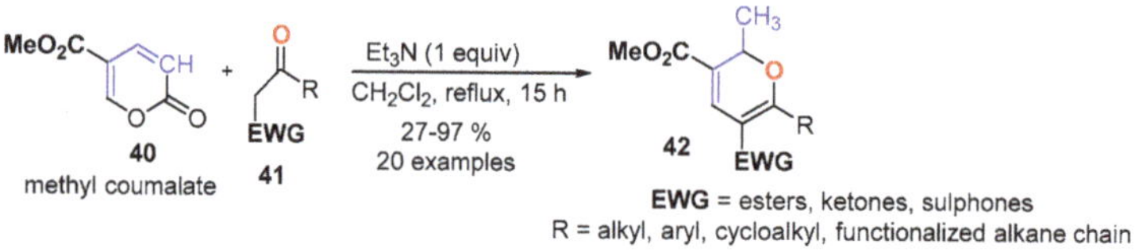

Scheme 15. Domino synthesis of tetrasubstituted 2HPs **42** from methyl coumalate **40**.

A one pot synthesis of 2,2,4,6-tetrasubstituted 2HPs **46** has been developed using Bayllis–Hillman carbonates **43** and β,γ-unsaturated α-oxo-esters **44** (Scheme 16) [48]. The one-pot reaction involved a phosphine-catalyzed condensation of **43** and **44** to give intermediate 4,5-dihydrofuran **45**, which, in the presence of pyrrolidine and heat, rearranged to the 2HP **46**. Authors gave a tentative mechanism for this pyrrolidine-catalyzed rearrangement. All the examples incorporated an aromatic (heterocyclic) substituent at R^1, but the authors do not explain if this was a mandatory property of this substituent.

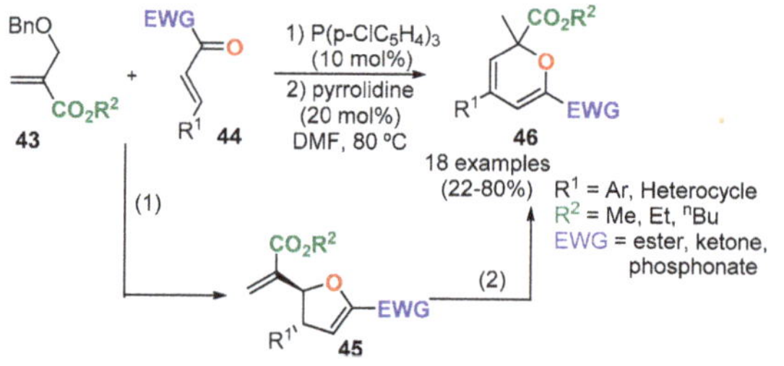

Scheme 16. Synthesis of tetrasubstituted 2HPs **46** from 4,5-dihydrofurane **45**.

3.3. From Allenolates

Stable 2,4,5,6-tetrasubstituted 2HPs **49** have been synthesized by the phosphine-catalyzed [3 + 3] annulation of ethyl 5-acetoxypenta-2,3-dienoate **47** and 1,3-dicarbonyl compounds **48** (Scheme 17) [49]. The scope of the reaction was wide, tolerating a good variety of the substituents. The presence of the ester group at the C_5 position of the ring was fundamental for the stability of the 2HP **49**.

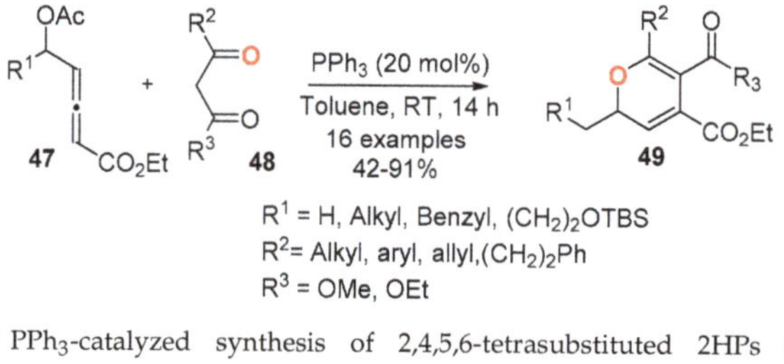

Scheme 17. PPh$_3$-catalyzed synthesis of 2,4,5,6-tetrasubstituted 2HPs **49** from ethyl 5-acetoxypent-2,3-dienoate **47** and 1,3-dicarbonyl compounds **48**.

3.4. From Alkynes

3.4.1. From Propargyl Vinyl Ethers

Propargyl vinyl ethers **50** have been successfully rearranged into stable 2,4,5,6-tetrasubstituted 2HPs **51** through a one pot procedure involving a Ag(I)-catalyzed propargyl Claisen rearrangement followed by a tandem DBU-catalyzed isomerization/6π-oxa-electrocyclization reaction (Scheme 18) [50]. The protocol used secondary propargyl vinyl ethers (they bear only one substituent at the propargylic position; R^3), and it required the installation of an ester group at the C_5 position of the ring and substitution at the C_6 position to give stability to the monocyclic 2HP **51**.

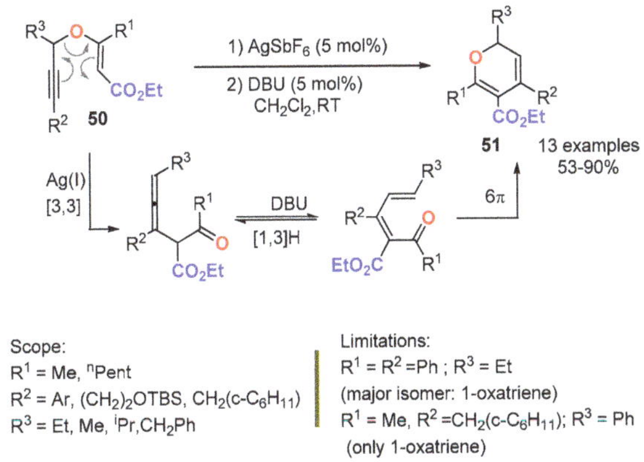

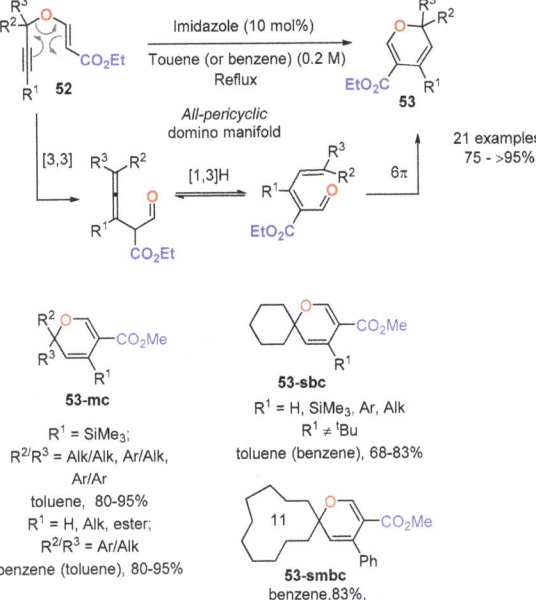

Scheme 18. One pot synthesis of 2,3,4,6-tetrasubstituted 2HPs **51** from propargyl vinyl ethers **50**.

More recently, a metal-free domino strategy has been developed for the synthesis of 2,2,4,5-tetrasubstituted 2HPs **53** from propargyl vinyl esters **52** (Scheme 19) [51–53]. The strategy made use of an imidazole-catalyzed all-pericyclic domino manifold entailing a sequential propargyl Claisen rearrangement/[1,3]-H shift/oxa-6π-electrocyclization set of reactions. Again, the presence of an ester functionality at the position C_5 of the ring was mandatory to stabilize the final 2HP **53**. The double substitution at C_2 favored the 2HP formation (steric information) and offered a wide range of optional substitution patterns at the ring (Alk/Alk, Ar/Alk, Ar/Ar). The protocol delivered 2HP structures endowed with different topologies, including monocycles (**53-mc**), 2,2-spiro-bicycles (**53-sbc**), and 2,2-spiro-macrobicycles (**53-smbc**). The main limitation arose from the presence of a ^{t}Bu substituent at the alkyne position (R^1): In this case, the reaction followed a different pathway through a sequential [1,7]-H shift/6π-electrocyclization/MeOH elimination set of reactions [54].

Scheme 19. All pericyclic domino synthesis of 2,2,4,5-tetrasubstituted 2HPs **53** from propargyl vinyl ethers **52**.

An alternative protocol using propargyl alcohols **54** and alkyl ethylendicarboxylates **55** has been developed (Scheme 20) [24,55]. The protocol generated 2,3,4,5,6-pentasubstituted 2HPs **56**, incorporating two identical ester functionalities at C_5 and C_6, and a halogen atom at C_3. The protocol used DABCO as the catalyst and N-iodosuccinimide (NIS) or N-bromosuccinimide (NBS) as the halogenation agent to generate 2HPs **56** in moderate to good yields. In all the conditions explored in Scheme 20, the substituents at the propargyl alcohol were aromatics (R^1/R^2 = Ar). The authors did not specify if this was a limitation to the procedure, or was just an inconvenient choice of starting materials.

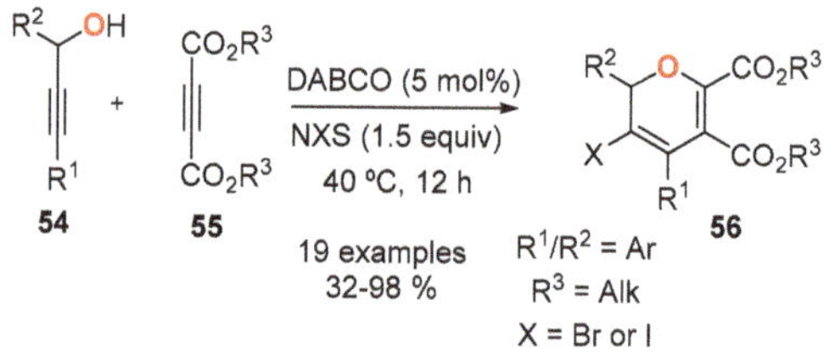

Scheme 20. Synthesis of 2,3,4,5,6-pentasubstituted 2HP **56** from propargyl alcohols **54** and dialkyl acetelynedicarboxylates **55**.

3.4.2. From Diynes

The Ni(0)-catalyzed cycloaddition of diynes **57** and aldehydes **58** has been explored in the construction of bicyclic 2HPs **59** (Scheme 21) [26,56]. Authors found that the structure of the diyne **57**, mainly the substitution at the terminal positions ($R^1 \neq H$), and the length of the chain connecting the alkyne units, exerted a great influence on the bicyclic 2HP formation reaction. The worst yield was observed when acetaldehyde was used as the aldehyde (28%), whereas the best was observed with *n*-butanal (90%).

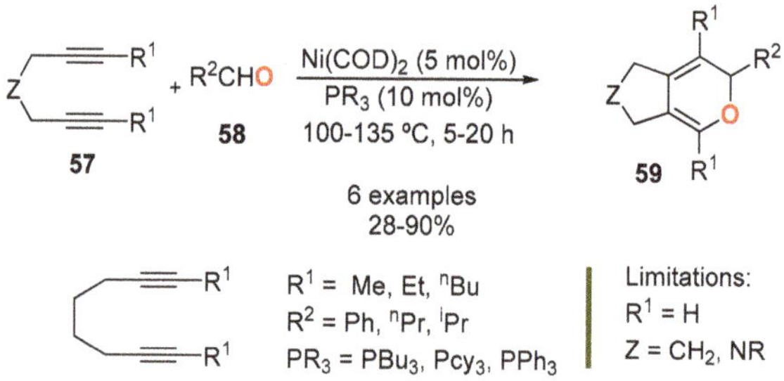

Scheme 21. Ni(0)-catalyzed synthesis of bicyclic 2HP **59** from diynes **57** aldehydes **58**.

A transition metal-free, cycloisomerization of diynols **60** to generate bicyclic 2HPs **61** has been reported (Scheme 22) [26]. The reaction was catalyzed by a cooperative catalytic system entailing Ca^{2+} catalyst (5 mol%) and camphorsulphonic acid (10 mol%), in the presence of benzaldehyde as a mild Lewis basic electron donor. The reaction was carried out without exclusion of air and moisture, and it tolerated a wide range of functionalities on the electron rich 2HP ring. The main limitations arose from substituents R^2/R^3 at the propargylic terminal position, which only allowed the alkyl/alkyl combination, and from R^1, which had to be aromatic. The only limitation for the aromatic substituent at R^1 was the presence of a free hydroxyl group at the *ortho* position of the ring. As long these restrictions were kept, excellent yields of 2HPs were obtained. The mechanistic proposal involved the formation of a propargylic tertiary cation **61**, which afforded the cyclic 1-oxa-2,4,5-triene intermediate **62**, which isomerized to **63** and rearranged into 2HP **64**.

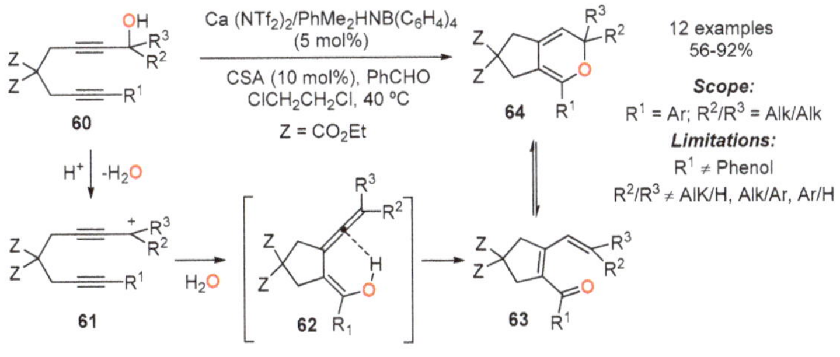

Scheme 22. Ca^{2+}/H^+-catalyzed synthesis of bicyclic 2HP **64** from diynols **60**.

3.4.3. From Alkenes: Tandem Stille-Oxa-Electrocyclization Reaction

Highly substituted bicyclic 2HPs **67** have been synthesized by a palladium-catalyzed tandem Stille-oxa-electrocyclization reaction between vinyl stannanes **65** and vinyl iodides **66** (Scheme 23) [57,58]. The strategy was a convergent alternative to the known methods for constructing similar bicyclic 2HPs, and it has been used in the total synthesis of natural products [59–61]. Although it required the prior stereoselective construction of both vinyl derivatives, the strategy had several advantages: It was convergent, highly diastereoselective, and required mild reaction conditions with low catalyst loadings (5 mol%). In the pattern of construction depicted in Scheme 23, the main restriction came from the nature of the vinyl iodide **66**, which had to have substituents at the vinyl ($R^3 \neq H$) and allylic positions ($R/R' \neq H$) to stabilize the 2HP ring form by steric destabilization of the 1-oxatriene form.

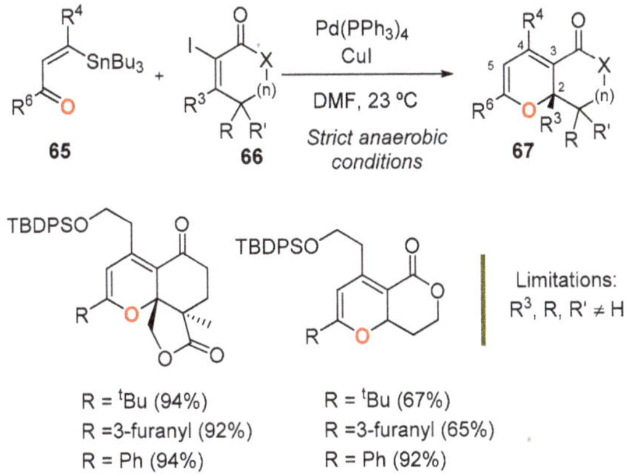

Scheme 23. Stille-oxa-electrocyclization strategy to access bicyclic 2HP **67**.

4. Summary and Conclusions

We have discussed the structural and electronic factors controlling the valence isomerism of 2HPs and how they can be harnessed to design effective synthesis of 2HPs. The most common routes to access these heterocycles relies on the 6π-electrocyclization of the corresponding 1-oxatriene isomers; thus, the synthetic challenge translates into the synthesis of the 1-oxatriene precursor. We have gathered the most transited routes to these species, including the proper Knoevenagel reaction, the tandem propargyl Claisen rearrangement/[1,3]-H shift reactions hosted by propargyl vinyl ethers, the cycloisomerization of diynes, and the Stille coupling of vinyl iodides and vinyl stannanes. From the large number of methods reported in the literature to access these heterocycles, we have selected only those able to generate stable rings with a convenient amount of structural/functional diversity. We hope that this review has filled the existing gap in literature regarding the reactivity and synthesis of these heterocycles, and that it finds use in future applications of these heterocycles.

Acknowledgments: The authors thank the Spanish Ministries of Economy and Competitiveness (MINECO) and Science, Innovation and Universities (MICINN), and the European Regional Development Funds (ERDF) for financial support (CTQ2015-63894-P and PGC2018-094503-B-C21). S.D.H. thanks La Laguna University and Cajasiete for a pre-doctoral contract.

Conflicts of Interest: The authors declare no conflict of interest. The funders had no role in the design of the study; in the collection, analyses, or interpretation of data; in the writing of the manuscript, or in the decision to publish the results.

References

1. Fravel, B.W. Pyrans and their Benzo Derivatives: Applications. In *Comprehensive Heterocyclic Chemistry III*; Katritzky, A.R., Ramsdem, C.A., Scriven, E.F.V., Taylor, R.J.K., Eds.; Elsevier: Oxford, UK, 2008; Volume 7, pp. 701–726.
2. Hsung, R.P.; Kurdyumov, A.V.; Sydorenko, N. A formal [3 + 3] cycloaddition approach to Natural-Product synthesis. *Eur. J. Org. Chem.* **2005**, *2005*, 23–44. [CrossRef]
3. Beaudry, C.M.; Malerich, J.P.; Trauner, D. Biosynthetic and biomimetic electrocyclizations. *Chem. Rev.* **2005**, *105*, 4757–4778. [CrossRef]
4. Drygina, O.V.; Garnovskii, A.D.; Kazantsev, A.V. 2H-pyrans (review). *Chem. Heterocycl. Compd.* **1985**, *21*, 239–253. [CrossRef]
5. Brimble, M.A.; Gibson, J.S.; Sperry, J. Pyrans and their Benzo Derivatives: Synthesis. In *Comprehensive Heterocycles Chemistry III*; Katritzky, A.R., Ramsdem, C.A., Scriven, E.F.V., Taylor, R.J.K., Eds.; Elsevier: Oxford, UK, 2008; Volume 5, pp. 419–699.
6. Hepworth, J.D.; Gabbutt, C.D.; Heron, B.M. Pyrans and their Benzo Derivatives: Synthesis. In *Comprehensive Heterocycles Chemistry II*; Katritzky, A.R., Rees, C.W., Scriven, R.J.K., Eds.; Elsevier: Oxford, UK, 1996; Volume 7, pp. 301–468.
7. Hepworth, J.D. Pyrans and Fused Pyrans: (iii) Synthesis and Applications. In *Comprehensive Heterocycles Chemistry*; Katritzky, A.R., Rees, C.W., Eds.; Elsevier: Oxford, UK, 1984; Volume 3, pp. 737–883.
8. Hepworth, J.D.; Heron, B.M. Six-membered ring systems: With O and/or S atoms. In *Progress in Heterocyclic Chemistry*; Gribble, G.W., Joule, J.A., Eds.; Elsevier: Oxford, UK, 2009; Volume 20, pp. 365–398.
9. Kuthan, J.; Šebek, P.; Böhm, S. New Developments in the Chemistry of Pyrans. *Adv. Heterocyclic. Chem.* **1995**, *62*, 19–135.
10. Pratap, R.; Ram, V.J. Natural and Synthetic Chromenes, Fused Chromenes, and Versatility of Dihydrobenzo[*h*]chromenes in Organic Synthesis. *Chem. Rev.* **2014**, *114*, 10476–10526. [CrossRef]
11. Majumdar, N.; Paul, N.P.; Mandal, S.; de Bruin, B.; Wulff, W.D. Catalytic Synthesis of 2H-Chromenes. *ACS Catal.* **2015**, *5*, 2329–2366. [CrossRef]
12. Ellis, G.P.; Lockhart, I.M. *The Chemistry of Heterocyclic Compounds: Chromenes, Chromanones, and Chromones*; Ellis, G.P., Ed.; Wiley-VCH: New York, NY, USA, 2009; Volume 31, pp. 1–1196.
13. Ellis, G.P. Pyrans and fused pyrans: (ii) reactivity. In *Comprehensive Heterocycles Chemistry*; Katritzky, A.R., Rees, C.W., Eds.; Elsevier: Oxford, UK, 1984; Volume 3, pp. 647–736.
14. Schweizer, E.E.; Meeder-Nycz, O. *Chromenes, Chromanes, Chromones*; Ellis, G.P., Ed.; Wiley-Interscience: New York, NY, USA, 1977; Volume 31, pp. 11–139.
15. Hepworth, J.D.; Heron, B.M. Synthesis and photochromic properties of naphthopyrans. In *Progress in Heterocyclic Chemistry*; Gribble, G.W., Joule, J.A., Eds.; Elsevier: Oxford, UK, 2005; Volume 17, pp. 33–62.
16. Vogel, E. Valence Isomerizations in compounds with strained rings. *Angew. Chem. Int. Ed. Engl.* **1963**, *2*, 1–11. [CrossRef]
17. Rodriguez-Otero, J. Study of the electrocyclization of (Z)-hexa-1,3,5-triene and its heterosubstituted analogues based on ab initio and DFT calculations. *J. Org. Chem.* **1999**, *64*, 6842–6848. [CrossRef]
18. Marvell, E.N.; Caple, G.; Gosink, T.A.; Zimmer, G. Valence isomerization of a *cis*-dienone to an α-pyran. *J. Am. Chem. Soc.* **1966**, *88*, 619–620. [CrossRef]
19. Marvell, E.N.; Gosink, T.A. Valence isomerization of 2,4,6-trimethyl-2H-pyran. *J. Org. Chem.* **1972**, *37*, 3036–3037. [CrossRef]

20. Lillya, C.P.; Kluge, A.F. Molecular spectra and conformations of conjugated dienones. *J. Org. Chem.* **1971**, *36*, 1977–1988. [CrossRef]
21. Gosink, T.A. Valence isomers. Substituent effects on the equilibrium between 2H-pyrans and cis dienones. *J. Org. Chem.* **1974**, *39*, 1942–1943. [CrossRef]
22. Krasnaya, Z.A. Dienone ⇄ 2H-pyran valence isomerization. *Chem. Heterocycl. Comp.* **1999**, *35*, 1255–1271. [CrossRef]
23. Adams, R.D.; Chen, L. Coupling of alkynes to carbon monoxide at a dimanganese center. A new route to carboxylate-functionalized pyrans. *J. Am. Chem. Soc.* **1994**, *116*, 4467–4468. [CrossRef]
24. Fan, M.; Yan, Z.; Liu, W.; Liang, Y. DABCO-catalyzed reaction of α-halo carbonyl compounds with dimethyl acetylenedicarboxylate: A novel method for the preparation of polysubstituted furans and highly functionalized 2H-pyrans. *J. Org. Chem.* **2005**, *70*, 8204–8207. [CrossRef] [PubMed]
25. Qing, F.L.; Gao, W.Z. The first synthesis of 4-trifluoromethyl-2H-pyrans by palladium-catalyzed cyclization of (E)-3-alkynyl-3-trifluoromethyl allylic alcohols. *Tetrahedron Lett.* **2000**, *41*, 7727–7730. [CrossRef]
26. Rauser, M.; Schroeder, S.; Niggemann, M. Cooperative catalysis: Calcium and camphorsulfonic acid catalyzed cycloisomerization of diynols. *Chem. Eur. J.* **2015**, *21*, 15929–15933. [CrossRef] [PubMed]
27. Tsuda, T.; Kiyoi, T.; Miyane, T.; Saegusa, T. Nickel(0)-catalyzed reaction of diynes with aldehydes. *J. Am. Chem. Soc.* **1988**, *110*, 8570–8572. [CrossRef]
28. Voskressensky, L.G.; Festa, A.A.; Varlamov, A.V. Domino reactions based on Knoevenagel condensation in the synthesis of heterocyclic compounds. Recent advances. *Tetrahedron* **2014**, *70*, 551–572. [CrossRef]
29. Tietze, L.F.; Beifuss, U. The Knoevenagel reaction. In *Comprehensive Organic Synthesis*; Trost, B.M., Ed.; Pergamon Press: Oxford, UK, 1992; Volume 2, pp. 341–394.
30. de Groot, A.; Jansen, B.J.M. A simple synthesis of 2H-pyrans; a one-step synthesis of flindersine. *Tetrahedron Lett.* **1975**, *39*, 3407–3410. [CrossRef]
31. Lelais, G.; MacMillan, D.W.C. Modern Strategies in organic catalysis: The advent and development of iminium activation. *Aldrichimica Acta* **2006**, *39*, 79–87.
32. Erkkila, A.; Majander, I.; Pihko, P.M. Iminium catalysis. *Chem. Rev.* **2007**, *107*, 5416–5470. [CrossRef] [PubMed]
33. Shen, H.C.; Wang, J.; Cole, K.P.; McLaughlin, M.J.; Morgan, C.D.; Douglas, C.J.; Hsung, R.P.; Coverdale, H.A.; Gerasyuto, A.I.; Hahn, J.M.; et al. A formal [3 + 3] cycloaddition reaction. improved reactivity using α,β-unsaturated iminium salts and evidence for reversibility of 6π-electron electrocyclic ring closure of 1-oxatrienes. *J. Org. Chem.* **2003**, *68*, 1729–1735. [CrossRef] [PubMed]
34. Hu, H.; Harrison, T.J.; Wilson, P.D. A modular and concise total synthesis of (±)-daurichromenic acid and analogues. *J. Org. Chem.* **2004**, *69*, 3782–3786. [CrossRef] [PubMed]
35. Kurdyumov, A.V.; Lin, N.; Hsung, R.P.; Gullickson, G.C.; Cole, K.P.; Sydorenko, N.; Swidorski, J.J. A Lewis acid-catalyzed formal [3 + 3] cycloaddition of α,β-unsaturated aldehydes with 4-hydroxy-2-pyrone, diketones, and vinylogous esters. *Org. Lett.* **2006**, *8*, 191–193. [CrossRef] [PubMed]
36. Hubert, C.; Moreau, J.; Batany, J.; Duboc, A.; Hurvois, J.P.; Renaud, J.L. Brønsted acid-catalyzed synthesis of pyrans via a formal [3 + 3] cycloaddition. *Adv. Synth. Catal.* **2008**, *350*, 40–42. [CrossRef]
37. Lee, Y.R.; Kim, D.H.; Shim, J.J.; Kim, S.K.; Park, J.H.; Cha, J.S.; Lee, C.S. One-pot synthesis of 2H-pyrans by indium(iii) chloride-catalyzed reactions. Efficient synthesis of pyranocoumarins, pyranophalenones, and pyranoquinolinones. *Bull. Korean Chem. Soc.* **2002**, *23*, 998–1002. [CrossRef]
38. Jung, E.J.; Lee, Y.R.; Lee, H.J. Iodine-catalyzed one-pot synthesis of 2H-pyrans by domino Knoevenagel/6π-electrocyclization. *Bull. Korean Chem. Soc.* **2009**, *30*, 2833–2836.
39. Burke, M.D.; Schreiber, S.L. A planning strategy for diversity-oriented synthesis. *Angew. Chem. Int. Ed.* **2004**, *43*, 46–58. [CrossRef] [PubMed]
40. Leutbecher, H.; Williams, L.A.D.; Rösner, H.; Beifuss, U. Efficient synthesis of substituted 7-methyl-2H,5H-pyrano[4,3-b]pyran-5-ones and evaluation of their in vitro antiproliferative/cytotoxic activities. *Bioorg. Med. Chem. Lett.* **2007**, *17*, 978–982. [CrossRef] [PubMed]
41. Peng, W.; Hirabaru, T.; Kawafuchi, H.; Inokuchi, T. Substituent-controlled electrocyclization of 2,4-dienones: Synthesis of 2,3,6-trisubstituted 2H-pyran-5-carboxylates and their transformations. *Eur. J. Org. Chem.* **2011**, *2011*, 5469–5474. [CrossRef]
42. Kumari, P.; Narayana, C.; Dubey, S.; Gupta, A.; Sagar, R. Stereoselective synthesis of natural product inspired carbohydrate fused pyrano[3,2-c]quinolones as antiproliferative agents. *Org. Biomol. Chem.* **2018**, *16*, 2049–2059. [CrossRef] [PubMed]

43. Jung, E.J.; Park, B.H.; Lee, Y.R. Environmentally benign, one-pot synthesis of pyrans by domino Knoevenagel/6π-electrocyclization in water and application to natural products. *Green Chem.* **2010**, *12*, 2003–2011. [CrossRef]
44. Narayan, S.; Muldoon, J.; Finn, M.G.; Fokin, V.V.; Kolb, H.C.; Sharpless, K.B. "On water": Unique reactivity of organic compounds in aqueous suspension. *Angew. Chem. Int. Ed.* **2005**, *44*, 3275–3279. [CrossRef] [PubMed]
45. Riviera, M.J.; Mischine, M.P. Green one-pot synthesis of 2H-pyrans under solvent-free conditions catalyzed by ethylenediammonium diacetate. *Synth. Commun.* **2013**, *43*, 208–220. [CrossRef]
46. Peña, J.; Moro, R.F.; Basabe, P.; Marcos, I.S.; Díez, D. Solvent free L-proline-catalysed domino Knoevenagel/6π-electrocyclization for the synthesis of highly functionalised 2H-pyrans. *RSC Adv.* **2012**, *2*, 8041–8049. [CrossRef]
47. Chang, L.; Plevová, K.; Thorimbert, S.; Dechoux, L. Preparation of substituted 2H-pyrans via a cascade reaction from methyl coumalate and activated methylene nucleophiles. *J. Org. Chem.* **2017**, *82*, 5499–5505. [CrossRef]
48. Xie, P.; Yang, J.; Zheng, J.; Huang, Y. Sequential catalyst phosphine/secondary amine promoted [1 + 4]/rearrangement domino reaction for the construction of (2H)-pyrans and 2-oxabicyclo[2,2,2]oct-5-ene skeletons. *Eur. J. Org. Chem.* **2014**, *2014*, 1189–1194. [CrossRef]
49. Hu, J.; Dong, W.; Wu, X.Y.; Tong, X. PPh$_3$-catalyzed [3 + 3] annulations of 5-acetoxypenta-2,3-dienoate with 1C,3O-bisnucleophiles: Facile entry to stable monocyclic 2H-pyrans. *Org. Lett.* **2012**, *14*, 5530–5533. [CrossRef]
50. Menz, H.; Kirsch, S.F. Synthesis of stable 2H-pyran-5-carboxylates via a catalyzed propargyl-Claisen rearrangement/oxa-6π electrocyclization strategy. *Org. Lett.* **2006**, *8*, 4795–4797. [CrossRef]
51. Tejedor, D.; Delgado-Hernández, S.; Diana-Rivero, R.; Díaz-Díaz, A.; García-Tellado, F. A domino strategy for the synthesis of 2H-pyrans from propargyl vinyl ethers. *Eur. J. Org. Chem.* **2019**, *2019*, 1784–1790. [CrossRef]
52. Tejedor, D.; Díaz-Díaz, A.; Diana-Rivero, R.; Delgado-Hernández, S.; García-Tellado, F. Synthesis and utility of 2,2-dimethyl-2H-pyrans: Dienes for sequential Diels–Alder/Retro-Diels–Alder reactions. *Org. Lett.* **2018**, *20*, 7987–7990. [CrossRef] [PubMed]
53. Tejedor, D.; Delgado-Hernández, S.; Peyrac, J.; González-Platas, J.; García-Tellado, F. Integrative pericyclic cascade: An atom economic, multi C-C bond-forming strategy for the construction of molecular complexity. *Chem. Eur. J.* **2017**, *23*, 10048–10052. [CrossRef] [PubMed]
54. Tejedor, D.; Cotos, L.; Márquez-Arce, D.; Odriozola-Gimeno, M.; Torrent-Sucarrat, M.; Cossío, F.P.; García-Tellado, F. Microwave-assisted organocatalyzed rearrangement of propargyl vinyl ethers to salicylaldehyde derivatives: An experimental and theoretical study. *Chem. Eur. J.* **2015**, *21*, 18280–18289. [CrossRef] [PubMed]
55. Chong, Q.; Wang, C.; Wang, D.; Wang, H.; Wu, F.; Xin, X.; Wan, B. DABCO-catalyzed synthesis of 3-bromo-/3-iodo-2H-pyrans from propargyl alcohols, dialkyl acetylene dicarboxylates, and N-bromo-/N-iodosuccinimides. *Tetrahedron Lett.* **2015**, *56*, 401–403. [CrossRef]
56. Yamamoto, Y.; Okude, Y.; Mori, S.; Shibuya, M. Combined experimental and computational study on ruthenium(II)-catalyzed reactions of diynes with aldehydes and N,N-dimethylformamide. *J. Org. Chem.* **2017**, *82*, 7964–7973. [CrossRef]
57. Tambar, U.K.; Kano, V.; Zepernick, J.F.; Stoltz, B.M. The development and scope of a versatile tandem Stille-oxa-electrocyclization reaction. *Tetrahedron Lett.* **2007**, *48*, 345–350. [CrossRef]
58. Tambar, U.K.; Kano, T.; Stoltz, B.M. Progress toward the total synthesis of saudin: Development of a tandem Stille-oxa-electrocyclization reaction. *Org. Lett.* **2005**, *7*, 2413–2416. [CrossRef]
59. Li, C.; Johnson, R.P.; Porco, J.A. Total synthesis of the quinone epoxide dimer (+)-torreyanic acid: Application of a biomimetic oxidation/electrocyclization/Diels-Alder dimerization cascade. *J. Am. Chem. Soc.* **2003**, *125*, 5095–5106. [CrossRef]
60. Shoji, M.; Yamaguchi, J.; Kakeya, H.; Osada, H.; Hayashi, Y. Total synthesis of (+)-epoxyquinols A and B. *Angew. Chem. Int. Ed.* **2002**, *41*, 3192–3194. [CrossRef]
61. Li, C.; Lobkovsky, E.; Porco, J.A. Total synthesis of (±)-torreyanic acid. *J. Am. Chem. Soc.* **2000**, *122*, 10484–10485. [CrossRef]

© 2019 by the authors. Licensee MDPI, Basel, Switzerland. This article is an open access article distributed under the terms and conditions of the Creative Commons Attribution (CC BY) license (http://creativecommons.org/licenses/by/4.0/).

Review

Recent Advances on Metal-Free, Visible-Light-Induced Catalysis for Assembling Nitrogen- and Oxygen-Based Heterocyclic Scaffolds

Robert Pawlowski, Filip Stanek and Maciej Stodulski *

Institute of Organic Chemistry Polish Academy of Sciences, Kasprzaka 44/52, 01-224 Warsaw, Poland;
robert.pawlowski@icho.edu.pl (R.P.); filip.stanek@icho.edu.pl (F.S.)
* Correspondence: maciej.stodulski@icho.edu.pl; Tel.: +48-22-343-21-30

Academic Editor: Gianfranco Favi
Received: 27 March 2019; Accepted: 15 April 2019; Published: 18 April 2019

Abstract: Heterocycles are important class of structures, which occupy a major space in the domain of natural and bioactive compounds. For this reason, development of new synthetic strategies for their controllable synthesis became of special interests. The development of novel photoredox systems with wide-range application in organic synthesis is particularly interesting. Organic dyes have been widely applied as photoredox catalysts in organic synthesis. Their low costs compared to the typical photocatalysts based on transition metals make them an excellent alternative. This review describes proceedings since 2015 in the area of application of metal-free, visible-light-mediated catalysis for assembling various heterocyclic scaffolds containing five- and six-membered rings bearing nitrogen and oxygen heteroatoms.

Keywords: photocatalysis; photoredox; visible-light-induced catalysis; photoredox cyclization; organic dyes; heterocycles

---

1. Introduction

Heterocycles are a very important class of structural motifs that can be found in many pharmaceuticals and natural products [1–3]. Numerous of them contain five- or six-membered rings bearing different heteroatom, like nitrogen or oxygen. Drimentine G, for example, is composed of three heterocyclic rings and indicates anticancer and antibacterial activities; others like Captopril is an ACE inhibitor used for the treatment of hypertension (Figure 1) [4–7]. As a consequence of potent bioactivity of many heterocycles, significant efforts have been devoted toward the development of new synthetic strategies for their preparation [3].

During the last years, application and development of visible-light-mediated catalysis in organic chemistry became a topic of immense importance [8,9]. It is due to the fact that opposite to many reactions, a light source is used to generate reactive species without the need to apply stoichiometric activators [10,11]. To fulfill the needs of the solar energy usage, structurally diverse compounds based on transition metals have been developed [12]. To pursue the ideal chemical transformation, light-emitting diodes (LEDs) or compact fluorescent lamps (CFLs) are used as a light source because of their low costs, reasonable energy consumption, and general availability.

To create heterocyclic scaffolds, comprehensive photocatalytic strategies based on various metal complexes have been already disclosed [13–20]. However, their metal-free analogues play a significant role in terms of green chemistry [21] and medicine because of the high toxicity of many transition-metal complexes [22–24]. Organic dyes capable of visible-light-spectra absorption are generally less expensive and less toxic compared to the classical iridium or ruthenium catalysts. Thanks to high air stability, they are also easy to handle. For this reason, organic dyes act as an attractive alternative to transition-metal complexes [25–27]. Figure 2 shows the examples of most commonly used metal-free catalysts in photoredox chemistry.

This article presents the coverage of the recent advances in the application of metal-free, visible-light-mediated catalysis for assembling five- and six-member heterocyclic scaffolds containing nitrogen and oxygen heteroatoms. We are mainly focusing on the new metal-free photochemical reactions discovered after the year 2015 in the synthesis of aromatic and non-aromatic heterocycles. The article is organized into three chapters, which describe synthetic strategies for the preparation of nitrogen-containing heterocycles, oxygen-containing heterocycles, and heterocycles containing more than one heteroatom.

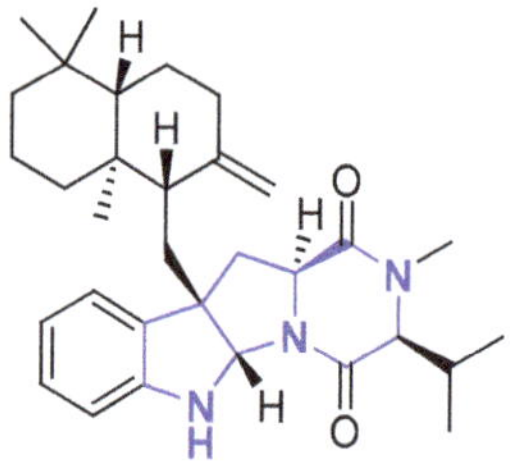

1 Drimentine G

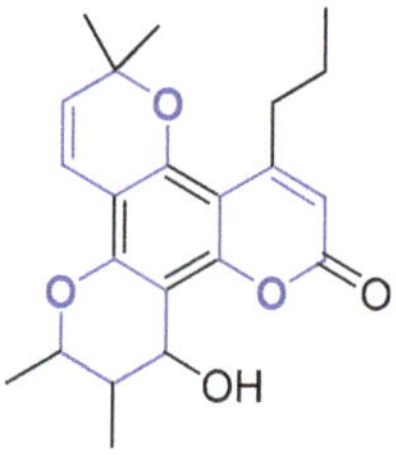

2 Calanolides A-B

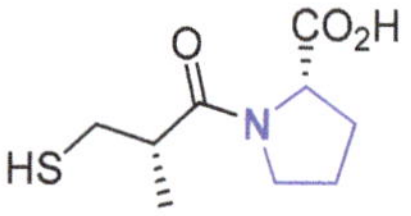

3 Captopril

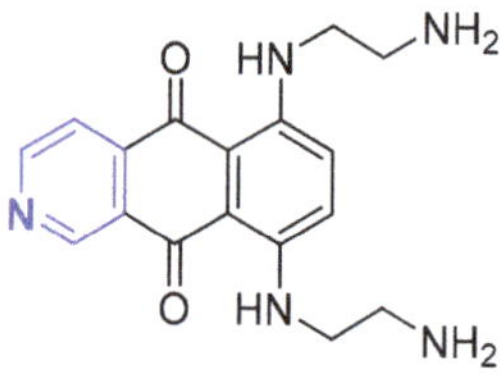

4 Pixantrone

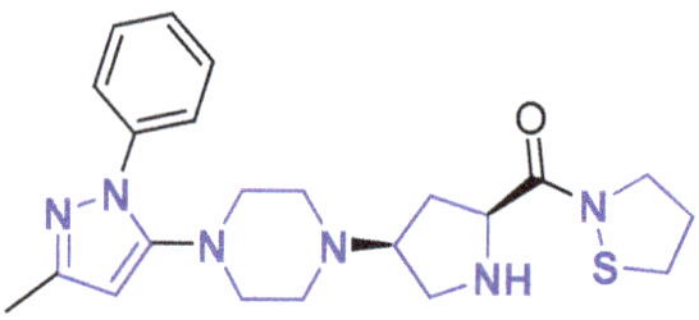

5 Teneligliptin

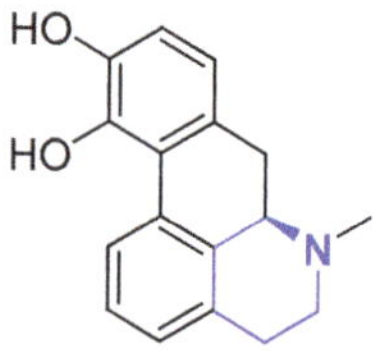

6 Apomorphine

Figure 1. Selected example of current drugs containing heterocyclic scaffold.

2. Application of Visible-Light-Mediated Catalysis in Synthesis of Heterocyclic Compounds

Many organic compounds isolated from nature contain five- or six-membered heterocycle scaffolds. These compounds are very interesting due to their potent bioactivities and applications as therapeutic drugs. For example, many heterocyclic derivatives, like emetine, papaverine, theophylline, etc. containing heterocyclic scaffold are used as life-saving pharmaceuticals [28]. Moreover, heterocyclic systems very often act as suitable intermediates in the synthesis of more complex molecules [29]. As part of their well-defined structure, they may contain *N*-, *O*-, *S*-, or other heteroatoms. Among the reaction strategies that lead to heterocycles, many of them use heavy metals, or require harsh reaction conditions [30]. Therefore, development of more sustainable approaches for constructing highly functionalized heterocycles have gained a considerable interest.

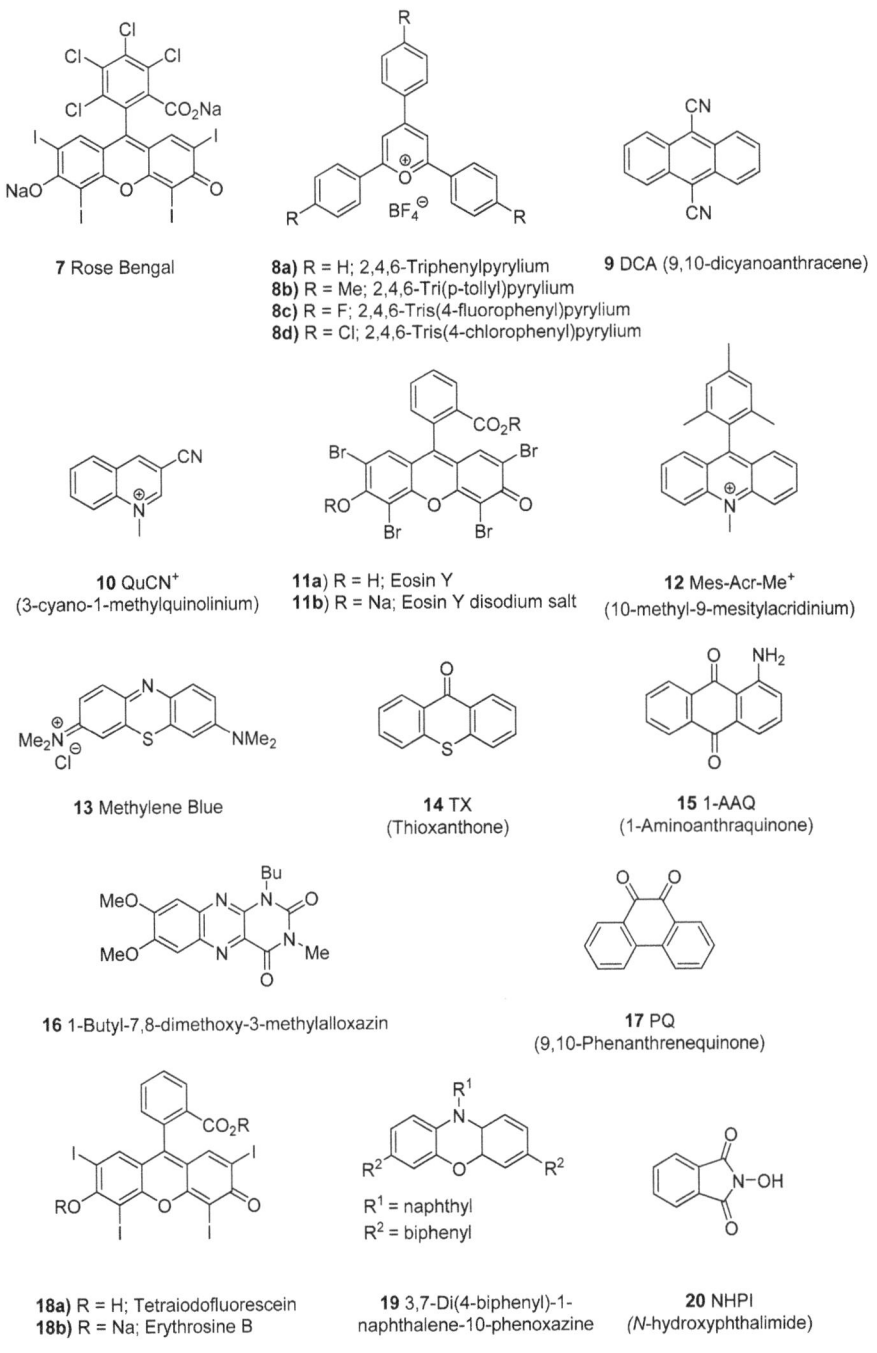

Figure 2. Examples of metal-free photoredox catalysts.

2.1. N-containing Heterocycles

2.1.1. Aromatic Heterocycles

Nitrogen bearing heterocycles constitute the majority in the field of heterocyclic chemistry. Various biologically active natural or synthetic products prevail the substituted indole motif. For example, some of the 2-substituted-indoles have been found as the successful leukotriene-modifier drugs in cardiovascular disease [31], others gain therapeutic interest in the treatment of cancer, HIV, heart disease, allergies, etc. [32]. To construct functionalized indoles, Kshirsagar and co-workers disclosed a photoredox-catalyzed vicinal thioamination of alkynes [33]. In this context, Eosin Y **11a** catalyzed radical cascade annulation generates various 3-sulfenylindoles **22** in up to 86% yield (Scheme 1a). This metal and strong-oxidant-free synthesis is based on the proposed mechanism presented in Scheme 1b. Eosin Y **11a** (EY) is first excited by the blue LED irradiation to EY*, which is then reduced to EY$^{\bullet-}$ by oxidizing thiophenol to the radical species **24**. Re-oxidation of EY$^{\bullet-}$ to its ground state takes place by air oxygen producing $O_2^{\bullet-}$ species. Deprotonation of thiophenol cation radical **24** by the $O_2^{\bullet-}$ gives hydroperoxyl radical **28** and radical **25**, which undergoes addition to the triple bond of **21** and produces another radical **26**. Subsequent intramolecular cyclization followed by N_2 release delivers intermediate radical **27**. Finally, hydrogen atom transfer (HAT) from hydrogen source present in the reaction mixture gives product **22**.

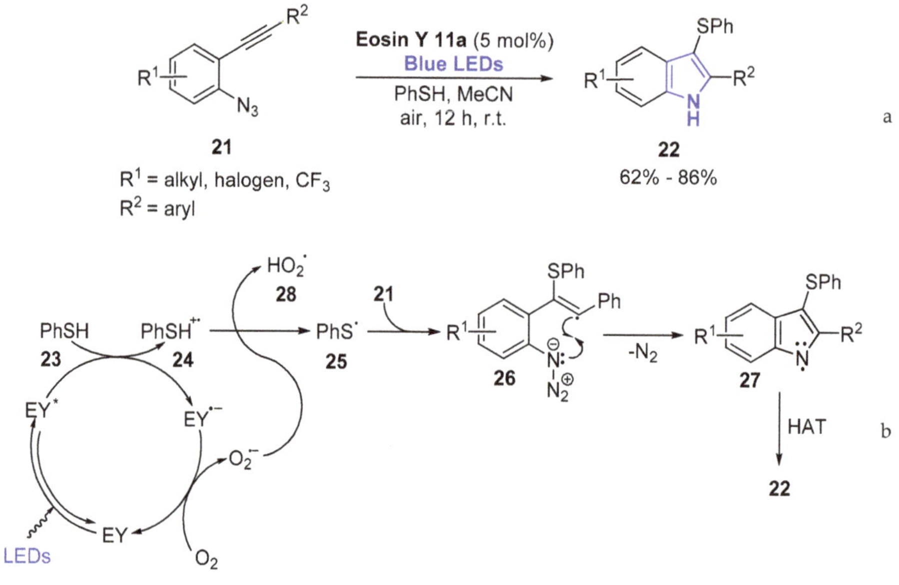

Scheme 1. (a) Vicinal thioamination of alkynes, mediated by Eosin Y; (b) Proposed mechanism for vicinal thioamination of alkynes, mediated by Eosin Y.

2,3-Disubstituted indoles can be also obtained by cyclization of arylsulfonyl chlorides with *ortho*-azidoarylalkenes [34]. This transition-metal free process described by Gu group, emerges as an efficient protocol, exhibiting high functional-group tolerance. Various indoles **31** can be prepared in up to 84% yield (Scheme 2a). Mechanism of this annulation is shown in Scheme 2b. Excited Eosin Y **11a** (EY) is responsible for reduction by SET of benzenesulfonyl chloride **30** to aryl radical **32** and EY$^{\bullet+}$. Formed phenyl radical **32** undergoes addition to the triple bond of **29** forming radical intermediate **33**, which subsequently undergoes intramolecular cyclization with the azide group to N-radical intermediate **34** with extrusion of nitrogen. Finally, H-atom abstraction from cyclohexadiene

(CHD) by **34** forms desired product **31** and CHD radical **35**, which closes the photoredox cycle by SET process regenerating Eosin catalyst in its ground state (Scheme 2b).

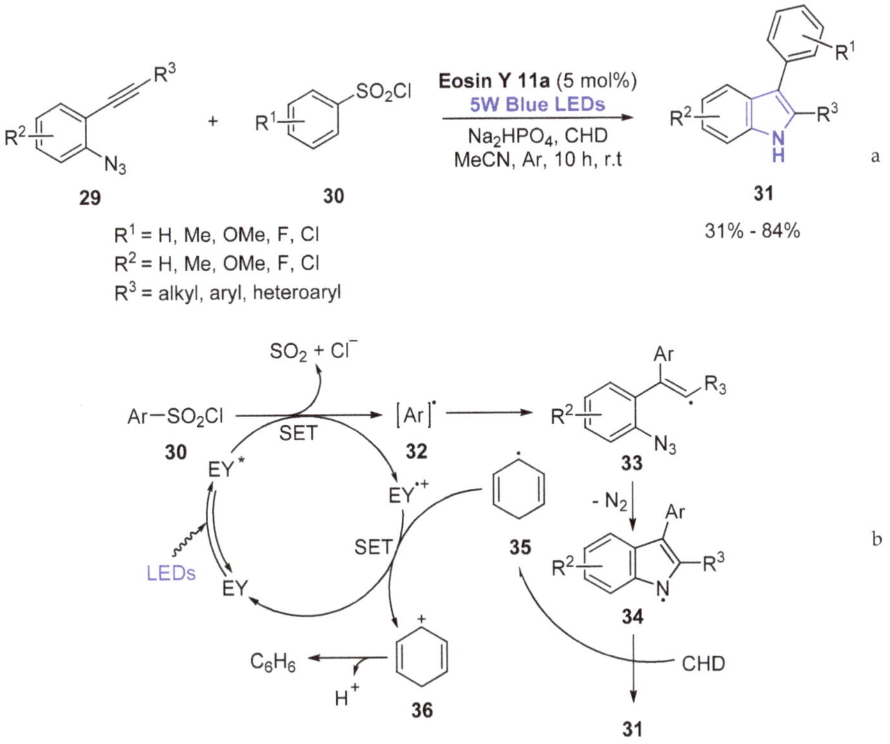

Scheme 2. (a) Visible-light-induced cyclization of arylsulfonyl chlorides; (b) Proposed mechanism.

Fluorine-containing compounds can often be found in medicinal chemistry due to improvement bioavailability, lipophilicity, or the metabolic stability compared to their non-fluorinated analogs [35]. For this reason, the synthesis of fluorinated heterocycles is particularly interesting.

The intramolecular cyclization of various 1,6-enynes **37** can be found as a general and powerful strategy in the synthesis of indole derivatives. In this context, Kumar and co-workers proposed metal and oxidant-free visible-light-induced trifluoromethylation of alkynes **37** (Scheme 3), using 9,10-phenanthrequinone catalyst **17** under compact fluorescent lamp (CFL) irradiation [36].

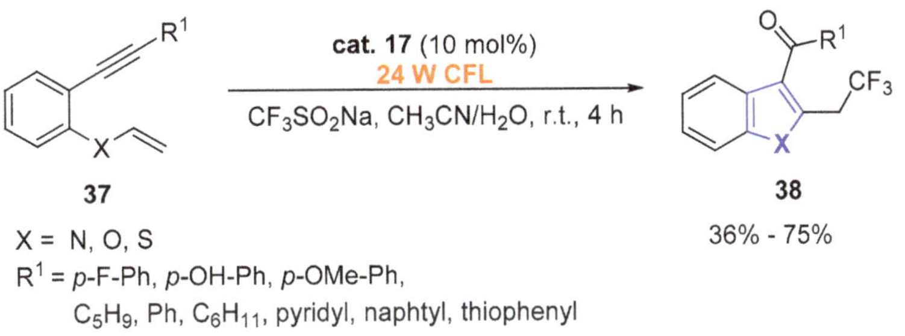

Scheme 3. Dehydrogenative cascade trifluoromethylation and oxidation of 1,6-enynes.

In addition, the described protocol is also effective for constructing benzofuran or thiophene frameworks. CFL irradiation of the CF$_3$SO$_2$Na derivative generates the initiating CF$_3$ radical, which undergoes selective addition to the double bond acceptor, while subsequent intramolecular cyclization gives final aromatic derivatives **38** in up to 75% yield.

The cascade-type reactions have been established as a general strategy for the synthesis of highly functionalized compounds [37]. In this context, Brasholz and co-workers described a visible-light-induced photocatalytic cascade reaction in the synthesis of indoloisoquinolines **40** using amino-substituted anthraquinone **15** as photocatalyst [38]. Formation of new heterocyclic ring is based on the dehydrogenation–cyclization–oxidation cascade that converts tetrahydroisoquinolines **39** into substituted tetracyclic heterocycles **40** (Scheme 4a). A plausible reaction mechanism starts on electron transfer between amine **39** and the photoinduced anthraquinone catalyst **15** (AQ*) with the formation of amine radical cation **41**. Hydrogen atom abstraction from **41** by superoxide radical anion converts amine **41** into iminium ion **42**. Subsequent deprotonation of **42** gives nitroenamine **43**, which undergoes electrocyclic ring closure to ylide **44**. Finally, rearomatization followed by photoinduced catalytic oxidation leads to appropriate 12-nitroindoloisoquinoline **40** in up to 69% yield (Scheme 4b).

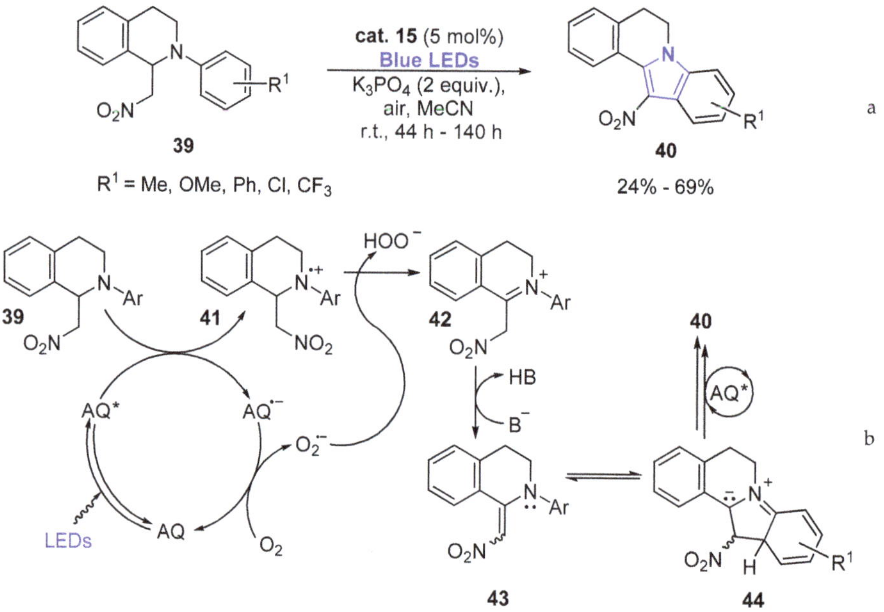

Scheme 4. (a) A visible-light-induced cross-dehydrogenative cascade in the synthesis of tetrahydroisoquinoline derivatives; (b) Proposed mechanism.

Pyridine core is widespread in many bioactive compounds [1–3]. The versatile methodology employing visible-light-promoted [2 + 2 + 2] cyclization of alkynes with nitriles to pyridines recently has been described by Wang and Meng group [39]. Using pyrylium salt **8d** as photoredox catalysts, various 2,3,6-trisubstituted pyridines **47** can be prepared under blue LED irradiation. A wide range of functional group tolerance makes various pyridines **47** accessible in up to 79% yield (Scheme 5a). In Scheme 5b, a reasonable mechanism for this transformation is suggested. Blue LED irradiation of the pyrylium salt **8d** (PC) in the presence of phenyl acetylene **45** generates the radical cation **51**, which reacts with the nitrile **46** and generates intermediate **48**. Subsequent addition of **48** to another phenyl acetylene molecule **45** affords intermediate **49**, which undergoes intramolecular cyclization to pyridine radical cation **50**. Finally, reduction of **50** by pyrylium anion (PC⁻) via a SET process provides the product **47** and regenerated catalyst (PC) in its ground state.

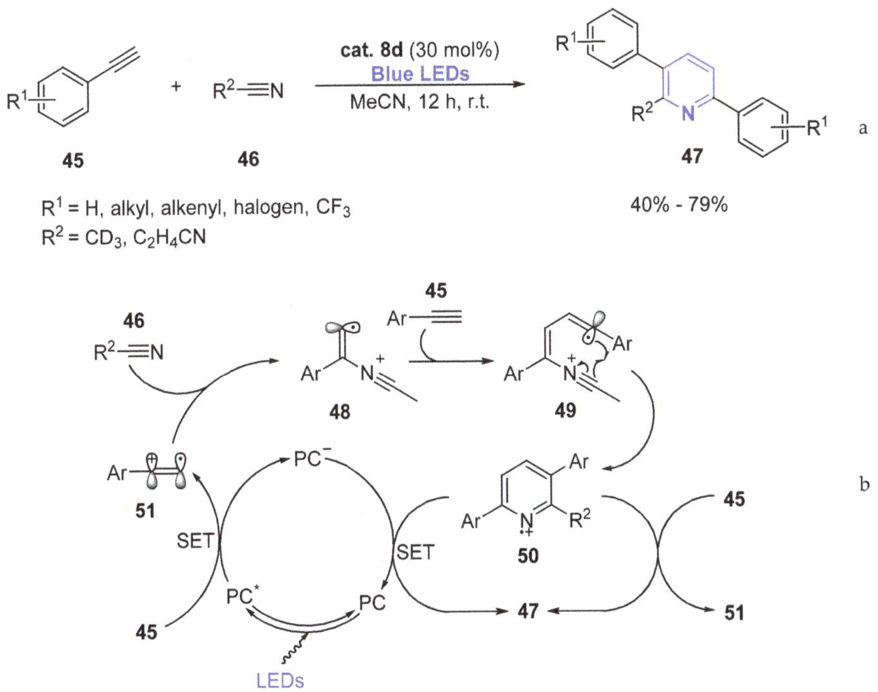

Scheme 5. (a) [2 + 2 + 2] Cyclization of alkynes with nitriles; (b) Proposed mechanism.

Recently, visible-light-photoredox catalysis has been used for the synthesis of various quinolones as well. The Zhang group developed an impressive example of N-propargyl aromatic amine application in the synthesis of 3-arylsulfonylquinolines **55** (Scheme 6a). The authors demonstrated a visible-light-induced multicomponent cascade cycloaddition of **52** with diaryliodonium salts **53** and sulfur dioxide [40]. This three-component reaction can be achieved using Eosin Y **11a** catalysts under irradiation by green LEDs in the presence of DABSO (DABCO·(SO$_2$)$_2$) as sulfur dioxide source. This transformation performs effectively with a wide functional group tolerance in up to 84% yield.

On the basis of a series of control experiments, the authors proposed a plausible catalytic cycle (Scheme 6b). First, irradiation of Eosin Y **11a** by green LEDs generates excited Eosin Y (EY*), which undergoes oxidative quenching with iodonium salt **53** to appropriate aryl radical **61** and EY•⁺. Subsequent reaction of aryl radical **60** with DABSO generates sulfonyl radical **56**, which react regioselectively with the triple bond of **52** forming alkenyl radical **57**. Intramolecular cyclization of **57** produce annulated aryl radical **58**. Deprotonation of **58** generates another radical **59**, which is oxidized to **60** by EY•⁺ thus regenerating eosin catalyst in its ground state. Alternatively, radical **59** can be oxidized by **53** leading to the dihydroquinoline derivative **60** (path II). Finally, sulfonated 1,2-dihydroquinoline **60** undergoes dehydroaromatization to the product **55**.

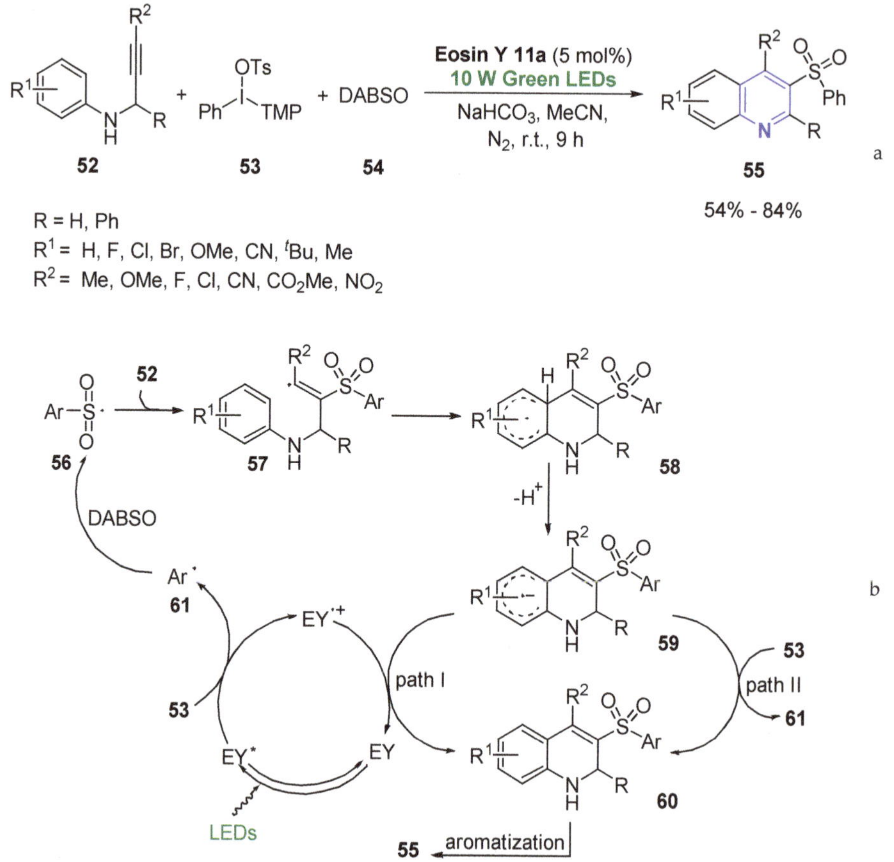

Scheme 6. (a) Cycloaddition of N-propargyl aromatic amines, diaryliodonium salts, and sulfur dioxide; (b) Proposed mechanism.

As an efficient strategy for the construction of phenanthridine skeleton, the radical cyclization involving an iminyl radical is often invoked [41]. To develop more efficient and greener approach, Xie group reported visible-light-mediated cyclization of O-2,4-dinitrophenyl oximes **62** to phenanthridine **63** using Eosin Y **11a** under CFL irradiation (Scheme 7a) [42]. The authors proposed a plausible catalytic cycle for this valuable process (Scheme 7b). Firstly, excited Eosin EY* is reduced by (i-Pr)$_2$NEt to Eosin EY$^{\bullet-}$, which undergoes single electron transfer (SET) with **62** and generate radical anion **64**.

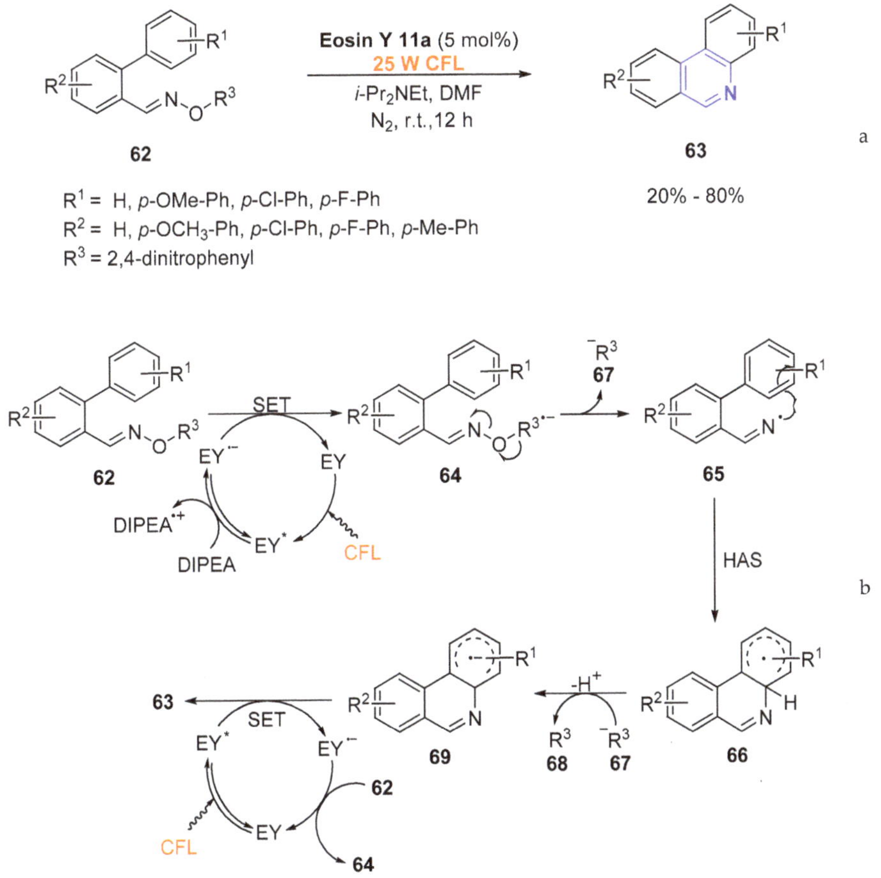

Scheme 7. (a) Cyclization of O-(2,4-dinitrophenyl)oximes to phenanthridine; (b) Proposed mechanism.

Fragmentation of the intermediate **64** followed by intramolecular cyclization gives radical **66**, which is then deprotonated by phenoxyl anion **67** to radical anion **69**. Formed intermediate **69** undergoes another photocatalytic cycle with the excited state of eosin by SET leading to the final product **63** in moderate to good yields (up to 80%).

In a recent investigation, Guo applied decarboxylative cyclization of N-acyloxylphthalimides **71** with vinyl azides **70** in the synthesis of various phenanthridine **72** [43]. This tandem radical addition/cyclization process affords substituted products in up to 79% yield (Scheme 8a). The authors proposed a possible mechanism for this transformation (Scheme 8b). Firstly, irradiation of Eosin Y **11a** by CFL affords the excited Eosin Y catalyst (EY*). Subsequent single electron transfer with **71** generates radical anion **73**, which undergoes fragmentation to CO_2, phtalimide anion **74**, and radical **76**. Addition of **76** to the double bond of **75** followed by extrusion of nitrogen produce iminyl radical **77**, which undergoes series of transformations: intramolecular cyclization, oxidation by Eosin Y$^{•+}$, and deprotonation to the final product **72**.

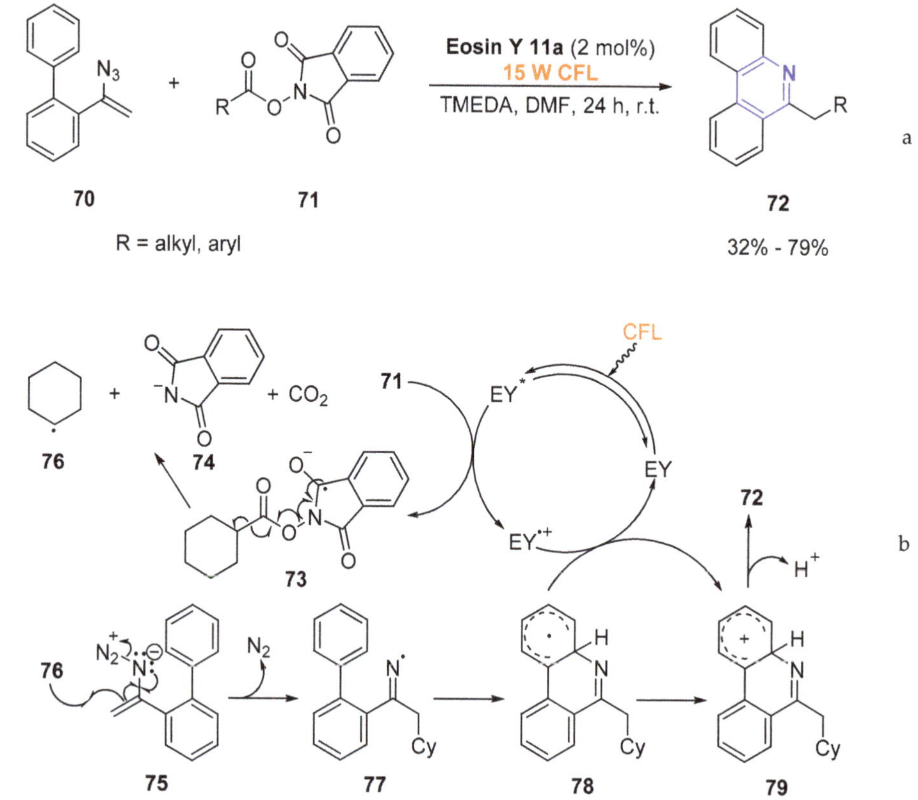

Scheme 8. (a) Decarboxylative radical cyclization of vinyl azides with N-acyloxyphthalimides; (b) Proposed mechanism.

2.1.2. Non-Aromatic Heterocycles

Non-aromatic heterocycles are an integral part of heterocycles' family. Same as their aromatic analogues they are widely distributed in nature, therefore significant efforts have been devoted to their synthesis [1–3].

In a recent investigation, Leonori reported a novel visible-light-mediated generation of 5-membered cyclic imines **81** [44]. Reaction is based on the hydroimination and iminohydroxylation cyclization of olefins **80** using Eosin Y **11a** as a photoredox catalyst (Scheme 9a). In this process, non-aromatic five heterocyclic ring is formed in up to 84% yield. Based on experimental findings, a plausible reaction mechanism is depicted in Scheme 9b. The suggested mechanism is based on single electron transfer from the enhanced Eosin catalyst (EY*) to the starting material **80** with production of anion radical **82**. Radical **82** undergoes fragmentation to phenoxide **85** and radical intermediate **83**. Next, radical **83** undergoes 5-*exo*-trig cyclization to the five-membered N-heterocycle radical **84**. Finally, hydrogen abstraction from **84** by cyclohexadiene (CHD) leads to the desired product **81** and CHD radical **35**, which closes the photoredox cycle by SET process regenerating Eosin catalyst in its ground state.

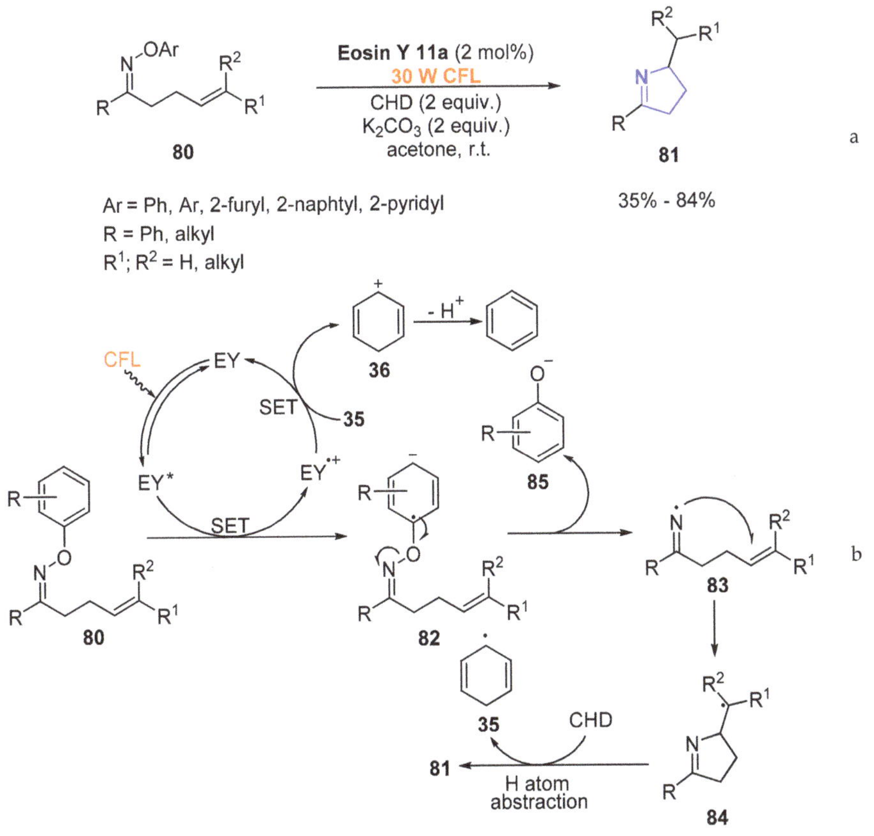

Scheme 9. (a) Hydroimination and iminohydroxylation cyclization of olefins; (b) Proposed mechanism.

Cyclization of N-arylacrylamides **86** have been found as a useful strategy in the synthesis of indolin-2-one compounds **88**. Novák and Tóth group developed cyclization of trifluoromethylated acrylamides under visible-light conditions catalyzed by Erythrosine B **18b** (Scheme 10) [45]. The elaborated arylation/cyclization sequence is initiated by aryl radical generation from diazonium salt **87**, which undergoes addition to the double bond of **86** followed by radical cyclization. In this process, structurally different heterocyclic units **88** can be prepared in up to 83% yield.

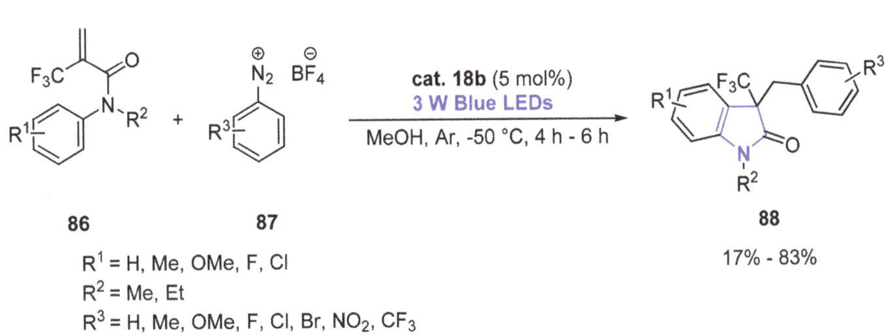

Scheme 10. Synthesis of 3-(trifluoromethyl)indolin-2-one derivatives under photoredox conditions.

Various dihydroisoquinolin derivatives can be obtained by using Eosin Y **11b** catalyst under visible-light conditions [46]. Huang group showed that substituted 3,4-dihydroisoquinolinones **90** can be received in a very good yields by intramolecular cyclization of appropriate amides **89** (Scheme 11a). Mechanistic pathway for this transformation is presented in Scheme 11b. It starts by the formation of excited state of Eosin catalyst (EY*) and deprotonation of **89** by DBU with anion **91** formation. Single electron transfer between anion **91** and EY* leads to the radical **92**. Subsequent intramolecular cyclization followed by another SET cycle recovers ground state of Eosin catalyst together with the anion **94** formation, which undergoes protonation to the final product **90**.

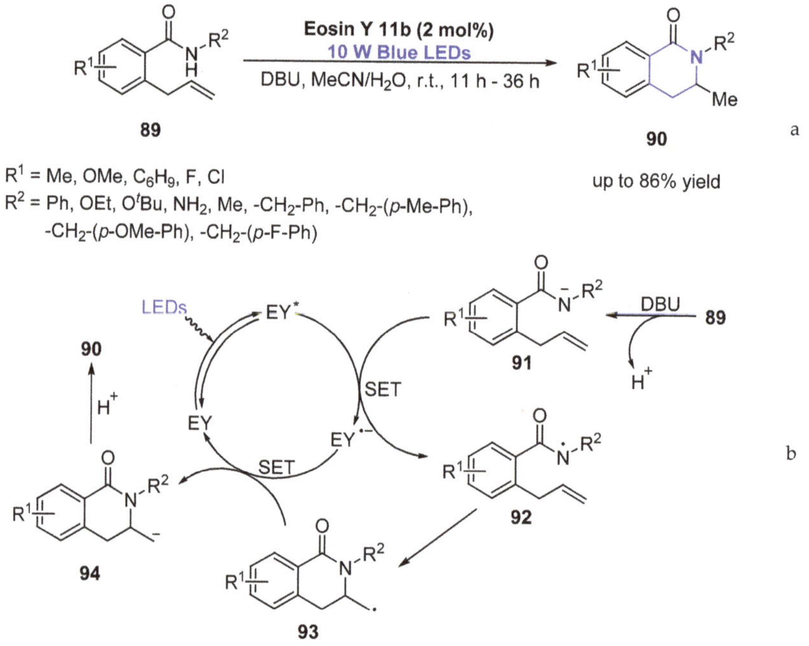

Scheme 11. (a) Synthesis of 3,4-dihydroisoquinolinones; (b) Proposed mechanism.

Throughout the past decades, stereoselective synthesis became of special interest, both by academia and industry scientists [47]. Recently, Sivaguru group [48] demonstrated that heterocycles possessing multiple stereocenters can be prepared using photochemical chemistry by intramolecular [2 + 2] photocycloaddition (Scheme 12). By incorporating axial chirality into the system, cycloaddition process can be effectively controlled leading to the appropriate stereoisomer **96** in up to 99% yield.

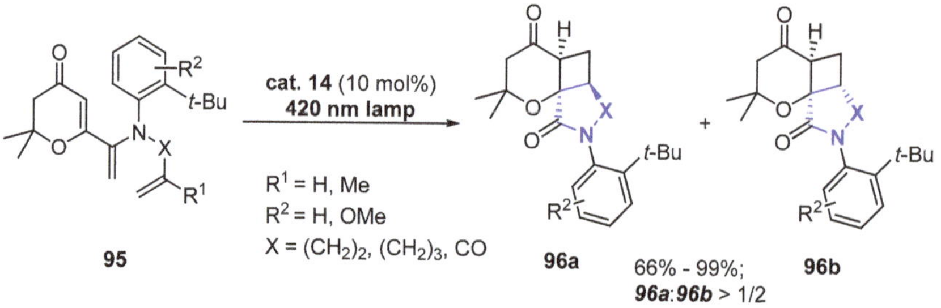

Scheme 12. [2 + 2] Photocycloaddition of enones.

Tetrahydroquinolines (THQ) are an important class of compounds being significant synthetic target for chemists. In this context, Guan and co-workers reported synthesis of tetrahydroquinoline derivatives 99 utilizing Rose Bengal 7 (RB) as a catalyst under irradiation by the compact fluorescent lamp (CFL) [49]. This newer approach to the previously reported by Rueping and coworkes [50] proceeds via tandem radical cyclization of N,N-dimethylanilines 97 with 2-benzylidenemalononitriles 98 to six-membered non-aromatic heterocycles 99 in up to 74% yield (Scheme 13a). The mechanism of this transformation can be rationalized on the basis of $^1O_2$ formation by interaction of O_2 with the excited-state of Rose Bengal catalyst (RB*) (Scheme 13b). Generated singlet oxygen proceeds single electron transfer with N,N-dimethylaniline 97 with the formation of cation radical 100. Absorbing proton by dioxygen radical anion results in the formation of amine radical 101, which react with 2-benzylidenepropanedinitrile 98 and produce the alkyl radical 102. Finally, intramolecular cyclization of 102 followed by rearomatization by the second electron transfer/proton elimination step gives the final product 99.

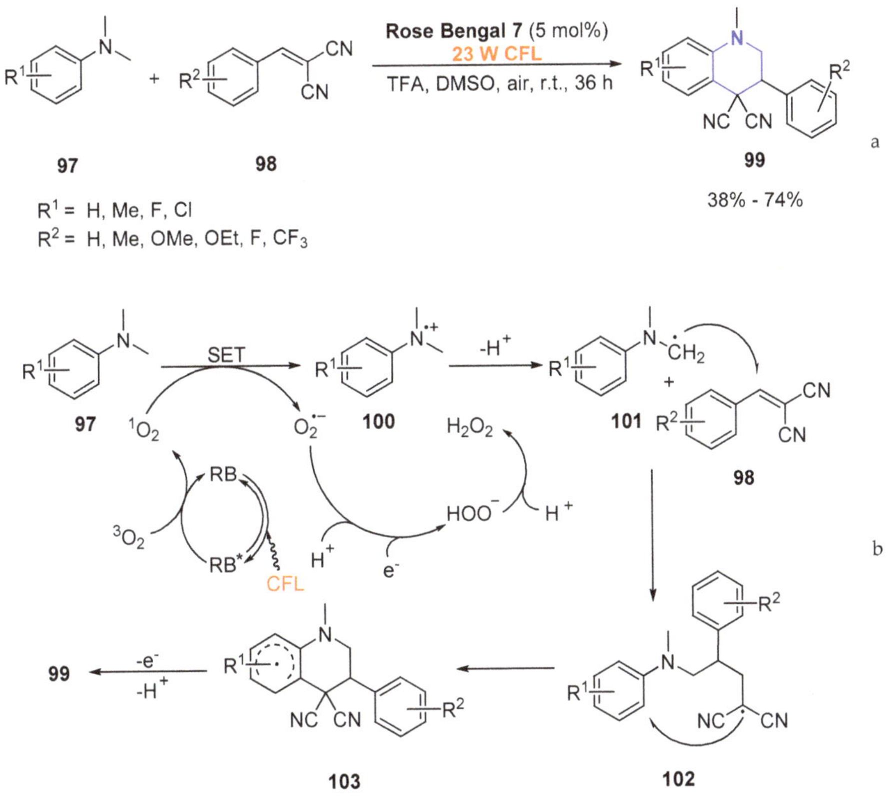

Scheme 13. (a) Synthesis of tetrahydroquinoline derivatives; (b) Proposed mechanism for synthesis of tetrahydroquinoline derivatives.

Similarly, using (N,N-dimethyl)anilines 104 for the formation of α–amino radicals, Zhang [51] and Yadav [52] group revealed an efficient [4 + 2] cyclization to the THQ analogues 106. In this process, N-hydroxyphthalimide 20 was used as an organophotoredox catalyst under white LEDs irradiation (Scheme 14a). Elaborated protocol involves C(sp^3)-H activation of N-methylanilines 104 for the formation of α–amino radical 108, which undergoes [4 + 2] cyclization with maleimide 105 to form intermediate **109**. Finally, aromatization of **109** leads to the desired product **106** in up to 94% yield (Scheme 14b).

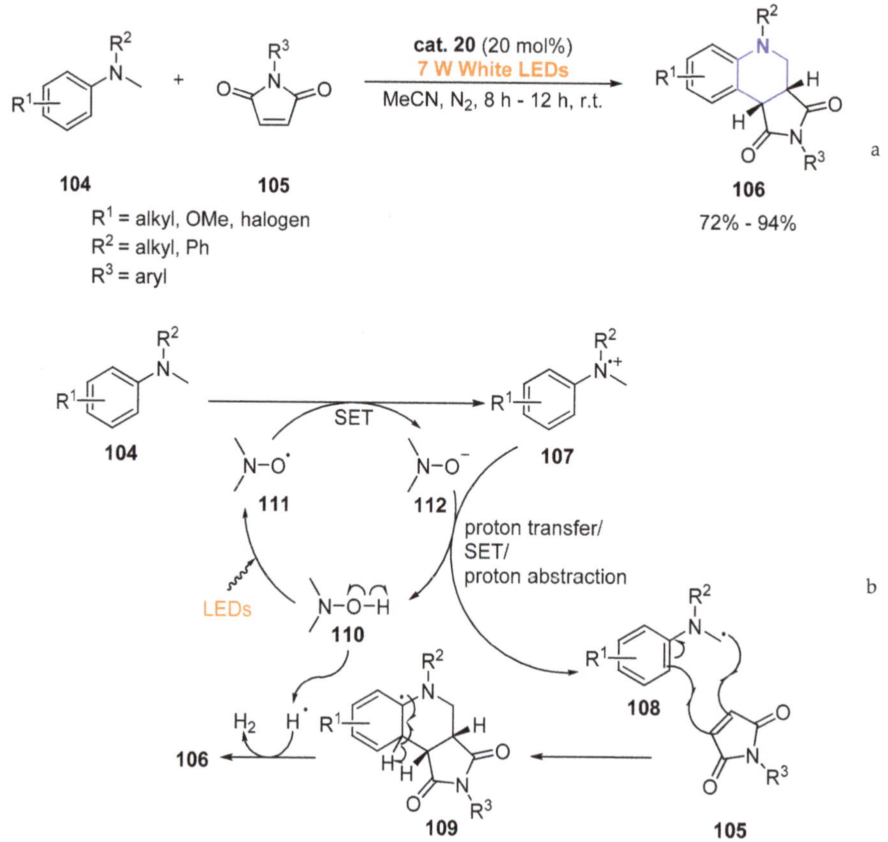

Scheme 14. (**a**) [4 + 2] Radical cyclization of N-methylanilines with maleimides in the synthesis of tetrahydroquinolines; (**b**) Proposed mechanism.

By utilizing sulfonyl radicals, many sulfonyl containing compounds can be prepared using photoredox catalysis. The use of cheap, non-irritating sodium sulfinate opens the way to the possibility to generate sulfonyl radicals by means of visible-light-induced catalysis. Zuo et al. disclosed a tandem radical addition/cyclization concept for the synthesis of isoquinolinediones [53]. A mild radical cascade reaction is based on sulfonyl radical generation from the appropriate sodium sulfinate **114** and Eosin Y **11a** under blue LED irradiation (Scheme 15a). In this tandem approach, a broad range of substrates and functional group are tolerated leading to the desired products in up to 82% yield. The reaction starts on generation of excited state of Eosin (EY*) by blue LEDs irradiation which is reductively quenched by sulfinate **114** to sulfinate radical **116** and EY$^{•-}$ catalyst (Scheme 15b). More stable form of radical **116**—radical **117** undergoes subsequent addition to the double bond of **113** generating radical **118**, which yields radical **119** by intramolecular cyclization. Finally, the single electron and proton transfer between EY$^{•-}$ and **119** delivers the desired product **115**. The photocatalytic cycle is completed by the reaction of Eosin hydride with proton, which leads to the release of hydrogen from the reaction mixture.

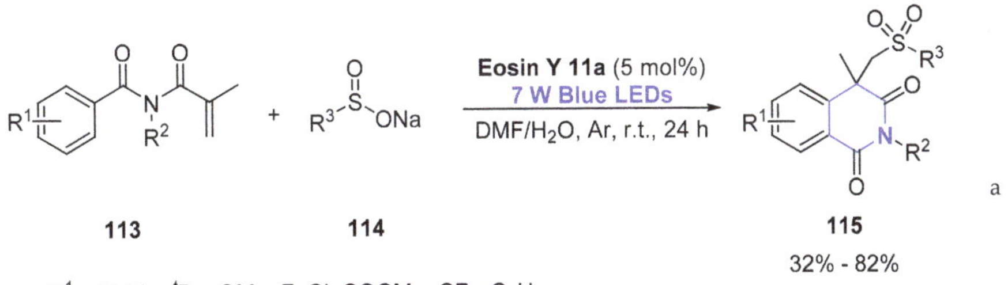

R¹ = H, Me, ᵗBu, OMe, F, Cl, COOMe, CF₃, C₄H₄
R² = H, Me, Et, ⁿPr, ⁱPr, CH₂Ph, C₃H₅
R³ = Et, Me, Ph, p-Me-Ph, p-Cl-Ph, p-F-Ph

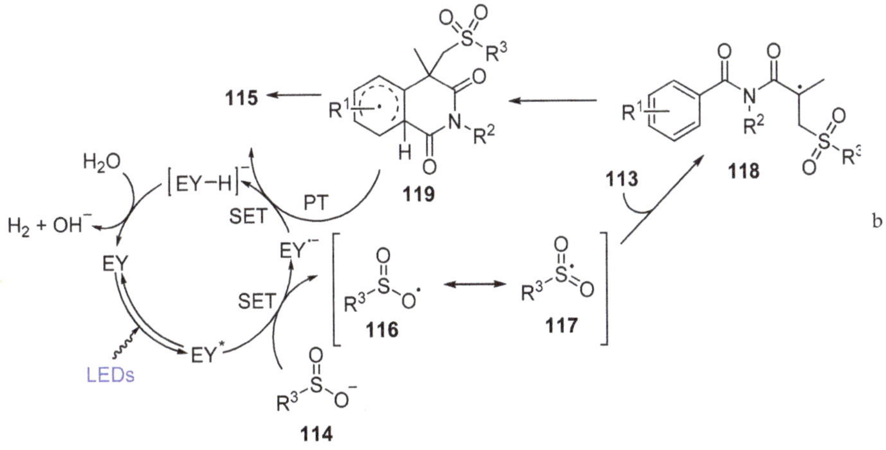

Scheme 15. (a) Preparation of isoquinolinediones; (b) Proposed mechanism for preparation of isoquinolinediones.

2.2. O-Containing Heterocycles

2.2.1. Aromatic Heterocycles

Similar to N-containing heterocycles, their oxygen analogues can be also found in many natural products and pharmaceuticals [54,55], attracting broad interests by scientific community. A new route toward the synthesis of a functionalized furan ring has been recently disclosed by Lei and co-workers [56]. Lei elaborated the synthesis of this useful scaffold through the visible-light-mediated oxidative [3 + 2] cycloaddition of enols and alkynes **121** (Scheme 16a). This highly selective method of constructing substituted furans is based on the usage of Methylene Blue **13** as catalyst in combination with $(NH_4)_2S_2O_8$ as a terminal oxidant. General mechanism for this transformation is disclosed in Scheme 16b. First, visible-light irradiation of Methylene Blue (MB) leads to the formation of its excited-state MB*. The persulfate is then reduced by MB* to form the sulfate radical anion. Subsequent hydrogen atom transfer between **120** and sulfate radical anion forms intermediate **123**, which undergoes intermolecular addition to the triple bond of alkyne **121** with the formation of radical **124**. Next, intramolecular radical addition to the carbonyl oxygen forms a five-membered O-heterocycle **125** radical. Finally, oxidation of **125** by MB•+ followed by deprotonation leads to the final product **122**.

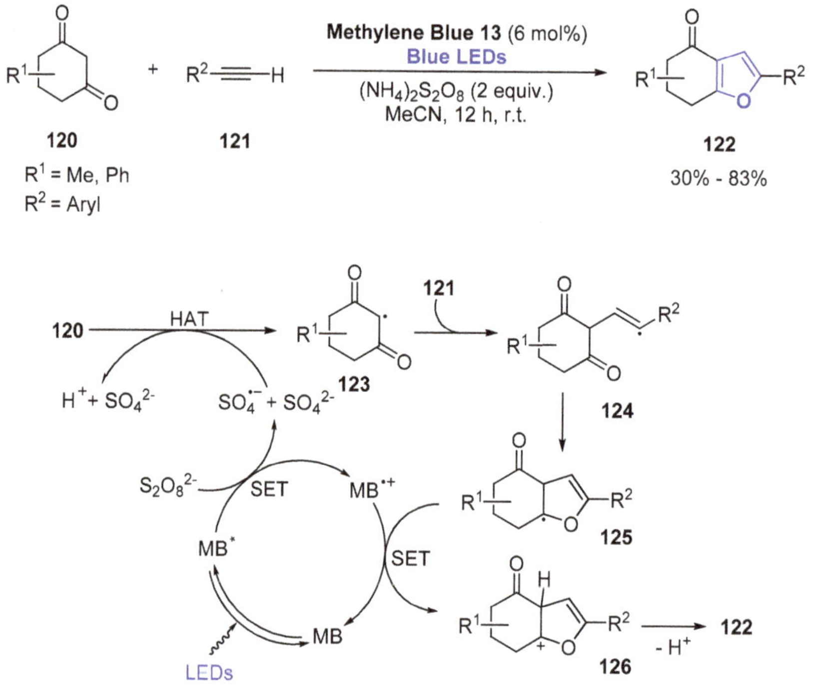

Scheme 16. (a) Visible-light-mediated oxidative [3 + 2] cycloaddition in the synthesis of furans; (b) Proposed mechanism.

2.2.2. Non-Aromatic Heterocycles

In addition to heteroaromatic compounds, their non-aromatic oxygen-containing analogues are equally important. Butyrolactones are a class of heterocycles highly prevalent in nature, indicating many interesting bioactivities [57,58]. Given their importance, many synthetic strategies for their preparations have been elaborated, including photoredox catalysis. In 2015, Nicewicz group presented a synthesis of α-benzyloamino-γ-butyrolactones **129** via polar radical crossover cycloaddition reaction [59]. The photoredox reaction is carried out using the Fukuzimi acridinium **12** photooxidant, easily accessible oxime acids **128** and alkenes **127**. Using given methodology, various lactones **129** with high functional group variations are accessible in very good yields (Scheme 17a). On the basis of the author's previous work, they proposed a plausible reaction pathway for this interesting transformation (Scheme 17b) [60]. First, a single electron oxidation of the alkene **127** by the excited state of catalyst (PC^{+*}) affords cation radical **137**. Next, the radical **137** reacts with the oxime acid **128** resulting in the formation of the radical **130**. Subsequent deprotonation of compound **130**, followed by intramolecular 5-*exo*-trig radical cyclization furnishes N-centered radical **131**. Finally, hydrogen atom transfer between **131** and **133** affords final product **129** and regenerates the acridinium catalyst **12** in its ground state. Authors indicated, however, that another mechanistic pathway without the use of H-atom donor co-catalyst might be involved.

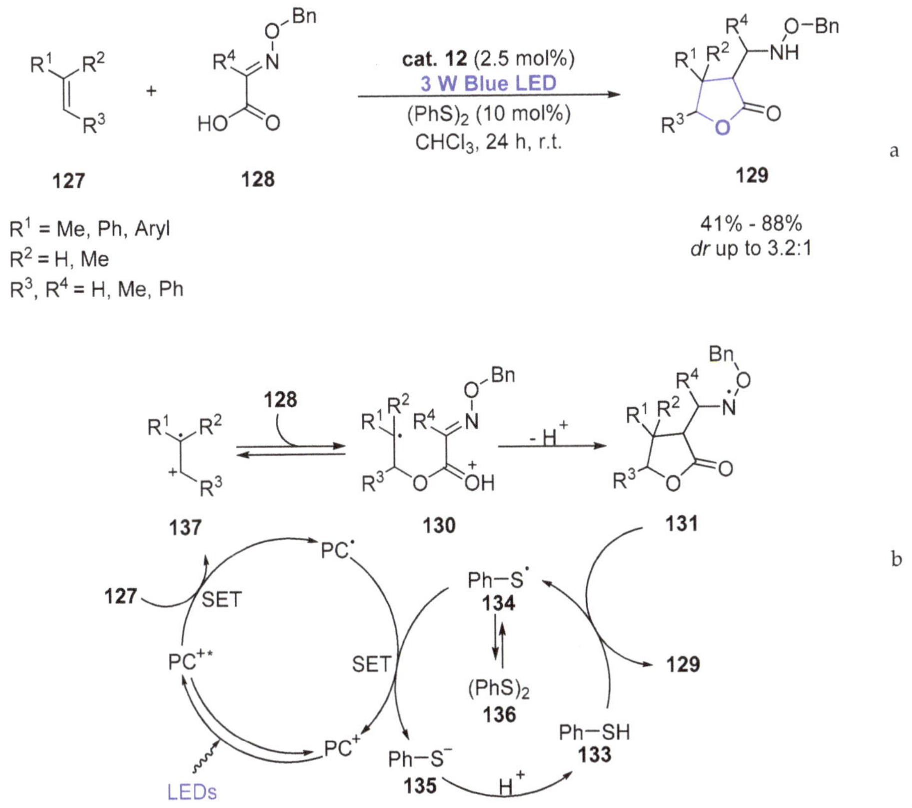

Scheme 17. (a) Synthesis of 3,4-di-substituted mercaptolactones; (b) Proposed mechanism.

The synthesis of structurally similar butyrolactones by metal-free visible-light-mediated catalysis has been also described by Liu [61] and Shah groups [62]. Moreover, by using amide analogues of **128** various γ-lactams or pyrrolidines can be prepared as well [63].

An impressive example of a chiral ion-pair photoredox organocatalyst [64] in enantioselective hydroetherification of alkenols in stereoselective synthesis of tetrahydrofuran analogs has been recently disclosed by Luo and co-workers [65]. The described work showed that ion pair catalysts indicate improved activity compared to Fukuzimi catalysts **12** and proves that visible-light-mediated catalysis is feasible for stereoselective transformations (Scheme 18a). This reaction is based on the three stage process: single electron transfer, cyclization, and hydrogen transfer and has been described already in detail by Nicewicz group [60]. However, it has been disclosed by the authors that the origin of the higher activity observed with an ion pair catalyst is related with cyclization/H-transfer sequence. Phosphate anion in ion pair catalyst endows a longer lifetime of the chiral photocatalysts triplet state, introduces chirality, and assists in the H-shift (Scheme 18b).

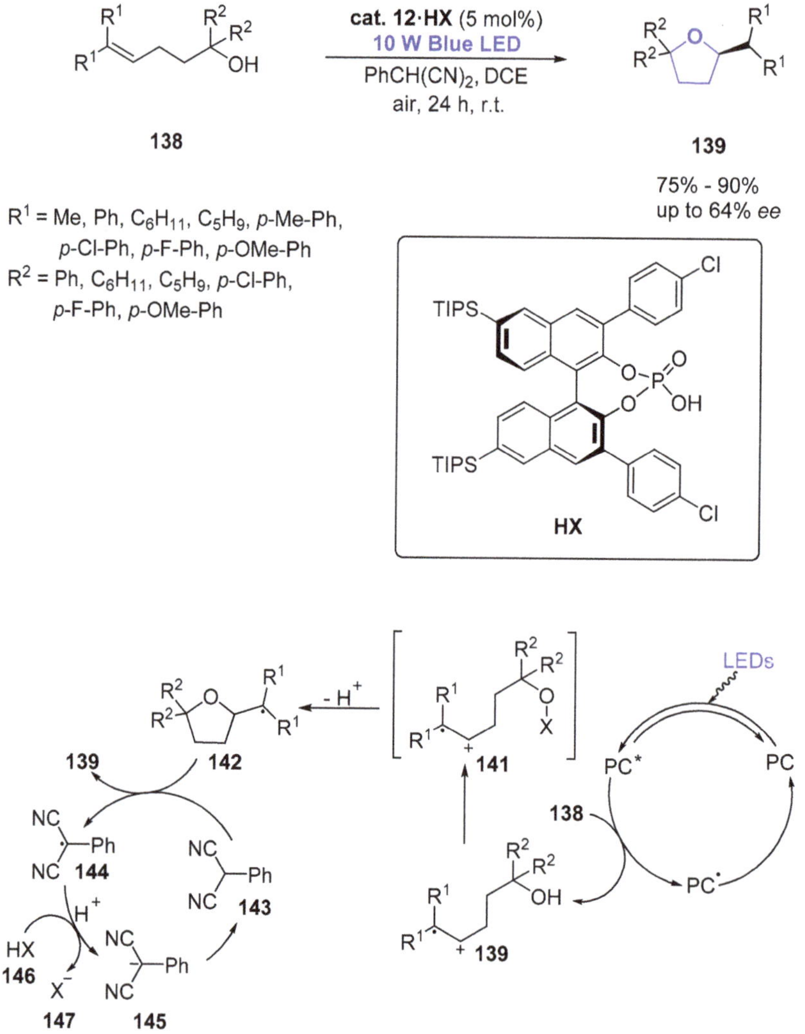

Scheme 18. (a) Hydroetheryfication of alkenols; (b) Proposed mechanism.

Cibulka group revealed that Flavin derivatives 16 can be a successful organic catalyst, able to perform intramolecular [2 + 2] cycloaddition of 148 under 400 nm violet LEDs irradiation [66–68] (Scheme 19a). 1-Butyl-7,8-dimethoxy-3-methylalloxazine 148 under irradiation undergoes intramolecular [2 + 2] cycloaddition of both styrene dienes forming bicyclic unit 149 (Scheme 19b). This photo-induced cycloaddition takes place with a broad spectrum of dienes leading to the formation of desired products 149 in good to excellent yields (up to 97%).

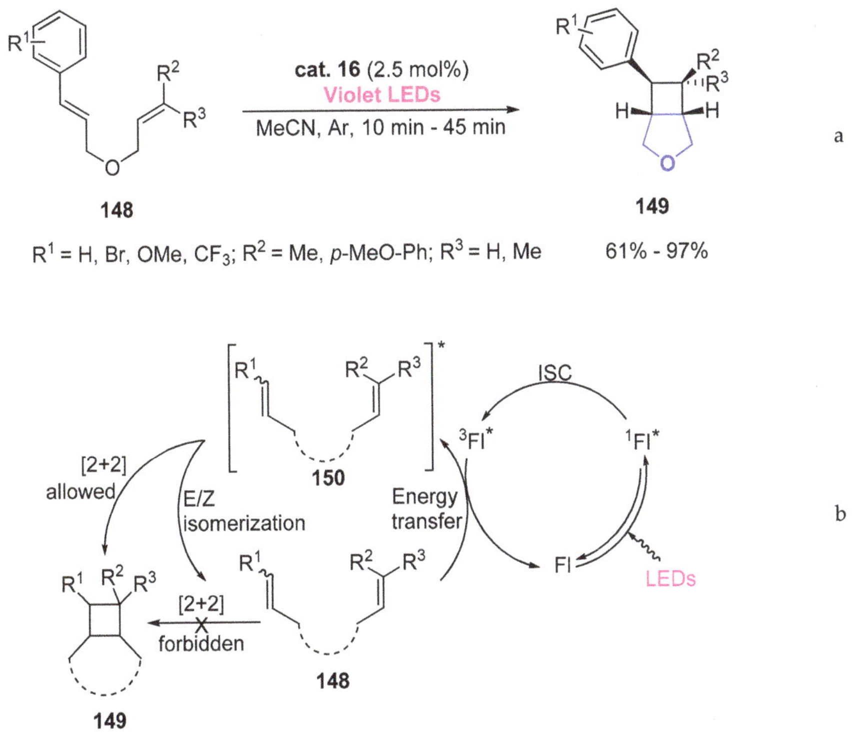

Scheme 19. (a) Flavin-mediated visible-light [2 + 2] cycloaddition of dienes; (b) Proposed mechanism.

Ionic approach for polyene cyclization has been extensively explored, thus constituting the main synthetic strategy of polyene ring synthesis [69]. However, stereoselective radical cyclization can also have significant benefits [70]. In this context, Zhang and Luo group [71] demonstrated visible-light-induced cyclization of polyenes in the synthesis of various oxygen-containing heterocycles (Scheme 20a). The reported protocol is based on radical cascade cyclization using Eosin Y **11a** as a catalyst and green LEDs. Desired products **152** are prepared in moderate to excellent yields (up to 90%) and high stereoselectivities (d.r. > 19:1). In Scheme 20b, a reasonable mechanism for this transformation is suggested. First, excited state of Eosin catalyst (EY*) is formed upon green LEDs' irradiation. Next, intersystem crossing (ISC) forms triplet ^{3}EY*, which allows for electron transfer from the substrate **151** to the ^{3}EY* resulting in the formation of the radical cation **153** and EY radical anion (EY$^{•-}$). The generated radical **153** undergoes intramolecular radical cascade cyclization together with the hydrogen shift to the intermediate **154**. Finally, intermediate **154** is reduced to the desired product **152**, whereas Eosin radical anion (EY$^{•-}$) is oxidized, regenerating catalyst and completing the photocatalytic cycle.

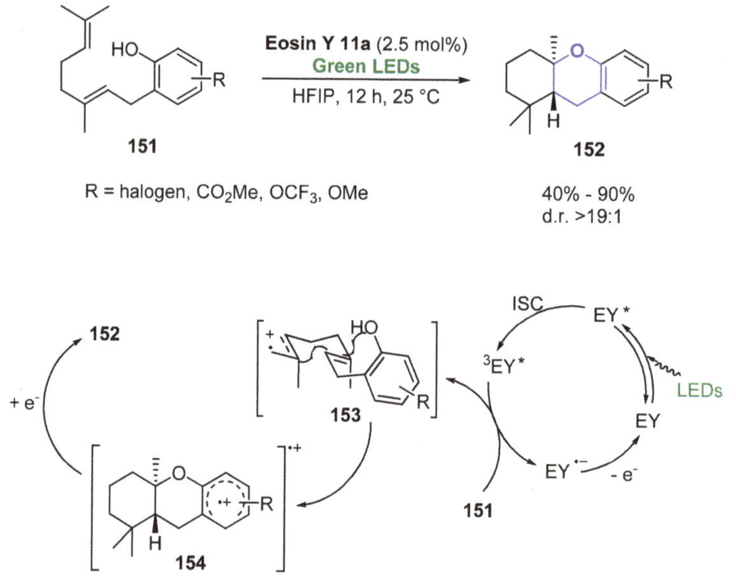

Scheme 20. (a) Visible-light-mediated radical cascade cyclization of polyenes; (b) Proposed mechanism.

In addition to the sulfonated quinolines (Scheme 6), its oxygen containing analogs—cumarines can be also obtained. In this context, various alkynes undergoe arylsulfonylation with arylsulfinic acid and TBHP using Eosin Y **11a** as a catalyst. It is noteworthy that appropriate cumarines can be obtained in good yields with a wide functional group tolerance [72].

2.3. Heterocycles Bearing More Than One Heteroatom

An efficient and straightforward strategy for building functionalized, N- or O-bearing heteroatoms is the 1,3-dipolar cycloaddition [73]. Xia, Yang and collaborators [74] described visible-light-mediated *anti*-regioselective 1,3-dipolar cycloaddition of nitrones **155** with alkenes **156** (Scheme 21a). The nitrones are cyclized with styrenes and aliphatic alkenes via a polar radical crossover cycloaddition reaction using triphenylpyrrylium catalyst **8b** under blue light irradiation. This transformation is based on the mechanism presented in Scheme 21b. Initially, upon blue LEDs irradiation excited state of catalysts is formed (PC*), which then oxidize double bond of **156** via the reductive quenching process. Electrophilic addition of intermediate **158** to nitrone **155**, followed by a radical cyclization leads to the intermediate **159**. This intermediate act as an oxidant enabling regeneration of the catalyst by electron transfer alongside with the product formation. The authors also suggest a possible radical chain propagation between **158** and alkene **155**.

Woo group [75] unveiled that nitrones can be also generated in situ from oxaziridines in photoredox conditions. They described synthetic method for the preparation of 4-isoxazolines **162** in a visible-light photoredox-catalyzed [3 + 2] cycloaddition of oxaziridines **160** with alkynes **161** (Scheme 22a). Described methodology relies on the in situ generation of nitrones from oxaziridines **160** by the single electron transfer process, which undergoes [3 + 2] cycloaddition with various alkynes. On the basis of a series of experiments, the following mechanism is proposed by the authors (Scheme 22b). Excited (Acr⁺-Mes*) catalyst formed upon blue LEDs irradiation allows for the single electron oxidation process of aziridine **160** alongside with the reduced photocatalyst (Acr•-Mes) formation. The radical cation **163** is then converted into the nitrone radical **164** through the ring opening process. Single-electron reduction of nitrone radical **164** by Acr•-Mes regenerates catalyst and forms nitrone **165**. Finally, [3 + 2] cycloaddition of nitrone **165** with alkyne **161** provides the product **162**.

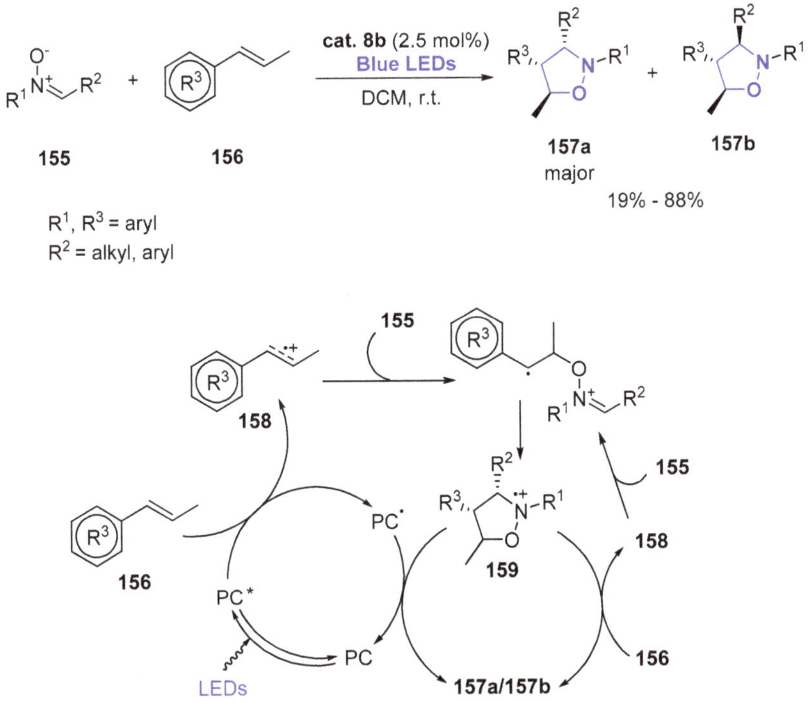

a

b

Scheme 21. (**a**) Visible-light-mediated nitrone 1,3-dipolar cycloaddition; (**b**) Proposed mechanism.

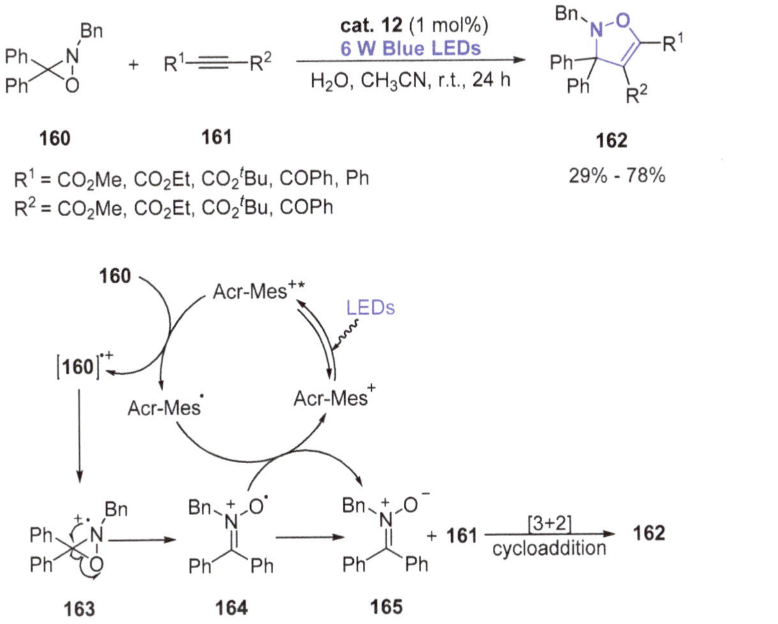

a

b

Scheme 22. (**a**) [3 + 2] Cycloaddition of oxaziridines with alkynes; (**b**) Proposed mechanism.

1,2,4-Oxadiazole derivatives besides exhibiting biological activities have also found application in light-emitting diodes (OLEDs) [76]. Due to the importance of this group, Cho and co-workers developed oxidative cyclization of amidoximes under visible-light conditions [77]. Described protocol involves the intramolecular oxidative cyclization of amidoximes **166** in the presence of the triphenylpyrylium **8c** catalyst and molecular oxygen as the oxidant, promoted by compact fluorescent lamp (Scheme 23a). Opposite to all previous examples, in this particular transformation T(p-F)PPT act as both an electrophilic catalyst and a photocatalyst due to the fact that reaction works also in the dark, although less efficient. The authors proposed a possible mechanism for this transformation (Scheme 23b). Reaction starts by the nucleophilic addition of **166** to the triphenylpyrrylium ion **168** and generates intermediate **169**, which undergoes homolytic dissociation and produce two radicals: **171** and **170**. Additionally, this process is accelerated by the visible-light irradiation. The catalyst **168** is then regenerated by the molecular oxygen oxidation of **170** intermediate. Beside catalyst regeneration cycle, the radical **171** undergoes an intramolecular 1,5-hydrogen atom transfer (HAT) and produces the radical **172**. Subsequent oxidation of the compound **172** to the iminium ion by molecular oxygen or excited catalyst followed by intramolecular cyclization yields the final product **167**.

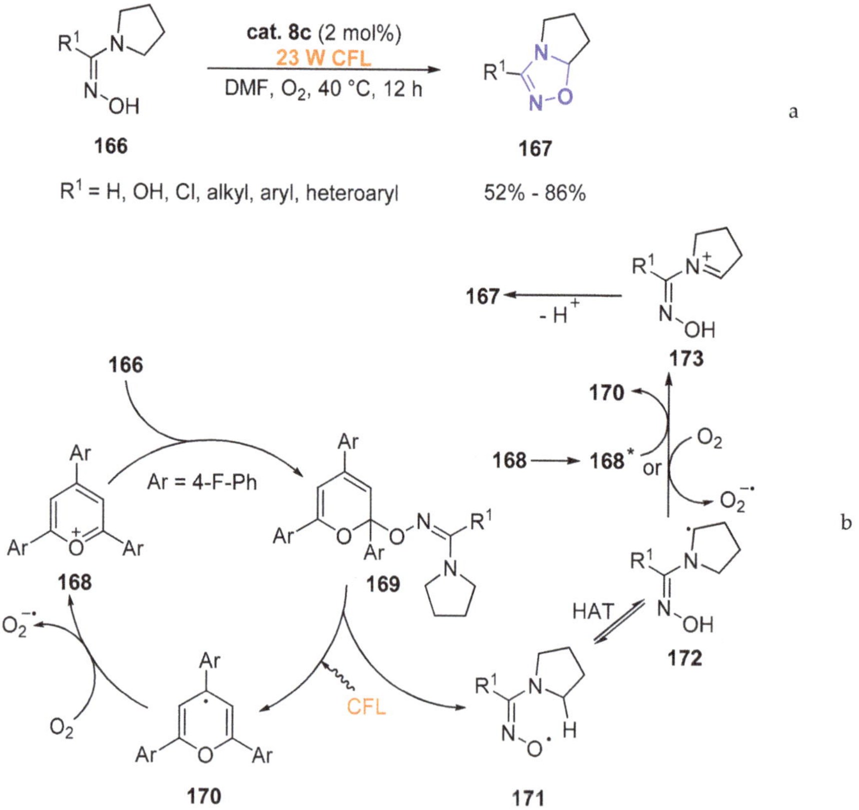

Scheme 23. (a) Application of amidoximes in visible-light-driven synthesis of oxadiazolines; (b) Proposed mechanism.

Many natural or synthetic compounds containing oxazoline scaffold have been found to possess some level of interesting bioactivities [78]. Moreover, these structures are important building blocks in the syntheses of many chiral ligands in stereoselective synthesis [79]. Therefore, novel methodologies for the synthesis of these structural motifs are highly desirable. In this context, Nicewicz and co-workers disclosed photoredox-catalyzed hydrofunctionalization of unsaturated amides and thioamides in the synthesis of 2-oxazolines and 2-thiazolines [80]. This intramolecular functionalization is based on dual catalytic system comprised from **174** and phenyl disulphide (Scheme 24a).

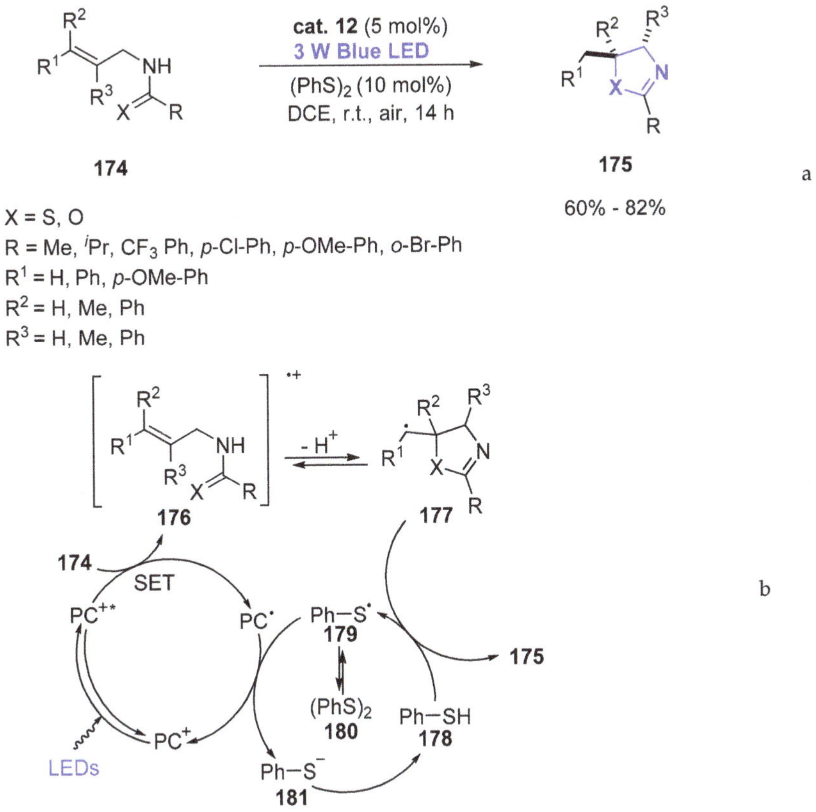

X = S, O
R = Me, ^{i}Pr, CF$_3$ Ph, p-Cl-Ph, p-OMe-Ph, o-Br-Ph
R^1 = H, Ph, p-OMe-Ph
R^2 = H, Me, Ph
R^3 = H, Me, Ph

Scheme 24. (a) Hydrofunctionalization of unsaturated amides; (b) Proposed mechanism.

This mild and efficient protocol leading to the products **175** in good yields (up to 82%) tolerates also many functional groups. In Scheme 24b, a reasonable mechanism for this transformation is suggested. Excitation of acridinium catalyst **12** by blue LED irradiation generates the active oxidant (PC⁺*), which accepts an electron from substrate **174**, generating cation radical intermediate **176**. Subsequent cyclization and proton loss affords cyclic radical intermediate **177**. Finally, hydrogen atom transfer from thiophenol **178** generates product **175** and a thiyl radical **179**. The thiyl radical **179** is presumed to re-oxidize the acridine radical catalyst (PC•), regenerating catalyst in its ground state.

An impressive example of visible-light-mediated generation and utilization of fluorophosgene was developed by König group [81]. Presented procedure is based on the cleavage of an aryl trifluoromethoxy ether **183** and in situ conversion of formed fluorophosgene for the synthesis of carbamates, carbonates, and urea derivatives **184** (Scheme 25a). The reported protocol is based on the single-electron reduction of 4-(trifluoromethoxy)benzonitrile **183** and its fragmentation into fluorophosgene **183a**, which undergoes further intramolecular cyclization with 1,2-diamines, aminoalcohols or diols **182** (Scheme 25b). In this way, carbonates, various carbamates, and urea derivatives **184** can be prepared in moderate to excellent yields.

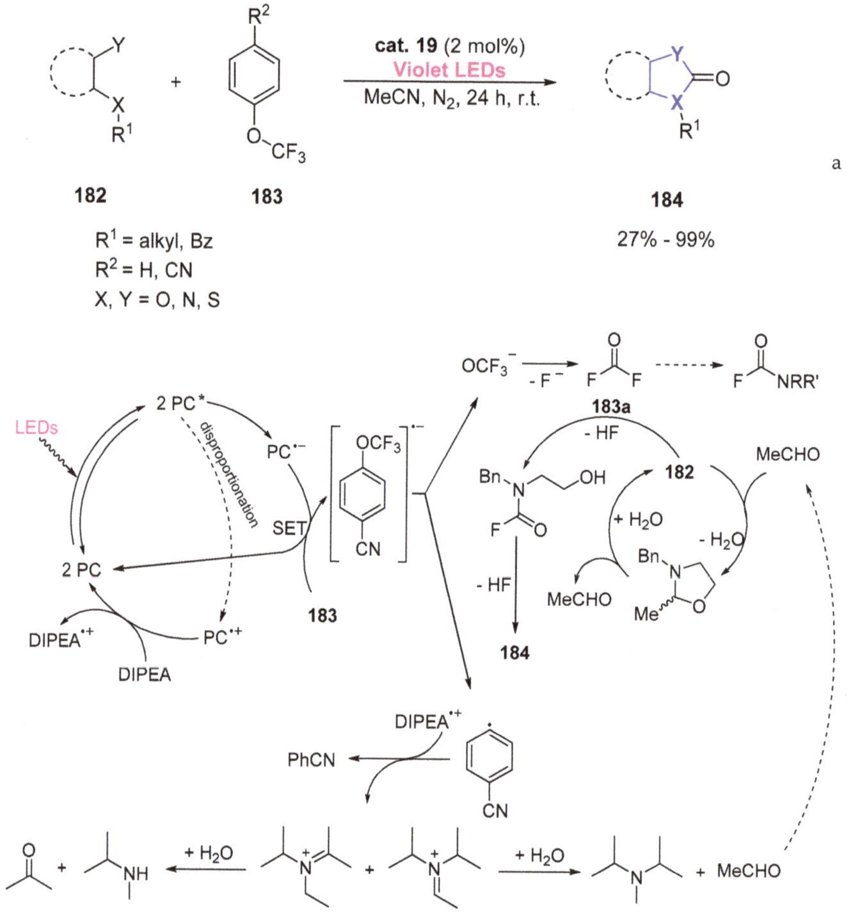

Scheme 25. (a) Visible-light-mediated in situ generation and conversion of fluorophosgene; (b) Proposed mechanism.

Quinazolines are another class of heterocyclic compounds extensively studied. Itoh et al. described the synthesis of quinazolines **187** under visible-light conditions by employing coupling of primary amines **185** and aldehydes **186** (Scheme 26a) [82]. A possible mechanism as proposed by the authors is disclosed in Scheme 26b. Initially, cyclization between the diamine **185** and the aldehyde **186** takes place leading to the diamine **188**. In the same time, Rose Bengal **7** (RB) is excited under irradiation followed by intersystem crossing (ISC) providing ^{3}RB*. Subsequent energy transfer from triplet RB catalyst to oxygen produces reactive singlet oxygen $^1O_2$ responsible for the oxidation of the compound **188** into the final product **187**.

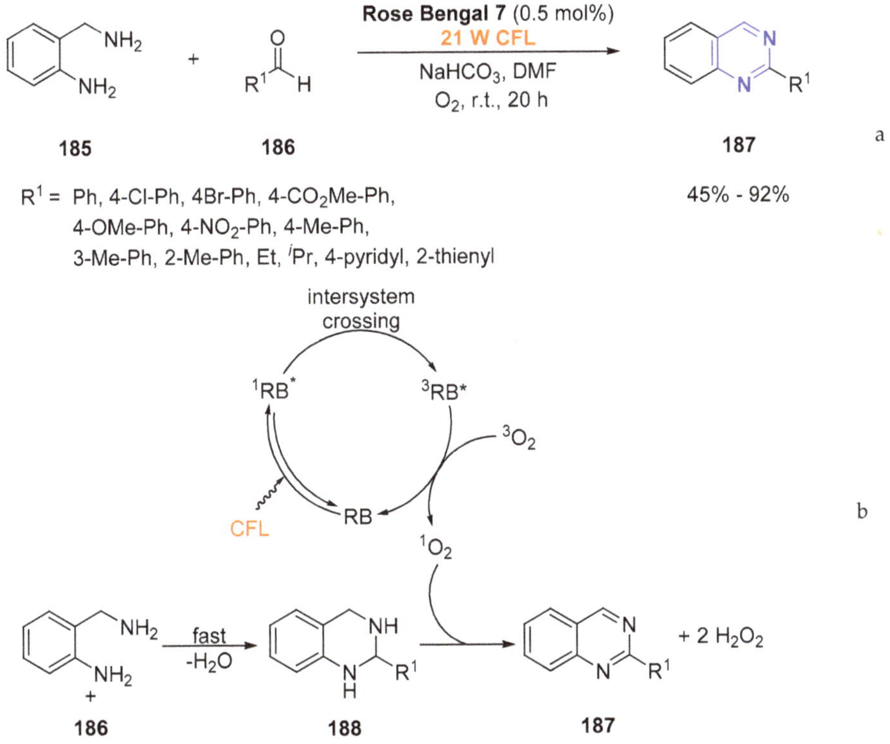

Scheme 26. (a) Synthesis of quinazolines; (b) Proposed mechanism.

More recently, Das group also demonstrated cyclization of the diamines **189** and aldehydes **190** into heterocycles **191**, but in different reaction conditions [83] (Scheme 27). Opposite to the previous example, this protocol is based on CO_2-catalyzed dehydrogenation/aromatization of amines under photoredox conditions. In this context, a corresponding amine formed upon cyclization of diamine **189** and aldehyde **190** is dehydrogenated and aromatized to the product **191**. A broad range of functional group can be used and desired products can be prepared in up to 91% yield. In addition to the diamine and aldehydes coupling, multisubstituted quinazolines and benzimidazoles can also be obtained by the radical cyclization of N-alkyl-N'-arylamidines catalyzed by Rose Bengal **7** in the presence of CBr_4 and K_2CO_3. Tang group demonstrated that the given methodology tolerates a wide range of functional groups with a higher reaction efficiency [84].

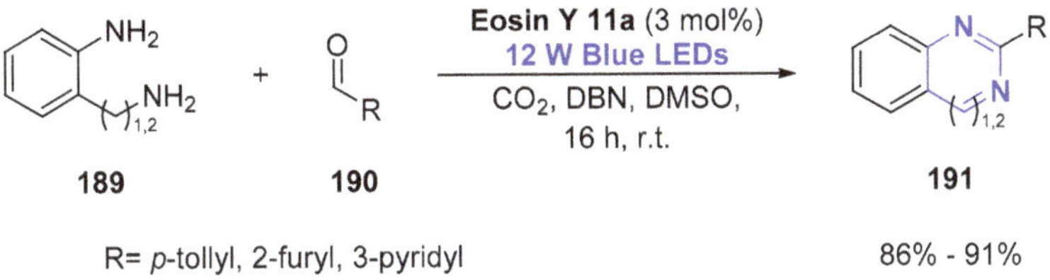

Scheme 27. Dehydrogenation of amines to imines.

The cyano moiety is widely used in organic synthesis because it undergoes many important transformations [85]. Sun et al. described alkoxycarbonylation–addition–cyclization sequence triggered by Eosin Y **11a** under white LEDs irradiation in the synthesis of polyheterocycles **194** [86].

This visible-light-induced cascade reaction is initiated by the intermolecular radical addition of alkyl carbazate **193** to a double bond of N-arylacrylamide **192** followed by cyano-mediated cyclization. Desired poly nitrogen containing heterocycles **194** can be prepared in up to 81% yield (Scheme 28a).

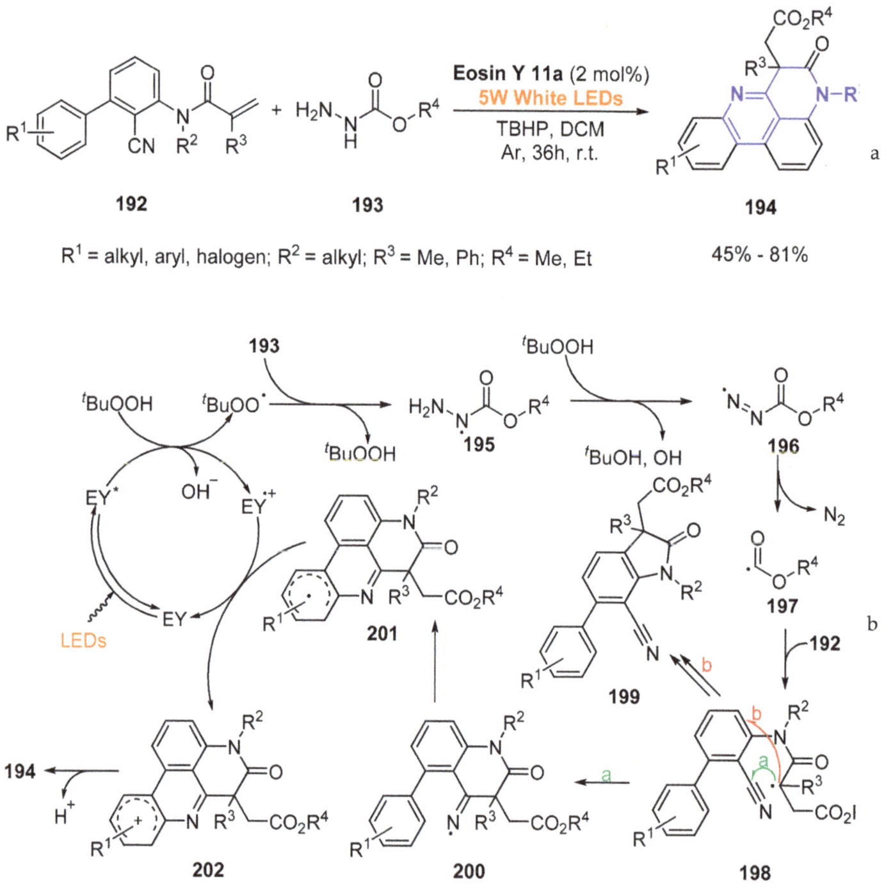

Scheme 28. (a) Photoredox alkoxycarbonylation–addition–cyclization sequence; (b) Proposed mechanism.

The suggested mechanism is depicted in Scheme 28b. Excited state of Eosin EY* catalyst formed upon white LEDs' irradiation undergoes single electron transfer with TBHP producing t-butoxy radical, which detach hydrogen from carbazate **193** leading to the radical intermediate **195**. Sequential dehydrogenation of **195** followed by nitrogen release provides alkoxycarbonyl radical **197**. At this moment, generated radical **197** reacts with double bond of the starting material **192** leading to the intermediate **198**, which undergoes intramolecular cyclization with cyano group to the radical intermediate **200** (path a). Alternatively, the radical **198** can undergo intramolecular cyclization with the aromatic ring (path b) leading to the undesired product **199** in trace amount. The radical **200** undergoes another intramolecular cyclization producing conjugated radical **201**, which after oxidation by the Eosin Y cation radical (EY•+) forms intermediate **202**. Final product **194** is obtained by the deprotonation of intermediate **202**.

In the context of synthesis of THIQ analogues, Baruah et al. elaborated cross dehydrogenative approach toward 1,3-oxazines using visible-light-induced catalysis [87]. 1-Aminoalkyl-2-naphthols

203 easily undergoes intramolecular cyclization using Eosin Y **11a** as photoredox catalyst and green light source (Scheme 29a). On the basis of a series of control experiments, the authors proposed a plausible catalytic cycle (Scheme 29b). The excited Eosin Y (EY*) accepts an electron from the starting material **203** to form a cation radical **205**. Next, the generated EY$^{•-}$ radical anion undergoes electron transfer with molecular oxygen regenerating the ground state Eosin Y **11a** and forming superoxide radical anion. Hydrogen removal from **205** and subsequent intramolecular addition of hydroxyl to the imine group provides the final product **204** in up to 86% yield.

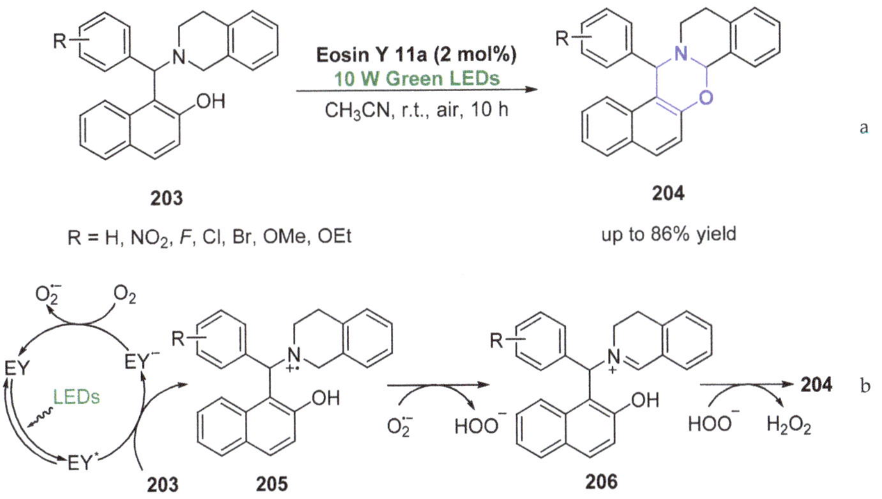

Scheme 29. (a) Synthesis of 1,3-oxazines under photoredox conditions; (b) Proposed mechanism.

Quinazolinone structure is another important structural unit in heterocyclic chemistry [1–3]. Idrolone, Hydromox, and sildenafil citrate are the representative drugs containing quinazolinone motif. Pan and co-workers [88] demonstrated synthesis of sulfonated quinazolinones **209** in visible-light-induced oxidative/reductive cyclization of N-cyanamide alkenes **207** (Scheme 30a). Presented strategy features mild reaction conditions and broad substrate suitability. Various complex N-heterocycles **209** can be accessed in up to 98% yield. On the basis of their experimental observations and literature studies, authors proposed the mechanism depicted in Scheme 30b. Firstly, Eosin Y **11b** is irradiated by green LEDs to generate the excited-state Eosin Y (EY*). Secondly, electron transfer from Eosin to tert-butyl hydroperoxide (TBHP) generates t-butyloxy radical and OH$^-$. Subsequent hydrogen atom removal from **214** produces sulfur-centered radical **215**, which undergoes addition to the double bond of **207** and produces radical intermediate **210**. Intramolecular cyclization cascade delivers radical **212**, which is oxidized by EY$^{•+}$ to the cationic intermediate **213** through a single-electron-transfer route regenerating the same EY in its ground state. Finally, deprotonation of **213** gives the expected product **209**.

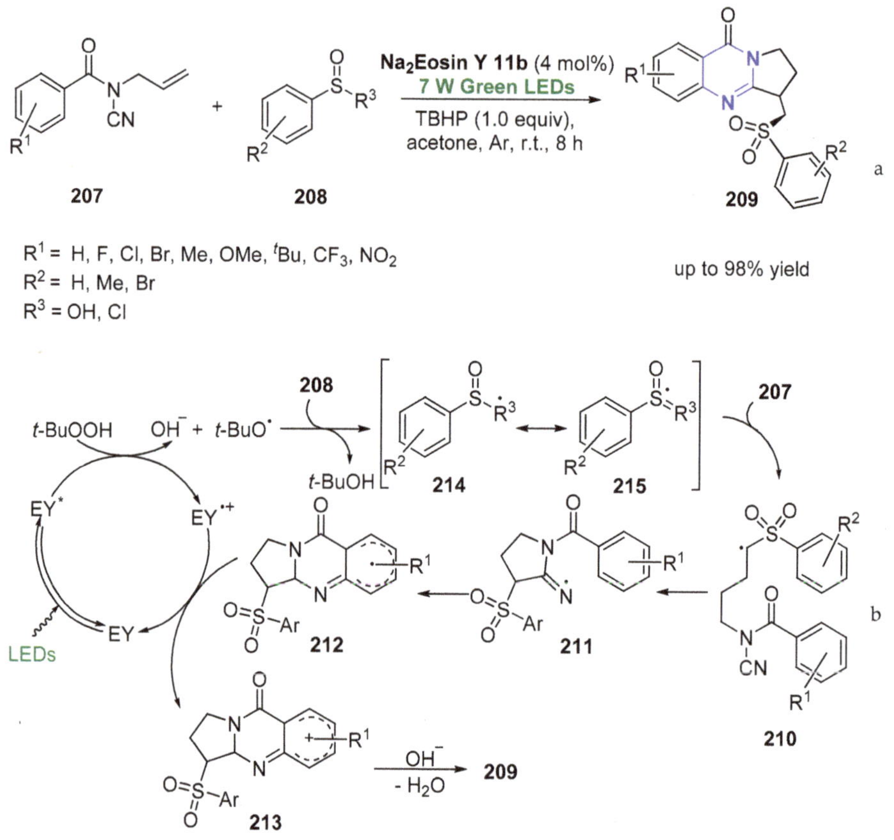

Scheme 30. (a) Synthesis sulfonated quinazolinones via oxidative/reductive cyclization; (b) Proposed mechanism.

In addition to the given examples, other heterocycles containing multiple heteroatoms like imidazopyridines [89] or oxadiazoles [90] can be also obtained using metal-free visible-light-mediated catalysis.

3. Conclusions

In this review, recent advances in the synthesis of nitrogen and oxygen heterocyclic compounds via a metal-free visible-light-induced catalysis have been discussed. It has been shown that various metal-free organic dyes are effective photo-redox catalysts in the synthesis of many structurally different heterocycles containing one or more heteroatoms. Both non-aromatic and aromatic heterocyclic units are readily accessible by using visible-light-mediated catalysis in a straightforward modular way. Reported works indicate that organic dyes capable of visible-light-spectra absorption act as an attractive alternative to the transition-metal complexes. The variety of the presented examples indicates the potential use of the described methodologies in organic synthesis, drug discovery, or materials science. Moreover, the use of inexpensive organic, transition metal-free catalysts makes described protocols very practical and hold promise for broader application in industry and scientific community.

Author Contributions: Conceptualization, M.S.; Writing-Original Draft Preparation, M.S.; Writing-Review & Editing, M.S., R.P. and F.S.; Visualization, R.P. and F.S.

Funding: The authors acknowledge the Foundation for Polish Science (TEAM/2017-4/38). R.P. also acknowledges the Ministry of Science and Higher Education (Diamond Grant 0019/DIA/2018/47).

Conflicts of Interest: The authors declare no conflict of interest.

References

1. Joule, J.A.; Mills, K. *Heterocyclic Chemistry*, 5th ed.; Wiley-Blackwell: Hoboken, NJ, USA, 2010.
2. Pozharskii, A.F.; Soldatenkov, A.T.; Katritzky, A.R. *Heterocycles in Life and Society: An Introduction to Heterocyclic Chemistry, Biochemistry and Applications*, 2nd ed.; Wiley: Hoboken, NJ, USA, 2011.
3. Eicher, T.; Hauptmann, S.; Speicher, A. *The Chemistry of Heterocycles: Structures, Reactions, Synthesis, and Applications*; 3rd Completely Revised and Enlarged ed.; Wiley-VCH: Weinheim, Germany, 2013.
4. Creagh, T.; Ruckle, J.L.; Tolbert, D.T.; Giltner, J.; Eiznhamer, D.A.; Dutta, B.; Flavin, M.T.; Xu, Z.Q. Safety and pharmacokinetics of single doses of (+)-Calanolide a, a novel, naturally occurring nonnucleoside reverse transcriptase inhibitor, in healthy, human immunodeficiency virus-negative human subjects. *Antimicrob. Agents Chemother.* **2001**, *45*, 1379–1386. [CrossRef] [PubMed]
5. Newman, D.J.; Cragg, G.M. Natural Products as Sources of New Drugs from 1981 to 2014. *J. Nat. Prod.* **2016**, *79*, 629–661. [CrossRef]
6. Sun, Y.; Li, R.; Zhang, W.; Li, A. Total synthesis of Indotertine A and Drimentines A, F, and G. *Angew. Chem. Int. Ed.* **2013**, *52*, 9201–9204. [CrossRef]
7. Yajima, T.; Yajima, K.; Hayashi, M.; Takahashi, H.; Yasuda, K. Efficacy and safety of teneligliptin in addition to insulin therapy in type 2 diabetes mellitus patients on hemodialysis evaluated by continuous glucose monitoring. *Diabetes Res. Clin. Pract.* **2016**, *122*, 78–83. [CrossRef] [PubMed]
8. Ghosh, S. *Visible-Light-Active Photocatalysis: Nanostructured Catalyst Design, Mechanisms, and Applications*; Wiley-VCH: Weinheim, Germany, 2018.
9. Stephenson, C.R.J.; Yoon, T.P.; MacMillan, D.W.C. *Visible Light Photocatalysis in Organic Chemistry*; Wiley-VCH: Weinheim, Germany, 2018.
10. Parsons, A.F. *An Introduction to Free Radical Chemistry*; Wiley-Blackwell: Hoboken, NJ, USA, 2000.
11. Togo, H. *Advanced Free Radical Reactions for Organic Synthesis*; Elsevier Science: Amsterdam, The Netherlands, 2004.
12. Romero, N.A.; Nicewicz, D.A. Organic Photoredox Catalysis. *Chem. Rev.* **2016**, *116*, 10075–10166. [CrossRef] [PubMed]
13. Zhang, B.; Studer, A. Recent advances in the synthesis of nitrogen heterocycles via radical cascade reactions using isonitriles as radical acceptors. *Chem. Soc. Rev.* **2015**, *44*, 3505–3521. [CrossRef]
14. Chen, J.R.; Hu, X.Q.; Lu, L.Q.; Xiao, W.J. Exploration of Visible-Light Photocatalysis in Heterocycle Synthesis and Functionalization: Reaction Design and Beyond. *Acc. Chem. Res.* **2016**, *49*, 1911–1923. [CrossRef]
15. Tanoury, G. Photochemical Synthesis of Azaheterocycles. *Synthesis* **2016**, *48*, 2009–2025. [CrossRef]
16. Zhou, L.; Lokman Hossain, M.; Xiao, T. Synthesis of N-Containing Heterocyclic Compounds Using Visible-light Photoredox Catalysis. *Chem. Rec.* **2016**, *16*, 319–334. [CrossRef] [PubMed]
17. Budén, M.E.; Bardagi, J.I.; Rossi, R.A. Constructing Heterocycles by Visible Light Photocatalysis. *Curr. Org. Synth.* **2017**, *14*, 398–429.
18. Xuan, J.; Studer, A. Radical cascade cyclization of 1,n-enynes and diynes for the synthesis of carbocycles and heterocycles. *Chem. Soc. Rev.* **2017**, *46*, 4329–4346. [CrossRef] [PubMed]
19. Bogdos, M.K.; Pinard, E.; Murphy, J.A. Applications of organocatalysed visible-light photoredox reactions for medicinal chemistry. *Beilstein J. Org. Chem.* **2018**, *14*, 2035–2064. [CrossRef] [PubMed]
20. Xuan, J.; Lu, L.-Q.; Chen, J.-R.; Xiao, W.-J. Visible-Light-Driven Photoredox Catalysis in the Construction of Carbocyclic and Heterocyclic Ring Systems. *Eur. J. Org. Chem.* **2013**, *2013*, 6755–6770. [CrossRef]
21. Zhang, W.; Cue, B.W. *Green Techniques for Organic Synthesis and Medicinal Chemistry*; Wiley: Hoboken, NJ, USA, 2012.
22. Richter, G.W.; Solez, K. *Transition Metal Toxicity*; Academic Press: Cambridge, MA, USA, 1990; Volume 31.
23. Wang, S.; Shi, X. Molecular Mechanisms of Metal Toxicity and Carcinogenesis. *Mol. Cell. Biochem.* **2001**, *222*, 3–9. [CrossRef] [PubMed]
24. Egorova, K.S.; Ananikov, V.P. Toxicity of Metal Compounds: Knowledge and Myths. *Organometallics* **2017**, *36*, 4071–4090. [CrossRef]

25. Hari, D.P.; Konig, B. Synthetic applications of eosin Y in photoredox catalysis. *Chem. Commun.* **2014**, *50*, 6688–6699. [CrossRef]
26. Srivastava, V.; Singh, P.P. Eosin Y catalysed photoredox synthesis: A review. *RSC Adv.* **2017**, *7*, 31377–31392. [CrossRef]
27. Ray, S.; Samanta, P.K.; Biswas, P. *Visible-Light-Active Photocatalysis: Nanostrucured Catalyst Design, Mechanisms, and Applicatios*; Wilech-VCH: Weinheim, Germany, 2018; pp. 393–415.
28. Taylor, A.P.; Robinson, R.P.; Fobian, Y.M.; Blakemore, D.C.; Jones, L.H.; Fadeyi, O. Modern advances in heterocyclic chemistry in drug discovery. *Org. Biomol. Chem.* **2016**, *14*, 6611–6637. [CrossRef]
29. Baumann, M.; Baxendale, I.R. An overview of the synthetic routes to the best selling drugs containing 6-membered heterocycles. *Beilstein J. Org. Chem.* **2013**, *9*, 2265–2319. [CrossRef]
30. Minozzi, M.; Nanni, D.; Spagnolo, P. From azides to nitrogen-centered radicals: Applications of azide radical chemistry to organic synthesis. *Chem. Eur. J.* **2009**, *15*, 7830–7840. [CrossRef]
31. Funk, C.D. Leukotriene modifiers as potential therapeutics for cardiovascular disease. *Nat. Rev. Drug Discov.* **2005**, *4*, 664–672. [CrossRef] [PubMed]
32. Vitaku, E.; Smith, D.T.; Njardarson, J.T. Analysis of the structural diversity, substitution patterns, and frequency of nitrogen heterocycles among U.S. FDA approved pharmaceuticals. *J. Med. Chem.* **2014**, *57*, 10257–10274. [CrossRef] [PubMed]
33. Tambe, S.D.; Rohokale, R.S.; Kshirsagar, U.A. Visible-Light-Mediated Eosin Y Photoredox-Catalyzed Vicinal Thioamination of Alkynes: Radical Cascade Annulation Strategy for 2-Substituted-3-sulfenylindoles. *Eur. J. Org. Chem.* **2018**, 2117–2121. [CrossRef]
34. Gu, L.; Jin, C.; Wang, W.; He, Y.; Yang, G.; Li, G. Transition-metal-free, visible-light induced cyclization of arylsulfonyl chlorides with o-azidoarylalkynes: A regiospecific route to unsymmetrical 2,3-disubstituted indoles. *Chem. Commun.* **2017**, *53*, 4203–4206. [CrossRef]
35. Ojima, I. *Fluorine in Medicinal Chemistry and Chemical Biology*; Wiley-Blackwell: Chichester, UK, 2009.
36. Jana, S.; Verma, A.; Kadu, R.; Kumar, S. Visible-light-induced oxidant and metal-free dehydrogenative cascade trifluoromethylation and oxidation of 1,6-enynes with water. *Chem. Sci.* **2017**, *8*, 6633–6644. [CrossRef]
37. Nicolaou, K.C.; Chen, J.S. The art of total synthesis through cascade reactions. *Chem. Soc. Rev.* **2009**, *38*, 2993–3009. [CrossRef] [PubMed]
38. Rusch, F.; Unkel, L.N.; Alpers, D.; Hoffmann, F.; Brasholz, M. A visible light photocatalytic cross-dehydrogenative coupling/dehydrogenation/6pi-cyclization/oxidation cascade: Synthesis of 12-nitroindoloisoquinolines from 2-aryltetrahydroisoquinolines. *Chem. Eur. J.* **2015**, *21*, 8336–8340. [CrossRef]
39. Wang, K.; Meng, L.G.; Wang, L. Visible-Light-Promoted [2+2+2] Cyclization of Alkynes with Nitriles to Pyridines Using Pyrylium Salts as Photoredox Catalysts. *Org. Lett.* **2017**, *19*, 1958–1961. [CrossRef]
40. Sun, D.; Yin, K.; Zhang, R. Visible-light-induced multicomponent cascade cycloaddition involving N-propargyl aromatic amines, diaryliodonium salts and sulfur dioxide: Rapid access to 3-arylsulfonylquinolines. *Chem. Commun.* **2018**, *54*, 1335–1338. [CrossRef]
41. Pan, C.; Han, J.; Zhang, H.; Zhu, C. Radical arylalkoxycarbonylation of 2-isocyanobiphenyl with carbazates: Dual C-C bond formation toward phenanthridine-6-carboxylates. *J. Org. Chem.* **2014**, *79*, 5374–5378. [CrossRef]
42. Liu, X.; Qing, Z.; Cheng, P.; Zheng, X.; Zeng, J.; Xie, H. Metal-Free Photoredox Catalyzed Cyclization of O-(2,4-Dinitrophenyl)oximes to Phenanthridines. *Molecules* **2016**, *21*, 1690. [CrossRef] [PubMed]
43. Yang, J.C.; Zhang, J.Y.; Zhang, J.J.; Duan, X.H.; Guo, L.N. Metal-Free, Visible-Light-Promoted Decarboxylative Radical Cyclization of Vinyl Azides with N-Acyloxyphthalimides. *J. Org. Chem.* **2018**, *83*, 1598–1605. [CrossRef]
44. Davies, J.; Booth, S.G.; Essafi, S.; Dryfe, R.A.; Leonori, D. Visible-Light-Mediated Generation of Nitrogen-Centered Radicals: Metal-Free Hydroimination and Iminohydroxylation Cyclization Reactions. *Angew. Chem. Int. Ed.* **2015**, *54*, 14017–14021. [CrossRef] [PubMed]
45. Gonda, Z.; Béke, F.; Tischler, O.; Petró, M.; Novák, Z.; Tóth, B.L. Erythrosine B Catalyzed Visible-Light Photoredox Arylation-Cyclization of N-Alkyl-N-aryl-2-(trifluoromethyl)acrylamides to 3-(Trifluoromethyl)indolin-2-one Derivatives. *Eur. J. Org. Chem.* **2017**, *2017*, 2112–2117. [CrossRef]
46. Zou, S.; Geng, S.; Chen, L.; Wang, H.; Huang, F. Visible light driven metal-free intramolecular cyclization: A facile synthesis of 3-position substituted 3,4-dihydroisoquinolin-1(2H)-one. *Org. Biomol. Chem.* **2019**, *17*, 380–387. [CrossRef] [PubMed]

47. Carreira, E.M.; Kvaerno, L. *Classics in Stereoselective Synthesis*; Wiley-VCH: Weinheim, Germany, 2009.
48. Clay, A.; Vallavoju, N.; Krishnan, R.; Ugrinov, A.; Sivaguru, J. Metal-Free Visible Light-Mediated Photocatalysis: Controlling Intramolecular [2+2] Photocycloaddition of Enones through Axial Chirality. *J. Org. Chem.* **2016**, *81*, 7191–7200. [CrossRef] [PubMed]
49. Xin, J.-R.; Guo, J.-T.; Vigliaturo, D.; He, Y.-H.; Guan, Z. Metal-free visible light driven synthesis of tetrahydroquinoline derivatives utilizing Rose Bengal. *Tetrahedron* **2017**, *73*, 4627–4633. [CrossRef]
50. Zhu, S.; Das, A.; Bui, L.; Zhou, H.; Curran, D.P.; Rueping, M. Oxygen Switch in Visible-Light Photoredox Catalysis: Radical Additions and Cyclizations and Unexpected C–C-Bond Cleavage Reactions. *J. Am. Chem. Soc.* **2013**, *135*, 1823–1829. [CrossRef]
51. Liang, Z.; Xu, S.; Tian, W.; Zhang, R. Eosin Y-catalyzed visible-light-mediated aerobic oxidative cyclization of N,N-dimethylanilines with maleimides. *Beilstein J. Org. Chem.* **2015**, *11*, 425–430. [CrossRef] [PubMed]
52. Yadav, A.K.; Yadav, L.D.S. Visible light photoredox catalysis with N-hydroxyphthalimide for [4+2] cyclization between N-methylanilines and maleimides. *Tetrahedron Lett.* **2017**, *58*, 552–555. [CrossRef]
53. Zuo, K.-L.; He, Y.-H.; Guan, Z. Metal-Free Visible-Light Photocatalytic Tandem Radical Addition-Cyclization Strategy for the Synthesis of Sulfonyl-Containing Isoquinolinediones. *Eur. J. Org. Chem.* **2019**, *2019*, 939–948. [CrossRef]
54. Martins, P.; Jesus, J.; Santos, S.; Raposo, L.R.; Roma-Rodrigues, C.; Baptista, P.V.; Fernandes, A.R. Heterocyclic Anticancer Compounds: Recent Advances and the Paradigm Shift towards the Use of Nanomedicine's Tool Box. *Molecules* **2015**, *20*, 16852–16891. [CrossRef] [PubMed]
55. Cossy, J.; Guérinot, A. Advances in Heterocyclic Chemistry. Natural Products Containing Oxygen Heterocycles—Synthetic Advances Between 1990 and 2015. *Adv. Heterocycl. Chem.* **2016**, *119*, 107–142.
56. Shao, A.; Luo, X.; Chiang, C.W.; Gao, M.; Lei, A. Furans Accessed through Visible-Light-Mediated Oxidative [3+2] Cycloaddition of Enols and Alkynes. *Chemistry* **2017**, *23*, 17874–17878. [CrossRef] [PubMed]
57. Chen, M.; Wang, K.L.; Liu, M.; She, Z.G.; Wang, C.Y. Bioactive steroid derivatives and butyrolactone derivatives from a gorgonian-derived *Aspergillus* sp. fungus. *Chem. Biodivers.* **2015**, *12*, 1398–1406. [CrossRef] [PubMed]
58. Niu, X.; Dahse, H.M.; Menzel, K.D.; Lozach, O.; Walther, G.; Meijer, L.; Grabley, S.; Sattler, I. Butyrolactone I derivatives from *Aspergillus terreus* carrying an unusual sulfate moiety. *J. Nat. Prod.* **2008**, *71*, 689–692. [CrossRef]
59. Cavanaugh, C.L.; Nicewicz, D.A. Synthesis of alpha-Benzyloxyamino-gamma-butyrolactones via a Polar Radical Crossover Cycloaddition Reaction. *Org. Lett.* **2015**, *17*, 6082–6085. [CrossRef]
60. Romero, N.A.; Nicewicz, D.A. Mechanistic insight into the photoredox catalysis of anti-markovnikov alkene hydrofunctionalization reactions. *J. Am. Chem. Soc.* **2014**, *136*, 17024–17035. [CrossRef]
61. Zhang, Q.B.; Ban, Y.L.; Zhou, D.G.; Zhou, P.P.; Wu, L.Z.; Liu, Q. Preparation of alpha-Acyloxy Ketones via Visible-Light-Driven Aerobic Oxo-Acyloxylation of Olefins with Carboxylic Acids. *Org. Lett.* **2016**, *18*, 5256–5259. [CrossRef]
62. Kouser, F.; Sharma, V.K.; Rizvi, M.; Sultan, S.; Chalotra, N.; Gupta, V.K.; Nandi, U.; Shah, B.A. Stereoselective synthesis of 3,4-di-substituted mercaptolactones via photoredox-catalyzed radical addition of thiophenols. *Tetrahedron Lett.* **2018**, *59*, 2161–2166. [CrossRef]
63. Gesmundo, N.J.; Grandjean, J.M.; Nicewicz, D.A. Amide and amine nucleophiles in polar radical crossover cycloadditions: Synthesis of gamma-lactams and pyrrolidines. *Org. Lett.* **2015**, *17*, 1316–1319. [CrossRef] [PubMed]
64. Morse, P.D.; Nguyen, T.M.; Cruz, C.L.; Nicewicz, D.A. Enantioselective counter-anions in photoredox catalysis: The asymmetric cation radical Diels-Alder reaction. *Tetrahedron* **2018**, *74*, 3266–3272. [CrossRef] [PubMed]
65. Yang, Z.; Li, H.; Li, S.; Zhang, M.-T.; Luo, S. A chiral ion-pair photoredox organocatalyst: Enantioselective anti-Markovnikov hydroetherification of alkenols. *Organ. Chem. Front.* **2017**, *4*, 1037–1041. [CrossRef]
66. Jirásek, M.; Straková, K.; Neveselý, T.; Svobodová, E.; Rottnerová, Z.; Cibulka, R. Flavin-Mediated Visible-Light [2+2] Photocycloaddition of Nitrogen- and Sulfur-Containing Dienes. *Eur. J. Org. Chem.* **2017**, *2017*, 2139–2146. [CrossRef]
67. Mojr, V.; Pitrová, G.; Straková, K.; Prukała, D.; Brazevic, S.; Svobodová, E.; Hoskovcová, I.; Burdziński, G.; Slanina, T.; Sikorski, M.; et al. Flavin Photocatalysts for Visible-Light [2+2] Cycloadditions: Structure, Reactivity and Reaction Mechanism. *ChemCatChem* **2018**, *10*, 849–858. [CrossRef]

68. Mojr, V.; Svobodova, E.; Strakova, K.; Nevesely, T.; Chudoba, J.; Dvorakova, H.; Cibulka, R. Tailoring flavins for visible light photocatalysis: Organocatalytic [2+2] cycloadditions mediated by a flavin derivative and visible light. *Chem. Commun.* **2015**, *51*, 12036–12039. [CrossRef] [PubMed]
69. Yoder, R.A.; Johnston, J.N. A Case Study in Biomimetic Total Synthesis: Polyolefin Carbocyclizations to Terpenes and Steroids. *Chem. Rev.* **2005**, *105*, 4730–4756. [CrossRef]
70. Justicia, J.; Alvarez de Cienfuegos, L.; Campana, A.G.; Miguel, D.; Jakoby, V.; Gansauer, A.; Cuerva, J.M. Bioinspired terpene synthesis: A radical approach. *Chem. Soc. Rev.* **2011**, *40*, 3525–3537. [CrossRef]
71. Yang, Z.; Li, H.; Zhang, L.; Zhang, M.T.; Cheng, J.P.; Luo, S. Organic Photocatalytic Cyclization of Polyenes: A Visible-Light-Mediated Radical Cascade Approach. *Chem. Eur. J.* **2015**, *21*, 14723–14727. [CrossRef]
72. Yang, W.; Yang, S.; Li, P.; Wang, L. Visible-light initiated oxidative cyclization of phenyl propiolates with sulfinic acids to coumarin derivatives under metal-free conditions. *Chem. Commun.* **2015**, *51*, 7520–7523. [CrossRef]
73. Hashimoto, T.; Maruoka, K. Recent advances of catalytic asymmetric 1,3-dipolar cycloadditions. *Chem. Rev.* **2015**, *115*, 5366–5412. [CrossRef] [PubMed]
74. Zheng, L.; Gao, F.; Yang, C.; Gao, G.L.; Zhao, Y.; Gao, Y.; Xia, W. Visible-Light-Mediated Anti-Regioselective Nitrone 1,3-Dipolar Cycloaddition Reaction and Synthesis of Bisindolylmethanes. *Org. Lett.* **2017**, *19*, 5086–5089. [CrossRef]
75. Jang, G.S.; Lee, J.; Seo, J.; Woo, S.K. Synthesis of 4-Isoxazolines via Visible-Light Photoredox-Catalyzed [3+2] Cycloaddition of Oxaziridines with Alkynes. *Org. Lett.* **2017**, *19*, 6448–6451. [CrossRef] [PubMed]
76. Li, Q.; Cui, L.S.; Zhong, C.; Jiang, Z.Q.; Liao, L.S. Asymmetric design of bipolar host materials with novel 1,2,4-oxadiazole unit in blue phosphorescent device. *Org. Lett.* **2014**, *16*, 1622–1625. [CrossRef]
77. Soni, V.K.; Kim, J.; Cho, E.J. Organocatalytic Oxidative Cyclization of Amidoximes for the Synthesis of 1,2,4-Oxadiazolines. *Adv. Synth. Catal.* **2018**, *360*, 2626–2631. [CrossRef]
78. Tilvi, S.; Singh, K.S. Synthesis of Oxazole, Oxazoline and Isoxazoline Derived Marine Natural Products: A Review. *Curr. Organ. Chem.* **2016**, *20*, 898–929. [CrossRef]
79. Hargaden, G.C.; Guiry, P.J. Recent applications of oxazoline-containing ligands in asymmetric catalysis. *Chem. Rev.* **2009**, *109*, 2505–2550. [CrossRef] [PubMed]
80. Morse, P.D.; Nicewicz, D.A. Divergent Regioselectivity in Photoredox-Catalyzed Hydrofunctionalization Reactions of Unsaturated Amides and Thioamides. *Chem. Sci.* **2015**, *6*, 270–274. [CrossRef]
81. Petzold, D.; Nitschke, P.; Brandl, F.; Scheidler, V.; Dick, B.; Gschwind, R.M.; Konig, B. Visible-Light-Mediated Liberation and In Situ Conversion of Fluorophosgene. *Chem. Eur. J.* **2019**, *25*, 361–366. [CrossRef]
82. Yamaguchi, T.; Sugiura, Y.; Yamaguchi, E.; Tada, N.; Itoh, A. Synthetic Method for the Preparation of Quinazolines by the Oxidation of Amines Using Singlet Oxygen. *Asian J. Org. Chem.* **2017**, *6*, 432–435. [CrossRef]
83. Riemer, D.; Schilling, W.; Goetz, A.; Zhang, Y.; Gehrke, S.; Tkach, I.; Hollóczki, O.; Das, S. CO_2-Catalyzed Efficient Dehydrogenation of Amines with Detailed Mechanistic and Kinetic Studies. *ACS Catal.* **2018**, *8*, 11679–11687. [CrossRef]
84. Shen, Z.C.; Yang, P.; Tang, Y. Transition Metal-Free Visible Light-Driven Photoredox Oxidative Annulation of Arylamidines. *J. Org. Chem.* **2016**, *81*, 309–317. [CrossRef] [PubMed]
85. Wang, M.X. Enantioselective biotransformations of nitriles in organic synthesis. *Acc. Chem. Res.* **2015**, *48*, 602–611. [CrossRef] [PubMed]
86. Li, X.; Fang, X.; Zhuang, S.; Liu, P.; Sun, P. Photoredox Catalysis: Construction of Polyheterocycles via Alkoxycarbonylation/Addition/Cyclization Sequence. *Org. Lett.* **2017**, *19*, 3580–3583. [CrossRef] [PubMed]
87. Borpatra, P.J.; Deb, M.L.; Baruah, P.K. Visible light-promoted metal-free intramolecular cross dehydrogenative coupling approach to 1,3-oxazines. *Tetrahedron Lett.* **2017**, *58*, 4006–4010. [CrossRef]
88. Qian, P.; Deng, Y.; Mei, H.; Han, J.; Zhou, J.; Pan, Y. Visible-Light Photoredox Catalyzed Oxidative/Reductive Cyclization Reaction of N-Cyanamide Alkenes for the Synthesis of Sulfonated Quinazolinones. *Org. Lett.* **2017**, *19*, 4798–4801. [CrossRef] [PubMed]
89. Yadav, S.; Srivastava, M.; Rai, P.; Tripathi, B.P.; Mishra, A.; Singh, J.; Singh, J. Oxidative organophotoredox catalysis: A regioselective synthesis of 2-nitro substituted imidazopyridines and 3-substituted indoles, initiated by visible light. *New J. Chem.* **2016**, *40*, 9694–9701. [CrossRef]

90. Yadav, L.; Kapoorr, R.; Singh, S.; Tripathi, S. Photocatalytic Oxidative Heterocyclization of Semicarbazones: An Efficient Approach for the Synthesis of 1,3,4-Oxadiazoles. *Synlett* **2015**, *26*, 1201–1206. [CrossRef]

Sample Availability: Samples of the compounds are not available from the authors.

© 2019 by the authors. Licensee MDPI, Basel, Switzerland. This article is an open access article distributed under the terms and conditions of the Creative Commons Attribution (CC BY) license (http://creativecommons.org/licenses/by/4.0/).

MDPI
St. Alban-Anlage 66
4052 Basel
Switzerland
Tel. +41 61 683 77 34
Fax +41 61 302 89 18
www.mdpi.com

Molecules Editorial Office
E-mail: molecules@mdpi.com
www.mdpi.com/journal/molecules